Molecular and Cellular Biology of Neuroprotection in the CNS

ADVANCES IN EXPERIMENTAL MEDICINE AND BIOLOGY

Recent Volumes in this Series

Volume 506
LACRIMAL GLAND, TEAR FILM, AND DRY EYE SYNDROME 3: BASIC SCIENCE AND CLINICAL RELEVANCE
Edited by David A. Sullivan, Michael E. Stern, Kazuo Tsubota, Darlene A. Dartt, Rose M. Sullivan, and B. Britt Bromberg

Volume 507
EICOSANOIDS AND OTHER BIOACTIVE LIPIDS IN CANCER, INFLAMMATION, AND RADIATION INJURY 5
Edited by Kenneth V. Honn, Lawrence J. Marnett, Santosh Nigam, and Charles Serhan

Volume 508
SENSORIMOTOR CONTROL OF MOVEMENT AND POSTURE
Edited by Simon C. Gandevia, Uwe Proske and Douglas G. Stuart

Volume 509
IRON CHELATION THERAPY
Edited by Chiam Hershko

Volume 510
OXYGEN TRANSPORT TO TISSUE XXIII: OXYGEN MEASUREMENTS IN THE 21ST CENTURY: BASIC TECHNIQUES AND CLINICAL RELEVANCE
Edited by David F. Wilson, John Biaglow and Anna Pastuszko

Volume 511
PEDIATRIC GENDER ASSIGNMENT: A CRITICAL REAPPRAISAL
Edited by Stephen A. Zderic, Douglas A. Canning, Michael C. Carr and Howard McC. Snyder III

Volume 512
LYMPHOCYTE ACTIVATION AND IMMUNE REGULATION IX: LYMPHOCYTE TRAFFIC AND HOMEOSTASIS
Edited by Sudhir Gupta, Eugene Butcher and William Paul

Volume 513
MOLECULAR AND CELLULAR BIOLOGY OF NEUROPROTECTION IN THE CNS
Edited by Christian Alzheimer

Volume 514
PHOTORECEPTORS AND CALCIUM
Edited by Wolfgang Baehr and Krzysztof Palczewski

Volume 515
NEUROPILIN: FROM NERVOUS SYSTEM TO VASCULAR AND TUMOR BIOLOGY
Edited by Dominique Bagnard

Volume 516
TRIPLE REPEAT DISORDERS OF THE NERVOUS SYSTEM
Edited by Lubov T. Timchenko

Volume 517
DOPAMINERGIC NEURON TRANSPLANTATION IN THE WEAVER MOUSE MODEL OF PARKINSON S DISEASE
Edited by Lazaros C. Triarhou

Volume 518
ADVANCES IN MALE MEDIATED DEVELOPMENTAL TOXICITY
Edited by Bernard Robaire and Barbara F. Hales

A Continuation Order Plan is available for this series. A continuation order will bring delivery of each new volume immediately upon publication. Volumes are billed only upon actual shipment. For further information please contact the publisher.

Molecular and Cellular Biology of Neuroprotection in the CNS

Edited by
Christian Alzheimer
Institute of Physiology
University of Kiel
Kiel, Germany

Kluwer Academic / Plenum Publishers
New York, Boston, Dordrecht, London, Moscow
Landes Bioscience / Eurekah.com
Georgetown, Texas, U.S.A.

Library of Congress Cataloging-in-Publication Data

CIP applied for but not received at time of publication.

Molecular and Cellular Biology of Neuroprotection in the CNS
Edited by Christian Alzheimer
ISBN 0-306-47414-X
AEMB volume number: 513

Kluwer Academic / Plenum Publishers
233 Spring Street, New York, NY 10013
http://www.wkap.nl

Landes Bioscience
810 S. Church Street, Georgetown, TX 78626
http://www.landesbioscience.com; http://www.eurekah.com
Landes tracking number: 1-58706-104-X

10 9 8 7 6 5 4 3 2 1

A C.I.P. record for this book is available from the Library of Congress.

Printed in the United States of America.

PREFACE

The adult mammalian brain is not well equipped for self-repair. Although neuronal loss reinstalls parts of the molecular machinery that is essential for neuronal development, other factors and processes actively impede regeneration of the damaged brain. Many therapeutic efforts thus aim to promote or inhibit these endogenous pathways. In addition, more radical approaches appear on the horizon, such as replacement of lost neurons with grafted tissue.

Neurorepair, however, is not the topic of this book. Here, we go one step back in the sequence of events that lead eventually to the demise of a neuronal population. This book focuses on the precious period when an initial damaging event evolves into a vast loss of neurons. The time frame might be hours to days in acute brain injury or months to years in chronic neurodegenerative diseases.

Given the limited capacity of regeneration, protecting neurons that are on the brink of death is a major challenge for basic and clinical neuroscience, with implications for a broad spectrum of neurological and psychiatric diseases, ranging from stroke and brain trauma to Parkinson«s and Alzheimer«s disease. In recent years, rapid progress has been made in unravelling many of the cellular and molecular players in neuronal death and survival. However, as the field develops into more and more specialized branches, the notion of common pathogenic pathways of neuronal loss might get buried under the wealth of novel data.

Thus it seems a timely endeavor to provide an overview on the most exciting recent developments in neuroprotective signaling and experimental neuroprotection. This book brings together experts from cellular and molecular neurobiology, neurophysiology, neuroanatomy, neuropharmacology, neuroimmunology and neurology. It is my hope that the book serves as a reference text for both basic neuroscientists and clinicians, offering a fresh look at many (certainly not all) of the highly intertwined processes that determine the fate of CNS neurons in the face of acute or chronic insults.

The book is written mostly from the viewpoint of the basic scientist who works at the cellular and molecular level, but who also develops and tests new hypotheses using animal models of acute and chronic brain injury. Although many of the new findings hold promise for therapeutic interventions, their translation into clinically relevant neuroprotective strategies is still in its infancy. If this book helps to bridge this gap, it will certainly be worth the effort.

I thank my publisher, Ron Landes, for his support and the opportunity to put this volume together. It was a pleasure working with Cynthia Dworaczyk, who coordinated the production of this book in a most skillful fashion. Finally, I am greatly indebted to the authors for their time and their valuable contributions.

Christian Alzheimer
Munich, February 2002

PARTICIPANTS

Dr. Christian Alzheimer
Institute of Physiology
University of Munich
Pettenkoferstrasse 12
D-80336 Munich
Germany
Present Address:
Institute of Physiology
University of Kiel
Olshausenstrasse 40
D-24098 Kiel
Germany

Dr. Guiseppe Battaglia
Istituto Neurologico Mediterraneo
Neuromed
86077 Pozzilli
Italy

Dr. Christian Behl
Max Planck Institute of Psychiatry
Kraepelinstrasse 2-10
80804 Munich
Germany

Dr. Martin Berry
King«s College
GKT School of Biomedical Research,
Neuronal Damage and Repair
Centre for Neuroscience, Hodgkin
Building
Guy«s Campus London Bridge
London SE1 1UL
UK

Dr. Ulrike Bl mer
Department of Neurosurgery
University of Kiel
Weimarer Str. 8
D-24106 Kiel
Germany

Dr. Valeria Bruno
Istituto Neurologico Mediterraneo
Neuromed
86077 Pozzilli
Italy

Dr. Samantha L. Budd
Astra Zeneca
S dert lje Bioscience
14157 Huddinge
Sweden

Dr. Georg Dechant
Max-Planck-Institute of Neurobiology
Am Klopferspitz 18a
D-82152 Martinsried
Germany

Dr. Ulrich Dirnagl
Department of Neurology
Experimental Neurology
Charit
D-10098 Berlin
Germany

Dr. Matthias Endres
Department of Neurology
Experimental Neurology
Charit
D-10098 Berlin
Germany

Dr. Peter J. Flor
Nervous System Research
Novartis Pharma AG
CH-4002 Basel
Switzerland

Dr. Gwenn Garden
Department of Neurology
UW School of Medicine
Seattle, WA 98195
U.S.A.

Dr. Arnold Ganser
Department of Hematology
and Oncology
Medical School Hannover
Carl-Neuberg-Strasse 1
D-30625 Hannover
Germany

Dr. Fabrizio Gasparini
Nervous System Research
Novartis Pharma AG
CH-4002 Basel
Switzerland

Dr. Saadi Ghatan
Department of Neurological Surgery
UW School of Medicine
Seattle, WA 98195
U.S.A.

Dr. Thomas Gillessen
Max-Planck-Institute of Psychiatry
Kraepelinstrasse 2 - 10
D-80804 Munich
Germany

Dr. Joseph T. Ho
Department of Neurological Surgery
UW School of Medicine
Seattle, WA 98195
U.S.A.

Dr. Yoshito Kinoshita
Department of Neurological Surgery
UW School of Medicine
Seattle, WA 98195
U.S.A.

Dr. Kerstin Krieglstein
Department of Anatomy
and Neuroanatomy
University of G ttingen
Kreuzbergring 36
D-37075 G ttingen
Germany

Dr. Sebastian Jander
Department of Neurology
University of D sseldorf
Moorenstrasse 5
D-40225 D sseldorf
Germany

Dr. Mark D. Johnson
Department of Neurological Surgery
UW School of Medicine
Seattle, WA 98195
U.S.A.

Dr. Stuart A. Lipton
The Burnham Institute
10901 North Torrey Pines Road
La Jolla, California 92037
U.S.A.

Dr. Ann Logan
Molecular Neuroscience
Department of Medicine
Wolfson Research Laboratories
Queen Elizabeth Hospital
Edgbaston, Birmingham B15 2TH
United Kingdom

Dr. Hugo Marti
Institute of Physiology
University of Z rich
Winterthurerstrasse 190
CH-8057 Z rich
Switzerland

Dr. Richard S. Morrison
Department of Neurological Surgery
UW School of Medicine
Seattle, Washington 98195-6470
U.S.A.

Dr. Ferdinando Nicoletti
Istituto Neurologico Mediterraneo
Neuromed
86077 Pozzilli
and
Department of Human Physiology and
Pharmacology
University La Sapienza
Rome
Italy

Dr. Harald Neumann
Neuroimmunology
European Neuroscience Institute
Waldweg 33
D-37073 G ttingen
Germany

Dr. William M. Pardridge
Department of Medicine
UCLA School of Medicine
Los Angeles, California 90095-1682
U.S.A.

Dr. Rod J. Sayer
Department of Physiology
University of Otago
P.O. Box 913
Dunedin
New Zealand

Dr. Michaela Scherr
Department of Hematology and
Oncology
Medical School Hannover
Carl-Neuberg-Strasse 1
D-30625 Hannover
Germany

Dr. Michael Schroeter
Department of Neurology
University of D sseldorf
Moorenstrasse 5
D-40225 D sseldorf
Germany

Dr. Guido Stoll
Department of Neurology
University of W rzburg
Josef-Schneider-Strasse 11
D-97080 W rzburg
Germany

Dr. Trevor W. Stone
Division of Neuroscience
and Biomedical Systems
West Medical Building
University of Glasgow
Glasgow G12 8QQ
Scotland

Dr. Davide Trotti
Department of Neurology
Cecil B. Day Laboratory for
Neuromuscular Research
Massachusetts General Hospital,
Harvard Medical School
Charlestown, MA 02129
U.S.A.

Dr. Klaus Unsicker
Institute of Anatomy and Cell Biology
University of Heidelberg
Im Neuenheimer Feld 307
D-69120 Heidelberg
Germany

Dr. Sabine Werner
Institute of Cell Biology
ETH H nggerberg
CH-8093 Zurich
Switzerland

Dr. Midori A. Yenari
Department of Neurosurgery
Stanford University
1201 Welch Road
MSLS Building P304
Stanford, California 94305-5487
U.S.A.

CONTENTS

I. NEURONAL CELL DEATH
OVERVIEW OF BASIC MECHANISMS 1

1. EXCITATORY AMINO ACID NEUROTOXICITY 3

Thomas Gillessen, Samantha L. Budd and Stuart A. Lipton

Historical Perspective 3
Clinical Relevance of Excitatory Amino Acid Neurotoxicity 5
Epilepsy 5
Traumatic Brain Injury 7
Hypoxia/Ischemia 7
Neurodegenerative Diseases 9
Intoxication with Exogenous Excitatory Amino Acids 9
Implication of Distinct Glutamate Receptor Classes in Excitotoxicity 10
Ionic Dependence of Excitotoxic Cell Damage 12
Mitochondrial Dysfunction 14
The Role of Reactive Oxygen Species
in Excitotoxicity 16
Role of Nitric oxide and other Reactive Nitrogen Species in Excitotoxicity 19
Excitotoxicity, Calcium Loading
and Apoptosis 21
Key Signaling Players in Neuronal Apoptosis 22
Conclusions 25

2. NEURONAL SURVIVAL AND CELL DEATH
SIGNALING PATHWAYS 41

Richard S. Morrison, Yoshito Kinoshita, Mark D. Johnson, Saadi Ghatan,
Joseph T. Ho and Gwenn Garden

Abstract 41
Introduction 41
Death receptor-Mediated Neuronal Apoptosis 42
Signal Transduction Pathways 48
Nuclear Signaling Pathways 53
p53-Mediated Cell Death Signaling Pathways 56
Bcl-2 Family Members and Mitochondrial Integrity 60
Proteolytic Enzymes 64

Calpains 66
Abbreviations 66

3. DETRIMENTAL AND BENEFICIAL EFFECTS OF INJURY-INDUCED INFLAMMATION AND CYTOKINE EXPRESSION IN THE NERVOUS SYSTEM 87

Guido Stoll, Sebastian Jander and Michael Schroeter

Abstract 87
Introduction 88
Glial Cell Populations in the Central Nervous System 88
Entry of Inflammatory Cells into the CNS: The Blood-Brain-Barrier and Immunological Cell adhesion Molecules 89
Inflammation and CNS Injury 90
Beneficial Effects of Neuroinflammation 99
Acknowledgement 104

4. CELLULAR AND MOLECULAR DETERMINANTS OF GLIAL SCAR FORMATION 115

Ann Logan and Martin Berry

Introduction 115
Development of the Scar 118
Axon Regeneration and Scarring 122
Cytokines and Scarring 124
Trophic Regulation of the Scar 133
Protease Regulation of Scarring 134
Therapeutic Modulation of the Scar 135
Conclusions 137

II. ION CHANNELS, RECEPTORS AND SIGNALING PATHWAYS 159

5. Na^+ CHANNELS AND Ca^{2+} CHANNELS OF THE CELL MEMBRANE AS TARGETS OF NEUROPROTECTIVE SUBSTANCES 161

Christian Alzheimer

Introduction 161
Na^+ Channels 162
Na^+ Channel Blockers as Neuroprotective Agents 168
Ca^{2+} Channels 173
Acknowledgements 177

6. INTRACELLULAR Ca^{2+} HANDLING ... 183

Rod J. Sayer

Abstract ... 183
Introduction ... 183
Cytosolic Ca^{2+} Buffering ... 184
Ca^{2+} Buffering and Neuroprotection ... 185
The Endoplasmic Reticulum Ca^{2+} Store ... 187
The Endoplasmic Reticulum and Neuroprotection ... 188
Mitochondria and Ca^{2+} Homeostasis ... 189
Mitochondria, Neurotoxicity and Neuroprotection ... 190
Future Challenges ... 191

7. NEUROPROTECTIVE ACTIVITY OF METABOTROPIC GLUTAMATE RECEPTOR LIGANDS ... 197

Peter J. Flor, Giuseppe Battaglia, Ferdinando Nicoletti, Fabrizio Gasparini and Valeria Bruno

Abstract ... 197
Introduction ... 198
Chemical Structures and Receptor Profile of Neuroprotective mGluR Ligands ... 199
Physico-Chemical and Pharmacokinetic Properties of Neuroprotective mGluR Ligands ... 202
Group-I mGluRs as Targets for Neuroprotective Drugs ... 204
Neuroprotection mediated by Group-II mGluRs ... 207
Role of Group-III mGluRs in Neuroprotection ... 211
Conclusions and Outlook ... 214

8. A ROLE FOR GLUTAMATE TRANSPORTERS IN NEURODEGENERATIVE DISEASES ... 225

Davide Trotti

Introduction ... 225
The Na^+/K^+-Dependent High Affinity Glutamate Transporters ... 226
Localization of Glutamate Transporters ... 227
Functional Properties, Stoichiometry and Kinetics of the Glutamate Transporters ... 227
Glutamate Transporter Topology ... 230
Role of Glutamate Transporters in the Excitatory Neurotransmission ... 230
Regulation of Glutamate Transporters ... 234
Glutamate Transporters in Disease States ... 236
Concluding Remarks ... 242

9. PURINES AND NEUROPROTECTION 249

Trevor W. Stone

Abstract 249
Adenosine 250
Adenosine Receptors 250
A2A Receptors 257
Abbreviations 270

10. HEAT SHOCK PROTEINS AND NEUROPROTECTION 281

Midori A. Yenari

Abstract 281
Introduction 281
Where and When is Hsp70 Expressed? 283
Correlative Evidence for a Neuroprotective Role 285
Neuroprotection with Hsp70 Overexpression 286
Potential Mechanisms of Protection 291
Conclusion 295
Acknowledgments 295

III. NEUROTROPHIC FACTORS AS NEUROPROTECTIVE AGENTS 301

11. NEUROTROPHINS 303

Georg Dechant and Harald Neumann

Summary 303
From Neurotrophin Physiology to Therapy 303
Structure and Physiological Functions of Neurotrophins and Their Receptors 304
Neurotrophins in Animal Models of Pathological Situations and Clinical Trials 314
Non-Neuroprotective and Side Effects of Neurotrophins 319
Potential Improvements of Neurotrophin Therapy 320
Acknowledgments 324

12. FIBROBLAST GROWTH FACTORS AND NEUROPROTECTION 335

Christian Alzheimer and Sabine Werner

Introduction Fibr oblast Growth Factors (FGFs) and FGF Receptors 335
Distribution of FGFs in Adult Brain 336
FGFs and Neuronal Development 337
Upregulation of FGFs after Brain Injury 338
Neuroprotective Effects of FGF2 339
FGFs and Glia 340

Neuroprotective Mechanisms of FGF2 341
Summary and Conclusions 343
Acknowledgements 345

13. TGF-βS AND THEIR ROLES IN THE REGULATION OF NEURON SURVIVAL 353

Klaus Unsicker and Kerstin Krieglstein

Abstract 353
Introduction 353
TGF-βs: A Brief Overview of Their Molecular Biology, Biochemistry and Signaling 354
Expression of TGF-βs and TβRs in the Nervous System 358
TGF-βs and the Regulation of Proliferation, Survival and Differentiation of Neurons 360
Regulation of Neuron Survival and Maintenance by Members of the TGF-β Superfamily Other Than TGF-βs Proper 364
Conclusions 366
Acknowledgements 366

14. VASCULAR ENDOTHELIAL GROWTH FACTOR 375

Hugo H. Marti

The Members of the VEGF Family 375
Regulation of VEGF and VEGF Receptor Expression 378
Pleiotropic Action of VEGF in the CNS 380
Intracellular Signaling Events 387
Conclusion 387
Acknowledgments 388

IV. ADVANCES IN DRUG DELIVERY TO CNS NEURONS 395

15. BLOOD-BRAIN BARRIER DRUG TARGETING ENABLES NEUROPROTECTION IN BRAIN ISCHEMIA FOLLOWING DELAYED INTRAVENOUS ADMINISTRATION OF NEUROTROPHINS 397

William M. Pardridge

Abstract 397
Blood-Brain Barrier, Neurotrophins and Neurological Disease 398
Chimeric Peptide Technology 401
Targeting Chimeric Neurotrophic Factors to the Brain 406
Targeting Gene Therapeutics to the Brain 423
Drug Targeting to the Human Brain 426
Acknowledgments 427

16. INVASIVE DRUG DELIVERY 431

Ulrike Blömer , Arnold Ganser and Michaela Scherr

Abstract 431
Vectors in Gene Therapy 432
Direct DNA Delivery and Synthetic Nonviral Vectors 433
Viral Vectors and Gene Delivery 434
Herpes Simplex Viral Vectors 435
Adenoviral Vectors 436
Adeno-Associated Viral Vector (AAV) 437
Retrovirus 437
Biology of Retroviruses 438
Moloney Murine Leukemia Virus and Lenitvirus Based Vectors 440
Retroviral Gene Transfer Ex Vivo 441
Retroviral Gene Transfer In Vivo 443
Retroviral Vector Safety 444
Conclusion 445

V. NEUROPROTECTIVE STRATEGIES IN ANIMAL AND IN VITRO MODELS OF NEURONAL DAMAGE 453

17. ISCHEMIA AND STROKE 455

Matthias Endres and Ulrich Dirnagl

Abstract 455
Epidemiological Data 455
Introduction 455
Global vs. Focal Ischemia 456
Animal Models of Cerebral ischemia 456
In vitro Models of Cerebral ischemia 458
Importance of Physiologic Parameters for Stroke Outcome 458
Pathophysiological Cascades Following Cerebral ischemia 459
Glutamate Receptors and Excitotoxicity 462
Tissue Acidosis 464
Protein Synthesis and Early Gene Expression 464
Molecular Mechanisms 466
Experimental Evidence for Caspase-Mediated Cell Death Following Cerebral Ischemia 467
Caspase Inhibition Protects From Cerebral ischemia 467
Conclusion 468

18. NEUROPROTECTIVE STRATEGIES IN ALZHEIMER'S DISEASE 475

Christian Behl

Introduction 475
What is the Cause of AD? 476
AD Genetics and Biochemistry 479

Neurotoxicology of Aβ 482
AD Risk Factors 484
Mouse Models of AD 485
Clinical AD Therapy 486
Experimental AD Therapies 487
AD Prevention 489
Final Remarks 492

INDEX 497

I. NEURONAL CELL DEATH

Overview of Basic Mechanisms

EXCITATORY AMINO ACID NEUROTOXICITY

Thomas Gillessen[1], Samantha L. Budd[2] and Stuart A. Lipton[3]

HISTORICAL PERSPECTIVE

The excitatory potency of the acidic amino acids glutamate and aspartate in various regions of the central nervous system (CNS) has been recognized since the 1960 s.[1,2] Nevertheless, the earlier findings that these amino acids are (1) constituents of intermediary metabolism and are (2) located in the brain ubiquitously in high concentrations rendered them unlikely candidates as neurotransmitters. These findings fueled a sustained debate about their physiological role as neurotransmitters in the 1970s. Today, L-glutamate is accepted as the predominant fast excitatory neurotransmitter in the vertebrate brain.

In parallel to studies on the physiological role of these amino acids, it has been observed since the 1950s that glutamate can exert toxic effects on the nervous system under certain conditions. Following the systemic application of glutamate to mice, toxic effects on retinal neurons were described.[3] Further studies in the 1970s corroborated these toxic effects and extended this view by showing neuronal cell death following oral intake of glutamate or aspartate in brain regions devoid of the blood-brain barrier in mice and nonhuman primates (Fig. 1).[4-9]

Thus, L-glutamate is the primary excitatory transmitter in the mammalian CNS but is cytotoxic under certain conditions. This relation between the physiological function as excitatory amino acid (EAA) and the pathological effect is reflected by the term excitotoxicity introduced in the 1970s by Olney et al. [10]

With the introduction of structural transmitter analogues, local injections of the glutamate agonist kainate were shown in the late 1970s and 1980s to induce cell death with a similar pattern of damage in different brain regions, thus confirming

[1]Institut fuer Pharmakologie und Toxikologie, Bereich Studien und Wissenschaft, Neuherbergstrasse 11, 80937 Muenchen, Germany; Present address: Max-Planck-Institute of Psychiatry, Kraepelinstrasse 2-10, 80804 Muenchen, Germany, [2]Astra Zeneca R&D S dert lje, Bioscience, 141 57 Huddinge, Sweden and [3]The Burnham Institute, 10901 Torrey Pines Road, La Jolla, CA 92037.

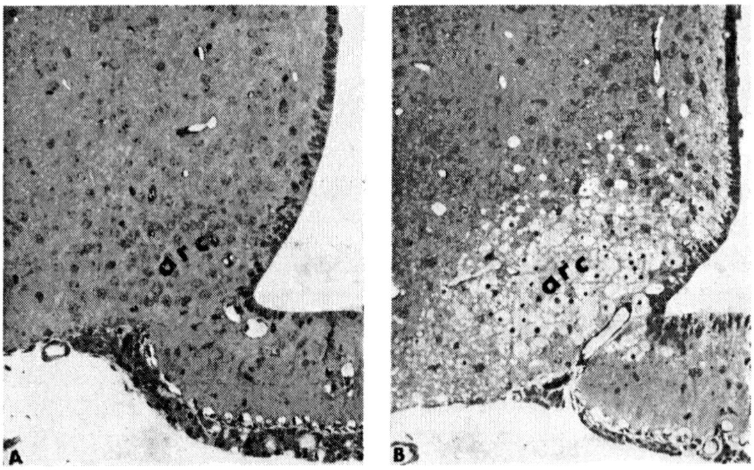

Figure 1. A, Tissue section through arcuate nucleus (arc) of hypothalamus from a 10 day old mouse (control; x150). No signs of pathology are present. B, Section trough arcuate nucleus (arc) from a 10 day old mouse treated orally with 1 g/kg sodium glutamate (x150). There is a considerable number of necrotic cells within the arcuate region. Reprinted with permission from Olney JW, Ho OL. Brain damage in infant mice following oral intake of glutamate, aspartate or cysteine. Nature 1970; 227:609-611, copyright © 1970 Macmillan Publishers Ltd.

the neurotoxic effect.[11-14] Today, it is well recognized, that exogenous or endogenous agonists of EAA receptors can induce cell death in CNS neurons.

Our knowledge about the role of glutamate as an excitatory neurotransmitter and its cytotoxic effects increased in parallel during the 1980s, and notably, the development of substances that antagonized the excitatory function[15] also stimulated studies on mechanisms underlying the toxic effects.[16-18] The discovery of different EAA receptor subtypes in conjunction with the introduction of selective receptor antagonists revealed that the glutamate-induced cell death was induced by excessive ionotropic glutamate receptor activation.[19-26]

The view that activation of different ionotropic EAA receptors can induce excitotoxic cell death was supported by subsequent studies on the ionic mechanisms underlying excitotoxicity. It was demonstrated, that excessive calcium loading plays a pivotal role in neuronal cell death following the intense stimulation of ionotropic glutamate receptors,[27,28] and since then the implication of ion homeostasis dysregulation and dysfunction in calcium signaling have been studied extensively (Fig. 2).[29-33]

Regarding the mode of cell death, glutamate-induced neuronal cell death has been judged originally as necrotic from the morphological appearance.[6,34-36] However, the observation of a delayed neuronal cell death in the penumbra of ischemic lesions[37] and after EAA exposure[38,39] has stimulated studies in the 1990s focussing on the mode of cell death. Today, there is compelling evidence that failure in extracellular glutamate homeostasis can result in different modes of cell death

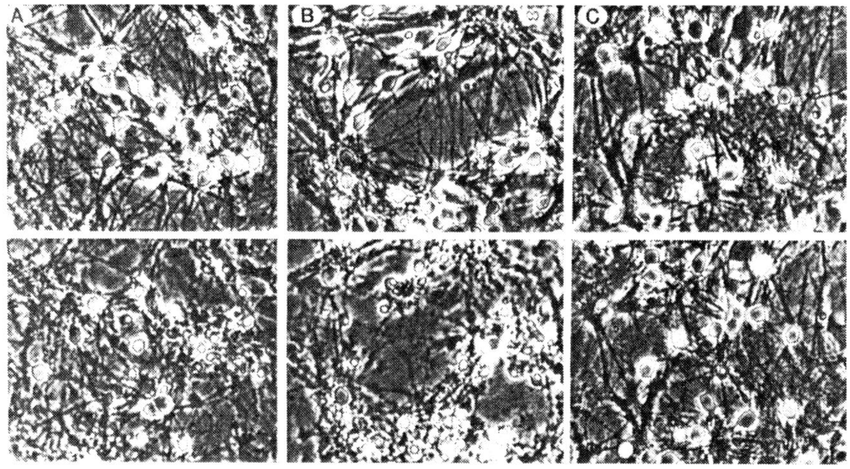

Figure 2. Cortical neurons in cell culture before (top row) and 1 day after (bottom row) a 5 min incubation in 500 μM glutamate. A, Na^+ ions replaced with equimolar choline. B, 1 μM tetrodotoxin added. C, Ca^{2+} ions omitted. Considerable cell death occurred even under replacement of Na^+ or addition of 1 μM tetrodotoxin. Omission of Ca^{2+} resulted in a substantial decrease in neuronal cell loss. Reprinted with permission from Choi D W. Glutamate neurotoxicity in cortical cell culture is calcium dependent. Neurosci Lett 1985; 58:293-297, © 1985 Elsevier Science.

with morphological and biochemical features of either apoptosis or necrosis depending on the severity of the insult, with more fulminant insults causing rapid energy failure because of lack of ionic homeostasis and thus necrosis.[40-44]

CLINICAL RELEVANCE OF EXCITATORY AMINO ACID NEUROTOXICITY

There is evidence that excitotoxicity is involved in acute brain damage under pathophysiological conditions following status epilepticus, mechanical trauma or ischemia (Fig. 3).[45] Moreover, glutamate cytotoxicity seems to be partly involved in many neurodegenerative diseases.

EPILEPSY

Histopathological studies on the brains of patients suffering from chronic epilepsy have revealed that certain brain regions show structural alterations with severe loss of neurons and reactive gliosis.[46,47] Brain pathology in epilepsy is described best for human temporal lobe epilepsy, resulting in sclerosis of the hippocampus that extends into the amygdala and the parahippocampal gyrus and is termed mesial temporal sclerosis , hippocampal sclerosis or Ammon s horn sclerosis . Pronounced brain damage has been observed following sustained epileptiform activity with seizures lasting more than 30 min, called status

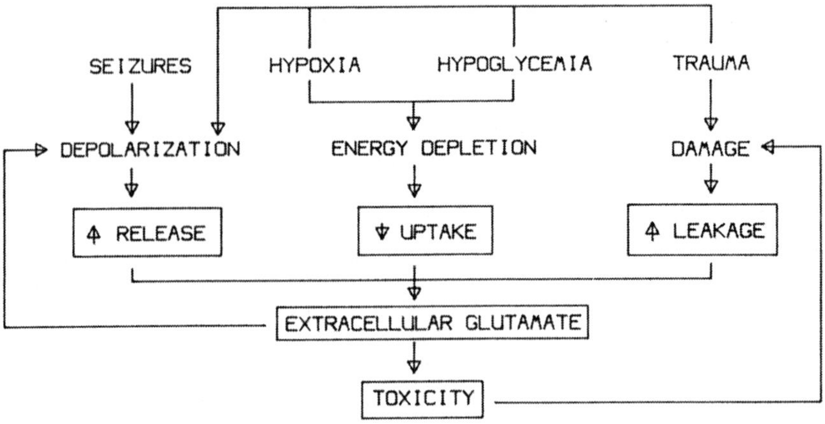

Figure 3. Glutamate induces acute CNS injury. This historical schematic summarizes that seizures, hypoxia, hypoglycemia and trauma share common mechanisms of acute injury. The excitatory activity of the transmitter glutamate is linked to its toxic effects (excitotoxicity). Reprinted with permission from Choi D W. Glutamate neurotoxicity and diseases of the nervous system. Neuron 1988; 1:623-634, copyright © 1988 Elsevier Science.

epilepticus (SE).[48-50] Importantly, neuronal cell loss after SE is not distributed equally across all hippocampal subfields and the extent of damage is in the order CA1 > CA4 > CA3 > CA2, thus indicating different susceptibility to SE-induced cell death.[12,50]

In animal models of epilepsy using chemoconvulsant-induced or electrical stimulation-induced SE, similar patterns of brain damage were observed.[51-54] Experiments aimed at the observation of ultrastructural changes have demonstrated that certain features of cell damage, such as swelling of dendrites and soma, are independent of the mechanism used to induce SE.[51-56]

Evidence for the implication of EAA-mediated excitotoxicity in SE-induced neuronal cell death arises from several experiments. First, the morphological appearance of cell damage following SE is similar to damage following systemic or local application of the glutamate receptor agonists L-glutamate, NMDA or kainate.[6,36,51,53,55,56] Second, the administration of ionotropic glutamate receptor antagonists that inhibit excitotoxic cell death in cultured neurons can prevent cell death induced by epileptiform activity.[57-60]

Apart from the above-mentioned evidence for excitotoxicity in epilepsy, there has been a considerable debate regarding the mode of cell death. There is now a large body of evidence suggesting the implication of both, necrotic and apoptotic cell death following pathologically relevant EAA receptor activation. Recently, several reports have added evidence for the implication of apoptotic pathways in epilepsy-associated neuronal cell death.[61-66] In conclusion, the mode of cell death in SE-induced brain pathology may be the result of intensity and duration of glutamate receptor activation, with a shift from apoptotic death to necrotic death with increasing intensity and duration of receptor activation.[40,43]

TRAUMATIC BRAIN INJURY

Following head trauma, mechanical brain injury can be accompanied by secondary changes as metabolic glutamate leaks uncontrollably from the neuronal cytoplasm[67] causing subsequent excitotoxic damage of surrounding neurons. However, compared to research on mechanisms underlying the cell damage in epilepsy, studies on traumatic injury have been sparse and there is less evidence for the implicated modes of cell death. Some studies have suggested the sudden release of excitatory amino acids from the cytoplasm into the extracellular space and subsequent bioenergetic failure as well as ultrastructural damage that can be diminished by NMDA antagonists, all indicative of excitotoxicity,[67-69] whereas others have demonstrated activation of caspase enzymes, internucleosomal DNA fragmentation and induction of immediate early genes, indicative of apoptosis.[69-71] Most recently, in neonatal models of traumatic brain injury (TBI) (mortality and morbidity from head trauma is highest in children), the resulting excitotoxicity has been shown to elicit both apoptosis and necrosis. Necrosis occurs localized to the site of impact, and within 4 hr of the insult, whereas a secondary apoptotic damage occurs between 6 — 24 hr and is found in the areas surrounding the primary necrosis.[72] In this model, the secondary damage was more severe than the primary damage suggesting a preponderance of apoptosis over necrosis.

HYPOXIA/ISCHEMIA

Under conditions of local or global ischemia, neurons are deprived of glucose and oxygen, resulting in bioenergetic failure and subsequent decrease of ion gradients across the plasma membrane.[73] The resulting plasma membrane depolarization leads to increased synaptic release of glutamate, and the diminished Na^+ gradient is followed by attenuated Na^+-dependent glutamate uptake or even reversed glutamate transport in terms of transporter-mediated release of glutamate from neurons and astrocytes.[74-76] Moreover, osmotic cell swelling following the influx of Na^+, Cl^- and H_2O can result in plasma membrane rupture and further release of cytoplasmic glutamate into the extracellular space. In summary, hypoxia/ischemia results in a secondary net increase in the extracellular glutamate concentration.[77-79] As in the other above-mentioned acute CNS insults, this increase in extracellular glutamate results in excitotoxic damage.[80-83] The implication of excitotoxicity in hypoxia/ ischemia is corroborated by the observation that NMDA-type glutamate receptor antagonists can reduce neuronal death in animal models of cerebral ischemia.[84-87] However, the actual set of circumstances in humans appears to differ from the defined animal models.[88] In addition to the well-accepted concept of excitotoxicity-associated acute necrotic cell death, there are several lines of evidence, that neurons can undergo

Table 1. Neurodegenerative diseases thought to be mediated at least in part through stimulation of glutamate receptors

Huntington s disease (pathologic process mimicked by injection of the endogenous NMDA agonist quinolinate; mitochondrial inhibitors, which make neurons more susceptible to glutamate toxicity, can reproduce this process)
AIDS dementia complex (human immunodeficiency virus-associated cognitive-motor complex) (evidence that neuronal loss is ameliorated by NMDA antagonists in vitro and in animal models)
Neuropathic pain syndromes (e.g., causalgia or painful peripheral neuropathies with a central component blocked by NMDA-receptor antagonists or inhibitors of nitric oxide synthase)
Olivopontocerebellar atrophy (some recessive forms associated with glutamate dehydrogenase deficiency)
Parkinsonism (mimicked by impaired mitochondrial metabolism, which renders neurons more susceptible to glutamate-induced toxicity)
Amyotrophic lateral sclerosis (primary defect may be a mutation in superoxide dismutase gene, which may render motor neurons more vulnerable to glutamate-induced toxicity; there is also evidence for decreased glutamate reuptake)
Mitochondrial abnormalities and other inherited or acquired biochemical disorders (partial listing)
MELAS syndrome (mitochondrial myopathy, encephalopathy, lactic acidosis, and stroke-like episodes due to a point mutation in mitochondrial DNA)
MERRF (myoclonus epilepsy with ragged-red fibers, signifying mitochondrial DNA mutation; also frequently accompanied by ataxia, weakness, dementia, and hearing loss)
Leber s disease (point mutation in mitochondrial DNA, presenting with delayed-onset optic neuropathy and occasionally degeneration of basal ganglia, with dystonia, dysarthria, ataxia, tremors, and decreased vibratory and position sense)
Wernicke s encephalopathy (thiamine deficiency)
Rett syndrome (disease of young girls, presenting with seizures, dementia, autism, stereotypical hand wringing, and gait disorder)
Hyperhomocysteinemia and homocysteinuria (L-homocysteine has been shown to be a weak NMDA-like agonist as is L-homocysteic acid)
Hyperprolinemia (L-proline is a weak NMDA-like agonist)
Nonketotic hyperglycinemia (a case report of some improvement after treatment with an NMDA antagonist)
Hydroxybutyric aminoaciduria
Sulfite oxidase deficiency
Combined systems disease (vitamin B12 deficiency, which may result in accumulation of homocysteine)
Lead encephalopathy
Alzheimer s disease (data that the vulnerability of neurons to glutamate can be increased by §-amyloid protein)
Hepatic encephalopathy (perhaps a component, although inhibitory neurotransmitters are more clearly involved)
Tourette s syndrome (deficits in basal ganglia have been proposed to be mediated by glutamate or glutamate-like toxins)
Drug addiction, tolerance, and dependency (animal modes suggest that NMDA antagonists may be helpful in treatment)
Multiple sclerosis
Depression/anxiety
Glaucoma

Table 2. Exogenous excitotoxins

	Kainic acid	**Domoic acid**	**BOAA**
Structure	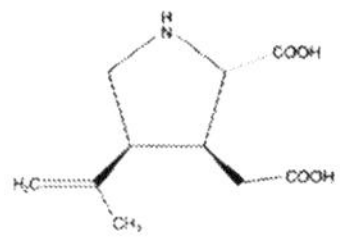	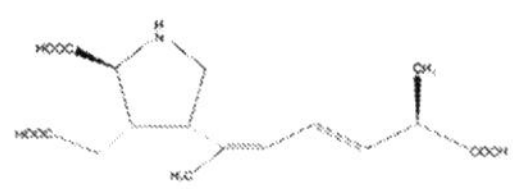	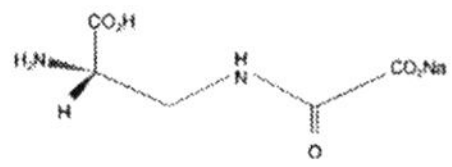
Biological source	*Digenea simplex*	*Chondria armata*	*Lathyrus sativus*
Consumption	Ascaricide	Blue mussels	Chick pea (staple food)
Binding site Ki	21 nM (kainate)	13 nM (kainate)	760 nM (AMPA)
Toxicity in culture	67 μM		10-100 μM
Acute toxic dose (rodents)	9-15 mg/kg i.p. (pathology in hippocampus)	4 mg/kg i.p.	500 mg/kg i.p.
Neurological syndrome	None reported	Limbic seizures; amnesia	Neurolathyrism

Reprinted with permission from Meldrum B, Garthwaite J. Excitatory amino acid neurotoxicity and neurodegenerative diseases. Trends Pharmacol Sci 1990; 11:379-387, copyright © 1990 Elsevier Science.

apoptosis following ischemia.[41,89] Among various biochemical markers of apoptosis, DNA fragmentation[90,91] and activation of caspase-3 have been observed[63,92] in models for cerebral ischemia. Moreover, application of caspase-inhibitors results in reduced infarct size[42,93,94] and decreases cell death in cultured neurons following ischemia.[95]

NEURODEGENERATIVE DISEASES

In addition to acute neurological disorders, many chronic neurodegenerative diseases may exhibit a component of glutamate-dependent neuronal damage, including apoptosis or injury to dendrites and axons (Table 1). This arises when the primary disease causes neuronal injury which in turn may cause the leak or release of excessive glutamate. Additionally, elevated inflammatory responses in many of these diseases can also contribute to excessive glutamate release or decreased glutamate clearance from the synaptic cleft.[96,97]

INTOXICATION WITH EXOGENOUS EXCITATORY AMINO ACIDS

Several structural analogues of endogenous EAAs have been introduced to the neurosciences during the 1980s, which exhibit similar or even higher excitatory and neurotoxic potency compared to the endogenous EAAs. In 1987, an outbreak of domoic acid poisoning following ingestion of mussels occurred in Canada. Patients suffered from acute headache, seizures, sensory dysfunctions and showed motor signs. In four fatal cases, neuropathological studies revealed lesions predominantly in the hippocampus and amygdala, resembling the lesion pattern after application of the exogenous excitotoxin kainate.[98] It was reconstructed that the mussels had accumulated domoic acid, synthesized by the phytoplankton *Nitzchia pungens.*[98] Domoic acid is a structural analogue of kainic acid, which is also synthesized by seaweed, but compared to kainate, domoic acid has a higher excitatory potency.[99,100] In parallel to its high excitatory potency, domoic acid exerts potent excitotoxic effects on CNS neurons (**Table** 2). Experimental administration of domoic acid to rodents and monkeys has resulted in brain damage with ultrastructural features resembling L-glutamate excitotoxicity.[101-103]

Ingestion of the chick pea *Lathyrus sativus*, which contains another EAA structural analogue results in acute spastic motor signs following the consumption. The clinical features of this motor disorder called lathyrism were known even by the ancient Greeks but the toxic component, the amino acid β-N-oxalyl-L-alanine (BOAA) was identified only some decades ago (Table 2).[104] Experimental administration of BOAA is known to induce the features of lathyrism in nonhuman primates[105] and application of BOAA to cell cultures results in cell death with structural features of excitotoxicity[106] that can be attenuated by non-NMDA receptor antagonists.[107,108] Importantly, the high excitotoxic potency of this glutamate receptor ligand is in accordance with its high excitatory potency as agonist at AMPA receptors.[109]

IMPLICATION OF DISTINCT GLUTAMATE RECEPTOR CLASSES IN EXCITOTOXICITY

L-glutamate, the most abundant excitatory transmitter in the brain, binds to different classes of receptors comprising different types of ionotropic receptors as well as metabotropic receptors. Ionotropic glutamate receptors are ligand-gated ion channels, which are named after agonists that bind preferentially to these receptor subtypes. They include α-amino-3-hydroxy-5-methyl-isoxalole-4-propionate (AMPA), kainate and N-methyl-D-aspartate (NMDA) receptors.

AMPA receptors are widely distributed in the CNS and are expressed on many different types of neurons. They control a cation channel that is permeable to Na^+ and K^+ ions with a single channel conductance < 20 pS.[110] Activation of AMPA receptors results in the fast onset of an excitatory postsynaptic current (EPSC) with rapid desensitization.[111] This current shapes the fast component of glutamatergic EPSCs in CNS neurons.[112] Importantly, certain AMPA receptor subtypes can exhibit substantial Ca^{2+} permeability and thereby contribute to the Ca^{2+}-dependent form of excitotoxic cell damage.[113-116] Ca^{2+}-permeable channels are formed from the receptor subunits GluR1 or GluR3, whereas coassembly with GluR2 results in only poor Ca^{2+} permeability.[117-119]

Another type of ionotropic glutamate receptor termed kainate receptor is less well understood because of lack of sufficiently selective agonists and antagonists. As with AMPA receptors, kainate receptors control a cation channel that desensitizes rapidly. Since the agonist kainate activates kainate receptors but also AMPA receptors, their physiological and pathophysiological role remains elusive. Recently, results based on gene-targeted rodents lacking certain kainate receptor subtypes have extended our knowledge of localization and function of kainate receptors.[120,121] From these studies one can conclude that kainate receptors are not only localized to postsynaptic sites but also appear to be localized presynaptically, suggesting a role for modulation of synaptic strength (for review see refs. 122,123).

The third type of ionotropic glutamate receptor, the NMDA receptor differs fundamentally from AMPA and kainate receptors in several ways. First, the pore is significantly permeable to Ca^{2+} ions,[124,125] but also to K^+ and Na^+ ions.[126] In contrast to AMPA receptors, NMDA receptors exhibit a high single channel conductance (50 pS) and desensitize much slower (for review see refs. 127,128). NMDA receptors are widely distributed in different types of CNS neurons and shape the late component of glutamatergic EPSCs. Second, the opening of the ligand-gated cation channel does not only depend on binding of agonist but is voltage-dependent, since the channel is blocked by Mg^{2+}ions at resting membrane potentials and a depolarization of the plasma membrane is required to relieve the Mg^{2+}-dependent block. Third, activation of NMDA receptors requires binding of a coagonist to the so-called glycine-binding site of the NMDA receptor. Very recently, the amino acid D-serine has been suggested to be an endogenous ligand for the glycine-binding site.[129] This is supported by the findings that (1) D-serine has a high potency to potentiate NMDAR-mediated neurotransmission, (2) D-serine is colocalized with NMDARs in the forebrain and (3) enzymatic degradation of the amino acid attenuates NMDAR-mediated neurotransmission.

As discussed at the start of this chapter, experimental application of the endogenous nonselective agonist glutamate has been shown to induce excitotoxic cell damage.[4-8] As with L-glutamate, administration of the agonists kainate[11,12,14,20] or NMDA[19] induced brain damage, thus confirming the implication of different glutamate receptor classes.

The implication of different ionotropic glutamate receptor types in induction of brain damage has been further studied by the use of selective receptor antagonists. Experimental administration of competitive or uncompetitive NMDA antagonists[21,22,85,86,130] or application of kainate antagonists attenuates cell death in cell cultures as well as brain damage.[20] Unfortunately, the clinical use of these receptor antagonists has been hampered by the inhibition of physiological NMDAR-mediated neurotransmission, resulting in various adverse effects.[131] However, the uncompetitive NMDA antagonist memantine (1-amino-3,5-dimethyladamantane hydrochloride), which has already been used for years for treatment of Parkinson s disease, is sufficiently tolerated. Memantine is an open-channel blocker, but has faster kinetics than MK-801. This results in substantial inhibitory drug action under conditions of prolonged exposure to glutamate but much less inhibition under millisecond exposure to glutamate.[132] This raises the hypothesis, that memantine should be suited to treat disease conditions associated with elevated glutamate concentrations for prolonged periods. Indeed, several groups have demonstrated its neuroprotective effectiveness combined with only few adverse effects.[133-135]

Whereas the role of ionotropic glutamate receptors in excitotoxicity has been studied extensively, the significance of metabotropic glutamate receptors (mGluRs) in excitotoxicity is less well understood. Importantly, there is no indication, that activation or inhibition of mGluRs by itself exerts excitotoxicity.[136] Instead, in view of the literature it appears more likely, that specific metabotropic receptor subtypes can be involved in modulation of ionotropic receptor-mediated excitotoxicity. This may be due to the coupling of different mGluR subtypes to different signal transduction pathways and effectors. Group I mGluRs comprise the subtypes $mGluR_1$ and $mGluR_5$ which in heterologous expression systems couple to phospholipid hydrolysis through phospholipase C. Among other effects, this results in second messenger-mediated activation of protein kinase C and Ca^{2+} release from IP_3-sensitive Ca^{2+} stores.[137-139] Application of recently developed selective receptor ligands suggests, that activation of group I, and particularly mGluR1 receptors amplifies NMDA-mediated excitotoxicity.[136,140-144]

Group II receptors comprise the subtypes $mGluR_2$ and $mGluR_3$ that are negatively coupled to the adenylate cyclase pathway. Recent pharmacological studies using selective agonists indicate that group II receptor activation results in protection against NMDA-mediated excitotoxicity.[141,145-147] Likewise, group III receptors ($mGluR_4$ and $mGluR_{6-8}$) are negatively coupled to adenylyl cyclase, and activation of mGluR4 or mGluR8 group III receptors results in attenuation of NMDA-mediated neurotoxicity.[147-149]

IONIC DEPENDENCE OF EXCITOTOXIC CELL DAMAGE

Activation of different ionotropic receptor types is linked to excitotoxic cell damage through the underlying ion currents. Depending on the predominance of either Na^+ or Ca^{2+} influx, two different components of excitotoxicity have been suggested. Na^+ ion influx mediated by activation of NMDA-type and non-NMDA-type glutamate receptors is followed by secondary influx of Cl^- and H_2O and results in swelling of neurons.[28,34,150,151] This acute form of cell damage depends on the transmembrane Na^+ and Cl^- gradients[28] and can be prevented by extracellular substitution of Na^+ and Cl^- with impermeant ions.[152,153] In contrast to the acute, primarily Na^+-dependent osmotic damage, a more delayed mode of cell death has been attributed to Ca^{2+} influx.[28,153] This delayed Ca^{2+}-dependent cell death can be largely attenuated by inhibition of NMDA receptors,[19,29,84,154] removal of extracellular Ca^{2+} ions[23,155,156] or buffering of cytoplasmic Ca^{2+} by membrane permeable chelators.[157,158] Therefore, the Ca^{2+}-dependent component was traditionally believed to be induced exclusively by NMDA receptor activation.

The finding that the late component can be mimicked by calcium ionophores in presence of Ca^{2+} ions has corroborated the Ca^{2+}-dependence of the delayed cell death.[28] However, the discovery of Ca^{2+}-permeable AMPA/kainate receptors, has lead to the view that neurons expressing these Ca^{2+}-permeable non-NMDA receptors can also undergo Ca^{2+}- dependent delayed cell death.[32,159] In summary, different routes of Ca^{2+} influx such as through NMDA receptors, Ca^{2+}-permeable AMPA/kainate receptors and voltage-dependent Ca^{2+} channels, which are all involved in physiological signaling, are implicated in excitotoxic cell death.

The suggested key role of Ca^{2+} in excitotoxicity has been subsequently confirmed by the finding, that the extent of excitotoxic cell death correlates with the total amount of Ca^{2+} uptake and is independent of the route of entry. In some cases Zn^{2+} can substitute for Ca^{2+} as the cation inducing excitotoxic damage.[29,159,160]

The Ca^{2+} overload observed following sustained stimulation of NMDA receptors results from the inability of cellular Ca^{2+} homeostasis, such as extrusion of Ca^{2+} across the plasma membrane by Na^{+}/Ca^{2+} antiporter[161-163] and Ca^{2+} ATPase[164] or Ca^{2+} sequestration by the endoplasmic reticulum and mitochondria[165,166] to remove the large influx of Ca^{2+}.

Interestingly, during sustained exposure to glutamate, the intracellular calcium concentration $[Ca^{2+}]_c$rises rapidly to a peak value and thereafter slowly recovers to an elevated plateau level.[167] However, the $[Ca^{2+}]_c$ can also manifest a second, delayed increase,[30,168] immediately preceding cell death (Fig. 4). This delayed $[Ca^{2+}]_c$ increase, termed delayed calcium deregulation (DCD) is irreversible and reflects irreversible loss of cellular Ca^{2+} homeostasis. Although the interval between the initial Ca^{2+} spike and the DCD varies within a cell population exposed to NMDA receptor agonists, once DCD occurs within a neuron, it invariably precedes cell death. The temporal relation between neuronal Ca^{2+} levels and delayed cell death suggests, that Ca^{2+}-dependent effector mechanisms are involved, which do not need sustained high Ca^{2+} levels but are triggered by transient changes in $[Ca^{2+}]_c$.

Since many enzymes are activated by transient or sustained $[Ca^{2+}]$ elevation various effector mechanisms may be implicated in Ca^{2+}-mediated excitotoxic cell death. Consequently, a variety of Ca^{2+}-dependent hydrolytic enzymes, including lipases and proteases, have been suggested to be involved in excitotoxic neuronal damage. Activation of the Ca^{2+}-dependent phospholipase A2 has been observed following NMDA receptor activation[169] and the subsequent catabolism of released arachidonic acid by lipoxygenases and cyclooxygenases (COX), is also associated with concomitant production of reactive oxygen species (ROS).[169-171] In addition, activation of phospholiase A2 and subsequent release of arachidonic acid may inhibit transporter-mediated glutamate uptake from the extracellular space.[33,172,173]

Among several Ca^{2+}-activated proteases, the activity of the Ca^{2+}-dependent cysteine protease calpain is increased following glutamate receptor-mediated Ca^{2+} loading. [174-177] Calpain activation results in proteolysis of structural proteins and degradation of the neuronal cytoskeleton.[177] Furthermore, calpain may direct the mode of cell death to necrosis by preventing the cytochrome c-mediated activation of caspases (see following).[174]

Ca^{2+}-mediated activation of nitric oxide synthase may be another pathway involved in excitotoxic cell death since neurons constitutively express the isoenzyme called neuronal nitric oxide synthase (nNOS). nNOS is activated by glutamate receptor-mediated $[Ca^{2+}]_c$ increases[178,179] and there is compelling evidence, that nNOS activation is linked to excitotoxic damage[31] since inhibition of NO formation results in protection of neurons from glutamate receptor-mediated cell death.[179-182] Studies on nNOS-deficient neuronal cultures confirmed the role of NO in glutamate receptor-mediated neurotoxicity since in nNOS-deficient cortical cultures the toxic effects elicited by administration of NMDA are markedly attenuated.[183] The role of NO in excitotoxicity is mediated at least partly through the reaction of NO with superoxide anions (O_2^-) to form peroxynitrite ($ONOO^-$).[184]

Other classes of enzymes are also thought to be involved in the Ca^{2+}-mediated cell death but in less direct ways. Calcineurin is a Ca^{2+}/calmodulin-activated phosphatase which can dephosphorylate nNOS, thereby increasing its activity[185] and potentially increasing excitotoxic damage. Calcineurin has been convincingly demonstrated to be involved in neuronal cell death since pharmacological experiments using the calcineurin inhibitors cyclosporin A and FK-506 have revealed attenuation of excitotoxic cell death.[186,187] Accordingly, inhibition of calmodulin can also be shown to decrease excitotoxic cell death.[182,188]

MITOCHONDRIAL DYSFUNCTION

Studies on the time course of glutamate receptor-induced $[Ca^{2+}]_c$ increase have demonstrated that if the neuronal cell does not succumb to the insult, $[Ca^{2+}]_c$ slowly recovers after the termination of agonist application.[30,167] This recovery is due to several mechanisms that ensure the cellular Ca^{2+} homeostasis under physiological conditions (Fig. 5) such as extrusion across the plasma membrane by the Na^+/Ca^{2+} exchanger[189] and the Ca^{2+}-ATPase, or sequestration of Ca^{2+} into the endoplasmic reticulum[189,190] and mitochondria.[191,192]

In theory, the driving force for Ca^{2+} ions to enter the mitochondrial matrix in energized mitochondria is formed by the strong electrochemical gradient for Ca^{2+} across the inner mitochondrial membrane, mainly due to the mitochondrial membrane potential $\Delta\psi_m$ ($\Delta\psi_m \approx -180$ mV). Therefore, normal energized mitochondria can take up substantial amounts of calcium.[193] Net uptake of calcium occurs in isolated mitochondria whenever $[Ca^{2+}]_c$ rises above a set point in the high nanomolar range (>0.5 μM), indicating the dynamic equilibrium of mitochondrial uptake and extrusion mechanisms.[194] $[Ca^{2+}]_c$ can be shown to increase above this set point[195] during excessive cytoplasmic Ca^{2+} loading following ionotropic glutamate receptor activation and results in net mitochondrial Ca^{2+} uptake.[191,192]

It has been proposed that mitochondrial Ca^{2+} uptake can shape the time course of cytoplasmic $[Ca^{2+}]_c$.[196] During excessive Ca^{2+} loading into neurons, mitochondrial Ca^{2+} uptake appears to blunt the $[Ca^{2+}]_c$ increase since mitochondrial depolarization by protonophores prior to Ca^{2+} loading results in increased $[Ca^{2+}]_c$.[163,196-199] After termination of the Ca^{2+} influx, $[Ca^{2+}]_c$ slowly decreases and Ca^{2+} redistribution from mitochondria into the cytoplasm begins when $[Ca^{2+}]_c$ decreases below the set point.[192,196]

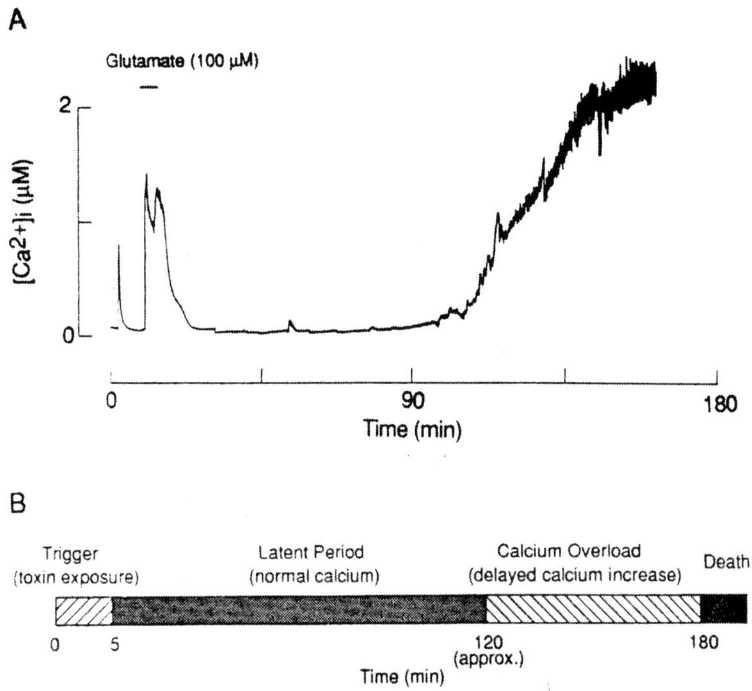

Figure 4. Time course of glutamate-induced Ca^{2+} overload. A, intracellular free Ca^{2+} concentration $[Ca^{2+}]_c$ of a hippocampal neuron. $[Ca^{2+}]_c$ was measured using the calcium-sensitive fluorochrome indo-1. Glutamate (100 μM) was applied for 5 min by superfusion (see bar). With a delay of about 90 min following glutamate application an irreversible calcium increase occurred. B, Schematic presentation of the $[Ca^{2+}]_c$ time course. Reprinted with permission from Randall RD, Thayer SA. Glutamate-induced calcium transient triggers delayed calcium overload and neurotoxicity in rat hippocampal neurons. J Neurosci 1992; 12:1882-1895, © 1992 Society for Neuroscience.

Ca^{2+} uptake into mitochondria results in a decrease in $\Delta\psi_m$ that can be monitored in isolated energized mitochondria[200-202] as well as in situ.[203-207] This depolarization is due to the balancing of the charge transfer carried by Ca^{2+} ions by re-entry of protons into the mitochondrial matrix. Under normal conditions, Ca^{2+}-induced depolarization is transient and serves to activate mitochondrial dehydrogenases[208-214] and the mitochondrial ATP synthase,[215] which in turn results in increased electron transport through the respiratory chain and consequently increased outward flux of protons.

However, studies on in situ mitochondrial membrane potential using cationic fluorochromes have demonstrated convincingly that the time course of $\Delta\psi_m$ depolarization can vary considerably, and depends on the time of Ca^{2+} loading. Short pulses of NMDA result in at least partial recovery of $\Delta\psi_m$[203] whereas prolonged NMDA exposure or higher agonist concentrations result in sustained depolarization with negligible recovery.[40,203,206] This sustained depolarization indicates that excessive Ca^{2+} uptake triggers severe mitochondrial bioenergetic dysfunction. Sustained depolarization has been ascribed to various underlying mechanisms, and ROS seem to play a key role in several of the suggested hypotheses.

In neurons as in other tissues, mitochondria are a major source of ROS;[216,217] nonetheless, mitochondria are themselves susceptible to oxidative damage. Excessive Ca^{2+} uptake into isolated mitochondria is known to result in increased formation of ROS (Fig. 6),[218,219] which in turn inhibit pyruvate dehydrogenase[220] and tricarboxylic acid cycle enzymes[221,222] as well as complex I of the respiratory chain.[223-225] In intact neurons, ROS formation has been monitored with redox-sensitive fluorochromes, which are oxidized by ROS to form fluorescent molecules.[226-228] These studies indicate that ionotropic glutamate receptor-induced Ca^{2+} loading enhances the production of O_2^-.[227,229-231]

The main site of ROS formation within mitochondria is within the respiratory chain. Complex I[232,233] as well as complex III[234-236] are thought to participate in one electron reduction of molecular oxygen, resulting in the generation of O_2^- which in turn can lead to other ROS. Nevertheless, the exact biophysical link between mitochondrial Ca^{2+} uptake and increased mitochondrial ROS production remains unclear.

THE ROLE OF REACTIVE OXYGEN SPECIES IN EXCITOTOXICITY

Evidence for the implication of ROS in excitotoxic cell damage arises from experiments showing enhanced production of O_2^- [227,230,231,237] following ionotropic glutamate receptor over-stimulation and from studies using radical scavengers or inhibitors of the formation of certain ROS.[227] These studies unequivocally demonstrate that removal of ROS results in attenuation of glutamate receptor-induced cell death.[238,239]

Several Ca^{2+}-dependent processes that increase the endogenous production of ROS have been described, which are all assumed to be implicated in excitotoxic cell death following NMDA receptor activation. As stated above, Ca^{2+} loading into mitochondria appears to be one important mechanism of ROS production under conditions associated with excitotoxicity.[227,229-231,237,240] Apart from this mechanism, activation of Ca^{2+}-dependent phospholipase A2 has been observed following NMDA receptor activation[169] and catabolism of released arachidonic acid by lipoxygenases and cyclooxygenases has been implicated in excitotoxicity through the concomitant production of ROS.[169-171]

Some ROS exhibit a high reactivity with correspondingly short half-lives and can undergo many different reactions. Generally speaking, ROS can exert multiple damaging reactions to proteins, lipids, carbohydrates and nucleic acids, thereby disrupting cellular functions. The increased production of ROS is therefore a potential threat to cellular homeostasis and neuronal survival if production is not balanced by

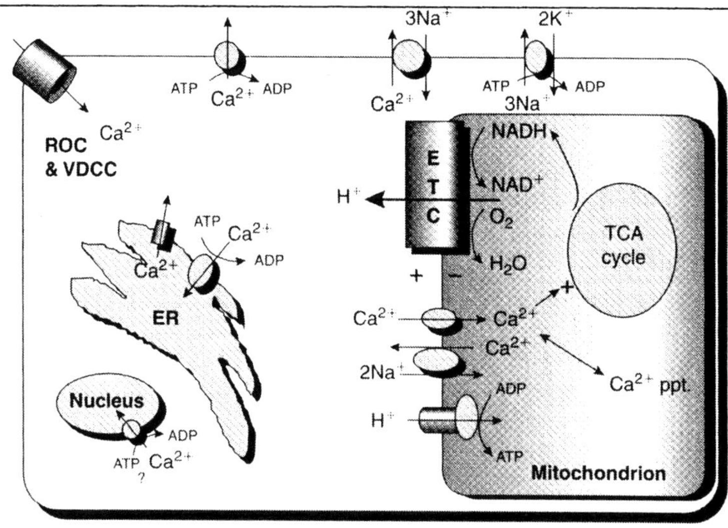

Figure 5. Calcium transport in neurons. Influx of Ca^{2+} ions through receptor operated channels (ROC) and voltage-dependent calcium channels (VDCC) activates extrusion mechanisms and calcium sequestration into mitochondria, endoplasmic reticulum (ER) and nucleus. Reproduced with permission from Murphy AN, Fiskum G. Bcl-2 and Ca^{2+}-mediated mitochondrial dysfunction in neuronal cell death. Biochem Soc Symp 1999; 66:33-41, copyright © 1999 the Biochemical Society.

the capacity of endogenous antioxidant mechanisms. Some basic characteristics of activated oxygen species are listed in Table 3.

From a multitude of ROS-mediated disturbances following NMDA receptor activation, only some examples will be discussed. Since mitochondria appear to be the major source of Ca^{2+}-induced increase in ROS formation and ROS are highly reactive, mitochondria are prone to damage by ROS. As discussed above, several mitochondrial enzymes, such as NADH:CoQ oxidoreductase, succinate dehydrogenase, ATP synthase, pyruvate dehydrogenase and the citric acid cycle enzyme aconitase, are inhibited by ROS including O_2^-, H_2O_2 or ¥OH.[220,221,223,241,242] The delayed deregulation of cellular Ca^{2+} homeostasis (DCD) that has been observed after NMDA exposure could be caused by ROS-dependent mechanisms. This hypothesis is supported by the finding that experimentally induced production of O_2^- by menadione results in enhancement of DCD,[238] whereas dismutation of O_2^- by the manganoporphyrin Mn-TBAP attenuates DCD.[238] Importantly, the redox-sensitive mechanism underlying DCD is still unresolved. Although oxidative stress-induced dysfunction of the plasma membrane Ca^{2+}-ATPase has been demonstrated[243] and this dysfunction has been proposed as a mechanism underlying the NMDA receptor-induced DCD,[238,244] the conclusion that mitochondrially generated ROS are involved in plasma membrane protein dysfunction should be judged cautiously because of the small reaction range.

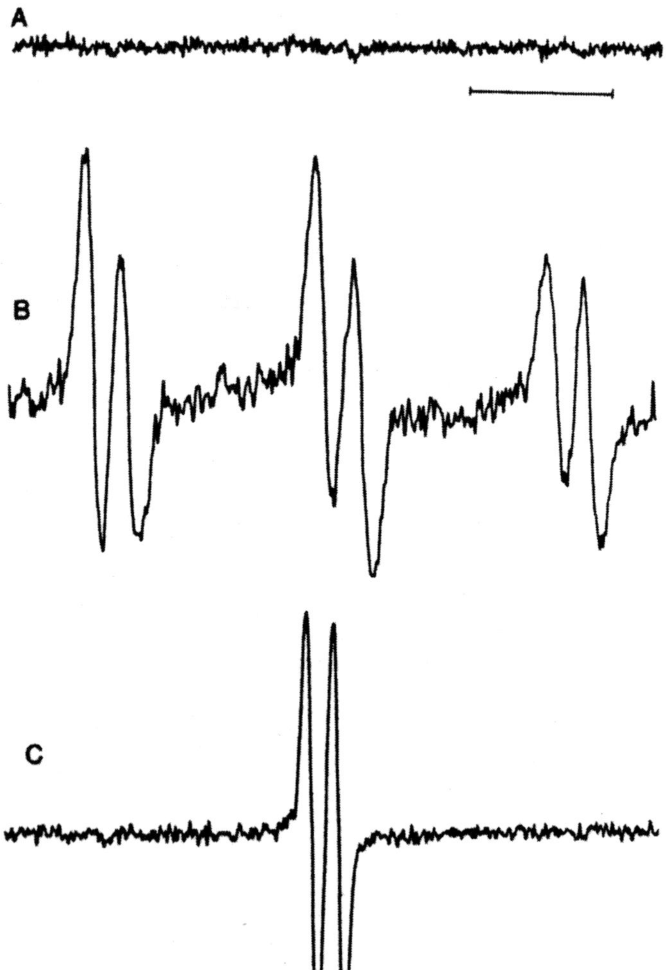

Figure 6. Electron paramagnetic resonance spectra from intact, coupled mitochondria isolated from rat cerebral cortex. A, no signals indicative of free radicals were detected in the absence of Na^+ and Ca^{2+}. B, following exposure to 14 mM Na^+ and 2.5 µM Ca^{2+} signals could be detected, indicating free radical production. C, incubation with ascorbate resulted in detection of an ascorbyl radical signal following incubation in 14 mM Na^+ and 2.5 µM Ca^{2+}, supporting the finding that free radicals were produced under this condition. Reprinted with permission from Dykens JA. Isolated cerebral and cerebellar mitochondria produce free radicals when exposed to elevated Ca^{2+} and Na^+: implications for neurodegeneration. J Neurochem 1994; 63:584-591, © 1994 Blackwell Science Ltd.

Ca^{2+}-induced increase in ROS can result in mitochondrial dysfunction including dissipation of $\Delta\psi_m$, bioenergetic failure and disturbance of cellular Ca^{2+} homeostasis. Apart from the inhibition of enzyme activity, the opening of a large nonselective pore in the inner mitochondrial membrane termed mitochondrial permeability transition (mPT) has been suggested as a key mechanism in neuronal excitotoxicity.[203,206] The basic concept of a pore-mediated permeabilization of the inner mitochondrial membrane rather than membrane damage had been suggested as early as the 1970s. The early studies by Haworth and Hunter[245] showed that various divalent cations could induce or prevent mitochondrial swelling and that the assumed pore is selective for the permeation of solutes below 1500 kDa. The idea of a large nonspecific channel was confirmed almost one decade later by electrophysiological experiments. Single-channel patch-clamp recordings from the inner mitochondrial membrane revealed a large conductance channel (1.3 nS),[246] which opened in response to the addition of Ca^{2+} to the mitochondrial preparation. Notably, the activation and inhibition characteristics of this mitochondrial megachannel are similar to the proposed mPT. Application of Ca^{2+}, ROS, inorganic phosphate and the adenine nucleotide translocator ligand atractylate induce permeability transition and opening of the megachannel, whereas Mg^{2+}, antioxidants, ADP, cyclosporin A and the adenine nucleotide translocator ligand bongkrekic acid (BA) inhibit permeability transition and channel opening.[219,247-256] This strikingly similar behavior has resulted in the conclusion that the mPT and the mitochondrial megachannel are virtually identical.[257]

Dissipation of $\Delta\psi_m$, bioenergetic failure and disturbance of mitochondrial Ca^{2+} homeostasis following activation of NMDA receptors all could be explained by opening of the mPT.[203,206] However, conclusions regarding the involvement of mPT in neuronal excitotoxicity until recently were based on pharmacological experiments using the mPT inhibitor cyclosporin A and should be treated cautiously since cyclosporin A also inhibits calcineurin and the multidrug-resistance channel.[258,259] Recently, however, our group succeeded in demonstrating that BA, a more specific inhibitor of mPT, could prevent NMDA-induced neuronal apoptosis, suggesting the involvement of mPT in this form of excitotoxic cell death.[260]

In summary, several ROS-dependent mechanisms have been suggested, mainly based on data from isolated mitochondria. In any event, the complex environment of in situ mitochondria, containing, for example, mPT enhancers as well as inhibitors, prevents a definitive interpretation of the physiological and pathophysiological relevance of the proposed mechanisms at this time.

ROLE OF NITRIC OXIDE AND OTHER REACTIVE NITROGEN SPECIES IN EXCITOTOXICITY

Nitric oxide (NO) is a well-recognized messenger molecule in the CNS that affects various cellular functions, e.g., neuronal transmitter release, synaptic plasticity and gene expression (see refs. 261-263). NO is produced by several types of cells expressing three distinct isoforms of the enzyme nitric oxide synthase (NOS, NADPH-diaphorase) that converts L-arginine into NO and citrulline: (1) neuronal NOS (nNOS), (2) inducible or immunologic NOS (iNOS) in microglia and astrocytes and (3) endothelial NOS (eNOS) predominantly in endothelial cells of brain blood vessels. Unlike classical neurotransmitters, NO is freely diffusible and therefore cannot be stored in synaptic vesicles. Once synthesized, NO diffuses across cell membranes and thus can reach (1) various compartments within the NO-producing cells and (2) different types of surrounding cells (Fig. 7). In contrast to conventional transmitters, the activity of the messenger molecule is not terminated by reuptake or enzymatic degradation but by chemical reactions with target molecules and its spontaneous formation of nitrite. The absence of control over its activity by release and reuptake mechanisms is functionally compensated by a tight regulation of its synthesis. However, the regulatory mechanisms involved in physiological signaling may also implicated under pathophysiological conditions. During the last decade, pharmacological experiments using NO donors, NOS inhibitors or exogenous substrates (reduced hemoglobin) for competitive reactions with NO suggested that NO is linked to neuronal cell death.[31,43,182] Even more important, cell death under conditions assumed to induce excitotoxicity could be largely attenuated by inhibition of NO formation or blockade of NO effects.[31,182,264] Subsequent studies on nNOS-deficient neuronal cultures confirmed the role of NO in glutamate receptor-mediated neurotoxicity since toxic effects elicited by administration of NMDA are markedly attenuated in nNOS-deficient cortical cultures compared with wild-type neurons.[183] Since nNOS activity is known to increase with increasing calcium concentrations, the NMDAR-mediated calcium uptake provides the link between conditions associated with excitotoxicity and NO-mediated cell damage.[178]

Although NO is a free radical, it is not as reactive as most ROS[265] and calculations based on its reactivity had estimated a reaction range of about 100 μm.[261,263]

The well-accepted implication of NO in excitotoxicity has been suggested to depend on increased NO formation and the concomitant production of O_2^-. Under this condition, O_2^- reacts with NO extremely fast to form peroxynitrite ($ONOO^-$).[266-268] $ONOO^-$ can undergo a multitude of different chemical reactions with various substrates, e.g., hydroxylation or nitration of tyrosine residues, lipid peroxidation or apparent decomposition into NO_2 and ¥OH and subsequently, this toxic molecule has been implicated in NO-induced cell damage.[43,182,184,269]

So far, several pathways of $ONOO^-$ toxicity have been suggested but with respect to its high reactivity, it appears unlikely that these are the only mechanisms involved in NO-mediated cell death.

Intense exposure to NO/O_2^- with resultant $ONOO^-$ formation has been demonstrated convincingly to result in neuronal necrosis because of energy failure, while mild insults lead to apoptosis.[43] This bioenergetic failure may be due to (1) increased ATP consumption under the condition of poly-(ADP ribose) polymerase (PARP) activation,[270] which occurs following $ONOO^-$ -mediated DNA damage and (2) decreased ATP synthesis through NO-mediated inhibition of the respiratory chain machinery.[271-273]

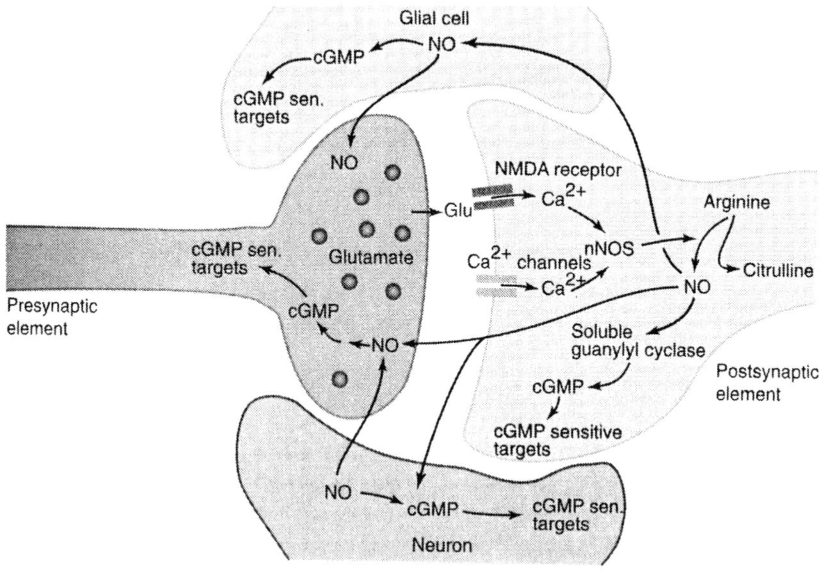

Figure 7. Schematic presentation of a nitric oxide (NO) producing neuron. NO formed from arginine by the enzyme nitric oxide synthase freely diffuses across membranes and thus can affect various organelles within the same neuron as well as other surrounding neurons and glia. Reprinted with permission from Deutch AY, Roth RH. Neurotransmitters. In: Fundamental Neuroscience 1999:193-234. Zigmond MJ, Bloom FE, Landis SC, Roberts JL, Squire LR, eds. San Diego: Academic press. © 1999 Academic Press.

Very recently, p38 mitogen-activated protein (MAP) kinase has been suggested as another pathway implicated in NO-induced cell death. Application of a p38 MAP kinase inhibitor significantly attenuated NO-induced caspase activation, Bax translocation, and neuronal cell death.[274]

EXCITOTOXICITY, CALCIUM LOADING AND APOPTOSIS

A great advance in excitotoxicity came with the understanding that neurons can die by apoptosis. In response to excitotoxicity the neuron activates apoptosis i.e., dismantling of itself including DNA, cytoskeleton, and production of ATP. While in acute neurological situations such as stroke the cells in the ischemic core die rapidly by necrosis, at least in animal models, the cells in the penumbra also go on to die but show markers of apoptosis, such as oligonucleosomal DNA damage.[275,276] Further evidence of apoptosis in excitotoxicity came with the examination of the activation of caspases in animal models of stroke.[277] Inhibition of caspases in general by intracerebroventricular injections of N-benzyloxycarbonyl-Val-Ala-Asp-fluoromethylketone (z-VAD.FMK), or selective inhibition of caspase-3 with N-benzyloxycarbonyl-Asp-Glu-Val-Asp-fluoromethylketone (z-DEVD.FMK) reduced the ischemic infarct volume by as much as 60%.[277] These caspase inhibitors remain effective even when applied hours after the insult, making them effective later than NMDA receptor blockade with MK-801.[94] While the morphology of dying ischemic cells does not precisely match that of classical apoptosis,[278] and in vivo excitotoxic injury results in neurodegeneration along an apoptosis-necrosis continuum, studies with caspase inhibitors clearly indicate a role for apoptosis in ischemia.

Apoptosis is also a feature of neurons dying in neurodegenerative disorders. DNA fragmentation has been detected in human post mortem samples of Parkinson s Disease,[279,280] Alzheimer s Disease,[281,282] Huntington s Disease,[282,283] and amyotrophic lateral sclerosis.[284] However, the presence of DNA fragmentation is not a guarantee that the cells are dying by apoptosis,[285] and DNA fragmentation can be influenced by ante mortem hypoxia.[286] More recently other recognized apoptotic molecules such as caspases,[287,288] and $BclX_L$[289] have been shown in Alzheimer s Disease. It is now largely accepted that neuron loss in chronic neurodegeneration is mediated by apoptosis.

KEY SIGNALING PLAYERS IN NEURONAL APOPTOSIS

Caspases are central to the induction of neuronal cell death.[94,277,290] Their actions on key intracellular substrates, including other protease zymogens, make caspases the primary executioners of the cell death program. The 13 known caspases can be separated into two functional groups; those that initiate apoptosis by receiving the initiating signals, and those that effect the dismantling of the cell. The caspases that receive or integrate apoptotic signals (caspases-8, -10, -2, -9) cleave and activate the downstream effector caspases (caspases-3, -7, -6). Perhaps the last reversible step in the death of neurons is the activation of the caspase-3,[291] although caspase-independent mechanisms almost certainly exist.[174,292] As discussed above, peptide inhibitors of caspase-3 block the death of neurons in many different situations (for review see ref. 293). The central importance of caspase-3 in neurons was clearly shown in caspase-3—deficient mice which have a doubling of brain size, correlated with decreased apoptosis and premature death.[294]

Mitochondria appear to provide a link between the initiator caspases and the downstream effector caspases (Fig. 8). In nonneuronal cells, mitochondria have been shown to accelerate activation of effector caspases by releasing proapoptotic molecules, such as cytochrome c,[295,296] the apoptosis-inducing factor,[297] and SMAC/diabolo.[298] Currently in neuronal systems, only the pathway utilized by cytochrome c has been fully elucidated. Cytochrome c triggers activation of caspase-9 which in turn cleaves caspases-3, -6, -7.[299] Release of cytochrome c from mitochondria has received much attention as a commitment to apoptosis.[295,296,300,301] Nevertheless, the release of cytochrome c from mitochondria has been shown to be not sufficient for neuronal apoptosis,[260,302] and microinjection of cytochrome c into sympathetic neurons does not lead to death in the absence of additional stimuli.[303] Therefore, cytochrome c

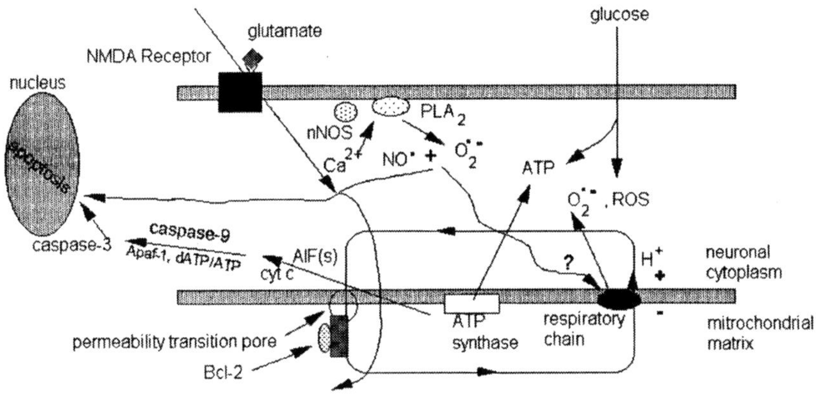

Figure 8. Model incorporating several features of apoptotic injury in neurons. After a relatively mild insult by glutamate at the NMDA receptor (initiating excessive Ca^{2+} influx and generating NO and O_2^- to form $ONOO^-$), the mitochondrial membrane potential transiently depolarizes and a drop in ATP transiently results, but this not a sufficient compromise in energy to severely disrupt the pumps and therefore does not result in necrosis. Rather, mitochondrial permeability transition occurs and a slight swelling of the mitochondria ensues. Caspases, Bcl-2 and $Bcl\text{-}X_L$, located at the mitochondrial membrane, may affect permeability transition. In neurons, as ATP synthesis recovers, cytochrome *c* (cyt *c*) is released from the mitochondria. In conjunction with apoptotic protease activating factor-1 (Apaf-1) and dATP/ATP, cytochrome *c* triggers activation of caspase-9 and in turn caspase 3, leading to apoptosis.

requires other factors to initiate the caspase cascade. These factors, Apaf-1, dATP and procaspase-9 along with cytochrome c are known collectively as the apoptosome.[304] The processed caspase-9 is then able to cleave and activate caspase-3.

Considerable uncertainty surrounds the mechanism of release of proapoptotic factors from the mitochondria. Possibilities include (i) transport by pore-forming Bcl-2 family proteins such as Bax,[305] (ii) opening of the permeability transition pore (PTP) in the inner mitochondrial membrane leading to rupture of the outer mitochondrial membrane,[306] or (iii) transport coupled to the voltage-dependent anion channel in the outer mitochondrial membrane.[301,307] This remains an important question to answer, since cytochrome c release from the mitochondria will eventually bring about mitochondrial dysfunction and an energetic deficit. Studies with nonneuronal cells have shown that mitochondrial membrane potential is maintained by ATP hydrolysis following the exit of cytochrome c.[308] The release of cytochrome c from mitochondria can be stimulated by some caspases and by the proapoptotic members of the Bcl-2 family Bid and Bax.[300,305,306] On the contrary, survival factors can act to prevent release of cytochrome c, e.g., channel activity of Bax is inhibited by anti-apoptotic Bcl-2 family members such as Bcl-2 or Bcl-xL, and ordinarily Apaf-1 is sequestered at the membrane by Bcl-xL, and in the absence of cytochrome c cannot interact with pro—caspase-9.[309] While the role of signaling molecules in neuronal apoptosis has been studied extensively in cell culture models, the in vivo data is sparse for many of the apoptotic pathways. However, several of the key

Table 3. Characteristics of reactive oxygen species

Reactive oxygen species	Symbol	Characteristics
Superoxide anion	$O_2^{\cdot}$	Good reductant, poor oxidant
Hydroxyl radical	¥OH	Highly reactive, small diffusion range
Hydrogen peroxide	H_2O_2	Oxidant, high diffusion range

signaling molecules in apoptosis have been detected following ischemia, e.g., the release of cytochrome c,[310,311] Fas/TNFR,[312] Bax,[313] phospho-JNK,[314] and caspase-9.[315]

The initiators of apoptosis in neurons are diverse including the withdrawal of trophic factors, and signaling through the transmembrane death receptors such as Fas/TNFR family as well as intrinsic signaling pathways that sense overload of ROS, Ca^{2+}, and H^+ (for review see ref. 290). In vitro, neuronal cell death can be induced by a variety of treatments: axotomy, excitotoxins, trophic deprivation, etc. These diverse stimuli have demonstrated that neurons indeed contain numerous apoptotic signaling pathways. Excitotoxicity has been achieved in cultures of primary neurons using either a direct insult such as activation of glutamate receptors, or with a secondary insult such as with Aβ (amyloid peptide), hypoxia, etc. These various paradigms which ultimately over-stimulate glutamate receptors all result in the activation of caspases. Upstream of the caspases however, excitotoxic insults can simultaneously activate more than one pathway.

As discussed in earlier sections, over-stimulation of NMDA receptors results in an influx of calcium. Calcium at high (> 500 nM) concentrations inappropriately activates many enzymes, is accumulated within the nucleus, endoplasmic reticulum and mitochondria, and as a general disturbance to the cell s ionic homeostasis causes a large energy drain as the cell tries to extrude the excess calcium (reviewed in ref. 316). All of these occurrences can and do trigger the neuron to apoptosis. From an extensive and rapidly unfolding literature, the following points demonstrate some of the pathways that are concurrently activated by excitotoxic entry of calcium.

Isolated brain mitochondria when exposed to high calcium release cytochrome c.[317] Accordingly, prolonged activation of NMDA receptors in cultured neurons and subsequent calcium loading results in release of cytochrome c from mitochondria followed by the activation of caspase-3.[260,318] Thus in excitotoxicity, cytochrome c release from mitochondria, in the presence of the other members of the apoptosome, is a direct signal to apoptosis. Apart from the mitochondrion as a key player in the apoptosis of neurons, the endoplasmic reticulum (ER) seems to participate in apoptotic signaling. The ER regulates cellular responses to stress and intracellular calcium levels. Treatment of cultured neurons with glutamate or thapsigargin induces the expression of caspase-12 on the ER outer membrane.[319,320] This results in cleavage and activation of cytosolic caspase-3.[319]

As discussed in an earlier section, glutamate receptor activation also activates the Ca^{2+}-dependent cysteine protease calpain.[174-177] Calpain inhibitors appear to be at least as effective as caspase inhibitors in preventing neuronal death and DNA fragmentation,[293] and have been demonstrated to reduce neuronal cell death caused by hypoxia.[321]

Other concomitantly activated pathways include the p38 MAP kinase and c-jun N-terminal kinase or stress-activated kinase (JNK/SAPK) which promote cell death.[322] SAPK/JNK is activated by tyrosine kinase and elevated intracellular calcium, and in primary striatal cultures, treatment with glutamate causes both increased phosphorylation of its substrate c-jun and increased activity of JNK.[323] Glutamate has been shown to activate p38 MAP kinase in rat cerebellar granule neurons,[324] and inhibition of p38 MAP kinase has been demonstrated to protect in vitro retinal ganglion neurons from NMDA-induced apoptosis.[325]

DNA damage is a common trigger for apoptosis in many different tissues, and in the mature nervous system, p53 is essential for neuronal death in response to DNA damage, ischemia and excitotoxicity.[326] Lately, signaling pathways that are responsible for triggering p53-dependent neuronal apoptosis are starting to be elucidated, and involve cell cycle deregulation and also activation of the JNK pathway.

CONCLUSIONS

A large body of evidence has accumulated showing that excessive activation of EAA receptors by the primary excitatory amino acid L-glutamate under pathophysiological conditions or intoxication with exogenous agonists results in excitotoxic cell death. Studies on the ion currents occurring under conditions of excitotoxicity have revealed that imbalanced Na^+ and Cl^- gradients as well as Ca^{2+} overloading are implicated in cell death. Calcium ions (and in some cases zinc ions) seem to play a pivotal role in EAA neurotoxicity, since this ubiquitous intracellular messenger can activate a multitude of enzymatic and nonenzymatic signaling and execution pathways implicated in neuronal cell death. There is compelling evidence that the mode of cell death that neurons undergo in excitotoxicity varies considerably along an apoptosis-necrosis continuum, depending on the severity of the insult. To date, numerous mechanisms involved in necrosis as well as various signaling and execution pathways that are implicated in apoptosis have been shown convincingly. However, while the role of signaling molecules in neuronal apoptosis has been studied extensively in in vitro models, the in vivo data is sparse for many of the pathways.

REFERENCES

1. Curtis DR, Watkins JC. The excitation and depression of spinal neurons by structurally related amino acids. J Neurochem 1960; 6117-141.
2. Krnjevic K, Phillis JW. Actions of certain amines on cerebral cortical neurones. Br J Pharmacol 1963; 20471-490.

3. Lucas DR, Newhouse JP. The toxic effect of sodium L-glutamate on the inner layers of the retina. Arch Ophtalmol 1957; 58193-201.
4. Olney JW, Sharpe LG. Brain lesions in an infant rhesus monkey treated with monsodium glutamate. Science 1969; 166(903):386-388.
5. Olney JW, Ho OL. Brain damage in infant mice following oral intake of glutamate, aspartate or cysteine. Nature 1970; 227(258):609-611.
6. Olney JW. Glutamate-induced neuronal necrosis in the infant mouse hypothalamus. An electron microscopic study. J Neuropathol Exp Neurol 1971; 30(1):75-90.
7. Olney JW, Sharpe LG, Feigin RD. Glutamate-induced brain damage in infant primates. J Neuropathol Exp Neurol 1972; 31(3):464-488.
8. Olney JW, Rhee V, Gubareff TD. Neurotoxic effects of glutamate on mouse area postrema. Brain Res 1977; 120(1):151-157.
9. Olney JW. Neurotoxicity of excitatory amino acids. In: McGeer EG, Olney JW, McGeer PL, eds. Kainic acid as a Tool in Neurobiology. New York: Raven, 1978:95-121.
10. Olney JW, Misra CH, de Gubareff T. Cysteine-S-sulfate: Brain damaging metabolite in sulfite oxidase deficiency. J Neuropathol Exp Neurol 1975; 34(2):167-177.
11. Coyle JT, Molliver ME, Kuhar MJ. In situ injection of kainic acid: A new method for selectively lesioning neural cell bodies while sparing axons of passage. J Comp Neurol 1978; 180(2):301-323.
12. Nadler JV, Perry BW, Cotman CW. Intraventricular kainic acid preferentially destroys hippocampal pyramidal cells. Nature 1978; 271(5646):676-677.
13. Schwarcz R, Coyle JT. Neurochemical sequelae of kainate injections in corpus striatum and substantia nigra of the rat. Life Sci 1977; 20(3):431-436.
14. Schwarcz R, Coyle JT. Kainic acid: Neurotoxic effects after intraocular injection. Invest Ophthalmol Vis Sci 1977; 16(2):141-148.
15. Watkins JC, Krogsgaard LP, Honore T. Structure-activity relationships in the development of excitatory amino acid receptor agonists and competitive antagonists. Trends Pharmacol Sci 1990; 11(1):25-33.
16. Greenamyre JT. The role of glutamate in neurotransmission and in neurologic disease. Arch Neurol 1986; 43(10):1058-1063.
17. Olney JW, Price MT, Fuller TA et al. The anti-excitotoxic effects of certain anesthetics, analgesics and sedative-hypnotics. Neurosci Lett 1986; 68(1):29-34.
18. Schwarcz R, Meldrum B. Excitatory aminoacid antagonists provide a therapeutic approach to neurological disorders. Lancet 1985; 2(8447):140-143.
19. Choi DW, Koh JY, Peters S. Pharmacology of glutamate neurotoxicity in cortical cell culture: attenuation by NMDA antagonists. J Neurosci 1988; 8(1):185-196.
20. Koh JY, Goldberg MP, Hartley DM et al. Non-NMDA receptor-mediated neurotoxicity in cortical culture. J Neurosci 1990; 10693-705.
21. Michaels RL, Rothman SM. Glutamate neurotoxicity in vitro: Antagonist pharmacology and intracellular calcium concentrations. J Neurosci 1990; 10(1):283-292.
22. Frandsen A, Drejer J, Schousboe A. Direct evidence that excitotoxicity in cultured neurons is mediated via N-methyl-D-aspartate (NMDA) as well as non-NMDA receptors. J Neurochem 1989; 53(1):297-299.
23. Rothman SM, Thurston JH, Hauhart RE. Delayed neurotoxicity of excitatory amino acids in vitro. Neuroscience 1987; 22(2):471-480.
24. Garthwaite G, Garthwaite J. Amino acid neurotoxicity: intracellular sites of calcium accumulation associated with the onset of irreversible damage to rat cerebellar neurones in vitro. Neurosci Lett 1986; 71(1):53-58.
25. Garthwaite G, Hajos F, Garthwaite J. Ionic requirements for neurotoxic effects of excitatory amino acid analogues in rat cerebellar slices. Neuroscience 1986; 18(2):437-447.
26. Garthwaite G, Garthwaite J. Differential dependence on Ca^{2+} of N-methyl-D-aspartate and quisqualate neurotoxicity in young rat hippocampal slices. Neurosci Lett 1989; 97(3):316-322.

27. Choi DW. Glutamate neurotoxicity in cortical cell culture is calcium dependent. Neurosci Lett 1985; 58(3):293-297.
28. Choi DW. Ionic dependence of glutamate neurotoxicity. J Neurosci 1987; 7(2):369-379.
29. Hartley DM, Kurth MC, Bjerkness L et al. Glutamate receptor-induced $45Ca^{2+}$ accumulation in cortical cell culture correlates with subsequent neuronal degeneration. J Neurosci 1993; 13(5):1993-2000.
30. Randall RD, Thayer SA. Glutamate-induced calcium transient triggers delayed calcium overload and neurotoxicity in rat hippocampal neurons. J Neurosci 1992; 12(5):1882-1895.
31. Dawson VL, Dawson TM, London ED et al. Nitric oxide mediates glutamate neurotoxicity in primary cortical cultures. Proc Natl Acad Sci USA 1991; 88(14):6368-6371.
32. Brorson JR, Manzolillo PA, Miller RJ. Ca^{2+} entry via AMPA/KA receptors and excitotoxicity in cultured cerebellar Purkinje cells. J Neurosci 1994; 14(1):187-197.
33. Volterra A, Trotti D, Cassutti P et al. High sensitivity of glutamate uptake to extracellular free arachidonic acid levels in rat cortical synaptosomes and astrocytes. J Neurochem 1992; 59(2):600-606.
34. Olney JW. Inciting excitotoxic cytocide among central neurons. Adv Exp Med Biol 1986; 203:631-645.
35. Ingvar M, Morgan PF, Auer RN. The nature and timing of excitotoxic neuronal necrosis in the cerebral cortex, hippocampus and thalamus due to flurothyl-induced status epilepticus. Acta Neuropathol Berl 1988; 75(4):362-369.
36. Olney JW. Glutamate-induced retinal degeneration in neonatal mice. Electron microscopy of the acutely evolving lesion. J Neuropathol Exp Neurol 1969; 28(3):455-474.
37. Siesjo BK. Pathophysiology and treatment of focal cerebral ischemia. Part I: Pathophysiology. J Neurosurg 1992; 77(2):169-184.
38. Choi DW, Rothman SM. The role of glutamate neurotoxicity in hypoxic-ischemic neuronal death. Annu Rev Neurosci 1990; 13171-182.
39. Hahn JS, Aizenman E, Lipton SA. Central mammalian neurons normally resistant to glutamate toxicity are made sensitive by elevated extracellular Ca^{2+}: Toxicity is blocked by the N-methyl-D-aspartate antagonist MK-801. Proc Natl Acad Sci USA 1988; 85(17):6556-6560.
40. Ankarcrona M, Dypbukt JM, Bonfoco E et al. Glutamate-induced neuronal death: A succession of necrosis or apoptosis depending on mitochondrial function. Neuron 1995; 15(4):961-973.
41. Choi DW. Ischemia-induced neuronal apoptosis. Curr Opin Neurobiol 1996; 6(5):667-672.
42. Gottron FJ, Ying HS, Choi DW. Caspase inhibition selectively reduces the apoptotic component of oxygen-glucose deprivation-induced cortical neuronal cell death. Mol Cell Neurosci 1997; 9(3):159-169.
43. Bonfoco E, Krainc D, Ankarcrona M et al. Apoptosis and necrosis: Two distinct events induced, respectively, by mild and intense insults with N-methyl-D-aspartate or nitric oxide/ superoxide in cortical cell cultures. Proc Natl Acad Sci U S A 1995; 92(16):7162-7166.
44. Kure S, Tominaga T, Yoshimoto T et al. Glutamate triggers internucleosomal DNA cleavage in neuronal cells. Biochem Biophys Res Commun 1991; 179(1):39-45.
45. Choi DW. Glutamate neurotoxicity and diseases of the nervous system. Neuron 1988; 1(8):623-634.
46. Margerison JH, Corsellis JA. Epilepsy and the temporal lobes. A clinical, electroencephalographic and neuropathological study of the brain in epilepsy, with particular reference to the temporal lobes. Brain 1966; 89(3):499-530.
47. Sagar HJ, Oxbury JM. Hippocampal neuron loss in temporal lobe epilepsy: correlation with early childhood convulsions. Ann Neurol 1987; 22(3):334-340.
48. Corsellis JA, Bruton CJ. Neuropathology of status epilepticus in humans. Adv Neurol 1983; 34:129-139.
49. Hauser WA. Status epilepticus: Frequency, etiology, and neurological sequelae. Adv Neurol 1983; 343-14.

50. DeGiorgio CM, Tomiyasu U, Gott PS et al. Hippocampal pyramidal cell loss in human status epilepticus. Epilepsia 1992; 33(1):23-27.
51. Olney JW, deGubareff T, Sloviter RS. Epileptic brain damage in rats induced by sustained electrical stimulation of the perforant path. II. Ultrastructural analysis of acute hippocampal pathology. Brain Res Bull 1983; 10(5):699-712.
52. Turski WA, Cavalheiro EA, Bortolotto ZA et al. Seizures produced by pilocarpine in mice: A behavioral, electroencephalographic and morphological analysis. Brain Res 1984; 321(2):237-253.
53. Clifford DB, Olney JW, Maniotis A et al. The functional anatomy and pathology of lithium-pilocarpine and high-dose pilocarpine seizures. Neuroscience 1987; 23(3):953-968.
54. Ben-Ari Y. Limbic seizure and brain damage produced by kainic acid: mechanisms and relevance to human temporal lobe epilepsy. Neuroscience 1985; 14(2):375-403.
55. Meldrum BS. Cell damage in epilepsy and the role of calcium in cytotoxicity. Adv Neurol 1986; 44849-855.
56. Olney JW, Collins RC, Sloviter RS. Excitotoxic mechanisms of epileptic brain damage. In: Delgado-Escueta AV, Ward AA, eds. Basic Mechanisms of the Epilepsies. New York: Raven Press, 1986:857-877.
57. Clifford DB, Olney JW, Benz AM et al. Ketamine, phencyclidine, and MK-801 protect against kainic acid-induced seizure-related brain damage. Epilepsia 1990; 31(4):382-390.
58. Clifford DB, Zorumski CF, Olney JW. Ketamine and MK-801 prevent degeneration of thalamic neurons induced by focal cortical seizures. Exp Neurol 1989; 105(3):272-279.
59. Fariello RG, Golden GT, Smith GG et al. Potentiation of kainic acid epileptogenicity and sparing from neuronal damage by an NMDA receptor antagonist. Epilepsy Res 1989; 3(3):206-213.
60. Sparenborg S, Brennecke LH, Jaax NK et al. Dizocilpine (MK-801) arrests status epilepticus and prevents brain damage induced by soman [published erratum appears in Neuropharmacology 1993 Mar;32(3):313]. Neuropharmacology 1992; 31(4):357-368.
61. Becker AJ, Gillardon F, Blumcke I et al. Differential regulation of apoptosis-related genes in resistant and vulnerable subfields of the rat epileptic hippocampus. Brain Res Mol Brain Res 1999; 67(1):172-176.
62. Charriaut MC, Aggoun ZD, Represa A et al. Apoptotic features of selective neuronal death in ischemia, epilepsy and gp 120 toxicity. Trends Neurosci 1996; 19(3):109-114.
63. Gillardon F, Bottiger B, Schmitz B et al. Activation of CPP-32 protease in hippocampal neurons following ischemia and epilepsy. Brain Res Mol Brain Res 1997; 50(1-2):16-22.
64. Henshall DC, Chen J, Simon RP. Involvement of caspase-3-like protease in the mechanism of cell death following focally evoked limbic seizures. J Neurochem 2000; 74(3):1215-1223.
65. Henshall DC, Clark RS, Adelson PD et al. Alterations in bcl-2 and caspase gene family protein expression in human temporal lobe epilepsy. Neurology 2000; 55(2):250-257.
66. Pollard H, Charriaut MC, Cantagrel S et al. Kainate-induced apoptotic cell death in hippocampal neurons. Neuroscience 1994; 63(1):7-18.
67. Faden AI, Demediuk P, Panter SS et al. The role of excitatory amino acids and NMDA receptors in traumatic brain injury. Science 1989; 244(4906):798-800.
68. Faden AI, Simon RP. A potential role for excitotoxins in the pathophysiology of spinal cord injury. Ann Neurol 1988; 23(6):623-626.
69. Bittigau P, Pohl D, Sifringer M et al. Modeling pediatric head trauma: Mechanisms of degeneration and potential strategies for neuroprotection. Restor Neurol Neurosci 1998; 13(1-2):11-23.
70. Shah PT, Yoon KW, Xu XM et al. Apoptosis mediates cell death following traumatic injury in rat hippocampal neurons. Neuroscience 1997; 79(4):999-1004.
71. Yakovlev AG, Knoblach SM, Fan L et al. Activation of CPP32-like caspases contributes to neuronal apoptosis and neurological dysfunction after traumatic brain injury. J Neurosci 1997; 17(19):7415-7424.

72. Pohl D, Bittigau P, Ishimaru MJ et al. N-Methyl-D-aspartate antagonists and apoptotic cell death triggered by head trauma in developing rat brain. Proc Natl Acad Sci U S A 1999; 96(5):2508-2513.
73. Hansen AJ. Effect of anoxia on ion distribution in the brain. Physiol Rev 1985; 65(1):101-148.
74. Hirsch JA, Gibson GE. Selective alteration of neurotransmitter release by low oxygen in vitro. Neurochem Res 1984; 9(8):1039-1049.
75. Nicholls D, Attwell D. The release and uptake of excitatory amino acids [see comments]. Trends Pharmacol Sci 1990; 11(11):462-468.
76. Szatkowski M, Attwell D. Triggering and execution of neuronal death in brain ischaemia: two phases of glutamate release by different mechanisms. Trends Neurosci 1994; 17(9):359-365.
77. Benveniste H, Drejer J, Schousboe A et al. Elevation of the extracellular concentrations of glutamate and aspartate in rat hippocampus during transient cerebral ischemia monitored by intracerebral microdialysis. J Neurochem 1984; 43(5):1369-1374.
78. Drejer J, Benveniste H, Diemer NH et al. Cellular origin of ischemia-induced glutamate release from brain tissue in vivo and in vitro. J Neurochem 1985; 45(1):145-151.
79. Obrenovitch TP, Richards DA. Extracellular neurotransmitter changes in cerebral ischaemia. Cerebrovasc Brain Metab Rev 1995; 7(1):1-54.
80. Choi DW. Calcium-mediated neurotoxicity: relationship to specific channel types and role in ischemic damage. Trends Neurosci 1988; 11(10):465-469.
81. Rothman SM, Olney JW. Glutamate and the pathophysiology of hypoxic Ischemic brain damage. Ann Neurol 1986; 19(2):105-111.
82. Rothman S. Synaptic release of excitatory amino acid neurotransmitter mediates anoxic neuronal death. J Neurosci 1984; 4(7):1884-1891.
83. Lee JM, Zipfel GJ, Choi DW. The changing landscape of ischaemic brain injury mechanisms. Nature 1999; 399(6738 Suppl):A7-14.
84. Albers GW, Goldberg MP, Choi DW. Do NMDA antagonists prevent neuronal injury? Yes. Arch Neurol 1992; 49(4):418-420.
85. Simon RP, Swan JH, Griffiths T et al. Blockade of N-methyl-D-aspartate receptors may protect against ischemic damage in the brain. Science 1984; 226(4676):850-852.
86. Wieloch T. Hypoglycemia-induced neuronal damage prevented by an N-methyl-D-aspartate antagonist. Science 1985; 230(4726):681-683.
87. Benveniste H. The excitotoxin hypothesis in relation to cerebral ischemia. Cerebrovasc Brain Metab Rev 1991; 3(3):213-245.
88. Dirnagl U, Iadecola C, Moskowitz MA. Pathobiology of ischaemic stroke: An integrated view. Trends Neurosci 1999; 22(9):391-397.
89. Martin LJ, Al Abdulla NA, Brambrink AM et al. Neurodegeneration in excitotoxicity, global cerebral ischemia, and target deprivation: A perspective on the contributions of apoptosis and necrosis. Brain Res Bull 1998; 46(4):281-309.
90. Linnik MD, Zobrist RH, Hatfield MD. Evidence supporting a role for programmed cell death in focal cerebral ischemia in rats. Stroke 1993; 24(12):2002-2008.
91. MacManus JP, Buchan AM, Hill IE et al. Global ischemia can cause DNA fragmentation indicative of apoptosis in rat brain. Neurosci Lett 1993; 164(1-2):89-92.
92. Namura S, Zhu J, Fink K et al. Activation and cleavage of caspase-3 in apoptosis induced by experimental cerebral ischemia. J Neurosci 1998; 18(10):3659-3668.
93. Fink K, Zhu J, Namura S et al. Prolonged therapeutic window for ischemic brain damage caused by delayed caspase activation. J Cereb Blood Flow Metab 1998; 18(10):1071-1076.
94. Endres M, Namura S, Shimizu SM et al. Attenuation of delayed neuronal death after mild focal ischemia in mice by inhibition of the caspase family. J Cereb Blood Flow Metab 1998; 18(3):238-247.
95. Chen J, Nagayama T, Jin K et al. Induction of caspase-3-like protease may mediate delayed neuronal death in the hippocampus after transient cerebral ischemia. J Neurosci 1998; 18(13):4914-4928.

96. Kaul M, Garden GA, Lipton SA. Pathways to neuronal injury and apoptosis in HIV-associated dementia. Nature 2001; 410(6831):988-994.
97. Bezzi P, Domercq M, Brambilla L et al. CXCR4-activated astrocyte glutamate release via TNFalpha: amplification by microglia triggers neurotoxicity. Nat Neurosci 2001; 4(7):702-710.
98. Teitelbaum JS, Zatorre RJ, Carpenter S et al. Neurologic sequelae of domoic acid intoxication due to the ingestion of contaminated mussels [see comments]. N Engl J Med 1990; 322(25):1781-1787.
99. Biscoe TJ, Evans RH, Headley PM et al. Domoic and quisqualic acids as potent amino acid excitants of frog and rat spinal neurones. Nature 1975; 255(5504):166-167.
100. Biscoe TJ, Evans RH, Headley PM et al. Structure-activity relations of excitatory amino acids on frog and rat spinal neurones. Br J Pharmacol 1976; 58(3):373-382.
101. Tryphonas L, Iverson F. Neuropathology of excitatory neurotoxins: the domoic acid model. Toxicol Pathol 1990; 18(1 Pt 2):165-169.
102. Tryphonas L, Truelove J, Nera E et al. Acute neurotoxicity of domoic acid in the rat. Toxicol Pathol 1990; 18(1 Pt 1):1-9.
103. Tryphonas L, Truelove J, Iverson F. Acute parenteral neurotoxicity of domoic acid in cynomolgus monkeys (*M. fascicularis*) [published erratum appears in Toxicol Pathol 1990; 18(3):431]. Toxicol Pathol 1990; 18(2):297-303.
104. Rao SLN, Adiga PR, Sarma PS. Biochemistry 1964; 145218-5220.
105. Spencer PS, Roy DN, Ludolph A et al. Lathyrism: Evidence for role of the neuroexcitatory aminoacid BOAA. Lancet 1986; 2(8515):1066-1067.
106. Nunn PB, Seelig M, Zagoren JC et al. Stereospecific acute neuronotoxicity of uncommon plant amino acids linked to human motor-system diseases. Brain Res 1987; 410(2):375-379.
107. Ross SM, Seelig M, Spencer PS. Specific antagonism of excitotoxic action of uncommon amino acids assayed in organotypic mouse cortical cultures. Brain Res 1987; 425(1):120-127.
108. Ross SM, Spencer PS. Specific antagonism of behavioral action of uncommon amino acids linked to motor-system diseases. Synapse 1987; 1(3):248-253.
109. Bridges RJ, Stevens DR, Kahle JS et al. Structure-function studies on N-oxalyl-diamino-dicarboxylic acids and excitatory amino acid receptors: evidence that beta-L-ODAP is a selective non-NMDA agonist. J Neurosci 1989; 9(6):2073-2079.
110. Ascher P, Nowak L. Quisqualate- and kainate-activated channels in mouse central neurones in culture. J Physiol Lond 1988; 399227-245.
111. Mosbacher J, Schoepfer R, Monyer H et al. A molecular determinant for submillisecond desensitization in glutamate receptors. Science 1994; 266(5187):1059-1062.
112. Hestrin S, Nicoll RA, Perkel DJ et al. Analysis of excitatory synaptic action in pyramidal cells using whole-cell recording from rat hippocampal slices. J Physiol Lond 1990; 422203-225.
113. Murphy SN, Miller RJ. Regulation of Ca^{++} influx into striatal neurons by kainic acid. J Pharmacol Exp Ther 1989; 249(1):184-193.
114. Gilbertson TA, Scobey R, Wilson M. Permeation of calcium ions through non-NMDA glutamate channels in retinal bipolar cells. Science 1991; 251(5001):1613-1615.
115. Ogura A, Akita K, Kudo Y. Non-NMDA receptor mediates cytoplasmic Ca^{2+} elevation in cultured hippocampal neurones. Neurosci Res N Y 1990; 9(2):103-113.
116. Pruss RM, Akeson RL, Racke MM et al. Agonist-activated cobalt uptake identifies divalent cation-permeable kainate receptors on neurons and glial cells. Neuron 1991; 7(3):509-518.
117. Jonas P, Racca C, Sakmann B et al. Differences in Ca^{2+} permeability of AMPA-type glutamate receptor channels in neocortical neurons caused by differential GluR-B subunit expression. Neuron 1994; 12(6):1281-1289.
118. Hollmann M, Hartley M, Heinemann S. Ca^{2+} permeability of KA-AMPA Gated glutamate receptor channels depends on subunit composition. Science 1991; 252(5007):851-853.
119. Nakanishi N, Shneider NA, Axel R. A family of glutamate receptor genes: evidence for the formation of heteromultimeric receptors with distinct channel properties. Neuron 1990; 5(5):569-581.

120. Contractor A, Swanson G, Heinemann SF. Kainate receptors are involved in short- and long-term plasticity at mossy fiber synapses in the hippocampus. Neuron 2001; 29(1):209-216.
121. Contractor A, Swanson GT, Sailer A et al. Identification of the kainate receptor subunits underlying modulation of excitatory synaptic transmission in the CA3 region of the hippocampus. J Neurosci 2000; 20(22):8269-8278.
122. Ben Ari Y, Cossart R. Kainate, a double agent that generates seizures: Two decades of progress. Trends Neurosci 2000; 23(11):580-587.
123. Frerking M, Nicoll RA. Synaptic kainate receptors. Curr Opin Neurobiol 2000; 10(3):342-351.
124. MacDermott AB, Mayer ML, Westbrook GL et al. NMDA-receptor activation increases cytoplasmic calcium concentration in cultured spinal cord neurones [published erratum appears in Nature 1986 Jun 26-Jul 2;321(6073):888]. Nature 1986; 321(6069):519-522.
125. Mayer ML, MacDermott AB, Westbrook GL et al. Agonist- and voltage-gated calcium entry in cultured mouse spinal cord neurons under voltage clamp measured using arsenazo III. J Neurosci 1987; 7(10):3230-3244.
126. Mayer ML, Westbrook GL. Permeation and block of N-methyl-D-aspartic acid receptor channels by divalent cations in mouse cultured central neurones. J Physiol Lond 1987; 394501-527.
127. Ascher P, Nowak L. Electrophysiological studies of NMDA receptors. Trends Neurosci 1987; 10(7):284-288.
128. Mayer ML, Westbrook GL. The physiology of excitatory amino acids in the vertebrate central nervous system. Prog Neurobiol 1987; 28(3):197-276.
129. Mothet JP, Parent AT, Wolosker H et al. D-serine is an endogenous ligand for the glycine site of the N-methyl-D-aspartate receptor. Proc Natl Acad Sci USA 2000; 97(9):4926-4931.
130. Rothman SM, Olney JW. Excitotoxicity and the NMDA receptor. Trends Neurosci 1987; 10(7):299-302.
131. Lees KR. Cerestat and other NMDA antagonists in ischemic stroke. Neurology 1997; 49(5 Suppl 4):S66-S69.
132. Chen HS, Lipton SA. Mechanism of memantine block of NMDA-activated channels in rat retinal ganglion cells: Uncompetitive antagonism. J Physiol 1997; 499(Pt 1):46.
133. Chen HS, Pellegrini JW, Aggarwal SK et al. Open-channel block of N-methyl-D-aspartate (NMDA) responses by memantine: therapeutic advantage against NMDA receptor-mediated neurotoxicity. J Neurosci 1992; 12(11):4427-4436.
134. Chen HS, Wang YF, Rayudu PV et al. Neuroprotective concentrations of the N-methyl-D-aspartate open-channel blocker memantine are effective without cytoplasmic vacuolation following post-ischemic administration and do not block maze learning or long-term potentiation. Neuroscience 1998; 86(4):1121-1132.
135. Seif eN, Peruche B, Rossberg C et al. Neuroprotective effect of memantine demonstrated in vivo and in vitro. Eur J Pharmacol 1990; 185(1):19-24.
136. Bruno V, Copani A, Knopfel T et al. Activation of metabotropic glutamate receptors coupled to inositol phospholipid hydrolysis amplifies NMDA-induced neuronal degeneration in cultured cortical cells. Neuropharmacology 1995; 34(8):1089-1098.
137. Sharp AH, McPherson PS, Dawson TM et al. Differential immunohistochemical localization of inositol 1,4,5-trisphosphate- and ryanodine-sensitive Ca^{2+} release channels in rat brain. J Neurosci 1993; 13(7):3051-3063.
138. Berridge MJ. Inositol trisphosphate and calcium signalling. Nature 1993; 361(6410):315-325.
139. Henzi V, MacDermott AB. Characteristics and function of $Ca(^{2+})$- and inositol 1,4,5-trisphosphate-releasable stores of Ca^{2+} in neurons. Neuroscience 1992; 46(2):251-273.
140. Bruno V, Battaglia G, Kingston A et al. Neuroprotective activity of the potent and selective mGlu1a metabotropic glutamate receptor antagonist, (+)-2-methyl-4 carboxyphenylglycine (LY367385): Comparison with LY357366, a broader spectrum antagonist with equal affinity for mGlu1a and mGlu5 receptors. Neuropharmacology 1999; 38(2):199-207.

141. Kingston AE, O Neill MJ, Bond A et al. Neuroprotective actions of novel and potent ligands of group I and group II metabotropic glutamate receptors. Ann NY Acad Sci 1999; 890:438-449.
142. Strasser U, Lobner D, Behrens MM et al. Antagonists for group I mGluRs attenuate excitotoxic neuronal death in cortical cultures. Eur J Neurosci 1998; 10(9):2848-2855.
143. Sagara Y, Schubert D. The activation of metabotropic glutamate receptors protects nerve cells from oxidative stress. J Neurosci 1998; 18(17):6662-6671.
144. Mukhin A, Fan L, Faden AI. Activation of metabotropic glutamate receptor subtype mGluR1 contributes to post-traumatic neuronal injury. J Neurosci 1996; 16(19):6012-6020.
145. Ambrosini A, Bresciani L, Fracchia S et al. Metabotropic glutamate receptors negatively coupled to adenylate cyclase inhibit N-methyl-D-aspartate receptor activity and prevent neurotoxicity in mesencephalic neurons in vitro. Mol Pharmacol 1995; 47(5):1057-1064.
146. Battaglia G, Bruno V, Ngomba RT et al. Selective activation of group-II metabotropic glutamate receptors is protective against excitotoxic neuronal death. Eur J Pharmacol 1998; 356(2-3):271-274.
147. Bruno V, Battaglia G, Copani A et al. Activation of class II or III metabotropic glutamate receptors protects cultured cortical neurons against excitotoxic degeneration. Eur J Neurosci 1995; 7(9):1906-1913.
148. Bruno V, Copani A, Bonanno L et al. Activation of group III metabotropic glutamate receptors is neuroprotective in cortical cultures. Eur J Pharmacol 1996; 310(1):61-66.
149. Bruno V, Battaglia G, Ksiazek I et al. Selective activation of mGlu4 metabotropic glutamate receptors is protective against excitotoxic neuronal death. J Neurosci 2000; 20(17):6413-6420.
150. Churchwell KB, Wright SH, Emma F et al. NMDA receptor activation inhibits neuronal volume regulation after swelling induced by veratridine-stimulated Na^+ influx in rat cortical cultures. J Neurosci 1996; 16(23):7447-7457.
151. Inglefield JR, Schwartz-Bloom RD. Activation of excitatory amino acid receptors in the rat hippocampal slice increases intracellular Cl^- and cell volume. J Neurochem 1998; 71(4):1396-1404.
152. Goldberg MP, Choi DW. Combined oxygen and glucose deprivation in cortical cell culture: calcium-dependent and calcium-independent mechanisms of neuronal injury. J Neurosci 1993; 13(8):3510-3524.
153. Dessi F, Charriaut MC, Ben Ari Y. Glutamate-induced neuronal death in cerebellar culture is mediated by two distinct components: A sodium-chloride component and a calcium component. Brain Res 1994; 650(1):49-55.
154. Hartley DM, Choi DW. Delayed rescue of N-methyl-D-aspartate receptor-mediated neuronal injury in cortical culture. J Pharmacol Exp Ther 1989; 250(2):752-758.
155. Murphy TH, Malouf AT, Sastre A et al. Calcium-dependent glutamate cytotoxicity in a neuronal cell line. Brain Res 1988; 444(2):325-332.
156. Garthwaite G, Garthwaite J. Neurotoxicity of excitatory amino acid receptor agonists in rat cerebellar slices: dependence on calcium concentration. Neurosci Lett 1986; 66(2):193-198.
157. Tymianski M, Wallace MC, Spigelman I et al. Cell-permeant $Ca(^{2+})$ chelators reduce early excitotoxic and ischemic neuronal injury in vitro and in vivo. Neuron 1993; 11(2):221-235.
158. Malcolm CS, Ritchie L, Grieve A et al. A prototypic intracellular calcium antagonist, TMB-8, protects cultured cerebellar granule cells against the delayed, calcium-dependent component of glutamate neurotoxicity. J Neurochem 1996; 66(6):2350-2360.
159. Lu YM, Yin HZ, Chiang J et al. $Ca(^{2+})$-permeable AMPA/kainate and NMDA channels: High rate of Ca^{2+} influx underlies potent induction of injury. J Neurosci 1996; 16(17):5457-5465.
160. Eimerl S, Schramm M. The quantity of calcium that appears to induce neuronal death. J Neurochem 1994; 62(3):1223-1226.
161. Andreeva N, Khodorov B, Stelmashook E et al. Inhibition of Na^+/Ca^{2+} exchange enhances delayed neuronal death elicited by glutamate in cerebellar granule cell cultures. Brain Res 1991; 548(1-2):322-325.

162. Mattson MP, Guthrie PB, Kater SB. A role for Na^+-dependent Ca^{2+} extrusion in protection against neuronal excitotoxicity. FASEB J 1989; 3(13):2519-2526.
163. White RJ, Reynolds IJ. Mitochondria and Na^+/Ca^{2+} exchange buffer glutamate-induced calcium loads in cultured cortical neurons. J Neurosci 1995; 15(2):1318-1328.
164. Carafoli E. Calcium pump of the plasma membrane. Physiol Rev 1991; 71(1):129-153.
165. Gunter TE, Gunter KK, Sheu SS et al. Mitochondrial calcium transport: Physiological and pathological relevance. Am J Physiol 1994; 267(2 Pt 1):C313-C339.
166. Pozzan T, Rizzuto R, Volpe P et al. Molecular and cellular physiology of intracellular calcium stores. Physiol Rev 1994; 74(3):595-636.
167. Dubinsky JM. Intracellular calcium levels during the period of delayed excitotoxicity. J Neurosci 1993; 13(2):623-631.
168. Tymianski M, Charlton MP, Carlen PL et al. Source specificity of early calcium neurotoxicity in cultured embryonic spinal neurons. J Neurosci 1993; 13(5):2085-2104.
169. Dumuis A, Sebben M, Haynes L et al. NMDA receptors activate the arachidonic acid cascade system in striatal neurons. Nature 1988; 336(6194):68-70.
170. Chan PH, Fishman RA. Transient formation of superoxide radicals in polyunsaturated fatty acid-induced brain swelling. J Neurochem 1980; 35(4):1004-1007.
171. Lazarewicz JW, Wroblewski JT, Palmer ME et al. Activation of N-methyl-D-aspartate-sensitive glutamate receptors stimulates arachidonic acid release in primary cultures of cerebellar granule cells. Neuropharmacology 1988; 27(7):765-769.
172. Williams JH, Errington ML, Lynch MA et al. Arachidonic acid induces a long-term activity-dependent enhancement of synaptic transmission in the hippocampus. Nature 1989; 341(6244):739-742.
173. Pellegrini-Giampietro DE, Cherici G, Alesiani M et al. Excitatory amino acid release from rat hippocampal slices as a consequence of free-radical formation. J Neurochem 1988; 51(6):1960-1963.
174. Lankiewicz S, Marc LC, Truc BN et al. Activation of calpain I converts excitotoxic neuron death into a caspase-independent cell death. J Biol Chem 2000; 275(22):17064-17071.
175. Lee KS, Frank S, Vanderklish P et al. Inhibition of proteolysis protects hippocampal neurons from ischemia. Proc Natl Acad Sci USA 1991; 88(16):7233-7237.
176. Siman R, Noszek JC, Kegerise C. Calpain I activation is specifically related to excitatory amino acid induction of hippocampal damage. J Neurosci 1989; 9(5):1579-1590.
177. Siman R, Noszek JC. Excitatory amino acids activate calpain I and induce structural protein breakdown in vivo. Neuron 1988; 1(4):279-287.
178. Garthwaite J, Garthwaite G, Palmer RM et al. NMDA receptor activation induces nitric oxide synthesis from arginine in rat brain slices. Eur J Pharmacol 1989; 172(4-5):413-416.
179. Gunasekar PG, Kanthasamy AG, Borowitz JL et al. NMDA receptor activation produces concurrent generation of nitric oxide and reactive oxygen species: implication for cell death. J Neurochem 1995; 65(5):2016-2021.
180. Dawson TM, Hung K, Dawson VL et al. Neuroprotective effects of gangliosides may involve inhibition of nitric oxide synthase. Ann Neurol 1995; 37(1):115-118.
181. Cazevieille C, Muller A, Meynier F et al. Superoxide and nitric oxide cooperation in hypoxia/reoxygenation-induced neuron injury. Free Radic Biol Med 1993; 14(4):389-395.
182. Dawson VL, Dawson TM, Bartley DA et al. Mechanisms of nitric oxide-mediated neurotoxicity in primary brain cultures. J Neurosci 1993; 13(6):2651-2661.
183. Dawson VL, Kizushi VM, Huang PL et al. Resistance to neurotoxicity in cortical cultures from neuronal nitric oxide synthase-deficient mice. J Neurosci 1996; 16(8):2479-2487.
184. Lipton SA, Choi YB, Pan ZH et al. A redox-based mechanism for the neuroprotective and neurodestructive effects of nitric oxide and related nitroso-compounds [see comments]. Nature 1993; 364(6438):626-632.
185. Dawson TM, Steiner JP, Dawson VL et al. Immunosuppressant FK506 enhances phosphorylation of nitric oxide synthase and protects against glutamate neurotoxicity [see comments]. Proc Natl Acad Sci USA 1993; 90(21):9808-9812.

186. Ankarcrona M, Dypbukt JM, Orrenius S et al. Calcineurin and mitochondrial function in glutamate-induced neuronal cell death. FEBS Lett 1996; 394(3):321-324.
187. Marcaida G, Kosenko E, Minana MD et al. Glutamate induces a calcineurin-mediated dephosphorylation of Na^+,$K(^+)$-ATPase that results in its activation in cerebellar neurons in culture. J Neurochem 1996; 66(1):99-104.
188. Marcaida G, Minana MD, Grisolia S et al. Lack of correlation between glutamate-induced depletion of ATP and neuronal death in primary cultures of cerebellum. Brain Res 1995; 695(2):146-150.
189. Blaustein MP. Calcium transport and buffering in neurons. Trends Neurosci 1988; 11(10):438-443.
190. Berridge MJ. Elementary and global aspects of calcium signalling. J Physiol Lond 1997; 499(Pt 2):291-306.
191. Peng TI, Greenamyre JT. Privileged access to mitochondria of calcium influx through N-methyl-D-aspartate receptors. Mol Pharmacol 1998; 53(6):974-980.
192. Peng TI, Jou MJ, Sheu SS et al. Visualization of NMDA receptor-induced mitochondrial calcium accumulation in striatal neurons. Exp Neurol 1998; 149(1):1-12.
193. Scarpa A, Azzone GF. The mechanism of ion translocation in mitochondria. 4. Coupling of K^+ efflux with Ca^{2+} uptake. Eur J Biochem 1970; 12(2):328-335.
194. Nicholls DG, Scott ID. The regulation of brain mitochondrial calcium-ion transport. The role of ATP in the discrimination between kinetic and membrane-potential-dependent calcium-ion efflux mechanisms. Biochem J 1980; 186(3):833-839.
195. Nicholls DG. The regulation of extramitochondrial free calcium ion concentration by rat liver mitochondria. Biochem J 1978; 176(2):463-474.
196. Werth JL, Thayer SA. Mitochondria buffer physiological calcium loads in cultured rat dorsal root ganglion neurons. J Neurosci 1994; 14(1):348-356.
197. Budd SL, Nicholls DG. A reevaluation of the role of mitochondria in neuronal Ca^{2+} homeostasis. J Neurochem 1996; 66(1):403-411.
198. Wang GJ, Thayer SA. Sequestration of glutamate-induced Ca^{2+} loads by mitochondria in cultured rat hippocampal neurons. J Neurophysiol 1996; 76(3):1611-1621.
199. Kiedrowski L, Costa E. Glutamate-induced destabilization of intracellular calcium concentration homeostasis in cultured cerebellar granule cells: role of mitochondria in calcium buffering. Mol Pharmacol 1995; 47(1):140-147.
200. Akerman KE. Changes in membrane potential during calcium ion influx and efflux across the mitochondrial membrane. Biochim Biophys Acta 1978; 502(2):359-366.
201. Nicholls DG. Calcium transport and porton electrochemical potential gradient in mitochondria from guinea-pig cerebral cortex and rat heart. Biochem J 1978; 170(3):511-522.
202. Heaton GM, Nicholls DG. The calcium conductance of the inner membrane of rat liver mitochondria and the determination of the calcium electrochemical gradient. Biochem J 1976; 156(3):635-646.
203. White RJ, Reynolds IJ. Mitochondrial depolarization in glutamate-stimulated neurons: an early signal specific to excitotoxin exposure. J Neurosci 1996; 16(18):5688-5697.
204. Budd SL, Nicholls DG. Mitochondria, calcium regulation, and acute glutamate excitotoxicity in cultured cerebellar granule cells. J Neurochem 1996; 67(6):2282-2291.
205. Khodorov B, Pinelis V, Vergun O et al. Mitochondrial deenergization underlies neuronal calcium overload following a prolonged glutamate challenge. FEBS Lett 1996; 397(2-3):230-234.
206. Schinder AF, Olson EC, Spitzer NC et al. Mitochondrial dysfunction is a primary event in glutamate neurotoxicity. J Neurosci 1996; 16(19):6125-6133.
207. Vergun O, Keelan J, Khodorov BI et al. Glutamate-induced mitochondrial depolarisation and perturbation of calcium homeostasis in cultured rat hippocampal neurones. J Physiol Lond 1999; 519 Pt 2451-466.

208. Denton RM, Richards DA, Chin JG. Calcium ions and the regulation of NAD^+-linked isocitrate dehydrogenase from the mitochondria of rat heart and other tissues. Biochem J 1978; 176(3):899-906.
209. Hansford RG. Relation between cytosolic free Ca^{2+} concentration and the control of pyruvate dehydrogenase in isolated cardiac myocytes. Biochem J 1987; 2411-51.
210. Hansford RG, Zorov D. Role of mitochondrial calcium transport in the control of substrate oxidation. Mol Cell Biochem 1998; 1841-2.
211. McCormack JG, Denton RM. The effects of calcium ions and adenine nucleotides on the activity of pig heart 2-oxoglutarate dehydrogenase complex. Biochem J 1979; 180(3):533-544.
212. McCormack JG, England PJ. Ruthenium Red inhibits the activation of pyruvate dehydrogenase caused by positive inotropic agents in the perfused rat heart. Biochem J 1983; 2142-5.
213. McCormack JG, Denton RM. Mitochondrial Ca^{2+} transport and the role of intramitochondrial Ca^{2+} in the regulation of energy metabolism. Dev Neurosci 1993; 15(3-5):165-173.
214. Robb GL, Burnett P, Rutter GA et al. Integrating cytosolic calcium signals into mitochondrial metabolic responses. EMBO J 1998; 17(17):4987-5000.
215. Moreno SR. Contribution of the translocator of adenine nucleotides and the ATP synthase to the control of oxidative phosphorylation and arsenylation in liver mitochondria. J Biol Chem 1985; 260(23):12554-12560.
216. Skulachev VP. Role of uncoupled and non-coupled oxidations in maintenance of safely low levels of oxygen and its one-electron reductants. Q Rev Biophys 1996; 29(2):169-202.
217. Dykens JA. Mitochondrial free radical production and oxidative pathophysiology: Implications for neurodegenerative disease. In: Beal MF, Howell N, Bodis-Wollner J, eds. Mitochondria and Free radicals in Neurodegenerative Diseases. New-York: Wiley-Liss, 1997:29-55.
218. Dykens JA. Isolated cerebral and cerebellar mitochondria produce free radicals when exposed to elevated CA^{2+} and Na^+: implications for neurodegeneration. J Neurochem 1994; 63(2):584-591.
219. Kowaltowski AJ, Castilho RF, Grijalba MT et al. Effect of inorganic phosphate concentration on the nature of inner mitochondrial membrane alterations mediated by Ca^{2+} ions. A proposed model for phosphate-stimulated lipid peroxidation. J Biol Chem 1996; 271(6):2929-2934.
220. Tabatabaie T, Potts JD, Floyd RA. Reactive oxygen species-mediated inactivation of pyruvate dehydrogenase. Arch Biochem Biophys 1996:3362-3366.
221. Patel M, Day BJ, Crapo JD et al. Requirement for superoxide in excitotoxic cell death. Neuron 1996; 16(2):345-355.
222. Sheu KF, Blass JP. The alpha-ketoglutarate dehydrogenase complex. Ann NY Acad Sci 1999; 89361-78.
223. Zhang Y, Marcillat O, Giulivi C et al. The oxidative inactivation of mitochondrial electron transport chain components and ATPase. J Biol Chem 1990; 265(27):16330-16336.
224. Hillered L, Ernster L. Respiratory activity of isolated rat brain mitochondria following in vitro exposure to oxygen radicals. J Cereb Blood Flow Metab 1983; 3(2):207-214.
225. Malis CD, Bonventre JV. Mechanism of calcium potentiation of oxygen free radical injury to renal mitochondria. A model for post-ischemic and toxic mitochondrial damage. J Biol Chem 1986; 261(30):14201-14208.
226. Bindokas VP, Jordan J, Lee CC et al. Superoxide production in rat hippocampal neurons: selective imaging with hydroethidine. J Neurosci 1996; 16(4):1324-1336.
227. Budd SL, Castilho RF, Nicholls DG. Mitochondrial membrane potential and hydroethidine-monitored superoxide generation in cultured cerebellar granule cells. FEBS Lett 1997; 415(1):21-24.
228. Rothe G, Valet G. Flow cytometric analysis of respiratory burst activity in phagocytes with hydroethidine and 2',7'-dichlorofluorescin. J Leukoc Biol 1990; 47(5):440-448.
229. Lafon CM, Pietri S, Culcasi M et al. NMDA-dependent superoxide production and neurotoxicity. Nature 1993; 364(6437):535-537.

230. Reynolds IJ, Hastings TG. Glutamate induces the production of reactive oxygen species in cultured forebrain neurons following NMDA receptor activation. J Neurosci 1995; 15(5 Pt 1):3318-3327.
231. Dugan LL, Sensi SL, Canzoniero LM et al. Mitochondrial production of reactive oxygen species in cortical neurons following exposure to N-methyl-D-aspartate. J Neurosci 1995; 15(10):6377-6388.
232. Herrero A, Barja G. ADP-regulation of mitochondrial free radical production is different with complex I- or complex II-linked substrates: Implications for the exercise paradox and brain hypermetabolism. J Bioenerg Biomembr 1997; 29(3):241-249.
233. Turrens JF, Boveris A. Generation of superoxide anion by the NADH dehydrogenase of bovine heart mitochondria. Biochem J 1980; 191(2):421-427.
234. Turrens JF, Alexandre A, Lehninger AL. Ubisemiquinone is the electron donor for superoxide formation by complex III of heart mitochondria. Arch Biochem Biophys 1985; 237(2):408-414.
235. Demin OV, Kholodenko BN, Skulachev VP. A model of O2.-generation in the complex III of the electron transport chain. Mol Cell Biochem 1998; 184(1-2):21-33.
236. Boveris A, Cadenas E, Stoppani AO. Role of ubiquinone in the mitochondrial generation of hydrogen peroxide. Biochem J 1976; 156(2):435-444.
237. Prehn JH. Mitochondrial transmembrane potential and free radical production in excitotoxic neurodegeneration. Naunyn Schmiedebergs Arch Pharmacol 1998; 357(3):316-322.
238. Castilho RF, Ward MW, Nicholls DG. Oxidative stress, mitochondrial function, and acute glutamate excitotoxicity in cultured cerebellar granule cells. J Neurochem 1999; 72(4):1394-1401.
239. Lafon CM, Culcasi M, Gaven F et al. Nitric oxide, superoxide and peroxynitrite: putative mediators of NMDA-induced cell death in cerebellar granule cells. Neuropharmacology 1993; 32(11):1259-1266.
240. Bondy SC, Lee DK. Oxidative stress induced by glutamate receptor agonists. Brain Res 1993; 610(2):229-233.
241. Huschenbett J, Zaidi A, Michaelis ML. Sensitivity of the synaptic membrane Na^+/Ca^{2+} exchanger and the expressed NCX1 isoform to reactive oxygen species. Biochim Biophys Acta 1998; 1374(1-2):34-46.
242. Yan LJ, Levine RL, Sohal RS. Oxidative damage during aging targets mitochondrial aconitase [published erratum appears in Proc Natl Acad Sci U S A 1998 Feb 17;95(4):1968]. Proc Natl Acad Sci U S A 1997; 94(21):11168-11172.
243. Hoyal CR, Thomas AP, Forman HJ. Hydroperoxide-induced increases in intracellular calcium due to annexin VI translocation and inactivation of plasma membrane Ca^{2+}-ATPase. J Biol Chem 1996; 271(46):29205-29210.
244. Castilho RF, Hansson O, Ward MW et al. Mitochondrial control of acute glutamate excitotoxicity in cultured cerebellar granule cells. J Neurosci 1998; 18(24):10277-10286.
245. Haworth RA, Hunter DR. The Ca^{2+}-induced membrane transition in mitochondria. II. Nature of the Ca^{2+} trigger site. Arch Biochem Biophys 1979; 195(2):460-467.
246. Petronilli V, Szabo I, Zoratti M. The inner mitochondrial membrane contains ion-conducting channels similar to those found in bacteria. FEBS Lett 1989; 259(1):137-143.
247. Szabo I, Zoratti M. The giant channel of the inner mitochondrial membrane is inhibited by cyclosporin A. J Biol Chem 1991; 266(6):3376-3379.
248. Bernardi P, Vassanelli S, Veronese P et al. Modulation of the mitochondrial permeability transition pore. Effect of protons and divalent cations. J Biol Chem 1992; 267(5):2934-2939.
249. Szabo I, Bernardi P, Zoratti M. Modulation of the mitochondrial megachannel by divalent cations and protons. J Biol Chem 1992; 267(5):2940-2946.
250. Broekemeier KM, Dempsey ME, Pfeiffer DR. Cyclosporin A is a potent inhibitor of the inner membrane permeability transition in liver mitochondria. J Biol Chem 1989; 264(14):7826-7830.

251. Brustovetsky N, Klingenberg M. Mitochondrial ADP/ATP carrier can be reversibly converted into a large channel by Ca^{2+}. Biochemistry 1996; 35(26):8483-8488.
252. Chernyak BV, Bernardi P. The mitochondrial permeability transition pore is modulated by oxidative agents through both pyridine nucleotides and glutathione at two separate sites. Eur J Biochem 1996; 238(3):623-630.
253. Kowaltowski AJ, Castilho RF, Vercesi AE. Ca($^{2+}$)-induced mitochondrial membrane permeabilization: role of coenzyme Q redox state. Am J Physiol 1995; 269(1 Pt 1):C141-C147.
254. Kowaltowski AJ, Castilho RF, Vercesi AE. Opening of the mitochondrial permeability transition pore by uncoupling or inorganic phosphate in the presence of Ca^{2+} is dependent on mitochondrial-generated reactive oxygen species. FEBS Lett 1996; 378(2):150-152.
255. Kowaltowski AJ, Netto LE, Vercesi AE. The thiol-specific antioxidant enzyme prevents mitochondrial permeability transition. Evidence for the participation of reactive oxygen species in this mechanism. J Biol Chem 1998; 273(21):12766-12769.
256. Vercesi AE, Kowaltowski AJ, Grijalba MT et al. The role of reactive oxygen species in mitochondrial permeability transition. Biosci Rep 1997; 17(1):43-52.
257. Szabo I, Zoratti M. The mitochondrial megachannel is the permeability transition pore. J Bioenerg Biomembr 1992; 24(1):111-117.
258. Friche E, Jensen PB, Nissen NI. Comparison of cyclosporin A and SDZ PSC833 as multidrug-resistance modulators in a daunorubicin-resistant Ehrlich ascites tumor. Cancer Chemother Pharmacol 1992; 30(3):235-237.
259. Tiberghien F, Loor F. Ranking of P-glycoprotein substrates and inhibitors by a calcein-AM fluorometry screening assay. Anticancer Drugs 1996; 7(5):568-578.
260. Budd SL, Tenneti L, Lishnak T et al. Mitochondrial and extramitochondrial apoptotic signaling pathways in cerebrocortical neurons. Proc Natl Acad Sci USA 2000; 97(11):6161-6166.
261. Garthwaite J, Boulton CL. Nitric oxide signaling in the central nervous system. Annu Rev Physiol 1995; 57683-706-706.
262. Zhang J, Snyder SH. Nitric oxide in the nervous system. Annu Rev Pharmacol Toxicol 1995; 35213-33-33.
263. Beckman JS, Koppenol WH. Nitric oxide, superoxide, and peroxynitrite: The good, the bad, and ugly. Am J Physiol 1996; 271(5 Pt 1):C1424-C1437.
264. Schulz JB, Matthews RT, Jenkins BG et al. Blockade of neuronal nitric oxide synthase protects against excitotoxicity in vivo. J Neurosci 1995; 15(12):8419-8429.
265. Butler AR, Flitney FW, Williams DL. NO, nitrosonium ions, nitroxide ions, nitrosothiols and iron-nitrosyls in biology: A chemist s perspective. Trends Pharmacol Sci 1995; 16(1):18-22.
266. Beckman JS, Beckman TW, Chen J et al. Apparent hydroxyl radical production by peroxynitrite: Implications for endothelial injury from nitric oxide and superoxide. Proc Natl Acad Sci USA 1990; 87(4):1620-1624.
267. Radi R, Beckman JS, Bush KM et al. Peroxynitrite oxidation of sulfhydryls. The cytotoxic potential of superoxide and nitric oxide. J Biol Chem 1991; 266(7):4244-4250.
268. Radi R, Beckman JS, Bush KM et al. Peroxynitrite-induced membrane lipid peroxidation: the cytotoxic potential of superoxide and nitric oxide. Arch Biochem Biophys 1991; 288(2):481-487.
269. Leist M, Fava E, Montecucco C et al. Peroxynitrite and nitric oxide donors induce neuronal apoptosis by eliciting autocrine excitotoxicity. Eur J Neurosci 1997; 9(7):1488-1498.
270. Zhang J, Dawson VL, Dawson TM et al. Nitric oxide activation of poly(ADP-ribose) synthetase in neurotoxicity. Science 1994; 263(5147):687-689.
271. Bolanos JP, Almeida A, Stewart V et al. Nitric oxide-mediated mitochondrial damage in the brain: mechanisms and implications for neurodegenerative diseases. J Neurochem 1997; 68(6):2227-2240.
272. Bolanos JP, Heales SJ, Land JM et al. Effect of peroxynitrite on the mitochondrial respiratory chain: differential susceptibility of neurones and astrocytes in primary culture. J Neurochem 1995; 64(5):1965-1972.

273. Brorson JR, Schumacker PT, Zhang H. Nitric oxide acutely inhibits neuronal energy production. J Neurosci 1999; 19(1):147-158.
274. Ghatan S, Larner S, Kinoshita Y et al. p38 MAP kinase mediates Bax translocation in nitric oxide-induced apoptosis in neurons. J Cell Biol 2000; 150(2):335-347.
275. Li Y, Chopp M, Jiang N et al. Induction of DNA fragmentation after 10 to 120 minutes of focal cerebral ischemia in rats. Stroke 1995; 26(7):1252-1257.
276. Linnik MD, Miller JA, Sprinkle-Cavallo J et al. Apoptotic DNA fragmentation in the rat cerebral cortex induced by permanent middle cerebral artery occlusion. Brain Res Mol Brain Res 1995; 32(1):116-124.
277. Hara H, Friedlander RM, Gagliardini V et al. Inhibition of interleukin 1beta converting enzyme family proteases reduces ischemic and excitotoxic neuronal damage. Proc Natl Acad Sci USA 1997; 94(5):2007-2012.
278. Lookeren-Campagne M, Gill R. Ultrastructural morphological changes are not characteristic of apoptotic cell death following focal cerebral ischaemia in the rat. Neurosci Lett 1996; 213(2):111-114.
279. Mochizuki H, Goto K, Mori H et al. Histochemical detection of apoptosis in Parkinson s disease. J Neurol Sci 1996; 137(2):120-123.
280. Tompkins MM, Basgall EJ, Zamrini E et al. Apoptotic-like changes in Lewy-body-associated disorders and normal aging in substantia nigral neurons. Am J Pathol 1997; 150(1):119-131.
281. Su JH, Anderson AJ, Cummings BJ et al. Immunohistochemical evidence for apoptosis in Alzheimer s disease. Neuroreport 1994; 5(18):2529-2533.
282. Dragunow M, Faull RL, Lawlor P et al. In situ evidence for DNA fragmentation in Huntington s disease striatum and Alzheimer s disease temporal lobes. Neuroreport 1995; 6(7):1053-1057.
283. Portera-Cailliau C, Hedreen JC, Price DL et al. Evidence for apoptotic cell death in Huntington disease and excitotoxic animal models. J Neurosci 1995; 15(5 Pt 2):3775-3787.
284. Martin LJ. Neuronal death in amyotrophic lateral sclerosis is apoptosis: possible contribution of a programmed cell death mechanism. J Neuropathol Exp Neurol 1999; 58(5):459-471.
285. Grasl-Kraupp B, Ruttkay-Nedecky B, Koudelka H et al. In situ detection of fragmented DNA (TUNEL assay) fails to discriminate among apoptosis, necrosis, and autolytic cell death: a cautionary note. Hepatology 1995; 21(5):1465-1468.
286. Kingsbury AE, Mardsen CD, Foster OJ. DNA fragmentation in human substantia nigra: apoptosis or perimortem effect? Mov Disord 1998; 13(6):877-884.
287. Uetsuki T, Takemoto K, Nishimura I et al. Activation of neuronal caspase-3 by intracellular accumulation of wild-type Alzheimer amyloid precursor protein. J Neurosci 1999; 19(16):6955-6964.
288. Selznick LA, Holtzman DM, Han BH et al. In situ immunodetection of neuronal caspase-3 activation in Alzheimer disease. J Neuropathol Exp Neurol 1999; 58(9):1020-1026.
289. Drache B, Diehl GE, Beyreuther K et al. Bcl-xl-specific antibody labels activated microglia associated with Alzheimer s disease and other pathological states. J Neurosci Res 1997; 47(1):98-108.
290. Pettmann B, Henderson CE. Neuronal cell death. Neuron 1998; 20(4):633-647.
291. Villa P, Kaufmann SH, Earnshaw WC. Caspases and caspase inhibitors. Trends Biochem Sci 1997; 22(10):388-393.
292. Oppenheim RW, Flavell RA, Vinsant S et al. Programmed cell death of developing mammalian neurons after genetic deletion of caspases. J Neurosci 2001; 21(13):4752-4760.
293. Villa PG, Henzel WJ, Sensenbrenner M et al. Calpain inhibitors, but not caspase inhibitors, prevent actin proteolysis and DNA fragmentation during apoptosis. J Cell Sci 1998; 111(Pt 6):22.
294. Kuida K, Zheng TS, Na S et al. Decreased apoptosis in the brain and premature lethality in CPP32-deficient mice. Nature 1996; 384(6607):368-372.
295. Kluck RM, Bossy-Wetzel E, Green DR et al. The release of cytochrome c from mitochondria: a primary site for Bcl-2 regulation of apoptosis. Science 1997; 275(5303):1132-1136.

296. Yang J, Liu X, Bhalla K et al. Prevention of apoptosis by Bcl-2: Release of cytochrome c from mitochondria blocked. Science 1997; 275(5303):1129-1132.
297. Susin SA, Zamzami N, Castedo M et al. Bcl-2 inhibits the mitochondrial release of an apoptogenic protease. J Exp Med 1996; 184(4):1331-1341.
298. Du C, Fang M, Li Y et al. Smac, a mitochondrial protein that promotes cytochrome c-dependent caspase activation by eliminating IAP inhibition. Cell 2000; 102(1):33-42.
299. Li P, Nijhawan D, Budihardjo I et al. Cytochrome c and dATP-dependent formation of Apaf-1/caspase-9 complex initiates an apoptotic protease cascade. Cell 1997; 91(4):479-489.
300. Bossy-Wetzel E, Green DR. Caspases induce cytochrome c release from mitochondria by activating cytosolic factors. J Biol Chem 1999; 274(25):17484-17490.
301. Shimizu S, Narita M, Tsujimoto Y. Bcl-2 family proteins regulate the release of apoptogenic cytochrome c by the mitochondrial channel VDAC [see comments]. Nature 1999; 399(6735):483-487.
302. Rosse T, Olivier R, Monney L et al. Bcl-2 prolongs cell survival after Bax-induced release of cytochrome c [see comments]. Nature 1998; 391(6666):496-499.
303. Deshmukh M, Johnson EM. Evidence of a novel event during neuronal death: development of competence-to-die in response to cytoplasmic cytochrome c. Neuron 1998; 21(4):695-705.
304. Zou H, Li Y, Liu X et al. An APAF-1.cytochrome c multimeric complex is a functional apoptosome that activates procaspase-9. J Biol Chem 1999; 274(17):11549-11556.
305. Marzo I, Brenner C, Zamzami N et al. Bax and adenine nucleotide translocator cooperate in the mitochondrial control of apoptosis. Science 1998; 2812027-2031.
306. Zamzami N, Susin SA, Marchetti P et al. Mitochondrial control of nuclear apoptosis. J Exp Med 1996; 183(4):1533-1544.
307. Vander-Heiden MG, Li XX, Gottleib E et al. Bcl-x_L promotes the open configuration of the voltage-dependent anion channel and metabolite passage through the outer mitochondrial membrane. J Biol Chem 2001; 276(22):19414-19419.
308. Rego AC, Vesce S, Nicholls DG. The mechanism of mitochondrial membrane potential retention following release of cytochrome c in apoptotic GT1-7 neural cells. Cell Death Differ 2001; 8(10):995-1003.
309. Hu Y, Benedict MA, Wu D et al. Bcl-x_L interacts with Apaf-1 and inhibits Apaf-1-dependent caspase-9 activation. Proc Natl Acad Sci USA 1998; 95(8):4386-4391.
310. Fujimura M, Morita FY, Murakami K et al. Cytosolic redistribution of cytochrome c after transient focal cerebral ischemia in rats. J Cereb Blood Flow Metab 1998; 18(11):1239-1247.
311. Ouyang YB, Tan Y, Comb M et al. Survival- and death-promoting events after transient cerebral ischemia: phosphorylation of Akt, release of cytochrome c and Activation of caspase-like proteases. J Cereb Blood Flow Metab 1999; 19(10):1126-1135.
312. Matsuyama T, Hata R, Yamamoto Y et al. Localization of Fas antigen mRNA induced in postischemic murine forebrain by in situ hybridization. Brain Res Mol Brain Res 1995; 34(1):166-172.
313. Cao G, Minami M, Pei W et al. Intracellular Bax translocation after transient cerebral ischemia: implications for a role of the mitochondrial apoptotic signaling pathway in ischemic neuronal death. J Cereb Blood Flow Metab 2001; 21(4):321-333.
314. Hayashi T, Sakai K, Sasaki C et al. c-Jun N-terminal kinase (JNK) and JNK interacting protein response in rat brain after transient middle cerebral artery occlusion. Neurosci Lett 2000; 284(3):195-199.
315. Krajewski S, Krajewska M, Ellerby LM et al. Release of caspase-9 from mitochondria during neuronal apoptosis and cerebral ischemia. Proc Natl Acad Sci USA 1999; 96(10):5752-5757.
316. Budd SL. Mechanisms of neuronal damage in brain hypoxia/ischemia: focus on the role of mitochondrial calcium accumulation. PHARMACOL THER 1998; 80(2):203-229.
317. Andreyev A, Fiskum G. Calcium induced release of mitochondrial cytochrome c by different mechanisms selective for brain versus liver. Cell Death Differ 1999; 6(9):825-832.

318. Luetjens CM, Bui NT, Sengpiel B et al. Delayed mitochondrial dysfunction in excitotoxic neuron death: cytochrome c release and a secondary increase in superoxide production. J Neurosci 2000; 20(15):5715-5723.
319. Nakagawa T, Zhu H, Morishima N et al. Caspase-12 mediates endoplasmic-reticulum-specific apoptosis and cytotoxicity by amyloid-beta. Nature 2000; 403(6765):98-103.
320. Rao RV, Hermel E, Castro-Obregon S et al. Coupling endoplasmic reticulum stress to the cell death program. Mechanism of caspase activation. J Biol Chem 2001; 276(36):33869-33874.
321. Wang KK, Nath R, Posner A et al. An alpha-mercaptoacrylic acid derivative is a selective nonpeptide cell-permeable calpain inhibitor and is neuroprotective. Proc Natl Acad Sci USA 1996; 93(13):6687-6692.
322. Harper SJ, LoGrasso P. Signalling for survival and death in neurones: the role of stress-activated kinases, JNK and p38. Cell Signal 2001; 13(5):299-310.
323. Schwarzschild MA, Cole RL, Hyman SE. Glutamate, but not dopamine, stimulates stress-activated protein kinase and AP-1-mediated transcription in striatal neurons. J Neurosci 1997; 17(10):3455-3466.
324. Kawasaki H, Morooka T, Shimohama S et al. Activation and involvement of p38 mitogen-activated protein kinase in glutamate-induced apoptosis in rat cerebellar granule cells. J Biol Chem 1997; 272(30):18518-18521.
325. Kikuchi M, Tenneti L, Lipton SA. Role of p38 mitogen-activated protein kinase in axotomy-induced apoptosis of rat retinal ganglion cells. J Neurosci 2000; 20(13):5037-5044.
326. Miller FD, Pozniak CD, Walsh GS. Neuronal life and death: An essential role for the p53 family. Cell Death Differ 2000; 7(10):880-888.

NEURONAL SURVIVAL AND CELL DEATH SIGNALING PATHWAYS

Richard S. Morrison[1], Yoshito Kinoshita[1], Mark D. Johnson[1], Saadi Ghatan[1], Joseph T. Ho[1] and Gwenn Garden[2]

ABSTRACT

Neuronal viability is maintained through a complex interacting network of signaling pathways that can be perturbed in response to a multitude of cellular stresses. A shift in the balance of signaling pathways after stress or in response to pathology can have drastic consequences for the function or the fate of a neuron. There is significant evidence that acutely injured and degenerating neurons may die by an active mechanism of cell death. This process involves the activation of discrete signaling pathways that ultimately compromise mitochondrial structure, energy metabolism and nuclear integrity. In this review we examine recent evidence pertaining to the presence and activation of anti- and pro-cell death regulatory pathways in nervous system injury and degeneration.

INTRODUCTION

Neuronal viability is maintained through a complex interacting network of signaling pathways that can be perturbed in response to a multitude of cellular stresses. A shift in one or more of these signaling pathways can alter the fate of a neuron resulting in cell death or continued survival. The nature of the stresses affecting neurons, the duration of the stresses, the developmental stage of the neuron and a variety of other factors influence the signaling pathways that are ultimately affected.

[1]Department of Neurological Surgery, University of Washington School of Medicine, Box 356470, Seattle, Washington 98195-6470 and [2]Department of Neurology, University of Washington School of Medicine, Seattle, Washington 98195.

These diverse parameters may also regulate the temporal response as well as the final disposition of the affected neurons.

Apoptosis is a mechanism of cell death that plays a fundamental role during the development of many tissues including the central nervous system. Apoptosis has traditionally been distinguished in developing tissues on the basis of specific morphological and biochemical criteria, including perinuclear chromatin condensation, cell shrinkage and endonuclease-mediated internucleosomal DNA fragmentation into a ladder pattern. More recently , the term apoptosis has been used to describe the programmed biochemical pathways of cell death that accompany development, tissue injury and degeneration. In the context of this review, apoptosis will refer to the biochemical pathways of cell death. There is accumulating evidence that acutely injured and degenerating neurons may die by a process of apoptosis, contributing to the loss of neurons observed under these conditions. The possibility that neurons may succumb under some circumstances by an active mechanism of cell death has raised interest in the regulatory pathways governing these processes.

In this review we examine recent evidence pertaining to the presence, activation and contribution of these regulatory pathways to nervous system development, injury and degeneration. The review begins with a description of cell death signals initiating through plasma membrane receptors followed by a description of relevant pro- and anti-apoptotic signal transduction pathways activated by various cellular stresses and trophic factor receptors. The review then considers signal transduction cascades activated in the nucleus and finally the critical mediators of viability that are related to mitochondrial function and disassembly of essential cellular components.

DEATH RECEPTOR-MEDIATED NEURONAL APOPTOSIS

Members of the tumor necrosis factor receptor (TNFR) family are involved in a number of physiological processes, including neuronal cell death during development and after injury. This family of receptors includes Fas/CD95/Apo1, TNFR1, TNFR2, DR3/TRAMP/Wsl-1, TRAIL-R1/DR4, TRAIL-R2/DR5/Killer/TRICK2, TRAIL-R3/DcR1/TRID, TRAIL-R4/DcR2/TRUNDD, DcR3, osteoprotegerin, DR6, $p75^{NTR}$ and others.[1] The prototypic members of the TNF receptor family are transmembrane proteins containing an extracellular cysteine-rich domain and an intracellular cytoplasmic protein-binding sequence called the death domain (DD). Activation of Fas/CD95, TNFR1, TRAIL-R1, TRAIL-R2 and DR6 receptors by their respective ligands leads to the recruitment of an intracellular protein complex known as the DISC (death-inducing signaling complex), followed by downstream caspase activation.

Fas/CD95 /Apo1

Fas receptors and Fas ligand (FasL) are expressed on both astrocytes and neurons in normal rat and human brain.[2] Exogenous application of FasL can cause neuronal apoptosis in vivo and in vitro, and this death can be blocked by a selective caspase 8 inhibitor.[3] Embryonic motoneurons co-express Fas and FasL at the time of programmed cell death, and their death may be reduced by blocking antibodies to the Fas receptor or by the use of a specific caspase 8 inhibitor.[4] Other studies have revealed a role for Fas in neuronal death after injury. In spinal cord ischemia, studies suggest the rapid formation of a complex containing the Fas receptor and procaspase 8, followed by activation of caspases 8 and 3 and neuronal apoptosis.[5] FasL is also upregulated in cortical neurons after cerebral ischemia.[6,7] In *lpr* mice expressing a dysfunctional Fas receptor, infarct volume after ischemia was reduced when compared to wild type control mice.[7] Nervous system infections may also be a stimulus for activation of the Fas/FasL pathway. Both Fas and FasL are upregulated in the brains of patients with HIV.[8] In vitro studies indicate that neuronal death in HIV dementia is not due to direct viral infection of neurons, but may be due to the release from activated microglia of other cytokines such as TNF and/or FasL.[9]

Most of the studies elucidating the mechanisms underlying Fas /FasL receptor signaling have been performed in non-neuronal cells (for review, see refs. 10,11). The binding of FasL to its receptor leads to trimerization of the receptor and results in the recruitment of an intracellular adapter protein, FADD (Fas-associated death domain). FADD binds to the DD of the Fas/CD95/Apo1 receptor via homophilic DD-DD interactions. FADD also contains a separate death effector domain (DED) at its N-terminal, which interacts directly with a homologous region in the prodomain of pro-caspase 8. The association of the trimerized Fas/CD95/Apo1 receptor, FADD and pro-caspase 8 forms the DISC. Pro-caspase 8 subsequently undergoes autocatalytic cleavage to yield its active form. Caspase 8 may then cleave and activate caspases 3, 6 and 7 directly, thereby leading to cell death.[12] Alternatively, caspase 8 may cleave Bid to form truncated Bid (t-Bid; see ref. 13). Some investigators have argued that there are at least two Fas/CD95/Apo1L signaling pathways that occur after DISC formation.[14] One pathway involves mitochondrial amplification of caspase activation, while the other results in mitochondrial dysfunction that occurs only after activation of caspases 8 and 3. Whereas the former apoptotic pathway is inhibitable by bcl-2 overexpression, the latter is not. The reasons for these cell-type specific differences in the order of caspase activation and mitochondrial dysfunction after Fas activation are not completely known. Nevertheless, it is apparent that caspase 8 is the primary apical caspase involved in signaling by Fas/CD95/Apo1L and other members of the TNFR superfamily. Caspase 10 also contains several DEDs, and can be recruited to the DISC after activation of TNF and TRAIL (TNF-related apoptosis-inducing ligand) receptors.[15] Thus, caspase 10 may also serve as an apical caspase in some cells, as has been demonstrated in lymphocytes.[16] However, the fact that caspase 8 gene deletion completely abrogates TNFR1 and Fas receptor-induced apoptosis indicates that caspase 8 is primarily responsible for death receptor signaling by the TNFR superfamily.[12]

TNF/TNFRs

Recent studies suggest that TNF may play an important role in the control of neuronal survival. For example, TNFα induced apoptosis in cultured primary rat cortical neurons and in differentiated PC12 cells.[17] Function-blocking antibodies to TNFα or TNFR1 partially protected embryonic mouse sympathetic and sensory neurons from apoptosis induced by NGF withdrawal.[18] Moreover, fewer sensory and sympathetic neurons died during the phase of naturally occurring cell death in TNFα-deficient mice than in wild type mice, and sympathetic neurons derived from such mice survived better than their wild type counterparts. Other investigators have found that TNFR1$^{-/-}$ and TNFR2$^{-/-}$ mice have an increased number of apoptotic cells in the injured mouse spinal cord.[19] The proposed explanation for this result was that activation of TNFRs in the wild type animals resulted in NF-κB activation and increased expression of c-IAP2, thereby inhibiting caspase-mediated apoptosis.

Like the Fas/CD95/Apo1L receptor, TNF receptors (TNFRs) also signal cell death via the formation of an intracellular protein complex.[15] After binding its ligand, TNFR1 trimerizes and recruits the intracellular DD-containing protein, TRADD (TNFR-associated death domain). TRADD, in turn, recruits FADD, pro-caspase 8 and the serine-threonine kinase RIP (receptor-interacting protein). TNFR1 also couples to a number of other intracellular signaling pathways. The serine-threonine kinase interacting protein (RIP) also interacts with the NF-κB pathway.[20] Activation of NF-κB is thought to oppose TNF-induced apoptosis in many cell types. TNFR1 can also associate with another factor, TRAF2 (TNF receptor-associated factor 2), which appears to couple TNFR1 to both NF-κB and c-Jun N-terminal kinase (JNK) activation.[21] Like TNFR1, DR6 has also been shown to activate NF-κB and JNK (Pan et al, 1998). Additional studies indicate that TNFRs may even couple to the nuclear transcription factor AP-1.[22]

TNFR2 lacks a cytoplasmic death domain. Thus, it was initially unclear how activation of the receptor leads to cell death. However, a recent study by Grell et al[23] has indicated that activation of TNFR2 leads to increased synthesis of TNF, which then acts in an autocrine fashion on TNFR1 receptors to initiate cell death.

TRAIL/TRAIL-Rs

TRAIL/Apo-2L is a type II transmembrane protein which, based on homology to TNF, was predicted (and subsequently shown) to be most effective in activating TRAIL receptors as a multimer.[24] The tissue distribution of TRAIL receptors and their ligand varies widely, with both being commonly expressed in the same tissues[24, 25]. There are currently five known TRAIL receptors, including TRAIL-R1/DR4, TRAIL-R2/DR5/Killer, TRAIL-R3/DcR1/LIT, TRAIL-R4/DcR2/TRUNDD, DcR3 and osteoprotegerin, a soluble receptor that also binds osteoclast differentiation factor.[15,24,26] TRAIL-R1 and TRAIL-R2 are 58% identical, and are type I transmembrane proteins that contain extracellular cysteine-rich domains and intracellular cytoplasmic domains containing a DD similar to that of the TNFRs.[24] TRAIL-R3/

DcR1 and TRAIL-R4/DcR2 have a somewhat different structure than that seen in TRAIL-R1 or TRAIL-R2. TRAIL-R3/DcR1 contains only a partial cytoplasmic DD, while TRAIL-R4/DcR2 lacks any transmembrane component, and is instead attached to the cell surface by a glycosylphosphatidylinositol linker.[1,24,27]

Interestingly, TRAIL preferentially induces apoptosis in tumorigenic cells and not in most normal tissues, including the brain.[28,29] However, this finding has recently been challenged by reports of TRAIL-induced apoptosis in normal human hepatocytes and in epileptic human brain.[30] Messenger RNA for TRAIL and its receptors have been localized in many tissues, including normal human brain tissue.[31] Exposure of epileptic human brain slices to TRAIL induces apoptosis,[32] and TRAIL induces cell death in primary mouse cortical neurons and neuroblastoma cells in vitro.[6,33]

The mechanisms underlying TRAIL/Apo-2L receptor signaling are not as well understood as those for CD95/Fas/Apo1 and TNFRs. As mentioned previously, TRAIL receptor activation leads to activation of caspase 8,[1,34] and caspase inhibition blocks TRAIL-induced apoptosis.[35] Studies have demonstrated inhibition of TRAIL signaling in FADD deficienT cells, as well as direct binding of FADD and TRADD to TRAIL-R1 and TRAIL-R2.[34,36] Thus, it seems likely that TRAIL receptor activation results in DISC formation, caspase 8 activation and cell death in a manner similar to that observed with TNFRs and the Fas receptor. There may be other signaling pathways coupled to TRAIL receptors as well, although their functional role remains to be determined. For example, TRAIL-R4 has been shown to activate NFκB, despite having an incomplete DD.[27]

The factors determining sensitivity to TRAIL are only now being elucidated. TRAIL itself is upregulated after cerebral ischemia,[6] and both Trail-R2/DR5 and TRAIL-R3/DcR1 can be induced in a variety of tissues in a p53-dependent or p53-independent manner.[37-39] These findings suggest that sensitivity to TRAIL-mediated apoptosis may be regulated in the nervous system, at least in part by injury-induced regulation of the expression of TRAIL and its receptors.

Both TRAIL-R3 and TRAIL-R4 have been called decoy receptors because their ligation fails to induce apoptosis. One hypothesis is that these non-functional decoy receptors bind the available TRAIL, thereby protecting TRAIL-sensitive cells from the ligand.[40,41] However, studies examining expression of mRNA and protein for TRAIL-R3/DcR1 and TRAIL-R4/DcR2 in a variety of cell types have not found a significant correlation with TRAIL sensitivity.[29,42,43] Thus, the function of TRAIL-R3 and TRAIL-R4 has not been conclusively demonstrated.

$p75^{NTR}$

Recent studies have provided evidence that the low affinity neurotrophin receptor, $p75^{NTR}$,has significant homology to other members of the TNFR superfamily, and that it may play a role as a death receptor under some circumstances. In addition to its well established role as a component of neurotrophin receptors, $p75^{NTR}$ may induce the death of selected neuronal populations when expressed in the absence of *Trk* receptors. For example, embryonic or neonatal motor neurons

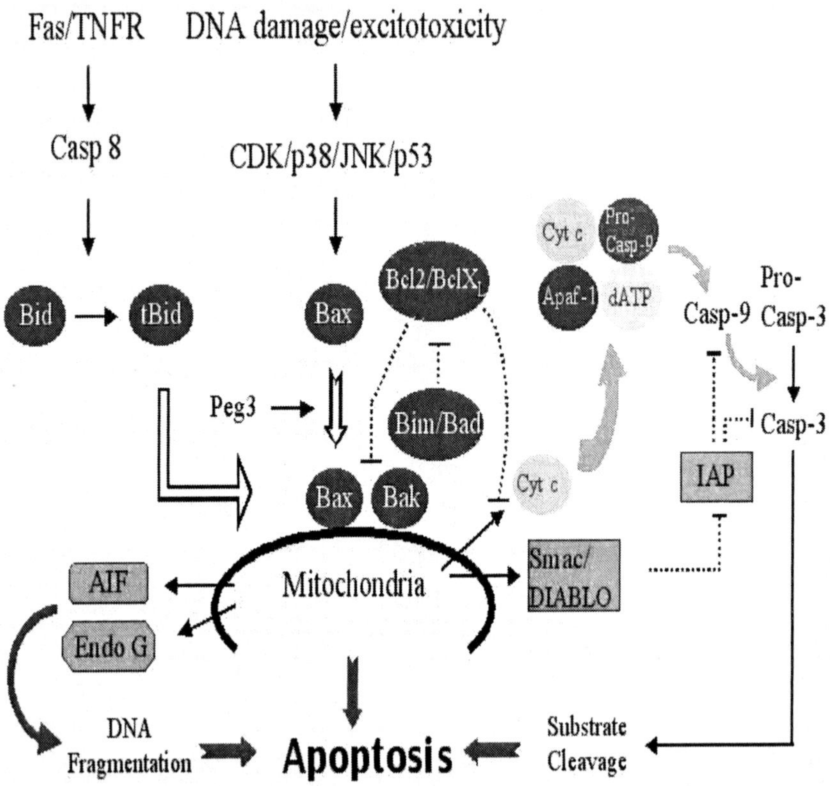

Figure 1. Cell death signaling pathways in neurons. Schematic representation of proposed cell death pathways initiated by excitotoxic and genotoxic damage in the nervous system. In this model, excitotoxicity and DNA damage are common activators of neuronal damage, although DNA damage may ensue from excitotoxicity and other stresses possibly as a result of increased production of reactive oxygen species. These cellular stresses are linked to the activation of numerous downstream signaling pathways such as the cyclin-dependent and stress activated kinases, DNA-dependent protein kinase (not pictured), ATM (not pictured) and p53. Cell death signals can also be activated through cell membrane receptors such as the Fas and TNF receptors. Both sets of pathways seem to have at least one common convergence point which revolves around members of the Bcl-2 family. The Bcl-2 family of apoptotic regulators influences mitochondrial integrity. If the anti-apoptotic members (Bcl-2/Bcl-X_L) are dominant, mitochondrial function and integrity are maintained. However, if the pro-apoptotic members (Bax, Bak, Bad, Bim Bok, Noxa, PUMA) are dominant, mitochondrial membrane permeability is compromised and numerous apoptotic mediators are released from the intermitochondrial space including cytochrome c (cyt c), endonuclease G (endo G) and apoptosis inducing factor (AIF). Cytochrome c is one of the factors required for the processing of pro-caspase 9, which in turn activates caspase-3. Caspase-3 cleaves a variety of cellular substrates, which culminates in the disassembly of the cell. Smac/DIABLO is another mitochondrial factor that promotes cell death by suppressing the action of the inhibitor of apoptosis proteins (IAP) which inhibit the action of caspases.

derived from wild type, but not p75^{NTR}-deficient mice undergo apoptosis after NGF exposure.[44,45] These cells also show increased survival after axotomy when compared to wild type motor neurons, suggesting that p75^{NTR} may promote neuronal cell death

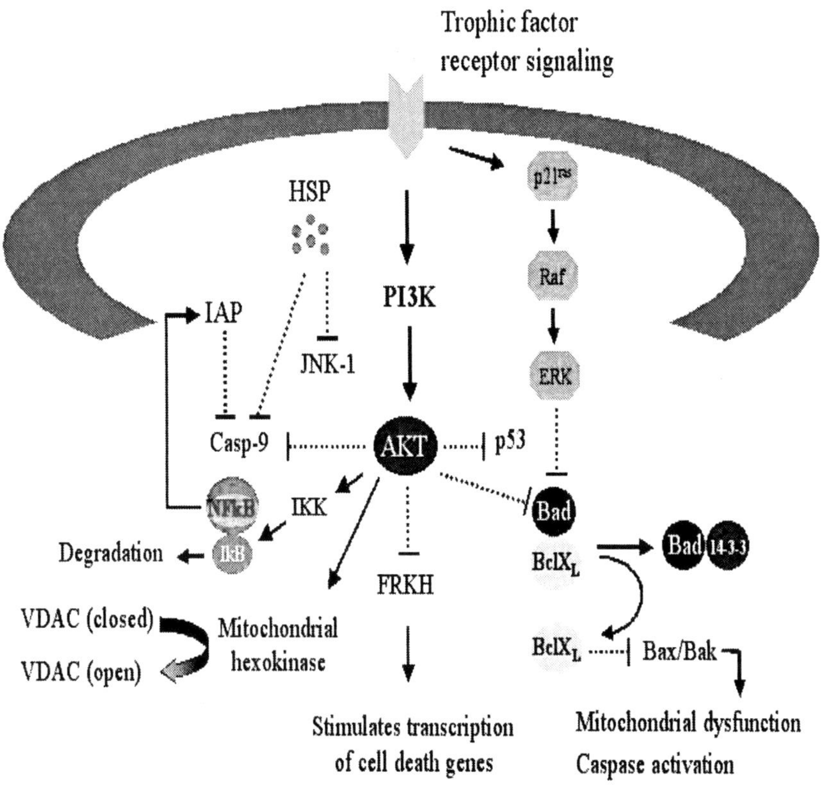

Figure 2. Survival signaling pathways in neurons. Schematic representation of pathways that promote neuronal survival. In this model, a major factor contributing to the survival of CNS neurons is the stimulation of the PI3K-Akt kinase pathway following trophic factor receptor activation. Akt is a central player as it phosphorylates a number of critical substrates. For example, phosphorylation of the forkhead transcription factor inhibits its ability to transcribe death-promoting genes, while the phosphorylation of Bad promotes its separation from Bcl-2/Bcl-X_L and its sequestration by the 14-3-3 protein. The sequestration of Bad allows Bcl-2/Bcl-X_L to suppress the death promoting action of Bax and Bak at the mitochondria. Akt has also been shown to antagonize the apoptotic action of p53 in hippocampal neurons, as well as caspase-9 activity, although the latter action has only been demonstrated for human caspase-9. Akt phosphorylates hexokinase, which in turn maintains the voltage dependent anion channel in an active state preventing the opening of the permeability transition pore. Akt has also been shown to enhance the degradation of I_kB thus helping to promote NF_kB-mediated transcriptional activation. NF_kB induces the caspase inhibitors IAP1 and IAP2 and pro-survival Bcl-2 family members Bcl-X_L and Bfl-1/A1 (not shown). The extracellular signal-regulated kinase, ERK is also activated in response to trophic factor signaling. ERK phosphorylates multiple substrates and its phosphorylation of Bad, acting in concert with Akt, promotes the sequestration of Bad by 14-3-3. Heat shock proteins (HSP) are induced by a wide array of insults and act at multiple levels to mitigate stress signals by inhibiting caspase activation or through interactions with JNK-1.

after injury.[46] $p75^{NTR}$ knockout mice also show decreased cell death of a subset of spinal cord interneurons, retinal neurons, and sympathetic neurons.[47,48]

The mechanism by which $p75^{NTR}$ signals cell death is poorly understood. It has a different death domain structure than that of other TNFRs, and there is conflicting evidence as to whether it signals apoptosis when occupied or unoccupied.[11] Studies have implicated NFκB activation, ceramide production and caspase activation in the $p75^{NTR}$ signaling pathway (for review, see ref. 49). In immortalized striatal neurons containing an inducible $p75^{NTR}$, expression of the receptor was found to activate caspase 9, 6 and 3 in an NGF-independent manner.[50] Cell death was inhibited by Bcl-X_L, by a dominant negative form of caspase 9 and by a viral FLIP, E8. The protection by viral FLIP suggests that DEDs are involved in the signaling cascade. However, FADD, TRADD and caspase 8 were not involved. Other proteins identified that interact with $p75^{NTR}$ and promote apoptosis include NRIF (neurotrophin receptor interacting factor, ref. 51), NADE (p75NTR-associated cell death executor, ref. 52), and NRAGE (neurotrophin receptor-interacting MAGE homolog, ref. 53), while RIP2[54] was shown to suppress apoptosis.

SIGNAL TRANSDUCTION PATHWAYS

Signaling Kinases- Pro- and Anti-Apoptotic Effectors in the Nervous System

The mitogen-activated protein (MAP) kinases and phosphatidylinositol-3 kinase (PI3K) are serine/threonine protein kinases that play critical roles in neuronal growth, differentiation and survival (Figs. 1 and 2). In general, activation of the extracellular signal-regulated kinase members of the MAP kinase family (ERK or p42/p44 MAP kinase) and the PI3K-Akt signaling pathway promote cell survival, while members of the MAP kinase family known as the stress activated protein kinases (SAPK s), c-Jun N-terminal kinases (JNK s) and the p38 MAP kinase (p38 MAPK), promote cell death. In this section, we will review the molecular pathways and contributions of these kinases based upon evidence from animal studies, as well as neuronal cell lines and various cultured neuronal populations.

PI3K-Akt Signaling

Observations over the past decade have identified the PI3K-Akt pathway s importance in mediating survival in PC12 cells[55] and cultured neurons from the peripheral[56] and central nervous systems.[57] A recent study by Kuruvilla et al[58] demonstrated the importance of PI3K activity under conditions that may more accurately simulate in vivo function. By applying NGF exclusively onto the distal axon compartment, the authors were able to demonstrate that PI3K activation in the axons and its retrograde signaling play a critical role in survival of both the axons and cell body.[58]

Neurotrophic factors such as NGF, brain-derived neurotrophic factor (BDNF), glial cell line-derived neurotrophic factor (GDNF), and insulin-like growth factor I (IGF-I), activate the PI3-Akt signaling cascade through corresponding receptor tyrosine kinases such as the high affinity neurotrophin receptors (Trk s) (reviewed in ref. 59). After receptor dimerization, PI3K is recruited to the plasma membrane where its catalytic subunit generates lipid second messengers, phosphoinositide phosphates (PIP_2, PIP_3) at the inner surface of the plasma membrane. Phosphoinositide-dependent protein kinase-1 (PDK1) then acts in concert with PIP_2 and PIP_3 to phosphorylate and activate Akt (a.k.a. protein kinase B, PKB), a protein kinase first identified in the AKT virus (reviewed in ref. 60). Alternately, Trk receptors may stimulate PI3K via the Ras G-protein, insulin receptor substrate (IRS) signaling, and Gab-1, an adaptor protein, which binds to Trk and directly stimulates PI3K.[61,62]

Dominant negative and constitutively active forms of Ras,[62] PI3K,[63,64] and Akt[65] have been used to study signaling through the PI3K-Akt pathway. These and other studies have demonstrated downstream signaling effects that regulate cellular survival, proliferation, and metabolism. For example, Akt phosphorylates and inactivates FKHRL1, a member of the family of Forkhead transcriptional regulators. Inactivated FKHRL1 is unable to induce the expression of death genes in cerebellar granule neurons.[66] In primary hippocampal neurons subjected to hypoxia or nitric oxide, p53 activation and p53-mediated Bax upregulation are also blocked by Akt signaling.[67] Akt activates the cAMP—responsive element binding protein (CREB) and nuclear factor-κB (NF-κB), additional transcriptional regulators that may promote neuronal survival.[68,69] In addition, Akt can directly inhibit the apoptotic machinery by phosphorylation at sites both upstream (BAD[70]) and downstream (Caspase 9[71]) of mitochondrial cytochrome c release. Finally, there is evidence to support the role of Akt in promoting neuronal survival through metabolic effects, by regulating glucose metabolism in neurons.[72]

ERK/MAP Kinase Pathways

The ERK cascade represents one of the evolutionarily conserved MAP kinase pathways, first found to be important in the regulation of neuronal survival in response to neurotrophin withdrawal.[73] A wide body of evidence has also been collected to demonstrate the role of ERK in neuronal plasticity and memory formation.[74,75] For the purposes of this discussion, we will focus on the data supporting ERK s role in promoting survival although ERK s can be activated by toxic stimuli as well and have been implicated in cell death.[76-78]

At least six ERK isoforms have been identified.[79] In general, the common pathway leading to ERK activation by trophic factor signaling involves engagement of the membrane receptor at the cell surface, activation of small ($p21^{ras,rap}$) GTPases, which in turn activate the protein kinase Raf. Raf, in turn, phosphorylates and activates MAP kinase kinase 1/2 (MKK 1/2), which phosphorylates and activates ERK (reviewed in ref. 80). The ERK pathway can also be stimulated and modulated by intracellular Ca^{++} levels, protein kinase A, diacylglycerol, and cAMP (reviewed in refs. 74,75,81).

Constitutively activated or dominant negative mutants directed at Ras, Raf, and MEK 1/2 (a.k.a. MKK1/2) have been used to determine the effects of the ERK pathway in the setting of neurotrophin withdrawal. Current evidence indicates a direct role for the ERK pathway in protecting neurons after injury. Activation of the ERK pathway is essential for BDNF-mediated protection of cortical neurons from DNA damage,[84] glutamate-induced excitotoxicity and ischemic preconditioning.[85] The role of ERK in suppressing neuronal apoptosis, however, has been controversial. The importance of this pathway as a neuroprotective mechanism appears related to the nature of the cell type and exposure. For example, a MKK1 protein kinase inhibitor protects against damage resulting from focal cerebral ischemia.[82] These conflicting findings may stem from the use of different experimental systems and the possibility that the regulation of ERK is complex and might also be controlled by MKK1-independent mechanisms.[83] While PI3K-Akt is thought to be more important in protecting cells from trophic factor withdrawal, ERK is thought to protect against injury-induced apoptosis.[84]

The mechanisms by which ERK mediates its protective effect are poorly understood. Downstream signaling may be directed through the serine/threonine kinase p90/rsk2 and CREB phosphorylation,[68] but a direct link between CREB and ERK has not yet been established in neurons. Alternately, ERK can phosphorylate Bad resulting in its sequestration by 14-3-3, and prevents Bad-mediated induction of apoptosis.[68]

JNKs

As their name implies, stress activated protein kinases, such as the c-Jun N-terminal kinase (JNK) are activated by numerous noxious stimuli including trophic factor withdrawal[73], excitotoxicity,[86,87] seizures,[88] irradiation,[86] hypoxia,[89] exposure to beta-Amyloid (Abeta),[90,91] arsenite toxicity[92] and axotomy[93] (reviewed in ref. 94). At least ten isoforms of JNK are expressed in the human adult brain, encoded by three separate genes through alternative splicing mechanisms.[95]

The relevance of JNK signaling to neuronal cell death has been demonstrated using JNK-deficient mice, JNK inhibitors and dominant negative forms of upstream activating kinases. Transfecting cortical neurons with various dominant negative forms of components comprising the JNK signaling pathway significantly decreased the number of neurons undergoing apoptosis in response to the β-amyloid peptide.[90] Similarly, treatment with an inhibitor of JNK activation, CEP-1347, effectively blocked increases in cellular JNK activity and protected PC12 cells, sympathetic neurons[91] and cortical neurons[96] from β-amyloid-induced-death. These results are consistent with the demonstration that JNK is activated and re-distributed in the brains of patients with Alzheimer s disease.[97] Moreover, CEP-1347 also inhibited MPTP-mediated MKK4 and JNK signaling and attenuated MPTP-induced dopaminergic cell loss[98] suggesting that the JNK pathway may be activated in the degenerative process in Parkinson s disease. Cultured rat sympathetic neurons and neuronally differentiated PC12 cells were protected from nerve growth factor withdrawal, exposure to ultraviolet irradiation, and oxidative stress following treatment with the JNK inhibitor CEP-1347[98]. Interestingly, CEP-1347 failed to protect undifferentiated PC12 cells induced to die by serum withdrawal or Jurkat T cells from Fas activation, even though each injury stimulus activated JNK and was inhibited by CEP-1347. Blocking JNK activation also protects neurons from cell death induced by the withdrawal of survival signals.[73,99,100] These results are consistent with the data obtained applying NGF withdrawal to JNK3-deficient sympathetic neurons.[101] Moreover, the deficiency in JNK3 activity[87] and the expression of a mutant form of jun that is resistant to phosphorylation by JNK[102] confers resistance to kainic acid-induced seizures and subsequent neuronal damage. Collectively, these results illustrate the importance of JNK signaling to the regulation of neuronal injury responses.

Upstream regulation of JNK activation is poorly understood. As in the ERK system, the immediate upstream activators of JNK are MKK s. While the ERK pathway is stimulated by MKK s 1/2, JNK appears to be activated by MKK4[103] or MKK7,[104] depending upon the cell type, developmental stage[105] and stressful stimulus.[106] Other kinases such as MLK3[107], MEKK1[108] and ASK1[109,110] lie further upstream of the MKKs, and probably activate both MKK s and JNK in a similar cell type and stimulus dependent manner.[106] ASK1 also activates p38 MAPK and is upregulated in PC12 cells following NGF withdrawal.[111] Overexpression of a constitutively active mutant of ASK1 activates JNK and induces apoptosis in differentiated PC12 cells and sympathetic neurons.[111] JNK activity is further modified by JNK interacting proteins (JIP s), which are scaffolding proteins thought to sequester all of the necessary kinases for JNK activation.[112]

The downstream effectors of the JNK pathway are numerous and include both cytoplasmic and nuclear targets. In the nucleus, JNK s phosphorylate and activate the transcription factor c-jun,[113] a component of the AP-1 transcription factor, which regulates genes involved in apoptosis. For example, in PC12 cells, JNK activation is followed by induction of Fas ligand (FasL) expression[100] (see Fas/CD95/Apo1 in the previous section). Cytoplasmic targets include p53[114] (see Nuclear Signaling Pathways in the following section), which is stabilized and activated by JNK,[115] and a cell death domain protein MADD, which co-translocates with JNK to the nucleus after hypoxic stress.[116]

p38 MAP Kinases

p38 MAPK s, like JNK s, are stress activated kinases that are highly conserved in evolution, exist in multiple isoforms, and are activated in response to numerous cellular stresses (reviewed in refs. 94,106). Like other members of the MAP kinase family, upstream regulation involves phosphorylation by MKK s, specifically MKK3, MKK4, and MKK6.[117-119] The downstream targets of the p38 MAPK pathway are similar to the targets of JNKs, and involve transcription factors, kinases, and pathways influencing pro-inflammatory cytokines (reviewed in ref. 106). However, little is known about the downstream targets of p38 MAPK signaling in the brain.

Inhibition of p38 MAPK has been shown to attenuate neuronal cell death in response to a variety of different cellular stresses. The p38 MAPK inhibitor SB203580 significantly reduced apoptosis in potassium-deprived cerebellar granule cells[120,121] and prevented the activation of downstream effectors such as caspase 3.[120] p38 MAPK inhibition also conferred protection against neuronal cell death induced by C2-ceramide,[122] oxidative damage[123] and trophic factor deprivation.[73,124] Focal ischemia induces p38 MAPK activity and treatment with a second-generation p38 MAPK inhibitor, SB 239063, significantly reduced infarct size and behavioral deficits.[125] Inhibiting p38 MAPK activity in the focal ischemia model was also associated with decreased expression of IL-1beta and TNFα, cytokines known to contribute to stroke-induced brain injury. Glutamate-mediated cell death, an important contributor to stroke-induced changes in viability, is also attenuated in cerebellar granule cells by p38 MAPK inhibition.[126]

Mechanistic aspects of the contribution of p38 MAPK activation to cell death were further revealed in a study from our laboratory.[127] We showed that nitric oxide (NO)-induced cell death of human neuroblastoma cells and murine cortical neurons in culture was dependent upon p38 MAPK activity and Bax. We further demonstrated that inhibition of p38 MAPK activity blocked Bax translocation from the cytosol to the mitochondria and thereby conferred protection from cell death. Cheng et al recently reported a similar result.[128] Thus, p38 MAPK represents a potential target for interrupting Bax translocation and preserving neuronal viability following neuronal injury involving NO toxicity. Collectively, the stress activated kinases, p38 MAPK and JNK, play important roles in the neuronal response to injury and newly developed inhibitors for these enzymes hold promise for therapeutic intervention.

Glycogen Synthase Kinase-3β

Another emerging kinase that has been implicated in neuronal cell death is glycogen synthase kinase-3β (GSK-3β). GSK-3β regulates a diverse array of proteins including many transcription factors (for review, see ref. 129) and is inhibited by the PI3K pathway.[130,131] Its function in the nervous system has been linked to neuropathology, such as Alzheimer s disease, based on its ability to promote tau hyperphosphorylation[132,133] although the significance of this finding is still controversial.[134] Several cellular stresses including serum deprivation stimulate GSK3β activity in cortical neurons prior to the induction of apoptosis.[72] Expression of an inhibitory GSK3β binding protein or a dominant interfering form of GSK3β reduced neuronal apoptosis, suggesting that GSK3beta contributes to trophic factor deprivation-induced apoptosis. Moreover, overexpressing GSK3β in neurons was sufficient to trigger neuronal cell death. Elucidating the intriguing role played by this kinase in neuronal cell survival/death is certainly warranted and will require the identification of specific substrates and their functions in the context of a specific survival or death promoting stimulus.

NUCLEAR SIGNALING PATHWAYS

Cyclins, Cyclin-Dependent Kinases, Rb and E2F1

What is seemingly a contradiction in our understanding of cell death in post-mitotic neurons, is the demonstration that neuronal damage results in the elaboration of events that are normally associated with proliferating cells. Neurons express a range of proteins after injury that normally function to control cell cycle progression (for review see ref. 135). Alterations in the levels of cyclin-dependent kinases (CDK s) and their respective cyclin partners have been observed in neurons in response to trophic factor deprivation,[136] ischemia and seizures,[137] stroke,[138] exposure to the β-amyloid peptide[139,140] and DNA damage.[141] In addition, abnormal expression of these cell cycle regulators has also been implicated in neurodegenerative diseases such as Alzheimer s disease[142-148] and amyotrophic lateral sclerosis (ALS).[149] Pharmacological and molecular manipulation of CDK activity has been used to evaluate the relationship between cell cycle regulatory proteins and neuronal cell death.

The use of CDK inhibitors or dominant negative forms of CDK 4/6 protects cultured postmitotic sympathetic neurons from death evoked by NGF withdrawal.[150] Pharmacological inhibitors of cyclin-dependent kinases, such as flavopiridol, olomoucine, and roscovitine, prevent cerebellar granule neuron apoptosis induced by non-depolarizing KCl.[151] Moreover, camptothecin-induced cell death of PC12 cells, sympathetic neurons, and cerebral cortical neurons is also suppressed by the CDK-inhibitors flavopiridol and olomoucine.[152,153] However, these cell cycle inhibitors did not prevent cell death induced by a strong oxidative stress (SOD1 depletion)[154] suggesting that cell cycle proteins are not involved in all forms of neuronal cell death.

The mechanism by which the CDK s facilitate neuronal cell death is under active investigation. Increased activity of cyclin-dependent kinase 5 (Cdk5) may contribute to neuronal death and cytoskeletal abnormalities in Alzheimer s disease[148] and ALS[149] through its ability to hyperphosphorylate tau, although the finding that cdk5 is a major protein tau kinase has been disputed.[155] However, Cdk5 may alter neuronal viability through its effects on other signaling cascades, consistent with the finding that it directly phosphorylates beta-catenin and regulates the latter s binding to presenilin-1.[156]

Cdk4/6 activation results in the phosphorylation of the Rb tumor suppressor protein in neurons following DNA damage.[141] One consequence of Rb phosphorylation is the activation of E2F family members. Overexpression of E2F1 is sufficient to induce neuronal cell death,[157-159] while E2F1 deficiency protects neurons from multiple cellular stresses including DNA damage,[160] ischemia,[161,162] staurosporine,[157] reduced potassium[158] and β-amyloid toxicity.[163] In addition, overexpression of dominant negative versions of DP-1, a binding partner for E2F family members, also protects neurons from cell death induced by DNA damage.[139,160] Interestingly, overexpression of E2F activity in dissociated sensory neurons from adult rats stimulated their entry into S-phase, although the authors found no evidence for subsequent mitotic events in the E2F-overexpressing cells.[159] These results suggest that other molecular signals are responsible for sister chromatid separation and continued cycling in neurons. The absence of these additional mitotic factors in damaged or degenerating neurons might provide an explanation for why the inappropriate activation of cell cycle regulatory genes promotes neuronal cell death.

The CDK-dependent inactivation of Rb is likely to have many other effects beyond the activation E2F family members. Rb has been shown to repress NFκB transcriptional activity[164] as well as JNK activity,[165] both important neuronal cell death mediators. The relevance of Rb phosphorylation to the process of neuronal cell death is illustrated by the finding that overexpressing a mutant form of Rb lacking critical phosphorylation sites protects neurons from cell death induced by DNA damage.[160,166] Moreover, deregulation of the Rb cell cycle pathway during development promotes ectopic cell cycle entry and elevated apoptosis which is, in part, p53-dependent.[167,168] The latter finding is consistent with a role for the Rb-regulated E2F1 protein as a specific inducer of apoptosis and p53 accumulation.[169]

p53 Tumor Suppressor Gene

The p53 Gene

The p53 tumor suppressor gene encodes a nuclear phosphoprotein that functions as a key regulator of DNA repair, cell cycle progression and apoptosis. The p53 protein is upregulated in response to a diverse array of cellular stresses, including DNA damage, hypoxia, oxidative stress, ribonucleotide depletion and oncogene activation.[170,171] p53 protein levels are primarily regulated by changes in protein degradation in response to injury. In response to cellular stress, p53 induces its biological response largely through the transactivation of specific target genes. These downstream effectors have been characterized with respect to p53-mediated growth arrest[172] but the pathways associated with p53-mediated apoptosis have not been completely elucidated.[173] In addition to its transactivating activity, p53 may also promote apoptosis by repressing the expression of select genes.[174,175] Moreover, p53-mediated apoptosis may also occur through transcription-independent pathways requiring direct protein-protein interactions.[176,177]

p53 Expression and Neuronal Injury

Alterations in p53 mRNA and protein expression have been associated with neuronal damage in a variety of model systems (for review see ref. 178). Neuronal damage resulting from stimulation by excitatory amino acids (excitotoxicity) has been strongly associated with p53 accumulation. The systemic injection of kainic acid, an excitotoxin that produces seizures, induced p53 expression in neurons exhibiting morphological evidence of damage.[179,180] Activation of glutamate receptors by intrastriatal infusion of either N-methyl-D-aspartate (NMDA), the NMDA receptor agonist quinolinic acid (QA) or kainic acid produced a significant elevation in p53 levels in striatal neurons.[181-183] Elevated expression of the p53 gene has also been observed in models of experimental traumatic brain injury.[184-186] As early as 6 h post-injury, p53 mRNA is induced predominantly in neurons that are vulnerable to traumatic brain injury. Transient or permanent occlusion of the middle cerebral artery causes ischemia-induced cell death in striatal and cerebral cortical neurons, which is associated with a significant increase in the expression of p53 mRNA[187] and protein.[188]

Elevated p53 immunoreactivity has also been detected in brain tissue derived from animal models of human neurodegenerative disease or from patients that have been diagnosed with a neurodegenerative disorder. Patients with Alzheimer s disease[189,190] show increased p53 immunoreactivity in morphologically damaged neurons, consistent with the presence of increased p53 immunoreactivity in neurons from mice overexpressing the beta-amyloid peptide (Aβ 1-42).[191] Mutation of the E6-AP ubiquitin ligase in a mouse model of Angelman syndrome results in increased cytoplasmic abundance of the p53 protein in hippocampal pyramidal neurons and cerebellar Purkinje neurons[192]. Animals expressing the Angelman mutation display motor dysfunction, inducible seizures and a deficiency in contextual learning. Thus, increased levels of the p53 protein in Angelman syndrome resulting from abnormalities in the ubiquitination process may contribute to neuronal dysfunction.

The brains of patients with Down s syndrome, a genetic disorder manifesting a similar pathology to Alzheimer s disease, have also been shown to express elevated levels of apoptosis effectors, including the p53 protein.[190,193,194] Motor neuron degeneration observed in amyotrophic lateral sclerosis has been associated with increased levels of p53 in motor neurons of the spinal cord and motor cortex.[195] Recent evidence also suggests that p53 could be involved in the pathogenesis of Huntington s disease. The Huntington s disease protein, huntingtin, was found to interact with p53 and the CREB-binding protein,[196] and to repress transcription of several p53-regulated promoters. These results raise the possibility that the mutant huntingtin protein may cause neuronal dysfunction and cell death by interacting with transcription factors and altering gene transcription.

The results obtained with in vitro models of neuronal injury are consistent with the data described above for the in vivo models. Excitotoxicity, which figured so prominently in the whole animal studies, is a potent inducer of p53 protein in cultured cerebellar granule neurons.[197,198] Another potent stimulus for elevating p53 expression in cultured neurons is DNA damage induced by cytosine arabinoside,[199,200] ionizing radiation,[166] camptothecin[153] and the absence of essential DNA repair proteins.[201] Hypoxia in culture, which models the ischemia produced by middle cerebral artery occlusion, increases p53 protein expression in rat embryonic cortical neurons.[202]

p53 Expression and Neuronal Cell Death

The physiological relevance of p53 expression to neuronal cell death has been evaluated in numerous models of injury and disease. p53-deficient mice or neurons derived from these mice have been used most often, but inhibitors of p53 expression or p53 function have also been used to evaluate the role of p53 in the context of neuronal injury. The absence of p53 has been shown to protect neurons from a wide variety of toxic insults including focal ischemia,[203] ionizing radiation[204-208] kainate-induced seizures,[209] MPTP-induced neurotoxicity,[210] methamphetamine-induced neurotoxicity,[211] adrenalectomy,[212] traumatic brain injury,[186] DNA damaging agents,[199,200,206,213-215] glutamate,[197,198] hypoxia[202,216] and NGF withdrawal.[217,218] A role for p53 has also been demonstrated for apoptosis associated with developmental neuronal death in certain subpopulations of neurons[217] as well as cell death occurring during abnormal development[167, 219]. In addition to the use of p53-deficient mice, the application of antisense oligonucleotides[197,200,202,220,221] or a synthetic inhibitor of p53[222] have been successfully used to protect neurons against cell death induced by a variety of cellular stresses.

However, the absence of p53 does not protect neurons against all forms of cellular stress. Cerebellar neurons lacking p53 die when transferred to a low potassium medium,[213] and postnatal cortical and hippocampal neurons also die after staurosporine exposure in a p53-independent manner.[204] Cerebellar granule neuron death induced by methylazoxymethanol is not alleviated in p53-null mice. Moreover, two separate reports failed to demonstrate that p53 is involved in the death of neurons in a mutant SOD1 transgenic mouse model of familial ALS.[223,224] Despite evidence that p53 plays an important role in mediating cell death after acute neuronal injury, additional studies will be required to fully evaluate the role of p53 in late onset neurodegenerative diseases.

P53-MEDIATED CELL DEATH SIGNALING PATHWAYS

The mechanism by which p53 specifies the neuronal response to injury is poorly understood. However, currently available data suggest that the Bcl-2 family member, Bax, is involved in p53-mediated neuronal death. Bax-deficient neurons are protected from cell death induced by DNA damaging agents[204,215,225] and adenovirus-mediated p53 over-expression.[215,226] Moreover, various forms of neuronal injury are associated with Bax translocation from the cytosol to the mitochondria.[127,153,227,228] The redistribution of Bax to the mitochondria has been associated with a reduction in mitochondrial membrane potential, mitochondrial release of cytochrome c and activation of caspases,[229-233] suggesting that caspases may also be a component of a p53-induced cell death pathway. Recent studies indeed demonstrated that p53 is required for caspase activation in response to genotoxic stress.[215,225,226,234] These findings suggest that some forms of neuronal injury invoke a common pathway involving signal transduction through p53, Bax, mitochondrial dysfunction, cytochrome c release and caspase activation.

Clearly, additional studies are required to elucidate the downstream effectors mediating neuronal cell death in response to p53 activation. The glyceraldehyde-3-phosphate dehydrogenase (GAPDH) gene has been identified as a p53-inducible gene in cultured cerebellar granule neurons subjected to DNA damage.[200] Apafl, a key element of the apoptosome, which regulates initiation of the caspase-9 cascade, is induced in neurons in response to p53 activation or DNA damage.[235] Other genes, such as DR5, Fas, Fas ligand,[173] PERP,[236] Noxa and PUMA[237] have been shown to be induced by apoptotic stimuli as a result of p53 activation in a variety of non-neuronal cell types, but the involvement of these genes in p53-dependent neuronal apoptosis is not known.

It is also critical that the pathways responsible for activating and suppressing p53 activity be identified. In this regard it is interesting to note that the important survival promoting protein, Akt, can protect neurons from cell death by inhibiting p53-dependent transcriptional activity.[67] These results demonstrate the interconnection that exists between pathways that govern cell death and viability, and serve to remind us that the response and the outcome of neurons to stress are exceedingly complex.

DNA Damage-Activated Enzymes

Excessive DNA damage may result from exposure to free radicals generated by oxidative stress and thereby may be a common initiating stimulus for neuronal apoptosis. Oxidative stress is associated with a variety of neuropathological conditions including acute excitotoxicity (stroke, traumatic brain injury) and progressive neurodegeneration (Alzheimer s, Parkinson s, Huntington s, ALS).[238,239] Several enzymes are activated in response to DNA damage (strand breaks), and are associated with the control of cell cycle checkpoints, DNA repair programs and the regulation of apoptosis. These include the gene product for ataxia telangiectasia (ataxia telangiectasia mutated, ATM), DNA-dependent protein kinase (DNA-PK) and poly(ADP-ribose) polymerase-1 (PARP-1). Interestingly, all three of these enzymes are substrates for caspase-3[240-244] implicating them in the regulation of apoptosis.

ATM

ATM (ataxia telangiectasia mutated), a protein kinase belonging to the PI3 Kinase family, is activated by DNA double-strand breaks secondary to ionizing radiation (IR) and radiomimetics. Multiple phosphorylation substrates have been identified including p53, MDM2, and BRCA1, proteins involved in DNA repair and cell cycle checkpoint control (for a review, see ref. 245). The generation of an ATM-deficient mouse verified the relationship of ATM function to DNA damage-induced responses. For example, $ATM^{-/-}$ mice display extreme sensitivity to IR due to its toxic effects on the gastrointestinal tract, and fibroblasts derived from the mice show compromised G1/S checkpoint function.[246] In contrast to these results from non-neuronal and proliferative cell types, developing CNS neurons in $ATM^{-/-}$ mice turn out to be radioresistant compared to those in wild-type mice which show massive cell death, indicating that IR-induced neuronal apoptosis is ATM-dependent.[208] IR-induced, ATM-mediated neuronal cell death also depends on p53, Bax and caspase-3.[225] ATM is known to phosphorylate p53 at Ser15 directly[247,248] and Ser20 through the activation of CHK2.[249] ATM also phosphorylates MDM2 at Ser395.[250] These phosphorylation events are thought to collectively contribute to p53 stabilization and transcriptional activation.[251,252] However, it is still not known precisely how these ATM-dependent modifications of p53 result in the induction of apoptotic mediators.

Interestingly, IR-induced death of neural progenitor cells is not dependent on ATM, while p53 is required for apoptosis in both neural progenitor and post-mitotic neuronal populations.[225,253] Thus, the linkage between ATM and p53 dependent cell death is not observed in undifferentiated, multipotent precursors, but appears to develop in newly formed post-mitotic neurons; it still remains to be established if ATM is also required for IR-induced apoptosis in fully differentiated neurons in adults. Indeed, adult neurons, irrespective of the ATM genotype, don t seem to be sensitive to IR.[246] This implies that a critical function of ATM in neuronal apoptotic signaling may be the elimination of neurons born with a substantial amount of DNA damage accumulated during development.[253] Consistent with this idea is the observation that ATM deficiency in Ataxia Telangiectasia (A-T) patients causes early-onset progressive neurodegeneration, which is apparently due to compromised genomic integrity based on the phenotypic similarities of A-T to pathological conditions caused by genetic defects in DNA repair and cell cycle checkpoints.[254] ATR, an ATM-related protein primarily involved in the response to UV-induced DNA damage,[255] has not been evaluated in the process of neuronal apoptosis and remains to be examined.

DNA-Dependent Protein Kinase

DNA-dependent protein kinase (DNA-PK) is a member of the PI3 kinase family and is activated in response to DNA strand breaks,[256] and is essential for DNA repair.[257] DNA-PK phosphorylates p53,[258] MDM2,[259] and other factors implicated in the regulation of cell death including c-abl[260] and IkB.[261] However, several recent reports suggest that DNA-PK is not required for neuronal apoptosis, but may instead be important for neuronal survival after genotoxic insult. Neurons derived from *scid* mice, which express a truncated form of DNA-PK with impaired kinase activity, show higher rates of spontaneous cell death and increased vulnerability to apoptotic insults in culture as well as in vivo.[262-264] Although an inhibitor of DNA-PK reportedly reduces the manifestation of apoptotic nuclear changes in staurosporine-treated neuroblastoma cells,[265] this finding may be due to the tumorigenic and proliferative properties of neuroblastoma cells. These results suggest that the compromise in

DNA repair activity associated with the *scid* mutation facilitates the accumulation of unrepaired DNA strand breaks, which eventually culminates in neuronal apoptosis.

Poly(ADP-Ribose) Polymerase

Poly(ADP-ribose) polymerase-1 (PARP-1) is the principal member of an expanding family of PARP enzymes,[266] which are activated in response to single strand DNA breaks. In response to DNA damage PARP-1 modifies a variety of nuclear proteins, including itself, by attaching poly(ADP-ribose) chains. The activated enzyme plays an important role in base excision repair and transcriptional regulation, accounting for its contribution to genomic stability.[267-269] Apart from its involvement in DNA repair, PARP-1 has been proposed to have a major impact on cell viability as excessive activation of the enzyme depletes cellular NAD^+ and ATP, causing necrotic cell death[270] (for a review, see ref. 271). In agreement with this idea, PARP-1 inhibitors attenuate neuronal cell death caused by excitotoxicity in culture.[272,273] More compelling results were obtained from studies involving PARP-1 knockout mice, in which PARP-1 deficiency resulted in substantially reduced areas of infarct in an animal model of stroke.[274,275] In addition, the absence of PARP-1 also dramatically reduces the size of lesions induced by intrastriatal NMDA injection[276] and the extent of MPTP-induced loss of dopaminergic neurons.[277] The protection conferred by genetic or pharmacological suppression of PARP-1 activity in models of stroke,[278] NMDA excitotoxicity[276] and traumatic brain injury[279] appears to be long-lasting based on anatomical and behavioral analyses.

The cleavage of PARP-1 by caspase-3 during the process of apoptosis is thought to serve as a switch between necrotic vs. apoptotic cell death, ensuring thaT cells maintain sufficient ATP levels to die by apoptosis. Consistent with this idea, PARP-1 inhibitors attenuate necrotic, but not apoptotic, neuronal death in an in vitro model of cerebral ischemia.[280] What is not clear in the stroke model, however, is whether, subsequent to PARP-1 inhibition, neurons in the core region of an infarct that would otherwise die by necrosis are rescued completely or die by apoptosis. Even though ATP depletion is prevented by PARP-1 inhibition, neurons in the core region still suffer from extensive oxidative damage, which should eventually activate apoptotic pathways. Apparently, however, even the apoptotic process is efficiently blocked since the protection afforded by PARP-1 inhibition is sustained for at least 2-3 weeks,[276,278,279] beyond the typical time course of apoptosis. Thus, these results suggest that PARP-1 inhibition may also work to suppress apoptotic neuronal cell death. This would be consistent with the finding that PARP-1 inhibition can partially suppress p53-dependent neuronal cell death[234] as well as p53 induction.[281]

A number of important issues remain to be examined with respect to the role of PARP in neuronal injury. For example, it is not clear whether NAD/ATP depletion can account for the entire cascade of PARP-mediated changes in neuronal viability. In addition, it is not clear if other members of the PARP family are involved in the neuronal response toxic insults. Nevertheless, data obtained thus far, suggests that PARP-1 plays an important role in several forms of neuronal apoptosis.

BCL-2 FAMILY MEMBERS AND MITOCHONDRIAL INTEGRITY

Properties of the Bcl-2 Family

Proteins belonging to the Bcl-2 family are key regulators of neuronal cell death and survival, and individual family members can serve to inhibit or promote apoptosis.[282] The prototypical anti-apoptotic member, Bcl-2, was first discovered because of its involvement in 95% of follicular B cell lymphomas as a result of a chromosomal translocation t(14;18). High levels of bcl-2 did not act to increase proliferation but rather increased cell survival, thus defining a new class of oncogenes.[283-285] Proteins in this family share a number of Bcl-2 homology domains (BH domains). In general, the anti-apoptotic members, Bcl-2 and Bcl-X_L, contain BH1-4, whereas the pro-apoptotic members, Bax and Bak, contain BH1-3. A third subset of family members including Bad, Bik, Blk, Hrk, Bim, Bid[286] and Noxa[287] contain only the BH3 domain. Many of the family members contain a hydrophobic C-terminal tail, which may serve to anchor these proteins into intracellular membranes.

Bcl-2 proteins regulate the apoptotic cascade mainly at the level of the mitochondria. Upon activation by an apoptotic signal, pro-apoptotic Bcl-2 proteins accumulate at the mitochondrial membrane.[288-291] This is associated with alterations in mitochondrial membrane potential and the release of several pro-apoptotic factors including, cytochrome c, which, in conjunction with Apaf-1, facilitates the activation of the caspase cascade (Fig. 1).[292,293]

The impact of pro-apoptotic Bcl-2 proteins on mitochondria can be inhibited by Bcl-2 and/or Bcl-X_L.[294,295] Although the mechanism underlying the regulation of mitochondrial membrane permeability is not completely understood, current hypotheses suggest that Bcl-2 family members regulate the outer mitochondrial membrane by interaction with the voltage-dependent anion channel (VDAC),[296] and other components of the mitochondrial permeability transition pore (PTP) complex.[297] Furthermore, there is evidence indicating that some Bcl-2 family members can form hetero- and homodimers with inherent ion channel-forming abilities.[298-300]

The ability of Bcl-2 family members to form dimers is essential to their function and regulation. Enforced homodimerization of Bax stimulates Bax translocation, caspase activation and cell death.[301] On the other hand, heterodimerization of anti-apoptotic members such as Bcl-2 or Bcl-X_L with pro-apoptotic members such as Bax can inhibit or activate apoptosis depending on the relative levels of each protein (Rheostat model[302]). Mutational analyses have shown that the BH domains are important for the dimerization of Bcl-2 family members. The BH1 and BH2 domains in Bcl-2 and Bcl-X_L are essential for their dimerization to Bax.[303] NMR and X-ray crystallographic analyses of Bcl-X_L show that the α-helical BH1-3 domains are closely juxtaposed to form a hydrophobic binding pocket.[304,305] Mutation of the

BH3 domain in Bax, Bad and Bid have shown that this domain is essential for both the binding ability of these proteins and their pro-apoptotic function.[306,307] These findings support the rheostat model for Bcl-2 proteins in apoptotic regulation. This hypothesis proposes that anti-apoptotic and pro-apoptotic Bcl-2 proteins exist in a delicate balance at the mitochondrial membrane. A shift in one direction or the other may determine the fate of the cell.

However, the rheostat model is perhaps too simple, as more recent evidence has shown the regulation of these proteins to be much more complex. Pro-apoptotic proteins such as Bad and Bid contain the critical amphipathic BH3 domain, but lack a C-terminal hydrophobic tail present in many of the other membrane-anchored family members. These proteins can move between the cytosol and membranes in a regulated fashion. For example, the pro-apoptotic activity of Bad is regulated by phosphorylation. When Bad is phosphorylated on two specific serine residues it is sequestered by binding to 14-3-3 and can no longer bind Bcl-2 or Bcl-X_L.[308] Phosphorylation also appears to play an important role in regulating other Bcl-2 family proteins. For example, Bcl-2 can be phosphorylated at serine 70 to activate Bcl-2 s anti-apoptotic activity or it can be hyperphosphorylated on multiple sites to inhibit Bcl-2 activity.[309,310] Furthermore, we have shown that p38 MAP kinase activity is involved in regulating Bax function.[127] These changes in phosphorylation state may very well affect the protein-protein interactions of these Bcl-2 family members.

The BH3 domain-only proteins function by modulating the activity of other Bcl-2 family members. Bcl-2 proteins such as Bad, Bim, Bid, Bik, Blk and Hrk can bind to the hydrophobic groove of the anti-apoptotic family members as well as pro-apoptotic family members such as Bax and Bak.[311,312] For example, Bid is a pro-apoptotic Bcl-2 homolog that, once cleaved by caspase-8, is targeted to mitochondria by cardiolipin.[313] Truncated Bid binds and activates pro-apoptotic members, Bax and Bak.[314-316] Other members, such as Bim, Bad and Noxa bind and inhibit the anti-apoptotic members, Bcl-2 and Bcl-X_L.[287,317,318] These interactions suggest that the BH3 domain-only proteins play an important role regulating the function of other Bcl-2 family members.

Bcl-2 Family Members in the Nervous System

Members of the Bcl-2 family are expressed in the nervous system during development and in the adult. Bcl-2 is widely expressed during development, whereas in the adult, its expression is low in the CNS but remains high in the PNS. Conversely, Bcl-X_L remains highly expressed in the nervous system throughout both development and adult life.[319] Bax is down regulated in the adult CNS but highly expressed in the nervous system during the period of natural cell death when the number of neurons is reduced by 20-80%, presumably to match the number of innervating neurons with the size of the target tissue.[320]

Studies with transgenic mice have demonstrated the importance of Bcl-2 family members in regulating neuronal cell death. Animals overexpressing Bcl-2 contained 30-40% more neurons than wild-type animals following the period of developmental neuronal death. Furthermore, neurons overexpressing Bcl-2 are protected from apoptosis after facial and sciatic nerve axotomy as well as optic nerve transection. Although the exact mechanism is not known, it may be related to the effects of Bcl-2 on intracellular Ca^{2+} homeostasis.[321] High levels of Bcl-2 expression have also proven protective in other models of neuronal cell death, for example, following ischemia and excitotoxic injury.[322]

Conversely, the absence of pro-apoptotic family members such as Bax tends to be protective. For example, Bax deficiency protects neurons from ionizing radiation, changes in extracellular ionic conditions, excitotoxicity and ischemia.[204,215,323] One of the important initiators of cell death following stroke is thought to be DNA damage. DNA damage can independently activate cyclin-dependent kinases and p53 pathways, and both mediate Bax activation in neurons.[153,215] Recently, a novel splice variant of Bax, Bax κ, has been shown to promote death following ischemia, and its mRNA was distributed in selectively vulnerable hippocampal CA1 neurons that are destined to die following global ischemia.[324]

Recent evidence has shown Bim to be important in regulating neuronal cell death. Upon withdrawal of nerve growth factor (NGF), neurons activate c-Jun and induce BIM_{EL} expression, and subsequently die in a Bax-dependent manner.[325] Furthermore, overexpression of BIM_{EL} induces cytochrome c release and apoptosis even in the presence of NGF. Finally, neurons treated with Bim antisense oligonucleotides and neurons from $Bim^{-/-}$ mice die more slowly following NGF withdrawal.[326,327] These studies suggest that Bim and perhaps other BH3 domain-only proteins play redundant roles upstream of Bax in the apoptotic pathway following growth factor withdrawal, and that these proteins may be regulated by kinases such as JNK.[328]

On the other hand, there are some Bcl-2 family members that act quite different in neuronal versus non-neuronal cells. Bid is an example of a Bcl-2 family member that seems to have cell type specific functions. One report suggests that, unlike in other cell types, Bid does not play an essential role in either naturally occurring or AraC-induced cell death in neurons.[329] Conversely, another report suggests that cleavage of Bid may amplify caspase-8-induced neuronal death in a seizure model.[330] Interestingly, another BH3 domain-only protein, N-Bak, plays opposite roles in neuronal versus non-neuronal cells. N-Bak is a neuron specific, BH3 only isoform of Bak that is anti-apoptotic in neurons, but pro-apoptotic in non-neuronal cells.[331] This was the first example of a neuron-specific Bcl-2 family member.

Bcl-2 family members have also been implicated in various neurodegenerative diseases such as Parkinson s Disease (PD), Alzheimer s Disease (AD) and motor neuron diseases such as amyotrophic lateral sclerosis (ALS) and spinal muscular atrophy (SMA). Overexpression of Bcl-2 and ablating Bax expression protected dopaminergic neurons from MPTP toxicity, a model of PD.[332,333] Presenilin-1 and 2, genes involved in AD, have also been shown to interact with Bcl-X_L.[334] Mutations in the superoxide dismutase-1 (SOD-1) gene are linked to familial ALS, and spinal cord motor neuron death in mutant SOD-1 mice is associated with a decrease in Bcl-2 expression and an increase in Bax expression.[335-337] Furthermore, in the spinal cords of these mice, Bax translocates from the cytosol to the mitochondria during the progression of the disease.[338] Overexpression of Bcl-2 delays caspase expression, increases motor neuron survival, and improves muscle function in mutant SOD-1 mice.[335,339] However, the disease eventually progresses and duration of the disease is not altered.[340] Survival of motor neuron protein (SMN), which has been implicated in SMA, has also been shown to interact with Bcl-2.[341] In fact, Bcl-2 acts synergistically with SMN to inhibit Bax induced or Fas mediated toxicity.[342] These studies suggest a possible role for Bcl-2 family members in a variety of neuropathological processes. Moreover, manipulating the expression and/or activity of Bcl-2 family members may prove instrumental in maintaining neuronal survival after injury and in response to disease.

Mitochondrial Pro-Apoptotic Factors

Mitochondria play a pivotal role in the amplification of an apoptotic signal since cytochrome c is normally sequestered in the mitochondrial intra-membrane space. In many cell types, apoptosis is associated with a loss in the normal electrochemical gradient employed by mitochondria to generate ATP (the proton motive force).[343] Loss of mitochondrial membrane potential leads to opening of a protein complex known as the permeability transition pore.[344] When this channel is opened, cytochrome c is released into the cytoplasm, leading to the formation of the apoptosome and activation of caspase-9.[343] In neurons mitochondrial cytochrome c release can be independent of permeability transition.[345] Cytochrome c release in this situation may be mediated by the action of pro-apoptotic bcl-2 family members[300] (discussed elsewhere).

In addition to cytochrome c, several other factors are released from mitochondria in response to apoptotic stimuli. One pro-apoptotic protein released by mitochondria is the apoptosis-inducing factor (AIF). AIF is released from mitochondria after permeability transition and translocates to the nucleus where it promotes chromatin condensation and DNA fragmentation in a caspase independent manner.[346] A second factor normally sequestered in mitochondria is known by the dual name of Smac/DIABLO. When released into the cytoplasm, this small protein upregulates the proteolytic activity of caspases by interacting with members of the IAP family. A third pro-apoptotic factor normally resident in mitochondria, endonuclease G, was recently identified as an important apoptosis initiator.[347] At this point, little is known about the role of the mitochondrial proteins Smac/DIABLO, AIF or endonuclease G in excitotoxicity or death receptor mediated neuronal apoptosis. Caspase-2 deficient neurons demonstrate increased levels of smac, which is thought to allow developmental neuronal death to proceed in the absence of caspase-2 by releasing caspase-9 from IAP inhibition.[348] A mutation in the *C. elegans* homologue for endonuclease G also prevents normal progression to neuronal apoptosis during development.[349]

PROTEOLYTIC ENZYMES

Caspases

Pioneering studies aimed at identifying genes required for programmed neuronal death during the course of *C. elegans* development led to the discovery that one of these genes shared strong molecular homology with interlukin-1 converting enzyme (ICE).[350] ICE turned out to be a member of a family of proteases later designated caspase enzymes.[351] Caspase stands for cysteine-dependent aspartate specific protease. To date, fourteen family members have been identified. Some family members (caspase-4 and 5) are thus far unique to humans and others, caspase-12 and 13, have not yet been identified in the human genome. All caspases are synthesized as inactive proenzymes and activated by proteolytic cleavage. The pro-enzyme is cleaved into an N-terminal peptide, a large subunit containing the catalytic portion of the enzyme and a small subunit at the C-terminal end of the peptide. The active enzyme consists of a tetrameric complex of two large subunits and two small subunits[352]. Caspase enzymes have been divided into several distinct groups based on protein structure and putative function. Caspases 1, 4, 5, 11 and 13 are classified as cytokine activators, caspases 2, 8, 9, 10 and 12 are believed to act as initiators of the apoptotic cascade, and caspases 3, 6 and 7 are designated executioners of apoptosis. Caspase 14 appears to have a unique role in supporting the terminal differentiation of keratinocytes in the epidermis and cornea without promoting the typical features of apoptosis.[353,354] Not all studies have supported these designations, and the biologic function associated with some of the less well studied members of this enzyme family have not been evaluated in neurons or neural tissues. In addition, some debate remains about the possibility that caspases known as cytokine activators, primarily caspase-1, may also participate as apoptotic initiators in neurons[355] or oligodendrocytes.[356,357]

The initiator caspases can be further subdivided into two distinct groups. One group is activated following ligand binding to members of the tumor necrosis factor alpha (TNF-α) receptor (TNF-R) family including FAS, TNF-R and TRAIL receptors. These caspases (8 and 10) contain death effector domains (DED) in the N-terminal region of the pro-enzyme. The DED confers homomeric binding ability to adaptor proteins in the death inducing signaling complex (DISC) at the cytoplasmic region of TNF-R family members know as death receptors. Ligand binding of death receptor family members promotes receptor clustering. The predominant theory of caspase activation is that death receptor clustering brings molecules of pro-caspase-8 or pro-caspase-10 in such close proximity that they can be activated by autocleavage.[352]

A second group of caspases (2 and 9) contain a domain know as the caspase recruitment domain (CARD). The CARD domain appears to confer homomeric binding capabilities between caspases and their specific regulatory complexes. Caspase-2 is recruited to the DISC through interactions between its CARD and the

CARD on an adapter protein called RAIDD. RAIDD contains both a CARD domain and a DED domain allowing it to interact with caspase-2 on one end and a DED containing serine/threonine kinase, RIP that is associated with TNF-R1.[358] Thus despite lacking a DED domain, caspase-2 may still be activated by proximity induced auto-processing. Procaspase-9 binds apoptotic protease activating factor-1 (Apaf-1) by CARD domain interaction and Apaf-1 oligomerization promotes proximity induced caspase-9 activation.[359,360] Apaf-1 is a required peptide co-factor for caspase-9 activation, but caspase-9 activation also requires two additional factors; cytochrome c and dATP.[361,362] The complex of these factors, designated the apoptosome, is under intense investigation and additional regulatory proteins continue to be identified.[363] In addition, there is some evidence that human caspase-9 activity is regulated by phosphorylation[71] although the putative phosphorylation sites are not conserved in several other species.[364]

Once initiator caspases are activated, they can proteolytically process executioner caspases (caspases 3, 6 and 7) into their active form. This group of caspases contain truncated N-terminal sequences without known regulatory function. Once activated, their substrates include nuclear and cytoskeletal components that must be proteolyzed for a cell to develop the phenotypic characteristics of apoptosis. The activation of executioner caspases has been demonstrated in variety of neuronal death paradigms. A large number of studies have demonstrated that caspase-3 is activated in the CNS in response to a variety of injurious stimuli including ischemia, excitotoxic insult, trauma and neurodegenerative diseases.[365-371] Caspase-3 is also required for normal developmental cell death in the CNS.[372] Caspase-6 may play a role in the pathogenesis of Alzheimer s and Huntington s disease.[373,374] Caspase-7 is activated during motoneuron degeneration in a mouse model of amyotrophic lateral sclerosis.[338] However, the inhibition or ablation of executioner caspases may not improve neuronal survival, even though the morphologic changes associated with apoptosis do not develop.[375,376]

Endogenous Caspase Inhibitors

Apoptosis appears to be a tightly regulated cellular process. In addition to specific patterns of caspase activation, several endogenous inhibitors of caspase enzyme activity have also been described. One class of caspase inhibitor acts by forming proteolytically inactive heterodimeric complexes with active caspases. Examples of this strategy include splice variants of caspase-2[377] and caspase-14.[354] Very little is known about the presence of caspase inhibitory splice variants in the central or peripheral nervous system. The FLICE (caspase-8) inhibitory protein (FLIP) shares a high degree of sequence homology with caspase-8 and inhibits caspase-8 activity by promoting the formation of a caspase-8/FLIP heterodimers devoid of proteolytic activity.[378] Motoneurons expressing high levels of FLIP are resistant to FasL induced apoptosis[4] and FLIP is expressed in neuroblasts in rat cortex during the early postnatal period.[379] The localization of FLIP in developing neural tissue suggests that FLIP may play a regulatory role in cell survival during nervous system development.

However, FLIP knockout mice have an embryonic lethal phenotype similar to that observed in caspase-8 deficient mice,[380] suggesting that endogenous caspase inhibition is also required for normal caspase function, perhaps by preventing proteolytic inactivation of other important apoptosis regulators.

The peptide family of inhibitor of apoptosis (IAP) genes[381] employs a second strategy for endogenous caspase inhibition. This family of proteins consists of at least seven mammalian members including X-chromosome-linked inhibitor of apoptosis (XIAP). All members of this family contain at least one BIR (baculovirus inhibitor of apoptosis repeat) domain that binds and inhibits the proteolytic activity of caspase-3 and 7. Some of the IAP proteins also inhibit caspase-9 activity (for review see ref. 382). Inhibition by IAP family members can be overcome by several different mechanisms. The c-terminal region of some IAPs contains a domain referred to as the RING (Really Interesting New Gene) domain, which contains ubiquitin E3 ligase activity that targets them for rapid proteasomal degradation.[383] Two other proteins that bind IAPs and prevent caspase inhibition have also been identified. The XIAP-associated factor 1 (XAF1) binds XIAP in the nucleus, potentially promoting release of caspase inhibition once the enzymes have been able to translocate into the nuclear compartment.[384] The second mitochondria-derived activator of caspase (Smac) or DIABLO protein is normally sequestered in mitochondria but will bind and inhibit the anti-apoptotic activities of IAPs when it is released into the cytoplasm.[385,386]

CALPAINS

In addition to caspases, there are additional classes of proteases associated with neuronal cell death. The calpains are calcium-activated cysteine proteases implicated in neuronal cell death following acute neurological insults. Administration of calpain inhibitors during the initial 24-h period following injury can attenuate injury-induced derangements of neuronal structure and function following traumatic brain injury and excitotoxicity.[387-393] At least one substrate, fodrin (spectrin), appears to be shared between the calpains and caspases, but it is not presently known if the caspases and calpains activate one another. In addition, a novel serine protease has been identified which is expressed at ten-fold higher concentrations in the CNS than in peripheral tissues of the rat and human.[394] The mRNA for this protease was found to be significantly elevated by excitotoxic injury. It seems clear that proteases can have profound effects on neuronal viability following injury although additional studies will be required to determine if they represent a rate-limiting step in the apoptotic process.

ABBREVIATIONS

AIF, apoptosis-inducing factor; ALS, amyotrophic lateral sclerosis; Apaf-1, apoptotic protease activating factor-1; ATM, ataxia telangiectasia mutated; BH domain, Bcl-2 homology domain; CARD, caspase recruitment domain; CDK, cyclin-dependent kinase; DD, death domain; DISC, death-inducing signaling complex; DNA-PK, DNA-dependent protein kinase; ERK, extracellular signal-regulated kinase; FADD, Fas-associated death domain; FasL, Fas ligand; FLIP, FLICE (caspase-8) inhibitory protein; IAP, inhibitor of apoptosis; IR, ionizing radiation; JNK, c-Jun N-terminal kinase; MKK, MAP kinase kinase; NMDA, N-methyl-D-aspartate; PARP, poly(ADP-ribose) polymerase; PI3K, phosphatidylinositol-3 kinase; RIP, receptor-interacting protein; TNFR, tumor necrosis factor receptor; TRADD, TNFR-associated death domain; TRAF2, TNF receptor-associated factor 2

REFERENCES

1. Walczak H, Krammer PH. The CD95 (APO-1/Fas) and the TRAIL (APO-2L) apoptosis systems. Exp Cell Res 2000; 256:58-66.
2. Bechmann I, Mor G, Nilsen J, Eliza M, Nitsch R, Naftolin F. FasL (CD95L, Apo1L) is expressed in the normal rat and human brain: Evidence for the existence of an immunological brain barrier. Glia 1999; 27:62-74.
3. Felderhoff-Mueser U, Taylor DL, Greenwood K, Kozma M, Stibenz D, Joashi UC et al. Fas/CD95/APO-1 can function as a death receptor for neuronal cells in vitro and in vivo and is upregulated following cerebral hypoxic-ischemic injury to the developing rat brain. Brain Pathol 2000; 10:17-29.
4. Raoul C, Henderson CE, Pettmann B. Programmed cell death of embryonic motoneurons triggered through the Fas death receptor. J Cell Biol 1999; 147:1049-1062.
5. Matsushita K, Wu Y, Qiu J, Lang-Lazdunski L, Hirt L, Waeber C et al. Fas receptor and neuronal cell death after spinal cord ischemia. J Neurosci 2000; 20:6879-6887.
6. Martin-Villalba A, Herr I, Jeremias I, Hahne M, Brandt R, Vogel J et al. CD95 ligand (Fas-L/APO-1L) and tumor necrosis factor-related apoptosis-inducing ligand mediate ischemia-induced apoptosis in neurons. J Neurosci 1999; 19:3809-3817.
7. Rosenbaum DM, Gupta G, D Amore J, Singh M, Weidenheim K, Zhang H et al. Fas (CD95/APO-1) plays a role in the pathophysiology of focal cerebral ischemia. J Neurosci Res 2000; 61:686-692.
8. Elovaara I, Sabri F, Gray F, Alafuzoff I, Chiodi F. Upregulated expression of Fas and Fas ligand in brain through the spectrum of HIV-1 infection. Acta Neuropathol (Berl) 1999; 98:355-362.
9. Kaul M, Garden GA, Lipton SA. Pathways to neuronal injury and apoptosis in HIV-associated dementia. Nature 2001; 410:988-994.
10. Sharma K, Wang RX, Zhang LY, Yin DL, Luo XY, Solomon JC et al. Death the Fas way: Regulation and pathophysiology of CD95 and its ligand. Pharmacol Ther 2000; 88:333-347.
11. Raoul C, Pettmann B, Henderson CE. Active killing of neurons during development and following stress: a role for p75(NTR) and Fas? Curr Opin Neurobiol 2000; 10:111-117.
12. Varfolomeev EE, Schuchmann M, Luria V, Chiannilkulchai N, Beckmann JS, Mett IL et al. Targeted disruption of the mouse Caspase 8 gene ablates cell death induction by the TNF receptors, Fas/Apo1, and DR3 and is lethal prenatally. Immunity 1998; 9:267-276.
13. Li H, Zhu H, Xu CJ, Yuan J. Cleavage of BID by caspase 8 mediates the mitochondrial damage in the Fas pathway of apoptosis. Cell 1998; 94:491-501.
14. Scaffidi C, Fulda S, Srinivasan A, Friesen C, Li F, Tomaselli KJ et al. Two CD95 (APO-1/Fas) signaling pathways. Embo J 1998; 17:1675-1687.
15. Bratton SB, MacFarlane M, Cain K, Cohen GM. Protein complexes activate distinct caspase cascades in death receptor and stress-induced apoptosis. Exp Cell Res 2000; 256:27-33.

16. Wang J, Zheng L, Lobito A, Chan FK, Dale J, Sneller M et al. Inherited human Caspase 10 mutations underlie defective lymphocyte and dendritic cell apoptosis in autoimmune lymphoproliferative syndrome type II. Cell 1999; 98:47-58.
17. Reimann-Philipp U, Ovase R, Weigel PH, Grammas P. Mechanisms of cell death in primary cortical neurons and PC12 cells. J Neurosci Res 2001; 64:654-660.
18. Barker V, Middleton G, Davey F, Davies AM. TNFalpha contributes to the death of NGF-dependent neurons during development. Nat Neurosci 2001; 4:1194-1198.
19. Kim GM, Xu J, Song SK, Yan P, Ku G, Xu XM et al. Tumor necrosis factor receptor deletion reduces nuclear factor-kappaB activation, cellular inhibitor of apoptosis protein 2 expression, and functional recovery after traumatic spinal cord injury. J Neurosci 2001; 21:6617-6625.
20. Lin Y, Devin A, Cook A, Keane MM, Kelliher M, Lipkowitz S et al. The death domain kinase RIP is essential for TRAIL (Apo2L)-induced activation of IkappaB kinase and c-Jun N-terminal kinase. Mol Cell Biol 2000; 20: 6638-6645.
21. Wajant H, Henkler F, Scheurich P. The TNF-receptor-associated factor family: Scaffold molecules for cytokine receptors, kinases and their regulators. Cell Signal 2001; 13:389-400.
22. Baud V, Karin M. Signal transduction by tumor necrosis factor and its relatives. Trends Cell Biol 2001; 11:372-377.
23. Grell M, Zimmermann G, Gottfried E, Chen CM, Grunwald U, Huang DC et al. Induction of cell death by tumour necrosis factor (TNF) receptor 2, CD40 and CD30: A role for TNF-R1 activation by endogenous membrane-anchored TNF. Embo J 1999; 18:3034-3043.
24. Griffith TS, Lynch DH. TRAIL: A molecule with multiple receptors and control mechanisms. Curr Opin Immunol 1998; 10:559-563.
25. Wiley SR, Schooley K, Smolak PJ, Din WS, Huang CP, Nicholl JK et al. Identification and characterization of a new member of the TNF family that induces apoptosis. Immunity 1995; 3:673-682.
26. Sheikh MS, Fornace AJ, Jr. Death and decoy receptors and p53-mediated apoptosis. Leukemia 2000; 14:1509-1513.
27. Degli-Esposti MA, Dougall WC, Smolak PJ, Waugh JY, Smith CA, Goodwin RG. The novel receptor TRAIL-R4 induces NF-kappaB and protects against TRAIL-mediated apoptosis, yet retains an incomplete death domain. Immunity 1997; 7:813-820.
28. Ashkenazi A, Pai RC, Fong S, Leung S, Lawrence DA, Marsters SA et al. Safety and antitumor activity of recombinant soluble Apo2 ligand. J Clin Invest 1999; 104:155-162.
29. Zhang XD, Nguyen T, Thomas WD, Sanders JE, Hersey P. Mechanisms of resistance of normal cells to TRAIL induced apoptosis vary between different cell types. FEBS Lett 2000; 482:193-199.
30. Jo M, Kim TH, Seol DW, Esplen JE, Dorko K, Billiar TR et al. Apoptosis induced in normal human hepatocytes by tumor necrosis factor- related apoptosis-inducing ligand. Nat Med 2000; 6:564-567.
31. Frank S, Kohler U, Schackert G, Schackert HK. Expression of TRAIL and its receptors in human brain tumors. Biochem Biophys Res Commun 1999; 257:454-459.
32. Nitsch R, Bechmann I, Deisz RA, Haas D, Lehmann TN, Wendling U et al. Human brain-cell death induced by tumour-necrosis-factor-related apoptosis-inducing ligand (TRAIL). Lancet 2000; 356:827-828.
33. Eggert A, Grotzer MA, Zuzak TJ, Wiewrodt BR, Ikegaki N, Brodeur GM. Resistance to TRAIL-induced apoptosis in neuroblastoma cells correlates with a loss of caspase-8 expression. Med Pediatr Oncol 2000; 35:603-607.
34. Sprick MR, Weigand MA, Rieser E, Rauch CT, Juo P, Blenis J et al. FADD/MORT1 and caspase-8 are recruited to TRAIL receptors 1 and 2 and are essential for apoptosis mediated by TRAIL receptor 2. Immunity 2000; 12:599-609.
35. Mariani SM, Matiba B, Armandola EA, Krammer PH. Interleukin 1 beta-converting enzyme related proteases/caspases are involved in TRAIL-induced apoptosis of myeloma and leukemia cells. J Cell Biol 1997; 137:221-229.

36. Kuang AA, Diehl GE, Zhang J, Winoto A. FADD is required for DR4- and DR5-mediated apoptosis: lack of trail- induced apoptosis in FADD-deficient mouse embryonic fibroblasts. J Biol Chem 2000; 275:25065-25068.
37. Meng RD, McDonald ER, 3rd, Sheikh MS, Fornace AJ, Jr., El-Deiry WS. The TRAIL decoy receptor TRUNDD (DcR2, TRAIL-R4) is induced by adenovirus-p53 overexpression and can delay TRAIL-, p53-, and KILLER/DR5-dependent colon cancer apoptosis. Mol Ther 2000; 1:130-144.
38. Meng RD, El-Deiry WS. p53-independent upregulation of KILLER/DR5 TRAIL receptor expression by glucocorticoids and interferon-gamma. Exp Cell Res 2001; 262:154-169.
39. Sheikh MS, Huang Y, Fernandez-Salas EA, El-Deiry WS, Friess H, Amundson S et al. The antiapoptotic decoy receptor TRID/TRAIL-R3 is a p53-regulated DNA damage-inducible gene that is overexpressed in primary tumors of the gastrointestinal tract. Oncogene 1999; 18:4153-4159.
40. Pan G, Ni J, Wei YF, Yu G, Gentz R, Dixit VM. An antagonist decoy receptor and a death domain-containing receptor for TRAIL. Science 1997; 277:815-818.
41. Sheridan JP, Marsters SA, Pitti RM, Gurney A, Skubatch M, Baldwin D et al. Control of TRAIL-induced apoptosis by a family of signaling and decoy receptors. Science 1997; 277:818-821.
42. Griffith TS, Rauch CT, Smolak PJ, Waugh JY, Boiani N, Lynch DH et al. Functional analysis of TRAIL receptors using monoclonal antibodies. J Immunol 1999; 162:2597-2605.
43. Zhang XD, Franco A, Myers K, Gray C, Nguyen T, Hersey P. Relation of TNF-related apoptosis-inducing ligand (TRAIL) receptor and FLICE-inhibitory protein expression to TRAIL-induced apoptosis of melanoma. Cancer Res 1999; 59:2747-2753.
44. Sedel F, Bechade C, Triller A. Nerve growth factor (NGF) induces motoneuron apoptosis in rat embryonic spinal cord in vitro. Eur J Neurosci 1999; 11:3904-3912.
45. Wiese S, Metzger F, Holtmann B, Sendtner M. The role of p75NTR in modulating neurotrophin survival effects in developing motoneurons. Eur J Neurosci 1999; 11:1668-1676.
46. Ferri CC, Moore FA, Bisby MA. Effects of facial nerve injury on mouse motoneurons lacking the p75 low- affinity neurotrophin receptor. J Neurobiol 1998; 34:1-9.
47. Bamji SX, Majdan M, Pozniak CD, Belliveau DJ, Aloyz R, Kohn et al. The p75 neurotrophin receptor mediates neuronal apoptosis and is essential for naturally occurring sympathetic neuron death. J Cell Biol 1998; 140:911-923.
48. Frade JM, Barde YA. Genetic evidence for cell death mediated by nerve growth factor and the neurotrophin receptor p75 in the developing mouse retina and spinal cord. Development 1999; 126:683-690.
49. Casaccia-Bonnefil P, Kong H, Chao MV. Neurotrophins: the biological paradox of survival factors eliciting apoptosis. Cell Death Differ 1998; 5:357-364.
50. Wang X, Bauer JH, Li Y, Shao Z, Zetoune FS, Cattaneo E et al. Characterization of a p75(NTR) apoptotic signaling pathway using a novel cellular model. J Biol Chem 2001; 276:33812-33820.
51. Casademunt E, Carter BD, Benzel I, Frade JM, Dechant G, Barde YA. The zinc finger protein NRIF interacts with the neurotrophin receptor p75(NTR) and participates in programmed cell death. Embo J 1999; 18:6050-6061.
52. Mukai J, Hachiya T, Shoji-Hoshino S, Kimura MT, Nadano D, Suvanto P et al. NADE, a p75NTR-associated cell death executor, is involved in signal transduction mediated by the common neurotrophin receptor p75NTR. J Biol Chem 2000; 275:17566-17570.
53. Salehi AH, Roux PP, Kubu CJ, Zeindler C, Bhakar A, Tannis LL et al. NRAGE, a novel MAGE protein, interacts with the p75 neurotrophin receptor and facilitates nerve growth factor-dependent apoptosis. Neuron 2000; 27:279-288.
54. Khursigara G, Bertin J, Yano H, Moffett H, DiStefano PS, Chao MV. A prosurvival function for the p75 receptor death domain mediated via the caspase recruitment domain receptor-interacting protein 2. J Neurosci 2001; 21:5854-5863.

55. Yao R, Cooper GM. Requirement for phosphatidylinositol-3 kinase in the prevention of apoptosis by nerve growth factor. Science 1995; 267:2003-2006.
56. Crowder RJ, Freeman RS. Phosphatidylinositol 3-kinase and Akt protein kinase are necessary and sufficient for the survival of nerve growth factor-dependent sympathetic neurons. J Neurosci 1998; 18:2933-2943.
57. Ghosh A, Greenberg ME. Distinct roles for bFGF and NT-3 in the regulation of cortical neurogenesis. Neuron 1995; 15:89-103.
58. Kuruvilla R, Ye H, Ginty DD. Spatially and functionally distinct roles of the PI3-K effector pathway during NGF signaling in sympathetic neurons. Neuron 2000; 27:499-512.
59. Segal RA, Greenberg ME. Intracellular signaling pathways activated by neurotrophic factors. Annu Rev Neurosci 1996; 19:463-489
60. Brunet A, Datta SR, Greenberg ME. Transcription-dependent and -independent control of neuronal survival by the PI3K-Akt signaling pathway. Curr Opin Neurobiol 2001; 11:297-305.
61. Holgado-Madruga M, Moscatello DK, Emlet DR, Dieterich R, Wong AJ. Grb2-associated binder-1 mediates phosphatidylinositol 3-kinase activation and the promotion of cell survival by nerve growth factor. Proc Natl Acad Sci USA 1997; 94:12419-12424.
62. Kauffmann-Zeh A, Rodriguez-Viciana P, Ulrich E, Gilbert C, Coffer P, Downward J et al. Suppression of c-Myc-induced apoptosis by Ras signalling through PI(3)K and PKB. Nature 1997; 385:544-548.
63. Dhand R, Hara K, Hiles I, Bax B, Gout I, Panayotou G et al. PI 3-kinase: structural and functional analysis of intersubunit interactions. Embo J 1994; 13:511-521.
64. Hu Q, Klippel A, Muslin AJ, Fantl WJ, Williams LT. Ras-dependent induction of cellular responses by constitutively active phosphatidylinositol-3 kinase. Science 1995; 268:100-102.
65. Dudek H, Datta SR, Franke TF, Birnbaum MJ, Yao R, Cooper GM et al. Regulation of neuronal survival by the serine-threonine protein kinase Akt. Science 1997; 275:661-665.
66. Brunet A, Bonni A, Zigmond MJ, Lin MZ, Juo P, Hu LS et al. Akt promotes cell survival by phosphorylating and inhibiting a Forkhead transcription factor. Cell 1999; 96:857-868.
67. Yamaguchi A, Tamatani M, Matsuzaki H, Namikawa K, Kiyama H, Vitek MP et al. Akt activation protects hippocampal neurons from apoptosis by inhibiting transcriptional activity of p53. J Biol Chem 2001; 276:5256-5264.
68. Bonni A, Brunet A, West AE, Datta SR, Takasu MA, Greenberg ME. Cell survival promoted by the Ras-MAPK signaling pathway by transcription-dependent and -independent mechanisms. Science 1999; 286:1358-1362.
69. Maggirwar SB, Sarmiere PD, Dewhurst S, Freeman RS. Nerve growth factor-dependent activation of NF-kappaB contributes to survival of sympathetic neurons. J Neurosci 1998; 18:10356-10365.
70. Datta SR, Dudek H, Tao X, Masters S, Fu H, Gotoh Y et al. Akt phosphorylation of BAD couples survival signals to the cell- intrinsic death machinery. Cell 1997; 91:231-241.
71. Cardone MH, Roy N, Stennicke HR, Salvesen GS, Franke TF, Stanbridge E et al. Regulation of cell death protease caspase-9 by phosphorylation. Science 1998; 282:1318-1321.
72. Hetman M, Cavanaugh JE, Kimelman D, Xia Z. Role of glycogen synthase kinase-3beta in neuronal apoptosis induced by trophic withdrawal. J Neurosci 2000; 20:2567-2574.
73. Xia Z, Dickens M, Raingeaud J, Davis RJ, Greenberg ME. Opposing effects of ERK and JNK-p38 MAP kinases on apoptosis. Science 1995; 270:1326-1331.
74. Impey S, Obrietan K, Storm DR. Making new connections: role of ERK/MAP kinase signaling in neuronal plasticity. Neuron 1999; 23:11-14.
75. Sweatt JD. The neuronal MAP kinase cascade: a biochemical signal integration system subserving synaptic plasticity and memory. J Neurochem 2001; 76:1-10.
76. Murray B, Alessandrini A, Cole AJ, Yee AG, Furshpan EJ. Inhibition of the p44/42 MAP kinase pathway protects hippocampal neurons in a cell-culture model of seizure activity. Proc Natl Acad Sci USA 1998; 95:11975-11980.
77. Kulich SM, Chu CT. Sustained extracellular signal-regulated kinase activation by 6-hydroxydopamine: Implications for Parkinson s disease. J Neurochem 2001; 77:1058-1066.

78. Stanciu M, DeFranco DB. Prolonged nuclear retention of activated ERK promotes cell death generated by oxidative toxicity or proteasome inhibition in a neuronal cell line. J Biol Chem 2002; 277:4010-4017.
79. Abe MK, Kahle KT, Saelzler MP, Orth K, Dixon JE, Rosner MR. ERK7 is an autoactivated member of the MAPK family. J Biol Chem 2001; 276:21272-21279.
80. Hetman M, Xia Z. Signaling pathways mediating anti-apoptotic action of neurotrophins. Acta Neurobiol Exp 2000; 60:531-545
81. Grewal SS, York RD, Stork PJ. Extracellular-signal-regulated kinase signalling in neurons. Curr Opin Neurobiol 1999; 9:544-553.
82. Alessandrini A, Namura S, Moskowitz MA, Bonventre JV. MEK1 protein kinase inhibition protects against damage resulting from focal cerebral ischemia. Proc Natl Acad Sci USA 1999; 96:12866-12869.
83. Grammer TC, Blenis J. Evidence for MEK-independent pathways regulating the prolonged activation of the ERK-MAP kinases. Oncogene 1997; 14:1635-1642.
84. Hetman M, Kanning K, Cavanaugh JE, Xia Z. Neuroprotection by brain-derived neurotrophic factor is mediated by extracellular signal-regulated kinase and phosphatidylinositol 3-kinase. J Biol Chem 1999; 274:22569-22580.
85. Gonzalez-Zulueta M, Feldman AB, Klesse LJ, Kalb RG, Dillman JF, Parada LF et al. Requirement for nitric oxide activation of p21(ras)/extracellular regulated kinase in neuronal ischemic preconditioning. Proc Natl Acad Sci USA 2000; 97:436-441.
86. Ferrer I, Ballabriga J, Pozas E. Transient forebrain ischemia in the adult gerbil is associated with a complex c-Jun response. Neuroreport 1997; 8:2483-2487.
87. Yang DD, Kuan CY, Whitmarsh AJ, Rincon M, Zheng TS, Davis RJ et al. Absence of excitotoxicity-induced apoptosis in the hippocampus of mice lacking the JNK3 gene. Nature 1997; 389:865-870.
88. Jeon SH, Kim YS, Bae CD, Park JB. Activation of JNK and p38 in rat hippocampus after kainic acid induced seizure. Exp Mol Med 2000; 32:227-230.
89. Chihab R, Ferry C, Koziel V, Monin P, Daval JL. Sequential activation of activator protein-1-related transcription factors and JNK protein kinases may contribute to apoptotic death induced by transient hypoxia in developing brain neurons. Brain Res Mol Brain Res 1998; 63:105-120.
90. Morishima Y, Gotoh Y, Zieg J, Barrett T, Takano H, Flavell R et al. Beta-amyloid induces neuronal apoptosis via a mechanism that involves the c-Jun N-terminal kinase pathway and the induction of Fas ligand. J Neurosci 2001; 21:7551-7560.
91. Troy CM, Rabacchi SA, Xu Z, Maroney AC, Connors TJ, Shelanski ML et al. beta-Amyloid-induced neuronal apoptosis requires c-Jun N-terminal kinase activation. J Neurochem 2001; 77:157-164.
92. Namgung U, Xia Z. Arsenite-induced apoptosis in cortical neurons is mediated by c-Jun N-terminal protein kinase 3 and p38 mitogen-activated protein kinase. J Neurosci 2000; 20:6442-6451.
93. Herdegen T, Claret FX, Kallunki T, Martin-Villalba A, Winter C, Hunter T et al. Lasting N-terminal phosphorylation of c-Jun and activation of c-Jun N-terminal kinases after neuronal injury. J Neurosci 1998; 18:5124-5135.
94. Mielke K, Herdegen T. JNK and p38 stresskinases Degenerative effectors of signal-transduction-cascades in the nervous system. Prog Neurobiol 2000; 61:45-60.
95. Gupta S, Barrett T, Whitmarsh AJ, Cavanagh J, Sluss HK, Derijard B et al. Selective interaction of JNK protein kinase isoforms with transcription factors. Embo J 1996; 15:2760-2770.
96. Bozyczko-Coyne D, O Kane TM, Wu ZL, Dobrzanski P, Murthy S, Vaught JL et al. CEP-1347/KT-7515, an inhibitor of SAPK/JNK pathway activation, promotes survival and blocks multiple events associated with Abeta-induced cortical neuron apoptosis. J Neurochem 2001; 77:849-863.

97. Zhu X, Raina AK, Rottkamp CA, Aliev G, Perry G, Boux H et al. Activation and redistribution of c-Jun N-terminal kinase/stress activated protein kinase in degenerating neurons in Alzheimer s disease. J Neurochem 2001; 76:435-441.
98. Saporito MS, Thomas BA, Scott RW. MPTP activates c-Jun NH(2)-terminal kinase (JNK) and its upstream regulatory kinase MKK4 in nigrostriatal neurons in vivo. J Neurochem 2000; 75:1200-1208.
99. Harding TC, Xue L, Bienemann A, Haywood D, Dickens M, Tolkovsky AM et al. Inhibition of JNK by overexpression of the JNK binding domain of JIP-1 prevents apoptosis in sympathetic neurons. J Biol Chem 2001; 276:4531-4534.
100. Le-Niculescu H, Bonfoco E, Kasuya Y, Claret FX, Green DR, Karin M Withdrawal of survival factors results in activation of the JNK pathway in neuronal cells leading to Fas ligand induction and cell death. Mol Cell Biol 1999; 19:751-763.
101. Bruckner SR, Tammariello SP, Kuan CY, Flavell RA, Rakic P, Estus S. JNK3 contributes to c-Jun activation and apoptosis but not oxidative stress in nerve growth factor-deprived sympathetic neurons. J Neurochem 2001; 78:298-303.
102. Behrens A, Sibilia M, Wagner EF. Amino-terminal phosphorylation of c-Jun regulates stress-induced apoptosis and cellular proliferation. Nat Genet 1999; 21:326-329.
103. Yang D, Tournier C, Wysk M, Lu HT, Xu J, Davis RJ et al. Targeted disruption of the MKK4 gene causes embryonic death, inhibition of c-Jun NH2-terminal kinase activation, and defects in AP-1 transcriptional activity. Proc Natl Acad Sci USA 1997; 94:3004-3009.
104. Holland PM, Suzanne M, Campbell JS, Noselli S, Cooper JA. MKK7 is a stress-activated mitogen-activated protein kinase kinase functionally related to hemipterous. J Biol Chem 1997; 272:24994-24998.
105. Ham J, Eilers A, Whitfield J, Neame SJ, Shah B. c-Jun and the transcriptional control of neuronal apoptosis. Biochem Pharmacol 2000; 60:1015-1021.
106. Harper SJ, LoGrasso P. Signalling for survival and death in neurones: The role of stress-activated kinases, JNK and p38. Cell Signal 2001; 13:299-310.
107. Tibbles LA, Ing YL, Kiefer F, Chan J, Iscove N, Woodgett JR et al. MLK-3 activates the SAPK/JNK and p38/RK pathways via SEK1 and MKK3/6. Embo J 1996; 15:7026-7035.
108. Lin A, Minden A, Martinetto H, Claret FX, Lange-Carter C, Mercurio F et al. Identification of a dual specificity kinase that activates the Jun kinases and p38-Mpk2. Science 1995; 268:286-290.
109. Ichijo H, Nishida E, Irie K, ten Dijke P, Saitoh M, Moriguchi T et al. Induction of apoptosis by ASK1, a mammalian MAPKKK that activates SAPK/JNK and p38 signaling pathways. Science 1997; 275:90-94.
110. Wang XS, Diener K, Jannuzzi D, Trollinger D, Tan TH, Lichenstein H et al. Molecular cloning and characterization of a novel protein kinase with a catalytic domain homologous to mitogen-activated protein kinase kinase kinase. J Biol Chem 1996; 271:31607-31611.
111. Kanamoto T, Mota M, Takeda K, Rubin LL, Miyazono K, Ichijo H et al. Role of apoptosis signal-regulating kinase in regulation of the c-Jun N-terminal kinase pathway and apoptosis in sympathetic neurons. Mol Cell Biol 2000; 20:196-204.
112. Whitmarsh AJ, Cavanagh J, Tournier C, Yasuda J, Davis RJ. A mammalian scaffold complex that selectively mediates MAP kinase activation. Science 1998; 281:1671-1674.
113. Derijard B, Hibi M, Wu IH, Barrett T, Su B, Deng T et al. JNK1: A protein kinase stimulated by UV light and Ha-Ras that binds and phosphorylates the c-Jun activation domain. Cell 1994; 76:1025-1037.
114. Hu MC, Qiu WR, Wang YP. JNK1, JNK2 and JNK3 are p53 N-terminal serine 34 kinases. Oncogene 1997; 15:2277-2287.
115. Fuchs SY, Adler V, Pincus MR, Ronai Z. MEKK1/JNK signaling stabilizes and activates p53. Proc Natl Acad Sci USA 1997; 95:10541-10546.
116. Zhang Y, Zhou L, Miller CA. A splicing variant of a death domain protein that is regulated by a mitogen-activated kinase is a substrate for c-Jun N-terminal kinase in the human central nervous system. Proc Natl Acad Sci USA 1998; 95:2586-2591.

117. Derijard B, Raingeaud J, Barrett T, Wu IH, Han J, Ulevitch RJ et al. Independent human MAP-kinase signal transduction pathways defined by MEK and MKK isoforms. Science 1995; 267:682-685.
118. Raingeaud J, Gupta S, Rogers JS, Dickens M, Han J, Ulevitch RJ et al. Pro-inflammatory cytokines and environmental stress cause p38 mitogen- activated protein kinase activation by dual phosphorylation on tyrosine and threonine. J Biol Chem 1995; 270:7420-7426.
119. Han J, Lee JD, Jiang Y, Li Z, Feng L, Ulevitch RJ. Characterization of the structure and function of a novel MAP kinase kinase (MKK6). J Biol Chem 1996; 271:2886-2891.
120. Nath R, McGinnis K, Dutta S, Shivers B, Wang KK. Inhibition of p38 kinase mimics survival signal-linked protection against apoptosis in rat cerebellar granule neurons. Cell Mol Biol Lett 2001; 6:173-184
121. Yamagishi S, Yamada M, Ishikawa Y, Matsumoto T, Ikeuchi T, Hatanaka H. p38 mitogen-activated protein kinase regulates low potassium-induced c-Jun phosphorylation and apoptosis in cultured cerebellar granule neurons. J Biol Chem 2001; 276:5129-5133.
122. Willaime S, Vanhoutte P, Caboche J, Lemaigre-Dubreuil Y, Mariani J, Brugg B. Ceramide-induced apoptosis in cortical neurons is mediated by an increase in p38 phosphorylation and not by the decrease in ERK phosphorylation. Eur J Neurosci 2011; 13:2037-2046.
123. McLaughlin B, Pal S, Tran MP, Parsons AA, Barone FC, Erhardt JA et al. p38 activation is required upstream of potassium current enhancement and caspase cleavage in thiol oxidant-induced neuronal apoptosis. J Neurosci 2001; 21:3303-3311.
124. De Zutter GS, Davis RJ. Pro-apoptotic gene expression mediated by the p38 mitogen-activated protein kinase signal transduction pathway. Proc Natl Acad Sci USA 2001; 98:6168-6173.
125. Barone FC, Irving EA, Ray AM, Lee JC, Kassis S, Kumar S et al. Inhibition of p38 mitogen-activated protein kinase provides neuroprotection in cerebral focal ischemia. Med Res Rev 2001; 21:129-145.
126. Kawasaki H, Morooka T, Shimohama S, Kimura J, Hirano T, Gotoh Y et al. Activation and involvement of p38 mitogen-activated protein kinase in glutamate-induced apoptosis in rat cerebellar granule cells. J Biol Chem 1997; 272:18518-18521.
127. Ghatan S, Larner S, Kinoshita Y, Hetman M, Patel L, Xia Z et al. p38 MAP kinase mediates Bax translocation in nitric oxide-induced apoptosis in neurons. J Cell Biol 2000; 150:335-347.
128. Cheng A, Chan SL, Milhavet O, Wang S, Mattson MP. p38 MAP kinase mediates nitric oxide-induced apoptosis of neural progenitor cells. J Biol Chem 2001; 276:43320-43327.
129. Grimes CA, Jope RS. The multifaceted roles of glycogen synthase kinase 3beta in cellular signaling. Prog Neurobiol 2001; 65:391-426.
130. Cross DA, Alessi DR, Cohen P, Andjelkovich M, Hemmings BA. Inhibition of glycogen synthase kinase-3 by insulin mediated by protein kinase B. Nature 1995; 378:785-789.
131. Delcommenne M, Tan C, Gray V, Rue L, Woodgett J, Dedhar S. Phosphoinositide-3-OH kinase-dependent regulation of glycogen synthase kinase 3 and protein kinase B/AKT by the integrin-linked kinase. Proc Natl Acad Sci USA 1998; 95:11211-11216.
132. Spittaels K, Van den Haute C, Van Dorpe J, Geerts H, Mercken M, Bruynseels K et al. Glycogen synthase kinase-3beta phosphorylates protein tau and rescues the axonopathy in the central nervous system of human four-repeat tau transgenic mice. J Biol Chem 2000; 275:41340-41349.
133. Lucas JJ, Hernandez F, Gomez-Ramos P, Moran MA, Hen R, Avila J. Decreased nuclear beta-catenin, tau hyperphosphorylation and neurodegeneration in GSK-3beta conditional transgenic mice. Embo J 2001; 20:27-39.
134. Mattson MP. Neuronal death and GSK-3beta: A tau fetish? Trends Neurosci 2001; 24:255-256.
135. Copani A, Uberti D, Sortino MA, Bruno V, Nicoletti F, Memo M. Activation of cell-cycle-associated proteins in neuronal death: a mandatory or dispensable path? Trends Neurosci 2001; 24:25-31.
136. Gao CY, Zelenka PS. Induction of cyclin B and H1 kinase activity in apoptotic PC12 cells. Exp Cell Res 1995; 219:612-618.

137. Timsit S, Rivera S, Ouaghi P, Guischard F, Tremblay E, Ben-Ari Y et al. Increased cyclin Dl in vulnerable neurons in the hippocampus after ischaemia and epilepsy: a modulator of in vivo programmed cell death? Eur J Neurosci 1999; 11:263-278.
138. Osuga H, Osuga S, Wang F, Fetni R, Hogan MJ, Slack RS et al. Cyclin-dependent kinases as a therapeutic target for stroke. Proc Natl Acad Sci USA 2000; 97:10254-10259.
139. Giovanni A, Wirtz-Brugger F, Keramaris E, Slack R, Park DS. Involvement of cell cycle elements, cyclin-dependent kinases, pRb, and E2F x DP, in B-amyloid-induced neuronal death. J Biol Chem 1999; 274:19011-19016.
140. Copani A, Condorelli F, Caruso A, Vancheri C, Sala A, Giuffrida Stella AM et al. Mitotic signaling by beta-amyloid causes neuronal death. Faseb J 1999; 13:2225-2234.
141. Park DS, Morris EJ, Padmanabhan J, Shelanski ML, Geller HM, Greene LA. Cyclin-dependent kinases participate in death of neurons evoked by DNA- damaging agents. J Cell Biol 1998; 143:457-467.
142. McShea A, Harris PL, Webster KR, Wahl AF, Smith MA. Abnormal expression of the cell cycle regulators P16 and CDK4 in Alzheimer s disease. Am J Pathol 1997; 150:1933-1939.
143. Vincent I, Jicha G, Rosado M, Dickson DW. Aberrant expression of mitotic cdc2/cyclin B1 kinase in degenerating neurons of Alzheimer s disease brain. J Neurosci 1997; 17:3588-3598.
144. Busser J, Geldmacher DS, Herrup K. Ectopic cell cycle proteins predict the sites of neuronal cell death in Alzheimer s disease brain. J Neurosci 1998; 18:2801-2807.
145. Ding XL, Husseman J, Tomashevski A, Nochlin D, Jin LW, Vincent I. The cell cycle Cdc25A tyrosine phosphatase is activated in degenerating postmitotic neurons in Alzheimer s disease. Am J Pathol 2000; 157:1983-1990.
146. Maccioni RB, Otth C, Concha, II, Munoz JP. The protein kinase Cdk5. Structural aspects, roles in neurogenesis and involvement in Alzheimer s pathology. Eur J Biochem 2001; 268:1518-1527.
147. Vincent I, Bu B, Hudson K, Husseman J, Nochlin D, Jin L. Constitutive Cdc25B tyrosine phosphatase activity in adult brain neurons with M phase-type alterations in Alzheimer s disease. Neuroscience 2001; 105:639-650
148. Patrick GN, Zukerberg L, Nikolic M, de la Monte S, Dikkes P, Tsai LH. Conversion of p35 to p25 deregulates Cdk5 activity and promotes neurodegeneration. Nature 1999; 402:615-622.
149. Nguyen MD, Lariviere RC, Julien JP. Deregulation of Cdk5 in a mouse model of ALS: Toxicity alleviated by perikaryal neurofilament inclusions. Neuron 2001; 30:135-147.
150. Park DS, Levine B, Ferrari G, Greene LA. Cyclin dependent kinase inhibitors and dominant negative cyclin dependent kinase 4 and 6 promote survival of NGF-deprived sympathetic neurons. J Neurosci 1997; 17:8975-8983.
151. Padmanabhan J, Park DS, Greene LA, Shelanski ML. Role of cell cycle regulatory proteins in cerebellar granule neuron apoptosis. J Neurosci 1999; 19:8747-8756.
152. Park DS, Morris EJ, Greene LA, Geller HM. Gl/S cell cycle blockers and inhibitors of cyclin-dependent kinases suppress camptothecin-induced neuronal apoptosis. J Neurosci 1997; 17:1256-1270.
153. Morris EJ, Keramaris E, Rideout HJ, Slack RS, Dyson NJ, Stefanis L et al. Cyclin-dependent kinases and P53 pathways are activated independently and mediate Bax activation in neurons after DNA damage. J Neurosci 2001; 21:5017-5026.
154. Park DS, Morris EJ, Stefanis L, Troy CM, Shelanski ML, Geller HM. Multiple pathways of neuronal death induced by DNA-damaging agents, NGF deprivation, and oxidative stress. J Neurosci 1998; 18:830-840.
155. Van den Haute C, Spittaels K, Van Dorpe J, Lasrado R, Vandezande K, Laenen I et al. Coexpression of human cdk5 and its activator p35 with human protein tau in neurons in brain of triple transgenic mice. Neurobiol Dis 2001; 8:32-44.
156. Kesavapany S, Lau KF, McLoughlin DM, Brownlees J, Ackerley S, Leigh PN et al. p35/cdk5 binds and phosphorylates beta-catenin and regulates beta-catenin/presenilin-1 interaction. Eur J Neurosci 2001; 13:241-247.

157. Hou ST, Callaghan D, Fournier MC, Hill I, Kang L, Massie B et al. The transcription factor E2F1 modulates apoptosis of neurons. J Neurochem 2000; 75:91-100.
158. O Hare MJ, Hou ST, Morris EJ, Cregan SP, Xu Q, Slack RS et al. Induction and modulation of cerebellar granule neuron death by E2F-1. J Biol Chem 2000; 275:25358-25364.
159. Smith DS, Leone G, DeGregori J, Ahmed MN, Qumsiyeh MB, Nevins JR. Induction of DNA replication in adult rat neurons by deregulation of the retinoblastoma/E2F G1 cell cycle pathway. Cell Growth Differ 2000; 11:625-633.
160. Park DS, Morris EJ, Bremner R, Keramaris E, Padmanabhan J, Rosenbaum M et al. Involvement of retinoblastoma family members and E2F/DP complexes in the death of neurons evoked by DNA damage. J Neurosci 2000; 20:3104-3114.
161. MacManus JP, Koch CJ, Jian M, Walker T, Zurakowski B. Decreased brain infarct following focal ischemia in mice lacking the transcription factor E2F1. Neuroreport 1999; 10:2711-2714.
162. Gendron TF, Mealing GA, Paris J, Lou A, Edwards A, Hou ST et al. Attenuation of neurotoxicity in cortical cultures and hippocampal slices from E2F1 knockout mice. J Neurochem 2001; 78:316-324..
163. Giovanni A, Keramaris E, Morris EJ, Hou ST, O Hare M, Dyson N et al. E2F1 mediates death of B-amyloid-treated cortical neurons in a manner independent of p53 and dependent on Bax and caspase 3. J Biol Chem 2000; 275:11553-11560.
164. Tamami M, Lindholm PF, Brady JN. The retinoblastoma gene product (Rb) induces binding of a conformationally inactive nuclear factor-kappaB. J Biol Chem 1996; 271:24551-24556.
165. Shim J, Park HS, Kim MJ, Park J, Park E, Cho SG et al. Rb protein down-regulates the stress-activated signals through inhibiting c-Jun N-terminal kinase/stress-activated protein kinase. J Biol Chem 2000; 275:14107-14111.
166. Jordan J, Galindo MF, Prehn JH, Weichselbaum RR, Beckett M, Ghadge GD et al. p53 expression induces apoptosis in hippocampal pyramidal neuron cultures. J Neurosci 1997; 17:1397-1405.
167. Macleod KF, Hu Y, Jacks T. Loss of Rb activates both p53-dependent and independent cell death pathways in the developing mouse nervous system. Embo J 1996; 15:6178-6188.
168. Lipinski MM, Macleod KF, Williams BO, Mullaney TL, Crowley D, Jacks T. Cell-autonomous and non-cell-autonomous functions of the Rb tumor suppressor in developing central nervous system. Embo J 2001; 20:3402-3413.
169. Nip J, Strom DK, Eischen CM, Cleveland JL, Zambetti GP, Hiebert SW. E2F-1 induces the stabilization of p53 but blocks p53-mediated transactivation. Oncogene 2001; 20:910-920.
170. Ko LJ, Prives C. p53: Puzzle and paradigm. Genes Dev 1996; 10:1054-1072.
171. Giaccia AJ, Kastan MB. The complexity of p53 modulation: Emerging patterns from divergent signals. Genes Dev 1998; 12:2973-2983.
172. Bates S, Vousden KH. p53 in signaling checkpoint arrest or apoptosis. Curr Opin Genet Dev 1996; 6:12-18.
173. Asker C, Wiman KG, Selivanova G. p53-induced apoptosis as a safeguard against cancer. Biochem Biophys Res Commun 1999; 265:1-6.
174. Miyashita T, Harigai M, Hanada M, Reed JC. Identification of a p53-dependent negative response element in the bcl- 2 gene. Cancer Res 1994; 54:3131-3135.
175. Roperch JP, Alvaro V, Prieur S, Tuynder M, Nemani M, Lethrosne F et al. Inhibition of presenilin 1 expression is promoted by p53 and p21WAF-1 and results in apoptosis and tumor suppression. Nat Med 1998; 4:835-838.
176. Ding HF, McGill G, Rowan S, Schmaltz C, Shimamura A, Fisher DE. Oncogene-dependent regulation of caspase activation by p53 protein in a cell-free system. J Biol Chem 273: 28378-28383.
177. Gottlieb E, and Oren M (1998) p53 facilitates pRb cleavage in IL-3-deprived cells: Novel pro- apoptotic activity of p53. Embo J 1998; 17:3587-3596.
178. Morrison RS, Kinoshita Y. The role of p53 in neuronal cell death. Cell Death Differ 2000; 7:868-879.

179. Sakhi S, Bruce A, Sun N, Tocco G, Baudry M, Schreiber SS. p53 induction is associated with neuronal damage in the central nervous system. Proc Natl Acad Sci USA 1994; 91:7525-7529.
180. Sakhi S, Sun N, Wing LL, Mehta P, Schreiber SS. Nuclear accumulation of p53 protein following kainic acid-induced seizures. Neuroreport 1996; 7:493-496.
181. Nakai M, Qin ZH, Chen JF, Wang Y, Chase TN. Kainic acid-induced apoptosis in rat striatum is associated with nuclear factor-kappaB activation. J Neurochem 2000; 74:647-658.
182. Qin ZH, Chen RW, Wang Y, Nakai M, Chuang DM, Chase TN. Nuclear factor kappaB nuclear translocation upregulates c-Myc and p53 expression during NMDA receptor-mediated apoptosis in rat striatum. J Neurosci 1999; 19:4023-4033.
183. Wang Y, Qin ZH, Nakai M, Chen RW, Chuang DM, Chase TN. Co-stimulation of cyclic-AMP-linked metabotropic glutamate receptors in rat striatum attenuates excitotoxin-induced nuclear factor-kappaB activation and apoptosis. Neuroscience 1999; 94:1153-1162.
184. Napieralski JA, Raghupathi R, McIntosh TK. The tumor-suppressor gene, p53, is induced in injured brain regions following experimental traumatic brain injury. Brain Res Mol Brain Res 1999; 71:78-86.
185. Lu J, Moochhala S, Kaur C, Ling E. Changes in apoptosis-related protein (p53, Bax, Bcl-2 and Fos) expression with DNA fragmentation in the central nervous system in rats after closed head injury. Neurosci Lett 2000; 290:89-92.
186. Martin LJ, Kaiser A, Yu JW, Natale JE, Al-Abdulla NA. Injury-induced apoptosis of neurons in adult brain is mediated by p53-dependent and p53-independent pathways and requires Bax. J Comp Neurol 2001; 433:299-311.
187. Chopp M, Li Y, Zhang ZG, Freytag SO. p53 expression in brain after middle cerebral artery occlusion in the rat. Biochem Biophys Res Commun 1992; 182:1201-1207.
188. Watanabe H, Ohta S, Kumon Y, Sakaki S, Sakanaka M. Increase in p53 protein expression following cortical infarction in the spontaneously hypertensive rat. Brain Res 1999; 837:38-45.
189. de la Monte SM, Sohn YK, Wands JR. Correlates of p53- and Fas (CD95)-mediated apoptosis in Alzheimer s disease. J Neurol Sci 1997; 152:73-83.
190. de la Monte SM, Sohn YK, Ganju N, Wands JR. P53- and CD95-associated apoptosis in neurodegenerative diseases. Lab Invest 1998; 78:401-411.
191. LaFerla FM, Hall CK, Ngo L, Jay G. Extracellular deposition of beta-amyloid upon p53-dependent neuronal cell death in transgenic mice. J Clin Invest 1996; 98:1626-1632.
192. Jiang YH, Armstrong D, Albrecht U, Atkins CM, Noebels JL, Eichele G et al. Mutation of the Angelman ubiquitin ligase in mice causes increased cytoplasmic p53 and deficits of contextual learning and long-term potentiation [see comments]. Neuron 1998; 21:799-811.
193. Sawa A. Neuronal cell death in Down s syndrome. J Neural Transm Suppl 1999; 57:87-97.
194. Seidl R, Fang-Kircher S, Bidmon B, Cairns N, Lubec G. Apoptosis-associated proteins p53 and APO-1/Fas (CD95) in brains of adult patients with Down syndrome. Neurosci Lett 1999; 260:9-12.
195. Martin LJ. p53 is abnormally elevated and active in the CNS of patients with amyotrophic lateral sclerosis. Neurobiol Dis 2000; 7:613-622.
196. Steffan JS, Kazantsev A, Spasic-Boskovic O, Greenwald M, Zhu YZ, Gohler H et al. The Huntington s disease protein interacts with p53 and CREB-binding protein and represses transcription. Proc Natl Acad Sci USA 2000; 97:6763-6768.
197. Uberti D, Belloni M, Grilli M, Spano P, Memo M. Induction of tumour-suppressor phosphoprotein p53 in the apoptosis of cultured rat cerebellar neurones triggered by excitatory amino acids. Eur J Neurosci 1998; 10:246-254.
198. Chen RW, Chuang DM. Long term lithium treatment suppresses p53 and Bax expression but increases Bcl-2 expression. A prominent role in neuroprotection against excitotoxicity. J Biol Chem 1999; 274:6039-6042.
199. Anderson CNG, Tolkovsky AM. A role for MAPK/ERK in sympathetic neuron survival: Protection against a p53-dependent, JNK-independent induction of apoptosis by cytosine arabinoside. J Neurosci 1999; 19:664-673.

200. Chen RW, Saunders PA, Wei H, Li Z, Seth P, Chuang DM. Involvement of glyceraldehyde-3-phosphate dehydrogenase (GAPDH) and p53 in neuronal apoptosis: evidence that GAPDH is upregulated by p53. J Neurosci 1999; 19:9654-9662.
201. Frank KM, Sharpless NE, Gao Y, Sekiguchi JM, Ferguson DO, Zhu C et al. DNA ligase IV deficiency in mice leads to defective neurogenesis and embryonic lethality via the p53 pathway. Mol Cell 2000; 5:993-1002.
202. Banasiak KJ, Haddad GG. Hypoxia-induced apoptosis: Effect of hypoxic severity and role of p53 in neuronal cell death. Brain Res 1998; 797:295-304.
203. Crumrine RC, Thomas AL, Morgan PF. Attenuation of p53 expression protects against focal ischemic damage in transgenic mice. J Cereb Blood Flow Metab 1994; 14:887-891.
204. Johnson MD, Xiang H, London S, Kinoshita Y, Knudson M, Mayberg M et al. Evidence for involvement of Bax and p53, but not caspases, in radiation-induced cell death of cultured postnatal hippocampal neurons. J Neurosci Res 1998; 54:721-733.
205. Wood KA, Youle RJ. The role of free radicals and p53 in neuron apoptosis in vivo. J Neurosci 1995; 15:5851-5857.
206. Enokido Y, Araki T, Tanaka K, Aizawa S, Hatanaka H. Involvement of p53 in DNA strand break-induced apoptosis in postmitotic CNS neurons. Eur J Neurosci 1996; 8:1812-1821.
207. D Sa-Eipper C, Leonard JR, Putcha G, Zheng TS, Flavell RA, Rakic P et al. DNA damage-induced neural precursor cell apoptosis requires p53 and caspase 9 but neither Bax nor caspase 3. Development 2001; 128:137-146.
208. Herzog KH, Chong MJ, Kapsetaki M, Morgan JI, McKinnon PJ. Requirement for Atm in ionizing radiation-induced cell death in the developing central nervous system. Science 1998; 280:1089-1091.
209. Morrison RS, Wenzel HJ, Kinoshita Y, Robbins CA, Donehower LA, Schwartzkroin PA. Loss of the p53 tumor suppressor gene protects neurons from kainate- induced cell death. J Neurosci 1996; 16:1337-1345.
210. Trimmer PA, Smith TS, Jung AB, Bennett JP, Jr. Dopamine neurons from transgenic mice with a knockout of the p53 gene resist MPTP neurotoxicity. Neurodegeneration 1996; 5:233-239.
211. Hirata H, Cadet JL. p53-knockout mice are protected against the long-term effects of methamphetamine on dopaminergic terminals and cell bodies. J Neurochem 1997; 69:780-790.
212. Sakhi S, Gilmore W, Tran ND, Schreiber SS. p53-deficient mice are protected against adrenalectomy-induced apoptosis. Neuroreport 1996; 8:233-235.
213. Enokido Y, Araki T, Aizawa S, Hatanaka H. p53 involves cytosine arabinoside-induced apoptosis in cultured cerebellar granule neurons. Neurosci Lett 1996; 203:1-4.
214. Araki T, Enokido Y, Inamura N, Aizawa S, Reed JC, Hatanaka H. Changes in c-Jun but not Bcl-2 family proteins in p53-dependent apoptosis of mouse cerebellar granule neurons induced by DNA damaging agent bleomycin. Brain Res 1998; 794:239-247.
215. Xiang H, Kinoshita Y, Knudson CM, Korsmeyer SJ, Schwartzkroin PA, Morrison RS. Bax involvement in p53-mediated neuronal cell death. J Neurosci 1998; 18:1363-1373.
216. Halterman MW, Miller CC, Federoff HJ. Hypoxia-inducible factor-1alpha mediates hypoxia-induced delayed neuronal death that involves p53. J Neurosci 1999; 19:6818-6824.
217. Aloyz RS, Bamji SX, Pozniak CD, Toma JG, Atwal J, Kaplan DR et al. p53 is essential for developmental neuron death as regulated by the TrkA and p75 neurotrophin receptors. J Cell Biol 1998; 143:1691-1703.
218. Vogel KS, Parada LF. Sympathetic neuron survival and proliferation are prolonged by loss of p53 and neurofibromin. Mol Cell Neurosci 1998; 11:19-28.
219. Lee EY, Chang CY, Hu N, Wang YC, Lai CC, Herrup K et al. Mice deficient for Rb are nonviable and show defects in neurogenesis and haematopoiesis [see comments]. Nature 1992; 359:288-294.
220. Yeung MC, Geertsma F, Liu J, Lau AS. Inhibition of HIV-1 gp120-induced apoptosis in neuroblastoma SK-N-SH cells by an antisense oligodeoxynucleotide against p53. AIDS 1998; 12:349-354.

221. Lakkaraju A, Dubinsky JM, Low WC, Rahman YE. Neurons are protected from excitotoxic death by p53 antisense oligonucleotides delivered in anionic liposomes. J Biol Chem 2001; 276:32000-32007.
222. Culmsee C, Zhu X, Yu QS, Chan SL, Camandola S, Guo Z et al. A synthetic inhibitor of p53 protects neurons against death induced by ischemic and excitotoxic insults, and amyloid beta-peptide. J Neurochem 2001; 77:220-228.
223. Kuntz C, Kinoshita Y, Beal F, Donehower LA, Morrison RS. The absence of p53 does not protect SOD1 mutant mice from onset of clincial symptoms or lethality. Exp Neurol 2000; 165:184-190.
224. Prudlo J, Koenig J, Graser J, Burckhardt E, Mestres P, Menger M et al. Motor neuron cell death in a mouse model of FALS is not mediated by the p53 cell survival regulator. Brain Res 2000; 879:183-187.
225. Chong MJ, Murray MR, Gosink EC, Russell HR, Srinivasan A, Kapsetaki M et al. Atm and Bax cooperate in ionizing radiation-induced apoptosis in the central nervous system. Proc Natl Acad Sci USA 2000; 97:889-894.
226. Cregan SP, MacLaurin JG, Craig CG, Robertson GS, Nicholson DW, Park DS et al. Bax-dependent caspase-3 activation is a key determinant in p53-induced apoptosis in neurons. J Neurosci 1999; 19:7860-7869.
227. McGinnis KM, Gnegy ME, Wang KK. Endogenous Bax translocation in SH-SY5Y human neuroblastoma cells and cerebellar granule neurons undergoing apoptosis. J Neurochem 1999; 72:1899-1906.
228. Putcha GV, Deshmukh M, Johnson EM, Jr. BAX translocation is a critical event in neuronal apoptosis: regulation by neuroprotectants, BCL-2, and caspases. J Neurosci 1999; 19:7476-7485.
229. Xiang J, Chao DT, Korsmeyer SJ. BAX-induced cell death may not require interleukin 1 beta-converting enzyme-like proteases. Proc Natl Acad Sci USA 1996; 93:14559-14563.
230. Vekrellis K, McCarthy MJ, Watson A, Whitfield J, Rubin LL, Ham J. Bax promotes neuronal cell death and is downregulated during the development of the nervous system. Development 1997; 124:1239-1249.
231. Martinou I, Missotten M, Fernandez PA, Sadoul R, Martinou JC. Bax and Bak proteins require caspase activity to trigger apoptosis in sympathetic neurons. Neuroreport 1998; 9:15-19.
232. Marzo I, Brenner C, Zamzami N, Jurgensmeier JM, Susin SA, Vieira HL et al. Bax and adenine nucleotide translocator cooperate in the mitochondrial control of apoptosis. Science 1998; 281:2027-2031.
233. Schuler M, Bossy-Wetzel E, Goldstein JC, Fitzgerald P, Green DR. p53 induces apoptosis by caspase activation through mitochondrial cytochrome c release. J Biol Chem 2000; 275:7337-7342.
234. Johnson MD, Kinoshita Y, Xiang H, Ghatan S, Morrison RS. Contribution of p53-dependent caspase activation to neuronal cell death declines with neuronal maturation. J Neurosci 1999; 19:2996-3006.
235. Fortin A, Cregan SP, MacLaurin JG, Kushwaha N, Hickman ES, Thompson CS et al. APAF1 is a key transcriptional target for p53 in the regulation of neuronal cell death. J Cell Biol 2001; 155:207-216.
236. Attardi LD, Reczek EE, Cosmas C, Demicco EG, McCurrach ME, Lowe SW et al. PERP, an apoptosis-associated target of p53, is a novel member of the PMP-22/gas3 family [In Process Citation]. Genes Dev 2000; 14:704-718.
237. Ferri KF, Kroemer G. Organelle-specific initiation of cell death pathways. Nat Cell Biol 2001; 3:E255-263.
238. Lewen A, Matz P, Chan PH. Free radical pathways in CNS injury. J Neurotrauma 2000; 17:871-890.
239. Mattson MP. Apoptosis in neurodegenerative disorders. Nat Rev Mol Cell Biol 2000; 1:120-129.

240. Tong X, Liu B, Dong Y, Sun Z. Cleavage of ATM during radiation-induced apoptosis: Caspase-3-like apoptotic protease as a candidate. Int J Radiat Biol 2000; 76:1387-1395.
241. Song Q, Lees-Miller SP, Kumar S, Zhang Z, Chan DW, Smith GC et al. DNA-dependent protein kinase catalytic subunit: A target for an ICE- like protease in apoptosis. Embo J 1996; 15:3238-3246.
242. Han Z, Malik N, Carter T, Reeves WH, Wyche JH, Hendrickson EA. DNA-dependent protein kinase is a target for a CPP32-like apoptotic protease. J Biol Chem 1996; 271:25035-25040.
243. Tewari M, Quan LT, O Rourke K, Desnoyers S, Zeng Z, Beidler DR et al. Yama/CPP32 beta, a mammalian homolog of CED-3, is a CrmA-inhibitable protease that cleaves the death substrate poly(ADP-ribose) polymerase. Cell 1995; 81:801-809.
244. Lazebnik YA, Kaufmann SH, Desnoyers S, Poirier GG, Earnshaw WC. Cleavage of poly(ADP-ribose) polymerase by a proteinase with properties like ICE. Nature 1994; 371:346-347.
245. Kastan MB, Lim DS. The many substrates and functions of ATM. Nat Rev Mol Cell Biol 2000; 1:179-186.
246. Barlow C, Hirotsune S, Paylor R, Liyanage M, Eckhaus M, Collins F et al. Atm-deficient mice: A paradigm of ataxia telangiectasia. Cell 1996; 86:159-171.
247. Canman CE, Lim DS, Cimprich KA, Taya Y, Tamai K, Sakaguchi K et al. Activation of the ATM kinase by ionizing radiation and phosphorylation of p53. Science 1998; 281:1677-1679.
248. Banin S, Moyal L, Shieh S, Taya Y, Anderson CW, Chessa L et al. Enhanced phosphorylation of p53 by ATM in response to DNA damage. Science 1998; 281:1674-1677.
249. Caspari T. How to activate p53. Curr Biol 2000; 10:R315-317.
250. Khosravi R, Maya R, Gottlieb T, Oren M, Shiloh Y, Shkedy D. Rapid ATM-dependent phosphorylation of MDM2 precedes p53 accumulation in response to DNA damage. Proc Natl Acad Sci USA 1999; 96:14973-14977.
251. Maya R, Balass M, Kim ST, Shkedy D, Leal JF, Shifman O et al. ATM-dependent phosphorylation of Mdm2 on serine 395: Role in p53 activation by DNA damage. Genes Dev 2001; 15:1067-1077.
252. Meek DW. Mechanisms of switching on p53: A role for covalent modification? Oncogene 1999; 18:7666-7675.
253. Lee Y, Chong MJ, McKinnon PJ. Ataxia telangiectasia mutated-dependent apoptosis after genotoxic stress in the developing nervous system is determined by cellular differentiation status. J Neurosci 2001; 21:6687-6693.
254. Rolig RL, McKinnon PJ. Linking DNA damage and neurodegeneration. Trends Neurosci 2000; 23:417-424.
255. Shiloh Y. ATM and ATR: networking cellular responses to DNA damage. Curr Opin Genet Dev 2001; 11:71-77.
256. Hammarsten O, DeFazio LG, Chu G. Activation of DNA-dependent protein kinase by single-stranded DNA ends. J Biol Chem 2000; 275:1541-1550.
257. Smith GC, Jackson SP. The DNA-dependent protein kinase. Genes Dev 1999; 13:916-934.
258. Lees-Miller SP, Sakaguchi K, Ullrich SJ, Appella E, Anderson CW. Human DNA-activated protein kinase phosphorylates serines 15 and 37 in the amino-terminal transactivation domain of human p53. Mol Cell Biol 1992; 12:5041-5049.
259. Mayo LD, Turchi JJ, Berberich SJ. Mdm-2 phosphorylation by DNA-dependent protein kinase prevents interaction with p53. Cancer Res 1997; 57:5013-5016.
260. Kharbanda S, Yuan ZM, Weichselbaum R, Kufe D. Determination of cell fate by c-Abl activation in the response to DNA damage. Oncogene 1998; 17:3309-3318.
261. Liu L, Kwak YT, Bex F, Garcia-Martinez LF, Li XH, Meek K et al. DNA-dependent protein kinase phosphorylation of IkappaB alpha and IkappaB beta regulates NF-kappaB DNA binding properties. Mol Cell Biol 1998; 18:4221-4234.
262. ChechlacZ M, Vemuri MC, Naegele JR. Role of DNA-dependent protein kinase in neuronal survival. J Neurochem 2001; 78:141-154.

263. Vemuri MC, Schiller E, Naegele JR. Elevated DNA double strand breaks and apoptosis in the CNS of scid mutant mice. Cell Death Differ 2001; 8:245-255.
264. Culmsee C, Bondada S, Mattson MP. Hippocampal neurons of mice deficient in DNA-dependent protein kinase exhibit increased vulnerability to DNA damage, oxidative stress and excitotoxicity. Brain Res Mol Brain Res 2001; 87:257-262.
265. Chakravarthy BR, Walker T, Rasquinha I, Hill IE, MacManus JP. Activation of DNA-dependent protein kinase may play a role in apoptosis of human neuroblastoma cells. J Neurochem 1999; 72:933-942.
266. Smith S. The world according to PARP. Trends Biochem Sci 2001; 26:174-179.
267. Herceg Z, Wang ZQ. Functions of poly(ADP-ribose) polymerase (PARP) in DNA repair, genomic integrity and cell death. Mutat Res 2001; 477:97-110.
268. Le Rhun Y, Kirkland JB, Shah GM. Cellular responses to DNA damage in the absence of Poly(ADP-ribose) polymerase. Biochem Biophys Res Commun 1998; 245:1-10.
269. Ziegler M, Oei SL. A cellular survival switch: Poly(ADP-ribosyl)ation stimulates DNA repair and silences transcription. Bioessays 2001; 23:543-548.
270. Ha HC, Snyder SH. Poly(ADP-ribose) polymerase is a mediator of necrotic cell death by ATP depletion. Proc Natl Acad Sci USA 1999; 96:13978-13982.
271. Ha HC, Snyder SH. Poly(ADP-ribose) polymerase-1 in the nervous system. Neurobiol Dis 2000; 7:225-239.
272. Cosi C, Suzuki H, Milani D, Facci L, Menegazzi M, Vantini G et al. Poly(ADP-ribose) polymerase: early involvement in glutamate-induced neurotoxicity in cultured cerebellar granule cells. J Neurosci Res 1994; 39:38-46.
273. Zhang J, Dawson VL, Dawson TM, Snyder SH. Nitric oxide activation of poly(ADP-ribose) synthetase in neurotoxicity. Science 1994; 263:687-689.
274. Eliasson MJ, Sampei K, Mandir AS, Hurn PD, Traystman RJ, Bao J et al. Poly(ADP-ribose) polymerase gene disruption renders mice resistant to cerebral ischemia. Nat Med 1997; 3:1089-1095.
275. Endres M, Wang ZQ, Namura S, Waeber C, Moskowitz MA. Ischemic brain injury is mediated by the activation of poly(ADP- ribose)polymerase. J Cereb Blood Flow Metab 1997; 17:1143-1151.
276. Mandir AS, Poitras MF, Berliner AR, Herring WJ, Guastella DB, Feldman A et al. NMDA but not non-NMDA excitotoxicity is mediated by Poly(ADP-ribose) polymerase. J Neurosci 2000; 20:8005-8011.
277. Mandir AS, Przedborski S, Jackson-Lewis V, Wang ZQ, Simbulan-Rosenthal CM, Smulson ME et al. Poly(ADP-ribose) polymerase activation mediates 1-methyl-4-phenyl-1, 2,3,6-tetrahydropyridine (MPTP)-induced parkinsonism. Proc Natl Acad Sci USA 1999; 96:5774-5779.
278. Ding Y, Zhou Y, Lai Q, Li J, Gordon V, Diaz FG. Long-term neuroprotective effect of inhibiting poly(ADP-ribose) polymerase in rats with middle cerebral artery occlusion using a behavioral assessment. Brain Res 2001; 915:210-217.
279. Whalen MJ, Clark RS, Dixon CE, Robichaud P, Marion DW, Vagni V et al. Traumatic brain injury in mice deficient in poly-ADP(ribose) polymerase: A preliminary report. Acta Neurochir Suppl 2000; 76:61-64
280. Moroni F, Meli E, Peruginelli F, Chiarugi A, Cozzi A, Picca R et al. Poly(ADP-ribose) polymerase inhibitors attenuate necrotic but not apoptotic neuronal death in experimental models of cerebral ischemia. Cell Death Differ 2001; 8:921-932.
281. Kruman, II, Culmsee C, Chan SL, Kruman Y, Guo Z, Penix L et al. Homocysteine elicits a DNA damage response in neurons that promotes apoptosis and hypersensitivity to excitotoxicity. J Neurosci 2000; 20:6920-6926.
282. Yang E, Korsmeyer SJ. Molecular thanatopsis: A discourse on the BCL2 family and cell death. Blood 1996; 88:386-401.

283. Bakhshi A, Jensen JP, Goldman P, Wright JJ, McBride OW, Epstein AL et al. Cloning the chromosomal breakpoint of t(14;18) human lymphomas: Clustering around JH on chromosome 14 and near a transcriptional unit on 18. Cell 1985; 41:899-906.
284. Cleary ML, Smith SD, Sklar J. Cloning and structural analysis of cDNAs for bcl-2 and a hybrid bcl- 2/immunoglobulin transcript resulting from the t(14;18) translocation. 1986; Cell 47:19-28.
285. Tsujimoto Y, Finger LR, Yunis J, Nowell PC, Croce CM. Cloning of the chromosome breakpoint of neoplastic B cells with the t(14;18) chromosome translocation. Science 1984; 226:1097-1099.
286. Huang DC, Strasser A. BH3-Only proteins-essential initiators of apoptotic cell death. Cell 2000; 103:839-842.
287. Oda E, Ohki R, Murasawa H, Nemoto J, Shibue T, Yamashita T et al. Noxa, a BH3-only member of the Bcl-2 family and candidate mediator of p53-induced apoptosis. Science 2000; 288:1053-1058.
288. Nechushtan A, Smith CL, Lamensdorf I, Yoon SH, Youle RJ. Bax and Bak coalesce into novel mitochondria-associated clusters during apoptosis. J Cell Biol 2001; 153:1265-1276.
289. Hsu YT, Wolter KG, Youle RJ. Cytosol-to-membrane redistribution of Bax and Bcl-X(L) during apoptosis. Proc Natl Acad Sci USA 1997; 94:3668-3672.
290. Wolter KG, Hsu YT, Smith CL, Nechushtan A, Xi XG et al. Movement of Bax from the cytosol to mitochondria during apoptosis. J Cell Biol 1997; 139:1281-1292.
291. Zha H, Fisk HA, Yaffe MP, Mahajan N, Herman B, Reed JC. Structure-function comparisons of the proapoptotic protein Bax in yeast and mammalian cells. Mol Cell Biol 1996; 16:6494-6508.
292. Thornberry NA, Lazebnik Y. Caspases: Enemies within. Science 1998; 281:1312-1316.
293. Green DR, Reed JC. Mitochondria and apoptosis. Science 1998; 281:1309-1312.
294. Kluck RM, Bossy-Wetzel E, Green DR, Newmeyer DD. The release of cytochrome c from mitochondria: A primary site for Bcl-2 regulation of apoptosis. Science 1997; 275:1132-1136.
295. Yang J, Liu X, Bhalla K, Kim CN, Ibrado AM, Cai J, Peng TI, Jones DP, and Wang X (1997) Prevention of apoptosis by Bcl-2: release of cytochrome c from mitochondria blocked. Science 275: 1129-1132.
296. Harris MH, Thompson CB. The role of the Bcl-2 family in the regulation of outer mitochondrial membrane permeability. Cell Death Differ 2000; 7:1182-1191.
297. Kowaltowski AJ, Smaili SS, Russell JT, Fiskum G. Elevation of resting mitochondrial membrane potential of neural cells by cyclosporin A, BAPTA-AM, and bcl-2. Am J Physiol Cell Physiol 2000; 279:C852-859.
298. Schendel SL, Montal M, Reed JC. Bcl-2 family proteins as ion-channels. Cell Death Differ 1998; 5:372-380.
299. Saito M, Korsmeyer SJ, Schlesinger PH. BAX-dependent transport of cytochrome c reconstituted in pure liposomes. Nat Cell Biol 2000; 2:553-555.
300. Martinou JC, Green DR. Breaking the mitochondrial barrier. Nat Rev Mol Cell Biol 2001; 2:63-67.
301. Gross A, Jockel J, Wei MC, Korsmeyer SJ. Enforced dimerization of BAX results in its translocation, mitochondrial dysfunction and apoptosis. Embo J 1998; 17:3878-3885.
302. Oltvai ZN, Milliman CL, Korsmeyer SJ. Bcl-2 heterodimerizes in vivo with a conserved homolog, Bax, that accelerates programmed cell death. Cell 1993; 74:609-619.
303. Yin XM, Oltvai ZN, Korsmeyer SJ. BH1 and BH2 domains of Bcl-2 are required for inhibition of apoptosis and heterodimerization with Bax. Nature 1994; 369:321-323.
304. Sattler M, Liang H, Nettesheim D, Meadows RP, Harlan JE, Eberstadt M et al. Structure of Bcl-xL-Bak peptide complex: recognition between regulators of apoptosis. Science 1997; 275:983-986.
305. Muchmore SW, Sattler M, Liang H, Meadows RP, Harlan JE, Yoon HS et al. X-ray and NMR structure of human Bcl-xL, an inhibitor of programmed cell death. Nature 1996; 381:335-341.

306. Wang K, Yin XM, Chao DT, Milliman CL, Korsmeyer SJ. BID: A novel BH3 domain-only death agonist. Genes Dev 1996; 10:2859-2869.
307. Zha J, Harada H, Osipov K, Jockel J, Waksman G, Korsmeyer SJ. BH3 domain of BAD is required for heterodimerization with BCL-XL and pro-apoptotic activity. J Biol Chem 1997; 272:24101-24104.
308. Zha J, Harada H, Yang E, Jockel J, Korsmeyer SJ. Serine phosphorylation of death agonist BAD in response to survival factor results in binding to 14-3-3 not BCL-X(L). Cell 1996; 87:619-628.
309. Blagosklonny MV. Unwinding the loop of Bcl-2 phosphorylation. Leukemia 2001; 15:869-874.
310. Ruvolo PP, Deng X, May WS. Phosphorylation of Bcl2 and regulation of apoptosis. Leukemia 2001; 15:515-522.
311. Wei MC, Lindsten T, Mootha VK, Weiler S, Gross A, Ashiya M et al. tBID, a membrane-targeted death ligand, oligomerizes BAK to release cytochrome c. Genes Dev 2000; 14:2060-2071.
312. Fesik SW. Insights into programmed cell death through structural biology. Cell 2000; 103:273-282.
313. Lutter M, Fang M, Luo X, Nishijima M, Xie X, Wang X. Cardiolipin provides specificity for targeting of tBid to mitochondria. Nat Cell Biol 2000; 2:754-761.
314. Korsmeyer SJ, Wei MC, Saito M, Weiler S, Oh KJ, Schlesinger PH. Pro-apoptotic cascade activates BID, which oligomerizes BAK or BAX into pores that result in the release of cytochrome c. Cell Death Differ 2000; 7:1166-1173.
315. Wei MC, Zong WX, Cheng EH, Lindsten T, Panoutsakopoulou V, Ross AJ et al. Proapoptotic BAX and BAK: A requisite gateway to mitochondrial dysfunction and death. Science 2001; 292:727-730.
316. Desagher S, Osen-Sand A, Nichols A, Eskes R, Montessuit S, Lauper S et al. Bid-induced conformational change of Bax is responsible for mitochondrial cytochrome c release during apoptosis. J Cell Biol 1999; 144:891-901.
317. O Connor L, Strasser A, O Reilly LA, Hausmann G, Adams JM, Cory S et al. Bim: A novel member of the Bcl-2 family that promotes apoptosis. Embo J 1998; 17:384-395.
318. Ottilie S, Diaz JL, Horne W, Chang J, Wang Y, Wilson G et al. Dimerization properties of human BAD. Identification of a BH-3 domain and analysis of its binding to mutant BCL-2 and BCL-XL proteins. J Biol Chem 1997; 272:30866-30872.
319. Merry DE, Korsmeyer SJ. Bcl-2 gene family in the nervous system. Annu Rev Neurosci 1997; 20:245-267
320. Oppenheim RW. Cell death during development of the nervous system. Annu Rev Neurosci 1991; 14:453-501
321. Sadoul R. Bcl-2 family members in the development and degenerative pathologies of the nervous system. Cell Death Differ 1998; 5:805-815.
322. Ay I, Sugimori H, Finklestein SP. Intravenous basic fibroblast growth factor (bFGF) decreases DNA fragmentation and prevents downregulation of Bcl-2 expression in the ischemic brain following middle cerebral artery occlusion in rats. Brain Res Mol Brain Res 2001; 87:71-80.
323. Miller TM, Moulder KL, Knudson CM, Creedon DJ, Deshmukh M, Korsmeyer SJ et al. Bax deletion further orders the cell death pathway in cerebellar granule cells and suggests a caspase-independent pathway to cell death. J Cell Biol 1997; 139:205-217.
324. Jin KL, Graham SH, Mao XO, He X, Nagayama T, Simon RP et al. Bax kappa, a novel Bax splice variant from ischemic rat brain lacking an ART domain, promotes neuronal cell death. J Neurochem 2001; 77:1508-1519.
325. Deckwerth TL, Elliott JL, Knudson CM, Johnson EM, Jr., Snider WD, Korsmeyer SJ. BAX is required for neuronal death after trophic factor deprivation and during development. Neuron 1996; 17:401-411.

326. Whitfield J, Neame SJ, Paquet L, Bernard O, Ham J. Dominant-negative c-Jun promotes neuronal survival by reducing BIM expression and inhibiting mitochondrial cytochrome c release. Neuron 2001; 29:629-643.
327. Putcha GV, Moulder KL, Golden JP, Bouillet P, Adams JA, Strasser A et al. Induction of BIM, a proapoptotic BH3-only BCL-2 family member, is critical for neuronal apoptosis. Neuron 2001; 29:615-628.
328. Harris CA, Johnson EM, Jr. Bh3-only bcl-2 family members are coordinately regulated by the JNK pathway and require Bax to induce apoptosis in neurons. J Biol Chem 2001; 276:37754-37760.
329. Leonard JR, D Sa C, Cahn BR, Korsmeyer SJ, Roth KA. Bid regulation of neuronal apoptosis. Brain Res Dev Brain Res 2001; 128:187-190.
330. Henshall DC, Bonislawski DP, Skradski SL, Lan JQ, Meller R, Simon RP. Cleavage of bid may amplify caspase-8-induced neuronal death following focally evoked limbic seizures. Neurobiol Dis 2001; 8:568-580.
331. Sun YF, Yu LY, Saarma M, Timmusk T, Arumae U. Neuron-specific Bcl-2 homology 3 domain-only splice variant of Bak is anti-apoptotic in neurons, but pro-apoptotic in non-neuronal cells. J Biol Chem 2001; 276:16240-16247.
332. Vila M, Jackson-Lewis V, Vukosavic S, Djaldetti R, Liberatore G, Offen D et al. Bax ablation prevents dopaminergic neurodegeneration in the 1-methyl-4-phenyl-1,2,3,6-tetrahydropyridine mouse model of Parkinson s disease. Proc Natl Acad Sci USA 2001; 98:2837-2842.
333. Offen D, Beart PM, Cheung NS, Pascoe CJ, Hochman A, Gorodin S et al. Transgenic mice expressing human Bcl-2 in their neurons are resistant to 6-hydroxydopamine and 1-methyl-4-phenyl-1,2,3,6-tetrahydropyridine neurotoxicity. Proc Natl Acad Sci USA 1998; 95:5789-5794.
334. Feinstein E, Kimchi A, Wallach D, Boldin M, Varfolomeev E. The death domain: A module shared by proteins with diverse cellular functions. Trends Biochem Sci 1995; 20:342-344.
335. Vukosavic S, Stefanis L, Jackson-Lewis V, Guegan C, Romero N, Chen C et al. Delaying caspase activation by Bcl-2: A clue to disease retardation in a transgenic mouse model of amyotrophic lateral sclerosis. J Neurosci 2000; 20:9119-9125.
336. Martin LJ. Neuronal death in amyotrophic lateral sclerosis is apoptosis: Possible contribution of a programmed cell death mechanism. J Neuropathol Exp Neurol 1999; 58:459-471.
337. Gonzalez de Aguilar JL, Gordon JW, Rene F, de Tapia M, Lutz-Bucher B, Gaiddon C et al. Alteration of the Bcl-x/Bax ratio in a transgenic mouse model of amyotrophic lateral sclerosis: Evidence for the implication of the p53 signaling pathway. Neurobiol Dis2000; 7:406-415.
338. Guegan C, Vila M, Rosoklija G, Hays AP, Przedborski S. Recruitment of the mitochondrial-dependent apoptotic pathway in amyotrophic lateral sclerosis. J Neurosci 2001; 21:6569-6576.
339. Azzouz M, Hottinger A, Paterna JC, Zurn AD, Aebischer P, Bueler H. Increased motoneuron survival and improved neuromuscular function in transgenic ALS mice after intraspinal injection of an adeno-associated virus encoding Bcl-2. Hum Mol Genet 2000; 9:803-811.
340. Kostic V, Jackson-Lewis V, de Bilbao F, Dubois-Dauphin M, Przedborski S. Bcl-2: Prolonging life in a transgenic mouse model of familial amyotrophic lateral sclerosis. Science 1997; 277:559-562.
341. Iwahashi H, Eguchi Y, Yasuhara N, Hanafusa T, Matsuzawa Y, Tsujimoto Y. Synergistic anti-apoptotic activity between Bcl-2 and SMN implicated in spinal muscular atrophy. Nature 1997; 390:413-417.
342. Coovert DD, Le TT, Morris GE, Man NT, Kralewski M, Sendtner M et al. Does the survival motor neuron protein (SMN) interact with Bcl-2? J Med Genet 2000; 37:536-539.
343. Bernardi P, Scorrano L, Colonna R, Petronilli V, Di Lisa F. Mitochondria and cell death. Mechanistic aspects and methodological issues. Eur J Biochem 1999; 264:687-701.
344. Zamzami N, Kroemer G. The mitochondrion in apoptosis: How Pandora s box opens. Nat Rev Mol Cell Biol 2001; 2:67-71.

345. Budd SL, Tenneti L, Lishnak T, Lipton SA. Mitochondrial and extramitochondrial apoptotic signaling pathways in cerebrocortical neurons. Proc Natl Acad Sci USA 2000; 97:6161-6166.
346. Susin SA, Lorenzo HK, Zamzami N, Marzo I, Snow BE, Brothers GM et al. Molecular characterization of mitochondrial apoptosis-inducing factor. Nature 1999; 397:441-446.
347. Li LY, Luo X, Wang X. Endonuclease G is an apoptotic DNase when released from mitochondria. Nature 2001; 412:95-99.
348. Troy CM, Rabacchi SA, Hohl JB, Angelastro JM, Greene LA, Shelanski ML. Death in the balance: alternative participation of the caspase-2 and -9 pathways in neuronal death induced by nerve growth factor deprivation. J Neurosci 2001; 21:5007-5016.
349. Parrish J, Li L, Klotz K, Ledwich D, Wang X, Xue D. Mitochondrial endonuclease G is important for apoptosis in *C. elegans*. Nature 2001; 412:90-94.
350. Yuan J, Shaham S, Ledoux S, Ellis HM, Horvitz HR. The *C. elegans* cell death gene ced-3 encodes a protein similar to mammalian interleukin-1 beta-converting enzyme. Cell 1993; 75:641-652.
351. Alnemri ES, Livingston DJ, Nicholson DW, Salvesen G, Thornberry NA, Wong WW et al. Human ICE/CED-3 protease nomenclature. Cell 1996; 87:171.
352. Stennicke HR, Salvesen GS. Caspases Controlling intracellular signals by protease zymogen activation. Biochim Biophys Acta 2000; 1477:299-306.
353. Eckhart L, Declercq W, Ban J, Rendl M, Lengauer B, Mayer C et al. Terminal differentiation of human keratinocytes and stratum corneum formation is associated with caspase-14 activation. J Invest Dermatol 2000; 115:1148-1151.
354. Kuechle MK, Predd HM, Fleckman P, Dale BA, Presland RB. Caspase-14, a keratinocyte specific caspase: mRNA splice variants and expression pattern in embryonic and adult mouse. Cell Death Differ 2001; 8:868-870.
355. Sanchez Mejia RO, Ona VO, Li M, Friedlander RM. Minocycline reduces traumatic brain injury-mediated caspase-1 activation, tissue damage, and neurological dysfunction. Neurosurgery 2001; 48:1393-1399; discussion 1399-1401.
356. Shibata M, Hisahara S, Hara H, Yamawaki T, Fukuuchi Y, Yuan J et al. Caspases determine the vulnerability of oligodendrocytes in the ischemic brain. J Clin Invest 2000; 106:643-653.
357. Hisahara S, Yuan J, Momoi T, Okano H, Miura M. Caspase-11 mediates oligodendrocyte cell death and pathogenesis of autoimmune-mediated demyelination. J Exp Med 2001; 193:111-122.
358. Grimm S, Stanger BZ, Leder P. RIP and FADD: two death domain -containing proteins can induce apoptosis by convergent, but dissociable, pathways. Proc Natl Acad Sci USA 1996; 93:10923-10927.
359. Zou H, Li Y, Liu X, Wang X. An APAF-1.cytochrome c multimeric complex is a functional apoptosome that activates procaspase-9. J Biol Chem 1999; 274:11549-11556.
360. Srinivasula SM, Ahmad M, Fernandes-Alnemri T, Alnemri ES. Autoactivation of procaspase-9 by Apaf-1-mediated oligomerization. Mol Cell 1998; 1:949-957.
361. Li P, Nijhawan D, Budihardjo I, Srinivasula SM, Ahmad M, Alnemri ES et al. Cytochrome c and dATP-dependent formation of Apaf-1/caspase-9 complex initiates an apoptotic protease cascade. Cell 1997; 91:479-489.
362. Hu Y, Benedict MA, Ding L, Nunez G. Role of cytochrome c and dATP/ATP hydrolysis in Apaf-1-mediated caspase-9 activation and apoptosis. Embo J 1999; 18:3586-3595.
363. Chu ZL, Pio F, Xie Z, Welsh K, Krajewska M, Krajewski S et al. A novel enhancer of the Apafl apoptosome involved in cytochrome c-dependent caspase activation and apoptosis. J Biol Chem 2001; 276:9239-9245.
364. Fujita E, Jinbo A, Matuzaki H, Konishi H, Kikkawa U, Momoi T. Akt phosphorylation site found in human caspase-9 is absent in mouse caspase-9. Biochem Biophys Res Commun 1999; 264:550-555.

365. Namura S, Zhu J, Fink K, Endres M, Srinivasan A, Tomaselli KJ et al. Activation and cleavage of caspase-3 in apoptosis induced by experimental cerebral ischemia. J Neurosci 1998; 18:3659-3668.
366. Hartmann A, Hunot S, Michel PP, Muriel MP, Vyas S, Faucheux BA et al. Caspase-3: A vulnerability factor and final effector in apoptotic death of dopaminergic neurons in Parkinson s disease. Proc Natl Acad Sci USA 2000; 97:2875-2880.
367. Su JH, Zhao M, Anderson AJ, Srinivasan A, Cotman CW. Activated caspase-3 expression in Alzheimer s and aged control brain: Correlation with Alzheimer pathology. Brain Res 2001; 898:350-357.
368. Li M, Ona VO, Chen M, Kaul M, Tenneti L, Zhang X et al. Functional role and therapeutic implications of neuronal caspase-1 and -3 in a mouse model of traumatic spinal cord injury. Neuroscience 2000; 99:333-342.
369. Tenneti L, Lipton SA. Involvement of activated caspase-3-like proteases in N-methyl-D-aspartate-induced apoptosis in cerebrocortical neurons. J Neurochem 2000; 74:134-142.
370. Benchoua A, Guegan C, Couriaud C, Hosseini H, Sampaio N, Morin D et al. Specific caspase pathways are activated in the two stages of cerebral infarction. J Neurosci 2001; 21:7127-7134.
371. Chen M, Ona VO, Li M, Ferrante RJ, Fink KB, Zhu S et al. Minocycline inhibits caspase-1 and caspase-3 expression and delays mortality in a transgenic mouse model of Huntington disease. Nat Med 2000; 6:797-801.
372. Kuida K, Zheng TS, Na S, Kuan C, Yang D, Karasuyama H et al. Decreased apoptosis in the brain and premature lethality in CPP32- deficient mice. Nature 2000; 384:368-372.
373. LeBlanc A, Liu H, Goodyer C, Bergeron C, Hammond J. Caspase-6 role in apoptosis of human neurons, amyloidogenesis, and Alzheimer s disease. J Biol Chem 1999; 274:23426-23436.
374. Wellington CL, Singaraja R, Ellerby L, Savill J, Roy S, Leavitt B et al. Inhibiting caspase cleavage of huntingtin reduces toxicity and aggregate formation in neuronal and nonneuronal cells. J Biol Chem 2000; 275:19831-19838.
375. Selznick LA, Zheng TS, Flavell RA, Rakic P, Roth KA. Amyloid beta-induced neuronal death is Bax-dependent but caspase- independent. J Neuropathol Exp Neurol 2000; 59:271-279.
376. D Mello SR, Kuan CY, Flavell RA, Rakic P. Caspase-3 is required for apoptosis-associated DNA fragmentation but not for cell death in neurons deprived of potassium. J Neurosci Res 2000; 59:24-31.
377. Droin N, Beauchemin M, Solary E, Bertrand R. Identification of a caspase-2 isoform that behaves as an endogenous inhibitor of the caspase cascade. Cancer Res 2000; 60:7039-7047.
378. Tschopp J, Irmler M, Thome M. Inhibition of fas death signals by FLIPs. Curr Opin Immunol 1998; 10:552-558.
379. Cheema ZF, Wade SB, Sata M, Walsh K, Sohrabji F, Miranda RC. Fas/Apo [apoptosis]-1 and associated proteins in the differentiating cerebral cortex: induction of caspase-dependent cell death and activation of NF-kappaB. J Neurosci 1999; 19:1754-1770.
380. Yeh WC, Itie A, Elia AJ, Ng M, Shu HB, Wakeham A et al. Requirement for Casper (c-FLIP) in regulation of death receptor-induced apoptosis and embryonic development. Immunity 2000; 12:633-642.
381. Yang YL, Li XM. The IAP family: Endogenous caspase inhibitors with multiple biological activities. Cell Res 2000; 10:169-177.
382. Holcik M, Korneluk RG. XIAP, the guardian angel. Nat Rev Mol Cell Biol 2001; 2:550-556.
383. Yang Y, Fang S, Jensen JP, Weissman AM, Ashwell JD. Ubiquitin protein ligase activity of IAPs and their degradation in proteasomes in response to apoptotic stimuli. Science 2000; 288:874-877.
384. Liston P, Fong WG, Kelly NL, Toji S, Miyazaki T, Conte D et al. Identification of XAF1 as an antagonist of XIAP anti-Caspase activity. Nat Cell Biol 2001; 3:128-133.

385. Verhagen AM, Ekert PG, Pakusch M, Silke J, Connolly LM, Reid GE et al. Identification of DIABLO, a mammalian protein that promotes apoptosis by binding to and antagonizing IAP proteins. Cell 2000; 102:43-53.
386. Du C, Fang M, Li Y, Li L, Wang X. Smac, a mitochondrial protein that promotes cytochrome c-dependent caspase activation by eliminating IAP inhibition. Cell 2000; 102:33-42.
387. Siman R, Noszek JC. Excitatory amino acids activate calpain I and induce structural protein breakdown in vivo. Neuron 1988; 1:279-287.
388. Siman R, Noszek JC, Kegerise C. Calpain I activation is specifically related to excitatory amino acid induction of hippocampal damage. J Neurosci 1989; 9:1579-1590.
389. Saatman KE, Bozyczko-Coyne D, Marcy V, Siman R, McIntosh TK. Prolonged calpain-mediated spectrin breakdown occurs regionally following experimental brain injury in the rat. J Neuropathol Exp Neurol 1996; 55:850-860.
390. Saatman KE, Murai H, Bartus RT, Smith DH, Hayward NJ, Perri BR et al. Calpain inhibitor AK295 attenuates motor and cognitive deficits following experimental brain injury in the rat. Proc Natl Acad Sci USA 1996; 93:3428-3433.
391. Faddis BT, Hasbani MJ, Goldberg MP. Calpain activation contributes to dendritic remodeling after brief excitotoxic injury in vitro. J Neurosci 1997; 17:951-959.
392. Kampfl A, Posmantur RM, Zhao X, Schmutzhard E, Clifton GL, Hayes RL. Mechanisms of calpain proteolysis following traumatic brain injury: Implications for pathology and therapy: Implications for pathology and therapy: a review and update. J Neurotrauma 1997; 14:121-134.
393. Rami A, Ferger D, Krieglstein J. Blockade of calpain proteolytic activity rescues neurons from glutamate excitotoxicity. Neurosci Res 1997; 27:93-97.
394. Scarisbrick IA, Towner MD, Isackson PJ. Nervous system-specific expression of a novel serine protease: Regulation in the adult rat spinal cord by excitotoxic injury. J Neurosci 1997; 17:8156-816

DETRIMENTAL AND BENEFICIAL EFFECTS OF INJURY-INDUCED INFLAMMATION AND CYTOKINE EXPRESSION IN THE NERVOUS SYSTEM

Guido Stoll[1,2], Sebastian Jander[1] and Michael Schroeter[1]

ABSTRACT

Lesions in the nervous system induce rapid activation of glial cells and under certain conditions additional recruitment of granulocytes, T-cells and monocytes/ macrophages from the blood stream triggered by upregulation of cell adhesion molecules, chemokines and cytokines. Hematogenous cell infiltration is not restricted to infectious or autoimmune disorders of the nervous system, but also occurs in response to cerebral ischemia and traumatic lesions. Neuroinflammation can cause neuronal damage, but also confers neuroprotection.

Granulocytes occlude vessels during reperfusion after transient focal ischemia, while the functional role of T-cells and macrophages in stroke development awaits further clarification. After focal cerebral ischemia neurotoxic mediators released by microglia such as the inducible nitric oxide synthase (leading to NO synthesis) and the cytokines interleukin-1β (IL-1β) and tumor necrosis factor-α (TNF-α) are upregulated prior to cellular inflammation in the evolving lesion and functionally contribute to secondary infarct growth as revealed by numerous pharmacological experiments and by use of transgenic animals. On the other hand, cytokine induction remote from ischemic lesions involves NMDA-mediated signalling pathways and confers neuroprotection. After nerve injury T cells can rescue CNS neurons. In the peripheral nervous system neuroinflammation is a prerequisite for successful regeneration that is impeded in the CNS. In conclusion, there is increasing evidence

[1]Department of Neurology, Heinrich-Heine-Universit t, Noorenstrasse 5, D-40225 D sseldorf, Germany and [2]Department of Neurology, Julius-Maximilians-Universit t, Josef-Schneider Str. 11, D-97080 W rzbur g, Germany.

that neuroinflammation represents a double edged sword. The opposing neurotoxic and neuroprotective properties of neuroinflammation during CNS injury provide a rich and currently unexplored set of research problems.

INTRODUCTION

Lesions in the nervous system induce a broad range of molecular and cellular responses involving activation of glial cells and under certain conditions additional recruitment of granulocytes, T-cells and monocytes/macrophages from the blood stream. Glial responses and leukocyte infiltration are triggered by local upregulation of cell adhesion molecules, chemokines and cytokines. In this Chapter the term neuroinflammation is used to describe the activation of microglia and astrocytes as well as hematogenous cell infiltration and induction of immune mediators. Neuroinflammation has variable net effects in disorders of the nervous system. In traumatic and ischemic brain injury secondary lesion expansion is accompanied by cytokine induction and leukocyte infiltration. Neurotoxic effects of inflammation, however, are counteracted by inflammation-associated neuroprotection. Likewise, in the peripheral nervous system (PNS), T-cell and macrophage infiltration is an essential part of the molecular programme facilitating successful regeneration after nerve injury. We here review the potential role of neuroinflammation in lesion development during traumatic and ischemic injury in the nervous system and its beneficial effects during repair processes and neuroprotection.

GLIAL CELL POPULATIONS IN THE CENTRAL NERVOUS SYSTEM

The central nervous system (CNS) contains a heterogenous population of non-neuronal cells, designated glia (astrocytes, oligodendrocytes and microglia). While oligodendrocytes enwrap axons with myelin sheaths facilitating fast nerve conduction, astrocytes are essential for the maintenance of the ion homeostasis and of the tightness of the blood-brain barrier (BBB). After injury and during other forms of activation astrocytes upregulate a large number of cell surface and structural molecules the most prominent of which represents glial fibrillary acidic protein (GFAP).[1] Astrocytes are an important cellular source of immune mediators such as the antiinflammatory cytokines interleukin-10 and transforming growth factors.[2,3] Thereby, astrocytes can profoundly downregulate microglial activity.[4,5] Astrocytes are critically involved in scar formation and repair processes after CNS injury.

Microglia, the resident macrophage population of the CNS, are strategically located in close vicinity to neurons in the grey matter and between fiber tracts in the white matter. Microglia quickly respond to virtually any functional disturbance or lesion in the CNS.[6,7] In the resting state microglia exhibit radially branched processes (ramified microglia) and constitutively express complement type-3 receptors (CR-3) and Fc receptors for binding of immunoglobulins. Microglial activation leads to retraction of processes followed by rounding of the cell body, in particular in conditions in which microglia finally act as phagocytes. This is associated with increased levels of CR-3 and major histocompatibility complex (MHC) class I and II antigen expression on the cellular surface which enable microglia to interact with T cells. Upon further transition into phagocytes microglia develop intracellular phagolysosomes that can be visualized by immunocytochemistry. All these molecules indicative of microglial activation are also present on hematogenous monocytes/ macrophages, which invade the CNS during pathological processes.[8] At the stage of phagocytic transformation microglia and macrophages become indistinguishable by morphological grounds and by immunocytochemistry. Thus a distinction between the relative contribution of activated local microglia and invading macrophages to the pathophysiology of CNS lesions is often difficult. Activated microglia and macrophages have the capacity to synthesize and secrete a large number of molecules, which potentially cause brain damage, induce repair processes or orchestrate immune responses. Molecules produced by microglia/macrophages include neurotoxins, cytokines, coagulation and complement factors, prostanoids, free radicals, extracellular matrix components, and enzymes,[9-16] but also neurotrophins.[17,18] In general, the knowledge about the circumstances under which each of these molecules is produced and the regulatory elements involved, is limited. This is mainly due to the complexity of possible interactions and the unknown contribution of other glial cells and infiltrating leukocytes in the in vivo situation.

ENTRY OF INFLAMMATORY CELLS INTO THE CNS: THE BLOOD-BRAIN-BARRIER AND IMMUNOLOGICAL CELL ADHESION MOLECULES

Infiltration of the CNS by hematogenous leukocytes requires a disturbance or molecular modification of the BBB. The BBB becomes passively permeable to hematogenous cells by necrotic tissue injury involving the vessel wall. Alternatively, activated leukocytes can pass the intact BBB by complex leukocyte/endothelium interactions via cell adhesion molecules.[19] On the other hand, in CNS lesion paradigms with preservation of the BBB such as brief episodes of focal or global ischemia which lead to selective neuronal death but not to tissue necrosis, leukocytes are virtually excluded from the evolving lesions. Similarly, white matter tract or nerve lesions (e.g., optic nerve) lack leukocyte infiltration in the CNS in contrast to the peripheral nervous system, where a robust inflammatory response occurs due to breakdown of the blood-nerve barrier.

CNS infiltration involves multiple cell adhesion steps that provide the traffic signal for leukocytes into the brain through the endothelial wall. Upregulation of P-selectin and E-selectin on endothelial cells reduces the velocity of leukocytes in the blood stream by binding to the ligand L-selectin constitutively expressed on the surface of leukocytes. The light binding of leukocytes to the vessel wall is strengthened under the influence of chemokines that are released from the site of injury. The ensuing adhesion process is mediated by the integrin family of adhesion molecules. Different molecules direct the adhesion of leukocyte subsets: lymphocytes constitutively bear the CD11a/CD18 (leukocyte function associated antigen-1 = LFA-1) and the very-late-antigen-4 complex (VLA-4) on their surface, monocytes LFA-1, VLA-4 and the CD11b/CD18 complex (corresponding to CR3) and granulocytes LFA-1 and the CD11b/CD18 complex. The corresponding endothelial counterreceptors that require active stimulation for expression are intercellular adhesion molecule-1 (ICAM-1) for LFA-1 and CD11b/CD18, and vascular cellular adhesion molecule-1 (VCAM-1) for VLA-4. Cell adhesion molecules on CNS endothelial cells are rapidly induced during autoimmune disorders, but also after focal cerebral ischemia.

INFLAMMATION AND CNS INJURY

Necrotic Brain Injury: Stab Wounds and Focal Ischemia

Stab Wounds

The most simple traumatic CNS lesion with destruction of the BBB is a stab wound. A device is penetrated into the cortex and causes a circumscribed lesion. The ensuing cellular response is limited to site of penetrating injury and the immediately surrounding tissue. Activation of microglia at the lesion edges and infiltration of the wound site by hematogenous macrophages occurs within hours after injury.[12] Microglia proliferate locally with a maximum at day 2-3 and macrophages rapidly clear trauma-induced debris. Microglia and macrophages express interleukin (IL)-1β, a cytokine probably involved in astrogliosis and neovascularisation in this setting.

Models of Focal Cerebral Ischemia

Cellular responses to cerebral ischemia are more complex. Focal ischemia can be induced in rodents by permanent or transient occlusion of the middle cerebral artery (MCAO) leading to cortical and subcortical infarcts of various size depending on the time of blood flow cessation.[20,21] In transient MCAO, lesions are smaller and inflammation is briefer than after permanent occlusion because tissue is salvaged by reperfusion.[22,23] Photochemically-induced cortical ischemia (photothrombosis, PT) is an alternative model.[24,25] After systemic injection of the dye rose bengal focal illumination of a brain area through the intact skull leads to endothelial alterations. These are followed by an early disruption of the blood-brain barrier, vasogenic edema, photochemically stimulated platelet aggregation with formation of clots, and thrombotic occlusion of small intracerebral vessels leading to infarctions that are highly reproducible in size and location. Regions of specific interest such as the necrotic infarct core, the immediate border zone exhibiting delayed neuronal death, and the non-ischemic ipsilateral cortex undergoing more discrete functional changes are difficult to differentiate in the MCAO models, but easily discernible in the photothrombosis model.

Immune Mediators and Secondary Infarct Growth: The Role of Interferon Regulatory Factor and Inducible NO Synthase

Focal impairment or cessation of blood flow to the brain restricts the delivery of substrates, most importantly oxygen and glucose, and thereby impairs maintenance of ionic gradients. This is followed by depolarization of neurons and glia which release excitatory amino acids into the extracellular space and accumulate Ca2+.[26] Ca2+ is a universal second messenger leading to production of proteolytic enzymes, free-radical species and further activates glutamate receptors. In the center of the ischemic territory, where the flow reduction is most severe, these processes induce rapid necrotic cell death. A significant proportion of neurons, however, die by an internal program of self-destruction, designated apoptosis or programmed cell death.[27] Apoptotic neurons are intermingled with necrotic neurons in the core of infarctions. In the boundary zone apoptotic cell death is ongoing during the first week after focal ischemia.[28-30] In experimental animals mediators of the immune system appear to play an essential role in this secondary infarct growth.

In two seminal papers Iadecola and colleagues formally proved that inflammatory mechanisms contribute to a significant infarct growth beyond 24 hr after focal cerebral ischemia.[31,32] Interferon regulatory factor 1 (IRF) is a transcription factor that can be activated by the proinflammatory cytokines tumor necrosis factor-α (TNF-α) and IL-1 β. IRF induces gene transcription of interferon-γ, inducible nitric oxide synthase (iNOS), and interleukin-1 converting enzyme (ICE; caspase 1) involved in the induction of apoptosis and in the cleavage of IL-1/IL-18 precursors into the mature active cytokine proteins.[33] After focal ischemia IRF-1 gene expression was markedly upregulated at 12hr and reached a peak at day 4. Knock-out mice lacking the IRF-1 gene were protected from ischemic brain damage and developed smaller infarctions accompanied by attenuation of neurological deficits. Although the molecular mechanisms underlying the beneficial effects in IRF-1 knock-out mice have not yet been identified, lack of nitric oxide (NO) induction is a likely candidate. NO is a small molecule that exerts pleiotropic actions. NO is synthesized by oxidation of L-arginine by the enzyme NO synthase (NOS) which exists in three isoforms, neuronal NOS (nNOS), endothelial NOS (eNOS) and inducible or immunological NOS (iNOS). NO production is enhanced at all stages of cerebral ischemia.[34] After MCAO in mice, iNOS mRNA expression in the post-ischemic brain began between 6- 12 hr, peaked at 96 hr and subsided after 7 days. Disruption of the iNOS gene in mice led to smaller infarcts and less motor deficit after focal ischemia. Most importantly, such reduction in ischemic damage and neurological deficit was observed 96 hr after ischemia, but not at 24 hr providing strong evidence

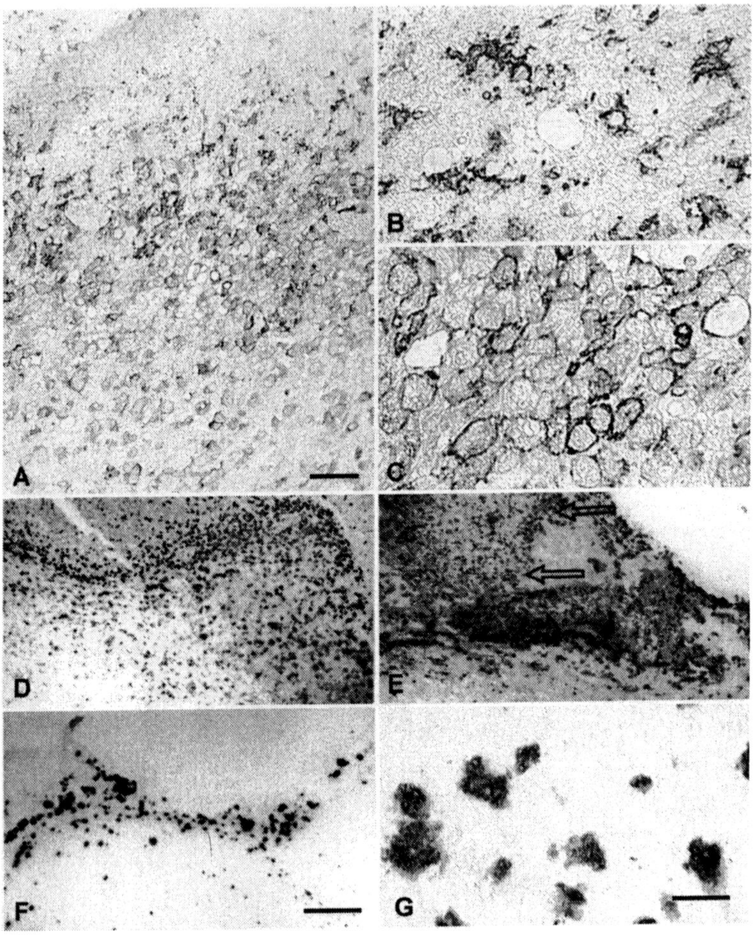

Figure 1. Microglial activation and macrophage infiltration at day 5 (A-C), 7 (D), 14 (E) and 28 (F,G) in the border zone after focal ischemia in humans (A-C) and after MCAO in the rat (D-G). Activated microglia strongly express MHC class II molecules (A-C). In (A) transition of microglia from a ramified phenotype to an ameboid phagocytic appearance can be followed from the periphery of the ischemic lesion (top, details shown in B and C) to the center (bottom). At the stage of full phagocytic transformation (C) the relative contribution of microglia and hematogenous macrophages to the pool of phagocytes is elusive because both cell types share identical immunocytochemical markers. (D-G) show the accumulation of phagocytes stained with the antibody ED1 in the border zone of focal ischemia in the rat. Note colliquation of the infarct and pseudocyste formation bordered by a dense wall of phagocytes (E, arrows). At day 28 most of the debris has been removed by phagocytes and the remaining cortex appears atrophic (F). (G) shows ED1+ phagocytes at higher magnification. Bar in (A) represents 250μm, 30μm in (G) which also applies to (B,C), and 500μm in (F) which also applies to (D, E)

that iNOS expression is one of the critical factors that contribute to the delayed expansion of brain damage. iNOS induces longlasting synthesis of large amounts of NO, which then reacts with superoxide to form peroxynitrite, a cytotoxic agent. On the cellular level iNOS could be localized in microglia, astrocytes and infiltrating leukocytes.[35]

Cell Infiltration: Granulocytes

In transient and permanent cerebral ischemia granulocytes are the first hematogenous cells that appear in the ischemic brain. After permanent occlusion of the MCA they accumulate in cerebral vessels within hours before they invade the infarct and its boundary zone.[36-38] This process peaks at 24 hrs after infarction, thereafter the number of granulocytes rapidly declines. At day 7 granulocytes are only occassionally seen. Granulocyte infiltration is preceded by an increased expression of cell adhesion molecules and chemokines.[39] Several groups demonstrated upregulation of ICAM-1 mRNA and protein as early as 3 hr on endothelial cells of intraparenchymal blood vessels in the ischemic cortex.[40-43] ICAM-1 expression peaked at 6 to 12 hr and persisted for several days. Similarly, mRNA for ELAM-1 and VCAM-1 was upregulated.[44,45] The functional relevance of cell adhesion processes in stroke development has been established in transient focal ischemia. Treatment with antibodies directed against the CD11b/CD18 complex on granulocytes and monocytes/macrophages of rats subjected to 2 hours of transient MCAO led to a significant reduction in infarct volume and to a decrease of apoptotic cells.[46] In parallel, infiltration by granulocytes was reduced. Similar results were obtained when a recombinant neutrophil inhibiting factor directed against the CD11b/CD18 complex was used.[47] Blocking of the corresponding ligand on endothelial cells, ICAM-1, had a stronger mitigating effect and reduced stroke volumes at day 2 by 80%. ICAM-1 knock out mice showed a fivefold decrease in infarct size.[48,49] In contrast, treatment with anti-leukocyte antibodies was ineffective in models of permanent MCAO.[50,51] The most likely explanation for this discrepancy is that granulocytes adhere to the microvascular endothelium via the ICAM-1/CD11b adhesion pathway thereby mechanically disconnecting dependent parenchyma from reperfusion (no reflow phenomenon) and inducing prolonged hypoxia which leads to extension of the infarct area into the penumbra zone. The pathophysiological consequences of the parenchymal granulocyte invasion has not been elucidated yet.

Cell Infiltration: T-Cells

In addition to granulocytes, a significant number of T cells identified by immunocyto-chemical surface markers invade the infarct region.[41,52] After both MCAO and photochemically-induced ischemia, CD5+ T cells started to infiltrate the infarct region from day 1 onwards. At day 3 their number had increased and reached a peak around day 7 followed by a substantial decrease within the next 7 days. T cells were preferentially located in the boundary zone of the infarctions often in close vicinity to blood vessels. The presence of T cells is surprising. T cell recruitment into the CNS is usually observed in autoimmune and inflammatory diseases such as experimental autoimmune encephalomyelitis (EAE) or multiple

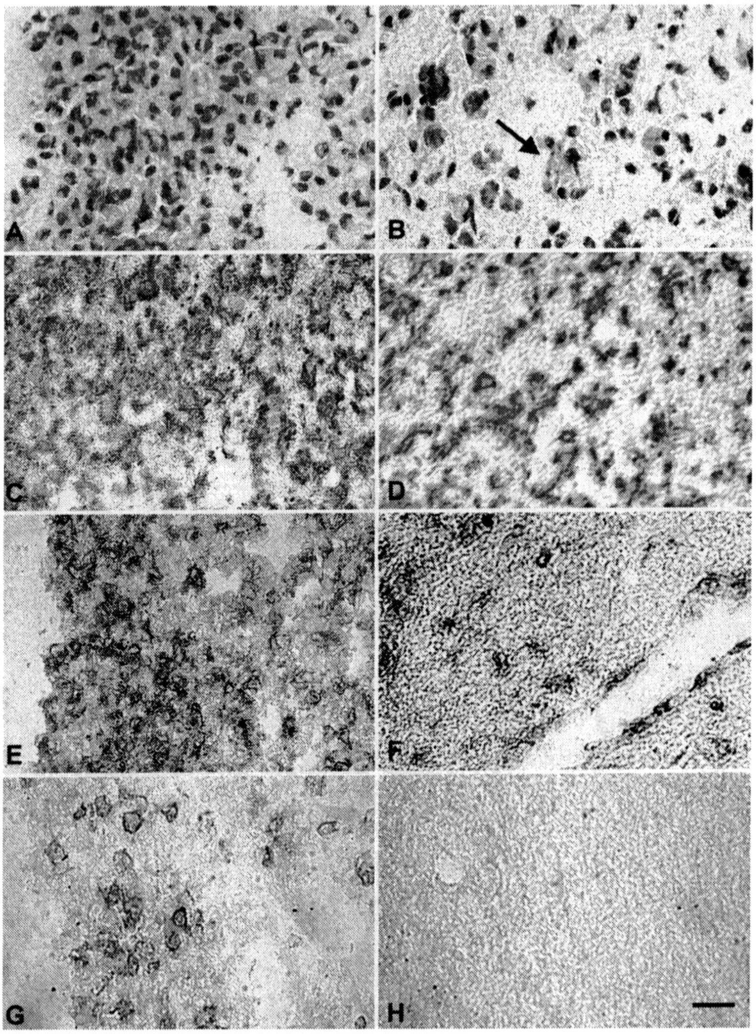

Figure 2. Differential microglial/macrophage responses at day 7 after transient MCAO in the rat. (A ,B) cresylviolet stain, (C, D) immunostaining for CR-3 (Ox 42), (E, F) for CD4, and (G, H) for CD8. (A) shows an area of ischemic pannecrosis with massive leukocyte infiltration (C, E , F). Note the appeareance of both CD4+ (E), and CD8+ (G) phagocytes. Contrastingly, in areas of selective neuronal death with preservation of glial structures (B) microglia still exhibit typical Ox 42+ processes (D) (arrow denotes degenerating [a]ghost neuron). In contrast to pannecrosis, microglia only partly transform into phagocytes, and upregulate CD4 (F), while CD8 expression is virtually absent (H). 50µm in A-H

sclerosis.[53] In EAE, systemic immunization with myelin proteins such as myelin basic protein (MBP) creates CD4+ helper/inducer T cells that are antigen-specific and enter the CNS after 10 to 12 days. As a second step, large numbers of nonspecific T cells and macrophages are recruited from the circulation and cause myelin destruction and clinical disease. In cerebral ischemia, the period between lesion induction and T cell infiltration is too short for generation of a systemic, antigen-specific immune response. Therefore, the T cell response is likely to be antigen-nonspecific. The functional consequences of T cell inflammation after focal cerebral ischemia is unclear at present (see also 3.2).

The Phagocyte Response: The Differential Contribution of Microglia and Macrophages

The most abundant leukocytes that enter the brain after focal cerebral ischemia are monocytes/macrophages (Fig. 1). They appear together with T cells but persist for a longer time period.[37,41,52,54] After permanent MCAO, monocytes/macrophages started to infiltrate the parenchyma at 12 hr and further increased in numbers up to day 14.[52,54,55] The entire infarct area was covered by macrophages at 3 days after MCAO.[52] Within the next four weeks the debris was completely cleared by these phagocytes. In the photothrombosis model macrophages behave differently. They were located in a ring-like fashion around the ischemic core in the first week and with a delay infiltrated the center of the lesion to remove debris.[41]

There is an intrinsic problem with the cellular identification of phagocytes after cerebral ischemia. Microglia, the resident macrophages of the brain, are activated after ischemia (Fig. 1A,B) and then become indistinguishable from hematogenous macrophages on morphological grounds and based on the expression of immunocytochemical markers.[8] Therefore, in the studies mentioned above, the pool of macrophages always contained activated microglia to an uncertain extent. To more specifically address the contribution of resident microglia to the pool of phagocytes hematogenous macrophages were temporarily depleted from the circulation by use of toxic liposomes.[56] When the phagocytic response between sham-treated and macrophage-depleted animals was assessed by immunocytochemistry, there was no difference in the number and distribution of ED1+ phagocytes at day 3 after photothrombosis.[57] Contrastingly, a dramatic difference was seen at day 6. In sham-treated rats ED1+ phagocytes largely outnumbered those in macrophage-depleted rats. These findings indicate that microglia are rapidly activated in the border zone of focal ischemic brain lesions and subsequently transform into phagocytes. With a certain delay which depends on the lesions paradigm additional monocytes/macrophages are recruited from the circulation probably due to local induction of chemokines such as the monocyte chemoattractant protein-1.[58-60]

Besides granulocytes, T cells and macrophages an unusual population of cells expressing the CD8 molecule appears in ischemic lesions of the rat cortex (Fig. 2).[41,52,61,62] These cells are abundant between days 3 and 6 and have almost disappeared at day 14. Usually, the CD8 molecule is restricted to T cells and natural killer (NK) cells.[63] However, most of the CD8+ cells in ischemic brain lesions are negative for T cell markers, but express microglia/macrophage markers.[61,62] Analysis of different ischemia paradigms revealed that the appearance of CD8+ microglia/macrophages in the CNS is indicative for a necrotic lesion (Fig. 2).[61] Ischemic brain lesions, however, are additionally infiltrated by an independent population of CD4+ macrophages representing the usual macrophage population in CNS injury and inflammation.[64,6] In contrast to the CD8+ population, the number of CD4+ macrophages gradually increased from day 2 on and peaked at day 14 when they covered the entire infarct region indicating a major role in debris removal. Hirji and coworkers independently reported on the expression of CD8 on peritoneal and pulmonary macrophages in vitro.[66,67] Upon stimulation of the CD8 molecule these macrophages produced NO in culture. NO derived from neurons and inflammatory cells has been implicated as a major pathogenic factor in ischemic brain damage (see above). The functional implications of the phenotypical diversity of microglia/macrophage responses in CNS injury in vivo, however, await further clarification.

Functional Aspects of the Microglia/Macrophage Response: The Neurotoxic Effects of the Cytokines IL-1β and TNF-α

Early microglial activation is involved in lesion development after cerebral ischemia. In support of this notion, nonspecific attenuation of microglial responses by minocycline was neuroprotective in both focal and global cerebral ischemia.[68,69] As stated in the introduction microglia are an important source of neurotoxic molecules which include proinflammatory cytokines.[70-76] Within one hour after focal ischemia increased IL-1mRNA levels can be measured in the ischemic cortex with peak expression during the first 24hr (Fig. 3).[77-79] Concomitantly, IL-1 receptors are upregulated.[80] In ischemic brain lesions IL-1β protein is mainly expressed by microglia/macrophages and endothelial cells,[73,81,82] while the contribution of neurons, astrocytes and oligodendrocytes is still controversial.[75] The most compelling evidence that IL-1β is involved in ischemic brain damage derives from pharmacological studies and from stroke induction in genetically manipulated laboratory animals.[83] Intracerebroventricular injection of IL-1β exacerbates brain damage after focal ischemia.[84,85] The effects of IL-1β can be antagonized by its natural counterplayer, IL1-receptor antagonist (IL-1ra). In ischemic lesions IL-1ra-mRNA levels are greatly increased at 12 hr after permanent MCAO and remain elevated for several days.[80] Currently, it appears that in vivo the agonistic IL-1β effects predominate over the antagonistic IL-1ra effects. Exogenous administration of IL-1ra either intracerebroventricularly or systemically, reduced focal ischemic brain damage.[83, 86-88] Interleukin-1 converting enzyme cleaves proIL-1 to generate biologically active IL-1β. In further support of a potentially noxious role of IL1β in ischemia, inhibition of ICE decreased infarct volumes.[89-91] Inhibition of ICE was accompanied by a decrease of IL-1β and TNF-α protein levels in ischemic lesions.[91] Although suggesting a pathological role for IL-1β, these neuroprotective effects might not solely be due to reduction of IL-1β production, since ICE additionally plays a critical role in activating apoptotic pathways and the activation of another proinflammatory cytokine, IL-18.

Similar to IL-1β, TNF-α is rapidly induced within the first three hours after MCAO and persists during the next few days (Fig. 3).[70,71,79,92,93] At the initial stage

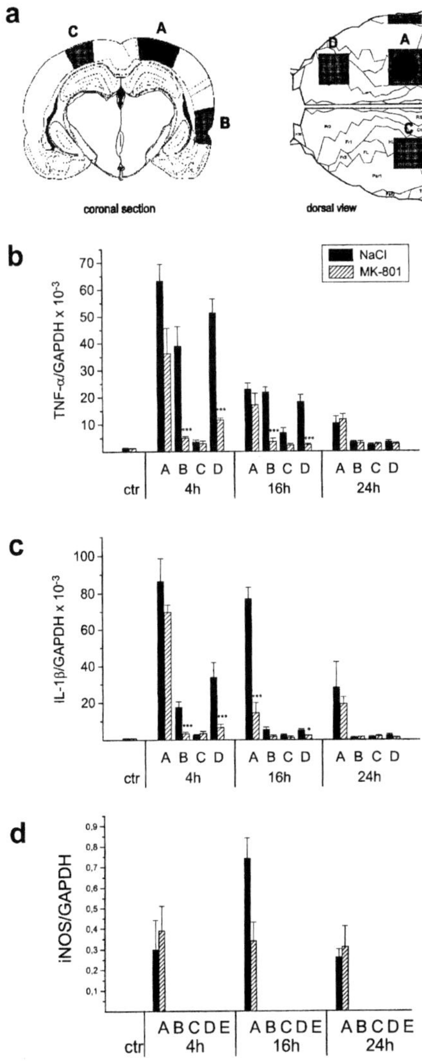

Figure 3. Expression of cytokine transcripts after focal ischemia induced by photothrombosis of the rat cortex (reverse transcriptase polymerase chain reaction). Black columns represent values after sham-treatment with i.v. NaCl, hatched columns expression levels after systemic application of the NMDA antagonist MK-801. Expression levels were normalized against the house keeping gene GAPDH. As shown in the schematic drawing (a) (A) denotes the ischemic lesion, (B, D) ipsilateral cortex areas remote from the lesion and (C) the homotypic cortex area of the contralateral hemisphere. Note rapid induction of TNF-α- (A in b) and IL-1β-mRNA (A in c) within the ischemic lesion, but also remote (B,D in b/c), while iNOS-mRNA expression (A in d) is restricted to the lesion. Remote cytokine induction is significantly suppressed after treatment MK-801 (hatched columns), while cytokine levels in the ischemic lesions are only marginally depressed.

of infarct development neurons appear to be the major source of TNF-α immunoreactivity.[92] In addition TNF-α has been localized in astrocytes, microglia, macrophages and granulocytes in ischemic brain lesions.[76,93] Concomitant with the expression of TNF-α, the TNF receptors p55 and p75 were upregulated within 6 and 24 hrs.[93] TNF receptor p55 has been implicated in transducing the cytotoxic signaling of TNF-α. The presence of TNF-α protein and TNF p55 receptors therefore suggested an injurious role. Intracerebroventricular injection of TNF-α 24 hr prior to MCAO significantly exacerbated the size of infarction probably by activating capillary endothelium to a proadhesive state.[70,71] TNF-α can induce apoptosis in a variety of targeT cells and thereby could contribute to delayed neuronal death.[94] Systemic treatment of rats within 1 to 2 hrs after onset of focal ischemia with the selective TNF synthesis inhibitor CNI-1493 led to a 70% redcution of the ensuing infarcts.[95] Similar neuroprotective effects due to TNF inhibiton were obtained with dexanabinol (HU-211), a nonpsychotropic cannabinoid analogue.[96] Neutralization of TNF by antibodies or application of a dimeric form of the type-1 TNF receptor had a similar mitigating effect on infarct development.[97,98] TNF-α, however, also protects neurons. As discussed below in detail the context of mediators at a given time after brain ischemia appears to determine whether the net effects of cytokines are neurotoxic or neuroprotective.[74,83]

Selective Neuronal Death

A different inflammatory response is seen in lesions in which frank necrosis is lacking and neurons mainly degenerate due to apoptosis. Such conditions can be induced by short global or focal ischemia.[99] Global cerebral ischemia refers to transient interruption of the blood flow to the brain by four vessel occlusion or bilateral occlusion of the carotid arteries. After brief periods of global cerebral ischemia selective loss of pyramidal cells occurs with a delay to 2-4 days in the CA1 region of the hippocampus with involvement of additional areas (CA3) after prolonged hypoxemia.[100] Transient forebrain ischemia leads to an early, but transient microglial reaction with upregulation of CR-3, MHC class I molecules, increased C1q biosynthesis, the recognition subcomponent of the classical complement activation pathway, and retraction of cellular processes in all hippocampal subfields including areas devoid of subsequent neuronal degeneration.[101-105] Increased staining of microglial cells was detected in the dentate hilus and CA1 area as early as 20 minutes after reperfusion and more intensely at 24 hrs. The strongest microglial reaction was observed 4-6 days after reperfusion. From days 4 to 7 the morphology and staining characteristics of microglial cells returned to normal in areas without neuronal death. In the CA1 region apoptotic bodies were observed between days 3 and 4 after ischemia, a period during which neuronal death is maximal.[33,106] Accordingly, the number of pyramidal neurons was significantly reduced at one week after ischemia.[107] Neuronal death was associated with the appearance of large numbers of transformed microglial cells which now showed an ameboid morphology.[104] Ameboid microglia expressed increased levels of NO synthase and of ICE.[108,109] In the dentate hilus, another hippocampal area undergoing neuronal cell loss, microglia clustered around degenerating neurons.[102,103] At day 21, CA1 was the only hippocampal region where these microglial activation markers were still upregulated. In contrast to focal cerebral ischemia, global ischemia does not appear to elicit a substantial recruitment of hematogenous inflammatory cells. T cells were scarce,[102] and macrophage and granulocyte infiltration was modest, although the endothelial adhesion molecule ICAM-1 was upregulated early in CA1.[38] Treatment with anti-ICAM-1 antibodies accordingly did not significantly affect appearance of macrophage-like cells at day 4.[38]

Areas of selective neuronal death which are characterized by loss of pyramidal neurons but preservation of endothelial cells and glia are also found after transient focal ischemia.[61,110] These regions show a similar restricted microglial response as CA1 in global ischemia.[61] Microglia are less ramified, upregulate CR3 and CD4 molecules and exhibit weak ED1 immunoreactivity as an indication of partial phagocytic transformation. Selective neuronal death in these regions can be attenuated by treatment with caspase inhibitors [111]. In contrast to pannecrotic ischemic lesions, CD8+ cells are virtually absent from areas of selective neuronal death (Fig. 2H).[61]

BENEFICIAL EFFECTS OF NEUROINFLAMMATION

Cytokines, Spreading Depression and Neuroprotection

Brief episodes of cerebral ischemia convey neuroprotection by making neurons less susceptible to a second ischemic event[112] (see also Chapter by Dirnagl and Endres in this book). There is evidence that neuroinflammation is involved in this process. Microglial and astrocytic responses are not restricted to the ischemic focus (Fig. 4). After focal ischemia microglia transiently upregulate CR3 and astrocytes express higher levels of GFAP in the entire ipsilateral cortex remote from the evolving lesion[113-115] in regions that lack histologically detectable tissue damage, leukocyte infiltration, and delayed apoptotic cell death.[29,30,114] Glial activation peaked between days 3 and 7. Remote astroglial activation could be completely blocked by treatment with the non-competitive N-methyl-D-aspartate (NMDA) receptor antagonist MK-801 while microglia only partly responded.[113,114] Morphological signs of glial activation were preceded by a rapid increase of IL-1β and TNF-α mRNA levels with peak levels within the first 24 hours (Fig. 3).[79] IL-1β protein was expressed by cortical microglia within 4 hours in the entire ipsilateral cortex and declined to baseline levels at 16 hours (Fig. 4H).[82] The cellular origin of TNF-α is not yet clear although some evidence suggests expression by neurons. Importantly, TNF-α and IL-1β expression in remote brain regions was not accompanied by iNOS induction in contrast to the coinduction in the ischemic focus (Fig. 3D). Since iNOS plays an essential role in NMDA-mediated neuronal injury,[116] this might explain why TNF-α and IL-1β expression in remote brain areas is probably neuroprotective rather than neurotoxic.

What are the molecular mechanisms underlying this widespread cytokine response? Remote cytokine induction after focal ischemia could be completely abolished by MK-801 treatment suggesting involvement of glutaminergic signalling through NMDA-receptors.[79] It has long been recognized that focal ischemic lesions in rodents provoke cortical spreading depression (CSD).[113,117] CSD is characterized by a transient suppression of all neuronal activity that repetitively extends from sites of increased extracellular potassium concentrations to the entire ipsilateral, but not contralateral hemisphere.[118] In the normal brain, CSD facilitates neuronal tolerance against a subsequent lethal ischemic challenge.[119,120] CSD can also be blocked by pretreatment with MK-801.[121] Taken together these observations suggested that CSD were responsible for the widespread ipsilateral TNF-α and IL-1β expression after focal cerebral ischemia. To formally prove this intriguing hypothesis we induced CSD by application of KCl to the cortical surface and found a similar glial activation and transient TNF-α and IL-1β expression restricted to the ipsilateral hemisphere as after focal ischemia (Fig. 4).[82]

Functionally, remote cytokine induction and widespread glial activation after focal ischemia and SD are probably part of a stress response leading to the development of ischemic tolerance and perilesional plasticity. TNF-α and IL-1β are involved in preconditioning leading to ischemic tolerance.[122-126] Intravenous pretreatment of rats with lipopolysaccharide, a potent TNF-α inducer, led to development of ischemic tolerance, an effect that could be prevented by coadministration of a TNF-α-binding protein.[127] Furthermore, intracisternal pretreatment of mice with TNF-α protected animals from ischemic injury after permanent MCAO.[98] In further support of a neuroprotective role of TNF-α in cerebral ischemia, mice lacking TNF receptors developed larger infarcts after focal ischemia.[128,129] In a model of brain percussion injury mice lacking TNF-α developed smaller memory deficits in the acute posttraumatic period at 48 hr postinjury, but showed persistent motor deficits after 4 weeks, when wild type animals had completely recovered suggesting a neuroprotective net effect on the long term.[130] The molecular mechanisms of TNF-α-mediated neuroprotection are not yet fully elucidated. Induction of a Ca^{2+}-scavenger proteins and mangane-superoxide dismutase are probably involved.[123] In vitro, hypoxic preconditioning protects neurons against hypoxic stress.[124,125] The neuroprotective effect was dependent on TNF-α induction. Ischemic tolerance that developed with a delay of 24-72 hrs was accompanied by TNF-α-mediated ceramide expression. Ceramide under certain conditons is a cytoprotectant.[131] Intraventricular administration of ceramide reduced brain damage after hypoxia-ischemia in one week old rats.[132] The therapeutic effects were associated with an upregulation of antiapoptotic bcl-2 molecules, and reduced numbers of apoptotic cells in the lesions. Taking into account the neurotoxicity of TNF-α as outlined above, present evidence suggests that TNF-α is deleterious in the acute phase of cerebral ischemia, but these negative effects are counterbalanced by an essential role of TNF-α in neuroprotection, recovery and repair. Similar to

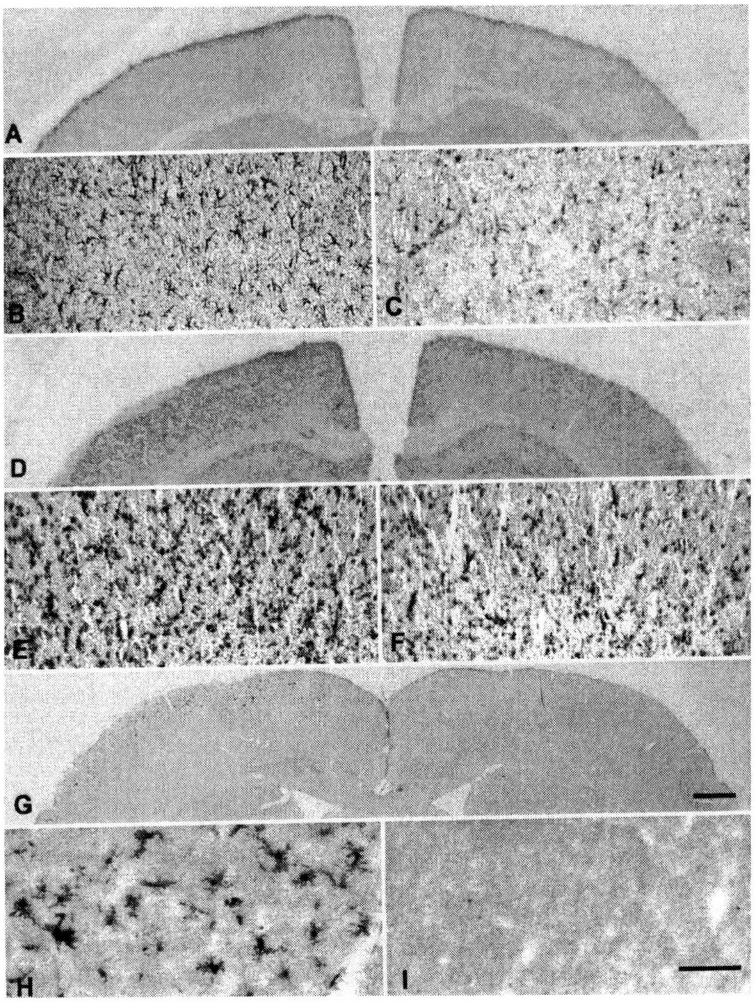

Figure 4. Glial activation and cytokine induction induced by cortical spreading depression (CSD). The findings depicted here are identical to the structural and molecular changes also seen in the remote ipsilateral cortex after focal ischemia. The left side of (A, D, G) represents an overview of the cerebral cortex ipsilateral to the CSD, while on the right side the contralateral hemisphere is shown. (B, E, H) and (C, F, I) exhibit morphological details at higher magnification from the corresponding area above. At 3 days after CSD astrocytes show marked upregulation of glial fibrillary acidic protein in the ipsilateral (B) compared to the contralateral hemisphere (C). Similarly microglia show stronger ipsilateral CR-3 expression (E) in comparison to the contralateral side (F). On the molecular level CSD induce strong microglial IL-1β expression as early as 8 hr (H) which again represents a strict ipsilateral response (H, I). Bar in (G) represents 1mm for (A, D, G), bar in (I) represents 30μm for (B, C, E, F, H, I).

TNF-α, IL-1β plays a dual role after cerebral ischemia. Despite its neurotoxic actions, IL-1β mediated induction of tolerance to global ischemia in gerbil hippocampal CA1 neurons.[122] Remote IL-1β induction after focal ischemia appears to be involved in postischemic brain plasticity. Schneider and colleagues[133] showed that IL-1β critically contributes to the maintenance of long-term potentiation (LTP) without affecting its induction. LTP is an important mechanism of brain plasticity and is facilitated in the vicinity of ischemic brain lesions.[134]

Neuroprotective Effects of T-Cell Inflammation

In autoimmune disorders of the CNS such as experimental autoimmune encephalomyelitis (EAE) a certain subset of T-cells, designated Th3-cells, can confer immunological tolerance. Animals previously tolerized to myelin basic protein (MBP) are resistant to induction of EAE by subsequent immunization with MBP or other myelin antigens.[135] This effect is due to the generation of immunosuppressive T cell subsets that express high levels of transforming growth factor-beta (TGF-β1).[136] Becker and colleagues[137] used a similar approach in focal cerebral ischemia. After immunization with MBP, tolerized animals developed significantly reduced infarct volumes when subjected to MCAO. Similar to EAE, T cells in the brain parenchyma of orally tolerized stroke animals exhibited TGF-β-immunoreactivity whereas TGF-β-positive T cells were absent in the control rats after MCAO. This study for the first time shows that antigen-specific tolerance can decrease infarct size in the Lewis rat MCAO model. The mechanisms of neuroprotection, however, have not yet been elucidated. The therapeutic effect of oral tolerization could be due to its immunosuppressive properties limiting inflammation. On the other hand, TGF-β1 directly promotes neuronal survival.[138] Focal cerebral ischemia leads to an increase in TGF-β expression by microglia and infiltrating macrophages.[139,140] Exogenous administration of TGF-β further limited injury and improved outcome in several models of cerebral ischemia.[141-143] The findings of Becker and colleagues point to immunomodulation as a new therapeutic strategy and, in particular, to a functional role of T cells in stroke development. Therefore, it is essential to know what population of lymphocytes are responsible for neuroprotection and whether this effect can be cellularly transferred to naive animals as shown in autoimmune inflammation.

Neuroprotective effects by T-cells have also been demonstrated in mechanical lesions to the nervous system. After a partial crush injury of the optic nerve, rats injected with activated anti-MBP T cells retained approximately three-fold more retinal ganglion cells with functionally intact axons than did rats injected with activated T cells specific for other antigens.[144] Similarly, active immunization of rats with MBP led to a robust survival-promoting effect on avulsed motoneurons which was attributed to the infiltration of the spinal cord by T-cells and natural killer lymphocytes expressing the neurotrophins brain-derived neurotrophic factor (BDNF) and neurotrophin-3 (NT-3).[145] Neurotrophin-positive T cells and macrophages have also been described in active EAE and MS lesions.[146] Taken together these findings suggest an important role of T cell and macrophage inflammation in the rescue of neurons during non-immune and autoimmune lesion paradigms of the nervous system.

Inflammation and Nerve Regeneration: Differences Between the Peripheral and Central Nervous System

Neuroinflammation is important for nerve regeneration. Transection/crush of a nerve or fiber tract leads to breakdown of the axoplasm within days, loss of axonal connectivity and to complete degeneration of the distal nerve stump. The injury-induced complex responses in the distal nerve or tract segment are referred to as Wallerian degeneration (WD).[147] In the peripheral nervous system axons promptly regenerate from the proximal stump into the degenerating distal nerve segment with a velocity of 1-2 mm per day and finally reach their target leading to functional recovery. In contrast, after CNS injury no such regrowth occurs. In the CNS regeneration appears to be suppressed by an adverse molecular and cellular environment. However, changing this environment facilitates a similar regenerative response within the CNS as revealed by the seminal experiments initially performed by Telo, a student of Ramon y Cajal,[148] and repeated with more success due to the availability of modern suture techniques by Aguayo and coworkers.[149] Both groups could show that transplantation of peripheral nerve tissue into transected optic nerves which belong to the CNS induced a similar regenerative response as in peripheral nerves. Conversely, peripheral nerves did not grow into CNS transplants. These findings indicated that the distal stump of a transected peripheral nerve provides a molecular milieu which facilitates regeneration in general in the nervous system. On the other hand, degenerating CNS fiber tracts contain growth-inhibitory factors that have been identified as the myelin and oligodendrocyte component Nogo and the myelin-associated glycoprotein (MAG).[150,151] Nogo which has been cloned recently[152,153] is expressed exclusively on CNS myelin sheaths and oligodendrocytes, while MAG is also present in myelin sheaths of peripheral nerves.

Neuroinflammation is an essential component of the growth promoting molecular microenvironment of injured peripheral nerves. Distal stumps of axotomised/crushed peripheral nerves are rapidly infiltrated by hematogenous T cells and macrophages.[154-156] Infiltration of nerves is accompanied by upregulation of proinflammatory cytokines[157,158] and leads to rapid removal of myelin debris containing growth-inhibitory MAG. In contrast, after injury to the optic nerve or to white matter tracts in the brain or spinal cord, hematogenous macrophages are largely excluded from the slowly degenerating distal stump probably because the blood-brain barrier remains intact.[154,159,160] Thereby myelin debris persists for months.[159] The functional importance of myelin/oligodendrocyte-associated growth-inhibition has been highlighted by neutralization experiments.[161,162] Application of anti-Nogo-antibodies allowed regeneration of lesioned fiber tracts to some extent along the original pathways, although most of the functional recovery is probably due to massive collateral sprouting from the corresponding contralateral uninjured fiber tracts. Most fibers of the orginal fiber tracts, however, were not able to pass the site of the lesion where a growth-impeding scar is rapidly formed in the CNS[148,163] (see below). In the CNS microglia could substitute hematogenous macrophages by phagocytic transformation and thereby compensate for the failure of macrophage entry into degenerating CNS fiber tracts. However, despite early signs of activation by upregulation of MHC class II and CD4 molecules, and the proinflammatory cytokine IL-18 after fiber tract injury, the microglial response does not mount into an efficient phagocytic transformation[154,158,159,164] which is regularly seen in the surround of focal cerebral infarcts.[57,114] It is tempting to speculate that further microglial transformation during Wallerian degeneration is inhibited by local, so far unknown factors. Accordingly, the phagocytic activity of both, macrophages and isolated brain-derived microglia was inhibited by exposure to optic nerve segments, but enhanced upon their exposure to sciatic nerve segments.[165] Moroever, peripheral-stimulated macrophages or in vitro activated microglia stimulated a regenerative response upon injection into transected optic nerve and spinal cord in support of a critical a role of neuroinflammation in nerve regeneration.[165-167]

Interestingly, a similar robust macrophage and T cell infiltration as seen along the entire degenerating nerve segments in the PNS occurs also in the CNS at the site of nerve or fiber tract injury probably because the blood-brain barrier is disrupted at this location.[156] Usually inflammation into lesioned tissue such as skin or heart is accompanied by scar formation during wound healing.[168] Similarly, at the site of injury and inflammation a scar is rapidly formed after optic nerve or fiber tract lesions which is nonpermissive for newly outgrowing nerve fibers.[163,169] The reason why this scar formation does not occur in PNS injury despite robust inflammation is unclear at present. These discrepancies between PNS and CNS injury make the important point that it is not the lack of inflammation alone that precludes nerve regeneration in the CNS, but the right or wrong molecular programming of the inflammatory response at the lesion site appears to be as important. The molecular mechanisms involved in scar formation are further described in detail in the Chapter by Logan and Berry. In conclusion, the insufficient activation or, alternatively, the active suppression of microglia, the lack of macrophage entry, and their inappropriate programming at the lesion site may account for regeneration failure in the CNS while, in contrast, neuroinflammation after PNS injury appears to be well designed. The variety of neuroinflammatory responses to injury of the nervous system provides a rich and currently unexplored set of research problems. Differentiation between beneficial and detrimental effects of inflammation is crucial for the development of treatment strategies selectively targeting harmful effects during disorders of the nervous system.

ACKNOWLEDGEMENTS

Work from the authors« laboratory was supported by the Deutsche Forschungsgemeinschaft and the Hermann- and Lilly-Schilling-Stiftung.

REFERENCES

1. Ridet JL, Malhotra SK, Privat A, Gage FH. Reactive astrocytes: Cellular and molecular cues to biological function. Trends Neurosci 1997; 20:570-577.

2. Constam DB, Philipp J, Malipiero UV, Ten Dijke P, Schachner M, Fontana A. Differential expression of transforming growth factor-beta 1, beta 2, and beta 3 by glioblastoma cells, astrocytes, and microglia. J Immunol 1992; 148:1404-1410.
3. Jander S, Pohl J, D«Urso D, Gillen C, Stoll G. Time course and cellular localisation of interleukin-10 mRNA and protein expression in autoimmune inflammation of the rat central nervous system. Am J Pathol 1998; 152:975-982.
4. Men ndez Iglesias B, Cerase J, Ceracchini C, Levi G, Aloisi F. Analysis of B7-1 and B7-2 costimulatory ligands in cultured mouse microglia: Upregulation by interferon-gamma and lipopolysaccharide and downregulation by interleukin-10, prostaglandin E2 and cAMP-elevating agents. J Neuroimmunol 1997; 72:83-93.
5. Letterio JJ, Roberts AB. Regulation of immune responses by TGF-β. Annu Rev Immunol 1998; 16:137-161
6. Kreutzberg GW. Microglia: A sensor for pathological events in the CNS. Trends Neurosci 1996; 19:312-318.
7. Stoll G, Jander S. The role of microglia and macrophages in the pathophysiology of the CNS. Prog Neurobiol 1999; 58:233-247.
8. Flaris NA, Densmore TL, Molleston MC, Hickey WF. Characterization of microglia and macrophages in the central nervous system of rats: Definition of the differential expression of molecules using standard and novel monoclonal antibodies in normal CNS and in four models of parenchymal reaction. Glia 1993; 7:34-40.
9. Piani D, Frei K, Do KQ, Cuenod M, Fontana A. Murine brain macrophages induce NMDA receptor mediated neurotoxicity in vitro by secreting glutamate. Neurosci Lett 1991; 133:159-162.
10. Chao CC, Hu S, Molitor TW, Shaskan EG, Peterson PK. Activated microglia mediate neuronal cell injury via a nitric oxide mechanism. J Immunol 1992; 149:2736-2741.
11. Banati RB, Gehrmann J, Schubert P, Kreutzberg GW. Cytotoxicity of microglia. Glia 1993; 7:111-118.
12. Giulian D, Corpuz M, Chapman S, Mansouri M, Robertson C. Reactive mononuclear phagocytes release neurotoxins after ischemic and traumatic injury to the central nervous system. J Neurosci Res 1993; 36:681-693.
13. Zielasek J, Hartung HP. Molecular mechanisms of microglial activation. Adv Neuroimmunol 1996; 6:191-222.
14. Flavin MP, Coughlin K, Ho LT. Soluble macrophage factors trigger apoptosis in cultured hippocampal neurons. Neurosci 1997; 80:437-448.
15. Minghetti L, Levi G. Microglia as effector cells in brain damage and repair: Focus on prostanoids and nitric oxide. Prog Neurobiol 1998; 54:99-125.
16. Ellison JA, Velier JJ, Spear P, Jonak ZL, Wang X, Barone FC, Feuerstein GZ. Osteopontin and its integrin receptor alpha(v)beta3 are upregulated during formation of the glial scar after focal stroke. Stroke 1998; 29:1698-1706.
17. Elkabes S, DiCicco-Bloom EM, Black IB. Brain microglia/macrophages express neurotrophins that selectively regulate microglial proliferation and function. J Neurosci 1996; 16:2508-2521.
18. Heese K, Hock C, Otten U. Inflammatory signals induce neurotrophin expression in human microglial cells. J Neurochem 1998; 70:699-707.
19. Springer TA. Traffic signals for lymphocyte recirculation and leukocyte emigration: The multistep paradigm. Cell 1994; 76:301-314.
20. McAuley MA. Rodent models of focal ischemia. Cerebrovasc Brain Metabol Rev 1995; 7:153-180.
21. Stoll G, Jander S, Schroeter M. Inflammation and glial responses in ischemic brain lesions. Prog Neurobiol 1998; 56:149-161.
22. Memezawa H, Smith ML, Siesj BK. Penumbral tissues salvaged by reperfusion following middle cerebral artery occlusion in rats. Stroke 1992; 23:552-559.
23. Zhang ZG, Chopp M, Powers C. Temporal profile of microglial response following transient (2h) middle cerebral artery occlusion. Brain Res 1997; 744:189-198.

24. Watson BD, Dietrich D, Busto R, Wachtel MS, Ginsberg MD. Induction of reproducible brain infarction by photochemically initiated thrombosis. Ann Neurol 1985; 17:497-504.
25. Schroeter M, Franke C, Stoll G, Hoehn M. Dynamic changes of MRI abnormalities in relation to inflammation and glial responses after photothrombotic cerebral infarction in the rat brain. Acta Neuropathol 2001; 101:114-122.
26. Dirnagl U, Iadecola C, Moskowitz MA. Pathobiology of ischaemic stroke: An integrated view. Trends Neurosci 1999; 22:391-397.
27. Bredesen DE. Neural apoptosis. Ann Neurol 1995; 38:839-851.
28. Li Y, Sharov VG', Jiang N, Zalonga C, Sabbah HN, Chopp M. Ultrastructural and light microscopic evidence of apoptosis after middle cerebral artery occlusion in the rat. Am J Pathol 1995; 146:1045-1051.
29. Braun JS, Jander S, Schroeter M, Witte OW, Stoll G. Spatiotemporal relationship of apoptotic cell death to lymphomonocytic infiltration in photochemically induced focal ischemia of the rat cerebral cortex. Acta Neuropath. 1996; 92:255-263.
30. Isenmann S, Stoll G, Schroeter M, Krajewski S, Reed JC, B hr M Differential regulation of Bax, bcl-2, and bcl-x proteins in focal cortical ischemia in the rat. Brain Pathol 1998; 8:49-63.
31. Iadecola C, Zhang F, Casey R, Nagayama M, Ross ME. Delayed reduction of ischemic brain injury and neurological deficits in mice lacking the inducible nitric oxide synthase gene. J Neurosci 1997; 17:9157-9164.
32. Iadecola C, Salkowski CA, Zhang F, Aber T, Nagayama M, Vogel SN, Ross ME. The transcription factor interferon regulatory factor 1 is expressed after cerebral ischemia and contributed to ischemic brain injury. J Exp Med 1999; 189:719-727.
33. Bhat RV, RiRocco R, Marcy VR, Flood DG, Zhu Y, Dobrzanski P, Siman R, Scott R, Contreras PC, Miller M. Increased expression of IL1-β converting enzyme in hippocampus after ischemia: Selective localization in microglia. J Neurosci 1996; 16:4146-4154.
34. Iadecola C. Bright and dark sides of nitric oxide in ischemic brain injury. Trends Neurosci 1997; 20:132-139.
35. Bidmon HJ, Wu J, Buchkremer-Raztmann I, Mayer B, Witte OW, Zilles K. Transient changes in the presence of nitric oxide synthases and nitrotyrosine immunoreactivity after focal cortical lesions. Neuroscience 1998; 82:377-395.
36. Kochanek PM, Hallenbeck JM. Polymorphonuclear leukocytes and monocytes/macrophages in the pathogenesis of cerebral ischemia and stroke. Stroke 1992; 23:1367-1379.
37. Garcia JH, Liu KF, Yoshida Y, Lian J, Chen S, del Zoppo G. Influx of leukocytes and platelets in an evolving brain infarct (Wistar rat). Am J Pathol 1994; 144:188-199.
38. Clark WM, Lauten JD, Lessov N, Woodward W, Coull BM. Time course of ICAM-1 expression and leukocyte subset infiltration in rat forebrain ischemia. Mol Chem Neuropathol 1995; 26:213-230.
39. Yamasaki Y, Matsuo Y, Matsuura N, Onodera H, Itoyama Y, Kogure K. Transient increase of cytokine-induced neutrophil chemoattractant, a member of the interleukin-8 family, in ischemic brain areas after focal ischemia in rats. Stroke 1995; 26:318-322.
40. Okada Y, Copeland BR, Mori E, Tung MM, Thomas WS, del Zoppo GJ. P-selektin and intercellular adhesion molecule-1 expression after focal brain ischemia and reperfusion. Stroke 1994; 25:202-211.
41. Jander S, Kraemer M, Schroeter M, Witte OW, Stoll G. Lymphocytic infiltration and expression of intercellular adhesion molecule-1 in photochemically induced ischemia of the rat cortex. J Cereb Blood Flow Metabol 1995; 15:42-51.
42. Zhang RL,Chopp M, Zalonga C, Jiang N, Gautam SC, Tang WX et al. The temporal profiles of ICAM-1 protein and mRNA expression after transient MCA occlusion in the rat. Brain Res 1995; 682:182-188.
43. Linsberg PJ, Carpen O, Paetau A, Karjalainen-Linsberg ML, Kaste M. Endothelial ICAM-1 expression associated with inflammatory cell response in human ischemic stroke. Circulation 1996; 94:939-945.

44. Wang X, Yue TL, Barone FC, Feuerstein GZ. Demonstration of increased endothelial-leukocyte adhesion molecule-1 mRNA expression in rat ischemic cortex. Stroke 1995; 26:1665-1669
45. Jander S, Pohl J, Gillen C, Schroeter M, Stoll G. Vascular cell adhesion molecule-1 mRNA is expressed in immune-mediated and ischemic injury of the rat nervous system. J Neuroimmunol 1996; 70:75-80.
46. Chopp M, Li Y, Jiang N, Zhang RL, Prostak J. Antibodies against adhesion molecules reduce apoptosis after middle cerebral artery occlusion in rat brain. J Cereb Blood Flow Metabol 1996; 16:578-584.
47. Jiang N, Moyle M, Soule HR, Rote WE, Chopp M. Neutrophile inhibitory factor is neuroprotective after focal ischemia in rats. Ann Neurol 1995; 38:935-942.
48. Connolly ES, Winfree CJ, Springer TA, Naka Y, Yan SD, Stern DM, Solomon RA, Gutlererez-Ramos JC, Pinski DJ. Cerebral protection in homozygous null ICAM-1 mice after middle cerebral artery occlusion: role of neutrophil adhesion in the pathogenesis of stroke. J Clin Invest 1996; 97:209-216.
49. Soriano SG, Lipton SA, Wang YF, Xiao M, Springer TA, Gutierrez-Ramos JC, Hickey PR. Intercellular adhesion molecule-1-deficient mice are less susceptible to cerebral ischemia-reperfusion injury. Ann.Neurol 1996; 39:618-624.
50. Zhang RL, Chopp M, Li Y, Zaloga C, Jiang N, Jones ML et al. Anti-ICAM-1 antibody reduces ischemic cell damage after transient but not permanent MCA occlusion in the Wistar rat. Stroke 1995; 26:1438-1443.
51. Garcia JH, Liu KR, Bree MP. Effects of CD11b/18 monoclonal antibody on rats with permanent cerebral artery occlusion. Am J Pathol 1996; 148:241-248.
52. Schroeter M, Jander S, Witte OW, Stoll G. Local immune responses in the rat cerebral cortex after middle cerebral artery occlusion. J Neuroimmunol 1994; 55: 195-203.
53. Wekerle H. Experimental autoimmune encephalomyelitis as a model of immune-mediated CNS disease. Curr Opin Neurobiol 1993; 3: 779-784.
54. Garcia JH, Liu KF, Yoshida Y, Lian J, Chen S, del Zoppo GJ. Influx of leukocytes and platelets in an evolving brain infarct (Wistar rat). Am J Pathol 1994;144:188-199.
54. Rupalla K, Allegrini PR, Sauer D, Wiesner C. Time course of microglia activation and apoptosis in various brain regions after permanent focal cerebral ischemia in mice. Acta Neuropath 1998; 96:172-178.
55. Clark RK, Lee EV, Fish CJ, White RF, Price WJ, Jonak ZL et al. Development of tissue damage, inflammation and resolution following stroke: An immunohistochemical and quantitative planimetric study. Brain Res Bull 1993; 31:565-572.
56. Van Roojien N. The liposome-mediated macrophage suicide« technique. J Immunol Meth 1989; 124:1-6.
57. Schroeter M, Jander S, Huitinga I, Witte OW, Stoll G. Phagocytic response in photochemically induced infarction of the rat cerebral cortex: the role of resident microglia. Stroke 1997; 28:382-386.
58. Kim JS, Gautam SC, Chopp M, Zalonga C, Jones ML, Ward PA, Welch KM. Expression of monocyte chemoattractant protein-1 and macrophage inflammatory protein-1 after focal cerebral ischemia in the rat. J Neuroimmunol 1995; 56:127-134.
59. Wang X, Yue TL, White RF, Barone FC, Feuerstein GZ. Monocyte chemoattractant protein-1 messenger RNA expression in rat ischemic cortex. Stroke 1995; 26:661-665.
60. Gourmala NG, Buttini M, Limonta S, Sauter A, Boddeke HW. Differential and time-dependent expression of monocyte chemoattractant protein-1 mRNA by astrocytes and macrophages in rat brain: effects of ischemia and peripheral lipopolysaccharide administration. J Neuroimmunol 1997; 74:35-44.
61. Schroeter M, Jander S, Huitinga I, Stoll G. CD8+ phagocytes in focal ischemia of the rat brain: Predominant origin from hematogenous macrophages and targeting to areas of pannecrosis. Acta Neuropathol 2001; 101:440-448.

62. Jander S, Schroeter M, D«Urso D, Gillen C, Witte OW, Stoll G. Focal cerebral ischemia of the rat brain elicits an unusual inflammatory response: early appearance of CD8+ macrophages/microglia. Eur J Neurosci 1998; 10:680-688.
63. Lawetzky A, Tiefenthaler G, Kubo R, H nig T. Identification and characterization or rat T cell populations expressing T cell receptors α/β and γ/δ. Eur J Immunol 1990; 20:343-349.
64. Perry VH, Gordon S. Modulation of CD4 antigen on macrophages and microglia in rat brain. J Exp Med 1987; 166:1138-1143.
65. Perry VH, Andersson PB, Gordon S. Macrophages and inflammation in the central nervous system. Trends Neurosci 1993; 16:268-273.
66. Hirji N, Lin TJ, Bissonnette E, Belosevic M, Befus AD. Mechanisms of macrophage stimulation through CD8: Macrophage CD8 alpha and CD8 beta induce nitric oxide production and associate killing of the parasite Leishmani major. J Immunol 1998;160:6004-6011.
67. Hirji N, Lin TJ, Gichrist M, Nault G, Nohara O, Grill BJ et al. Novel CD8 molecule on macrophages and masT cells: expression, function and signaling. Int Arch Allergy Immunol 1999; 118:180-182.
68. Yrjanheikki J, Keinanen R, Pellikka M, Hokfelt T, Koistinaho J. Tetracyclines inhibit microglial activation and are neuroprotective in global brain ischemia. Proc Natl Acad Sci USA 1998; 95:15769-15774.
69. Yrjanheikki J, Tikka T, Keinanen R, Goldsteins G, Chan PH, Koistinaho J. A tetracycline derivative, minocycline, reduces inflammation and protects against focal cerebral ischemia with a wide therapeutic window. Proc Natl Acad Sci USA 1999; 96:13496-13500.
70. Arvin B, Neville LF, Barone FC, Feuerstein GZ. The role of inflammation and cytokines in brain injury. Neurosci Biobehav Reviews 1996; 20:445-452.
71. Barone FC, Arvin B, White RF, Miller A, Web CL, Willette RN et al. Tumor necrosis factor-alpha. A mediator of focal ischemic brain injury. Stroke 1997; 28:1233-1244.
72. Peters O, Back T, Lindauer U, Busch C, Megow D, Dreier J et al. Increased formation of reactive oxygen species after permanent and reversible middle cerebral artery occlusion in the rat. J Cereb Blood Flow Metab 1998; 18:196-205.
73. Davis CA, Loddick SA, Toulmond S, Stroemer RP, Hunt J, Rothwell NJ. The progression and topographic distribution of interleukin-1????? expression after permanent middle cerebral artery occlusion in the rat. J Cereb Blood Flow Metab 1999; 19:87-98.
74. Shohami E, Ginis I, Hallenbeck JM. Dual role of tumor necrosis factor alpha in brain injury. Cytokine Growth Factor Rev 1999; 10:119-130.
75. Touzani O, Boutin H, Chuquet J, Rothwell N. Potential mechanisms of interleukin-1 involvement in cerebral ischemia. J Neuroimmunol 1999; 100:203-215.
76. Gregersen R, Lambertsen K, Finsen B. Microglia and macrophages are the major source of tumor necrosis factor in permanent middle cerebral artery occlusion in mice. J Cereb Blood Flow Metab 2000; 20:53-65.
77. Liu T, McDonnell PC, Young PR, White RF, Siren AL, Hallenbeck JM et al. Interleukin-1 mRNA expression in ischemic rat cortex. Stroke 1993; 24:1746-51
78. Buttini M, Sauter A, Boddeke HW. Induction of interleukin-1 beta mRNA after focal cerebral ischemia in the rat. Brain Res 1994; 23:126-134.
79. Jander S, Schroeter M, Stoll G. Role of NMDA receptor signalling in the regulation of inflammatory gene expression after focal brain ischemia. J Neuroimmunol 2000; 109:181-187.
80. Wang X, Barone FC, Aiyar NV et al. Interleukin-1 receptor and receptor antagonist gene expression after focal stroke in rats. Stroke 1997; 28:155-161
81. Zhang ZG, Chopp M, Goussev A. Cerebral vessels express interleukin 1 beta after focal ischemia. Brain Res 1998; 784:210-217.
82. Jander S, Schroeter M, Peters O, Witte OW, Stoll G. Cortical spreading expression induces proinflammatory cytokine gene expression in the rat brain. J Cerebr Blood Flow Metabol 2001; 21:218-225.
83. Rothwell NJ, Luheshi GN. Interleukin 1 in the brain: Biology, pathology and therapeutic target. Trends Neurosci 2000; 23:618-625.

84. Yamasaki Y, Matsuura N, Shozuhara H, Onodera H, Itoyama Y, Kogure K. Interleukin-1 as a pathogenic mediator of ischemic brain damage in rats. Stroke 1995; 26:676-680
85. Stroemer RP, Rothwell NJ. Exacerbation of ischemic brain damage by localized striatal injection of interlekin-1 beta in the rat. J Cereb Blood Flow Metabol 1998; 18:833-839.
86. Relton JK, Rothwell NJ. Interleukin-1 receptor antagonist inhibits ischemic and excitotoxic neuronal damage in the rat. Brain Res Bull 1992; 29:243-246.
87. Garcia JH, Liu KF, Relton JK. Interleukin-1 receptor antagonist decreases the number of necrotic neurons in rats with middle cerebral artery occlusion. Am J Pathol 1995; 147:1477-1486
88. Relton JK, Martin D, Thompson RC, Russell DA. Peripheral administration of interleukin-1 receptor antagonist inhibits brain damage after focal ischemica in the rat. Exp Neurol 1996; 138:206-213.
89. Loddick SA, MacKenzie A, Rothwell NJ. An ICE inhibitor, z-VAD-DCB attenuates ischemic brain damage in the rat. NeuroReport 1996; 7:1465-1468.
90. Hara H, Friedlander RM, Gagliardini V, Ayata C, Ayata G, Huang Z et al. Inhibition of ICE family proteases reduces ischemic and excitotoxic neuronal damage. Proc Natl Acad Sci USA 1997; 94:2007-2012.
91. Rabuffetti M, Sciorati C, Tarozzo G, Clementi E, Manfredi AA, Beltramo M. Inhibition of caspase-1-like acitivty by Ac-Tyr-Val-Ala-Asp-Chloromethyl ketone induces long-lasting neuroprotection in cerebral ischeami through appotosis reduction and decrease of proinflammatory cytokines. J Neurosci 2000; 20:4398-4404.
92. Liu T, Clark RK, McDonell PC, Young PR, White RF, Barone FC et al. Tumor necrosis factor-α expression in ischemic neurons. Stroke 1994; 25:1481-1488.
93. Botchina GI, Meistrell ME, Botchina IL, Tracey KJ. Expression of TNF and TNF receptors (p55 and p75) in the rat brain after focal cerebral ischemia. Mol Med 1997; 3:765-781.
94. Fehsel K, Kolb-Bachofen V, Kolb H. Analysis of TNF-α induced DNA strand breaks at the single cell level. Am J Pathol 1991; 139:251-254.
95. Meirstrell III ME, Botchkina GI, Wang. Tumor necrosis factor alpha is a brain damaging cytokine in cerebral ischemia. Shock 1997; 8:341-348.
96. Lecker R, Shohami E, Abramsky O. Daxanabiolol, a novel neuroprotective drug in experimental focal ischemia. J Neurol Sci 1999; 162:114-119.
97. Lavine SD, Hofman FM, Zlokovic BV. Circulating antibody against tumor necrosis factor alpha protects rat brain from reperfusion injury. J Cereb Blood Flow Metab 1998; 18:52-58.
98. Nawashiro H, Martin D, Hallenbeck JM. Inhibition of tumor necrosis factor and amelioration of brain infarction in mice. J Cereb Blood Flow mMetab1997; 17:229-232.
99. Pulsinelli WA, Brierly JB, Plum F. Temporal profile of neuronal damage in a model of transient forebrain ischemia. Ann Neurol 1982; 11:491-498.
100. Ordy JM, Wengenack TM, Bialobok P, Rodier P, Baggs RB, Dunlap WP et al. Selective vulnerability and early progression of hippocampal CA1 pyramidal cell degeneration and GFAP-positive astrocytes reactivity in the rat four-vessel occlusion model of transient global ischemia. Exp Neurol 1993; 119:128-139.
101. Morioka T, Kalehua AN, Streit WJ. The microglial reaction in the rat dorsal hippocampus following transient forebrain ischemia. J Cereb Blood Flow Metabol 1991; 11:966-973.
102. Morioka T, Kalehua AN, Streit WJ. Progressive expression of immunomolecules on microglial cells in the rat dorsal hippocampus following transient forebrain ischemia. Acta Neuropath (Berl) 1992; 83:149-157.
103. Finsen BR, Jorgensen MB, Diemer NH, Zimmer J. Microglial MHC antigen expression after ischemic and kainic acid lesions of the adult rat hippocampus. Glia 1993; 7:41-49.
104. Jorgensen MB, Finsen BR, Jensen MB, Castellano B, Diemer NH, Zimmer J. Microglial and astroglial reactions to ischemic and kainic acid-induced lesions of the adult rat hippocampus. Exp Neurol 1993; 120:70-88.

105. Sch fer MK, Schwaeble WJ, Post C, Salvati P, Calabresi M, Sim RB et al. Complement C1q is dramatically up-regulated in brain microglia in response to transient global cerebral ischemia. J Immunol 2000; 164:5444-5452.
106. Friedlander RM, Gagliardini V, Hara H, Fink KB, Li W, MacDonald G et al. Expression of a dominant negative mutant of interleukin-1 beta converting enzyme in transgenic mice prevents neuronal cell death induced by trophic factor withdrawl and ischemic brain injury. J Exp Med 1997; 185:933-940.
107. Lin B, Ginsberg MD, Busto R, Dietrich WD. Sequential analysis of subacute and chronic neuronal, astrocytic and microglial alterations after transient global ischemia in rats. Acta Neuropath 1998; 95:511-523.
108. Yamashita K, Kataoka Y, Yamashita YS, Himeno A, Tsutsumi K, Niwa M et al. Glial endothelia/nitric oxide system participates in hippocampus CA1 neuronal death of SHRPS following transient forebrain ischemia. Clin Exp Pharmacol Physiol 1995; 22(suppl. 1):S227-278.
109. Yamashita K, Kataoka Y, Sakurai-Yamashita Y, Shigematsu K, Himeno A, Niwa M et al. Involvement of glial endothelin/nitric oxide in delayed neuronal death of rat hippocampus after transient forebrain ischemia. Cell Mol Neurobiol 2000; 20:541-551.
110. Lehrmann E, Christensen T, Zimmer J, Diemer NH, Finsen B. Microglial and macrophage reactions mark progressive changes and define the penumbra in the rat neocortex and striatum after transient middle cerebral artery occlusion. J Comp Neurol 1997°; 386:461-476.
111. Endres M, Namura S, Shimizu-Sasamata M, Waeber C, Zhang L, Gomez-Isla T et al. Attenuation of delayed neuronal death after mild focal ischemia in mice by inhibition of the caspase family. J Cereb Blood Flow Metabol 1998; 18:238-247.
112. Chen J, Simon R. Ischemic tolerance in brain. Neurology 1997; 48:306-311.
113. Schroeter M, Schiene K, Kraemer M, Hagemann G, Weigel H, Eysel UT et al. Astroglial responses in photochemically induced focal ischemia of the rat cortex. Exp Brain Res 1995; 106:1-6.
114. Schroeter M, Jander S, Witte OW, Stoll G. Hetereogeneity of the microglial response in photochemically induced focal ischemia of the rat cerebral cortex. Neuroscience 1999; 89:1367-1377.
115. Yamashita K, Vogel P, Fritze K, Back T, Hossmann KA, Wiessner C. Monitoring the temporal and spatial activation pattern of astrocytes in focal cerebral ishcmeia using in situ hybridization to GFAP mRNA: comparison with sgp-2 and Hsp70 mRNA and the effect of glutamate receptor antagonists. Brain Res 1996; 735:285-297.
116. Hewett SJ, Csernansky CA, Choi DW. Selective potentiation of NMDA-induced neuronal injury following induction of astrocytic iNOS. Neuron 194; 13:487-494.
117. Dietrich D, Feng ZC, Leistra H, Watson BD, Rosenthal M. Photothrombotic infarction triggers multiple episodes of cortical spreading depression in distant brain regions. J Cereb Blood Flow Metab 1994; 14:20-28.
118. Martins-Ferreira H, Nedergaard M, Nicholson C. Perspectives on spreading depression. Brain Res Brain Res Rev 2000; 32:215-234.
119. Kobayashi S, Harris VA, Welsh FA. Spreading depression induces tolerance of cortical neurons to ischemia in rat brain. J Cereb Blood Flow Metab 1995; 15:721-727.
120. Yanamoto H, Hashimoto N, Nagata I, Kikuchi H. Infarct tolerance against temporary focal ischemia following spreading depression in rat brain. Brain Res 1998; 784:239-249.
121. Lauritzen M, Hansen AJ. The effect of glutamate receptor blockade on anoxic depolarization and cortical spreading depression. J Cereb Blood Flow Metab 1992; 12:223-229
122. Ohtsuki T, Ruetzeler CA, Tasaki K, Hallenbeck JM. Interleukin-1 mediates induction of tolerance to global ischemia in gerbil hippocampal CA1 neurons. J Cereb Blood Flow Metab 1996; 16:1137-1142.
123. Mattson MP. Neuroprotective signal transduction: relevance to stroke. Neurosci Biobehav Rev 1997; 21:193-206.

124. Ginis I, Schweizer U, Brenner M, Liu J, Azzam N, Spatz M et al. TNF-α pretreatment prevents subsequent activation of cultured brain cells with TNF-α and hypoxia via ceramide. Am J Physiol 1999; 45:C1171-C1183.
125. Liu J, Ginis I, Spatz M, Hallenbeck JM. Hypoxic preconditioning protects cultured neurons against hypoxic stress via TNF-α and ceramide. Am J Physiol Cell Pysiol 2000; 278:C144-C153.
126. Wang X, Li X, Erhardt JA, Barone FC, Feuerstein GZ. Detection of tumor necrosis factor-alpha mRNA induction in ischemic brain tolerance by means of real-time polymerase chain reaction. J Cereb Blood Flow Metab 2000; 20:15-20.
127. Tasaki K, Ruetzeler CA, Ohtsuki T, Martin D, Nawashiro H, Hallenbeck JM. Lipopolysaccharide pretreatment induces resistance against subsequent focal cerebral ischemic damage in spontaneously hypertensive rats. Brain Res 1997; 748:267-270.
128. Bruce AJ, Boling W, Kindy MS et al. Altered neuronal and microglial responses to excitotoxic and ischemic brain injury in mice lacking TNF receptors. Nat Med 1996; 2:788-794.
129. Gary DS, Bruce-Keller AJ, Kindy MS, Mattson MP. Ischemic and excitotoxic brain injury is enhanced in mice lacking the p55 tumor necrosis factor receptor. J Cereb Blood Flow Metabol 1998; 18:1283-1287.
130. Scherbel U, Raghupathi R, Nakamura M, Saatman KE, Trojanowski JQ, Neugebauer E et al. Differential acute and chronic responses of tumor necrosis factor-deficient mice to experimental brain injury. Proc Natl Acad Sci USA 1999; 96:8721-8726.
131. Goodmann Y, Mattson MP. Ceramide protects hippocampal neurons agianst excitotoxic and oxidative insults and amyloid-β-peptide toxicity. J Neruochem 1996; 66:869-872.
132. Chen Y, Ginis I, Halenbeck JM. The protective effect of ceramide in immature rat brain hypoxia-ischemia involves up-regulation of bcl-2 and reduction of TUNEL-positive cells. J Cereb Blood flow Metab 2001; 21:34-40
133. Schneider H, Pitossi F, Balschun D, Wagner A, del Rey A, Besedovsky HO. A neuromodulatory role of interleukin-1β in the hippocampus. Proc Natl Acad Sci USA 1998; 95:7778-7783.
134. Hagemann G, Redecker C, Neumann-Haefelin T, Freund HJ, Witte OW. Increased long-term potentiation in the surround of experimentally induced focal cortical infarction. Ann Neurol 1998; 44:255-258.
135. Miller A, Lider O, Roberts AB, Sporn MB, Weiner HL. Suppressor T cells generated by oral tolerization to myelin basic protein suppress both in vitro and in vivo immune responses by the release of transforming growth factor beta after antigen-specific triggering. Proc Natl Acad Sci USA 1992; 89:421-425.
136. Weiner HL, Friedman A, Miller A, Khoury SJ, al-Sabbagh A, Santos L et al. Oral tolerance: Immunologic mechanisms and treatment of animal and human organ-specific autoimmune diseases by oral administration of autoantigens. Annu Rev Immunol 1994; 12:809-837.
137. Becker KJ, McCarron RM, Ruetzler C, Laban O, Sternberg E, Flanders KJ et al. Immunologic tolerance to myelin basic protein decreases stroke size after transient focal cerebral ischemia. Proc Natl Acad Sci USA 1997; 94:10873-10878
138. Prehn JH, Backhauss C, Krieglstein J. Transforming growth factor-beta 1 prevents glutamate neurotoxicity in rat neocortical cultures and protects mouse neocortex from ischemic injury in vivo. J Cereb Blood Flow Metab 1993; 13:521-525.
139. Lehrmann E, Kiefer R, Finsen B, Diemer NH, Zimmer J, Hartung HP. Cytokines in cerebral ischemia: expression of transforming growth factor beta-1 mRNA in the postischemic adult hippocampus. Exp Neurol 1995; 131:1-10.
140. Lehrmann E, Kiefer R, Christensen T, Toyka KV, Zimmer J, Diemer NH et al. Microglia and macrophages are major sources of locally produced transforming growth factor-beta1 after transient middle cerebral artery occlusion in rats. Glia 1998; 24°:437-448.
141. Gross CE, Bednar MM, Howard DB, Sporn MB. Transforming growth factor-beta 1 reduces infarct size after experimental cerebral ischemia in a rabbit model. Stroke 1993; 24:558-562.

142. McNeill H, Williams C, Guan J, Dragunow M, Lawlor P, Sirimanne E et al. Neuronal rescue with transforming growth factor-beta 1 after hypoxic-ischemic brain injury. NeuroReport 1994; 14:901-904
143. Henrich-Noack P, Prehn JH, Krieglstein J. TGF-beta 1 protects hippocampal neurons against degeneration caused by transient global ischemia. Dose-response relationship and potential neuroprotective mechanisms. Stroke 1996; 27:1609-1614.
144. Moalem G, Leibowitz-Amit R, Yoles E, Mor F, Cohen I, Schwartz M. Autoimmune T cells protect neurons from secondary degeneration after central nervous system axotomy. Nat Med 1999; 5:49-55.
145. Hammarberg H, Lidman O, Lundberg C, Eltayeb SY, Gilen AW, Muhallab S et al. Neuroprotection by encephalomyelitis: rescue of mechanically injured neurons and neurotrophin production by CNS-infiltrating T and natural killer cells. J Neurosci 2000; 120:5283-5291.
146. Kerschensteiner M, Gallmeier E, Behrens L, Leal VV, Misgeld T, Klinkert WE et al. Activated human T cells, B cells, and monocytes produce brain-derived neurotrophic factor in vitro and in inflammatory brain lesions: A neuroprotective role of inflammation?. J Exp Med 1999; 189:865-870.
147. Stoll G, M ller HW. Nerve injury, axonal degeneration and neural regeneration: Basic insights. Brain Pathol 1999; 9:313-325.
148. Cajal RY. Cajal«s Degeneration & Regeneration of the Nervous System. Translated by R.M. May, DeFelipe. New York: Oxford University Press, 1991:589.
149. Aguayo AJ, David S, Richardson P, Bray G. Axon elongation in peripheral and central nervous system transplants. In: Fedoroff S, Hertz L, eds. Advances in Cellular Neurobiology. Vol 3. New York: Academic Press, 1982:215-234.
150. Bandtlow CE, Schwab ME. NI-35/250/Nogo-A: A neurite growth inhibitor restricting structural plasticity and regeneration of nerve fibers in the adult vertebrate CNS. Glia 2000; 29:175-181.
151. Qiu J, Cai D, Filbin MT. Glial inhibition of nerve regeneration in the mature mammalian CNS. Glia 2000; 29:166-174.
152. Chen MS, Huber AB, van der Haar ME, Frank M, Schnell L, Spillmann AA, Christ F, Schwab ME. Nogo-A is a myelin-assoicated neurite outgrowth inhibitor and an antigen for monoclonal antibody IN-1. Nature 2000; 403:434-439.
153. GranPr T, Nakamura F, Vartanian T, Strittmatter SM. Identification of the NOGO inhibitor of axona regeneration as a Reticulon protein. Nature 2000; 403:439-444.
154. Perry VH, Brown MC, Gordon S. The macrophage response to central and peripheral nerve injury. A possible role for macrophages in regeneration. J Exp Med 1987; 165:1218-1223.
155. Stoll G, Griffin JW, Li CY, Trapp BD. Wallerian degeneration of the peripheral nervous system: Participation of both Schwann cells and macrophages in myelin degradation. J Neurocytol 1989; 18:671-683.
156. Jander S, Lausberg F, Stoll G. Differential recruitment of CD8+ macrophages during Wallerian degeneration in the peripheral and central nervous system. Brain Pathol 2001; 11:27-38.
157. Gillen C, Jander S, Stoll G. Sequential expression of mRNA for proinflammatory cytokines and interleukin-10 in the rat peripheral nervous system: comparison between immune-mediated demyelination and Wallerian degeneration. J Neurosci Res 1998; 51:489-496.
158. Menge T, Jander S, Stoll G. Induction of the proinflammatory cytokine interleukin-18 by axonal injury. J Neurosci Res 2001; 65:332-339.
159. Stoll G, Trapp BD, Griffin JW. Macrophage function during Wallerian degeneration of the rat optic nerve: clearance of degenerating myelin and Ia expression. J Neurosci 1989b; 9:2327-2335.
160. George R, Griffin JW. Delayed macrophage responses and myelin clearance during Wallerian degeneration in the central nervous system: The dorsal radiculotomy model. Exp Neurol 1994; 129:225-236.

161. Bregman BR, Kunkel-Bagden E, Schnell L, Dai HN, Gao D, Schwab ME. Recovery from spinal cord injury mediated by antibodies to neurite growth inhibitors. Nature 1995; 378:498-501.
162. Z«Graggen W, Metz GAS, Kartje GL, Thallamir M, Schwab ME. Functional recovery and enhanced corticofugal plasticity after unilateral pyramidal tract lesion and blockade of myelin-associated neurite growth inhibitors in adult rats. J Neurosci 1998; 18:4744-4757.
163. Fields RD, Schwab ME, Silver J. Does CNS myelin inhibit axon regeneration? Neuroscientist 1999; 5:12-18.
164. Reichert F, Rotshenker S. Deficient activation of microglia during optic nerve degeneration. J Neuroimmunol 1996; 70:153-161.
165. Lazarov-Spiegeler O, Solomon AS, Schwartz M. Peripheral nerve stimulated macrophages simulate a peripheral nerve-like regenerative response in rat transected otpic nerve Glia 1998; 24:329-337.
166. Chamak B, Morandi V, Mallat M. Brain macrophages stimulate neurite outgrowth and re-generation by secreting thrombospondin. J Neurosci Res 1994; 38:221-233.
167. Rabchevsky AG, Streit WJ. Grafting of cultured microglial cells into the lesioned spinal cord of adult rats enhances neurite outgrowth. J Neurosci Res 1997; 47:34-48.
168. Singer AJ, Clark RAF. Cutaneous wound healing. New Engl J Med 1999; 341:738-746.
169. Stichel CC, Muller HW. Experimental strategies to promote axonal regeneration after traumatic central nervous system injury. Prog Neurobiol 1998; 56:119-148.

CELLULAR AND MOLECULAR DETERMINANTS OF GLIAL SCAR FORMATION

Ann Logan[1] and Martin Berry[2]

INTRODUCTION

Many axotomised neurons die and none of their severed axons regenerate after penetrant CNS injury in the adult. Debris and necrotic tissue is phagacytosed by hematogenous macrophages and microglia. The scar is formed by astrocytes interacting with fibrous tissue invading the wound from the meninges to re-establish the glia limitans around the margins of a central connective tissue core. Reactive astrocytes and synantocytes and invading meningeal fibroblasts all express axon growth inhibitory ligands within and about the lesion. Severed axons do however regenerate through the wound if stimulated by neurotrophic factors and prevent the deposition of scar tissue. The positive correlation between regeneration and scar failure suggests that neurotrophic factors act to (1) down-regulate receptors for growth inhibitory ligands allowing growth cones to proceed into the wound, and (2) stimulate secretion and release from growth cones of proteases, which actively disperse and inhibit fibrosis. All injury reactive cells in the wound, and the immigrating hematogenous elements, secrete a multitude of cytokines which trophically regulate the sequential development of the scar through acute, subacute and consolidation phases. TGF-β is probably the initiator of the cascade, and CTGF is its down stream mediator. Administration of recombinant anti-inflammatory compounds, CTGF antagonists, and competitive TGF-β receptor blockers, and also TGF-β neutralising antibodies, effectively reduces mesodermal scarring but has little effect on either the glial responses to injury, or axon regeneration. The use of the DNA of some of

[1]Molecular Neuroscience, Department of Medicine, Wolfson Research Laboratories, Queen Elizabeth Hospital, Edgbaston, Birmingham, B15 2TH, UK and [2]GKT School of Biomedical Research, Centre for Neuroscience, Guy's Campus-London Bridge, London SE1 1UL, UK.

these anti-scarring agents in gene therapies is also starting to be fruitful, and has advantages over recombinant protein administration.

No scar tissue is formed in the wounds of the fetal/neonatal CNS. Scarring is acquired in the neonatal period, in rats, for example, by about 12 days post natum (dpn). The factors accounting for scar failure and acquisition are unknown. Two possibilities seem plausible: (1) lower levels of endogenous fibrogenic cytokines in fetal/neonatal wounds; (2) presence of axons growing de novo through the wounds which inhibit scar formation -when axon growth ceases in the developing CNS, towards the end of the 2nd week, scarring is acquired.

Scar tissue (or a cicatrix) develops after penetrant, infarct and hemorrhagic lesions of the central nervous system (CNS). Scar acquisition in the injured CNS occurs neonatally, for example, in the rat, over a period of 8-12 (dpn). Before this time, no scar is formed in the injured developing rat brain.[1-3] A mature cicatrix is formed by 12-14 days by reconstitution of the glia limitans externa. After vascular accidents, a circumscribed island of fibrous tissue is deposited in the lesion, which becomes surrounded by a basal lamina. The cells that actively contribute to cicatrisation are derived from the blood, vasculature, CNS parenchyma, and the meninges. Platelets, leucocytes and macrophages are active acutely, whilst fibroblasts and astrocytes play a role sub-acutely.[4] Hematogenous cells initiate a cytokine/growth factor cascade [5] that mobilises endogenous fibroblasts and astrocytes to align along the lesion edge and to lay down a basal lamina at their interface. The clinical significance of scarring relates to: (1) the positive direct correlation between the failure of axon regeneration and cicatrisation, (2) the epileptogenic properties of the scar, and (3) subarachnoid fibrosis, precipitated after bleeding into the subarachnoid space, that leads to the development of post subarachnoid hemorrhage (SAH) chronic communicating hydrocephalus.[6] The recent discovery of axon growth inhibitory ligands both in the extracellular matrix and on the plasmalemma of fibroblasts in the scar,[7,8] has added poignancy to the former research. Anti-scarring therapeutic strategies for blocking the cytokine/growth factor cascade, by the administration of either neutralizing antibodies or competitive receptor blockers, are effective in reducing the density of the scar, whilst matrix enzymatic treatments reduce the axon growth inhibitory potency of the scar matrix. Anti-fibrotic treatment after SAH also has the potential to reduce post-SAH communicating hydrocephalus. Interestingly, recent work suggests that regenerating CNS fibers down-regulate receptors for axon growth inhibitory ligands and secrete metalloproteinases and plasminogen activators from growth cones, thereby controlling their immediate microenvironment. Accordingly, regenerating CNS axons traversing a wound inhibit scarring. Moreover, once a scar has become established, regenerating CNS axons can penetrate and partially disperse the mature scar.[9-12] These findings suggest that stimulation of robust axonal regeneration through a CNS lesion may either circumvent or enhance anti-fibrotic therapies.

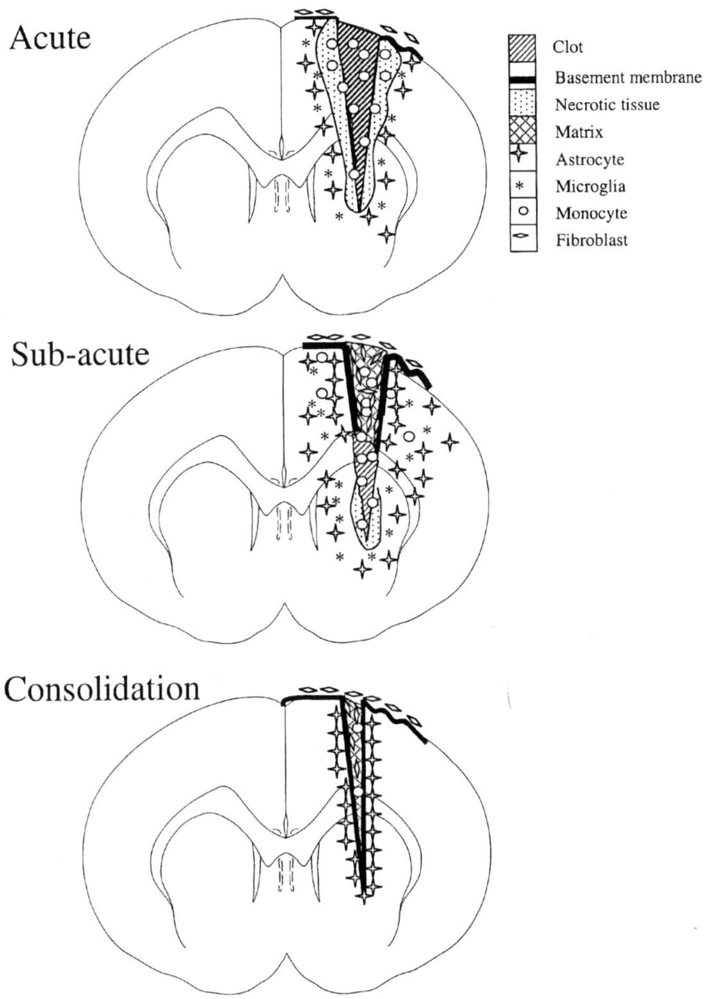

Figure 1. The cellular changes that occur after a penetrating injury to the cerebral hemisphere. Hematogenous macrophages delivered during the acute hemorrhagic phase (1-3 dpl), clear away the necrotic tissue and clot, whilst a reactive gliosis is initiated in the surrounding neuropil. In the sub-acute phase (4-8 dpl), scarring is initiated at the pial surface and progresses inwards: the clot is removed, the wound cavity infiltrated from above by meningeal fibroblasts, that start to produce fibrous ECM components including collagens, laminin and fibronectins in the core of the wound, and astrocyte processes accumulate at the superficial wound margins to begin formation of a glia limitans. Finally, during the consolidation phase (9-14 dpl), the scar contracts and matures. A declining reactive gliosis is seen at subsequent time points.

DEVELOPMENT OF THE SCAR

There are multiple CNS scarring paradigms currently in experimental use, all of which exhibit variants of a common pattern of cicatrisation exemplified by the reaction to a cerebral stab wound. Unless otherwise stated, the rat cerebral stab model has been adopted as the prototype of penetrant injury for documenting scarring in the CNS throughout this review (Fig. 1). Such lesions penetrate the glia limitans externa, brain parenchyma, and often the ependyma lining the intracerebral ventricles, and the choroid plexus within, severing blood vessels and neuronal and glial processes in the cortical and diencephalic neuropil.

Response to Injury of the Adult Brain

Scarring in the mature animal is characterised by three sequential overlapping but nonetheless discrete periods called the (i) acute (or hemorrhagic); (ii) subacute; and (iii) consolidation phases, occurring over the periods 0-3 days post lesion (dpl), 4-8 dpl, and 9-12 dpl, respectively. Commonly, the consolidation phase may be prolonged for up to 20 dpl.[4]

Acute phase (Fig. 1a): Massive bleeding immediately ensues after penetrant injury, pervading the lesion, subarachnoid space and ventricles, if the latter are impaled. Platelets in the clot probably initiate the cytokine cascade, augmented by leukocytes whose invasion is mediated by the expression of both endothelial addressins from the vasculature surrounding the wound and chemokines released from cells in the damaged parenchyma.[13,14] Pericytes and hematogenous monocytes and macrophages invade the infarcted rim of parenchyma that encloses the lesion cavity.[15] Neutrophils adhering to the peri-lesion vessels mediate a redistribution and loss of tight junction proteins, precipitating blood-brain barrier (BBB) breakdown, oedema and extended tissue damage.[16-22] Neutrophils may also damage neurons by secreting free radicals and proteases. Blockade of their recruitment into a wound reduces lesion size,[23] although BBB leakage occurs after brain damage in the absence of neutrophils.[24] Large numbers of macrophages accumulate in the clot, most of which are transformed monocytes, but some are resident macrophages that have migrated into the wound from the meninges and peri-vascular spaces.[25] Macrophages remove the clot and surrounding necrotic zone of parenchyma and also release cytokines and growth factors which in turn induce the release of endogenous trophic factors from glia and neurons in the viable peri-lesion neuropil.[5,26-28] Macrophages/ reactive microglia also express the chondroitin sulphate proteoglycan (CSPG), recognised by the NG2 antibody.[29] There are regional differences in the acute inflammatory response in the CNS, for example, significantly greater recruitment of inflammatory cells occurs in the cord compared with the cerebrum.[30]

Microglia in the neuropil about the lesion become reactive over the first 24 hrs.[31-37] There is hyperplasia, particularly about the wound; the new microglia probably derive from the resident population rather than from invading monocytes.[38-

[40] The reactive phenotype is seen within 30 minutes after injury, withdraws its processes, becomes amoeboid, up-regulates CD4, ED1, complement type 3 receptor (CR3 recognized by the OX-42 antibody), and nucleoside diphosphatase (NDPase), cytokines (e.g., TGFβs, IL-1, and IL-6), and NG2.[29, 41-46] There is also expression of the major histocompatibility antigens (MHC) I and II[47] and leukocyte common antigen (LCA).[48-51] Expression of some macrophage related proteins (MRP), e.g., MRP8 and MRP14, is delayed beyond 72 hours.[52] Microglia appear to detect changes in the micro-environment of the damaged neuropil[53] and become phagocytic removing neuronal debris[48] and stripping presynaptic boutons from axotomised neuronal somata.[54-57] In addition to this scavenger activity, microglia also have immune surveillance[31] and neuroprotective/regenerative functions after injury through the release of neurotrophins[58,59] and lipocortin-1, a putative neuroprotective agent.[60,61] Activated microglia persist in large numbers within the lesion and degenerating tracts through the sub-acute and consolidation phases and continue to scavenge axon debris.[61,62] Thereafter, numbers decrease, buT cells never disappear completely,[63] indelibly marking all sites of past CNS trauma. Since activated microglia and brain macrophages express increased numbers of peripheral benzodiazepine binding sites,[64] a radiolabelled ligand for the latter allows both recent and long standing brain lesions to be detected by PET-scanning.[65]

By the end of the acute phase, most of the clot and necrotic tissue has been removed from the lesion and replaced by masses of macrophages. The lesion cavity is reduced in size and the surrounding viable neuropil contains astrocytes which show the first signs of reactivity with an up-regulation of glial fibrillary protein (GPAP), S100 protein, vimentin, and the intermediate filament protein plectin,[66-69] most of which appear to be essential for normal scarring.[70] Both alpha and beta transcripts of GFAP are expressed.[71] Initiation of astrocyte reactivity is poorly understood[72] but could be mediated by: (1) blood borne factors,[73] including growth factors and hormone,[74-76] albumin,[77] thrombin,[78-80] angiotensin II[81] and cAMP;[82-84] (2) inflammatory cytokines released from activated macrophages and microglia;[85-89] and (3) activation of a neutral, Ca^{2+}-dependent protease, calpain-1, through Ca^{2+} influx.[90] The release of synaptic mediators from damaged neurons might also stimulate GFAP expression,[82,91,92] and such activated astrocytes may play a role in synaptic elimination and collateral re-innervation.[56] Blood in the ventricles is removed without fibrosis, but that in the subarachnoid space is replaced by fibrous tissue which obliterates the subarachnoid space and blocks the flow and drainage of cerebrospinal fluid through the arachnoid granulations into the dural venous sinuses.[6]

The necrosis developing at the lesion margin is generally attributed to ischemia, but inflammatory hematogenous cells and microglia probably contribute to this secondary brain damage.[93,94] A variety of apoptotic molecules are also expressed about the lesion, in the first 3 dpl, in neurons, microglia, astrocytes and oligodendroglia, including the tumour suppressor protein p35, Bcl-2, caspase 3, Bax and TUNEL.[95-98] There is a more sustained expression of these molecules in degenerating and demyelinating white matter tracts.[96,99-102] Blocking axolemmal

Na^+ channels with tetrodotoxin reduces the loss of axons in tracts but does not affect glial death.[103]

Sub-acute phase (Fig. 1): The structure of the scar becomes defined during this phase by the deposition of a sheet of basal lamina, undercoated with astrocytic end-feet, and overlain with a connective tissue down-growth from the pia mater. All three elements of the scar are contiguous with and similar, if not identical, to those of the glia limitans of the pia mater.[63, 67] Meningeal down-growth progressively transforms the macrophage-filled core of the lesion into a connective tissue matrix from the pial surface into the lesion depths.

GFAP reactivity in astrocytes reaches a peak in the sub-acute phase and masses of intensely GFAP+ astrocyte processes become stacked at the lesion edge. A more generalised GFAP up-regulation spreads more slowly through the astrocytes in the entire lesioned hemisphere and is complete by 20 dpl.[66,104] Some of the reactive astrocytes co-express clusterin,[105] nestin,[106] VEGF and the VEGF flt-1 receptor,[107] plectin[67] and the neurotensin receptor.[108] The processes of astrocytes are functionally coupled through gap junctions[109,110] and propagate Ca^{2+} waves intercellularly,[77,111,112] providing a mechanism for spread of GFAP reactivity, either by the release of ATP,[113-117] or diffusion of a soluble messenger.[113,118,119]

A basal lamina, rich in collagen types IV and V,[2] and laminin,[64,68] is laid down at the interface between the palisades of astrocyte processes and the mesodermal core of the lesion. Interactions between the meningeal fibroblasts and astrocyte processes leads to basal lamina production, since fibroblast invasion and genesis of the membrane occur pari passu, commencing sub-pially and progressing into the depths. By the end of the subacute phase, the entire lesion edge is invested in a basal lamina. Reactive astrocytes produce the extracellular matrix (ECM) molecules CSPG, neurocan, versican, and phosphacan/DSD-1, keratin sulphated proteoglycans, and tenascin-C,[120-128] all of which have axon growth inhibitory properties.[129,130] Astrocytes and neurons also produce the cell adhesion glycoprotein CD44 after injury, which binds to the ECM.[131] The protective roles of reactive astrocytes in the injured CNS have been reviewed by Eddleston and Mucke,[132] and include repair of the blood brain barrier and glia limitans externa, contributions to the scar, and the release of cytokines, neurotrophins and transporter molecules/enzymes for the metabolism of excitotoxic amino acids. Selective ablation of reactive astrocytes from the wound causes a 25-fold increase in hemotogenous cell influx, failure of BBB repair, and a substantial increase in the loss of neurons.[133]

A new injury reactive glial phenotype has been described and classified under the general terminology of NG2+ glia,[134,135] since they share an antigenic phenotype with oligodendrocyte progenitor cells.[135-140] One type of NG2+ glia in the adult CNS is a highly differentiated mature cell, morphologically distinct from the oligodendrocyte precursor, which we have called a synantocyte (from the Greek *synant*, meaning contact).[31] Synantocytes strongly express NG2 on their surface, but possess no other conventional glial epitopes. Morphologically, they are distinct differentiated stellate cells, with thin highly branched attenuated processes terminating, with astrocyte processes, on the axolemma of nodes of Ranvier[141,142] and, like astrocyte processes,[143] have pre-synaptic expansions,[144] and can themselves be post-synaptic.[145] Synantocytes express quisqualate sensitive receptors that promote raised intracellular Ca^{2+}.[146] Thus, synantocyte function may be modulated by glutamate released from both nodes of Ranvier, and synapses.[31] Butt et al [42] and Butt and Berry[142] also hypothesised that synantocytes might play a role during development in the nodal aggregation of sodium channels at incipient nodes of Ranvier. Synantocytes respond rapidly to CNS injury by an up-regulation of NG2 expression (an axon growth inhibitory ligand),[120,147] and outgrowth and thickening of processes.[29,148-153] Synantocytes appear capable of dedifferentiating after injury and of proliferating to augment the numbers of NG2+ oligodendrocyte precursor cells about the lesion and within demyelinated tracts.[154] They are, therefore, potentially able to replace oligodendrocytes lost after CNS trauma.[136,153,155,156]

Consolidation phase (Fig. 1): All the elements of the mature tripartite scar (a basal lamina with an undercoat of astrocyte processes and a covering of connective tissue) are in place at the end of the sub-acute phase of the injury response. During the consolidation phase, the core of the lesion contracts and is comprised of a collagen/fibronectin-rich extracellular matrix ECM as most but not all of the macrophages and fibroblasts disappear.[2] The density of GFAP+ intermediate filaments is reduced in the processes of perilesion astrocytes that become compacted below the basal lamina and bind together by multiple tight junctions. ED1+ reactive microglia persist in the neuropil immediately surrounding the lesion. Continuing diffuse retrograde and anterograde axon degeneration after tract severance is associated with sustained astrocyte and microglia activation in the absence of leukocyte infiltration,[62,157] and the expression of the S100A4 protein.[158] In cross section, the established scar is seen as two opposed membranes, comprising a rim of GFAP+ astrocyte processes abutting a laminin/collagen IV+ basal lamina, separated by a thin sheet of ECM containing a few scattered fibroblasts and ED1+ positive macrophages. In three dimensions, the mature scar is analogous to an invagination of the glia limitans externa, the connective tissue coats of which are fused around the line of invagination at the pial surface.

The fibrosis in the subarachnoid space after SAH continues through the sub-acute and into the consolidation phases if the bleeding, associated with a penetrant brain injury, is persistent and/or recurrent. The extent of the resulting arachnoid/pia adhesions is correlated with the severity of the bleed.[159-162] In the worst cases, adhesions extend widely within the subarachnoid space, basal cisterns, and arachnoid villi. The chronic post-SAH hydrocephalus that ensues is usually of the communicating type.[163-167]

Most deep penetrant wounds of the cerebral hemisphere impale the ventricles, damaging the ependymal lining. The ependyma does not regenerate after injury, probably because the epithelium is both non-mitotic and incapable of re-expressing fetal cytoskeletal and secretary proteins.[168,169] Damage to the ependyma exposes the underlying neuropil and subependymal cells, astrocyte processes herniate into the ventricles,[64,168,170,171] and ependymal rosettes form.[168] Ultimately, the aperture is sealed by a meshwork of subependymal astrocyte processes which never become epithelialized. Ependymal rupture associated with subependymal heterotopia may be an important post-traumatic aetiological epileptogenic factor.[172]

Para-lesion cystic cavities develop in the consolidation phase. They are more numerous in lesions of the brain stem and cord lesions than in those of the cerebral hemispheres.[173] The cysts are most frequently located in the subcortical white matter, lined by macrophages and a meshwork of CD81-positive astrocyte processes,[174] and may communicate with, or be isolated from the ventricle. The cavities are not epithelialised, but have a wall of reactive astrocyte processes. The aetiology of para-lesion cysts is unknown but may be caused by focal infarcts,[175] hemorrhage,[176,177] lysosome activity,[178] leakage of CSF from expanded hydrocephalic ventricles,[179-181] and/or microglial/macrophage activity,[182-185] particularly those expressing CD81.[174]

Response to Injury of the Fetal/Neonatal Brain

Before 8 dpn in the rat, no scar tissue is formed after a penetrant injury.[1,3] In the acute phase, the lesion is marked by hemorrhage, influx of inflammatory cells[186] and a transitory weak expression of GFAF by astrocytes in the peri-lesion neuropil.[187,188] Reactive astrocyte processes are short and exhibit no preferential orientation toward the lesion, as in the mature scar. As the clot is removed, the processes of neurons and glia in the walls of the lesion grow across the wound lumen to obliterate the original injury site. The only evidence of the lesion remaining at 20 dpl, for example, is a subpial and subventricular accumulation of astrocytes and an increased density of blood vessels. A relatively acellular strip of neuropil marks the line of the original lesion. Interestingly, although no scarring occurs, the glia limitans externa is repaired, demonstrating that astrocytes in these young animals are capable of forming a basal lamina when in contact with pial fibroblasts. This observation lead Maxwell et al[3] to suggest that a failure of migration of meningeal fibroblasts into the lesion may account for the absence of scarring. At the point where the ventricular surface is interrupted by the lesion, subventricular astrocytes plug the wound and attenuated ependymal cells only partially cover this surface.

From 8 dpn, mature scarring is gradually acquired as meningeal fibroblasts progressively invade the wound from the pial surface. By 12 dpn, invading fibroblasts reach the depths of the lesion, and thereafter a complete mature scar forms, associated with the development of para-lesion cysts[3] and the failure of axons to traverse the scar.

AXON REGENERATION AND SCARRING

Both scarring and the abortive regenerative response of axons to injury are correlated temporally and spatially, but a causal relationship in which, for example, scarring is envisaged to actively impede regeneration, although suggested,[189-191] has not been substantiated. Thus, after the administration of anti-inflammatory agents, like for example, pyromen,[189,192-194] glucocorticoids,[195,196] and ACTH,[197,198] which all reduce scarring, the claim that enhanced regeneration ensues has not been borne out,[199-201] although more recently, Stichel et al[202,203] have demonstrated impressive CNS axon regeneration after inhibition of collagen IV in the wound. However, the successful use of transforming growth factor β (TGF-β) antagonists to depress cicatrisation in the CNS has not resulted in concomitant regrowth of severed axons in the cerebrum.[204-206] Axons stop growing in the walls of the developing scar and rarely form neuromatous entanglements,[207] analogous to those found in peripheral nerve lesions. This observation ostensibly suggests that: (1) the scar does not act as a simple physical barrier to the growth of axons, and (2) that regeneration appears to be actively inhibited by the scar.

Several axon growth inhibitory ligands are recognised, e.g., injured membrane protein IMP; [208] astrocyte/synantocyte derived CSPGs,[209,210] including versican, brevican, neurocan, phosphacan, and NG2[120, 211-21]; myelin/oligodendrocyte derived inhibitors[214-217] including NOGO-A (NOGO-B and -C have yet to be characterised);[218-220] and meningeal fibroblast derived semaphorin-3A.[7,8,221-224] The growth cone collapse they all promote is mediated through receptors for the respective ligands e.g., NOGOR,[225] and the neuropilin-1/plexin-A1R complex.[226-228] In CNS lesions, NG2 is up-regulated in reactive astrocytes/synantocytes,[120] and the axon growth inhibitory potency of this CSPG is neutralised by removing the GAG side chains from the core protein.[147] Fibroblasts migrating into a developing scar express semaphorin-3A and arrest the growth of axons invading the wound, where neuropili-1 is also expressed.[7] Neutralising the growth arresting properties of both NOGO-A and CNS myelin,[229,230] and CSPG by chondroitinase ABC treatment[231] allows axons to regenerate in transected CNS white matter tracts. Cerebellar Purkinje cell axons begin to sprout into long-term white matter glial scars because either the inhibitory nature of the scar changes or permissive molecules become expressed.[232]

Retinal ganglion cell (RGC) axons regenerate through the inhibitory environment of the transected optic nerve of the adult rat, when RGC are stimulated both by Schwann cell derived neurotrophins applied intravitreally,[9-12] and also after injury to the lens [233, 234], without pre-neutralisation of putative axon growth inhibitory molecules. One correlate of this successful regeneration is the absence of scar tissue,[9-12,235] thought to be attributable to the secretion of metalloproteinases (MMPs) and plasminogen activators (PAs) from the growth cones of regenerating axons.[11,12,235] Thus, regenerating optic nerve fibers grow through inhibitory: reactive astrocytes,[236-243] NG2+ glia, and myelin debris and reactive oligodendrocytes.[217,218] In the chiasm, where the option to run in undamaged neuropil is available, axons preferentially regenerate in the degenerating decussating trajectories,[10] where abundant putative growth inhibitory molecules are most exposed.

Why are optic axons, under neurotrophin stimulation, able to regenerate unerringly through the growth inhibitory environments of an optic nerve lesion, distal optic nerve segment, and chiasm after optic nerve transection? One possibility is that neurotrophins not only promote the survival of axotomised RGC and stimulate the regrowth of their axons, but also: (1) down-regulate both NOGOR and the neuropilin-1/plexin-A1R complex, (2) secrete MMPs/PAs, and (3) regulate the expression of tissue inhibitors of MMPs (TIMPs) in the optic nerve.

Paradoxically, axons also fail to regenerate in CNS regions in which NOGO A is absent, e.g., (a) through grey matter; (b) either through the transected unmyelinated optic nerve of the BW mutant,[244-246] or over cryosections of unmyelinated neonatal optic nerve;[242, 243] and (c) through areas of the developing scar, where myelin and oligodendrocytes are entirely absent.[11,12,125,347] Conversely, CNS axons spontaneously regenerate through the myelinated adult: (a) anterior medullary velum (AMV);[248-250] (b) ventral funiculi of the cord, from the ventral horns to the ventral roots;[251-253] (c) adult CNS neuropil from implanted fetal CNS grafts;[254-256] (d) corpus callosum, fimbria,[257] and dorsal columns of the spinal cord[129] from adult DRG micro-implants; and (e) transected postcommissural fornix through reactive astrocytes/ microglia and chondroitin and keratin sulphate proteoglycans in the perilesion area.[202,203] These observations indicate that a regenerative outcome may not be correlated with either the presence or absence of growth inhibitory ligands, but with the availability of an exogenous neurotrophin stimulus. Axons will grow over growth inhibitory substrates if neurons are pre-treated with neurotrophins, providing evidence that priming with neurotrophins promotes the down-regulation of receptors for axon growth inhibitory ligands.[258]

The failure of a scar to form in the fetal/neonatal mammalian CNS is correlated with axon growth de novo through these early lesions (see above, and refs. 1, 3). Similarly, regenerating axons crossing the interface between fetal brain grafts and adult host mid-brain,[259,260] traversing the site of anastomosis between a peripheral nerve graft and the adult optic nerve,[261,262] and regenerating through optic nerve lesions[9-12] all inhibit scar formation. Growing axons express and secrete MMPs during development,[263-265] and a recapitulation of this activity during regeneration in the adult may explain why a scar is not formed in the presence of growing axons.

Thus, contrary to an old idea that scar tissue may inhibit regeneration,[236] evidence documented above suggests that regrowing axons regulate scarring in the lesion through which they pass. Thus, a scar is formed by default only if axon growth fails. Moreover, Berry et al[12] have shown that when the application of a trophic stimulus to RGC, which promotes optic nerve regeneration, is delayed until after the scar is formed, the subsequent surge of regrowing axons through the lesion disperses the established scar tissue. It is perhaps self evident that successful functional regeneration in the CNS is impossible unless neurotrophic stimulation not only supports growth cone advance, but also suppresses all impending interactions with putative growth inhibitory substrates within both the neuropil and the scar.

CYTOKINES AND SCARRING

After a penetrant CNS injury, there is release into the damaged neuropil of multiple hematogenous and endogenous cytokines, which interact to control the cellular changes described above. The up-regulation and orchestration of this complex

Table 1. Chemokines acutely released into CNS wounds

Chemokine	Reference
Growth-regulated oncogene (Gro)-α	McTigue et al[301]
Monocyte chemoattract proteins (MCP-1 and MCP-5)	Andjelkovic et al[298]; Flugel et al[302]; McTigue et al[303]; Muessel et al[266]; Galasso et al[304]
Macrophage inflammatory proteins (MIP)-1 α and β	Bona et al[271]; Ghirnikar et al[305-306]; Le et al[276]
Regulated on Activation, Normal T cell Expressed and Secreted (RANTES)	Bona et al[271]; Ghirnikar et al[305]
Platelet activating factor (PAF)	McClennan et al[279]
Macrophage migration inhibitory	Yoshimoto et al[293]; Lindsberg et al[307] factor (MIF)

cascade of trophic factors (see Fig. 2) must be exact to initiate the temporal sequence of tissue-specific effects observed (see Fig. 1).

Cytokines and the Response to Injury in the Mature Brain

Cytokines in the Acute Phase

The acute phase response is dominated by the actions of pro-inflammatory cytokines and chemokines (a contraction of *chemo*attractant and cyto*kine*) delivered into the wound from the blood and also rapidly produced in the damaged tissues. Chemokines are small, inducible, secreted members of the cytokine family primarily responsible for the chemoattraction of cells involved in the inflammatory process. They are classified into two main subfamilies: the CXC, or (a) chemokines that primarily recruit neutrophils, and the CC, or (b) chemokines which recruit mononuclear inflammatory cells.[266] Multiple members of both families are active in wounds after CNS injury.

Hemorrhage heralds the first trophic event of the injury response, with the delivery into the wound of platelet-derived growth factor (PDGF)[267] and TGF-β by platelet lysis.[5,26,27] In addition, insulin-like growth factors (IGFs) that circulate in the blood and cerebrospinal fluid (CSF) also appear rapidly in the hemorrhagic wound fluid.[268, 269] In vivo, thrombin enters the wound from the blood and activates microglia in the surrounding neuropil.[270] These circulating factors are supplemented almost immediately by cytokines, particularly interleukin (IL)-1α and β, IL-2, IL-6, IL-8

Table 2. Microglia-derived cytokines active acutely after CNS lesion

Cytokine	Reference
Epidermal growth factor (EGF)	Nieto-Sampedro et al[310]
Insulin-like growth factors (IGFs)	Walter et al[268,269]
Interleukins (IL-1β, IL-2 and IL-6)	Rothwell & Relton[311]; Lotan & Shwartz[312] ; Schiefer et al[313]; Gabellec et al[314]; Ali et al[315]; Hayashi et al[274]; Knoblach & Faden[316]; Streit et al[42]; Streit et al[317]; Bartholdi & Schwab[318]; Fassbender et al[319]; Hashizume et al[320]; Acarin et al[321]
Interferons	Rothwell et al[322]
Tumor necrosis factors (TNF-α and TNF-β)	Gehrmann et al[323]; Bartholdi & Schwab[318]; Streit et al[42]; Kamei et al[324]
Fibroblast growth factors (FGF-1 and FGF-2)	Logan et al[27]; Clarke et al[325]; Le et al[276]; Smith et al[326]
Transforming growth factor-βs (TGF-β1 and TGF-β2)	Logan et al; [27, 206, 327]; Logan & Berry[5]
Neuregulin	Erlich et al[328]
Fas ligand (FasL)	Beer et al[329]

and tumour necrosis factors,[271-276] released/transported from cerebral endothelial cells (CEC), from neutrophils entering the wound in the blood, and from monocyte-lineage cells already resident in the CNS parenchyma (reviewed by Fuerstein et al[277] and Brown & Khoshkbijar[278]). The results of impact injury on brain slices indicate that intrinsic rather than extrinsic factors are important, since microglial activation occurs in the absence of circulating factors.[45]

The adhesion molecules, cytokines and chemokines released/presented at the site of the BBB play an important role in mobilising peripheral inflammatory cells into the brain. Immediate post-hemorrhagic events are thus followed by a pro-inflammatory cascade over the next 48 hours, when the wound is rapidly invaded by polymorphs, monocytes and macrophages. Microglia in turn release a second wave of multiple cytokines and chemokines. Invasion is regulated by CEC actively engaged in microvascular stasis and leukocyte infiltration by producing a plethora of pro-inflammatory mediators.[275] When challenged by the hemorrhagic trophic stimuli and hypoxia, CEC release and/or express products of the arachidonic acid cascade

Table 3. Macrophage and leukocyte-derived cytokines active acutely after CNS lesion

Cytokine	Reference
Tumor necrosis factors (TNF-α and TNF-β)	Acarin et al[321]; Leskovar et al[332]; Le et al[276]; Nomoto et al[333]
Interleukins (IL-1β, IL-2 and IL-6)	Rothwell & Luheshi[334]; Leskovar et al[334]
Transforming growth factor-βs (TGF-β1 and TGF-β2)	Logan et al[27,205,327]
Fibroblast growth factors (FGF-1 and FGF-2)	Logan et al[27]; Clarke et al[325]; Smith et al[326]

with both vasoactive and pro-inflammatory properties, including prostaglandins, leukotrienes, and platelet-activating factor (PAF),[279-282] a reaction catalysed by cyclo-oxygenase-1 (COX-1) from macrophages and microglia.[283] These metabolites induce platelet and neutrophil activation and adhesion, change local cerebral blood flow and blood rheology, and increase BBB permeability. Vascular hyperpermeability, leading to tissue oedema and trophin influx, is probably reinforced by the local activity of MMP-3[284] and the rapid local up-regulation of vascular endothelial growth/permeability factor (VEGF), an event triggered by hypoxia.[285-289]

Ischemic CEC and glia up-regulate the expression and release of bioactive inflammatory cytokines and chemokines (see Table 1). Within hours these act together as potent neutrophil and monocyte chemoattractants, influencing the expression of integrins and cell adhesion molecules such as intracellular adhesion molecule (ICAM-1), neural (N)CAM, vascular (V)CAM and E- and P-selectins to amplify the immune response.[290-300]

Microglia are rapidly activated by released stimulatory factors, to up-regulate the expression of cytokines (see Table 2), some of which may, in turn, act on endothelial cells to increase their adhesion to circulating macrophages and leukocytes, supplementing the early chemokine response.[293,298,308,309]

Activated macrophages and leukocytes are the most persistent cell types in the wound. They reinforce the trophic effects of microglia, releasing multiple cytokines as detailed in Table 3. All of these may themselves be trophic or may stimulate the production of trophic substances from targeT cells, thereby establishing an endogenous supply of cytokines.[5,330,331]

It seems that acutely released PAF, TGF-β, TNF-α, ILs and other chemokines, as well as having beneficial effects on damaged neural tissue, are also directly detrimental, initiating edema,[311] and damaging glia[335] and neurons,[282,334,336-339] impending apoptosis that in compromised glia and neurons is signalled by their early expression of Fas and

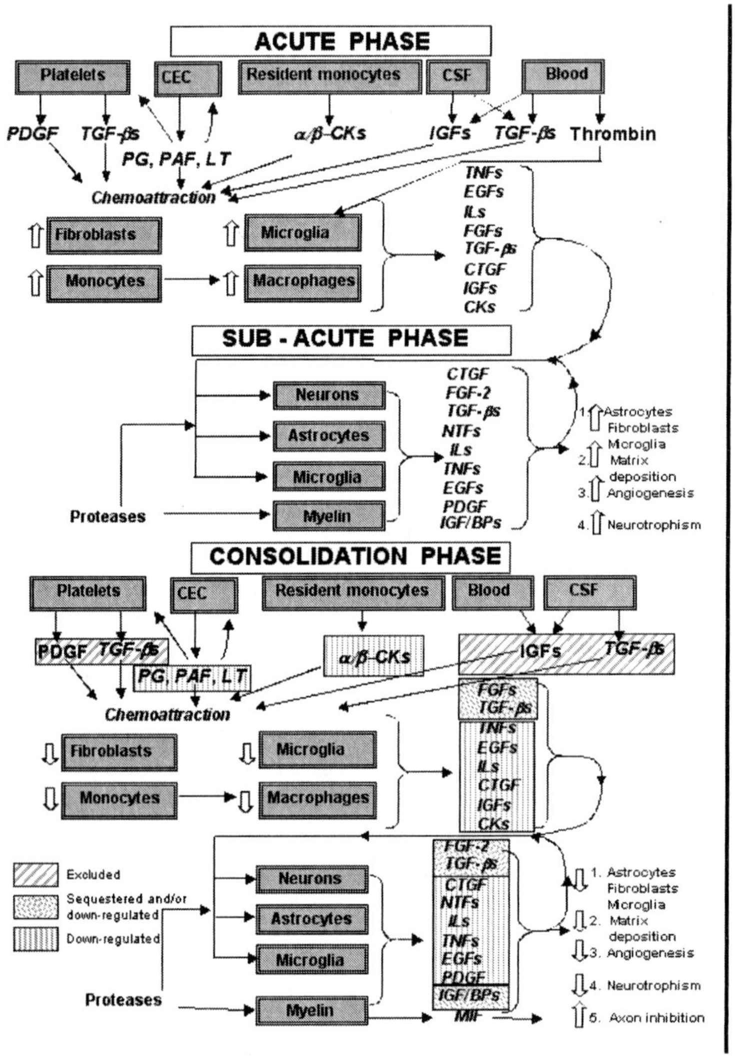

Figure 2. The up- and down-regulation of the cytokine cascade initiated by a penetrating CNS lesion. In the immediate acute phase (1-3 dpl), numerous hematogenously and locally derived cytokines are delivered into the wound to initiate the cellular and trophic injury responses which are maximal in the sub-acute period (4-8 dpl). In the consolidation phase (9-14 dpl), the hematogenously derived cytokines are denied access to the wound as the blood-brain-barrier repairs, and therefore become excluded from the wound fluid. Soluble cytokines that are matrix binding become sequestered within the deposited scar proteoglycans thereby losing bioactivity. The expression of locally derived cytokines becomes down-regulated.
CK, chemokine; CSF, cerebrospinal fluid; CTGF, connective tissue growth factor; EGF, epidermal growth factor; FGF, fibroblast growth factor; IGF/BP. Insulin-like growth factor/binding protein; IL, interleukin; LK, leukotrine; NIF, neurite growth inhibitory factors PAF, platelet activating factor; PDGF, platelet-derived growth factor; PG, prostaglandin; NTF, neurotrophic factors; TGF-β transforming growth factor-β; TNF, tumour necrosis factor.

Table 4. Cytokines active sub-acutely after CNS lesion

Cytokine	Reference
Tumor necrosis factors-α	Nomoto et al[333]; Acarin et al[321]
Interleukins (IL-1β, IL-6)	Acarin et al[321]
Fibroblast growth factors (FGF-1, FGF-2)	Logan et al[26]; Gomez-Pinella et al[351]; Gomide & Chadi[352]; Clarke et al[325]; Buytaert et al[353]; Smith et al[326]
Transforming growth factor-βs (TGF-β1 and TGF-β2)	Logan et al[27,206]; Hughes et al[354]; Jiang et al[355]; Martinez et al[356]; Acarin et al[321]
Connective tissue growth factor (CTGF)	Hertel et al[359]; Schwab et al[283]
Activin	Tretter et al[358,359]; Hughes et al[354]
Neuregulin	Tokita et al[360]
Insulin growth factors (IGF-1 and IGF-2)	Walter et al[268,269]; Hughes et al[354]
Platelet derived growth factor (PDGF)	Takayama et al[361]
Leukemia inhibitory factor (LIF)	Sugiura et al[362]
Ciliary neurotrophic factor (CNTF)	Park et al[363]
Glial derived neurotrophic factor (GDNF)	Widenfalk et al[364]; Bresjanac & Antaver[365]
Neurotrohins (NGF, BDNF and NT3)	Hughes et al[354]; Michael et al[366]; Dougherty et al[367]; Dreyfus et al[368]; Hayashi et al[274]; Krenz & Weaver[369]; Moalem et al[370]; Otten et al[371]; Widenfalk et al[364]; Liebl et al[372]

Fas ligand (FasL).[329] The first wave of cytokines, and in particular IL-β1, also mediates a systemic injury response, which is characterised by a rapid induction of changes in endocrine, metabolic and immune functions.[311,334,340]

Cytokines in the Sub-Acute Phase

The cytokine-mediated cytotoxic events of acute phase oedema and secondary tissue damage are reinforced in the sub-acute phase by the generation of nitric oxide (NO) by inducible NO synthase (iNOS) expressed by endothelial cells, microglia and astrocytes in the damaged tissues in response to cytokine stimulation.[319,341,342,343] NO contributes to the continuing breakdown of the BBB[343, 344] and is also thought to elicit cell membrane damage, mitochondrial dysfunction,[345] glial and axonal degeneration.[326, 341]

By 3-5 dpl, much of the hemorrhagic debris, including platelets, has been removed and most of the extravasated erythrocytes have disappeared along with their trophic contribution (Fig. 1). Chemokine activity has filled the core of the wound and its margins with cytokine-producing macrophages, monocytes, polymorphs and numerous activated microglia. They have also attracted matrix-depositing, connective tissue growth factor (CTGF)-expressing fibroblasts into the wound from the meninges.

Both microglia and macrophage-derived IL-1β, IL-2, IL-6, FGFs and TGF-βs are active through the acute and sub-acute phases and ensure that reactive astrocytes, neurons and activated T cells become primary sources of multiple neurotrophic growth factors and other cytokines[345-349] and, between 3-7 dpl in the rat, this expression reaches a peak to include those factors detailed in Table 4, together with their receptors. In turn, activated astrocytes may release various microglia stimulatory factors.[5,26,27,206] In this way, a positive trophic feedback is established which enables secreted glial growth factors to regulate neuronal survival well beyond the period of initial tissue insult. The ability of the released and bioactivated trophins to up-regulate the local expression of more endogenous cytokines and receptors is well documented and ensures propagation and expansion of the trophic cascade in the sub-acute wound environment.[350]

The many neurotrophic factors released during the acute and sub-acute phases, including NGF, BDNF, NT3, CNTF, GDNF, IGF, LIF, FGF and TGFβ, (see Table 4) may affect their own availability to damaged neurons by altering expression of their receptors. These include truncated extracellular domain forms of Trks that will, as they accumulate during the sub-acute phase, block neurotrophin activity by binding neurotrophins without activating the receptor signalling cascade. [372,373] Transiently increased supplies of neurotrophic factors and receptors are available sub-acutely after injury to enhance neuron survival by stimulating anti-apoptotic factors like the Bcls.[374] Axon regeneration is also promoted but later aborted, perhaps as the expression of these factors, and their receptors, becomes down-regulated.

During the sub-acute phase, neovascularisation occurs in the surrounding parenchymal tissue, probably in response to the local autocrine expression of endothelial cytokines and receptors such as VEGF[375] and angiopoietin,[376] and reinforced by FGF-2,[26] CTGF[377] and TGF-β,[27] which are all potent angiogenic factors.

The TGF-β isoforms released into the wound sub-acutely, acting with PDGF and FGF-2, probably initiate the trophic and cellular cascade that leads to scar formation, a response which is associated with abortive axonal growth.[5,26,204,206] Of these, the TGF-βs, probably acting via up-regulation of CTGF,[378-381] are primarily responsible for initiating the organisation of astrocytes into a glia limitans and for promoting ECM deposition by invading fibroblasts during the sub-acute phase.[5,204-

[206] The expression of TGF-β1 and TGF-β2 is rapidly up-regulated in the choroid plexus as well as in neurons and glia of damaged CNS tissue which supplements, through the CSF and neuropil, the acute delivery of the cytokine from blood.[26,204-206,301,382-384] Thus, between 1-7 dpl significant levels of TGF-β are sustained within wounds to orchestrate glial/collagen scar formation. Over this period, CTGF expression in damaged tissue is induced in glia, CEC, neurons, and invading meningeal fibroblasts in damaged tissue.[283,357,379] These fibrogenic molecules initiate the synthesis of collagens, fibronectins, proteoglycans, and elastin in the wound, and also modulate the activity of the MMPs and TIMPs, which remodel the injured tissues.[5,27,204,206] Hence, the raised levels of these cytokines induce the formation of a dense fibrous scar within the wound between 5-8 dpl.

Of relevance to the development of the trophic cascade is the sub-acute release, into the wound fluid, of extracellular matrix-degrading proteases, including the gelatinases MMP-2 and MMP-9 from macrophages, microglia and axonal growth cones,[385-387] membrane type (MT-) MMP1,[388] ADAM proteases,[389] and PAs,[384,390,391] synthesised in response to the released cytokines and neurotrophic factors. The proteases accelerate the tissue remodelling required during the axonal sprouting and cell migration phase and effect the rapid release and activation of depots of latent trophic factors that are normally sequestered within the tissue and in the tissue fluid. For example, significant amounts of inactive IGFs, bound to IGF binding proteins (IGFBPs), are associated with the myelin sheaths of most nerve tracts throughout the brain and spinal cord. The release of IGFs sub-acutely by protease activity, raises IGF levels in the CSF and interstitial fluid.[268,269] Similarly, released MMPs activate stores of TGF-β locally.[392] The solubilised and circulating IGFs in the wound fluid are supplemented by TGF-βs and FGFs that are synthesised by the choroid plexus and released into the CSF.[26,27,268,269] All these factors access the damaged neural tissues via circulating tissue fluids and have gliotrophic and neurotrophic effects. Production of TIMPs[385,388,393] and PA inhibitors[390,391] is increased late in the sub-acute phase, and this may add to the fibrotic build-up of ECM components which will be critical, during the subsequent consolidation phase, in dampening the cytokine cascade.

Thus, a cascade of trophic events is initiated acutely and is expanded and maintained through the sub-acute period, which ensures a rapid and sustained supply of functionally active cytokines locally within CNS wounds which act co-operatively to promote the range of post-injury cellular events described previously.

Cytokines in the Consolidation Phase

The endogenous expression of fibrogenic cytokines and neurotrophins is transient and not maintained through the consolidation phase of the injury response, hence limiting the scope for axon regeneration and dampening the inflammatory and fibrogenic cascade. The maturing glial scar presents a physical and probably also a biochemical barrier to regrowing axons. Reconstruction of the BBB blocks the supply of hematogenously-derived cytokines and chemokines, thereby removing the primary driving force of the trophic cascade. TGF-βs sub-acutely increase ECM synthesis and, in the later phases, inhibit protease release in the wound by down-regulating MMP expression and up-regulating the expression of TIMPs. The wound microenvironment is thus changed to one which now both sequesters and deactivates matrix-binding growth factors, such as the FGFs and TGF-βs, which are primary activators of the trophic cascade. As cellular trafficking within the wound ceases, the physical/chemical balance of the wound environment is tipped against regeneration, and ensures that the axonal growth response is not sustained. Hence, the whole trophic cascade becomes down-regulated, halting the associated dynamic cellular responses.

Cytokines and the Injury Response of the Immature Brain

We have described earlier that embryonic and perinatal neuronal tissues do not scar after injury and this failure has been attributed to differences of the trophic wound environment between immature and mature animals. Developmental changes in the trophic environment of wounds are ill-defined in immature animals, but it is clear that there are differences in the inflammatory cytokine/chemokine responses.[186,394]

For example, an IL-6 dependent decreased expression of growth inhibitory factor (GIF a member of the metallothionein family) in neurons, with concomitant increased GIF expression in astrocytes, after damage to the immature brain,[395] has been linked to muted gliosis, efficient wound repair and neuronal survival in neonates.[396] Some studies have also suggested that fetal wound healing involves far lower endogenous levels of the fibrogenic cytokines FGF-2 and TGF-β than the adult.[326, 397] It is reported that absence of scarring in lesioned fetal murine dermal tissue reflects a relative deficiency of TGF-β1 and FGF-2, but not PDGF within such wounds,[397,398] and the same seems to be true of CNS lesions.[326] There is a sparse inflammatory response in fetal/neonatal lesions, although there is some up-regulation of pro-inflammatory cytokines, like IL-1b, IL-6, and TNFα.[309,399] Titres of endogenous levels of TGF-β and FGF-2 proteins are high in the developing murine CNS,[400,401] and TGF-β levels increase rapidly, but only locally, in damaged tissue after injury, and are very quickly cleared from the wound.[402] Of significance is the observation that this suppressed TGF-β up-regulation is accompanied by a very muted increase in FGF-2 expression.[326]

Notwithstanding the muted cytokine/chemokine response, it may be that the key difference between fetal and adult injury responses is not the absolute levels of cytokines present acutely in the wound site, but the persistence of expression of fibrotic cytokines after wound closure. However, more recent experiments by ourselves and others seem to suggest that fetal/perinatal scarless healing involves different cellular and connective tissue events than adult repair, and that the repair process here may be dependent on more than the trophic environment of the wound.[5,398,403,404] For example, we suggest that axons growing de novo through neonatal CNS wounds inhibit scarring by the secretion of MMPs (see earlier).

TROPHIC REGULATION OF THE SCAR

The TGF-βs are potent fibrogenic proteins that have been shown to have a broad diversity of actions that relate to wound healing in many tissues, including the CNS. The well documented activities include stimulation of ECM synthesis, modulation of cell proliferation, and cell infiltration, immunosuppression, and neuroprotection (see reviews by Cui & Ackhurst[405]; McPherron & Lee[406]; Pratt & McPherson[407] for extensive references). The ECM is not only a physical supporting scaffold for tissue restructuring but, more importantly, a complex and dynamic molecular substratum which can modify a wide range of cellular activities. Since the ECM regulates cell interactions, proliferation, differentiation, morphology, adhesion, and migration, TGF-βs probably control cell activity in CNS wounds both directly and indirectly via the ECM. In general, TGF-βs promote ECM formation and can enhance the response of cells to ECM through alterations in integrin expression.[408,409] TGF-βs stimulate the deposition of several ECM components, including fibronectins, interstitial collagens (I, III, IV and V), thrombospondin, tenascin, laminin, and chondroitin/dermatan sulphate proteoglycans.[410,411] The enhancement of ECM protein synthesis increases transcription of ECM genes and, in some cases, stabilises their mRNAs. In addition, TGF-βs potentiate ECM production by inhibiting matrix protein degradation, by modulating the expression of genes encoding proteases and protease inhibitors. Thus, TGF-βs decrease synthesis and secretion of several proteases, including tissue-PA, urokinase-type PA, thiol protease, collagenase and transin, but increase the expression of protease inhibitors, PA inhibitor type 1(PAI-1), urokinase, and the TIMPs from macrophages, microglia and axonal growth cones.[410,411] The net effect of these activities is to promote ECM deposition. TGF-βs also regulate cell-substratum interactions by selectively altering integrin expression, a major class of cell-adhesion receptor, that bind fibronectin, collagen and other ECM proteins.[408,409] This modulation of the adhesion cascade regulates cell migration, thereby influencing fibroblast migration into the wound from the meninges.[412]

A role for TGF-§s has been implicated in numerous CNS pathologies in which fibrosis and neural dysfunction are causally associated. For example, in post-traumatic brain and spinal cord scarring,[27,204-206,382,384,413] post-surgical adhesions,[414] hemorrhagic stroke,[415-418] and SAH.[6,419,420] TGF-βs may also promote plaque development in Alzheimer s disease and Down s syndrome.[421-423] In all of these conditions, the levels of TGF-β are raised in both the CSF and also locally in damaged neural tissue. For example, whilst TGF-β1 and TGF-β2 are present in the intact brain and spinal cord at discrete locations and at relatively low levels, after CNS traumatic injury there is a steep elevation of titres of both isoforms. As described previously, initially titres are derived from hematogenous cells, but these are later supplemented by endogenous local synthesis by neurons and glia in the damaged neuropil, and by choroid plexus cells, leading to raised cytokine levels in the CSF.[27,204,206,384] The most direct evidence for a fibrogenic role for TGF-βs in the pathophysiology of CNS fibrosis comes from experiments in the lesioned brain. On the one hand, raised levels of TGF-βs are correlated with the increased deposition of scar material in such lesions, whilst immunoneutralisation with TGF-β antibodies markedly inhibits fibrogenic scarring.[204-206]

Since its identification,[424] the secreted protein of the immediate early gene CTGF has been implicated in several physiological and pathophysiological processes. The repertoire of biological activities of CTGF, including stimulation of cell proliferation and survival, and DNA synthesis, ECM production and angiogenesis[380] implicates the peptide as a regulator of fibrotic scarring in CNS and peripheral tissues. Most importantly for fibrogenesis, CTGF stimulates mitosis, chemotaxis via α5-integrin expression, adhesion, and matrix production, including collagen 1 and fibronectin, by fibroblasts. All these activities are strongly reminiscent of those of TGF-β (reviewed by Grotendorst[378]; Moussad & Brigstock[381]). But, whilst CTGF shares the biological actions of TGF-β on fibroblasts, it does not seem to share TGF-βs broader growth inhibitory actions, or to modulate immune/inflammatory cells.[378] CTGF also augments the activity of other fibrogenic growth factors like FGF-2,[425] as well as auto-inducing its own expression.[426]

Sustained CTGF expression is induced rapidly and robustly by TGF-βs.[427] Evidence for a functional link between CTGF and TGF-β has come from experiments blocking TGF-β effects with CTGF neutralizing antibodies or antisense oligonucleotides[428], so that it is now accepted that CTGF can act as a mediator of TGF-β s fibrogenic activities, although TGF-β-independent fibrogenic activities are also reported.[379,381,429,430] The potential relevance of CTGF to CNS scarring has recently been demonstrated in the intact adult CNS,[283,357,377] where CTGF mRNA and protein are expressed in astrocytes and in specific populations of neurons in the brain and spinal cord.[357,431,432] Levels of mRNA and protein are strongly induced in both cell types in parallel with TGF-β elevation after CNS injury.[283,357,377] The restricted expression and accumulation of CTGF in CNS wounds implicates the peptide as a key molecule regulating glial scarring, but direct confirmation of this awaits experiments with CTGF antagonists.

PROTEASE REGULATION OF SCARRING

Berry et al[12] have shown that RGC axons regenerating through optic nerve lesions regulate newly developing and established scar material in the lesion through which they pass and have proposed that this may be due to scar dissolution by MMPs released by axonal growth cones.[433] The observation of this naturally occurring phenomenon emphasises the potential importance of protease regulation of scarring, a biological activity that can be exploited therapeutically. We have reviewed the evidence for production of axon growth inhibitory molecules, including CSPG, in the scar (see above). Recently, experiments have suggested that CNS axon regeneration can be enhanced by degradation of CSPG in the scar after the local application of chondroitinase ABC.[231] It seems that selective degradation of scar components with proteases may be a useful therapeutic strategy in some situations. However, anti-fibrotic therapies, including those employing anti-TGF-β neutralizing antibodies and TGF-β receptor blockade with decorin, provide no evidence for enhancement of axon regeneration despite the reduction in scar material, including CSPG, at CNS lesion sites.[205,206,327]

THERAPEUTIC MODULATION OF THE SCAR

Targeting the restricted number of key primary activators which initiate the scarring cascade provides an opportunity for therapeutic intervention aimed at preventing scar formation. A number of trophic antagonists are potentially useful as scar attenuating agents. Those that block the activities of the fibrogenic cytokines TGF-β and CTGF are primary candidates. Logan et al[327] demonstrated that immunoneutralisation of TGF- β1 in a cerebral wound with an isoform-specific polyclonal TGF-β1 antibody dramatically inhibited cerebral fibrogenic scarring, although the inflammatory response was exacerbated. These findings established the principle of inhibition of fibrogenesis by TGF-β immunoneutralisation, but the therapeutic potential for the clinical use of animal-derived polyclonal antibodies for the treatment of patients remains limited. Recombinant monoclonal isoform-specific human antibodies are being developed to immunoneutralise TGF-β2 within wounds, and they markedly reduce CNS scar components, including the inhibitory molecule CSPG, and inflammation.[206] A similar inhibition of scar formation is obtained after treatment of CNS wounds with decorin, a naturally occurring proteoglycan that is a pan-TGF-β antagonist.[205] In each case of antagonist-induced scar inhibition, the glia limitans externa is reconstructed in the outer layers of the lesioned cortex, so that wound closure occurs normally at the pial membrane. However, whilst widespread reactive gliosis is apparent in the neuropil around the immunoneutralised wound, the numerous activated glia neither organized into the expected limiting membrane nor laid down a laminin-rich basal lamina in the wound; fibroblasts do not migrate into the wound and a glial/matrix scar is not formed in the lesion. Unlike the results of Moon et al,[231] the removal of the putative inhibitory scar did not enhance axonal regeneration through the lesion site, indicating the need for a simultaneously delivered neurotrophic stimulus. Nevertheless, these experiments have established the principle that therapeutic manipulation of CNS wounds effectively inhibits scar formation.

A number of TGF-β antagonists, other than neutralising antibodies, are becoming available to reduce tissue fibrosis. Transfecting the epithelium of the airway with DNA, encoding antifibrogenic decorin,[205] inhibits bleomycin-induced lung fibrosis[434] demonstrating the potential of the delivery of therapeutic genes encoding anti-fibrotics into injury responsive cells in CNS wounds. Strategies to block TGF-β intracellular signalling may also prove useful. For example, adenovirus-mediated gene transfer of dominant-negative Smad4 (a central mediator of the TGF-β signalling pathway) has been used to block TGF-β activity in pancreatic acinar cells.[435] Similarly, antisense-oligonucleotides inhibit TGF-β expression in human endothelial cells.[436] The targeted delivery of DNA encoding anti-fibrotic factors like decorin to injury responsive cells such as astrocytes in CNS wound sites is thus an attractive therapeutic strategy that is now being piloted by us in experimental studies.[437-438]

In the future, CTGF- rather than TGF-β-antagonists may ultimately prove more effective as anti-fibrotic agents since they have the added advantage of interfering with connective tissue formation but not affecting the potentially useful neurotrophic activities of the TGF-βs. Several anti-CTGF neutralising antibodies and antisense oligonucleotides have been used with some success to limit fibrosis in peripheral tissues (reviewed by Moussad & Brigstock[381])but have not so far been used in the injured CNS.

It has been suggested that activation of the renal TGF-β system in diabetes may be mediated through activation of the renin-angiotensin(AT) system (RAS).[439-442] Accordingly, AT converting enzyme (ACE) inhibitors and AT receptor antagonists have been used successfully to inhibit TGF-β receptor expression and collagen/ proteoglycan expression in the diabetic kidney.[442,443] Whether a similar strategy will prove useful in the CNS remains to be established, but angiotensinogen and at receptors are expressed in astrocytes and neurons in the intact CNS.[444] Furthermore, angiotensinogen mRNA, together with AT II immunoreactivity and AT1 and 2 receptor mRNA and protein are locally up-regulated in the spinal cord (Logan et al, unpublished observations) and brain[445] after mechanical injury. Hence, the acute systemic delivery of either ACE inhibitors or AT receptor antagonists after penetrant CNS injury may help suppress scar formation since the perturbed BBB may allow access of the drugs to the wound site. Moreover, AT II, acting through the AT2 receptor, promotes the regeneration of RGC axons in the transected optic nerve, after systemic administration.[446] These observations encourage the further investigation of anti-fibrotic properties of ACE inhibitors and AT receptor antagonists, and the neurotrophic affects of AT in the injured CNS.

A number of anti-inflammatory agents have been used during the acute phase of the CNS injury response to limit oedema, tissue damage and scarring. Of these, the corticosteroid methylprednisolone is perhaps the best known, but must be given within 30 minutes of injury to be effective.[447-450] More recently, other non-steroidal anti-inflammatory agents have been tested in the CNS with varying degrees of success: e.g., (1) nitroflurbiprofen and nitro-asparin reduces the brain inflammation associated with lipopolysaccharide infusion;[451] (2) the chemokine antagonist vMIPII decreases the number of hematogenous macrophages invading a cord contusion wound and reduces neuronal loss and gliosis;[452] (3) the anti-inflammatory cytokine, IL-10, down-regulates TNF-α expression after traumatic spinal cord injury, reducing lesion volume and improving functional recovery;[453] and (4) the thrombin inhibitor argatroban suppresses the infiltration of inflammatory cells and gliosis.[454] Others have used a monoclonal antibody to neutralise the expression of the aD sub-unit of β2, integrin, reducing the post-injury trafficking of hematogenous cells into spinal cord wounds by blocking their interaction with VCAM-1.[455] A similar monoclonal antibody strategy targeting the leukocyte adhesion molecule, P-selectin, reduces both leukocyte infiltration and lesion size after brain injury.[456] Complement depletion also seems to be an effective means of reducing infiltration of inflammatory cells.[457-

[458] Cyclosporin A is neuroprotective if given within the first 24 hours after injury.[459] A neutrophil elastase inhibitor, ONO-5046, is reported to contract lesion size by blocking the damaging effects of elastase released from activated neutrophils.[460] How anti-inflammatory therapeutic strategies fit with the recently reported neurotrophic effects of post-injury inflammation[367,369,461-465] remains to be assessed.

CONCLUSIONS

Presently, there is no treatment available for correcting the devastating affects of CNS injury because lost neurons are not replaced, axons do not regrow, and scarring cannot be controlled. This review summarises some of the data relevant to these sequelae, and highlights possible areas where therapy might be targeted. Like McGraw et al,[466] we believe future therapeutic success rests with a combinatorial strategy aimed at depressing inflammation, ECM deposition and fibroblast invasion, coupled with limiting gliosis and promoting axon regeneration, mindful of the possibility that some activities of injury responsive cells in the CNS may be beneficial.

REFERENCES

1. Berry M, Maxwell WL, Logan A et al. Deposition of scar tissue in the central nervous system. Acta Neurochir Suppl1983; 32:31-53.
2. Maxwell WI, Duance VC, Lehto M et al. The distribution of types I, III, IV and V collagens in penetrant lesions of the central nervous system of the rat. Histochem J 1984; 16:1219-1229.
3. Maxwell WL, Follows R, Ashhurst DE et al. The response of the cerebral hemisphere of the rat to injury. II. The neonatal rat. In: Phil Trans Roy Soc Lond Series B 328, 1990:501-513.
4. Berry M, Butt A, Logan A. Cellular response to penetrating CNS injury. In: Berry M, Logan A, eds. CNS Injuries: Cellular Responses and Pharmacological strategies. Boca Raton: CRC Press, 1999:1-18.
5. Logan A, Berry M. Transforming growth factor-β and CNS scarring. In: Berry M, Logan A, eds. CNS Injuries: Cellular Responses and Pharmacological strategies. Boca Raton: CRC Press, 1999:1-18.
6. Flood C, Akinwunmi J, Lagord C et al. Transforming growth factor-β1 in the cerebrospinal fluid of patients with subarachnoid hemorrhage: Titres derived from exogenous and endogenous sources. J Cereb Blood Flow Metab 2001; 21:157-162.
7. Pasterkamp RJ, Giger RJ, Ruitenberg MJ et al. Expression of the gene encoding the chemorepellent semaphorin III is induced in the fibroblast component of neural scar tissue formed following injury of the adult but not the neonatal CNS. Mol Cell Neurosci 1999; 13:146-166.
8. Pasterkamp RJ, Anderson PN, Verhaagen J. Peripheral nerve injury fails to induce growth of lesioned ascending dorsal column axons into spinal cord scar tissue expressing the axon repellent Semaphorin3A. Euro J Neurosci 2001; 13:457-471.
9. Berry M, Carlile J, Hunter A. Peripheral nerve explants grafted into the vitreous body of the eye promote the regeneration of retinal ganglion cell axons severed in the optic nerve. J Neurocytol 1996; 25: 147-170.
10. Berry M, Carlile J, Hunter A et al. Optic nerve regeneration after intravitreal peripheral nerve implants: Trajectories of axons regrowing through the optic chiasm into the optic tracts. J Neurocytol 1999; 28:721-741.

11. Berry M, Carlile L, Hunter AS et al. Optic nerve regeneration after intravitreal peripheral nerve implants: Early growth of axons through the wound. J Comp Neurol 2001. In review.
12. Berry M, Carlile L, Hunter AS et al. Optic nerve regeneration after intravitreal peripheral nerve implants: Axons growing through the wound site both inhibit the formation of new, and penetrate and disperse established scar tissue. J Neurosci 2001. In review.
13. Schall TJ, Bacon KB. Chemokines, leukocyte trafficking and inflammation. Cur Opin Immunol 1994; 6:865-873.
14. Pan LC, Lloyd C, Zhou H et al. Neurotactin, a membrane-anchored chemokine up-regulated in brain inflammation. Nature 1997; 387:611-617.
15. Dore-Duffy P, Owen C, Balabanov R et al. Pericyte migration from the vascular wall in response to traumatic brain injury. Microvasc Res 2000; 60:55-69.
16. Perry VH, Bell MD, Brown HC et al. Inflammation in the nervous system. Cur Opin Neurobiol 1995; 5:636-641.
17. Bell MD, Taub DD, Kunkel SJ et al. Recombinant human adenovirus with rat MIP-2 gene insertion causes prolonged PMN recruitment to the murine brain. Eur J Neurosci 1996; 8:1803-1811.
18. Bell MD, Taub DD, Perry VH. Overriding the brain s intrinsic resistance to leukocyte recruitment with intraparenchymal injections of recombinant chemokines. Neurosci 1996; 74:283-292.
19. Bolton SJ, Perry VH. Differential blood-brain barrier breakdown and leucocyte recruitment following excitotoxic lesions in juvenile and adult rats. Exp Neurol 1998; 154:231-240.
20. Barone FC, Feuerstein GZ. Inflammatory mediators and stroke: New opportunities for novel therapeutics. J Cereb Blood Flow Metab 1999; 19:819-834.
21. Roitbak T, Sykova E. Diffusion barriers evoked in the rat cortex by reactive astrogliosis. Glia 1999; 28:40-48.
22. Preston E, Webster J, Small D. Characteristics of sustained blood-brain barrier opening and tissue injury in a model for focal trauma in the rat. J Neurotrauma 2001; 18:83-92.
23. Zhang RL, Chopp M, Li Y et al. Anti-ICAM-1 antibody reduces ischemic cell damage after transient middle cerebral artery occlusion in the rat. Nature 1994; 44:1747-1751.
24. Whelen MJ, Carlos TM, Kochanel PM et al. Neutrophils do not mediate blood-brain barrier permeability early after controlled cortical impact in rats. J Neurotrauma 1999; 16:583-594.
25. Graeber MB, Streit WJ, Keifer R et al. New expression of myelomonocytic antigens by microglia and perivascular cells following lethal motor neuron injury. J Neuroimmunol 1990; 17:121-132.
26. Logan A, Frautschy SA, Gonzalex A-M et al. A time course of the focal elevation of synthesis of basic fibroblast factor and one of its high affinity receptors (flg) following a localised cortical brain injury. J Neurosci 1992; 12:3628-3837.
27. Logan A, Frautschy SA, Gonzalex A-M et al. Enhanced expression of transforming growth factor β in the rat brain after a localised cerebral injury. Brain Res 1992; 587:216-225.
28. Baird A, Mormede P, Bohlen P. Immunoreactive fibroblast growth factor in cells of peritoneal exudate suggests its identity with macrophage-derived growth factor. Biochem Biophys Res Commun 1985; 126:358-364.
29. Bu J, Akhtar N, Nishiyama A. Transient expression of the NG2 proteoglycan by a subpopulation of activated macrophages in an excitotoxic hippocampal lesion. Glia 2001; 34:296-310.
30. Schnell L, Fearn S, Klassen H et al. Acute inflammatory responses to mechanical lesions in the CNS: Differences between brain and spinal cord. Eur J Neurosci 1999; 11:3648-3658.
31. Berry M, Butt AM, Wilkin G et al. Structure and function of glia in the central nervous system. In: Lantos PH, Graham DI, eds. Greenfield s Neuropathology. Sydney, Auckland: Arnold, 2001. 7th Ed.
32. Perry VH, Gordan S. Macrophages in the nervous system. Int Rev Cytol 1991; 125:203-244.
33. Perry VH. Macrophages and the nervous system. In: Austin, Tx: RG Landes Company, 1994.

34. Kreutzberg CW, Blakemore WF, Graeber MB et al. Tenascin in the injured rat optic nerve and in non-neuronal cells in vitro: potential role in neural repair. J Comp Neurol 1994; 340:233-242.
35. Streit WJ. Microglial response to brain injury: A brief synopsis. Toxicol Path 2000; 28:28-30.
36. Streit WJ, Walter SA, Pennell NA. Reactive microgliosis. Prog Neurobiol 1999; 57:563-581.
37. Norton WT. Cell reactions following acute brain injury: a review. Neurochem Res 1999; 24:213-218.
38. Morshead CM, van der Kooy D. Separate blood and brain origins of proliferating cells during gliosis in adult brains. Brain Res 1990; 535:237-244.
39. Lassmann H, Schmied M, Vass K et al. Bone marrow derived elements and resident microglia in brain inflammation. Glia 1993; 7:19-24.
40. Matsumoto Y, Fugiwara M. Absence of donor-type major histocompatibility complex class I antigen bearing microglia in the rat central nervous system of radiation bone marrow chimeras. J Neuroimmunol 1992; 17:71-82.
41. Kreutzberg GW. Microglia: a sensor for pathological events in the CNS. Trends Neurosci 1996; 19:312-318.
42. Streit WJ, Semple-Rowland SL, Hurley SD et al. Cellular pathology of the central nervous system. In: Graham DI, Lantos PL, eds. Greenfield s Neuropathology. 6th ed. London, Sydney, Auckland: Arnold, 1997:85-1546.
43. Fitch MT, Doller C, Combs CK et al. Cellular and molecular mechanisms of glial scarring and progressive cavitation: In vivo and in vitro analysis of inflammation-induced secondary injury after CNS trauma. J Neurosci 1999; 19:8182-8198.
44. Jensen MB, Hegelund IV, Poulsen FR et al. Microglia reactivity correlates to the density and the myelination of the anterogradely degenerating axons and terminals following perforant path denervation of the mouse fascia dentata. Neurosci 1999; 93:507-518.
45. Koshinaga M, Katayama Y, Fukushima M et al. Rapid and widespread microglial activation induced by traumatic brain injury in rat brain slices. J Neurotrauma 2000; 17:185-192.
46. Cenci di Bello IM, Dawson RL, Levine JM et al. Generation of oligodendroglial progenitors in acute inflammation demyelinating lesions of the rat brain stem is associated with demyelination rather than inflammation. J Neurocytol 1999; 28:365-381.
47. Streit WJ, Graeber MB, Kreutzberg GW. Expression of Ia antigen on perivascular and microglial cells after sublethal and lethal motor neuron injury. Exp Neurol 1989; 105:115-126.
48. Carbonell AL, Boya J, Calvo JL et al. Ultrastructural study of the neuroglial and macrophagic reaction to Wallerian degeneration in the adult rat optic nerve. Histol Pathol 1991; 6:443-451.
49. Finsen BR, Jorgensen MB, Diemer NH et al. Microglia MHC antigen expression after ischemic and kainic acid lesions of the adult hippocampus. Glia 1993; 7:41-49.
50. Perry VH, Gordan S. Macrophages in the nervous system. Trends Neurosci 1988; 11:273-279.
51. Perry VH, Gordan S. Macrophages in the nervous system. Int Rev Cytol 1991; 125:203-244.
52. Engel S, Schluesener H, Mittelbronn M et al. Dynamics of microglial activation after human traumatic brain injury are revealed by delayed expression of macrophage-related proteins MRP8 and MRP14. Acta Neuropath 2000; 100:313-322.
53. Ng YK, Ling EA. Microglial reaction in focal cerebral ischemia induced by intra-carotid homologous clot injection. Histol Histopath 2001; 16:167-174.
54. Blinzinger K, Kreutzberg G. Displacement of synaptic terminals from regenerating motor neurons by microglial cells. Z Zellforsch Mikrosk Anat 1968; 85:145-157.
55. Kreutzberg GW, Barron KD. 5 -nucleotidase of microglial cells in the facial nucleus during axonal reaction. J Neurocytol 1978; 7:601-610.
56. Aldskogius H, Liu L, Svensson M. Glial responses to synaptic damage and plasticity. J Neurosci Res 1999; 58:33-41.
57. Bruce-Keller AJ. Microglial-neuronal interactions in synaptic damage and recovery. J Neurosci Res 1999; 58:191-201.
58. Banati RB, Graeber MB. Surveillance, intervention and cytotoxicity: Is there a protective role of microglia? Dev Neurosci 1994; 16:114-127.

59. Kreutzberg CW, Blakemore WF, Graeber MB. Cellular pathology of the central nervous system. In: Graham DI, Lantos PL, eds. Greenfield s Neuropathology. Vol. 1, 6th ed. Sydney Australia: Edward Arnold, 1997:Chapter 3.
60. Young KA, Hirst WD, Solito E et al. De novo expression of lipocortin-1 in reactive microglia and astrocytes in kainic acid lesioned rat cerebellum. Glia 1999; 26(4):333-343.
61. Oehmichen M, Theuerkauf I, Meissner C. Is traumatic axonal injury (AI) associated with an early microglial activation: Application of a double-labeling technique for simultaneous detection of microglia and AI. Acta Neuropath 1999; 97:491-494.
62. Csuka E, Hans VH, Ammann E et al. Cell activation and inflammatory response following traumatic axonal injury in the rat. Neuroreport 2000; 11:2587-90.
63. Maxwell WL, Follows R, Ashhurst DE et al. The response of the cerebral hemisphere of the rat to injury. I. The mature rat. Phil Trans Roy Soc Lond 1990; 328:479-500.
64. Dubois A, B navid s J, Peny B et al. Imaging of primary and remote ischemic and excitotoxic brain lesions. An autoradiographic study of peripheral type benzodiazepine binding sites in the rat and cat. Brain Res 1988; 445:77-90.
65. Ramsay SC, Weiller C, Myers K et al. Monitoring by PET of macrophage accumulation in brain after ischemic stroke. Lancet 1992; 339:1054-1055.
66. Mathewson AJ, Berry M. Observations on the astrocyte response in a cerebral stab wound in adult rats. Brain Res 1985; 327:61-69.
67. Reier PJ, Houle JO. The glia scar: Its bearing on axonal elongation and transplantation approaches in CNS repair. Adv Neurol 1988; 47:87-138.
68. Kalman M, Szabo A. Plectin immonopositivity appears in the astrocytes in the white matter but not in the grey matter after stab wounds. Brain Res 2000; 857:291-294.
69. Cerutti SM, Chadi G. S100 immunoreactivity is incresed in reactive astrocytes of the visual pathways following a mechanical lesion of the rat occipital cortex. Cell Biol Int 2000; 24:35-49.
70. Pekny M, Johansson CB, Eliasson C et al. Abnormal reaction to central nervous system injury in mice lacking glial fibrillary acidic protein and vimentin. J Cell Biol 1999; 145:503-514.
71. Condorelli DF, Nicoletti VG, Dell Alvani P et al. GFAP beta mRNA expression in the normal rat brain and after neuronal injury. Neurochem Res 1999; 24:709-714.
72. Eng LF. Regulation of glial intermediate filaments in astrogliosis. In: Dorenberg HD, Hertz L, Sohousboa A, eds. The Biochemical Pathology of Astrocytes. New York: Alan R. Liss, Inc., 1988:79-90.
73. Klatzo I. Neuropathological aspects of brain edema. J Neuropath Exp Neurol 1963; 26:1-14.
74. Morrison RS, DeVellis J, Lee Y-I et al. Hormone and growth factor induced synthesis of glial fibrillary acidic protein in rat astrocytes. J Neurosci Res 1985; 14:167-176.
75. Winter CG, Saotome Y, Levison SW et al. A role for ciliary neurotrophic factor as an inducer of reactive gliosis, the glial response to central nervous system injury. Proc Natl Acad Sci USA 1995; 92:5865-5869.
76. Eclancher F, Perraud F, Faltin J et al. Reactive astrogliosis after basic fibroblast growth factor (bFGF) injection in injured neonatal rat brain. Glia 1990; 3:502.
77. Nadal A, Fuentes E, Pastor J et al. Plasma albumin induces calcium waves in rat cortical astrocytes. Glia 1997; 19:343-351.
78. Nelson RB, Siman R. Rhrombin and its inhibitors regulate morphological and biochemical differentiation of astrocytes in vitro. Dev Brain Res 1990; 54:93-104.
79. Grabham P, Cunningham DB. Thrombin receptor activation stimulates astrocyte proliferation and reversal of stellation by distinct pathways: Involvement of tyrosine phosphorylation. J Neurochem 1995; 64:583-591.
80. Pindon A, Berry M, Hantai D. Thrombomodulin as a new marker of lesion-induced astrogliosis. Involvement of thrombin through the G protein coupled receptor PAR-1. J Neurosci 1997; 20:2543-2550.

81. Tallant EA, Higson JT. Angiotensin II activates distinct signal transduction pathways in astrocytes isolated from neonatal rat brain. Glia 1997; 19:333-342.
82. Tardy M, Le Prince G, Fages C et al. Neuron-glia interaction, effect of serotonin and DBcAMP on the expression of GFAP and its encoding message. Ann N Y Acad Sci 1991; 633:630-632.
83. Le Prince G, Fages C, Rolland B et al. DBcAMP effect on the expression of GFAP and of its encoding mRNA in astroglial primary cultures. Glia 1991; 4:322-326.
84. Kaneko R, Hagiwara N, Leader K et al. Glial-specific cAMP response of the glial fibrillary acidic protein gene in the RT5 cell lines. Proc Natl Acad Sci USA 1994; 91:4529-4533.
85. Giulian D, Woodward J, Young DG et al. Interleukin-1 injected into mammalian brain stimulates astrogliosis and neovascularisation. J Neurosci 1988; 8:2485.
86. Guilian D. Microglia and tissue damage in the cental nervous system. In: Levi G, ed. Differentiation and Functioning of Glial Cells. New York: Alan R. Liss, 1990:379.
87. Balasingam V, Yong VW. Attenuation of astroglial reactivity by interleukin-10. J Neurosci 1996; 16:2945-2955.
88. Yong VW, Tejeda-Berges T, Goodyer GG et al. Differential proliferative response of human and mouse astrocytes to gamma-interferon. Glia 1992; 6:269-280.
89. Yong VW. Cytokines, astrogliosis and neurotrophism following CNS trauma. In: Ransohoff R, Beneviste E, eds. Cytokines and the CNS. Boca Raton: CRC, 1996:309-327.
90. Du S, Rubin A, Klepper S et al. Calcium influx and activation of calpain I mediate acute reactive gliosis in injured spinal cord. Exp Neurol 1999; 157:96-105.
91. Le Prince C, Copin MC, Hardin H et al. Neuronal-glia interactions: Effects of serotonin on the astroglial expression of GFAP and its encoding message. Dev Brain Res 1990; 51:295-2968.
92. Bardakdjian J, Tardy M, Pimoule C et al. GABA metabolism in cultured glial cells. Neurochem Res 1979; 4:517-527.
93. Gong C, Hoff JT, Keep RF. Acute inflammatory reaction following experimental intracerebral hemorrhage in rat. Brain Res 2000; 87:57-65.
94. Grossman SD, Rosenberg LJ, Wrathal JR. Temporal-spatial pattern of actue neuronal and glial loss after spinal cord contusion. Exp Neurol 2001; 168:273-282.
95. Saito N, Yamamoto T, Watanabe T et al. Implications of p53 protein expression in experimental spinal cord injury. J Neurotrauma 2000; 17:173-182.
96. Raghupathi R, Graham DI, McIntosh TK. Apoptosis after traumatic brain injury. J Neurotrauma 2000; 17:927-938.
97. Citron BA, Arnold PM, Sebastian C et al. Rapid upregulation of caspase-3 in rat spinal cord after injury: mRNA, protein, and cellular localisation correlates with apoptotic cell death. Exp Neurol 2000; 166:213-226.
98. Beattie MS, Farooqui AA, Bresnahan JC. Review of current evidence for apoptosis after spinal cord injury. J Neurotrauma 2000; 17:915-925.
99. Casha S, Yu WR, Fehlings MG. Oligodendroglial apoptosis occurs along degenerating axons and is associated with FAS and p75 expression following spinal cord injury in the rat. Neurosci 2001; 103:203-218.
100. Li Y, Field PM, Raisman G. Death of oligodendrocytes and microglial phagocytoses of myelin precede immigration of Schwann cells into the spinal cord. J Neurocytol 1999; 28:417-427.
101. Li GL, Farooque M, Holtz A et al. Apoptosis of oligodendrocytes occurs for long distances away from the primary injury after compression trauma to rat spinal cord. Acta Neuropath 1999; 98:473-480.
102. Abe Y, Yamamoto T, Sugiyama Y et al. Apoptotic cells associated with Wallerian degeneration after experimental spinal cord injury: a possible mechanism of oligodendroglial death. J Neurotrauma 1999; 16:945-952.

103. Rosenberg LJ, Teng YD, Wrathall JR. Effects of the sodium channel blocker tetrodotoxin on acute white matter pathology after experimental contusive spinal cord injury. J Neurosci 1999; 19:6122-6133.
104. Monzon-Mayor M, Alvarex M, Arbelo-Galvan J et al. Long-term evolution of local, proximal and remote astrocyte responses after diverse nucleus basalis lesioning (an experimental Alzheimer model): GFAP immunocytochemical study. Brain Res 2000; 865:245-258.
105. Liu L, Svensson M, Aldskogius H. Clusterin upregulation following rubrospinal tract lesion in the adult rat. Exp Neurol 1999; 157:69-76.
106. Kaya SS, Mahmood A, Li Y et al. Expression of nestin after traumatic brain injury in rat brain. Brain Res 1999; 840:153-157.
107. Krum JM, Rosenstein JM. Transient coexpression of nestin, GFAP, and vascular endothelial growth factor in mature reactive astroglia following neural grafting or brain wounds. Exp Neurol 1999; 160:348-360.
108. Nouel D, Sarret P, Vincent JP et al. Pharmacological, molecular and functional characterization of glial neurotensin receptors. Neurosci 1999; 94:1189-1197.
109. Dermietzel R, Spray DC. Gap junctions in the brain: Where, what type, how many and why? Trends Neurosci 1993; 16:186-192.
110. Giaume C, Fromaget C, Aoumari AE et al. Gap junctions in cultured astrocytes: Single-channel currents and characterisation of channel forming protein. Neuron 1991; 6:133-143.
111. Cornell-Bell AH, Finkbeiner SM. Ca2+ waves in astrocytes. Cell Calcium 1991; 12:829-204.
112. Verkhratsky A, Kettenmann H. Calcium signalling in glia cells. Trends Neurosci 1996; 19:346-352.
113. Newman EA. Propagation of intercellular calcium waves in retinal astrocytes and M ller cells. J Neurosci 2001; 21:2215-2223.
114. Fields RD, Stevens B. ATP: An extracellular signaling molecule between neurons and glia. Trends Neurosci 2000; 23:625-633.
115. James G, Butt AM. P2X and P2Y purinoceptors mediate ATP-evoked calcium signalling in optic nerve glia in situ. Cell Calcium 2001. In press.
116. James G, Butt AM. Changes in P2Y and P2X purinoceptor expression in reactive glia following axonal degeneration in the rat optic nerve. Neurosci Lett 2001. In press.
117. James G, Butt AM. Adenosine 5 triphosphate evoked mobilization of intracellular calcium in central nervous system white matter of adult mouse optic nerve. Neurosci Lett 1999; 268:53-56.
118. Sanderson MJ, Charles AC, Boitano S et al. Mechanisms and function of intercellular calcium signalling. Mol Cell Endocrinol 1994; 98:173-187.
119. Sanderson MJ. Intercellular waves of communication. News Physiol Sci 1996; 11:262-269.
120. Fawcett JW, Asher RA. The glial scar and central nervous system repair. Brain Res Bull 1999; 49:377-391.
121. McKeon RJ, Schreiber RC, Dudge JS et al. Reduction of neurite outgrowth in a model of glial scarring following CNS injury is correlated with the expression of inhibitory molecules on reactive astrocytes. J Neurosci 1991; 1:3398-3398.
122. Bartach U, Bartsch S, Dorries U et al. Immunohistochemical localisation of tenascin in the developing and lesioned adult mouse optic nerve. Eur J Neurosci 1992; 4:338.
123. Laywell ED, Dorried U, Bartsch U et al. Enhanced expression of the developmentally regulated extracellular matrix molecule tenascin following adult brain injury. Proc Natl Acad Sci USA 1992; 89:2634-2638.
124. Ajemian A, Ness R, David S. Tenascin in the injured rat optic nerve and in non-neuronal cells in vitro: potential role in neural repair. J Comp Neurol 1994; 340:233-242.
125. Lips K, Stichel CC, Muller H-W. Restricted appearance of tenascin and chondroitin sulphate proteoglycans after transection and sprouting of adult rat post-commissural fornix. J Neurocytol 1995; 24:449-464.

126. Gates MA, Ladywell BO, Fillmore H et al. Astrocytes and extrcellular matrix following intracerebral transplantation of embryonic ventral mesencephalon or lateral ganglionic eminence. Neurosci 1996; 74:579-597.
127. Zhang Y, Winterbottom JK, Schachner M et al. Tenascin-C expression and axonal sprouting following CNS injury to the spinal dorsal columns in the adult rat. J Neurosci Res 2001; 49:433-450.
128. Zhang Y, Tohyama K, Winterbottom JK et al. Correlation between putative inhibitory molecules at the dorsal root entry zone and failure of dorsal root axon regeneration. Mol Cell Neurosci 1997; 17:444-459.
129. Davies SJ, Goucher DR, Doller C et al. Robust regeneration of adult sensory axons in degenerating white matter of the adult rat spinal cord. J Neurosci 1999; 19:5810-5822.
130. Taylor J, Pesheva P, Schachner M. Influence of janusin and tenascin on growth cone behavior in vitro. J Neurosci Res 1993; 35:347-362.
131. Jones LL, Liu Z, Shen J et al. Regulation of the cell adhesion molecule CD44 after nerve transection and direct trauma to the mouse brain. J Comp Neurol 2000; 426:468-492.
132. Eddleston M, Mucke L. Molecular profile of reactive astrocytes Implications for their role in neurological disease. Neurosci 1993; 54:15-36.
133. Bush TG, Puvanachandra N, Horner CH et al. Leukocyte infiltration, neuronal degeneration, and neurite outgrowth after ablation of scar-forming, reactive astrocytes in adult transgenic mice. Neuron 1999; 23:297-308.
134. Levine JM, Card JP. Light and electron microscopic localisation of a cell surface antigen (NG2) in the rat cerebellum: Association with smooth protoplasmic astrocytes. J Neurosci 1987; 7:2711-2720.
135. Levine JM, Stincone F, Lee YS. Development and differentiation of glial precursor cells in the rat cerebellum. Glia 1993; 7:307-321.
136. Stallcup WB, Beasley L. Bipotential glial precursor cells of the optic nerve express the NG2 proteoglycan. J Neurosci 1987; 7:2737-2744.
137. Levine JM, Stallcup WB. Plasticity of developing cerebellar cells in vitro studied with antibodies against the NG2 antigen. J Neurosci 1987; 7:2721-2731.
138. Nashiyama A, Lin X-H, Giese N et al. Co-localization of NG2 proteoglycan and PDGF-alpha receptor on O2A progenitor cells in the developing brain. J Neurosci Res 1996; 43:299-314.
139. Nashiyama A, Lin X-H, Giese N et al. Interaction between NG2: Proteoglycan and PDGF-alpha receptor on O2A cells is required for optimal response to PDGF. J Neurosci Res 1996; 43:315-330.
140. Reynolds R, Hardy R. Oligondendroglial progenitors labelled with the O4 antiody persist in the adult rat cerebral cortex in vivo. J Neurosci Res 1997; 47:455-470.
141. Butt AM, Duncan A, Hornby MF et al. Cells expressing the NG2 antigen contact nodes of Ranvier in adult CNS white matter. Glia 1999; 26:84-91.
142. Butt AM, Berry M. Oligodendrocytes and the control of myelination in vivo: New insights from the rat anterior medullary velum. J Neurosci Res 2000; 59:477-488.
143. Araque A, Purpara V, Sanzgiri RP et al. Tripartite synapses: Glia, the acknowledged partner. Trends Neurosci 1999; 22:208-215.
144. Ong WY, Levine JM. A light and electron microscopic study of NG2 chondroitin sulphate proteoglycan-positive oligodendrocyte precursor cells in the normal and kainate-lesioned rat hippocampus. Neurosci 1999; 92:83-95.
145. Bergles DE, Roberts JD, Somogyi P et al. Glutamatergic synapses on oligodendrocyte precursor cells in the hippocampus. Nature 2000; 405:187-191.
146. Fulton BP, Burne JF, Raff MC. Visualization of O-2A progenitor cells in the developing and adult rat optic nerve by quisqualate-stimulated cobalt uptake. J Neurosci 1992; 12:4816-4833.

147. Fidler PS, Schuette K, Asher RA et al. Comparing astrocytic cell lines that are inhibitory or permissive for axon growth: The major axon-inhibitory proteoglycan is NG2. J Neurosci 1999; 19:8778-8788.
148. Levine JM. Increased expression of the NG2 chondroitin sulphate proteoglycan after brain injury. J Neurosci 1994; 14:4716-4730.
149. Nishiyama A, Yu M, Drazba JA et al. Normal and reactive NG2+ glial cells are distinct from resting and activated microglia. J Neurosci Res 1997; 48: 299-312.
150. Redwine JM, Armstrong RC. In vivo proliferation of oligodendrocyte progenitors expressing PDGF alpha R during early remyelination. J Neurobiol 1998; 37:413-428.
151. Kierstad HS, Levine JM, Blakemore WF. Response of the oligodendrocyte progenitor cell population (defined by NG2 labelling) to demyelination of the adult spinal cord. Glia 1998; 22:161-170.
152. Ong WY, Levine JM. A light and electron microscopic study of NG2 chondroitin sulphate proteoglycan-positive oligodendrocyte precursor cells in the normal and kainate lesioned rat hippocampus. Neurosci 1999; 92:83-95.
153. Levine JM, Reynolds R, Fawcett JW. The oligodendrocyte precursor cell in health and disease. Trends Neurosci 2001; 24:39-47.
154. McTigue DM, Wei P, Stokes BT. Proliferation of NG2+ cells and altered oligodendrocyte numbers in the contused rat spinal cord. J Neurosci 2001; 21:3392-3400.
155. Nishiyama A, Chang AS, Trapp BD. NG2+ glia cells: A novel glial cell population in the adult brain. J Neuropath Exp Neurol 1999; 58:1113-1124.
156. Dawson MR, Levine JM, Reynolds R. NG2-expressing cells in the central nervous system: Are they oligodendroglial progenitors? J Neurosci Res 2000; 61:471-479.
157. Csuka E, Hans VH, Ammann E et al. Cell activation and inflammatory response following traumatic axonal injury in the rat. Neuroreport 2000; 11:2587-2590.
158. Kozlova EN, Lukanidin E. Metastasis-associated mtsl (s100A4) protein is selectively expressed in white matter astrocytes and is up-regulated after peripheral nerve or dorsal root injury. Glia 1999; 27:249-258.
159. Fisher CM, Kistler JP, Davis JM. Relation of cerebral vasospasm to subarachnoid hemorrhage visualised by computerised tomographic scanning. Neurosurg 1980; 6:1-9.
160. Graff-Radford NR, Torner J, Adams HP et al. Factors associated with hydrocephalus after subarachnoid hemorrhage. A report of the co-operative aneurysm study. Arch Neurol 1989; 46:744-752.
161. Tapaninaho A, Hernesniemi J, Vapalahti M et al. Shunt-dependent hydrocephalus after subarachnoid hemorrhage and aneurysm surgery: timing of surgery is not a risk factor. Act Neurochir 1993; 123:118-124.
162. Vale FL, Bradley EL, Fisher WS. The relationship of subarachnoid hemorrhage and the need for postoperative shunting. J Neurosurg 1997; 86:462-466.
163. Bagley C Jr. Blood in the cerebrospinal fluid. Resultant functional and organic alterations in the central nervous system. Arch Surg 1928; 17:18-38.
164. Hammes EM. Reaction of the meninges to blood. Arch Neurol Psychiat 1944; 52:505-514.
165. Suzuki S, Asia M, Ottomo M et al. Change in the subarachnoid space after experimental subarachnoid hemorrhage in the dog: scanning electron microscopic observation. Acta Neurochir 1977; 39:1-14.
166. Jackowski A, Crockard A, Burnstock G et al. Time course of intracranial pathophysiological changes following experimental subarachnoid hemorrhage in the rat. J Cereb Blood Flow Metab 1990; 10:835-849.
167. Motohashi O, Suzuki M, Yanai N et al. Thrombin and TGF-β promote human leptomeningeal cell proliferation in vitro. Neurosci Lett 1995; 90:105-108.
168. Sarnat HB. Ependymal reactions in injury: A review. J Neuropath Exp Neurol 1995; 54:1-15.
169. Bruni JE. Ependymal development, proliferation, and functions: A review. Microsc Res Tech 1998; 41:2-13.

170. Collins P, Fairman S. Repair of the ependyma in hydrocephalic brains. Neuropath Appl Neurobiol 1990; 16:45-46.
171. Berry M, Hunter AS, Duncan A et al. Axon-glial relations during regeneration of axons in the adult rat anterior medullary velum. J Neurocytol 1998; 27:915-937.
172. Raymond AA, Fish DR, Stevens JM et al. Subependymal heterotopia: A distinct neural migration disorder associated with epilepsy. J Neurol Neurosurg Psychiat 1994; 57:1195-1202.
173. Milhorat TH, Capocelli AL, Anzil AP et al. Pathological basis of spinal cord cavitation in syringomyelia: analysis of 105 autopsy cases. J Neurosurg 1995; 82:802-812.
174. Dijkstra S, Geisert EE JR, Gispen WH et al. Up-regulation of CD81 (target of the antiproliferative antibody; TAPA) by reactive microglia and astrocytes after spinal cord injury in the rat. J Comp Neurol 2000; 428:266-277.
175. Balentine JD. Pathology of experimental spinal cord trauma. I. The necrotic lesion as a function of vascular injury. Lab Invest 1978; 39:236-253.
176. Ducker TB, Kindt GW, Kempe LG. Pathological findings in acute experimental spinal cord trauma. J Neurosurg 1971; 35:700-707
177. Wallace MC, Tator CH, Lewis AJ. Chronic regenerative changes in the spinal cord after cord compression injury in rats. Surg Neurol 1987; 27:209-219.
178. Kao CC, Chang LW, Blookworth JJ. Axonal regeneration across transected mammalian spinal cords: An electron microscopic study of delayed microsurgical nerve grafting. Exp Neurol 1977; 54:591-615.
179. Gardner WJ, Goodall RJ. Surgical treatment of Arnold-Chiari malformation in adults. Explanation of its mechanism and importance of encephalography in diagnosis. J Neurosurg 1950; 7:199-206.
180. Williams B. On the pathogenesis of syringomyelia: A review. J R Soc Med 1980; 73: 798-806.
181. Williams B, Terry AF, Jones F et al. Syringomyelia as a sequel to traumatic paraplegia. Paraplegia 1981; 19:67-80.
182. Blight AR. Morphometric analysis of a model of spinal cord injury in guinea pigs, with behavioral evidence of delayed secondary pathology. J Neurol Sci 1991; 103:156-171.
183. Szczepanik AM, Fishkin RJ, Rush DK. Effects of chronic intrahippocampal infusion of lipopolysaccharide in the rat. Neurosci 1996; 70:57-65.
184. Fitch MT, Silver J. Activated macrophages and the blood brain barrier: Inflammation after CNS injury leads to increases in putative inhibitory molecules. Exp Neurol 1997; 148:587-603.
185. Fitch MT, Doller C, Combs CK et al. Cellular and molecular mechanisms of glial scarring and progressive cavitation: In vivo and in vitro analysis of inflammation-induced secondary injury after CNS trauma. J Neurosci 1999; 19:8182-8198.
186. Benjelloun N, Renolleau S, Represa A et al. Inflammatory responses in the cerebral cortex after ischemia in the P7 neonatal rat. Stoke 1999; 30:1916-1923.
187. Kalman M, Ajtai BM, Sommernes JH. Characteristics of glial reaction in the perinatal rat cortex: Effect of lesion size in the critical period . Neural Plast 2000; 7:147-165.
188. Ajtai BM, Kalman M. Axon growth failure following corpus callosum lesions precedes glial reaction in perinatal rats. Anat Embryol 2000; 202:313-321.
189. Clemente CD. Structural regeneration in the mammalian central nervous system and the role of neuroglia and connective tissue. In: Windle WF, ed. Regeneration in the Central Nervous System. SpriNGField: Thomas, 1995:147-161.
190. Clemente CD. Regeneration in the vertebrate nervous system. Int Rev Biol 1964; 6:257-301.
191. Windle WF. Regeneration of axons in the vertebrate central nervous system. Physiol Rev 1956; 36:427-440.
192. Windle WF, Clemente CD, Chambers WW. Inhibition of formation of a glial barrier as a means of permitting a peripheral nerve to grow into the brain. J Comp Neurol 1952; 96:359-369.
193. Windle WF, Clemente CD, Scott D et al. Induction of neural regeneration in the central nervous system of animals. Trans Amer Neurol Ass 77th Ann Meeting 1952:164-170.

194. Clemente CD. The regeneration of peripheral nerves inserted in the cerebral cortex and the healing of cerebral lesion. J Comp Neurol 1958; 109:123-152.
195. Caso A. Effectos del tratamiento de las lesiones cerebrales con ACTH. Bol Inst Estud Med Biol 1953; 11:61-74.
196. Clemente CD, Windle WF. Regeneration of severed nerve fibers in the spinal cord of the adult cat. J Comp Neurol 1954; 101:691-731.
197. McMasters RE. Regeneration of the spinal cord in the rat: Effects of Pyromen and ACTH upon the regeneration capacity. J Comp Neurol 1962; 119:113-125.
198. Fertig A, Kiernan JA, Seyan SS. Enhancement of axonal regeneration in the brain of the rat by corticotrophin and triiodothyronine. Exp Neurol 1971; 33:372-385.
199. Flint G, Berry M. A quantitative investigation of the response to injury of the central nervous system of rats treated with ACTH and triiodothyronine. Experientia 1973; 29:566-568.
200. Knowles JF, Berry M. The effects of deoxycorticosterone acetate on regeneration of axons in the mammalian central nervous system. Exp Neurol 1978; 62:1-15.
201. Berry M, Knowles J, Willis P et al. A reappraisal of the effects of ACTH on the response of the central nervous system to injury. J Anat 1979; 128:859-871.
202. Stichel CC, Hermanns S, Luhmann HJ et al. Inhibition of collagen IV deposition promotes regeneration of injured CNS axons. Eur J Neurosci 1999; 11:632-646.
203. Stichel CC, Niermann H, D Urso D et al. Basal membrane-depleted scar in lesioned CNS: Characteristics and relationships with regenerating axons. Neurosci 1999; 93:321-333.
204. Logan A, Berry M, Harper JR et al. Decorin, a natural inhibitor of transforming growth factor-β inhibits scar formation in the CNS. Proc. 76th Ann Meet Endocrinol Soc Abstr 1663, 1994.
205. Logan A, Baird A, Berry M. Decorin attenuates gliotic scar formation in the rat cerebral hemisphere. Exp Neurol 1999; 159:504-510.
206. Logan A, Green J, Hunter A et al. Inhibition of glial scarring in the injured rat brain by a recombinant human monoclonal antibody to transforming growth factor-β2. Eur J Neurosci 1999; 11:2367-2374.
207. Sung JH. Tangled masses of regenerating central nerve fibers (non-myelinated central neuromas) in the central nervous system. J Neuropath Exp Neurol 1964; 40:645-657.
208. Nieto-Sampedro M. Neurite outgrowth inhibitors in gliotic tissue. Adv Exp Med Biol 1999; 468:207-224.
209. Lemons ML, Howland DR, Anderson DK. Chondroitin sulfate proteoglycan immunoreactivity increases following spinal cord injury and transplantation. Exp Neurol 1999; 160:51-65.
210. Probstmeier R, Stichel CC, Muller HW et al. Chondroitin sulfates expressed on oligodendrocyte-derived tenascin-R are involved in neural cell recognition. Functional implications during CNS development and regeneration. J Neurosci Res 2000; 60:21-36.
211. McKeon RJ, Jurynec MJ, Buck CR. The chondroitin sulfate proteoglycans neurocan and phosphacan are expressed by reactive astrocytes in the chronic CNS glial scar. J Neurosci 1999; 19:10778-10788.
212. Haas CA, Rauch U, Thon N et al. Entorhinal cortex lesion in adult rats induces the expression of the neuronal chondroitin sulphate proteoglycan neurocan in reactive astrocytes. J Neurosci 1999; 19:9953-9963.
213. Thon N, Haas CA, Rauch U et al. The chondroitin sulphate proteoglycan brevican is upregulated by astrocytes after entorhinal cortex lesions in adult rats. Eur J Neurosci 2000; 12:2547-2558.
214. Berry M. Post-injury myelin-breakdown products inhibit axonal growth: An hypothesis to explain the failure of axonal regeneration in the mammalian central nervous system. Biblio Anat 1982; 23:1-11.
215. Schwab ME, Thoenen H. Dissociated neurons regenerate into sciatic but not optic nerve explants in culture irrespective of neurotrophic factors. J Neurosci 1985; 5:2415-2423.

216. Schwab ME, Caroni P. Rat CNS myelin and a subtype of oligodendrocytes in culture represent a non-permissive substrate for neurite growth and fibroblast spreading. J Neurosci 1988; 8:2381-2393.
217. Caroni P, Schwab ME. Antibody against myelin associated inhibitor of neurite growth neutralises non-permissive substrate properties of CNS white matter. Neuron 1988; 1:85-96.
218. Chen MS, Huber AB, van der Haar ME et al. Nogo-A is a myelin-associated neurite outgrowth inhibitor and an antigen for monoclonal antibody IN-1. Nature 2000; 403:434-439.
219. GrandPr T, Nakamura F, Vartanian SM et al. Identification of the Nogo inhibitor of axon regeneration as a reticulon protein. Nature 2000; 403:439-444.
220. Prinjha R, Moore T, Vinson M et al. Inhibitor of neurite outgrowth in humans. Nature 2000; 403:383-384.
221. Nakamura F, Tanaka M, Takahashi T et al. Neuropilin-1 extracellular domains mediate semaphorin D/III-induced growth cone collapse. Neuron 1998; 21:1093-1100.
222. Takahashi T, Nakamura F, Jin Z et al. Semaphorins A and E act as antagonists of neuropilin-1 and agonists of neuropilin-2 receptors. Nature Neurosci 1998; 1:487-493.
223. Chen H, He Z, Bagri A et al. Semaphorin-neuropilin interactions underlying sympathetic axon responses to class III semaphorins. Neuron 1998; 21:1283-1290.
224. Pasterkamp RJ, Verhaagen J. Emerging roles for semaphorins in neural regeneration. Brain Res Rev 2001; 35:36-54.
225. Fournier AE, GrandPr T, Strittmatter S. Identification of a receptor mediating Nogo-66 inhibition of axonal regeneration. Nature 2001; 409:341-346.
226. Rohm B, Ottemeyer A, Lohrum M et al. Plexin-neuropilin complexes mediate repulsion by the axon guidance signal semaphorin 3A. Mech Dev 2000; 93:95-104.
227. Takahashi T, Fournier A, Nakamura Y et al. Plexin-neuropilin-1 complexes from functional semaphorin-3A receptors. Cell 1999; 99:59-69.
228. Goodman ML, Poo CS, Tessier-Lavigne M-M et al. Plexins are a large family of receptors for transmembrane, secreted, and GPI-anchored semaphorins in vertebrates. Cell 1999; 99:71-80.
229. Schnell L, Schwab ME. Axonal regeneration in the rat spinal cord produced by an antibody against myelin-associated neurite growth inhibitors. Nature 1990; 343:269.
230. Huang DW, McKerracher L, Braun PE et al. A therapeutic vaccine approach to stimulate axon regeneration in the adult mammalian spinal cord. Neuron 1999; 24:639-647.
231. Moon LD, Asher RA, Rhodes KE et al. Regeneration of CNS axons back to their target following treatment of adult rat brain with chondroitinase ABC. Nat Neurosci 2001; 4(5):465-466.
232. Dusart I, Morel MP, Wehrle R et al. Late axonal sprouting of injured Purkinje cells and its temporal correlation with permissive changes in the glial scar. J Comp Neurol 1999; 408:399-418.
233. Leon S, Yin Y, Nguyen J et al. Lens injury stimulates axon regeneration in the mature rat optic nerve. J Neurosci 2000; 20:4615-4626.
234. Fischer D, Mitrofanis P, Thanos S. Cataractogenic lens injury prevents traumatic ganglion cell death and promotes axonal regeneration both in vivo and in culture. Invest Ophth Vis Res 2000; 41:3934-3954.
235. Berry M, Logan A. Response of the central nervous system to penetrant injury: Scarring and regeneration of axons. In: Crockard A, Hayward R, Hoff J, eds. Neurosurgery. Vol 1. London: Blackwell Science, 2000:479-498.
236. Wells MK, Bernstein JJ. Scar formation and the barrier hypothesis in failure of mammalian central nervous system regeneration. In: Dacy RG, Rimel RW, Jane JA, eds. Trauma in the Central Nervous System. New York: Raven Press, 1985:245-247.
237. Lindsay RM. Reactive gliosis. In: Federoff S, Vernakadis A, eds. Astrocytes. Vol. 3. New York: Academic Press, 1986:231-262.

238. Reier PJ. Gliosis following CNS injury: The anatomy of astrocytic scars and their influence on axonal elongation. In: Federoff S, Vernakadis A, eds. Astrocytes. Vol. 3. New York: Academic Press, 1986:263-324.
239. Liuzzi JL, Lasek RJ. Astrocytes block axonal regeneration in mammals by activation of the physiological stop pathway. Science 1987; 237:449-464.
240. Mansour M, Asher R, Dahl D et al. Permissive and non-permissive reactive astrocytes: Immunofluorescence study with antibodies to the glial hyaluronate-binding proteins. J Neurosci Res 1990; 25:300-311.
241. Snow DM, Lemmon V, Carrino DA et al. Sulphated proteoglycans in astroglial barriers inhibit neurite outgrowth in vitro. Exp Neurol 1990; 109:111-130.
242. Shewan D, Berry M, Bedi K et al. Embryonic optic nerve tissue fails to support neurite outgrowth by central and peripheral neurons in vitro. Eur J Neurosci 1993; 5:809-817.
243. Shewan D, Berry M, Cohen J. Extensive regeneration in vitro by early embryonic neurons on immature and adult tissue. J Neurosci 1995; 15:2057-2062
244. Berry M, Hall S, Follows R et al. Defective myelination in the optic nerve of the Brown-Wyse (BW) mutant rat. J Neurocytol 1989; 18:141-159.
245. Berry M, Hall S, Rees L et al. Regeneration of axons in the optic nerve of the Brown-Wyse (BW) mutant rat. J Neurocytol 1992; 21:426-448.
246. Hall S, Berry M, Wyse JPH. Regrowth of PNS axons through grafts of the optic nerve of the BW mutant rat. J Neurocytol 1992; 21:402-412.
247. Stichel CC, Wunderlich G, Schwab ME et al. Clearance of myelin constituents and axonal sprouting in the transected postcommissural fornix. Eur J Neurosci 1995; 7:401-411.
248. McConnell P, Berry M, Rees EL et al. The injury response of nerve fibers in the anterior medullary velum of the adult rat. Brain Res 1984; 323:257-268.
249. Derouiche A, Berry M, Sievers J. Regeneration of axons into the trochlear rootlet after anterior medullary lesions in the rat is specific for ipsilateral IVth nerve motoneurons. J Comp Neurol 1994; 341:340-350.
250. Berry M, Hunter AS, Duncan A et al. Axon-glial relations during regeneration of axons in the adult rat anterior medullary velum. J Neurocytol 1998; 27:915-937.
251. Cajal R. Degeneration and Regeneration in the Nervous System. London:Oxford University Press, 1928.
252. Risling M, Cullheim S, Hildebrand C. Reinnervation of the ventral root L7 from ventral horn neurons following intramedullary axotomy in adult cats. Brain Research 1983; 208:15-23.
253. Linda H, Cullheim S, Risling M. A light and electron microscopic study of intracellular HRP-labelled lumbar motoneurons after intramedullary axotomy in the adult cat. J Comp Neurol 1992; 318:188-208.
254. Wictorin K, Brundin P, Sauer H et al. Long distance directed axon growth from human dopaminergic neuroblasts implanted along the nigrostriatal pathway in 6-hydroxydopamine lesioned adult rats. J Comp Neurol 1992; 323:475-494.
255. Davies SJA, Field PM, Raisman G. Long fiber growth of axons of embryonic mouse hippocampal neurons microtransplanted into the adult fimbria. Eur J Neurosci 1993; 5:95-106.
256. Davies SJA, Field PM, Raisman G. Long interfasicular axon growth from embryonic neurons transplanted into adult myelinated tracts. J Neurosci 1994; 14:1596-1612.
257. Davies SJ, Fitch MT, Memberg SP et al. Regeneration of adult axons in white matter tracts of the central nervous system. Nature 1997; 390:680-683.
258. Cai D, Shen Y, De Lellard M et al. Prior exposure to neurotrophins blocks inhibition of axonal regeneration by MAG and myelin via a cAMP-dependent mechanism. Neuron 1999; 22:89-101.
259. Sievers J, Kruger S, Hansen CH et al. Integration of fetal brain transplants into adult brain: Morphological study of the host/graft interface. In: Bjorklund A, Steveni U, eds. Neural Grafting in the Mammalian CNS. Amsterdam: Elsevier Science Publishers, B.V., 1985:159-167.

260. Kruger S, Sievers J, Hanson C et al. Three morphologically distinct types of interface develop between adult host and fetal brain transplants; implications for scar formation in the adult central nervous system. J Comp Neurol 1986; 249:103-116.
261. Hall S, Berry M. Electron microscope study of the interaction of axons and glia at the site of anastomosis between the optic nerve and cellular and acellular sciatic nerve grafts. J Neurocytol 1989; 18:171-184.
262. Berry M, Rees L, Hall S et al. Optic axons regenerate into sciatic nerve isografts only in the presence of Schwann cells. Brain Res Bull 1992; 20:223-231.
263. Monard D. Cell-derived proteases and protease inhibitors as regulators of neurite outgrowth. Trends Neurosci 1988; 11:541-545.
264. Fawcett JW, Housdon E. The effects of protease inhibitors on axon growth through astrocytes. Dev 1990; 108:59-66.
265. Romanic AM, Madri JA. Extracellular matrix-degrading proteinase in the nervous system. Brain Path 1994; 4:145-156.
266. Muessel MJ, Berman NE, Klein RM. Early and specific expression of monocyte chemoattractant protein-1 in the thalamus induced by cortical injury. Brain Res 2000; 870:211-221.
267. Antoniades HN, Galanopoulos T, Neville-Golden J et al. Injury induces in vivo expression of PDGF and PDGF-receptor mRNAs in skin epithelial cells and PDGF mRNA in connective tissue fibroblasts. Proc Natl Acad Sci USA 1991; 88:565-569.
268. Walter H, Berry M, Hill DJ et al. Spatial and temporal changes in the insulin-like growth factor (IGF) axis indicate autocrine actions of IGF-1 within wounds of the rat brain. Endocrinol 1997; 138:3024-3034.
269. Walter HJ, Berry M, Hill DJ et al. Distinct sites of insulin-like growth factor (IGF)-II expression and localisation in lesioned rat brain: Possible roles of IGF binding proteins (IGFBPs) in the mediation of IGF-II activity. Endocrinol 1999; 140:520-532.
270. Moller T, Hanisch UK, Ransom BR. Thrombin-induced activation of cultured rodent microglia. J Neurochem 2000; 75:1539-1547.
271. Bona E, Andersson AL, Blomgren K et al. Chemokine and inflammatory cell response to hypoxia-ischemia in immature rats. Pediat Res 1999; 45:500-509.
272. Tonai T, Taketani Y, Ueda N et al. Possible involvement of interleukin-1 in cyclooxygenase-2 induction after spinal cord injury in rats. J Neurochem 1999; 72:302-309.
273. Pan W, Kastin AJ, Bell RL et al. Upregulation of tumor necrosis factor alpha transport across the blood-brain barrier after acute compressive spinal cord injury. J Neurosci 1999; 19:3649-3655.
274. Hayashi M, Ueyama T, Nemoto K et al. Sequential mRNA expression for immediate early genes, cytokines, and neurotrophins in spinal cord injury. J Neurotrauma 2000; 17:203-218.
275. Stanimirovic D, Satoh K. Inflammatory mediators of cerebral endothelium: A role in ischemic brain inflammation. Brain Pathol 2000; 10:113-126.
276. Le YL, Shih K, Bao P et al. Cytokine chemokine expression in contused rat spinal cord. Neurochem Int 2000; 36:417-425.
277. Fuerstein GZ, Wang X, Barone FC. The role of cytokines in the neuropathology of stroke and neurotrauma. Neuroimmunomod 1998; 5:143-159.
278. Brown KS, Khoshkbijar BA. Hematogenous cell responses to CNS injury. In: Berry M, Logan A, eds. CNS Injuries: Cellular Responses and Pharmocological Strategies. Baco Raton: CRC Press Inc., 1998:63-80.
279. McClennan KM, Smith PF, Darlington CL. Platelet-activating factor in the CNS. Prog Neurobiol 1996: 50:585-596.
280. Uhl E, Pickelmann S, Rohrich F et al. Influence of platelet-activating factor on cerebral microcirculation in rats: part 2. Local application. Stroke 1999; 30:880-886.
281. Wahl M, Schilling L, Unterberg A et al. Mediators of vascular and parenchymal mechanisms in secondary brain damage. Acta Neurochir Suppl 1993; 57:64-72.

282. Liu D, Li L, Augustus L. Prostaglandin release by spinal cord injury mediates production of hydrozyl radical, malondialdehyde and cell death: A site of the neuroprotective action of methylprednisolone. J Neurochem 2001; 77:1036-1047.
283. Schwab JM, Postler E, Nguyen TD et al. Connective tissue growth factor is expressed by a subset of reactive astrocytes in human cerebral infarction. Neuropath Appl Neurobiol 2000; 26:434-440.
284. Rosenberg GA, Cunningham LA, Wallace J et al. Immunohistochemistry of matrix metalloproteinases in reperfusion injury to rat brain: Activation of MMP-9 linked to stromelysin-1 and microglia in cell cultures. Brain Res 2001; 893:104-112.
285. Mayhan WG. VEGF increases permeability of the blood-brain barrier via a nitric oxide synthase/cGMP-dependent pathway. Am J Physiol 1999; 276:C1148-1153
286. Tsao MN, Li YQ, Lu G et al. Upregulation of vascular endothelial growth factor is associated with radiation-induced blood-spinal cord barrier breakdown. J Neuropath Exp Neurol 1999; 58:1051-1060.
287. van Bruggen N, Thibodeaux H, Palmer JT et al. VEGF antagonism reduces oedema formation and tissue damage after ischemia/reperfusion injury in the mouse brain. J Clin Invest 1999; 104:1613-1620.
288. Skold M, Culheim S, Hammarberg H et al. Induction of VEGF and VEGF receptors in the spinal cord after mechanical spinal injury and prostaglandin administration. Eur J Neurosci 2000; 12:3675-3686.
289. Vaquero J, Zurita M, de Oya S et al. Vascular endothelial growth/permeability factor in spinal cord injury. J Neurosurg 1999; 90:220-223.
290. Wahl SM, Wong H, McCartney-Francis N. Role of growth factors in inflammation and repair. J Cell Biochem 1989; 40:193-199.
291. Heino J, Ignotz RA, Hemler ME et al. Regulation of cell adhesion receptors by transforming growth factor-beta. Concomitant regulation of integrins that share a common beta 1 subunit. J Biol Chem 1989; 264:380-388.
292. Wang L, Goillot E, Tepper RI. IL-10 inhibits alloreactive cytotoxic T lymphocyte generation in vivo. Cell Immunol 1994; 159:152-169.
293. Yoshimoto T, Houkin K, Tada M et al. Induction of cytokines, chemokines and adhesion molecule mRNA in a rat forebrain reperfusion model. Acta Neuropath 1997; 93:154-158.
294. Whalen MJ, Carlos TM, Kochanek PM et al. Blood-brain barrier permeability, neutrophil accumulation and vascular adhesion molecule expression after controlled cortical impact in rats: A preliminary study. Acta Neurochir Suppl 1998; 71:212-214.
295. Whalen MJ, Carlos TM, Dixon CE et al. Effect of traumatic brain injury in mice deficient in intercellular adhesion molecule-1: Assessment of histopathologic and functional outcome. J Neurotrauma 1999; 16:299-309.
296. Kyrkanides S, Olschowka JA, Williams JP et al. TNF alpha and IL-1beta mediate intercellular adhesion molecule-1 induction via microglia-astrocyte interaction in CNS radiation injury. J Neuroimmunol 1999; 95:95-106.
297. Isaksson J, Farooque M, Holtz A et al. Expression of ICAM-1 and CD11b after experimental spinal cord injury in rats. J Neurotrauma 1999; 16:165-173.
298. Andjelkovic AV, Kerkovich D, Pachter JS. Monocyte: Astrocyte interactions regulate MCP-1 expression in both cell types. J Leuk Biol 2000; 68:545-552.
299. Brook GA, Houweling DA, Gieling RG et al. Attempted endogenous tissue repair following experimental spinal cord injury in the rat: Involvement of cell adhesion molecules L1 and NCAM? Eur J Neurosci 2000; 12:3224-3238.
300. Tzeng SF, Cheng H, Lee YS et al. Expression of neural cell adhesion molecule in spinal cords following a complete transection. Life Sci 2001; 68:1005-1012.
301. McTigue DM, Popovich PG, Morgan TE et al. Localisation of transforming growth factor-β1 and receptor mRNA after experimental spinal cord injury. Exp Neurol 2000; 163:220-230.
302. Flugel A, Hager G, Horvat A et al. Neuronal MCP-1 expression in response to remote nerve injury. J Cereb Blood Flow Metab 2001; 21:69-76.

303. McTigue DM, Tani M, Krivacic K et al. Selective chemokine mRNA accumulation in the rat spinal cord after contusion injury. J Neurosci Res 1998; 53:368-376.
304. Galasso JM, Liu Y, Szaflarski J et al. Monocyte chemoattractant protein-1 is a mediator of acute excitotoxic injury in neonatal rat brain. Neurosci 2000; 101:737-744.
305. Ghirnikar RS, Lee YL, He TR et al. Chemokine expression rat stab wound brain injury. J Neurosci Res 1996; 46:727-733.
306. Ghirnikar RS, Lee YL, Eng LF. Inflammation in traumatic brain injury: Role of cytokines and chemokines. Neurochem Res 1998; 23:329-340.
307. Lindsberg PJ, Hallenbeck JM, Feuerstein G. Platelet-activating factor in stroke and brain injury. Ann Neurol 1991; 30:117-129.
308. Pober JS, Doukas J, Hughes CW et al. The potential roles of vascular endothelium in immune-reactions. Hum Immunol 1990; 28:258-262.
309. Saliba E, Henrot A. Inflammatory mediators and neonatal brain damage. Biol Neo 2001; 79:224-227.
310. Nieto-Sampedro M, Saneto RP, de Vellis J et al. The control of glial populations in brain: Changes in astrocyte mitogenic and morphogenic factors in response to injury. Brain Res 1985; 343:320-328.
311. Rothwell NJ, Relton JK. Involvement of interleukin-1 and lipocortin-1 in ischemic brain-damage. Cereb Brain Metab Rev 1993; 5:178-198.
312. Lotan M, Shwartz M. Cross talk between the immune system and the nervous system in response to injury: implications for regeneration. FASEB 1994; 8:1026-1033.
313. Schiefer J, Topper R, Schmidt W et al. Expression of interleukin 6 in the rat striatum following stereotaxic injection of quinolinic acid. J Neuroimmunol 1998; 89:168-176.
314. Gabellec MM, Crumeyrolle-Adrias M, Le Saux F et al. Expression of interleukin-1 genes and interleukin-1 receptors in the mouse brain after hippocampal injury. Neurosci Res 1999; 33:251-260.
315. Ali C, Nicole O, Docagne F et al. Ischemia-induced interleukin-6 as a potential endogenous neuroprotective cytokine against NMDA receptor-mediated excitotoxicity in the brain. J Cereb Blood Flow Metab 2000; 20:956-966.
316. Knoblach SM, Faden AI. Cortical interleukin-1 beta elevation after traumatic brain injury in the rat: no effect of two selective antagonists on motor recovery. Neurosci Lett 2000; 289:5-8.
317. Streit WJ, Hurley SD, McGraw TS et al. Comparative evaluation of cytokine profiles and reactive gliosis supports a critical role for interleukin-6 and neuron-glia signalling during regeneration. J Neurosci Res 2000; 61:10-20.
318. Bartholdi D, Schwab ME. Expression of pro-inflammatory cytokine and chemokine mRNA upon experimental spinal cord injury in mouse: An in situ hybridization study. Eur J Neurosci 1997; 9:1422-38.
319. Fassbender K, Schneider S, Bertsch T et al. Temporal profile of release of interleukin-1β in neurotrauma. Neurosci Lett 2000; 284:135-138.
320. Hashizume H, DeLeo JA, Colburn RW et al. Spinal glial activation and cytokine expression after lumbar root injury in the rat. Spine 2000; 25:1206-1217.
321. Acarin L, Gonzalez B, Castellano B. Neuronal, astroglial and microglial cytokine expression after an excitotoxic lesion in the immature rat brain. Eur J Neurosci 2000; 12:3505-3520.
322. Rothwell N, Allan S, Toulmond S. The role of interleukin 1 in acute neurodegeneration and stroke: pathophysiological and therapeutic implications. J Clin Invest 1997; 100:2648-2652.
323. Gehrmann J, Matsumoto Y, Kreutzberg GW. Microglia Intrinsic immuneffector cell of the brain. Brain Res Rev 1995; 20:269-287.
324. Kamei H, Yoshida S, Yamasaki K et al. Severity of trauma changes expression of TNF-alpha mRNA in the brain of mice. J Surg Res 2000; 89:20-25.
325. Clarke WE, Berry M, Smith C et al. Coordination of fibroblast growth factor receptor 1 (FGFR1) and fibroblast growth factor-2 (FGF-2) trafficking to nuclei of reactive astrocytes around cerebral lesions in adult rats. Mol Cell Neurosci 2000; 17:17-30.

326. Smith C, Berry M, Clarke WE et al. Differential expression of fibroblast growth factor-2 and fibroblast growth receptor 1 in a scarring and nonscarring model of CNS injury in the rat. Eur J Neurosci 2001; 13:443-456.
327. Logan A, Berry M, Gonzalez AM et al. Effects of transforming growth factor β1 on scar production in the injured central nervous system. Eur J Neurosci 1994; 6:355-363.
328. Erlich S, Shohami E, Pinkas-Kramarski R. Closed head injury induces up-regulation of ErbB-4 receptor at the site of injury. Mol Cell Neurosci 2000; 16:597-608.
329. Beer R, Franz G, Schopf M et al. Expression of Fas and Fas ligand after experimental traumatic brain injury in the rat. J Cereb Blood Flow Metab 2000; 20:669-677.
330. Baird A, Mormede P, Bohlen P. Immunoreactive fibroblast growth factor in cells of peritoneal exudate suggests it identify with macrophage-derived growth factor. Biochem Biophys Res Com 1985; 126:358-364.
331. Friedman WJ. Cytokines regulate expression of the type 1 interleukin-1 receptor in rat hippocampal neurons and glia. Exp Neurol 2001; 168:23-31.
332. Leskovar A, Moriaty LJ, Turek JJ et al. The macrophage in acute neural injury: Changes in cell numbers over time and levels of cytokine production in mammalian central and peripheral nervous systems. J Exp Biol 2000; 203:1783-1795.
333. Nomoto Y, Yamamoto M, Fukishima T et al. Expression of nuclear factor kappa B and tumor necrosis factor alpha in the mouse brain after experimental thermal ablation injury. Neurosurg 2001; 48:158-166.
334. Rothwell NJ, Luheshi GN. Interleukin 1 in the brain: Biology, pathology and therapeutic target. Trends Neurosci 2000; 23:618-625.
335. Selmaj KW, Raine CS. Tumor necrosis factor mediates myelin and oligodendrocyte damage in vitro. Ann of Neurol 1988; 23:339-346.
336. Relton JK, Rothwell NJ. Involvement of cytokines in acute neurodegeneration in the CNS. Neurosci Biobehav Rev 1992; 17:217-227.
337. Araujo DM, Cotman CW. Basic FGF in astroglial, microglial and neuronal cultures: Characterisation of binding sites and modulation of release by lymphokines and trophic factors. J Neurosci 1992; 12:1668-1678.
338. Knoblach SM, Fan L, Faden AI. Early neuronal expression of tumor necrosis factor-alpha after experimental brain injury contributes to neurological impairment. J Neuroimmunol 1999; 95:115-125.
339. Allan SM. The role of pro- and anti-inflammatory cytokines in neurodegeneration. Ann NY Acad Sci 2000; 917:84-93.
340. Cartmell T, Luheshi GN, Hopkins SJ et al. Role of endogenous interleukin-1 receptor antagonist in regulating fever induced by localised inflammation in the rat. J Physiol 2001; 531:171-180.
341. Yick LW, Wu W, So KF et al. Time course of NOS expression and neuronal death in Clarke s nucleus following traumatic injury in adult rat spinal cord. Neurosci Lett 1998; 241:155-158.
342. Knerlich F, Schilling L, Gorlach C et al. Temporal profile of expression and cellular localisation of inducible nitric oxide synthase, interleukin-1β and interleukin converting enzyme after cryogenic lesion of the rat parietal cortex. Brain Res Mol Brain Res 1999; 68:73-87.
343. Thiel VE, Audus KL. Nitric oxide and blood-brain barrier integrity. Antiox Redox Signal 2001; 3:273-278.
344. Giovannoni G, Miller RF, Heales SJ et al. Elevated cerebrospinal fluid and serum nitrate and nitrite levels in patients with central nervous system complications of HIV-1 infection: A correlation with blood-brain-barrier dysfunction. J Neurol Sci 1998; 156:53-58.
345. Moss DW, Bates TE. Activation of murine microglial cell lines by lipopolysaccharide and interferon-gamma causes NO-mediated decreases in mitochondrial and cellular function. Eur J Neurosci. 2001; 13:529-38.

346. DeKosky ST, Styren SD, O Malley ME et al. Interleukin-1 receptor antagonist suppresses neurotrophin response in injured rat brain. Ann Neurol 1996; 39:123-127.
347. Fagan AM, Gage FH. Cholinergic sprouting in the hippocampus: A proposed role for IL-1. Exp Neurol 1990; 110:105-120.
348. Korsching S. The neurotrophic factor concept: A re-examination. J Neurosci 1993; 13:2739-2748.
349. Herx LM, Rivest S, Yong VW. Central nervous system-initiated inflammation and neurotrophism in trauma: IL-1 beta is required for the production of ciliary neurotrophic factor. J Immunol 2000; 165:2232-2239.
350. Phillips AO, Topley N, Morrisey K et al. Basic fibroblast growth factor stimulates the release of preformed transforming growth factor beta 1 from human proximal tubular cells in the absence of de novo gene transcription or mRNA translation. Lab Invest 1997; 76:591-600.
351. Gomez-Pinella F, Lee JW, Cotman CW. Basic FGF in adult rat brain: cellular distribution and response to entorhinal lesion and fimbria-fornix transection. J Neurosci 1992; 12:345-355.
352. Gomide VC, Chadi G. The trophic factors S-100b and basic fibroblast growth factor are increased in the forebrain reactive astrocytes of adult callosotomized rat. Brain Res 1999; 835:162-174.
353. Buytaert KA, Kline AE, Montanez S et al. The temporal patterns of c-Fos and basic fibroblast growth factor expression following a unilateral anteromedial cortex lesion. Brain Res 2001; 894:121-130.
354. Hughes PE, Alexi T, Walton M et al. Activity and injury-dependent expression of inducible transcription factors, growth factors and apoptosis-related genes within the central nervous system. Prog Neurobiol 1999; 57:421-450.
355. Jiang Y, McLennan IS, Koishi K et al. Transforming growth factor-beta 2 is anterogradely and retrogradely transported in motoneurons and up-regulated after nerve injury. Neurosci 2000; 97:735-742.
356. Martinez G, Carnazza ML, Di Giacomo C et al. Expression of bone morphogenetic protein-6 and transforming growth factor-β1 in the rat brain after a mild and reversible ischemic damage. Brain Res 2001; 894:1-11.
357. Hertel M, Yvonne T, Alzheimer C et al. Connective tissue growth factor: A novel player in tissue reorganisation after brain injury. Eur J Neurosci 2000; 12:376-380.
358. Tretter YP, Munz B, Hubner G et al. Strong induction of activin expression after hippocampal lesion. Neuroreport 1996; 7:1819-1823.
359. Tretter YP, Hertel M, Munz B et al. Induction of activin A is essential for the neuroprotective action of basic fibroblast growth factor in vivo. Nat Med 2000; 6:812-815.
360. Tokita Y, Keino H, Matsui F et al. Regulation of neuregulin expression in the injured rat brain and cultured astrocytes. J Neurosci 2001; 21:1257-1264.
361. Takayama S, Sasahara M, Iihara K et al. Platelet-derived growth factor B-chain-like immunoreactivity in injured rat brain. Brain Res 1994; 653:131-140.
362. Sugiura S, Lahav R, Han J et al. Leukaemia inhibitory factor is required for normal inflammatory responses to injury in the peripheral and central nervous systems in vivo and is chemotactic for macrophages in vitro. Eur J Neurosci 2000; 12:457-466.
363. Park KI. Transplantation of neural stem cells: Cellular and gene therapy for hypoxic-ischemic brain injury. Yonsei Med J 2000; 41:825-835.
364. Widenfalk J, Lundstromer K, Jubran M et al. Neurotrophic factors and receptors in the immature and adult spinal cord after mechanical injury or kainic acid. J Neurosci 2001; 21:3457-3475.
365. Bresjanac M, Antauer G. Reactive astrocytes of the quinolinic acid-lesioned rat striatum express GFRalpha1 as well as GDNF in vivo. Exp Neurol 2000; 164:53-59.
366. Michael GJ, Averill S, Shortland PJ et al. Axotomy results in major changes in BDNF expression by dorsal root ganglion cells: BDNF expression in large TrkB and TrkC cells, in pericellular baskets, and in projections to deep dorsal horn and dorsal column nuclei. Eur J Neurosci 1999; 11:3539-3551.

367. Dougherty KD, Dreyfus CF, Black IB. Brain-derived neurotrophic factor in astrocytes, oligodendrocytes, and microglia-macrophages after spinal cord injury. Neurobiol Dis 2000; 7:574-585.
368. Dreyfus CF, Dai X, Lercher LD et al. Expression of neurotrophins in the adult spinal cord in vivo. J Neurosci Res 1999; 56:1-7.
369. Krenz NR, Weaver LC. Nerve growth factor in glia and inflammatory cells of the injured rat spinal cord. J Neurochem 2000; 74:730-739.
370. Moalem G, Gdalyahu A, Shani Y et al. Production of neurotrophins by activated T cells: Implications for neuroprotective autoimmunity. J Autoimmunol 2000; 15:331-345.
371. Otten U, Marz P, Heese K et al. Cytokines and neurotrophins interact in normal and diseased states. Ann NY Acad Sci 2000; 917:322-330.
372. Liebl DJ, Huang W, Young W et al. Regulation of Trk receptors following contusion of the rat spinal cord. Exp Neurol 2001; 167:15-26.
373. King VR, Bradbury EJ, McMahon SB et al. Changes in truncated TrkB and p75 receptor expression in the rat spinal cord following spinal cord hemisection and spinal cord hemisection plus neurotrophin treatment. Exp Neurol 2000; 165:327-341.
374. Yuan J, Yankner BA. Apoptosis in the nervous system. Nature 2000; 407:802-809.
375. Skold M, Cullheim S, Hammarberg H et al. Induction of VEGF and VEGF receptors in the spinal cord after mechanical spinal injury and prostaglandin administration. Eur J Neurosci 2000; 12:3675-3686.
376. Lin TN, Wang CK, Cheung WM et al. Induction of angiopoietin and Tie receptor mRNA expression after cerebral ischemia-reperfusion. J Cereb Blood Flow Metab 2000; 20:387-395.
377. Schwab JM, Beschorner R, Nguyen TD et al. Differential cellular accumulation of connective tissue growth factor defines a subset of reactive astrocytes, invading fibroblasts, and endothelial cells following central nervous system injury in rats and humans. J Neurotrauma 2001; 18:377-388.
378. Grotendorst GR. Connective tissue growth factor: a mediator of TGF-beta action on fibroblasts. Cyt Growth Fact Rev 1997; 8:171-179.
379. Kothapalli D, Hayashi N, Grotendorst GR. Inhibition of TGF-beta-stimulated CTGF gene expression and anchorage-independent growth by cAMP identifies a CTGF-dependent restriction point in the cell cycle. FASEB J 1998; 12:1151-1161.
380. Brigstock DR. The connective tissue growth factor/cysteine-rich 61/nephroblastoma over expressed (CCN) family. Endocrinol Rev 1999; 20:189-206.
381. Moussad EEA, Brigstock DR. Connective tissue growth factor: What s in a name? Mol Genet Metab 2000; 71:276-292.
382. Lindholm D, Castren E, Kiefer R et al. Transforming growth factor-b1 in the rat brain: Increase after injury and inhibition of astrocyte proliferation. J Cell Biol 1992; 117:395-400.
383. Klempt ND, Sirimanne E, Gunn AJ et al. Hypoxia-ischemia induces transforming growth factor β1mRNA in the infant rat brain. Mol Brain Res 1992; 13:93-101.
384. Lagord C, Berry M, Logan A. Expression of TGFβ2 but not TGFβ1 correlates with the deposition of scar tissue in the lesioned spinal cord. Eur J Neurosci 2001. In press.
385. Lukes A, Mun-Bryce S, Lukes M et al. Extracellular matrix degradation by metalloproteinases and central nervous system diseases. Mol Neurobiol 1999; 19:267-284.
386. Vecil GG, Larsen PH, Corley SM et al. Interleukin-1 is a key regulator of matrix metalloproteinase-9 expression in human neurons in culture and following mouse brain trauma in vivo. J Neurosci Res 2000; 61:212-224.
387. de Castro RC Jr, Burns CL, McAdoo DJ et al. Metalloproteinase increases in the injured rat spinal cord. Neuroreport 2000; 11:3551-3554.
388. Rathke-Hartlieb S, Budde P, Ewert S et al. Elevated expression of membrane type 1 metalloproteinase (MT1-MMP) in reactive astrocytes following neurodegeneration in mouse central nervous system. FEBS Lett 2000; 481:227-234.

389. Schlomann U, Rathke-Hartlieb S, Yamamoto S et al. Tumor necrosis factor alpha induces a metalloprotease-disintegrin ADAM8 (CD 156): Implications for neuron-glia interactions during neurodegeneration. J Neurosci 2000; 20:7964-7971.
390. Dietzmann K, von Bossanyi P, Krause D et al. Expression of the plasminogen activator system and the inhibitors PAI-1 and PAI-2 in posttraumatic lesions of the CNS and brain injuries following dramatic circulatory arrests: An immunohistochemical study. Path Res Pract 2000; 196:15-21.
391. Tang H, Fu WY, Ip NY. Altered expression of tissue-type plasminogen activator and type 1 inhibitor in astrocytes of mouse cortex following scratch injury in culture. Neurosci Lett 2000; 285:143-146.
392. D Angelo M, Billings PC, Pacifici M et al. Authentic matrix vesicles contain active metalloproteases (MMP). A role for matrix vesicle-associated MMP-13 in activation of transforming growth factor-beta. J Biol Chem 2001; 276:11347-11353.
393. Jaworski DM. Differential regulation of tissue inhibitor of metalloproteinase mRNA expression in response to intracranial injury. Glia 2000; 30:199-208.
394. Balasingham V, Yong VW. Attenuation of astroglial reactivity by interleukin-10. J Neurosci 1996; 16:2945.
395. Acarin L, Carrasco J, Gonzalez B et al. Expression of growth inhibitory factor (metallothionein-III) mRNA and protein following excitotoxic immature brain injury. J Neuropath Exp Neurol 1999; 58:389-397.
396. Penkowa M, Moos T, Carrasco J et al. Strongly compromised inflammatory response to brain injury in interleukin-6-deficient mice. Glia 1999; 25:343-357.
397. Whitby DJ, Ferguson MWJ. Immunohistochemical localisation of growth factors in fetal wound healing. Develop Biol 1991; 147:207-215.
398. Adzick NS, Lorenz HP. Cells, matrix, growth factors, and the surgeon. The biology of scarless fetal wound repair. Ann Surg 1994; 220:10-18.
399. Silverstein FS, Barks JD, Hagan P et al. Cytokines and perinatal brain injury. Neurochem Int 1997; 30:375-383.
400. Flanders KC, Ludecke G, Engels S et al. Localisation and actions of transforming growth factor-βs in the embryonic nervous system. Develop 1991; 113:183-191.
401. Gonzalez A-M, Buscaglia M, Long M et al. Distribution of fibroblast growth factor in the 18 day rat foetus: Localisation in the basement membranes of diverse tissues. J Cell Biol 1990; 110:753-765.
402. Martin P, Dickson MC, Millan FA et al. Rapid induction and clearance of TGF-β1 is an early response to wounding in the mouse embryo. Develop Gen 1993; 14:225-238.
403. Martin P, Nobes C, McCluskey J et al. Repair of excisional wounds in the embryo. Eye 1994; 8:155-160.
404. Hopkinson-Woolley J, Hughes D, Gordon S et al. Macrophage recruitment during limb development and wound healing in the embryonic and fetal mouse. J Cell Sci 1994; 107:1159-1167.
405. Cui W, Akhurst RJ. Transforming growth factor βs: Biochemistry and biological actions in vitro and in vivo. In: LeRoith D, Bondy C, eds. Growth Factors in Health and Disease. Vol 1B. Greenwich: JAI Press, 1996:319-356.
406. McPherron AC, Lee S. Tre transforming growth factor β superfamily. In: LeRoith D, Bondy C, eds. Growth Factors in Health and Disease. Greenwich: JA1 Press, 1996:357-393.
407. Pratt BM, McPherson JM. TGF-beta in the central nervous system: potential roles in ischemic injury and neurodegenerative diseases. Cyt Growth Fac Rev 1997; 8:267-292.
408. Ignotz RA, Heino J, Massague J. Regulation of cell adhesion receptors by transforming growth factor-beta. Regulation of vitronectin receptor and LFA-1. J Biol Chem 1989; 264:389-392.
409. Heino J, Massague J. Transforming growth factor-beta switches the pattern of integrins expressed in MG-63 human osteosarcoma cells and causes a selective loss of cell adhesion to laminin. J Biol Chem 1989; 264:21806-21811.

410. Roberts AB, Sporn MB. The transforming growth factor-βs. In: Sporn MB, Roberts AB, eds. Peptide Growth Factors and their Receptors I: Handbook of Experimental Pharmacology. Berlin/Heidelberg: Springer-Verlag, 1990:418-472.
411. Roberts AB. Molecular and cell biology of TGF-beta. Min Electrol Metab 1998; 24:111-119.
412. Mita T, Yamashita H, Kaji Y et al. Effects of transforming growth factor β on corneal epithelial and stromal cell function in a rat wound healing model after excimer laser keratectomy. Graefes Arch Clin Exp Ophthalmol 1998; 236:824-843.
413. Struckhoff G. Transforming growth factor beta 1 and parathyroid hormone-related protein control the secretion of dipeptidyl peptidase II by rat astrocytes. Neurosci Lett 1995; 189:117-120.
414. Fukui N, Fukuda A, Kojima K et al. Suppression of fibrous adhesion by proteoglycan decorin. J Orthop Res 2001; 19:456-462.
415. Krupinski J, Kumar P, Kumar S et al. Increased expression of TGF-β in brain tissue after ischemic stroke in humans. Stoke 1996; 27:852.
416. Krupinski J, Vodovotz Y, Li C et al. Inducible nitric oxide production and expression of transforming growth factor-beta1 in serum and CSF after cerebral ischemic stroke in man. Nitric oxide 1998; 2:442-453.
417. Lehrmann E, Kiefer R, Christensen T et al. Microglia and macrophages are major sources of locally produced transforming growth factor-beta1 after transient middle cerebral artery occlusion in rats. Glia 1998; 24:437-448.
418. Slevin M, Krupinski J, Slowik A et al. Serial measurement of vascular endothelial growth factor and transforming growth factor-beta1 in serum of patients with acute ischemic stroke. Stroke 2000; 31:1863-1870.
419. Kitazawa K, Tada T. Elevation of transforming growth factor-beta 1 level in cerebrospinal fluid of patients with communicating hydrocephalus after subarachnoid hemorrhage. Stroke 1994; 25:1400-1404.
420. Daniel M, Lagord C, Akinwunmi J et al. Measurement of transforming growth factor-β1 levels in cerebrospinal fluid from patients with CNS injury. J Endocrinol 1997; 152:126.
421. Wyss-Coray T, Masliah E, Mallory M et al. Amyloidogenic role of cytokine TGF-beta1 in transgenic mice and in Alzheimer s disease. Nature 1997; 389:603-606.
422. Wyss-Coray T, Lin C, Sanan DA et al. Chronic overproduction of transforming growth factor-beta1 by astrocytes promotes Alzheimer s disease-like microvascular generation in transgenic mice. Am J Path 2000; 156:139-150.
423. Wyss-Coray T, Lin C, Yan F et al. TGF-beta1 promotes microglial amyloid-beta clearance and reduces plaque burden in transgenic mice. Nat Med 2001; 7:612-618.
424. Bradham DM, Igarashi A, Potter RL et al. Connective tissue growth factor: A cysteine-rich mitogen secreted by human vascular endothelial cells is related to the SRC-induced immediate early gene product CEF-10. J Cell Biol 1991; 114:1285-1294.
425. Kireeva ML, Latinkic BV, Kolesnikova TV et al. Cyr61 and Fisp1 are both ECM-associated signalling molecules: Activities, metabolism and localisation during development. Exp Cell Res 1997; 233:63-77.
426. Riser BL, Denichilo M, Cortes P et al. Regulation of connective tissue growth factor activity in cultured rat mesangial cells and its expression in experimental diabetic glomerulosclerosis. J Am Soc Nephrol 2000; 11:25-38.
427. Igarashi A, Bradham DM, Okochi H et al. Connective tissue growth factor. J Dermatol 1992; 19:642-643.
428. Duncan MR, Frazier KS, Abramson S et al. Connective tissue growth factor mediates transforming growth factor beta-induced collagen synthesis: Down-regulation by cAMP. FASEB J 1999; 13:1774-1786.
429. Kothapalli D, Frazier KS, Welply A et al. Transforming growth factor beta induces anchorage-independent growth of NRK fibroblasts via a connective tissue growth factor-dependent signaling pathway. Cell Growth Differ 1997; 8:61-68.

430. Kothapalli D, Grotendorst GR. CTGF modulates cell cycle progression in cAMP-arrested NRK fibroblasts. J Cell Physiol 2000; 182:119-126.
431. Kondo Y, Nakanishi T, Takigawa M et al. Immunohistochemical localisation of connective tissue growth factor in the rat central nervous system. Brain Res 1999; 834:146-151.
432. Albrecht C, von Der Kammer H, Mayhaus M et al. Muscarinicacetylcholine receptors induce the expression of the immediate early growth regulatory gene CYR61. J Biol Chem 2000; 275:28929-28936.
433. Berry M, Logan A. The response of the CNS to penetrant injury: scarring and regeneration of axons. Neurosurg 2000.
434. Kolb M, Margetts PJ, Galt T et al. Transient transgene expression of decorin in the lung reduces the fibrotic response to bleomycin. Am J Resp Crit Care Med 2001; 163:770-777.
435. Zhang L, Graziano K, Pham T et al. Adenovirus-mediated gene transfer of dominant-negative Smad4 blocks TGF-beta signaling in pancreatic acinar cells. Am J Physiol Gastrointest Liver Physiol 2001; 280:G1247-1253.
436. Hallscheidt P, Wallich R, Kauffmann GW et al. Inhibition of TGF-beta and VEGF expression in human endothelium cell lines with antisense-oligonucleotides: potential tools for gene therapy of chronic renal transplant rejection. Transplant Proc 2001; 33:2157-2159.
437. Berry M, Gonzalez AM, Clarke W et al. Sustained effects of gene-activated matrices after CNS injury. Mol Cell Neurosci 2001; 17:706-716.
438. Gonzalez AM, Baird A, Greenlees L et al. Delivery of DNA to neuronal cells after spinal cord injury. Proc Soc Neurosci 2000; 26:1888.
439. Kagami S, Border WA, Miller DE et al. Angiotensin II stimulates extracellular matrix protein synthesis through induction of transforming growth factor-β expression in rat glomerular mesangial cells. J Clin Invest 1994; 93:2431-2437.
440. Guh JY, Tang ML, Tang YL et al. Captopril reverses high-glucose-induced growth effects on LLC-PK_1 cells partly by decreasing transforming growth factor-β receptor protein expression. J Am Soc Nephrol 1996; 7:1207-1215.
441. Klahr S, Morrissey JJ. Comparative study ofACE inhibitors and angiotensin II receptor antagonists in interstitial scarring. Kidney Int Suppl 1997; 63:S111-S114.
442. Hill C, Logan A, Smith C et al. Angiotensin converting enzyme inhibitor suppresses glomerular transforming growth factor beta receptor expression in experimental diabetes in rats. Diabetologia 2001; 44:495-500.
443. Sasamura H, Shimizu-Hirota R, Nakaya H et al. Effects of AT1 receptor antagonist on proteoglycan gene expression in hypertensive rats. Hyperten Res 2001; 24:165-172.
444. Yang G, Gray TS, Sigmund CD et al. The angiotensinogen gene is expressed in both astrocytes and neurons in murine central nervous system. Brain Res 1999; 817:123-131.
445. Lippoldt A, Bunnemann B, Ueki A et al. On the plasticity of the cerebellar renin-angiotensin system: localisation of components and effects of mechanical perturbation. Brain Res 1994; 668:144-159.
446. Lucius R, Gallinat S, Rosenstiel P et al. The angiotensin II type 2 (AT2) receptor promotes axonal regeneration in the optic nerve of adult rats. J Exp Med 1998; 188:661-670.
447. Bracken MB, Shapard MJ, Holford TR et al. Methylprednisolone or tirilazad mesylate administration after acute spinal cord injury: 1-year follow up. Results of the third National Acute Spinal Cord Injury randomized controlled trial. J Neurosurg 1998; 89:699-706.
448. Oudega M, Vargas CG, Weber AB et al. Long-term effects of methylprednisolone following transection of adult rat spinal cord. Eur J Neurosci 1999; 11:2453-2464.
449. Yoon DH, Kim YS, Young W. Therapeutic time window for methylprednisolone in spinal cord injured rat. Yonse Med J 1999; 40:313-320.
450. Taoka Y, Okajima K, Uchiba M et al. Methylprednisolone reduces spinal cord injury in rats without affecting tumor necrosis factor-alpha production. J Neurotrauma 2001; 18:533-543.
451. Hauss-Wegrzyniak B, Willard LB, Del Soldato P et al. Peripheral administration of novel anti-inflammatories can attenuate the effects of chronic inflammation with in CNS. Brain Res 1999; 815:36-43.

452. Ghirnikar RS, Lee YL, End LF. Chemokine antagonist infusion attenuates cellular infiltration following spinal cord contusion injury in rat. J Neurosci Res 2000; 59:63-73.
453. Bethea JR, Nagashima H, Acosta MC et al. Systemically administered interleukin-10 reduces tumor necrosis factor-alpha production and significantly improves functional recovery following traumatic spinal cord injury in rats. J Neurotrauma 1999; 16:851-863.
454. Kudo A, Suzuki M, Kubo Y et al. Intrathecal administration of thrombin inhibitor ameliorates cerebral vasospasm. Use of a drug delivery system releasing hirudin. Cerebrovasc Dis 2000; 10:424-430.
455. Mabon PJ, Weaver LC, Dekaban GA. Inhibition of monocyte/macrophage migration to a spinal cord injury site by an antibody to the integrin alpha D: A potential new anti-inflammatory treatment. Exp Neurol 2000; 166:52-64.
456. Suzuki H, Abe K, Tojo SJ et al. Reduction of ischemic brain injury by anti-P-selectin monoclonal antibody after permanent middle cerebral artery occlusion in rat. Neurol Res 1999; 21:269-276.
457. Huang J, Kim LJ, Mealey R et al. Neuronal protection in stroke by an sLex-glycosylated complement inhibitory protein. Science 1999; 285:595-599.
458. Lew SM, Gross CE, Bednar MM et al. Complement depletion does not reduce brain injury in a rabbit model of thromboembolic stroke. Brain Res Bull 1999; 48:325-331.
459. Sullivan PG, Rabchevsky AG, Hicks RR et al. Dose-response curve and optimal dosing regimen of cyclosporin A after traumatic brain injury in rats. Neurosci 2000; 101:289-295.
460. Shimakura A, Kamanaka Y, Ikeda Y et al. Neutrophil elastase inhibition reduces cerebral ischemic damage in the middle cerebral artery occlusion. Brain Res 2000; 858:55-60.
461. Cohen IR, Schwartz M. Autoimmune maintenance and neuroprotection of the central nervous system. J Neuroimmunol 1999; 100:111-114.
462. Leon S, Yin Y, Nguyen J et al. Lens injury stimulates axon regeneration in the mature rat optic nerve. J Neurosci 2000; 20:4615-4626.
463. Hauben E, Nevo U, Yoles E et al. Autoimmune T cells as potential neuroprotective therapy for spinal cord injury. Lancet 2000; 355:286-287.
464. Schwartz M, Moalem G. Beneficial immune activity after CNS injury: Prospects for vaccination. J Neuroimmunol 2001; 113:185-192.
465. Yoles E, Hauben E, Algi O et al. Protective autoimmunity is a physiological response to CNS trauma. J Neurosci 2001; 21:3740-3748.
466. McGraw J, Hiebert GW, Steeves JD. Modulating astrogliosis after neurotrauma. J Neurosci Res 2001; 63:109-115.

II. ION CHANNELS, RECEPTORS AND SIGNALING PATHWAYS

Na^+ CHANNELS AND Ca^{2+} CHANNELS OF THE CELL MEMBRANE AS TARGETS OF NEUROPROTECTIVE SUBSTANCES

Christian Alzheimer*

INTRODUCTION

Loss of ion homeostasis is the cardinal event in the early neuronal response to acute ischemic and traumatic brain injury.[1-3] At short notice, excessive influx of Na^+ and Ca^{2+} overwhelming the compensatory mechanisms of the neuron will cause osmotic volume expansion (swelling), which might be reversible if the impact of the acute injury is limited and small, and appropriate therapeutic actions are taken. However, pathologically elevated Na^+ and Ca^{2+} levels in the cytosol are likely to trigger a cascade of molecular events that eventually lead to neuronal death (e.g., formation of reactive oxygen species, lipid peroxidation, mitochondrial dysfunction, activation of caspases etc.). Owing to its double nature as charge carrier and ubiquitous second messenger, Ca^{2+} has received much more attention than Na^+. Deviating from this usual bias, this Chapter will first take a closer look at the pathways of pathological Na^+ influx, at the multiple sequelae of excessive intracellular Na^+ accumulation, and at attempts to prevent Na^+ overload as a means to fight neuronal loss in acute injury.

*Department of Physiology, University of Munich, Pettenkoferstr. 12, D-80336 Munich, Germany; Present address: Institute of Physiology, University of Kiel, 24098 Kiel, Germany.

NA^+ CHANNELS

Pathophysiology of Na^+ Loading

The schematic drawing of Figure 1 illustrates mechanisms of Na^+ entry and the pathological events produced by elevated cytosolic Na^+. Two major and closely interrelated pathways can be identified through which Na^+ enters the cells, one being voltage-dependent Na^+ channels, the other being ionotropic glutamate receptors. With ATP levels drastically falling during ischemia and metabolic dysfunction, the efficacy of the Na^+/K^+ exchanger declines. The gradual breakdown of ion gradients leads to membrane depolarization, thereby opening voltage-dependent Na^+ channels in a regenerative fashion. The concomitant opening of voltage-dependent Ca^{2+} channels contributes to the detrimental rise in cytosolic Ca^{2+} and promotes vesicular glutamate release. At the same time, enhanced glutamate levels of synaptic and mainly nonsynaptic origin (see below) cause further Na^+ and Ca^{2+} influx through ionotropic glutamate receptors of the AMPA/kainate and NMDA subtype and aggravate membrane depolarization. The elevated Na^+ level in the cytosol has several fatal consequences for the survival of the cell: Firstly, the transmembrane 3 Na^+/2 Ca^{2+} exchanger, which normally uses the Na^+ gradient to extrude Ca^{2+} from the cell, operates now in reverse mode transporting Ca^{2+} into the cell. Secondly, a rise in cytosolic Na^+ activates the mitochondrial 2 Na^+ /Ca^{2+} exchanger causing Ca^{2+} release from the mitochondria. It has been estimated that about 50% of the sustained rise in cytosolic Ca^{2+} during ischemia is attributable to this mechanism.[4] Thirdly, the direction of the Na^+-dependent glutamate transporter, which recovers glutamate from the extracelluar space under physiological conditions, is reversed, leading to the release of glutamate. It is noteworthy that ATP depletion also impairs vesicle filling thus raising cytosolic glutamate and promoting the reversed operation of the glutamate carrier. Underscoring the significance of this pathway for glutamate excitotoxicity, recent evidence indicates that, compared with other mechanisms of nonsynaptic glutamate release in severe brain ischemia, reversed uptake is the predominant pathway.[5]

Spreading Depression

An electrophysiological hallmark of acute severe hypoxia in forebrain gray matter is the development of a self-regenerating, nearly complete membrane depolarization that is associated with a loss of neuronal activity and a collapse of ion gradients. This phenomenon closely resembles the spreading depression (SD) originally described by Le o [6] (for review see ref. 7). Recurrent waves of SD are thought to contribute to the growth of a focal lesion by extending the initial damage of cerebrovascular infarcts into marginal zones. SD-like episodes can be triggered in brain slice preparations by acute oxygen and/or glucose deprivation or by elevation

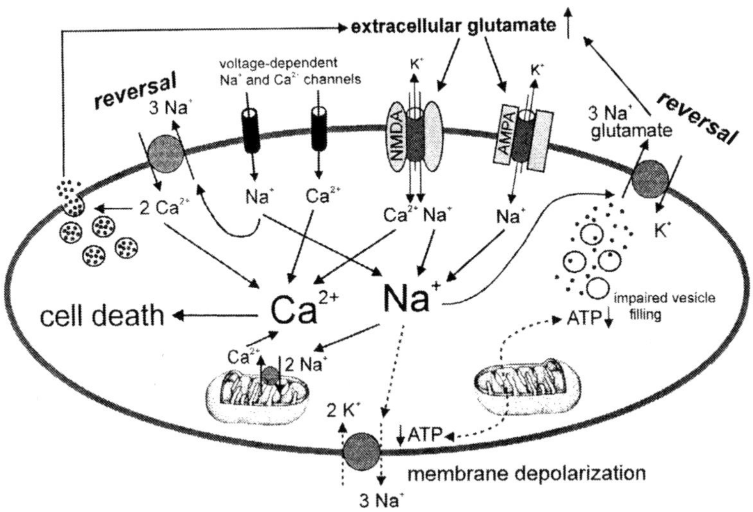

Figure 1. Mechanisms and consequences of Na^+ overload during acute brain injury.

of extracellular K^+. It is noteworthy that SD-like episodes evoked in vitro are electrophysiologically fully reversible, suggesting that the severe ionic and metabolic disturbances that completely silence electrical activity do not necessarily cause lasting neuronal damage at this early stage of a (time-limited) energy deprivation. The typical course of a SP-like episode in a CA1 pyramidal cell of the hippocampal slice is illustrated in Figure 2A (all experiments of this figure from M ller and Somjen, Ref. 8). Initially, hypoxia induces a small hyperpolarization with a concomitant decrease in input resistance resulting from the opening of ATP-dependent K^+ channels, Ca^{2+}-dependent K^+ channels and/or two-pore domain (2P) K^+ channels of the TREK/TRAAK subgroup.[9,10] The voltage trajectory then gradually depolarizes until a threshold potential is reached (b), from where the membrane potential rapidly decays to about -20 mV (c). This fast segment of the hypoxic depolarization is followed by a slow decline to almost 0 mV (d). With the restoration of oxygenation, the neuron gradually recovers. The subsequently illustrated experiments of Figure 2 support the above outlined scheme and underscore the importance of Na^+ influx in the early pathogenesis of acute brain injury: Firstly, reduction of extracellular Na^+ to 90 mM substantially increased the delay until the threshold for the rapid depolarization was reached (Fig. 2B). Secondly, blockade of AMPA/kainate and NMDA receptors by DNQX and CPP, respectively, also slowed, but to a lesser extent, the gradual depolarization towards the threshold where the fast depolarization takes off (Fig. 2C). Thirdly, additional blockade of voltage-dependent Na^+ channels with tetrodotoxin (TTX) abrogated a self-regenerative SD-like episode in about 40% of neurons tested.

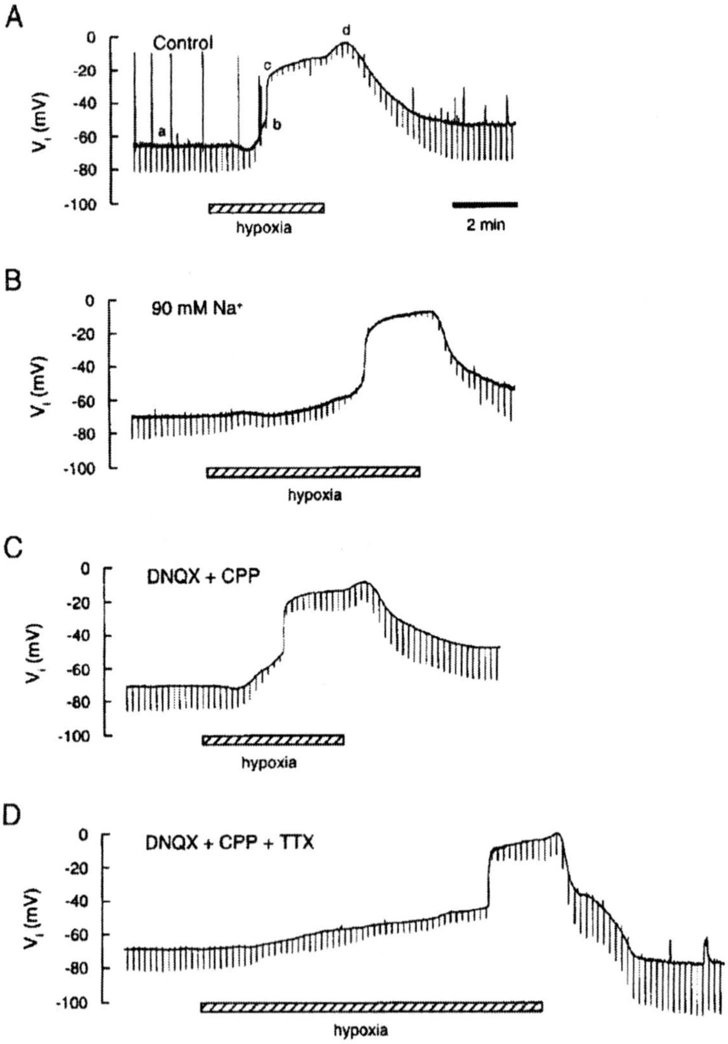

Figure 2. Contribution of voltage-dependent Na^+ channels and ionotropic glutamate receptors to hypoxia-induced depolarization of membrane potential of a hippocampal CA1 neuron in vitro. Voltage traces show hypoxic spreading depression (SD) under control conditions (A), in low external Na^+ solution (B), in the presence of the non-NMDA and NMDA receptor antagonists DNQX and CPP, respectively (C), and after further addition of the Na^+ channel blocker TTX (D). For explanation, see text. Reprinted with permission from M ller M & Somjen GG. J Neurophysiol 2000; 84:1869-1880.

In the remaining neurons, the onset of the rapid depolarization was more than sixfold postponed and its threshold was substantially increased (Fig. 2D). These experiments clearly indicate that Na^+ influx through voltage-gated Na^+ channels and ionotropic glutamate receptors is the essential pathophysiological mechanism triggering SD-like episodes in hypoxic neurons. Once the threshold is reached, however, other yet unknown processes are set in motion that proceed in an all-or-none fashion. These data suggest that the therapeutic window where Na^+ channel blockers will prove beneficial after the onset of stroke is relatively short. We should remember, however, that recurrent waves of spreading depression in vivo are considered a main route along which secondary brain damage migrates into marginal zones of the initial damage. Thus, inhibition of Na^+ loading of metabolically compromised neurons in the penumbra of the lesion should help to confine the extent of secondary brain damage.

Persistent Na^+ Current (I_{NaP})

Since a detailed account of the role of ionotropic glutamate receptors in acute brain damage is given elsewhere in this book (Chapter 1), I will focus this Chapter on voltage-dependent Na^+ channels as a pathway for pathological Na^+ accumulation in the cell. Direct measurements of intracellular ion distributions with the use of electron probe X-ray microanalysis in hippocampal slices exposed to combined oxygen/glucose deprivation strongly implicated Na^+ loading of neurons via voltage-gated, TTX-sensitive Na^+ channels as a critical event in ischemia-induced neuronal damage.[11] Given that the fast Na^+ current underlying the upstroke of the action potential typically inactivates within a millisecond or less upon depolarization, how can Na^+ channels generate a sustained Na^+ influx? In addition to producing the fast Na^+ current, voltage-dependent Na^+ channels of CNS neurons have been known since the early 1980«s to give rise to a small,TTX-sensitive Na^+ current, which does not (or only very slowly) inactivate. Because this sustained or persistent Na^+ current (I_{NaP}) is activated 10-15 mV below firing threshold, it adds depolarizing current to excitatory synaptic potentials, facilitates action potential generation, shortens interspike intervals, contributes to subthreshold membrane potential oscillations, and promotes burst discharges (reviewed in ref. 12). By now, I_{NaP} has been found in neurons of various CNS regions, including neocortex,[13-17] hippocampus,[18] entorhinal cortex,[19-21] thalamus,[22,23] striatum,[24,25] and cerebellum.[26,27]

The nature and single-channel mechanism of I_{NaP} has remained controversial since the current was first described.[12,28] The fact that I_{NaP} is activated at a more negative voltage range than the fast Na current was long thought to argue in favor of the two currents being mediated by different Na^+ channel subtypes. However, experiments using enzymatic removal of fast inactivation showed that the different voltage dependence of transient Na^+ current and I_{NaP} is well compatible with the hypothesis that the two currents arise from the same population of Na^+ channels.[29] In fact, single-channel recordings from pyramidal neurons of rat neocortex demonstrated that Na^+ channels display modal gating giving rise to both early and

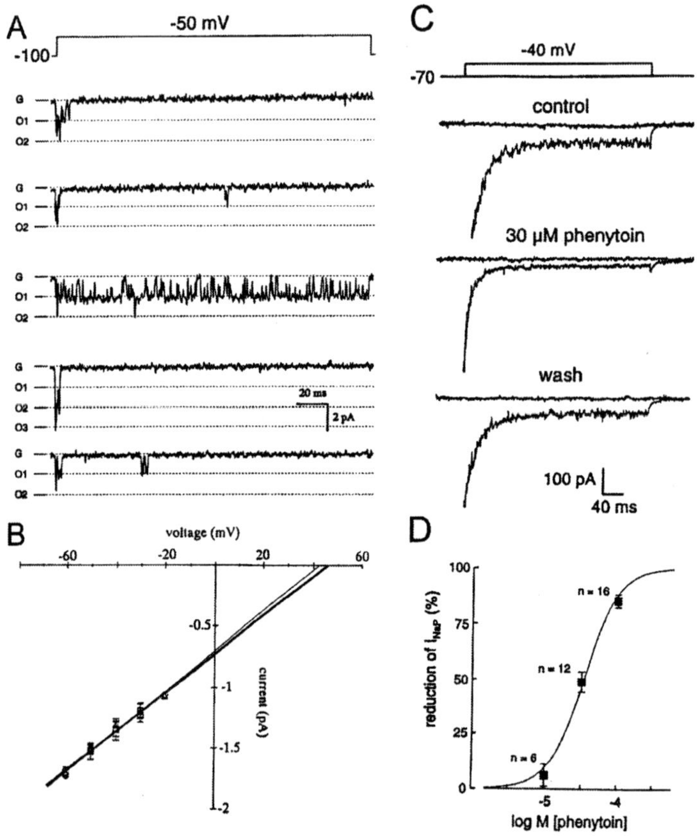

Figure 3. Ionic mechanism and pharmacological manipulation of persistent Na current (I_{NaP}). A: Modal gating of single Na^+ channels during sustained depolarization from -100 to -50 mV. The gating behavior of single Na^+ channels was recorded in the cell-attached configuration of the patch-clamp technique from an acutely isolated pyramidal cell of rat neocortex. The current traces illustrate three types of openings: (i) Transient, partially overlapping openings immediately following the onset of the depolarizing voltage step (all traces). (ii) Brief late openings (second and fifth trace). (iii) Sustained burst of openings resulting from temporary failure of regular inactivation (third trace). B: I-V relations for Na^+ channels mediating early transient openings (triangles), brief late openings (circles) and sustained burst openings (squares) suggest that all events arise from an electrophysiologically uniform population of Na^+ channels. C: Whole-cell recording of I_{NaP} in the same preparation after pharmacological suppression of voltage-dependent K^+ and Ca^{2+} channels. The sustained inward current (I_{NaP}) that remained after inactivation of fast Na^+ current is reversibly reduced by 30 μM phenytoin. D: Dose-response curve for inhibition by phenytoin of I_{NaP} (EC_{50} 34 μM). Reprinted with permission from Alzheimer C, Schwindt PC & Crill WE. J Neurosci 1993; 13:660-673 [copyright 1933 by the Society for Neuroscience] (B), and Chao TI & Alzheimer C. NeuroReport 1995; 6:1778-1780 (C, D).

(rare) late openings.[30] As shown in Figure 3A (middle trace), individual Na^+ channels recorded from the soma of a neocortical pyramidal cell occasionally fail to inactivate, producing a burst of openings during sustained depolarization, which give rise to I_{NaP} on the whole-cell level. Since the Na^+ channels underlying the early transient and the late openings did not differ in single channel conductance or ion selectivity (Fig. 3B), both currents are likely to be generated by an electrophysiologically (but not necessarily molecular) uniform population of Na^+ channels. The situation might be different in entorhinal cortex, where layer II principal cells appear to express Na^+ channel subtypes with a particular high probability of late openings.[31] With the advent of molecular cloning techniques and the use of heterologous expression systems to study the properties of the various Na^+ channel subunits (reviewed in ref. 32), a picture emerges that might reconcile apparently conflicting data on the nature of I_{NaP} in native preparations. It seems that all Na^+ channel subtypes are capable of modal gating with transient and late gating modes. Some subtypes, however, appear to be particularly prone to sustained late openings. Thus the pattern of Na^+ channel subtype expression in a particular neuron might be crucial in determining the size of its I_{NaP}. In the context of this review, an interesting question would be whether larger I_{NaP} confers higher susceptibility to hypoxic damage.

It is still controversial whether the properties of I_{NaP} are altered during ischemia. In hippocampal CA1 pyramidal cells, both hypoxia and inhibition of oxidative metabolism by sodium cyanide produced a significant increase of I_{NaP}, while having little effect on the fast Na^+ current.[33] This upregulation of I_{NaP} appears to result from the enhanced production of nitric oxide (NO) during ischemia.[34] Single-channel recordings suggest that persistent Na^+ channels can directly sense O_2 levels, with the redox reaction involved in increasing Na^+ channel activity during hypoxia presumably occurring in an associated regulatory protein.[35] Hypoxic augmentation of I_{NaP} (and of fast Na^+ current) was also reported from rat caudal hypothalamic neurons.[36] Additional evidence for the presumed hypoxia-induced upregulation of I_{NaP} comes from experiments in which elevation of K^+ in the bath solution (without concomitant hypoxia) reversibly augmented I_{NaP} in hippocampal neurons.[37] Remember that a rise in extracellular K^+ is the earliest ionic deviation after the onset of hypoxia.[2] Whereas these findings suggest that enhancement of I_{NaP} exacerbates Na^+ overload under hypoxic conditions, this concept is not generally accepted. In marked contrast to the above studies, hypoxia was found to affect neither I_{NaP} (Astman et al, Soc Neurosci Abstr 25, 288.12, 1999) nor the fast Na^+ current in layer V neurons from rat neocortex.[38] Interestingly, hypoxia did enhance I_{NaP} only when fluoride was the main anion in the pipette solution, but failed to do so when chloride or gluconate were substituted for fluoride, which might explain some of the divergent findings between the different laboratories (Astman et al, Soc Neurosci Abstr, 25,288.12, 1999). Finally, oxygen deprivation was reported to decrease fast Na^+ currents in CA1 neurons of rat hippocampus.[39] This inhibitory effect of hypoxia on Na^+ channel activity was ascribed to an upregulation of PKC. In that study, the effect of hypoxia on I_{NaP} was not directly investigated, but since PKC activation is known to reduce I_{NaP},[40] one would expect a parallel decrease in I_{NaP}.

In summary, there is ample evidence that TTX-sensitive, voltage-dependent Na^+ channels giving rise to I_{NaP} represent a significant pathway for pathological Na^+ loading. It remains to be determined, however, whether ischemia exerts a direct modulatory action on I_{NaP}, be it an upregulation in a vicious cycle-like fashion, or a downregulation as a compensatory mechanism.

Na^+ CHANNEL BLOCKERS AS NEUROPROTECTIVE AGENTS

From the above, it is obvious that a blockade of Na^+ channels underlying I_{NaP} should represent a promising target to prevent (or at least slow) the collapse of the cellular ion gradients. The major beneficial actions of suppressing Na^+ influx during ischemia can be summarized as follows.

1. Reducing the Na^+ load will decrease the energy demand and attenuate the metabolic stress imposed by the impaired energy supply, allowing the neuron to use the remaining ATP molecules for neuroprotective mechanisms other than fueling the Na^+/ K^+ antiporter system.

2. With the transmembrane Na^+ gradient preserved, the driving force for the Na^+/Ca^{2+} exchanger and the Na^+-dependent glutamate uptake system are maintained and both systems can operate in normal mode. Thus the neuron remains capable of extruding Ca^{2+} instead of becoming loaded with Ca^{2+}, and a nonsynaptic release of glutamate into the extracellular space is prevented.

3. With less Na^+ entering the cell, release of Ca^{2+} from mitochondria via the mitochondrial Na^+ /Ca^{2+} exchanger is diminished, which abates the ischemia-induced rise in cytosolic Ca^{2+}.

4. Inhibition of Na^+ influx attenuates the ischemic depolarization, thereby promoting the voltage-sensitive block of NMDA receptor channels by Mg^{2+} and reducing Ca^{2+} entry through voltage-dependent Ca^{2+} channels.

5. On the presynaptic side, partial block of axonal Na^+ channels seems to preferentially diminish the release of the excitatory neurotransmitter glutamate, without appreciably affecting GABAergic synaptic inhibition. This important and novel finding will be discussed in more detail in the following paragraph.

Pharmacological Features of Na^+ Channel Inhibitors Used for Neuroprotection

Before I come to in vitro experiments, animal studies and clinical trials with patients that were inspired by the above assumptions, I shall elaborate on the pharmacological properties of the drugs administered in this neuroprotective strategy.

Phenytoin

Since late Na^+ channel openings producing I_{NaP} on the whole-cell level have been advanced as the predominant pathway of pathological Na^+ influx during ischemia and energy deprivation, the ideal Na^+ channel blocker should preferentially diminish I_{NaP} without interfering with fast Na^+ current. Fortunately, a number of Na^+ channel blockers do indeed affect Na^+ channel gating in a manner that closely meets this requirement. I will use phenytoin as the prototype drug to explain how the voltage-dependence and kinetics of drug binding to Na^+ channels accounts for its therapeutical benefits.

Phenytoin is widely used in the antiepileptic treatment of patients with generalized tonic-clonic seizures and partial seizures. It is generally believed that phenytoin exerts its therapeutic effects primarily by blocking voltage-dependent Na^+ channels. Clinically most relevant, this block occurs in a frequency- and voltage-dependent fashion that allows phenytoin to suppress epileptiform discharges without disrupting regular spiking activity. A detailed electrophysiological analysis of the action of phenytoin on Na^+ currents of hippocampal neurons showed that the drug binds tightly but slowly to fast inactivated states of the Na^+ channel.[41] According to that study, the action of phenytoin on Na^+ channel gating resembles that of the normal inactivation particle, the major difference being that phenytoin binding and unbinding are much slower than the development and recovery from fast inactivation.[41] In functional terms, phenytoin helps to stabilize Na^+ channels in inactivated states from which the probability of opening is small.[42,43]

Since I_{NaP} is produced by Na^+ channels that temporarily fail to inactivate during sustained depolarization, phenytoin would be expected to diminish the occurrence of such failures and hence decrease I_{NaP}. As shown in Figure 3C, phenytoin indeed produced a significant and reversible reduction of I_{NaP} responses in neocortical pyramidal cells, which were evoked by prolonged depolarizing voltage steps.[44] Half-maximal suppression of I_{NaP} was achieved at 34 μM phenytoin (Fig. 3D). Therapeutic concentrations of orally administered phenytoin in epileptic patients might be well in the range where a considerable reduction of I_{NaP} would be expected. In contrast, tonic (use-independent) blockade of the fast Na^+ current requires several-fold higher concentrations when examined at the neurons« resting potential. Owing to the slow binding kinetics of phenytoin, the strong, but very brief depolarizations during normal action potential firing are insufficient for appreciable drug binding. During ischemia-induced depolarization, however, binding of phenytoin to Na^+ channels is strongly promoted. Consequently, the probability of late Na^+ channel openings that are thought to reflect incomplete inactivation is substantially reduced,[45] and I_{NaP} declines.

Riluzole and Other Novel Na^+ Channel Inhibitors

A similar mode of action was described for the relatively novel neuroprotective drug riluzole, which has both anti-epileptic and anti-ischemic properties.[46] Riluzole was found to bind selectively to inactivated rat brain IIA Na^+ channels ($Na_v1.2$) expressed in Xenopus oocytes and to inactivated Na^+ channels in myelinated nerve fibers, with only the latter effect occurring in a use-independent fashion.[47,48] As predicted by its mode of action, riluzole was found to produce a dose-dependent inhibition of I_{NaP}.[49] As for phenytoin, the effect on I_{NaP} was observed in a concentration range substantially lower than that required for inhibition of the fast Na^+ current. Since the recommended therapeutic concentration of riluzole in patients falls into the range where riluzole inhibits I_{NaP} but not fast Na^+ current, it seems reasonable to assume that suppression of I_{NaP} contributes to the neuroprotective action of riluzole.

In addition to its action on Na^+ channels, riluzole was also found to interfere with glutamate release.[50] Its anti-glutamate effects are believed to account for the reported efficacy of riluzole in treating amyotrophic lateral sclerosis (ALS, Lou Gehrig disease), a disorder where the loss of motoneurons might be causally linked to glutamate-mediated slow neurodegeneration. For a long time, the issue of whether riluzole directly affects glutamate release remained controversial. A recent study provides evidence that inhibition by riluzole of Na^+ channels fully accounts for its anti-glutamate effects.[51] Interestingly, the same concentration of riluzole (20 μM) that almost completely suppressed excitatory postsynaptic currents (EPSCs), barely affected GABAergic inhibitory postsynaptic currents (IPSCs). That this surprising effect did indeed result from inhibition of Na^+ channels was confirmed by experiments, in which a low concentration of TTX (30 nM) closely mimicked the differential action of riluzole on the two types of synaptic currents. A preferential suppression of EPSCs over IPSCs was also observed for the Na^+ channel blockers phenytoin (2.5 μM) and lamotrigine (10 μM), although their effects on EPSCs were substantially smaller. The partial block of axonal Na^+ channels most likely produces an altered action potential waveform, giving rise to less Ca^{2+} influx in the presynaptic terminal. Why would this only happen at terminals of excitatory, but not inhibitory fibers? The findings of the above study[51] are compatible with the notion that inhibitory synapses have a higher safety factor than excitatory synapses, possibly resulting from a higher density of axonal Na^+ channels in inhibitory neurons compared to excitatory neurons.

While the exact mechanisms rendering excitatory synapses more susceptible to partial block of Na^+ channels await further study, the selective attenuation of glutamatergic synaptic transmission should significantly contribute to both the anticonvulsant and the neuroprotective efficacy of Na^+ channel inhibitors. This property should prove particularly beneficial in slow neurodegenerative diseases like ALS, which are proposed to involve prolonged glutamatergic overstimulation of motoneurons. When extrapolating the data of the above study [51] on in vivo conditions, one should bear in mind, however, that these experiments were performed on solitary microculture neurons that make synapses with themselves, so-called autapses. In acute preparations with a largely preserved neuronal network such as the hippocampal slice and even more so in the intact brain, the various factors controlling transmitter release might be weighted in a different fashion. For example, in the hippocampal slice, effective neuroprotection by phenytoin was achieved at concentrations that did not alter excitatory synaptic transmission,[52] suggesting that the selective depression by phenytoin of presynaptic glutamate release described in microcultured neurons might not be a prerequisite for the drug«s neuroprotective action in other systems.

To complete this section, I should mention that several novel compounds such as lamotrigine, the lamotrigine derivative BW619C89, and lifarizine, which all displayed a neuroprotective profile in animal models of acute stroke (see below), are believed to exert their therapeutic action mainly through suppression of Na^+ channels (reviewed in ref. 53). Like riluzole, all three drugs block Na^+ channels by a preferential interaction with inactivated states of the Na^+ channels.[54-56] Although this issue has not been investigated yet, I would hence predict that these compounds all act as potent inhibitors of I_{NaP}.

Neuroprotective Potential of Na^+ Channel Inhibitors

Given the central role of pathological Na^+ influx in initiating subsequent neuronal damage, the ability of various Na^+ channel blockers to prevent Na^+ loading of energy-deprived neurons without disrupting electrical signaling in intact brain regions makes these compounds promising candidates for therapeutic interventions. It is hence not surprising that the last decade witnessed a surge of studies probing the neuroprotective potential of Na^+ channel inhibitors in various in vitro and in vivo models of acute brain damage. Since excellent and comprehensive overviews of these studies have been published recently,[53,57-59] I ask the interested reader to consult these reviews for a detailed account and a complete list of references. In the following, I want to highlight some, as I believe, representative findings.

In cultured neocortical neurons exposed to combined oxygen/glucose deprivation, TTX alone had only a limited neuroprotective effect.[60] In contrast, ionotropic glutamate receptor antagonists reduced neuronal death to a much larger extent than TTX, with TTX displaying mainly an additive neuroprotective effect if applied together with the glutamate antagonists. This bias towards glutamate antagonists might be attributed to the small amplitude of I_{NaP} in these cultures, which predicts that activation of I_{NaP} should represent only a minor route of pathological Na^+ influx. By contrast, in organotypic slice cultures of rat hippocampus, the noncompetitive NMDA receptor antagonist MK-801 and TTX afforded quite similar degrees of protection against energy deprivation, if applied during the insult. If administered during the recovery period, however, only MK-801 was capable of reducing neuronal damage, although to a lesser extent, whereas TTX was ineffective.[61,62] This implicates Na^+ channels in the early phase of neuronal damage, whereas excitotoxic NMDA receptor activation seems to extend well into the recovery period. In support of this notion, phenytoin was found to delay ischemic depolarization in the rat hippocampal slice, but did not allow for recovery of synaptic function once ischemic depolarization had fully developed.[63] However, when hippocampal slices were subjected to short periods (10-15 min) of hypoxia or combined hypoxia/hypoglycemia, phenytoin did afford effective protection against the consequences of energy deprivation.[64,65]

A vast number of animal studies strengthened the link between Na^+ channel inhibition and neuroprotection. In gerbils subjected to 5 min of forebrain ischemia, phenytoin pretreatment provided significant protection of hippocampal CA1 neurons.[64-66] In a clinically more relevant experimental setting, phenytoin (100 mg/kg i.p.) and lamotrigine (50 mg/kg i.p.) were administered twice to rats 30 min and 24.5 h after middle cerebral artery occlusion. Under these conditions, the two drugs reduced infarct size by 40% and 28%, respectively.[64,67] Very similar results were reported from rats subjected to tandem occlusion of the middle cerebral and the ipsilateral common carotid arteries.[68] Phenytoin (28 mg/kg i.v.) and the noncompetitive NMDA-receptor antagonist MK-801 (0.1 mg/kg i.v.) injected 30 min and 24 h after arterial occlusion significantly reduced infarct volume by 45% and 32%, respectively (notably, higher concentrations of MK-801 failed to decrease infarct size). Consistent with data from in vitro studies, a single injection of phenytoin (28 mg/kg) 30 min after occlusion was neuroprotective, whereas a delayed administration (>2 h after occlusion) was ineffective. I had already mentioned that riluzole was approved for the treatment of ALS. Findings from several animal models of stroke demonstrate that riluzole has also a significant anti-ischemic action.[46,69] Whereas the anticonvulsant lamotrigine displays only a modest neuroprotective potency against ischemia, its derivative BW619C89 emerged as an efficient anti-ischemic agent in several models of global and focal ischemia. Intravenous application of this compound (20 mg/kg) 5 min after occlusion of middle cerebral artery reduced total infarct volume by 57%.[70] BW619C89 (30 mg/kg) was also found to attenuate the damage produced in a rat model of traumatic brain injury (lateral fluid percussion).[71] Similarly, the related Na^+ channel blocker, BW1003C87 (10 mg/kg, i.v. infusion for 15 min), applied 15 min after a fluid percussion injury of moderate severity over the left parietal cortex, produced a significant reduction of focal brain edema.[72] Assuming that traumatic and ischemic forms of brain damage share common pathogenetic mechanisms at the cellular level, these data would encourage the administration of Na^+ channel inhibitors in traumatic brain injury.

In summary, the in vitro studies and the animal experiments lend support to the idea that voltage-gated Na^+ channels represent a significant and promising target in the early treatment of stroke and brain trauma. Compared with other treatment strategies to prevent excessive cation influx, i.e., application of ionotropic glutamate receptor antagonists or Ca^{2+} channel blockers, Na^+ channel inhibitors have two distinct advantages: Firstly, unlike the former, many of the Na^+ channel inhibitors afford neuroprotection at plasma levels that do not cause appreciable side effects in animals, as estimated from spontaneous locomotor patterns and cardiovascular responses.[68,73] Secondly, in contrast to glutamate antagonists, Na^+ channel blockers are also capable of rescuing axons from hypoxic damage in vitro.[74] Protection of white matter tracts might thus emerge as an additional benefit of Na^+ channel inhibitors in stroke and trauma.

Although Na^+ channel blockers hence appear to meet all criteria for successful use in patients, it is disappointing to learn that so far, clinical trials have not yet demonstrated convincing benefits in stroke patients. In a pilot study of the lifarizine study group,[75] lifarazine modestly decreased mortality from stroke in the treated group and slightly improved neurological outcome at 3 months. A phase II clinical trial of BW619C89 (sipatrigine) by continuous infusion in acute stroke was recently completed. No positive effects on outcome measures at 30 days or 3 months were demonstrated. None of the placebo patients, but 16 of 21 patients receiving sipatrigine suffered from adverse neuropsychiatric effects so that drug infusion had to be discontinued in 7 patients.[76] The results of a European phase II trial in acute stroke of sipatrigine scheduled to begin late 1999 have not yet been published. For phenytoin, no clinical development in stroke is currently pursued. A phase III trial in patients with acute ischemic stroke has been completed with no significant differences between placebo and fosphenytoin at 3 months. These results have not been published as of January 2001. The above pieces of information were obtained from an excellent website on stroke and related issues maintained by the Washington University at St. Louis (www.strokecenter.org). The interested reader should consult this site for a comprehensive and up-to-date coverage of ongoing clinical trials in stroke.

It is not clear why the clinical trials conducted so far have failed to detect beneficial actions of Na^+ channel inhibitors. Possibly, the delay between the ischemic insult and the onset of infusion, which was typically 6-12 h in the above trials, renders the drugs largely ineffective, as predicted by the pathophysiology of Na^+-dependent neuronal damage and the animal studies mentioned above.

Ca^{2+} CHANNELS

In a variety of pathological settings including stroke, cardiac arrest and traumatic brain injury, Ca^{2+} overload of neurons is generally considered the critical event triggering the various Ca^{2+}-dependent processes that eventually lead to neuronal death.[77-79] As already illustrated in the schematic drawing of Figure 1, several of the pathways producing elevated cytosolic Ca^{2+} levels are directly or indirectly coupled to excessive Na^+ influx, such as reversal of the transmembrane Na^+/Ca^{2+} exchanger, activation of the mitochondrial Na^+/Ca^{2+} antiporter, and Na^+-dependent membrane depolarization with subsequent opening of voltage-gated Ca^{2+} channels. Additional sources of cytosolic Ca^{2+} rises include Ca^{2+} entry through NMDA receptors and non-NMDA receptors lacking the GluR2 subunit, as well as IP_3-mediated release of Ca^{2+} from intracellular stores following activation of group I metabotropic glutamate receptors. Finally, the declining activity of Ca^{2+}-ATPases following ATP depletion impairs the removal of the accumulated cytosolic Ca^{2+}. The various mechanisms of Ca^{2+} overload offer a number of potential targets for neuroprotection, which are covered in several Chapters of this book. Here, I will discuss whether inhibition of voltage-gated Ca^{2+} channels is a promising strategy to abate the consequences of ischemic insult and traumatic brain injury.

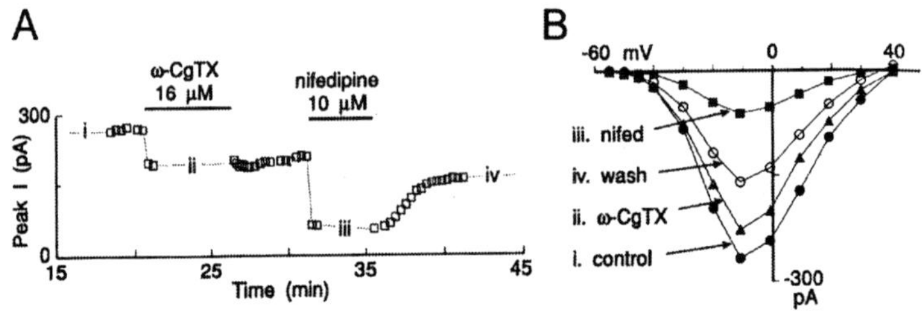

Figure 4. Whole-cell recording from acutely isolated human neocortical pyramidal cell reveals ω-CgTX-sensitive and nifedipine-sensitive components of high threshold Ca^{2+} current. A: High threshold currents were evoked by step depolarizations from -50 to -13 mV and the amplitude of peak current was plotted vs. time in the absence and presence of the channel inhibitors as indicated. B: I-V relations for the high threshold currents were determined at the time points indicated by like-numbers in A. Reprinted with permission from Sayer RJ, Brown AM, Schwindt PC & Crill WE. J Neurophysiol 1993; 69:1596-1606.

Ca^{2+} Channel Blockers and Neuroprotection

Once cytosolic Ca^{2+} had been pinpointed as the pivotal trigger of neuronal death in various forms of acute brain damage, the availability of organic Ca^{2+} channel blockers already approved for cardiovascular diseases made these drugs prime tools of neuroprotection and spawned a series of preclinical studies and clinical trials. A number of in vitro and in vivo studies provided indeed encouraging results, but, as a whole, the findings were often controversial and inconsistent across species and lesion models.[80-82] Even worse, a recent meta-analysis of published studies revealed that clinical trials with organic Ca^{2+} channel blockers all failed to produce positive effects on several outcome measures in stroke patients.[83]

Diversity of Ca^{2+} Channels

To better understand the pitfalls and potential benefits of Ca^{2+} channel inhibition in neuroprotection, the diversity of voltage-gated Ca^{2+} channels has to be taken into account.

In their basic structure closely resembling voltage-gated Na^+ channels (Na_v), voltage-gated Ca^{2+} channels are large heteromultimeric complexes of four or five distinct subunits.[84] They are composed of a_1 as the major subunit, and α_2, β, δ and (specifically for skeletal muscle) γ as auxiliary subunits. The a_1 subunit is the central functional element of the channel incorporating the conduction pore, the voltage sensor and gating apparatus, and binding sites for several drugs, toxins and regulatory molecules. While the accessory subunits may influence amplitude and kinetics of the current and, in some cases, confer sensitivity to channel modulators, the basic electrophysiological and pharmacological properties of Ca^{2+} channels are determined by the a_1 subunit. The diversity of native Ca^{2+} channels in excitable tissue is hence mainly attributable to the multiplicity of α_1 subunits, which are, in mammals, encoded by at least 10 distinct genes. A panel of expert Ca^{2+} channel researchers has recently proposed that the old nomenclatures should be replaced with one based on the structural relationships among the various α_1 subunits.[85] According to this novel nomenclature, the Ca_v1 family ($Ca_v1.1$-1.4) comprises the subunits α_{1S}, α_{1C}, α_{1D}, α_{1F}, which give rise to L-type currents, the Ca_v2 family ($Ca_v2.1$-2.3) comprises the subunits α_{1A}, α_{1B}, α_{1E}, which give rise to P/Q-type, N-type and R-type currents, respectively. Finally, the Ca_v3 family ($Ca_v3.1$.-3.3) includes subunits a_{1G}, a_{1H}, a_{1I}, which mediate T-type currents. With the exception of the T-type current, which is transient and requires only small depolarization from rest for activation (low-threshold Ca^{2+} current), the remaining Ca^{2+} currents share the common characteristics that they require stronger depolarization from rest for activation (high-threshold Ca^{2+} currents) and inactivate, apart from R-type channels, with much slower kinetics.

Only L-type channels are blocked by organic Ca^{2+} channel blockers such as dihydropyridines, phenylalkylamines and benzothiazepines, whereas Ca^{2+} channels of the N-, P/Q-, and R-type are blocked by specific polypeptide toxins isolated from spider and snail venoms. T-type channels are insensitive to L-type blockers and venom toxins, but have been reported to be selectively inhibited by the benzimidazole mibefradil.[86] L-type channels are expressed in various types of non-neuronal tissues as well as in neurons, where they contribute predominantly to somatodendritic Ca^{2+} influx. By contrast, N-, P/Q- and R-type channels are expressed primarily in neurons, where they are mainly involved in neurotransmitter release, but also participate in postsynaptic Ca^{2+} influx. Figure 4 illustrates the relative contribution of nifedipine-sensitive L-type current and ω-conotoxin GVIA-sensitive N-type current to the total high-threshold Ca^{2+} current of a pyramidal neuron soma acutely isolated from human neocortex.[87] These data are in agreement with immunohistochemical data showing expression of both L-type and N-type Ca^{2+} channels in the soma and dendrites of pyramidal neurons, with L-type channels clustered at the base of dendrites and N-type channels expressed more diffusely.[88-90]

Selective Blockers of Neuronal Ca^{2+} Channels

As already mentioned above, clinical trials consistently failed to detect positive effects of organic (L-type channel) blockers in stroke patients. Owing to the almost ubiquitous expression of L-type Ca^{2+} channels, a major shortcoming of L-type current blockers in neuroprotection is their lack of selectivity for neuronal Ca^{2+} channels. Cardiovascular liabilities might thus account, at least in part, for the inconsistent and sometimes even adverse effects of these compounds in acute brain damage. Whereas there seems no indication for L-type Ca^{2+} channel blockers in stroke and brain injury, the development of inhibitors specific for neuronal Ca^{2+} channels inspired new hope. In fact, SB201823 and SB206284, two compounds which target neuronal Ca^{2+} channels without displaying any particular subtype selectively, showed substantial efficacy in a number of ischemia models in vivo, lending credence to the idea that neuroprotective effects of Ca^{2+} channel antagonists occur independent of effects on hemodynamic parameters.[80]

An even more promising candidate is ziconotide (SNX-111), a synthetic ω-conotoxin MVIIA, which selectively inhibits N-type channels. Unlike most polypeptide toxins, which act as rather irreversible Ca^{2+} blockers, ziconotide shows slow unbinding in vitro. Several studies demonstrated considerable neuroprotective effects of ziconotide in vitro and in vivo. For example, in organotypic hippocampal slice cultures subjected to 3 h of oxygen deprivation, ziconotide prevented neuronal damage not only if preincubated, but also if application was delayed until after the hypoxic episode.[91] By contrast, neither nifedipine, a dihydropyridine, nor the mixed neuronal Ca^{2+} channel blocker SB201823A were protective under these experimental conditions. Sending a note of caution, however, ziconotide failed to protect neurons of the same preparation, if hypoxia was combined with glucose deprivation, a paradigm that is more likely to mimic ischemia than hypoxia alone. A neuroprotective action of ziconotide in animal models of stroke was first reported by Valentino et al (ref. 92), who found that the N-type channel blocker rescued hippocampal CA1 neurons from forebrain ischemia in the rat model of four-vessel occlusion. Most notably, a single bolus injection (i.v.) of ziconotide provided neuroprotection up to 24 h after the ischemic insult, a result confirmed in an independent study.[93] The mechanisms underlying ziconotide«s neuroprotective effect have remained controversial. In the study by Valentino et al (ref. 92), the neuroprotective efficacy of ziconotide did not correlate with its ability to inhibit glutamate release. By contrast, a study using rat focal cerebral ischemia attributed the reduction in infarct volume produced by ziconotide to the concomitant reduction in glutamate release.[94] Demonstrating its therapeutic potential in acute CNS trauma, ziconotide (again administered as a single bolus injection i.v.) produced a significant improvement of mitochondrial function in a rat model of traumatic brain injury when given at time intervals ranging from 15 min before injury up to 10h after injury, with the highest improvement between 2 and 6 h post injury.[95,96] The unexpectedly long window of opportunity for therapeutic intervention after stroke and brain trauma was a consistent finding in all animal studies with ziconotide, bearing particular significance for its prospective use in a clinical setting. This is in marked contrast to the much shorter time window (1 h or less) offered by Na^{+} channel inhibitors (see above). As of July 2001, ziconotide was FDA-approved for intrathecal application in chronic pain, but the status of trials for stroke and traumatic brain injury is unclear (www.strokecenter.org).

It is noteworthy that ziconotide penetrates the blood brain barrier (BBB) relatively poorly, thus requiring intravenous levels several orders of magnitude higher than its K_d-value at N-type channels. At these high plasma levels, the compound might cause side effects resulting from sympathetic blockade and histamine release from mast cells.[97] Nevertheless, the neuroprotective efficacy and, in particular, the extended therapeutic window make selective N-type channel blockers promising candidates to prevent neuronal loss in stroke and brain trauma, provided that compounds are developed which pass the BBB more easily.

ACKNOWLEDGEMENTS

I thank Dr. R. Sayer for comments on the manuscript. Support of my work by the Deutsche Forschungsgemeinschaft (DFG) is gratefully acknowledged.

REFERENCES

1. Friedman JE, Haddad GG. Removal of extracellular sodium prevents anoxia-induced injury in freshly dissociated rat CA1 hippocampal neurons. Brain Res 1994; 641:57-64.
2. Martin RL, Lloyd HG, Cowan AI. The early events of oxygen and glucose deprivation: Setting the scene for neuronal death? Trends Neurosci 1994; 17:251-257.
3. Taylor CP, Weber ML, Gaughan CL, Lehning EJ, LoPachin RM. Oxygen/glucose deprivation in hippocampal slices: Altered intraneuronal elemental composition predicts structural and functional damage. J Neurosci 1999; 19:619-629.
4. Zhang Y, Lipton P. Cytosolic Ca2+ changes during in vitro ischemia in rat hippocampal slices: major roles for glutamate and Na+-dependent Ca2+ release from mitochondria. J Neurosci 1999; 19:3307-3315.
5. Rossi DJ, Oshima T, Attwell D. Glutamate release in severe brain ischaemia is mainly by reversed uptake. Nature 2000; 403:316-321.
6. Leao. Further observations on the spreading depression of activity in the cerebral cortex. J Neurophysiol 1947; 10:409-414.
7. Somjen GG. Mechanisms of spreading depression and hypoxic spreading depression- like depolarization. Physiol Rev 2001; 81:1065-1096.
8. M ller M, Somjen GG. Na^+ dependence and the role of glutamate receptors and Na^+ channels in ion fluxes during hypoxia of rat hippocampal slices. J Neurophysiol 2000; 84:1869-1880.
9. Yamamoto S, Tanaka E, Higashi H. Mediation by intracellular calcium-dependent signals of hypoxic hyperpolarization in rat hippocampal CA1 neurons in vitro. J Neurophysiol 1997; 77:386-392.
10. Patel AJ, Honore E. Properties and modulation of mammalian 2P domain K^+ channels. Trends Neurosci 2001; 24:339-346.
11. LoPachin RM. Intraneuronal ion distribution during experimental oxygen/glucose deprivation. Routes of ion flux as targets of neuroprotective strategies. Ann NY Acad Sci 1999; 890:191-203.
12. Crill WE. Persistent sodium current in mammalian central neurons. Annu Rev Physiol 1996; 58:349-362.
13. Stafstrom CE, Schwindt PC, Crill WE. Negative slope conductance due to a persistent subthreshold sodium current in cat neocortical neurons in vitro. Brain Res 1982; 236:221-226.
14. Stafstrom CE, Schwindt PC, Chubb MC, Crill WE. Properties of persistent sodium conductance and calcium conductance of layer V neurons from cat sensorimotor cortex in vitro. J Neurophysiol 1985; 53:153-170.
15. Connors BW, Gutnick MJ, Prince DA. Electrophysiological properties of neocortical neurons in vitro. J Neurophysiol 1982; 48:1302-1320.
16. Alzheimer C, Schwindt PC, Crill WE. Postnatal development of a persistent Na^+ current in pyramidal neurons from rat sensorimotor cortex. J Neurophysiol 1993; 69:290-292.
17. Fleidervish IA, Gutnick MJ. Kinetics of slow inactivation of persistent sodium current in layer V neurons of mouse neocortical slices. J Neurophysiol 1996; 76:2125-2130.
18. French CR, Sah P, Buckett KJ, Gage PW. A voltage-dependent persistent sodium current in mammalian hippocampal neurons. J Gen Physiol 1990; 95:1139-1157.
19. Alonso A, Llinas RR. Subthreshold Na^+-dependent theta-like rhythmicity in stellate cells of entorhinal cortex layer II. Nature 1989; 342:175-177.

20. Magistretti J, Alonso A. Biophysical properties and slow voltage-dependent inactivation of a sustained sodium current in entorhinal cortex layer-II principal neurons: A whole-cell and single-channel study. J Gen Physiol 1999; 114:491-509.
21. Agrawal N, Hamam BN, Magistretti J, Alonso A, Ragsdale DS. Persistent sodium channel activity mediates subthreshold membrane potential oscillations and low-threshold spikes in rat entorhinal cortex layer V neurons. Neuroscience 2001; 102:53-64.
22. Jahnsen H, Llinas R. Ionic basis for the electro-responsiveness and oscillatory properties of guinea-pig thalamic neurones in vitro. J Physiol (Lond) 1984; 349:227-247.
23. Parri HR, Crunelli V. Sodium current in rat and cat thalamocortical neurons: Role of a non-inactivating component in tonic and burst firing. J Neurosci 1998; 18:854-867.
24. Chao TI, Alzheimer C. Do neurons from rat neostriatum express both a TTX-sensitive and a TTX- insensitive slow Na^+ current? J Neurophysiol 1995; 74:934-941.
25. Cepeda C, Chandler SH, Shumate LW, Levine MS. Persistent Na^+ conductance in medium-sized neostriatal neurons: Characterization using infrared videomicroscopy and whole cell patch- clamp recordings. J Neurophysiol 1995; 74:1343-1348.
26. Llinas R, Sugimori M. Electrophysiological properties of in vitro Purkinje cell somata in mammalian cerebellar slices. J Physiol (Lond)1980; 305:171-195.
27. Kay AR, Sugimori M, Llin s R. Kinetic and stochastic properties of a persistent sodium current in mature guinea pig cerebellar Purkinje cells. J Neurophysiol 1998; 80:1167-1179.
28. Taylor CP. Na^+ currents that fail to inactivate. Trends Neurosci 1993; 16:455-460.
29. Brown AM, Schwindt PC, Crill WE. Voltage dependence and activation kinetics of pharmacologically defined components of the high-threshold calcium current in rat neocortical neurons. J Neurophysiol 1993; 70:1530-1543.
30. Alzheimer C, Schwindt PC, Crill WE. Modal gating of Na^+ channels as a mechanism of persistent Na^+ current in pyramidal neurons from rat and cat sensorimotor cortex. J Neurosci 1993; 13:660-673.
31. Magistretti J, Ragsdale DS, Alonso A. High conductance sustained single-channel activity responsible for the low-threshold persistent Na^+ current in entorhinal cortex neurons. J Neurosci 1999; 19:7334-7341.
32. Catterall WA. From ionic currents to molecular mechanisms: The structure and function of voltage-gated sodium channels. Neuron 2000; 26:13-25.
33. Hammarstrom AK, Gage PW. Inhibition of oxidative metabolism increases persistent sodium current in rat CA1 hippocampal neurons. J Physiol (Lond) 1998; 510 (Pt 3):735-741.
34. Hammarstrom AK, Gage PW. Nitric oxide increases persistent sodium current in rat hippocampal neurons. J Physiol (Lond) 1999; 520 Pt 2:451-461.
35. Hammarstrom AK, Gage PW. Oxygen-sensing persistent sodium channels in rat hippocampus. J Physiol (Lond) 2000; 529 (Pt 1):107-118.
36. Horn EM, Waldrop TG. Hypoxic augmentation of fast-inactivating and persistent sodium currents in rat caudal hypothalamic neurons. J Neurophysiol 2000; 84:2572-2581.
37. Somjen GG, Muller M. Potassium-induced enhancement of persistent inward current in hippocampal neurons in isolation and in tissue slices. Brain Res 2000; 885:102-110.
38. Fleidervish IA, Gebhardt C, Astman N, Gutnick MJ, Heinemann U. Enhanced spontaneous transmitter release is the earliest consequence of neocortical hypoxia that can explain the disruption of normal circuit function. J Neurosci 2001; 21:4600-4608.
39. O Reilly JP, Cummins TR, Haddad GG. Oxygen deprivation inhibits Na^+ current in rat hippocampal neurones via protein kinase C. J Physiol (Lond) 1997; 503 (Pt 3):479-488.
40. Alroy G, Su H, Yaari Y. Protein kinase C mediates muscarinic block of intrinsic bursting in rat hippocampal neurons. J Physiol (Lond) 1999; 518 (Pt 1):71-79.
41. Kuo CC, Bean BP. Slow binding of phenytoin to inactivated sodium channels in rat hippocampal neurons. Mol Pharmacol 1994; 46:716-725.
42. Quandt FN. Modification of slow inactivation of single sodium channels by phenytoin in neuroblastoma cells. Mol Pharmacol 1988; 34:557-565.

43. Ragsdale DS, Scheuer T, Catterall WA. Frequency and voltage-dependent inhibition of type IIA Na^+ channels, expressed in a mammalian cell line, by local anesthetic, antiarrhythmic, and anticonvulsant drugs. Mol Pharmacol 1991; 40:756-765.
44. Chao TI, Alzheimer C. Effects of phenytoin on the persistent Na^+ current of mammalian CNS neurones. NeuroReport 1995; 6:1778-1780.
45. Segal MM, Douglas AF. Late sodium channel openings underlying epileptiform activity are preferentially diminished by the anticonvulsant phenytoin. J Neurophysiol 1997; 77:3021-3034.
46. Malgouris C, Bardot F, Daniel M, Pellis F, Rataud J, Uzan A et al. Riluzole, a novel antiglutamate, prevents memory loss and hippocampal neuronal damage in ischemic gerbils. J Neurosci 1989; 9:3720-3727.
47. Hebert T, Drapeau P, Pradier L, Dunn RJ. Block of the rat brain IIA sodium channel alpha subunit by the neuroprotective drug riluzole. Mol Pharmacol 1994; 45:1055-1060.
48. Benoit E, Escande D. Riluzole specifically blocks inactivated Na channels in myelinated nerve fibre. Pfl gers Arch 1991; 419:603-609.
49. Urbani A, Belluzzi O. Riluzole inhibits the persistent sodium current in mammalian CNS neurons. Eur J Neurosci 2000; 12:3567-3574.
50. Martin D, Thompson MA, Nadler JV. The neuroprotective agent riluzole inhibits release of glutamate and aspartate from slices of hippocampal area CA1. Eur J Pharmacol 1993; 250:473-476.
51. Prakriya M, Mennerick S. Selective depression of low-release probability excitatory synapses by sodium channel blockers. Neuron 2000; 26:671-682.
52. Weber ML, Taylor CP. Damage from oxygen and glucose deprivation in hippocampal slices is prevented by tetrodotoxin, lidocaine and phenytoin without blockade of action potentials. Brain Res 1994; 664:167-177.
53. Taylor CP, Meldrum BS. Na^+ channels as targets for neuroprotective drugs. Trends Pharmacol Sci 1995; 16:309-316.
54. McGivern JG, Patmore L, Sheridan RD. Actions of the novel neuroprotective agent, lifarizine (RS-87476), on voltage-dependent sodium currents in the neuroblastoma cell line, N1E-115. Br J Pharmacol 1995; 114:1738-1744.
55. Xie X, Lancaster B, Peakman T, Garthwaite J. Interaction of the antiepileptic drug lamotrigine with recombinant rat brain type IIA Na^+ channels and with native Na^+ channels in rat hippocampal neurones. Pfl gers Arch 1995; 430:437-446.
56. Xie XM, Garthwaite J. State-dependent inhibition of Na^+ currents by the neuroprotective agent 619C89 in rat hippocampal neurons and in a mammalian cell line expressing rat brain type IIA Na^+ channels. Neuroscience 1996; 73:951-962.
57. Obrenovitch TP. Sodium and potassium channel modulators: their role in neuroprotection. Int Rev Neurobiol 1997; 40:109-135.
58. Obrenovitch TP. Neuroprotective strategies: Voltage-gated Na^+-channel down-modulation versus presynaptic glutamate release inhibition. Rev Neurosci 1998; 9:203-211.
59. Urenjak J, Obrenovitch TP. Pharmacological modulation of voltage-gated Na^+ channels: A rational and effective strategy against ischemic drain damage. Pharmacol Rev 1996; 48:21-67.
60. Lynch JJ, III, Yu SP, Canzoniero LM, Sensi SL, Choi DW. Sodium channel blockers reduce oxygen-glucose deprivation-induced cortical neuronal injury when combined with glutamate receptor antagonists. J Pharmacol Exp Ther 1995; 273:554-560.
61. Breder J, Sabelhaus CF, Opitz T, Reymann KG, Schroder UH. Inhibition of different pathways influencing Na^+ homeostasis protects organotypic hippocampal slice cultures from hypoxic/hypoglycemic injury. Neuropharmacology 2000; 39:1779-1787.
62. Tasker RC, Coyle JT, Vornov JJ. The regional vulnerability to hypoglycemia-induced neurotoxicity in organotypic hippocampal culture: Protection by early tetrodotoxin or delayed MK-801. J Neurosci 1992; 12:4298-4308.
63. Watson GB, Lanthorn TH. Phenytoin delays ischemic depolarization, but cannot block its long- term consequences, in the rat hippocampal slice. Neuropharmacology 1995; 34:553-558.

64. Amagasa M, Mizoi K, Ogawa A, Yoshimoto T. Actions of brain-protecting substances against both oxygen and glucose deprivation in the guinea pig hippocampal neurons studied in vitro. Brain Res 1989; 504:87-93.
65. Stanton PK, Moskal JR. Diphenylhydantoin protects against hypoxia-induced impairment of hippocampal synaptic transmission. Brain Res 1991; 546:351-354.
66. Taft WC, Clifton GL, Blair RE, DeLorenzo RJ. Phenytoin protects against ischemia-produced neuronal cell death. Brain Res 1989; 483:143-148.
67. Rataud J, Debarnot F, Mary V, Pratt J, Stutzmann JM. Comparative study of voltage-sensitive sodium channel blockers in focal ischaemia and electric convulsions in rodents. Neurosci Lett 1994; 172:19-23.
68. Boxer PA, Cordon JJ, Mann ME, Rodolosi LC, Vartanian MG, Rock DM et al. Comparison of phenytoin with noncompetitive N-methyl-D-aspartate antagonists in a model of focal brain ischemia in rat. Stroke 1990; 21:III47-III51.
69. Pratt J, Rataud J, Bardot F, Roux M, Blanchard JC, Laduron PM et al. Neuroprotective actions of riluzole in rodent models of global and focal cerebral ischaemia. Neurosci Lett 1992; 140:225-230.
70. Leach MJ, Swan JH, Eisenthal D, Dopson M, Nobbs M. BW619C89, a glutamate release inhibitor, protects against focal cerebral ischemic damage. Stroke 1993; 24:1063-1067.
71. Sun FY, Faden AI. Neuroprotective effects of 619C89, a use-dependent sodium channel blocker, in rat traumatic brain injury. Brain Res 1995; 673:133-140.
72. Okiyama K, Smith DH, Gennarelli TA, Simon RP, Leach M, McIntosh TK. The sodium channel blocker and glutamate release inhibitor BW1003C87 and magnesium attenuate regional cerebral edema following experimental brain injury in the rat. J Neurochem 1995; 64:802-809.
73. Smith SE, Meldrum BS. Cerebroprotective effect of lamotrigine after focal ischemia in rats. Stroke 1995; 26:117-121.
74. Stys PK, Sontheimer H, Ransom BR, Waxman SG. Noninactivating, tetrodotoxin-sensitive Na^+ conductance in rat optic nerve axons. Proc Natl Acad Sci USA 1993; 90:6976-6980.
75. Squire IB, Lees KR, Pryse-Phillips W, Kertesz A, Bamford J. Efficacy and tolerability of lifarizine in acute ischemic stroke. A pilot study. Lifarizine Study Group. Ann NY Acad Sci 1995; 765:317-318.
76. Muir KW, Holzapfel L, Lees KR. Phase II clinical trial of sipatrigine (619C89) by continuous infusion in acute stroke. Cerebrovasc Dis 2000; 10:431-436.
77. Choi DW. Calcium-mediated neurotoxicity: Relationship to specific channel types and role in ischemic damage. Trends Neurosci 1988; 11:465-469.
78. Mattson MP. Calcium and free radicals: Mediators of neurotrophic factor and excitatory transmitter-regulated developmental plasticity and cell death. Perspect Dev Neurobiol 1996; 3:79-91.
79. Xiong Y, Gu Q, Peterson PL, Muizelaar JP, Lee CP. Mitochondrial dysfunction and calcium perturbation induced by traumatic brain injury. J Neurotrauma 1997; 14:23-34.
80. Hunter AJ. Calcium antagonists: Their role in neuroprotection. Int Rev Neurobiol 1997; 40:95-108.
81. Scriabine A. Calcium channel antagonists as neuroprotective agents. In: B r PR, Flint Beal M, eds. Neuroprotection in CNS Diseases. New York: Marcel Dekker Inc., 1997:27-51.
82. Kobayashi T, Mori Y. Ca^{2+} channel antagonists and neuroprotection from cerebral ischemia. Eur J Pharmacol 1998; 363:1-15.
83. Horn J, Limburg M. Calcium antagonists for ischemic stroke: A systematic review. Stroke 2001; 32:570-576.
84. Catterall WA. Structure and regulation of voltage-gated Ca^{2+} channels. Annu Rev Cell Dev Biol 2000; 16:521-555.
85. Ertel EA, Campbell KP, Harpold MM, Hofmann F, Mori Y, Perez-Reyes E et al. Nomenclature of voltage-gated calcium channels. Neuron 2000; 25:533-535.

86. Ertel SI, Ertel EA, Clozel JP. T-type Ca^{2+} channels and pharmacological blockade: Potential pathophysiological relevance. Cardiovasc Drugs Ther 1997; 11:723-739.
87. Sayer RJ, Brown AM, Schwindt PC, Crill WE. Calcium currents in acutely isolated human neocortical neurons. J Neurophysiol 1993; 69:1596-1606.
88. Westenbroek RE, Ahlijanian MK, Catterall WA. Clustering of L-type Ca^{2+} channels at the base of major dendrites in hippocampal pyramidal neurons. Nature 1990; 347:281-284.
89. Westenbroek RE, Hell JW, Warner C, Dubel SJ, Snutch TP, Catterall WA. Biochemical properties and subcellular distribution of an N-type calcium channel alpha 1 subunit. Neuron 1992; 9:1099-1115.
90. Mills LR, Niesen CE, So AP, Carlen PL, Spigelman I, Jones OT. N-type Ca^{2+} channels are located on somata, dendrites, and a subpopulation of dendritic spines on live hippocampal pyramidal neurons. J Neurosci 1994; 14:6815-6824.
91. Pringle AK, Benham CD, Sim L, Kennedy J, Iannotti F, Sundstrom LE. Selective N-type calcium channel antagonist omega conotoxin MVIIA is neuroprotective against hypoxic neurodegeneration in organotypic hippocampal-slice cultures. Stroke 1996; 27:2124-2130.
92. Valentino K, Newcomb R, Gadbois T, Singh T, Bowersox S, Bitner S et al. A selective N-type calcium channel antagonist protects against neuronal loss after global cerebral ischemia. Proc Natl Acad Sci USA 1993; 90:7894-7897.
93. Buchan AM, Gertler SZ, Li H, Xue D, Huang ZG, Chaundy KE et al. A selective N-type Ca^{2+}-channel blocker prevents CA1 injury 24 h following severe forebrain ischemia and reduces infarction following focal ischemia. J Cereb Blood Flow Metab 1994; 14:903-910.
94. Takizawa S, Matsushima K, Fujita H, Nanri K, Ogawa S, Shinohara Y. A selective N-type calcium channel antagonist reduces extracellular glutamate release and infarct volume in focal cerebral ischemia. J Cereb Blood Flow Metab 1995; 15:611-618.
95. Verweij BH, Muizelaar JP, Vinas FC, Peterson PL, Xiong Y, Lee CP. Mitochondrial dysfunction after experimental and human brain injury and its possible reversal with a selective N-type calcium channel antagonist (SNX-111). Neurol Res 1997; 19:334-339.
96. Verweij BH, Muizelaar JP, Vinas FC, Peterson PL, Xiong Y, Lee CP. Improvement in mitochondrial dysfunction as a new surrogate efficiency measure for preclinical trials: Dose-response and time-window profiles for administration of the calcium channel blocker Ziconotide in experimental brain injury. J Neurosurg 2000; 93:829-834.
97. Miljanich GP, Ramachandran J. Antagonists of neuronal calcium channels: Structure, function, and therapeutic implications. Annu Rev Pharmacol Toxicol 1995; 35:707-734.

INTRACELLULAR Ca^{2+} HANDLING

Rod J. Sayer*

ABSTRACT

Intracellular Ca^{2+} is regulated within three major compartments: the cytosol, the endoplasmic reticulum and mitochondria. This Chapter reviews the mechanisms involved in handling of Ca^{2+} within these compartments with reference to potential strategies for neuroprotection. In the cytosol, Ca^{2+} buffering has a major influence on Ca^{2+} signals. Cytosolic Ca^{2+}-binding proteins such as CB28 participate in Ca^{2+} buffering and may have a role in resistance to neurotoxicity. In the endoplasmic reticulum, a number of proteins are involved in Ca^{2+} uptake, lumenal buffering or release, and these may be of value as potential targets for therapeutic intervention. Mitochondria are receiving increasing attention for their role in Ca^{2+} storage and signaling, and as key players in the processes leading to cell death following Ca^{2+} overload. An improved understanding of how Ca^{2+} is controlled within these intracellular compartments, and how these compartments interact, will be important for neuroprotective strategies.

INTRODUCTION

The management of Ca^{2+} inside cells is sometimes referred to as Ca^{2+} homeostasis, but this understates the dynamic role of Ca^{2+} as a signaling molecule. Ca^{2+} transients encode information about neuronal activity via their amplitude, timing and subcellular location. This signaling system exploits a steep gradient of Ca^{2+}concentration ([Ca^{2+}]) between the cytosol (resting level ~100 nM) and the extracellular fluid and endoplasmic reticulum (ER), in which [Ca^{2+}] is three or four orders of magnitude higher. The gradients allow generation of rapid rises of cystolic

*Department of Physiology, University of Otago, P.O. Box 913, Dunedin, New Zealand.

$[Ca^{2+}]$, but have energetic costs and carry risks of Ca^{2+} overload. A complex set of mechanisms has evolved to allow active Ca^{2+} signaling while avoiding toxicity.

The impact of Ca^{2+} influx through the plasma membrane on the resulting rise in free cytosolic Ca^{2+} concentration ($[Ca^{2+}]_c$). Depends upon cytosolic buffering, the rates of removal from cytosol by sequestration into organelles (mainly ER and mitochondria), and extrusion from the cell via the plasma membrane. In addition, Ca^{2+} that enters may influence subsequent Ca^{2+} influx by inactivating voltage- or receptor-operated channels in the plasma membrane, or by activating K^+ channels that reduce membrane excitability, or may affect Ca^{2+} release from intracellular stores.

CYTOSOLIC Ca^{2+} BUFFERING

A consistent finding across many cell types, including neurons, is that the transient rises of free cytosolic Ca^{2+} concentration ($[Ca^{2+}]_c$) during Ca^{2+} influx represent only a tiny proportion of Ca^{2+} that enters the cell. The remainder does not contribute to the signal because it is bound to high-affinity Ca^{2+} binding sites (buffers) within the cytosol. Typical estimates suggest that for neurons only one in every 100-400 molecules of added Ca^{2+} remains free.[1,2] In some neurons such as Purkinje cells of the cerebellum that are adapted for high Ca^{2+} fluxes, the proportion that remains free may be as low as one in several thousand.[3]

Cytosolic Ca^{2+} buffering has a major impact on the spatial distribution of Ca^{2+} signals. Ca^{2+} diffuses much more slowly in cytosol than in unbuffered media, because of frequent encounters with Ca^{2+}—binding proteins that are either fixed or of lower mobility. Its range of messenger action is therefore likely to be short relative to unbuffered molecules such as IP_3.[4] Much of the signaling action of Ca^{2+} probably occurs via highly localized and steep rises in $[Ca^{2+}]$, for example immediately under the plasma membrane around Ca^{2+} channels, or at close associations between ER and mitochondria.

The cellular constituents that contribute the bulk of the cytosolic Ca^{2+} buffering have not been conclusively identified, but a large group of proteins belonging to the EF-hand family are known to be specialized for Ca^{2+}-binding.[5-8] Some members of this family, such as calmodulin, are ubiquitous across cell types and operate primarily as Ca^{2+} sensors. These undergo conformational change on binding Ca^{2+} and thereby regulate the activity of target proteins. Other EF-hand proteins, such as calbindin$_{28kDa}$ (CB28), parvalbumin (PV) and calretinin, are found in specific types of neurons and there is evidence that these function (at least in part) as Ca^{2+} buffers. In rat dorsal root ganglion neurons, introduction of CB28 or PV reduces the rate of rise and the peak of the $[Ca^{2+}]_c$ transient.[9] Pituitary tumor (GH3) cells transfected with CB28 show smaller increases in $[Ca^{2+}]_c$ for a given Ca^{2+} influx.[10] In Purkinje neurons, which normally contain both CB28 and PV, a fast component of the cytosolic Ca^{2+} transient is increased in CB28-null mutant mice.[11] CB28, PV and calretinin are generally regarded as mobile cytosolic buffers, but about one-third of the CB28 and calretinin in neurons is associated with insoluble cellular structures.[12,13]

Ca^{2+} BUFFERING AND NEUROPROTECTION

The possibility that certain Ca^{2+}-binding proteins might confer neuroprotection by buffering intracellular Ca^{2+} has generated considerable interest. CB28 and PV have received the most attention because they are expressed only in specific subtypes of neuron. Questions therefore arose about whether their presence or level of expression might be associated with resistance to excitotoxic or degenerative processes.

Several correlative studies indicated that neurons containing CB28 or PV were spared in some pathological conditions. In the hippocampal formation, the neurons most susceptible to damage from experimental seizures are the hilar interneurons and CA3 pyramidal cells, which lack CB28 or PV.[14] In Parkinson s disease, the subgroup of dopamine neurons in the substantia nigra that express CB28 are relatively spared.[15] The same neurons are resistant to MPTP toxicity in animal studies.[16] However, other findings have suggested a more complex picture. In neurodegenerative conditions such as Parkinson s, Alzheimer s and Pick s diseases, the loss of neurons containing Ca^{2+}—binding proteins varies across different brain areas (for a review see ref. 17). Furthermore, within a given region, vulnerability may depend more on the type of insult than the presence of Ca^{2+}—buffering proteins. For example, CA3 pyramidal cells do not express CB28 or PV but are more resistant to ischemia than CB28-containing neurons elsewhere in the hippocampus.[18]

Findings from investigations in CB28-transgenic mice[11] have tended to downplay or even contradict a neuroprotective role for CB28. Surprisingly, in models of ischemia CB28 null mutants suffer less hippocampal damage than wild types.[19] The CB28 null mutants are not more vulnerable than wild types to kainate-induced excitotoxicity in the hippocampus (although heterozygosity confers some resistance[20]). Substantia nigra dopamine neurons in the CB28-deficient mice are not more susceptible to degeneration following MPTP administration; the same population of cells is lost regardless of genotype.[21] Similarly, transgenic mice deficient in PV, or double knockouts for PV/CB28 or PV/calretinin are no more vulnerable than wild types to kainate excitotoxicity in the hippocampus.[22]

As a result of these studies, it has been suggested that a *downregulation* of CB28 may be neuroprotective. The loss of CB28 from dentate granule cells that is associated with seizures may reflect an adaptive response by the neuron to reduce its vulnerability to further threat.[23-25] Indeed a recent electrophysiological study of human dentate granule cells removed from patients with hippocampal sclerosis confirmed that loss of CB28 is associated with a reduced Ca^{2+} influx during trains of action potentials,[25] which would be consistent with a neuroprotective response.

The data from transgenic animals remains controversial, in part because the life-long lack of the deficient protein may allow the opportunity for developmental compensation. Therefore a substantial body of work continues to be based on the premise that some advantage may be gained by increasing the expression of CB28 in vulnerable neurons. Sapolsky s group, in particular, is reporting success in using herpes simplex viral vectors to express (or over-express) CB28. In cultured hippocampal neurons this confers resistance to hypoglycemic and glutaminergic insults.[26-28] In the whole animal, injections into the hippocampus lead to increased survival of neurons following kainic acid or 3-acetylpyridine neurotoxicity,[29] while striatal injections promote neuronal survival after middle cerebral artery occlusion.[30] Other groups have shown that cell culture lines transfected with CB28 have improved survival when challenged with NMDA[31,32] or that cultured motor neurons rendered particularly sensitive to glutamate by expression of a mutant Cu/Zn-superoxide dismutase are protected by coexpression of CB28.[33] On the other hand, there have been conflicting reports about whether hippocampal neurons that naturally express CB28 in cultures are more robust,[34,35] and in one study the over-expression of PV in mouse neocortical neurons was found to enhance neurotoxicity at some concentrations of NMDA.[36]

An alternative approach to the manipulation of intracellular Ca^{2+} buffering involves the use of exogenous Ca^{2+} chelators, e.g., BAPTA and its analogs. An early study of interneurons in the dentate hilus showed that these neurons, which usually deteriorated under prolonged stimulation, were protected by injection of BAPTA via the recording electrode.[37] This prompted further investigation into whether cell-permeant forms of BAPTA and its analogs might have efficacy as neuroprotective agents. BAPTA-AM and Quin-2 were shown to be taken up by spinal neurons in vitro and to reduce or delay neuronal death.[38,39] In vivo intravenous infusion of BAPTA-AM resulted in delivery of the chelator to the brain, affected the electrophysiology of hippocampal neurons as expected for a Ca^{2+} buffer, and reduced cortical infarct volume in a model of focal cerebral ischemia.[38] DP-b99 (DP-BAPTA), a variant of BAPTA designed to be preferentially active in lipid environments, is claimed to show efficacy in animal models and is under commercial development as a neuroprotective drug for cerebral ischemia (D-Pharm Ltd.; and see http://www.strokecenter.org/trials/ints/intPage95.htm). However it should be noted that not all studies have reported positive findings with exogenous chelators. In cultured hippocampal neurons, BAPTA-AM reduced the peak magnitudes of $[Ca^{2+}]_c$ transients evoked by brief applications of excitatory amino acids, but also slowed the recovery to baseline. With longer exposure to agonists, the peak $[Ca^{2+}]_c$ responses were not reduced, and excitotoxicity was either increased or similar to controls.[40,41]

The inconsistency of the above studies indicates that the actions of both endogenous and exogenous Ca^{2+} buffers are complex and still poorly understood. Their neuroprotective efficacy will depend upon many factors that vary with cell type and circumstance. For example, their accessibility to different subcellular compartments may be variable so that they buffer Ca^{2+} entry by some pathways more than others; they likely influence the mobility of intracellular Ca^{2+}, thereby affecting the rate and distance of Ca^{2+} diffusion, and the reduction of transient peaks of $[Ca^{2+}]_c$ needs to be balanced against the consequences of longer durations of raised $[Ca^{2+}]_c$. Ca^{2+} buffers may have important actions via binding of other ions, e.g., Zn^{2+}, and the majority of Ca^{2+}—buffering proteins may well have additional Ca^{2+}—sensor propertiesAs the field evolves it will become clearer how these factors impact on the various animal models and their applicability to human neuroprotection.

THE ENDOPLASMIC RETICULUM Ca^{2+} STORE

The ER is an elaborate membrane-enclosed network that extends from the nuclear envelope to the periphery of the cell, which in neurons includes axons and synaptic terminals, dendrites and spines. Its functions include the folding and assembly of newly-formed proteins, lipid synthesis and an important role in Ca^{2+} storage and signaling. The ER forms a connected network with a continuous lumen (also continuous with the nuclear envelope), but it does show some regional specialization; areas most prominently associated with ribosomes, and therefore protein synthesis, are termed rough ER and the remainder is known as smooth ER. Although the entire ER is believed to function as a Ca^{2+} store, some regional variation has been shown in the distribution of Ca^{2+} uptake pumps and release channels, and also in concentrations of stored Ca^{2+} (reviewed in ref. 42 and 43).

In neurons, most of the Ca^{2+} destined for uptake into the ER is probably admitted into the cell via voltage-gated Ca^{2+} channels, i.e., influx depends upon electrical activity of the plasma membrane. However, the importance of another route of Ca^{2+} entry, studied extensively in non-excitable cells, is not yet apparent for neurons. Termed *capacitative* or *store-operated* Ca^{2+} entry, this process is activated by a fall in $[Ca^{2+}]$ within the ER. It is uncertain whether the signal from the ER involves a diffusible messenger (putative calcium influx factor) or a mechanical linkage at regions where the ER is closely apposed to the plasma membrane.[44] The plasma membrane channels responsible for Ca^{2+} entry via this route have been characterized electrophysiologically in a number of non-excitable cells, and a strong candidate for the molecular identity of the channel has been reported recently.[45] This putative store-operated Ca^{2+} channel, CaT1, is a member of the TRP family[46] and is expressed in a number of tissues including brain.[47,48] It seems likely that a number of closely-related channels have similar roles.

Ca^{2+} is taken up from the cytosol into the ER by sarcoplasmic-endoplasmic reticulum Ca^{2+} ATPases (SERCAs). The SERCA family originates from three genes giving rise to multiple splice variants, with SERCA2b and SERCA3a being expressed in brain.[49] SCRCAa hydrolyze one ATP molecule for every two Ca^{2+} ions they pump into the ER lumen, with the different isoforms varying in Ca^{2+} affinity and rate of Ca^{2+} uptake.

Within the ER the majority of Ca^{2+} is buffered by Ca^{2+}—binding proteinsThe most prominent of these in neurons is calreticulin which binds Ca^{2+} with low affinity (K_d ~2 mM) but high capacity (>20 mol Ca^{2+} per mol protein[50,51]). Other proteins that contribute to Ca^{2+} buffering in the ER lumen include endoplasmin, BiP and protein disulfide isomerase. These molecules, including calreticulin, have a dual role in that they also assist with protein folding and assembly. Estimates suggest that when the ER stores are filled their total $[Ca^{2+}]$ is about 5-10 mM and their free $[Ca^{2+}]$ is 100-700 μM.[52]

The two main pathways for Ca^{2+} release from the ER are via IP_3 receptors (IP_3R) and ryanodine receptors (RyR). Both are Ca^{2+} channels composed of four subunits. IP_3R subunits are encoded by at least three genes (with splice variants), and the complete IP_3R is activated by IP_3 and also modulated by cytosolic Ca^{2+}. RyR originate from three genes and are gated by Ca^{2+} and in some cells by cADP-ribose. For both IP_3R and RyR the sensitivity to $[Ca^{2+}]_c$ is bell-shaped so that channel activity is promoted within a certain range; the actual values and degree of sensitivity vary with the multiple ribosome receptor isoforms. IP_3R and RyR show some differences of distribution within the brain. For example, IP_3R predominate over RyR in cerebellar Purkinje neurons and hippocampal CA1 pyramidal cells, whereas the reverse is true in the dentate gyrus and CA3 regions of the hippocampus.[53]

The role of the ER in Ca^{2+} homeostasis and Ca^{2+} signaling in neurons clearly involves much more than Ca^{2+}storage or uptake so as to maintain low $[Ca^{2+}]_c$. The properties of the IP_3R and RyR endow them with the ability to act as a source of Ca^{2+} transients (waves or spikes) within the cytosol. Ca^{2+} release from the ER can be triggered by the diffusion of IP_3 from distant sites of generation at the plasma membrane, in response to the activation of phospholipase C by extracellular stimuli. Alternatively, where the ER is near enough to a source of elevated $[Ca^{2+}]_c$, e.g., where it is closely apposed to plasma membrane that contains Ca^{2+} channels, RyR or IP_3 channel opening is triggered or enhanced in a process known as Ca^{2+}—induced Ca^{2+} release (CICR). The Ca^{2+} released from a localized area of ER may then promote further release from neighboring regions to form a propagating wave that spreads along the ER. Thus the ER in neurons should be thought of as an active excitable organelle that promotes the integration and communication of Ca^{2+} signals with the cell.[54]

Given that the neuron has ample machinery at the plasma membrane to generate Ca^{2+} transients, the ER must provide some advantages for Ca^{2+} signaling. As a source of Ca^{2+} signals, it has been suggested that the ER facilitates rises of free cytosolic Ca^{2+} deep within the cell, out of the range of the limited diffusion of Ca^{2+} entering from the plasma membrane. In addition, the extracellular space immediately surrounding many neurons may be of such limited volume as to be a poor source of Ca^{2+}.[56] As a means of terminating Ca^{2+} signals efficiently, the ER has in its favor a larger surface area than the plasma membrane and lower energetic costs because Ca^{2+} uptake is not opposed by a membrane potential.[55]

THE ENDOPLASMIC RETICULUM AND NEUROPROTECTION

A potential link between disturbances of ER homeostasis and pathological conditions is shown by experiments in which ER Ca^{2+} stores are artificially depleted. Agents such as thapsigargin and cyclopiazonic acid, which inhibit SERCA, promote store emptying. A number of studies have reported that this leads to suppression of protein synthesis, activation of stress genes, and ultimately apoptosis, a picture seen in various ischemic/neurotoxic conditions (reviewed in refs. 57-59). It is likely that both the low $[Ca^{2+}]$ within the ER and the high $[Ca^{2+}]$ associated with excessive or prolonged store emptying have deleterious effects, which might contribute to a pathological scenario. In the ER insufficient levels of intraluminal Ca^{2+} interfere with protein processing, at least in part because the chaperone activity of calreticulin and other ER proteins is Ca^{2+}—dependent.[51,60] In the cytoplasm potential targets for released Ca^{2+} include proteases, endonucleases, phospholipases and other Ca^{2+}—sensitive proteins that have been implicated in neurotoxicity.[61] Adequate ER functioning presumably requires an appropriate environment of cytoplasmic Ca^{2+} within a certain range, which might explain why cellular vulnerability seems to increase at both high and low extremes of $[Ca^{2+}]_c$.[62]

Drugs that influence the Ca^{2+}—handling machinery of the ER might provide opportunities for therapeutic intervention. For example, dantrolene, which blocks release of Ca^{2+} from the ER via RyR, has shown some neuroprotective efficacy in experimental models. It was reported to reduce glutamate-induced cytotoxicity in cultured cerebral cortical neurons,[63] to prevent seizure-induced cell death in rat hippocampal slices,[64] and to protect against death of hippocampal and cerebral cortical neurons following brain ischemia in gerbils and rats.[65,66]

Another approach is to identify proteins that would be potential targets for drugs because of their role in ER Ca^{2+} homeostais. For example, calreticulin has been suggested to have anti-apoptotic properties because inhibition of its expression increases ionomycin-induced cell death.[67,68] Another protein with evidence for anti-apoptotic activities is Bcl-2. This oncogenic protein resides in the membranes of the ER and nuclear envelope, and in the outer mitochondrial membrane. Its normal function in the ER membrane is unknown at present, but alterations in ER Ca^{2+} homeostasis have been described in cultured cells over-expressing Bcl-2. Transfected cells, unlike controls, were reported to maintain ER Ca^{2+} uptake and protein processing in the presence of thapsigargin,[70] to upregulate SERCA[71] or (in contrast) to show reduced loading of ER Ca^{2+} stores.[72] Interpretation of these findings is currently difficult, and it is unclear how they relate to the actions of the Bcl-2 family of proteins in mitochondria (see below).

MITOCHONDRIA AND Ca^{2+} HOMEOSTASIS

Mitochondria receive attention in any consideration of cell death or neuroprotection because of their key role in energy metabolism and because of the potentially damaging consequences of the reactive oxygen species that are generated as a by-product. However in the last few years there has been a change in perspective so that mitochondria are now seen as increasingly important for their influence on intracellular Ca^{2+} homeostasis.[73-75]

Ca^{2+} is taken up into mitochondria via a uniporter in the inner membrane driven by a membrane potential of 150-200 mV (negative inside). The membrane potential is generated by H^+ extrusion associated with mitochondrial respiration. A rise of inner mitochondrial $[Ca^{2+}]$ activates dehydrogenases in the citric acid cycle, thereby linking energy metabolism with cellular activity. The relatively low affinity of the uniporter raised the conundrum that the rise in bulk $[Ca^{2+}]_c$ in stimulated cells, peaking at several μM, seemed insufficient to allow significant mitochondrial uptake.

Accumulating evidence now points to a close functional relationship between mitochondria and the ER. At regions where these are in close apposition, microdomains of high $[Ca^{2+}]_c$ are formed by release from the ER, and mitochondrial uptake becomes significant.[74] In addition, the pattern of Ca^{2+} signals may be important with uptake being more efficient when mitochondria are exposed to pulses of Ca^{2+}.[76]

An important question is whether mitochondrial Ca^{2+} handling goes beyond a signaling function for regulation of activity within the organelle itself, to additionally influence signals within the cytoplasm. There is now substantial evidence that this is the case, although with some variation between cells. Uptake of Ca^{2+} into mitochondria appears to be of sufficient rapidity and magnitude to act as a buffer so that recovery from Ca^{2+} transients is accelerated. Ca^{2+} is then released from mitochondria over a slower time course due to action of the mitochondrial Na^+-Ca^{2+} exchanger, resulting in a prolongation of the raised $[Ca^{2+}]_c$ after the initial recovery.[77] One example of the functional importance of this phenomenon is the contribution of persistently raised $[Ca^{2+}]_c$ in presynaptic terminals to post-tetanic potentiation.[78] Another consequence of mitochondrial Ca^{2+} release may be a feedback influence on nearby ER Ca^{2+} channels, thereby affecting the propagation of cytosolic Ca^{2+} waves.[74,75]

MITOCHONDRIA, NEUROTOXICITY AND NEUROPROTECTION

An emerging body of evidence implicates Ca^{2+} overload of mitochondria, perhaps coupled with other factors such as nitric oxide or reactive oxygen species production as a crucial factor in the progression towards neuronal death (for detailed reviews see refs. 75,79-81). It is hypothesized that overloading of the mitochondria with Ca^{2+} has two important consequences: (a) loss of the proton gradient across the inner mitochondrial membrane and therefore cessation of ATP production and (b) release into the cytoplasm of cytochrome *c* and other proactive substances, with subsequent activation of caspases. The degree to which these occur might tip the balance between necrosis and apoptosis. With intense insult the loss of ATP might be sufficient to cause rapid necrosis, whereas lesser injuries may allow survival of enough mitochondria to produce energy for the delayed apoptotic process.

A leading hypothesis states that a key step in the pathological loss of function in mitochondria involves formation of a permeability transition pore (PTP) across the inner and outer membranes.[79] An abrupt opening of this pore, promoted by Ca^{2+}, oxidative stress, falling ATP concentration or other factors, is believed to cause mitochondrial depolarization, loss of proton gradient, and therefore uncoupling of oxidative phosphorylation. An associated increase in permeability of the outer membrane allows release of cytochrome *c* and other substances, e.g., apoptosis-inducing factor, from the inter-membrane space into the cytoplasm, with subsequent initiation of the apoptotic cascade. The putative makeup of the PTP is a complex of at least three proteins that forms at regions of close contact between the inner and outer mitochondrial membranes. These are the voltage-dependent ion channel (VDAC) in the outer membrane, the adenine nucleotide translocase (ANT) in the inner membrane (which exchanges ADP and ATP), and cyclophilin-D in the mitochondrial matrix. Whether formation of these components into the PTP occurs only under pathological conditions or whether transient openings might be important contributors to Ca^{2+} signaling[73] is controversial.

The hypothesis that the PTP is involved in cell death indicates possible targets for neuroprotective agents. For example, cyclosporin A is one drug known to bind to cyclophilin-D and to inhibit PTP opening. Mitochondrial depolarization and cell death in cultured neurons exposed to glutamatergic agonists are ameliorated or prevented by cyclosporin A.[82-84] Cyclosporin A injected into rat hippocampus has been reported to reduce the necrosis caused by forebrain ischemia.[85] However, cyclosporin A also binds to cyclophilins in the cytoplasm and ER so definitive attribution of its site of action awaits analogs that are specific for mitochondrial cyclophilin-D.[75] Another strategy might be to act further upstream by preventing loading of the mitochondria with Ca^{2+}. In cultured neurons, Stout et al[86] showed that pre-emptive depolarization of mitochondrial membranes with FCCP inhibited mitochondrial Ca^{2+} uptake and reduced glutamate neurotoxicity. A further potential target is the Ca^{2+} uptake uniporter which still awaits molecular characterization and development of blockers.

The Bcl-2 family of proteins has emerged as important regulators of apoptosis, and a major part of their action is proposed to occur in mitochondria (for reviews see refs. 87-89). Of this group, Bcl-2, Bcl-x_L and several others are anti-apoptotic whereas a number of structurally similar relatives promote cell death. Most members of the family have a hydrophobic C-terminal segment which localizes them to the cytoplasmic aspect of organellar membranes. Bcl-2 and Bcl-x_L are of particular interest as potential drug targets because they appear to oppose processes that disrupt mitochondrial function. When Bcl-2 or Bcl-x_L are applied as recombinant protein to isolated mitochondria, or over-expressed in cultured cells, they inhibit cytochrome *c* release in the face of apoptotic stimuli.[90-94] Consistent with these findings, over-expression of Bcl-2 in neuronal cell lines reduce cell death induced by a variety of factors,[95,96] and when over-expressed in transgenic mice, cortical infarcts induced by middle cerebral artery occlusion are smaller than in wild-type animals.[97] The mechanism of action of the Bcl-2 family of proteins is still uncertain, but it has been proposed that the anti-apoptotic members, e.g., Bcl-2 and Bcl-x_L, associate with the mitochondrial pore complex and stabilize its normal function in a way that inhibits pathological PTP activity.

FUTURE CHALLENGES

This short review gives a simplified and rather linear overview of the cellular machinery involved in Ca^{2+} handling. By linear, I mean that the components (cytosolic buffers, ER and mitochondria) are considered in succession, whereas a major task for the future will be to understand how these function together in a working system.

The close relationship between ER and mitochondria, for example, appears to be crucial for mitochondrial Ca^{2+} uptake, and therefore requires investigation of still smaller spatial and temporal domains. This makes heavy demands on technological development and also on our ability to construct useful models of interacting systems. Further complicating the picture, the majority of components are multifunctional. For example, Ca^{2+}-binding proteins can have both buffer and sensor actions; ER is involved in Ca^{2+} signaling, protein and lipid processing; and Ca^{2+} mitochondrial uptake is important for enzyme modulation and for shaping cytosolic Ca^{2+} transients. Of course, none of these functions is independent. To add to the picture, there are other players in the Ca^{2+} game. Some, such as the Golgi apparatus[98] and nuclear. envelope, may have a minor but distinct role; others, e.g., a large fraction of the buffering capacity of the cytosol, have a major role but are still not clearly defined. It remains a challenging and worthwhile enterprise for the basic scientist to reveal and recognize those key components of the Ca^{2+} handling system that may ultimately become useful targets for therapeutic intervention.

REFERENCES

1. Neher E. The use of fura-2 for estimating Ca buffers and Ca fluxes. Neuropharmacol 1995; 34:1423-1442.
2. Helmchen F, Imoto K, Sakmann B. Ca^{2+} buffering and action potential-evoked Ca^{2+} signaling in dendrites of pyramidal neurons. Biophy J 1996; 70:1069-1081
3. Fierro L, Llano I. High endogenous calcium buffering in Purkinje cells from rat cerebellar slices. J Physiol (Lond) 1996; 496:617-625.
4. Allbritton NL, Meyer T. Stryer L. Range of messenger action of calcium ion and inositol 1,4,5 trisphosphate. Science 1992; 258:1812-1815.
5. Heizmann CW, Hunziker W. Intracellular calcium-binding proteins: more sites than insights. Trends Biochem Sci 1991; 16:98-103.
6. Baimbridge KG, Celio MR, Rogers JH. Calcium binding proteins in the nervous system. Trends Neurosci 1992; 15:303-308.
7. Ikura M. Calcium binding and conformational response in EF-hand proteins. Trends Biochem Sci 1996; 21:14-17.
8. Lewit-Bentley A, R ty S. EF-hand calcium-binding proteins. Curr Opin Struct Biol 2000; 10:637-643.
9. Chard PS, Bleakman D, Christakos S et al. Calcium buffering properties of calbindin D_{28K} and parvalbumin in rat sensory neurones. J Physiol (Lond) 1993; 472:341-357.
10. Lledo P-M, Somasundaram B, Morton AJ et al. Stable transfection of calbindin-D_{28K} into the GH_3 cell line alters calcium currents and intracellular calcium homeostasis. Neuron 1992; 9:943-954.
11. Airaksinen MS, Eilers J, Garaschuk O et al. Ataxia and altered dendritic calcium signaling in mice carrying a targeted null mutation of the calbindin$_{D28k}$ gene. Proc Natl Acad Sci USA 1997; 94:1488-1493.
12. Hubbard MJ, McHugh NJ. Calbindin$_{30kDa}$ and calbindin$_{28kDa}$ (calretinin) are substantially localised in the particulate fraction of rat brain. FEBS Lett 1995; 374:333-337.
13. Sayer RJ, Turnbull CI, Hubbard MJ. Calbindin$_{28kDa}$ is specifically associated with extranuclear constitutents of the dense particulate fraction. Cell Tissue Res 2000; 302:171-180.
14. Sloviter RS. Calcium binding protein (calbindin-D_{28K}) and parvalbumin immunocytochemistry: Localization in the rat hippocampus with specific reference to selective vulnerability of the hippocampal neurons to seizure activity. J Comp Neurol 1989; 280:183-196.

15. Yamada T, McGreer PL. Baimbridge KG et al. Relative sparing in Parkinson s disease of substantia nigra dopamine neuron containing calbindin-D_{28K}. Brain Res 1990; 526:303-307.
16. Iacopino A, Christakos S, German D et al. Calbindin-D_{28K}-containing neurons in animal models of neurodegeneration: possible protection for excitotoxicity. Brain Res Mol Brain Res 1992; 13:251-261.
17. Heizmann CW, Braun K. Changes in Ca^{2+}—binding proteins in human neurodegenerative disorders. Trends Neurosci 1992; 15:259-264.
18. Freund TF, Buzs ki G, Leon A et al. Relationship of neuronal vulnerability and calcium-binding protein immunoreactivity in ischemia. Exp Brain Res 1990; 83:55-66.
19. Klapstein GJ, Vietla S, Lieberman DN et al. Calbindin-D_{28K} fails to protect hippocampal neurons against ischemia in spite of its cytoplasmic calcium-buffering properties: Evidence from calbindin-D_{28K} knockout mice. Neuroscience 1998; 85:361-373.
20. Gary DS, Sooy K, Chan SL et al. Concentration- and cell-specific effects of calbindin D28K on vulnerability of hippocampal neurons to seizure-induced injury. Brain Res Mol Brain Res 2000; 75:89-95.
21. Airaksinen MS, Thoenen H, Meyer M. Vulnerability of midbrain dopaminergic neurons in calbindin-D28K-deficient mice: lack of evidence for a neuroprotective role of endogenous calbindin in MPTP-treated and weaver mice. Eur J Neurosci 1997; 9:120-127.
22. Bouillert V, Schwaller B, Schurmans S et al. Neurodegenerative and morphogenic changes in a mouse model of temporal lobe epilepsy do not depend on the expression of calcium-binding proteins parvalbumin, calbindin, or calretinin. Neuroscience 2000; 97:47-58.
23. Miller JJ, Baimbridge KG. Biochemical and immunohistochemical correlates of kindling-induced epilepsy: role of calcium-binding protein. Brain Res 1983; 278:322-326.
24. K hr H, Mody I. Endogenous intracellular calcium buffering and the activation/inactivation of HVA calcium currents in rat dentate gyrus granule cells. J Gen Physiol 1991; 98:941-967.
25. N gerl UV, Mody I, Jeub M et al. Surviving granule cells of the sclerotic human hippocampus have induced Ca^{2+} influx because of a loss of calbindin-D_{28K} in temporal lobe epilepsy. J Neurosci 2000; 20:1831-1836.
26. Meier TJ, Ho DY, Sapolsky RM. Increased expression of calbindin D_{28K} via herpes simplex virus amplicon vector decreases calcium ion mobilization and enhances neuronal survival after hypoglycemic challenge. J Neurochem 1997; 69:1039-1047.
27. Meier TJ, Ho DY, Park TS et al. Gene transfer of calbindin D_{28K} cDNA via herpes simplex virus amplicon vector decreases cytoplasmic calcium ion response and enhances neuronal survival following glutaminergic challenge but not following cyanide. J Neurochem 1998; 71:1013-1023.
28. Monje ML, Phillips R, Sapolsky R. Calbindin overexpression buffers hippocampal cultures from the energetic impairments caused by glutamate. Brain Res 2001; 911:37-42.
29. Phillips RG, Meier TJ, Giuli LC et al. Calbindin D_{28K} gene transfer via herpes simplex virus amplicon vector decreases hippocampal damage in vivo following neurotoxic insults. J Neurochem 1999; 73:1200-1205.
30. Yenari MA, Minami M, Sun GH et al. Calbindin D28K overexpression protects striatal neurons from transient focal cerebral ischemia. Stroke 2001; 32:1028-1035.
31. D Orlando C, Fellay B, Schwaller B et al. Calretinin and calbindin D-28k delay the onset of cell death after excitotoxic stimulation in transfected P19 cells. Brain Res 2001; 909:145-158.
32. Rintoul GL, Raymond LA, Baimbridge KG. Calcium buffering and protection from excitotoxic cell death by exogenous calbindin D-28k in HEK 293 cells. Cell Calcium 2001; 29:277-287.
33. Roy J, Minotti S, Dong L et al. Glutamate potentiates the toxicity of mutant Cu/Zn-superoxide dismutase in motor neurons by postsynaptic calcium-dependent mechanisms. J Neurosci 1998; 18:9673-9684.
34. Mattsom MP, Rychlik B, Chu C et al. Evidence for calcium-reducing and excito-protective roles for the calcium-binding protein calbindin-D_{28k} in cultured hippocampal neurons. Neuron 1991; 6:41-51.

35. M ckel V, Fischer G. Vulnerability to excitotoxic stimuli of cultured rat hippocampal neurons containing the calcium-binding proteins calretinin and calbindin D_{28k}. Brain Res 1994; 648:109-120.
36. Klaus W, Grzesiek S, Labhardt AM et al. NMR investigation and secondary structure of domains I and II of rat brain calbindin D_{28k} (1-93). Eur J Biochem 1999; 262:933-938.
37. Scharfman HE, Schwartzkroin PA. Protection of dentate hilar cells from prolonged stimulation by intracellular calcium chelation. Science 1989; 246:257-260.
38. Tymianski M, Wallace MC, Spigelman I et al. Cell permeant Ca^{2+} chelators reduce early excitotoxic and ischemic neuronal injury in vitro and in vivo. Neuron 1993; 11:221-235.
39. Kudo Y, Takeda K, Yamazaki K. Quin2 protects against neuronal cell death due to Ca^{2+} overload. Brain Res 1990; 528:48-54.
40. Dubinsky JM. Effects of calcium chelators on intracellular calcium and excitotoxicity. Neurosci Lett 1993; 150:129-132.
41. Abdel-Hamid KM, Baimbridge KG. The effects of artificial calcium buffers on calcium responses and glutamate-mediated excitotoxicity in cultured hippocampal neurons. Neuroscience 1997; 81:673-687.
42. Meldolesi J, Pozzan T. The heterogeneity of ER Ca^{2+} stores has a key role in nonmuscle cell signaling and function. J Cell Biol 1998; 142:1395-1398.
43. Petersen OH, Tepikin A, Park MK. The endoplasmic reticulum: One continuous or several Ca^{2+} stores? Trends Neurosci 2001; 24:271-276.
44. Putney JW. Channelling calcium. Nature 2001; 410:648-649.
45. Yue L, Peng J-B, Hediger MA et al. CaT1 manifests the pore properties of the calcium-release-activated calcium channel. Nature 2001; 410:705-709.
46. Harteneck C, Plant TD, Schultz G. From worm to man: Three families of TRP channels. Trends Neurosci 2000; 23:159-166.
47. Peng J-B, Chen X-Z, Berger UV et al. Molecular cloning and characterization of a channel-like transporter mediating intestinal calcium absorption. J Biol Chem 1999; 274:22739-22746.
48. Barley NF, Howard A, O Callaghan D et al. Epithelial calcium transporter expression in human duodenum. Am J Physiol 2001; 280:G285-G290.
49. Misquitta CM, Mack DP, Grover AK. Sarco/endoplasmic reticulum Ca^{2+} (SERCA)-pumps: link to heart beats and calcium waves. Cell Calcium 1999; 25:277-290.
50. Baksh S, Michalak M. Expression of calreticulin in *Escherichia coli* and identification of its Ca^{2+} binding domains. J Biol Chem 1991; 21458-21465.
51. Michalak M, Mariani P, Opas M. Calreticulin, a multifunctional Ca^{2+} binding chaperone of the endoplasmic reticulum. Biochem Cell Biol 1998; 76:779-785.
52. Meldolesi J. Pozzan T. The endoplasmic reticulum Ca^{2+} store: A view form the lumen. Trends Biochem Sci 1998; 23:10-14.
53. Sharp AH, McPherson PS, Dawson TM et al. Differential immunohistochemical localization of inositol 1,4,5-trisphosphate- and ryanodine-sensitive Ca^{2+} release channels in rat brain. J Neurosci 1993; 13:3051-3063.
54. Berridge MJ. Neuronal calcium signalling. Neuron 1998; 21:13-26.
55. Taylor CW. Why do hormones stimulate Ca^{2+} mobilization? Biochem Soc Trans 1995; 23:637-642.
56. Paschen W, Doutheil J. Disturbances of the functioning of endoplasmic reticulum: A key mechanism underlying neuronal cell injury? J Cereb Blood Flow Metab 1999; 19:1-18.
57. Kass GEN, Orrenius S. Calcium signaling and cytotoxicity. Environ Health Perstect 1999; 107 Suppl 1:25-35.
58. Mattson MP, LaFerla FM, Chan SL et al. Calcium signaling in the ER: Its role in neuronal plasticity and neurodegenerative disorders. Trends Neurosci 2000; 23:222-229.
59. Paschen W, Doutheil J. Disturbance of endoplasmic reticulum functions: A key mechanism underlying cell damage? Acta Neurochir 1999; 73[Suppl]:1-5.
60. Corbett EF, Michalak M. Calcium, a signaling molecule in the endoplasmic reticulum. Trends Biochem Sci 2000; 25:307-311.

61. Sattler R, Tymianski M. Molecular mechanisms of calcium-dependent cytotoxicity. J Mol Med 2000; 78:3-13.
62. Lee J-M, Zipfel GJ, Choi DW, The changing landscape of ischaemic brain injury mechanisms. Nature 199; 399 Supp:A7-A14.
63. Frandsen A, Schousboe A. Dantrolene prevents glutamate cytotoxicity and Ca^{2+} release from intracellular stores in cultured cerebral cortical neurons. J Neurochem 1991; 56:1075-1078.
64. Pelletier MR, Wadia JS, Mills LR et al. Seizure-induced cell death produced by repeated tetanic stimulation in vitro: possible role of endoplasmic reticulum calcium stores. J Neurophysiol 1999; 81:3054-3064.
65. Zhang L, Andou Y, Masuda S et al. Dantrolene protects against ischemic, delayed neuronal death death in gerbil brain. Neurosci Lett 1993; 158:105-108.
66. Wei H, Perry DC. Dantrolene is cytoprotective in two models of neuronal cell death. J Neurochem 1996; 67:2390-2398.
67. Liu N, Fine RE, Simons E et al. Decreasing calreticulin expression lowers the Ca^{2+} response to bradykinin and increases sensitivity to ionomycin in NG-108-15 cells. J Biol Chem 1994; 46:28635-28639.
68. Johnson RJ, Liu N, Shanmugaratnam J et al. Increased calreticulin stability in differentiated NG-108-15 cells correlates with resistance to apoptosis induced by antisense treatment. Brain Res Mol Brain Res 1998; 53:104-111.
69. Lithgow T, van Driel R, Bertram JF et al. The protein product of the oncogene bcl-2 is a component of the nuclear envelope, the endoplasmic reticulum and the outer mitochondrial membrane. Cell Growth Differ 1994; 5:411-417.
70. He H, Lam M, McCormick TS et al. Maintenance of calcium homeostasis in the endoplasmic reticulum by Bcl-2. J Cell Biol 1997; 138:1219-1228.
71. Kuo TH, Kim H-RC, Zhu L et al. Modulation of endoplasmic reticulum calcium pump by Bcl-2. Oncogene 1998; 17:1903-1910.
72. Pinton P, Ferrari D, Magalhaes P et al. Reduced loading of intracellular Ca^{2+} stores and downregulation of capacitative Ca^{2+} influx in Bcl-2-overexpressing cells. J Cell Biol 2000; 148:857-862.
73. Miller RJ. Mitochondria The Kraken wakes! Trends Neurosci 1998; 21:95-97.
74. Rizzuto P, Bernardi P, Pozzan T. Mitochondria as all-around players of the calcium game. J Physiol (Lond) 2000; 529:37-47.
75. Duchen MR. Mitochondria and calcium: From cell signaling to cell death. J Physiol (Lond) 2000; 529:57-68.
76. Sparagna GC, Gunter KK, Sheu S-S et al. Mitochondria calcium uptake from physiological-type pulses of calcium. A description of the rapid uptake mode. J Biol Chem 1995; 270:27510-27515.
77. White RJ, Reynolds IJ. Mitochondria accumulate Ca^{2+} following intense glutamate stimulation of cultured rat forebrain neurones. J Physiol (Lond) 1997; 498:31-47.
78. Tang Y-G, Zucker R-S. Mitochondrial involvement in post-tetanic potentiation of synaptic transmission. Neuron 1997; 18:483-491.
79. Crompton M. Mitochondrial intermembrane junctional complexes and their role in cell death. J Physiol (Lond) 2000; 529:11-21.
80. Crompton M. The mitochondrial permeability transition pore and its role in cell death. Biochem J 1999; 341:233-249.
81. Reynolds IJ. Mitochondrial membrane potential and the permeability transtion in excitotoxicity. Ann NY Acad Sci 1999; 893:33-41.
82. Nieminen A-L, Petrie TG, Lemasters JJ et al. Cyclosporin A delays mitochondrial depolarization induced by *N*-methyl-D-aspartate in cortical neurons: Evidence of the mitochondrial permeability transition. Neuroscience 1996; 75:993-997.
83. Schinder AF, Olson EC, Spitzer NC et al. Mitochondrial dysfunction is a primary event in glutamate neurotoxicity. J Neurosci 1996; 16:6125-6133.

84. Vergun O, Keelan J, Khodorov BI et al. Glutamate-induced mitochondrial depolarisation and perturbation of calcium homeostasis in cultured rat hippocampal neurons. J Physiol (Lond) 1999; 519:451-466.
85. Uchino H, Eln r E, Uchino K et al. Cyclosporin A dramatically ameliorates CA1 hippocampal damage following transient forebrain ischaemia in the rat. Acta Physiol Scand 1995; 155:469-471.
86. Stout AK, Raphael HM, Kanterewicz BI et al. Glutamate-induced neuron death requires mitochondrial calcium uptake. Nature Neurosci 1998; 1:366-373.
87. Adams JM, Cory S. Life-or-death decisions by the Bcl-2 protein family. Trends Biochem Sci 2001; 26:61-66.
88. Tsujimoto Y, Shimizu S. Bcl-2 family: Life-or-death switch. FEBS Lett 2000; 466:6-10.
89. Vander Heiden MG, Thompson CB. Bcl-2 proteins: Regulators of apoptosis or of mitochondrial homeostasis? Nature Cell Biol 1999; 1:E209-E216.
90. Yang J, Liu X, Bhalla K et al. Prevention of apoptosis by Bcl-2: Release of cytochrome c from mitochondria blocked. Science 1997; 275:1129-1132.
91. Kluck RM, Bossy-Wetzel E, Green DR et al. The release of cytochrome c from mitochondria: A primary site for Bcl-2 regulation of apoptosis. Science 1997; 275:1132-1136.
92. Vander Heiden MG, Chandel NS, Williamson EK et al. Bcl-x_L regulates the membrane potential and volume homeostasis of mitochondria. Cell 1997; 91:627-637.
93. Narita M, Shimizu S, Ito T et al. Bax interacts with the permeability transition pore to induce permeability transition and cytochrome c release in isolated mitochondria. Proc Natl Acad Sci USA 1998; 95:14681-14686.
94. J rgenmeister JM, Xie Z, Deveraux Q et al. Bax directly induces release of cytochrome c from isolated mitochondria. Proc Natl Acad Sci USA 1998; 95:4997-5002.
95. Garcia I, Martinou I, Tsujimoto Y et al. Prevention of programmed cell death of sympathetic neurons by *bcl-2* proto-oncogene. Science 1992; 258:302-304.
96. Zhong L-T, Sarafian T, Kane DJ et al. *bcl-2* inhibits death of central neural cells induced by multiple agents. Proc Natl Acad Sci USA 1993; 90:4533-4537.
97. Martinou J-C, Dubois-Dauphin M, Staple JK et al. Overexpression of Bcl-2 in transgenic mice protects neurons from naturally occurring cell death and experimental ischemia. Neuron 1994; 13:1017-1030.
98. Pinton P, Pozzan T, Rizzuto R. The Golgi apparatus is an inositol 1,4,5-trisphosphate-sensitive Ca^{2+} store, with functional properties distinct form those of the endoplasmic reticulum EMBO J 1998; 17:5298-5308.

NEUROPROTECTIVE ACTIVITY OF METABOTROPIC GLUTAMATE RECEPTOR LIGANDS

Peter J. Flor,[1] Giuseppe Battaglia,[2] Ferdinando Nicoletti,[2,3] Fabrizio Gasparini[1] and Valeria Bruno[2]

ABSTRACT

Metabotropic glutamate receptors form a family of currently eight subtypes (mGluR1-8), subdivided into three groups (I-III). Activation of group-II (mGluR2 and -3) or group-III metabotropic glutamate receptors (mGluR4, -6, -7 and -8) has been established to be neuroprotective in vitro and in vivo. In contrast, group-I mGluRs (mGluR1 and -5) need to be antagonized in order to evoke protection. Initially, all neuroprotective mGluR ligands were analogues of L-glutamate. Those compounds were valuable to demonstrate protection in vitro, but showed limited applicability in animal models, particularly in chronic tests, due to low blood-brain-barrier penetration. Recently, systemically active and more potent and selective ligands became available, e.g., the group-II mGluR agonists LY354740 and LY379268 or group-I antagonists like MPEP (mGluR5-selective) and BAY36-7620 (mGluR1-selective). This new generation of pharmacological agents allows a more stringent assessment of the role of individual mGluR-subtypes or groups of receptors in various nervous system disorders, including ischaemia-induced brain damage, traumatic brain injury, Huntington s and Parkinson s-like pathology or epilepsy. Moreover, the use of genetically modified animals (e.g., knock-out mice) is starting to shed light on specific functions of mGluR-subtypes in experimental neuropathologies.

[1]Novartis Pharma AG, Nervous System Research, Basel, Switzerland, [2]Istituto Neurologico Mediterraneo-Neuromed, Pozzilli, Italy and [3]Department of Human Physiology and Pharmacology, University of Rome, Italy.

INTRODUCTION

The neurotoxicity of excitatory amino acids such as glutamate and some of its analogs, e.g., kainate and NMDA is well established in the central nervous systems.[1] Glutamate, the transmitter of the vast majority of excitatory synapses in the mammalian brain, activates ionotropic and metabotropic receptors (iGluRs and mGluRs), which mediate the physiological and the toxic effects of the transmitter. Activation of iGluRs results in Fast excitatory synaptic transmission via cation channel gating while mGluRs play a modulatory role by controlling membrane enzymes and second messengers. Molecular and physiological diversity of iGluRs was reviewed recently.[2,3] The G protein-coupled mGluRs form a family of currently eight subtypes (mGluR1 to -8), subdivided into three groups (I-III) on the basis of their amino acid sequence identities, pharmacological profiles and signal transduction pathways.[4-6] Group-I mGluRs (mGluR1 and mGluR5) are positively coupled to the phosphoinositide/Ca^{2+} cascade. Group-II (mGluR2 and mGluR3) and group-III (mGluR4, mGluR6, mGluR7 and mGluR8) receptors are both negatively coupled to adenylate cyclase in heterologous expression systems. The mGluR diversity is further increased by variants generated by alternative splicing at the intracellular C-terminal region. Splice variants for mGluR1 (termed mGluR1a, -b, -c, -d, -e),[4,7-9] mGluR4 (a, b),[10] mGluR5 (a, b),[11,12] mGluR7 (a, b)[13] and mGluR8 (a, b)[14] have been cloned and pharmacologically characterized. Little pharmacological or physiological differences between such mGluR splice variants, most likely generated from the same receptor gene, are reported to date.

The three groups of mGluRs can be discriminated pharmacologically with the use of selective agonists. 3,5-DHPG selectively activates group-I mGluRs, while (2*R*,4*R*)-APDC and LY-354740 are examples for group-II selective agonists; L-AP4, (*R*,*S*)-PPG, L-SOP and close analogues are selective agonists for group-III mGluRs.[6,13,15-21]

The individual mGluR-subtypes show a wide but distinct regional distribution throughout the mammalian nervous system. Subtypes of all three groups are highly expressed in neocortical layers, hippocampus, basal ganglia, thalamus/hypothalamus, cerebellum and spinal cord. Only mGluR6 appears to be exclusively restricted to retinal ON bipolar cells, where it couples to a cGMP-phosphodiesterase and amplifies visual transmission.[5] The precise subcellular localization of each individual mGluR subtype (see Fig. 1) has been studied extensively throughout the mammalian nervous system, mostly in hippocampus, basal ganglia, cerebellum, spinal cord, and retina. Group-I mGluRs, as shown by numerous studies, are found postsynaptically with the highest receptor density at perisynaptic locations outside the postsynaptic membrane specializations.[22,23] Similarly, it was shown that postsynaptic mGluR2/3 staining was also concentrated on the periphery of synaptic specializations.[24] However, work by Shigemoto et al[23] and Lujan et al[25] demonstrated that mGluR2 immunoreactivity was located primarily in the presynaptic terminals; most of the receptors were found distant from the release sites. mGluR3 is highly expressed in glial cells,[26-28] but the role of these glial receptors is as yet undetermined. The group-

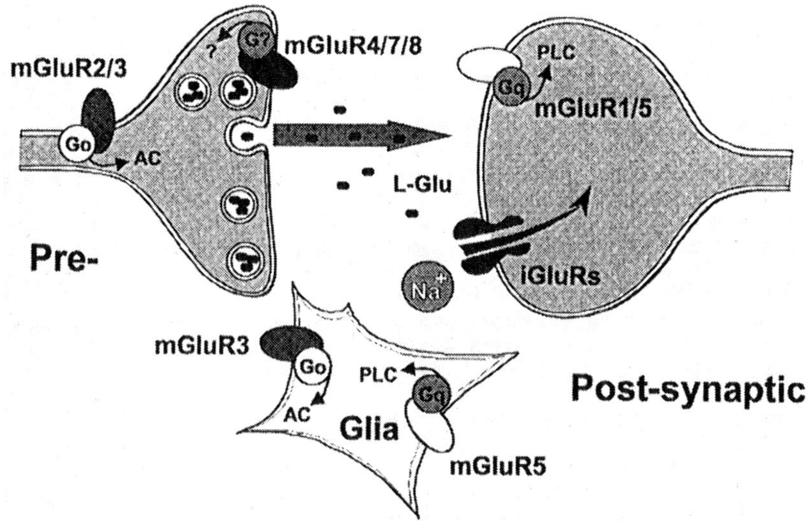

Figure 1. Glial, Pre-, and Postsynaptic localization of metabotropic glutamate receptor subtypes (mGluR1-8). Coupling of mGluR1 and -5 (group-I) to postsynaptic phospholipase C (PLC) via G_q protein is indicated. Group-II mGluRs (subtypes -2 and -3) are found at presynaptic and glial locations and coupling to adenylate cyclase is mediated by G_o and possibly G_i. Presynaptic expression of group-III mGluR subtypes -4, -7, and -8 is shown; specific G protein and second messenger coupling is still controversial. Due to restricted expression in retinal ON bipolar cells, mGluR6 is not shown.

III receptors mGluR4, -7 and -8 are localized presynaptically in, or near, the active zones and are thought to mediate presynaptic depression of glutamatergic synaptic potentials, most likely via inhibition of voltage-gated calcium entry and regulation of glutamate release.[6,23,29]

Here we discuss how activation or inhibition of distinct groups or subtypes of metabotropic glutamate receptors result in neuroprotective effects. Synaptic localization and signal transduction of the individual receptor subtypes will be considered to explain possible mechanisms of neuroprotection; and finally we will discuss the potential use of mGluR-selective compounds in the treatment of acute and chronic degenerative disorders of the nervous system.

CHEMICAL STRUCTURES AND RECEPTOR PROFILE OF NEUROPROTECTIVE MGLUR LIGANDS

Group-I Preferring Antagonists

The first generation of selective antagonists described for metabotropic glutamate receptors were phenyl glycine derivatives, e.g., 4CPG, MCPG, and 4C3HPG (Fig. 2). Those compounds were widely examined in model systems of neurodegeneration/

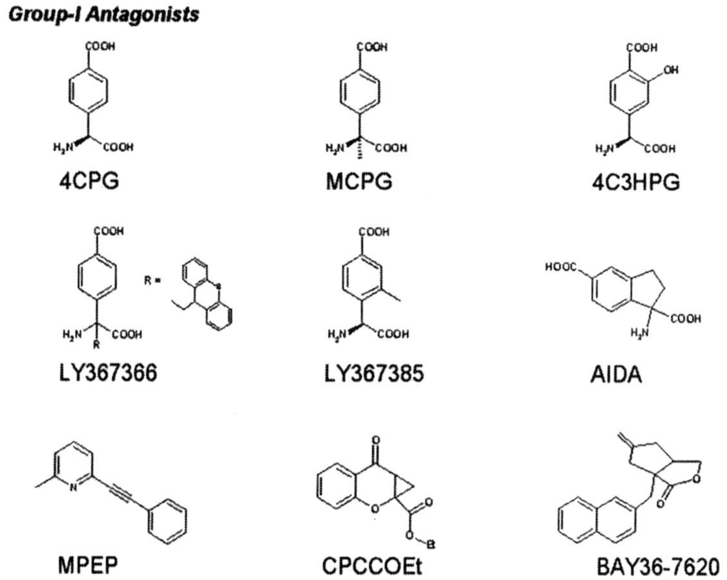

Figure 2. Chemical structures and abbreviated names of most commonly used group-I mGluR antagonists. For all compounds shown, neuroprotective effects were reported.

neuroprotection (see below). A very frequently used compound is MCPG which is an equally potent antagonist of mGluR1 and mGluR2 (IC_{50} = 20-500 µM).[9,30] MCPG is less potent at mGluR5 (IC_{50} = 200-1000 µM), and showed little antagonist activity at mGluR3, -4, -6, -7, and -8 (IC_{50} > 1000 µM).[9] 4CPG is a more potent antagonist for mGluR1 (IC_{50} = 20-80 µM) with little effects on mGluR2, -4, -5 (IC_{50} >500 µM), and no test results available for mGluR3, -6, -7, and -8;[9,30] 4C3HPG shows a very similar profile but in addition to the effects of 4CPG it is an agonist at mGluR2 (EC_{50} = 20 µM).[30]

The second generation of group-I mGluR antagonists showed higher potency and/or selectivity. LY367366 antagonizes mGluR1 and -5 (IC_{50} = 3-6 µM), interacts with group-II and -III to a lower extent (IC_{50} > 10 µM) and shows no activity at iGluRs; the compound LY367385 shows a very similar profile except, it doesn t antagonize mGluR5 (Fig. 2).[9,31] AIDA is a mGluR1-selective antagonist (IC_{50} = 4-200 µM) with little activity at mGluR2, -4, and -5 (IC_{50} > 1000 µM); tests at mGluR3, -6, -7 and -8 were not performed.[9,30]

More recently, amino acid-unrelated compounds with a high degree of selectivity for mGluR1 or mGluR5 were described. MPEP is a very potent and selective antagonist at mGluR5 with an IC_{50} of 32 nM, but it has no activity at all other mGluRs up to 10 µM; at this concentration it s also inactive at a broad selection of iGluRs (Fig. 2).[32] CPCCOEt is a very selective antagonist for mGluR1 (IC_{50} = 6.5 µM) which shows up to 100 µM no activity at mGluR2, -4, -5, -7, and -8 (mGluR3

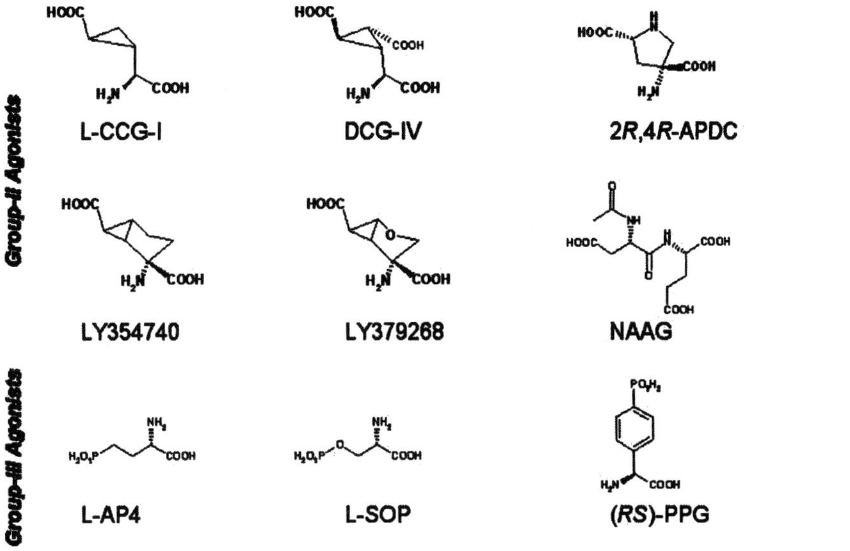

Figure 3. Chemical structures and abbreviated names of most commonly used group-II and -III mGluR agonists. For all compounds shown, neuroprotective effects were reported.

and -6 were not tested).[9,33] Similarly, BAY36-7620 was very recently described as a highly selective mGluR1 antagonist showing higher potency than CPCCOEt (IC_{50} = 0.16 μM; Fig. 2).[34]

Agonists for Metabotropic glutamate receptors

The first mGluR agonist used on neuronal tissue were quisqualate, ibotenate, and *trans*-ACPD or its active isomer 1*S*,3*R*-ACPD.[9] Although important insights into neuroprotective/neurodegenerative activity of mGluR agonists were gained (e.g., activation of group-I amplifies neuronal death),[35] more direct evidence for an involvement of specific groups/subtypes of mGluRs arose via the use of more selective agonists. The glutamate analogues L-CCG-I and DCG-IV (Fig. 3) were used to demonstrate group-II mGluRs as potential targets for neuroprotective drugs.[36] Both compounds show very potent agonist activity at mGluR2 and mGluR3 (EC_{50}-values = 0.1-0.9 μM), however at higher concentrations L-CCG-I is also an agonist at mGluR1, -4, -5, -6, -7, and -8 (EC_{50}-values = 2-9 μM), while DCG-IV antagonizes all group-III mGluRs (IC_{50} = 22-40 μM); DCG-IV is also a potent agonist of NMDA receptors.[9,37] Considerably improved compounds were published more recently. 2*R*,4*R*-APDC shows agonist EC_{50}-values at mGluR2 and mGluR3 of 0.3-0.4 μM.

The compound has no appreciable iGluR activity and doesn t activate or antagonize any group-I or -III mGluR up to 100 μM. Thus, 2*R*,4*R*-APDC is extremely specific for group-II receptors.[9] The compounds LY354740 and LY379268 show much higher potency and equally good specificity for mGluR2 and mGluR3 (EC_{50}-values = 3-20 nM), when compared to 2*R*,4*R*-APDC (Fig. 3).[9,20] NAAG is an interesting molecule, because it activates only mGluR3 (EC_{50} = 65 μM) but not mGluR2 (EC_{50} > 300 μM), and it doesn t show interaction with group-I and group-III receptors.[38]

An involvement of group-III mGluRs in neuroprotection has primarily been demonstrated by the use of the phosphono amino acids L-AP4 and (*R,S*)-PPG (Fig. 3). Both compounds are selective activators of group-III mGluRs with no appreciable activity at other mGluRs or iGluRs. The EC_{50}-values of L-AP4 at mGluR4 and -6 were reported to be 0.2-1 μM, 150-500 μM at mGluR7 and 0.06-0.9 μM at mGluR8. (*R,S*)-PPG is slightly less potent with EC_{50}-values at mGluR4, -6, -7, and -8 receptors of 5.2 μM, 4.7 μM, 185 μM and 0.2 μM, respectively.[5,9,18,21] The pharmacology of L-SOP at group-III mGluRs is very similar to (*R,S*)-PPG, but characterization of L-SOP at other receptors is not available to date.[9]

PHYSICO-CHEMICAL AND PHARMACOKINETIC PROPERTIES OF NEUROPROTECTIVE MGLUR LIGANDS

The investigation of the respective roles of each mGluR subtype in neurodegeneration and neuroprotection requires the use of ligands which are not only group- or subtype-selective but which also possess the adequate physico-chemical and pharmacokinetic properties needed for their use in the different experimental models.

General Considerations

Glutamate and aspartate are the two acidic representatives among the natural amino acids. The presence of a distal carboxylic acid enhances their polar properties in comparison to the other amino acids. These polar properties limit considerably their capacity to cross membranes by passive diffusion. Therefore, glutamate is actively transported through biological membranes by specific transporter proteins. Up to date, almost all mGluR-selective ligands are amino acid derivatives. The only non-amino acid compounds discussed in this report (see Figs. 2 and 3) are MPEP, CPCCOEt and BAY36-7620 which are all group-I antagonists.[32,34,39,40] All other structures shown in Figure 2 and 3 share the same functional groups as glutamate, namely the amino acid functionality and the distal carboxylic acid (or phosphate moiety in the case of group-III ligands). Despite the structural analogy to glutamic acid the majority of these molecules fails to be actively transported through the cell membranes by the specific glutamate transporter systems and as a consequence have a poor oral bioavailability and do not cross the blood-brain-barrier (BBB). The two reported exceptions, LY354740 and its derivative LY379268, are orally bioavailable, enter the brain and display potent agonist activity at mGluR2, and -3.[20,41] In addition, these two ligands have a good water solubility, which is generally the case for glutamate analogs and of advantage for the formulation and the use in *in vitro* neuroprotection assays. The recently published non-amino acid compounds MPEP and BAY 36-7620 also demonstrate oral bioavailability and good BBB penetration[32,39,42] in contrast to the earlier published mGluR1 antagonist CPCCOEt, which neither shows bioavailability nor BBB penetration.

In vivo Dosage and Side Effects

Most glutamate analogs such as MCPG, 4C3HPG and AIDA can only by studied in animal disease models after central administration (i.c.v.). The relatively low potency and lack of selectivity of such compounds usually doesn t allow conclusions on the side effect profile of antagonizing specific receptors. In contrast, MPEP, the mGluR5-selective antagonist at *in vitro* and *in vivo* conditions can easily be studied in animal models because it reaches the brain via oral and intravenous application (p.o. and i.v.).[32,42] Furthermore, MPEP shows a large window between the doses achieving the beneficial effects in animal models for inflammatory pain (10-100 mg/kg, p.o.) or anxiety (1-10 mg/kg, p.o.) and the dose inducing the first side effects (> 100 mg/kg, p.o.) such as inhibition of spontaneous locomotor activity.[32,42,43]

In the case of group-II mGluRs, two orally active compounds have been investigated: LY354740 and LY379268. Both compounds are nanomolar potent agonists with similar potency at mGluR2 and mGluR3 and the investigations *in vivo* show that both compounds penetrate the BBB. In animal models, LY354740 showed anxiolytic like effects after oral administration (0.5-10 mg/kg, p.o.) without the sedative or motor impairment side effects of the classical benzodiazepine anxiolytics. Other group-II agonists, discussed here, are not bioavailable and do not cross the BBB, therefore have to be centrally administered.

None of the currently known group-III-selective agonists, e.g., L-AP4, L-SOP or (*R,S*)-PPG, show oral bioavailability or BBB penetration. Therefore, *in vivo* experiments necessitate a central administration of the drugs. The presence of a very polar acidic moiety such as the phosphate/phosphonate group is certainly a key element for the poor bioavailability of these molecules. *In vivo* (i.c.v. administration) and *in vitro* investigations using (*R,S*)-PPG in mGluR4- and mGluR7-deficient mice showed that mobilization of mGluR4 occurred to achieve neuroprotective effects (10 nmol, i.c.v.)[21,44] whereas activation of mGluR7 produces anticonvulsive effects (Herman van der Putten et al, personal communication). Interestingly, the sedative and respiratory side-effects induced by central administration of higher doses of PPG (200-2000 nmol, i.c.v. in mice)[21,45] are also seen in mGluR4- or mGluR7-deficient animals, clearly indicating that these side-effects are mediated by an interaction of (*R,S*)-PPG with target(s) distinct from mGluR4 and mGluR7 (Herman van der Putten, personal communication).

GROUP-I MGLURS AS TARGETS FOR NEUROPROTECTIVE DRUGS

General and Introductory Considerations on Group-I Receptors

The role of mGluR1 and mGluR5 in neurodegeneration/neuroprotection has been debated for several years because of the contrasting effects of agonists (such as DHPG) in different *in vitro* models of neuronal toxicity.[46] However, the evidence that subtype-selective mGluR1 or -5 antagonists are consistently neuroprotective, suggests that endogenous activation of group-I receptors is permissive to cell death. mGluR1 and -5 are both coupled to polyphosphoinositide hydrolysis, but they differ for the kinetics of IP3-induced Ca^{2+} mobilization. In transfected cells, activation of mGluR1 induces a single-peaked Ca^{2+} response, whereas activation of mGluR5 promotes oscillatory increases in cytosolic free Ca^{2+} due to the PKC-dependent phosphorylation of a specific threonine residue that is just distal to the 7th trans-membrane domain of mGluR5.[47] Whether or not a similar difference exists under native conditions is unclear at present. Activation of both receptors regulates the activity of a variety of Ca^{2+} and K^+ channels and turns on a not-yet-defined voltage-gated inward current,[48,49] with the net effect of enhancing neuronal excitability. Particular attention deserves the description of a positive modulation of NMDA-gated ion currents by mGluR5, which provides the most likely substrate for the permissive action of mGluR5 activators on neuronal toxicity.[50-55] Interestingly, NMDA receptor activation can also potentiate mGluR5 responses by inducing a calcineurin-dependent dephosphorylation of a PKC phosphorylation site that participates in mGlu5 desensitization.[56,57] The Shank family of postsynaptic density proteins may cross-link Homer and PSD-95, thus allowing the functional coupling between mGluR5 and NMDA receptors in the same region of the dendritic spine.[58] Studies performed in transfected oocytes suggest that only NMDA receptors containing NR2A or -2B subunits are positively modulated by mGluR5, whereas NMDA receptors containing the NR2C (or -2D) subunit are not. This may help explain why in mature cerebellar granule cells (that predominantly express the NR2C subunit) activation of group-I mGluRs enhances NMDA responses only when the NR2C subunit is knocked down by antisense oligonucleotides.[59] A question that awaits a final answer is whether or not group-I mGluR subtypes are presynaptically localized and modulate neurotransmitter release. This issue is fundamental because the extent of excitotoxic degeneration is strictly related to glutamate and GABA release in a number of disorders, such as ischemia-induced neurodegeneration and temporal lobe epilepsy. A series of studies carried out in the laboratory of J. Sanchez-Prieto indicate that a first application of group-I mGluR agonists facilitates glutamate release in cortical or hippocampal synaptosomes,[60-65] whereas a second drug application inhibits release. This functional change in the modulation of glutamate release depends on the PKC-dependent phosphorylation of the receptor (presumably mGluR5), leading to a switch in the G-protein coupling. A similar scenario is observed in cultured cortical cells, where a first exposure to DHPG amplifies NMDA toxicity, whereas a second exposure is neuroprotective.[66] As a secondary release in endogenous glutamate contributes to the progression of NMDA toxicity, the functional state of presynaptic mGluR5 might be determinant for the final effect of group-I mGluR agonists on excitotoxic neuronal death. Electron microscopy studies have shown that group-I mGluRs are preferentially localized in the peripheral portion of postsynaptic densities rather than in presynaptic terminals.[22,23] However, the evidence that mGluR5 is found in the axon when co-transfected with Homer-1a raises the intriguing possibility that homer proteins regulate the dendritic or axon targeting of mGluR5.[67] Homer-1a is encoded by an early gene, which is switched on by neuronal hyperactivity. It will be interesting to examine whether changes in the subcellular distribution of mGluR5 are associated with synaptic hyperactivity that occurs during epilepsy or brain ischemia.

Effect of Group-I mGluR Agonists on Excitotoxic Neuronal Death

mGluR1/5 agonists may either amplify excitotoxic neuronal death or produce neuroprotection, depending on the neuronal type and the paradigm of toxicity. Possible explanations for contrasting data include (i) the absence or presence of the NR2C subunit in NMDA receptors; (ii) the existence of an activity-dependent switch between facilitatory and inhibitory mGluR subtypes (see above); and (iii) a role for glial cells expressing mGluR5. These aspects have been discussed in detail in a recent review.[46] It is noteworthy that excitotoxic neuronal death incorporates features of necrosis and apoptosis, and that apoptosis by trophic deprivation contributes to the overall neuronal death in brain ischemia.[68] One expects that, in experimental model of brain ischemia, pharmacological activation of group-I mGluRs amplifies necrotic death by further increasing cytosolic free Ca^{2+}, but supports the survival of neurons undergoing apoptosis by trophic deprivation through the same mechanism.[69] Therefore one can easily conclude that pharmacological activation of group-I mGluR subtypes is not valuable in the experimental treatment of acute or chronic neurodegenerative disorders.

Neuroprotective Activity of mGluR1 Antagonists

A battery of competitive or non-competitive mGluR1 antagonists has been tested in in vitro and in vivo models of excitotoxic neuronal death. AIDA, LY367385, 4CPG, 4C3HPG, and CPCCOEt protect mixed murine cortical cultures against NMDA toxicity,[31,70,71] and are also effective in in vivo models of excitotoxic death.[31,71,72] LY367385, 4C3HPG and AIDA are also neuroprotective in the gerbil model of global ischemia and in murine cortical cultures and rat organotypic hippocampal cultures subjected to oxygen-glucose deprivation.[31,73,74] mGluR1 is preferentially localized in GABAergic neurons and its activation depresses inhibitory synaptic transmission.[22,75-78] Hence, it has been hypothesized that mGluR1 antagonists are neuroprotective by enhancing GABAergic transmission. Initial evidence in this line has been provided by the observation that local infusion of AIDA increases the basal output of GABA in the gerbil hippocampus.[74] In murine cortical cultures, neuroprotection by LY367385 or CPCCOEt (but not by MPEP) is occluded by GABA or by SKF89976A (an inhibitor of GABA transporter), and is abolished by a cocktail of GABA-A and GABA-B receptor antagonists.[71] The same cocktail of antagonists prevents the protective activity of CPCCOEt against neurodegeneration induced by NMDA in in vivo studies.[71] A role for GABAergic transmission is strengthened by the evidence that LY367385 and CPCCOEt enable NMDA to enhance GABA release in the striatum, and that activation of mGluR1 inhibits GABAergic IPSCs in cortico-striatal slices.[71] Taken together, these data suggest that mGluR1 antagonists are neuroprotective by removing a tonic inhibitory control exerted by mGluR1 on GABA release. The finding that mGluR1 is localized also on GABAergic nerve terminals in the striatum is consistent with this hypothesis.[78] mGluR1 antagonists might be beneficial in CNS disorders characterized by an impairment of inhibitory synaptic transmission, such as Ammon s horn sclerosis.[79] Interestingly, the very recently introduced mGluR1 antagonist BAY36-7620 was shown to be protective in epilepsy, stroke and trauma models upon systemic administration.[34,39]

Neuroprotective Effects of mGluR5 Antagonists

The development of subtype-selective mGluR5-specific antagonists has allowed to establish that endogenous activation of mGluR5 facilitates cell death in a variety of models of neurodegeneration. MPEP and its ancestors SIB-1757 and SIB-1893 (which all behave as non-competitive mGluR5 antagonists) are potent and effective in attenuating NMDA toxicity in culture, and at least MPEP can also reduce excitotoxic neuronal death in the rat striatum.[80] As opposed to mGluR1 antagonists, these effects do no involve changes in GABAergic transmission[71] and might be explained with the ability of mGluR5 to positively modulate NMDA-gated ion currents (see above). In cortical cultures, neuroprotection by MPEP, SIB-1757 or SIB-1893 is observed at concentrations $<10\ \mu M$, and, therefore, cannot be ascribed to any non-specific interaction with NMDA receptors.[32,42,80,81] In addition, the neuroprotective action of MPEP is not mimicked by its isomer iso-MPEP, which fails to antagonize mGluR5 although it shares most of the structural and physico-chemical features of MPEP.[80] No data are yet published on the effect of MPEP in experimental models of hypoxic/ischemic neuronal death. The expectation is unclear, because activation of mGluR5 is known to support the survival of developing neurons,[80] and apoptosis by trophic deprivation contributes to the overall neuronal death in brain ischemia (see above).

A series of studies outlines a potential use of mGluR5 antagonists in the experimental therapy of chronic neurodegenerative disorders. In cultured cortical cells, MPEP and its analogs are neuroprotective against β-amyloid toxicity at concentrations lower than those required for neuroprotection against excitotoxic death. In these studies, β-amyloid peptide is applied to the cultures in the presence of MK-801 and DNQX, thus excluding any involvement of ionotropic glutamate receptors in the neuroprotective activity of mGlu5 receptor antagonists.[80] Interestingly, inhibition of mGluR1 by AIDA does not reduce but rather exacerbates §-amyloid toxicity in cortical cells.[83] Thus, at least in cultured neurons, endogenous activation of mGluR5 is required for the engagement of a death pathway in neurons exposed to §-amyloid peptide. This encourages the study of mGluR5 antagonists in animal models of Alzheimer s disease. Finally, mGluR5 antagonists may be promising in the treatment of Parkinson s disease (PD). An optimal antiparkinsonian drug should combine neuroprotective properties with the ability to relief the cardinal symptoms of the disease (i.e., bradykinesia, rigidity and tremor). Recent unpublished data show that systemically administered MPEP or SIB-1893 (both at 10 mg/kg, i.p.) protect nigro-striatal dopaminergic terminals against metamphetamine toxicity and reduce the amount of reactive oxygen species produced by metamphetamine in the striatum of freely moving animals (G. Battaglia et al, manuscript in preparation). On the other hand, activation of mGluR5 induces direct excitation, and selectively potentiates NMDA currents in neurons of the subthalamic nucleus.[53] As an increased activity of the subthalamic nucleus is implicated in the pathophysiology of bradykinesia, one can predict that mGluR5 antagonists can improve parkinsonian symptoms in experimental animals and humans. It will be interesting to test mGluR5 antagonists in classical experimental models of parkinsonism, such as the MPTP model in mice or monkeys.

NEUROPROTECTION MEDIATED BY GROUP-II MGLURS

mGluR2/3 Agonists Are Protective in Several Experimental Paradigms of Neurodegeneration

In the last decade several reports have shown that pharmacological activation of group-II mGluRs is neuroprotective in a variety of models of neuronal degeneration, including neuronal cultures, brain slices and *in vivo* models of excitotoxicity. The non-selective group-II mGluRs agonists, DCG-IV, 4C3HPG, L-CCG-I and 1S,3R-ACPD, protect neurons against NMDA toxicity in mixed cultures of mouse cerebral cortex.[36,66,84-89] This effect is substantially reduced by the group-II mGluR antagonists, PCCG-IV or MCCG-I, suggesting that activation of mGluR2/3 is responsible for neuroprotection and is supported by the finding that the highly selective mGluR2/3 agonist, 2*R*,4*R*-APDC, is neuroprotective in the same model.[90,91] DCG-IV or L-CCG-I are also neuroprotective against kainate toxicity in cultured cortical cells,[36] although Gottron et al[92] found that, in the same cultures, protection by DCG-IV is restricted to the small percentage of neurons that respond to kainate with an enhanced influx of Co^{2+}. Neuroprotection by group-II mGluR agonists has also been shown in primary cultures of cerebellar granule cells,[93] as well as in primary cultures of mesencephalic neurons,[94] challenged with excitotoxins. Recently, the potent, highly selective and systemically active group-II mGluR agonists, LY354740, LY379268 and LY389795, have been shown to prevent excitotoxicity in both rat and mouse cortical neuronal cultures (see Kingston et al and D Onofrio et al for support; but see also Behrens et al).[89,95-97] Interestingly, the effect of these compounds is enhanced by the presence of glial cells, suggesting an involvement of glial mGluR3 in the mechanism of neuroprotection.[96] Accordingly, NAAG, an endogenous selective mGluR3 agonist, protects cultured cortical neurons against NMDA toxicity.[38,86]

Selective and non selective group-II mGluR agonists exert neuroprotective activity also in further in vitro models of neuronal injury, i.e., oxygen-glucose deprivation- and staurosporine-induced neuronal death, in which neurons follow the apoptotic pathway of degeneration.[87,95,98,99]

Group-II mGluR agonists have also been shown to be neuroprotective in in vivo models of neurodegeneration. DCG-IV, infused i.c.v., protects vulnerable neurons against local or systemic injection of kainate.[100] In addition, LY379268, locally or systemically injected, protects against NMDA-induced striatal GABAergic neuronal loss in rats.[89] Intracerebroventricular injection of 4C3HPG protects hippocampal CA1 neurons in a gerbil model of global ischemia (BCAO), both when administered prior to and after the induction of ischemia.[101] The same degree of neuroprotection in hippocampal CA1 neurons, in the BCAO model, has been observed after systemic injection of LY354740 or LY379268.[102,103] In contrast, the infarct size following middle cerebral artery occlusion (MCAO), induced by endothelin-1 in rat, is unaffected by the drugs, suggesting that group-II mGluR agonists are likely to have more utility in global than in focal cerebral ischemia.[103,104] LY379268 exerts also neuroprotective activity in a neonatal rat model of hypoxia-ischemia.[105]

Recently, group-II mGluR agonists have been shown to possess anti-seizure activity. LY354740 is anticonvulsant in mice when systemically administered in both the ACPD-induced limbic seizure model and the pentetrazole- and picrotoxin-induced seizures.[20,106,107] Moreover, group-II mGluR agonists attenuate both in vitro and in vivo models of traumatic neuronal injury, showing an additive effect with NMDA- and group-I mGluR antagonists.[108] In addition, L-CCG-I and 4C3HPG inhibit excitotoxic phenomena mediated by kainate on spinal cord motor neurons, a model resembling amyotropic lateral sclerosis.[109]

Neuroprotection via mGluR2/3 Involves Glial-Neuronal Interactions

The mechanisms underlying group-II mGluR-mediated neuroprotection has been examined in cultured cortical cells exposed to a brief pulse with NMDA, a model of excitotoxicity in which the damage is induced by the influx of extracellular Ca^{2+} through the NMDA receptor channel during the pulse, and is amplified by the endogenous glutamate secondarily released after the pulse.[110] Group-II mGluRs are preferentially localized in the preterminal region of the axon, far from the active zone of neurotransmitter release, and their activation inhibits the release of glutamate, but only in response to high concentrations of glutamate that spread back to the most remote region of the axon. Moreover, mGluR3 is also expressed by astrocytes throughout the brain.

It has been initially suggested that activation of mGluR2/3 is neuroprotective by reducing the release of endogenous glutamate. However, an additional mechanism might be responsible for neuroprotection since the protein synthesis inhibitor, cycloheximide, prevents the neuroprotective activity of group-II mGluR agonists in cortical cell cultures.[85,95,96] Thus, it has been proposed that group-II mGluR activation triggers a specific program that requires new protein synthesis and that is likely to occur in glial cells, as the conditioned medium collected from pure cultures of astrocytes, transiently exposed to DCG-IV, L-CCG-I, 4C3HPG, LY379268 (which activate both mGluR2 and mGluR3) or to NAAG (a selective mGluR3 agonist), is highly neuroprotective when transferred to mixed cortical cultures already challenged with NMDA.[85] Moreover, the neuroprotective activity of group-II mGluR agonists is almost completely lost in the absence of glial cells.[95,96] This effect is likely to be mediated by the production and release from astrocytes of transforming growth factor-β (TGF-β), induced by the activation of glial mGluR3.[86]

In recombinanT cells, mGluR2/3 are negatively coupled to adenylate cyclase through a Gi protein. However, it has been shown that in mixed cortical cultures exposed to NMDA, neither forskolin nor dibutyril-cAMP counteract the neuroprotective activity of group-II mGluR agonists, excluding an involvement of the α subunits of the Gi protein coupled to group-II mGluRs. Therefore, to explain how activation of Gi-linked receptors leads to the production of neurotrophic factors, attention has been focused on the intracellular pathways activated by the βγ subunits released from the Gi protein (Fig. 4). A possible link is represented by the activation of mitogen activated protein (MAP) kinase and phosphatidylinositol (PI)-3-kinase which occurs both in primary cultures and in recombinanT cells (Fig. 4).[89] Accordingly, blockade of these pathways leads not only to a reduction of the TGF-β production induced by group-II mGluR activation, but also reverses the neuroprotective activity of mGluR2/3 agonists, an effect which also occurs in in vivo models of excitotoxicity.[89] MAP kinases, $P90^{rsk}$ and $P70^{S6k}$, which regulate gene expression and protein synthesis, could represent the downstream effectors of the de novo synthesis of TGF-β.

However, the proposed mechanism of glial-neuronal interaction mediated by TGF-_ production in response to mGluR2/3 activation, observed both in cortical cultures challenged with NMDA and in rat striatum infused with NMDA, is not found in hippocampus after BCAO.[111] As Gi-linked receptors, such as group-II mGluRs and A1 adeNOSine receptors, have been shown to induce in cultured astrocytes the release of other factors, such as nerve growth factor and S-100β protein, other than TGF-β,[112] it cannot be excluded that in different brain regions other factors modulate the neuroprotective activity of group-II mGluRs.

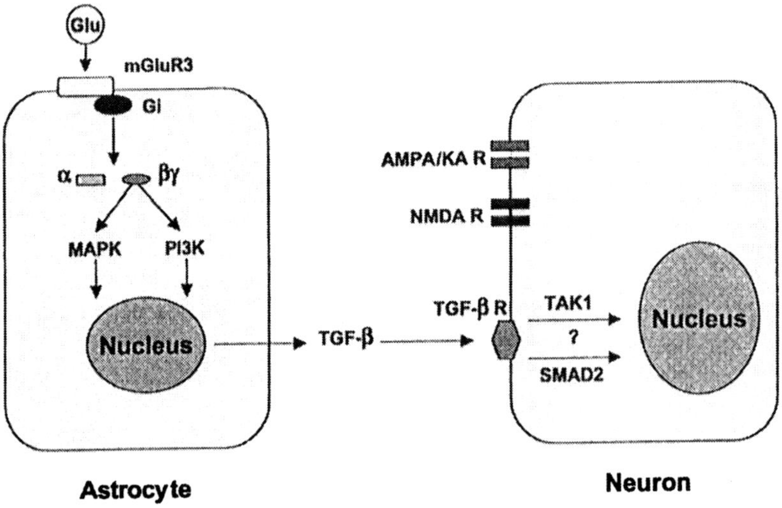

Figure 4. Neuroprotection via activation of mGluR3. Schematic representation of the cascade of events leading to TGF-§ secretion by astrocytes. Protein kinases (e.g., TAK1) and SMAD proteins are likely to transduce TGF-§ signals from cell surface to the nucleus of neurons.

mGluR2/3 Agonists and Potential Applicability in Acute and Chronic Neurodegeneration

Group-II mGluRs may be considered as a potential target for drugs aimed at reducing the progression of neuronal degeneration (Fig. 4). A safe neuroprotective drug with a favorable therapeutic window is particularly needed in the experimental therapy of ischemic brain damage. Group-II mGluR agonists might meet these criteria because: (i) they are neuroprotective against in vitro degeneration induced by oxygen-glucose deprivation and in in vivo models of global ischemia; (ii) they do not interfere with the normal excitatory synaptic transmission, in contrast to NMDA or AMPA receptor antagonists; (iii) they protect cultured neurons when applied after a toxic pulse with NMDA, at times at which NMDA antagonists are no longer protective, and they are even active in vivo when applied after the induction of an ischemic damage; (iiii) by locally increasing the production of neurotrophic factors such as TGF-β, they may provide a broad spectrum mechanism of protection, as TGF-β is known to exert neuroprotective activities against neuronal degeneration induced by excitotoxins, oxygen-glucose deprivation, β-amyloid peptide and the HIV capside protein gp120,[86,98,99,113-124] and (iiiii) mGluR3 is localized on the vascular side of astrocytes, in proximity of endothelial cells, so they can be easily reached by drugs present in the blood stream and able to cross the blood-brain-barrier.

Recently group-II mGluRs have been proposed as targets for the therapy of PD. Overactive glutamatergic afferents from the subthalamic nucleus could cause both an excitotoxic loss of dopaminergic neurons in the substantia nigra, and hyperactivation of GABAergic neurons in the globus pallidus, which leads to a reduction of motor activity.[125] Drugs, which can reduce glutamate release by acting at presynaptic group-II mGluRs might at one time delay the degeneration of substantia nigra neurons and improve motor activity.[126,127]

Furthermore, it s interesting to add that group-II mGluR agonists could interfere with the pathophysiology of neuropsychiatric disorders such as anxiety and schizophrenia, and they may also exert anti-addictive effects.[20,106,128-134]

ROLE OF GROUP-III MGLURS IN NEUROPROTECTION

Effects of Group-III mGluR Ligands in Experimental Paradigms of Neurodegeneration

There are numerous reports that describe activation of group-III mGluRs to be neuroprotective *in* vitro, and more recently, several in vivo observations support these findings. The agonists L-AP4, L-SOP and (*RS*)-PPG promote survival of rat cerebellar granule cells and protect cultured cortical and cerebellar neurons against toxic insults, such as prolonged β-amyloid peptide exposure, transient iGluR activation or mechanical damage.[21,44,98,99,136-138] In rat hippocampal slices exposed to a severe hypoxic/hypoglycaemic insult, (*RS*)-PPG improves the recovery of population spike amplitude in CA1, a parameter for functional synaptic transmission and neuronal viability. Such acutely isolated hippocampal slices are only viable for a few hours, and electrophysiogical recordings can only be performed in a limited time window after the damaging insult when most probably excitotoxicity is still the predominant component in the pathophysiology of hypoxic/hypoglycaemic neuronal damage.[45,139]

In a recent study we found (*RS*)-PPG to be neuroprotective against striatal lesions induced by local infusion of NMDA or quinolinic acid into the rat caudate nucleus and, to our knowledge, this provided the first in vivo evidence that activation of group-III mGluRs is neuroprotective in animal models.[21] The use of such in vivo excitotoxic injury models, to produce neuronal depletion, reactive gliosis and alterations of neurotransmitter levels, has been highly valuable for examining pathological patterns reminiscent of Huntington s disease (HD). Even if the primary cause of HD is unrelated, excitotoxic injury mediated by iGluR activation may play a role in progressive neuronal depletion.[140]

Extending the studies to further in vivo models of neurodegeneration, we and others confirmed neuroprotective effects of (*RS*)-PPG in rat and also mouse models of excitotoxicity (in vivo and in vitro), but, neither in focal cerebral ischaemia in mice, nor in global cerebral ischaemia in gerbils, nor in global cerebral ischaemia in rats (*RS*)-PPG had any significant influence on the extent of neuronal damage.[44,45]

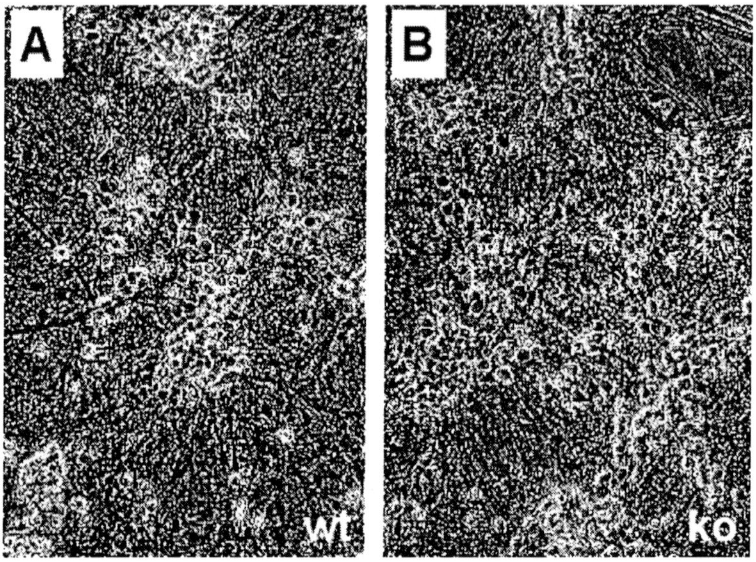

Figure 5. Phase-contrast photomicrographs of cultured cortical cells from CD1 (wild type, wt, A) and mGluR4-deficient (knock-out, ko, B) mice. Dissociated cortical neurons prepared from fetal mice at 14-16 days of gestation after 13 days of in vitro culture are shown. Morphology of individual neurons and synaptically connected networks are indistinguishable between wild type and knock-out cultures. The cortical cells were plated on a confluent layer of astrocytes (14 days in vitro, not visible on micrographs) raised from the same mouse genotypes as neurons.

Neuroprotection via Group-III mGluRs Is Mediated by mGluR4 and/or mGluR7 and Involves a Presynaptic Regulation of Transmitter Release

In contrast to group-II mGluRs that require glial components for neuroprotection, Group-III agonists are equally protective in mixed cortical cultures (containing astrocytes and neurons, see Fig. 5) and in pure cultures of neurons. Therefore, this neuroprotection is not expected to be mediated by glial factors, but rather involves one or more receptor subtype(s) expressed on neuronal structures. Recently, we have searched for the identity of the neuroprotective group-III mGluR subtype using mixed cortical cultures. This model is particularly suitable for the study because cortical neurons of mixed cultures express all currently known group-III mGluR subtypes.[137] A possible role in neuroprotection for one or more group-III mGluR subtype(s) was initially suggested by the use of racemic (*RS*)-PPG in in vitro neuroprotection paradigms.[21] More recently, the stereoisomers of (*RS*)-PPG were separated and it was found that all protective activity is harbored in the (+)-

isomer.[44,141] (+)-PPG is neuroprotective against NMDA toxicity with an EC_{50} value of about 5 μM. This value coincides with that found for the activation of recombinant mGluR4, but differs by at least 25-fold from the potency of PPG at mGluR7 and -8.[21] Furthermore, there is strong evidence for a critical role of mGluR4 in mediating neuroprotection via the use of cortical cultures prepared from mGluR4 subtype-deficient mice (-/-, Fig. 5B), where all group-III agonists [i.e., L-AP4, (*RS*)-PPG and L-SOP] failed to protect against NMDA toxicity. This was in contrast to the protective effects of group-III agonists in wild type (+/+) and heterozygous neurons and did not reflect a general refractoriness of -/- knockout neurons to mechanisms of protection, because the group-I/II mGluR ligands 4C3HPG, CPCCOEt and MPEP, retained their protective activity in those mGluR4 -/- cultures.[32,33,44,142]

We extended the study to the in vivo model of excitotoxic degeneration by unilaterally injecting NMDA ± (*RS*)-PPG into the caudate nucleus of wild type or mGluR4 -/- mice. This brain region has been selected because it receives an extensive glutamatergic innervation from the cerebral cortex. Low doses of (*RS*)-PPG (10 nmol), which should preferentially activate mGluR4 over mGluR7 receptors, were neuroprotective in +/+ mice, but were totally inactive in mGluR4 -/- mice. In contrast, the -/- mice were partially protected with higher doses of (*RS*)-PPG (100 nmol),[44] which are expected to recruit mGluR7 (unpublished observations by Herman van der Putten, based on experiments with mGluR7 -/- mice). The prominent role of mGluR4 and possibly mGluR7 in mediating group-III agonist-induced neuroprotection is also consistent with the high expression levels and broad distribution of mGluR4 and mGluR7 in many regions of mammalian brain, including basal ganglia, cortical areas and hippocampus.[23,143-145]

Inhibition of glutamate release by presynaptic mGluR4 and mGluR7 may represent a common mechanism of neuroprotection in vitro and in vivo. Accordingly, an enhanced release of endogenous glutamate has been shown to facilitate the progression of NMDA toxicity in cortical cultures,[110] and we have found that cortical cultures from mGluR4 -/- mice were more vulnerable to low concentrations of NMDA, and showed higher extracellular glutamate levels, as compared to wild type +/+ cultures. Moreover, in vivo microdialysis studies showed that intrastriatal infusion of NMDA increased extracellular glutamate levels to a greater extent in mGluR4 -/- than in +/+ mice, supporting the hypothesis that the mGluR4 subtype is necessary for the maintenance of the homeostasis of extracellular glutamate levels.[44] In addition to the regulation of presynaptic glutamate release, an inhibition of NMDA receptors by postsynaptic group-III mGluRs (hypothetical) via a protein phosphorylation cascade may also be involved in their neuroprotective effects.[146] Furthermore, group-III mGluRs also exert protection in primary hippocampal neurons by modulation of the free radical nitric oxide and the cascade of programmed cell death.[147] Thus, activation of group-III mGluRs could open several novel strategies to interfere with the progressive course of neurodegenerative disorders.

mGluR4/7 Activators May Become Applicable in Basal Ganglia Disorders and Other Chronic Degenerative Diseases

The broadest analysis of group-III mGluR activation in animal models of neuronal damage was recently preformed by Henrich-Noack et al.[45] The results, as summarized above, support the notion that group-III mGluR agonists are a quite valuable class of drugs against pathologies closely related to excitotoxic cell damage, but are most probably not effective enough when damaged brain tissue has progressed into a multifactorial pathology as it appears after an ischaemic challenge, e.g., in human stroke. This view is supported by several clinical stroke trials that have evaluated the potential of glutamate transmission-modifying drugs, and to date, the results of these attempts have been disappointing.[148,149]

In amyotropic lateral sclerosis (ALS) and Huntington s disease (HD), circumstantial evidence suggests that excitotoxicity may contribute to the pathogenic process. In fact, it was recently shown that an anti-glutamate drug, riluzole, provides some therapeutic benefit in the treatment of ALS and clinical trials aiming at neuroprotection in HD are in progress.[148] HD and Parkinson s disease (PD) are examples of neurodegenerative disorders where mitochondrial dysfunction may sensitize populations of basal ganglia neurons to excitotoxicity from synaptic glutamate, and recent evidence strongly suggests that the group-III receptors mGluR4 and mGluR7 modulate basal ganglia function at multiple sites.[143,144] Activation of mGluR4 and/or mGluR7 may increase viability of selective cell populations, e.g., GABAergic projecting neurons in the striatum which are selectively lost in HD. Alternatively, modulation of group-III mGluR activity may balance basal ganglia circuits under pathological conditions, for instance in PD where subthalamic nuclei are overactive.[144] Thus, reduction of glutamatergic transmission by group-III mGluR activation may well be a novel and viable mechanism for experimental therapy of ALS, HD, and PD.

Evidence for an involvement of group-III mGluRs, or glutamatergic transmission in general, in Alzheimer s disease (AD) is more indirect and speculative than for ALS, HD, and PD. Experimental evidence, however, shows that various group-III agonists are highly protective against programmed cell death in cultured cortical cells induced by prolonged β-amyloid peptide exposure.[80,98] On the other hand, no in vivo data from transgenic models of AD has been published to date that would support a role of group-III receptors in AD.

Finally, it s very interesting to mention the effectiveness of group-III agonists in various experimental animal models of epilepsy.[21,150-152]

CONCLUSIONS AND OUTLOOK

The early in vitro neuroprotection studies with the first mGluR-selective ligands such as MCPG, 4C3HPG, DCG-IV, and L-AP4 established a major principle: activation of group-II or group-III mGluRs is neuroprotective while antagonist activity at group-I mGluRs generally results in protection. Subsequent *in vivo* work was supportive for this principle and laid the ground to define possible target diseases for mGluR compounds. Based on our current knowledge, human stroke as target for mGluR ligands has to be considered unlikely due to no, or weak, activity of the so far tested compounds in focal ischemia models in rodents. On the other hand, a role for mGluRs in chronic neurodegenerative disorders, such as ALS, HD, PD or AD, receives increasingly more evidence. Clearly, the gold-standard animal models for those diseases are generally performed with transgenic mice expressing relevant disease genes, e.g., expanded trinucleotide repeat-containing huntingtin locus in the case of HD. Up to date, no studies with the improved and systemically active pharmacological tools such as LY354740, MPEP, or BAY36-7620 were reported in those models. Once available, those studies are likely to create excitement towards clinical neuroprotection trials with mGluR-selective compounds.

REFERENCES

1. Lipton SA, Rosenberg PA. Excitatory amino acids as a final common pathway for neurologic disorders. N Engl J Med 1994; 330:613-622.
2. Hollmann M, Heinemann S. Cloned glutamate receptors. Annu Rev Neurosci 1994; 17:31-108.
3. Dingledine R, Borges K, Bowie D et al. The glutamate receptor ion channels. Pharmacol Rev 1999; 51:7-61.
4. Tanabe Y, Masu M, Ishii T et al. A family of metabotropic glutamate receptors. Neuron 1992; 8:169-179.
5. Nakanishi S. Metabotropic glutamate receptors: Synaptic transmission, modulation, and plasticity. Neuron 1994; 13:1031-1037.
6. Conn PJ, Pin J-P. Pharmacology and functions of metabotropic glutamate receptors. Annu Rev Pharmacol Toxicol 1997; 37:205-237.
7. Laurie DJ, Boddeke HW, Hiltscher R et al. HmGlu1d, a novel splice variant of the human type I metabotropic glutamate receptor. Eur J Pharmacol 1996; 296:R1-R3.
8. Flor PJ, Gomeza J, Tones MA et al. The C-terminal domain of the mGluR1 metabotropic glutamate receptor affects sensitivity to agonists. J Neurochem 1996; 67:58-63.
9. Schoepp DD, Jane DE, Monn AJ. Pharmacological agents acting at subtypes of metabotropic glutamate receptors. Neuropharmacology 1999; 38:1431-1476.
10. Iversen L, Mulvihill E, Haldeman B et al. Changes in metabotropic glutamate receptor mRNA levels following global ischemia: increase of a putative presynaptic subtype (mGluR4) in highly vulnerable rat brain areas. J Neurochem 1994; 63:626-633.
11. Minakami R, Katsuki F, Yamamoto T et al. Molecular cloning and the functional expression of two isoforms of human metabotropic glutamate receptor subtype 5. Biochem Biophys Res Commun 1994; 199:1136-1143.
12. Joly C, Gomeza J, Brabet I et al. Molecular, functional, and pharmacological characterization of the metabotropic glutamate receptor type 5 splice variants: Comparison with mGluR1. J Neurosci 1995; 15:3970-81.
13. Flor PJ, van der Putten H, R egg D et al. A novel splice variant of a metabotropic glutamate receptor, human mGluR7b. Neuropharmacology 1997; 36:153-159.
14. Corti C, Rimland JM, Corsi M et al. Isolation and analysis of splice variants of the rat metabotropic glutamate receptors, mGluR7 and mGluR8. Soc Neurosci Abstr 1997; 23:939.
15. Okamoto N, Hori S, Akazawa C et al. Molecular characterization of a new metabotropic glutamate receptor mGluR7 coupled to inhibitory cyclic AMP signal transduction. J Biol Chem 1994; 269:1231-1236.

16. Duvoisin RM, Zhang C, Ramonell K. A novel metabotropic glutamate receptor expressed in the retina and olfactory bulb. J Neurosci 1995; 15:3075-3083.
17. Flor PJ, Lindauer K, P ttner I et al. Molecular cloning, functional expression and pharmacological characterization of the human metabotropic glutamate receptor type 2. Eur J Neurosci 1995; 7:622-629.
18. Flor PJ, Lukic S, R egg D et al. Molecular cloning, functional expression and pharmacological characterization of the human metabotropic glutamate receptor type 4. Neuropharmacology 1995; 34:149-155.
19. Johansen PA, Chase LA et al. Type 4a metabotropic glutamate receptor: identification of new potent agonists and differentiation from the L-(+)-2-amino-4-phosphonobutanoic acid-sensitive receptor in the lateral perforant pathway in rats. Mol Pharmacol 1995; 48:140-149.
20. Monn JA, Valli MJ, Massey SM et al. Design, synthesis, and pharmacological characterization of (+)-2-aminobicyclo[3.1.0]hexane-2,6-dicarboxylic acid (LY354740): A potent, selective, and orally active group 2 metabotropic glutamate receptor agonist possessing anticonvulsant and anxiolytic properties. J Med Chem 1997; 40:528-537.
21. Gasparini F, Bruno V, Battaglia G et al. (R,S)-4-phosphonophenylglycine, a potent and selective group III metabotropic glutamate receptor agonist is anticonvulsive and neuroprotective *in vivo*. J Pharmacol Exp Ther 1999; 290:1678-1687.
22. Baude A, Nusser Z, Roberts JD et al. The metabotropic glutamate receptor (mGluR1 alpha) is concentrated at perisynaptic membrane of neuronal subpopulations as detected by immunogold reaction. Neuron 1993; 11:771-787.
23. Shigemoto R, Kinoshita A, Wada E et al. Differential presynaptic localization of metabotropic glutamate receptor subtypes in the rat hippocampus. J Neurosci 1997; 17:7503-7522.
24. Petralia RS, Wang YX, Niedzielski AS et al. The metabotropic glutamate receptors, mGluR2 and mGluR3, show unique postsynaptic, presynaptic and glial localizations. Neuroscience 1996; 71:949-976.
25. Lujan R, Roberts JD, Shigemoto R et al. Differential plasma membrane distribution of metabotropic glutamate receptors mGluR1 alpha, mGluR2 and mGluR5, relative to neurotransmitter release sites. J Chem Neuroanat 1997; 13:219-241.
26. Ohishi H, Shigemoto R, Nakanishi S et al. Distribution of the mRNA for a metabotropic glutamate receptor (mGluR3) in the rat brain: An in situ hybridization study. J Comp Neurol 1993; 335:252-266.
27. Ohishi H, Ogawa-Meguro R, Shigemoto R et al. Immunohistochemical localization of metabotropic glutamate receptors, mGluR2 and mGluR3, in rat cerebellar cortex. Neuron 1994; 13:55-66.
28. Mineff E, Valtschanoff J. Metabotropic glutamate receptors 2 and 3 expressed by astrocytes in rat ventrobasal thalamus. Neurosci Lett 1999; 270:95-98.
29. Trombley PQ, Westbrook GL. L-AP4 inhibits calcium currents and synaptic transmission via a G-protein-coupled glutamate receptor. J Neurosci 1992; 12:2043-2050.
30. Br uner-Osborne H, Egebjerg J, Nielsen EO et al. Ligands for glutamate receptors: Design and therapeutic prospects. J Med Chem 2000; 43:2609-2645.
31. Bruno V, Battaglia G, Kingston A et al. Neuroprotective activity of the potent and selective mGlu1a metabotropic glutamate receptor antagonist, (+)-2-methyl-4-carboxyphenylglycine (LY367385): Comparison with LY367366, a broader spectrum antagonist with equal affinity for mGlu1a and mGlu5 receptors. Neuropharmacology 1999; 38:199-207.
32. Gasparini F, Lingenh hl K, Stoehr N, et al. 2-Methyl-6-(phenylethynyl)-pyridine (MPEP), a potent, selective and systemically active mGlu5 receptor antagonist. Neuropharmacology 1999; 38:1493-1503.
33. Litschig S, Gasparini F, Rueegg D, et al. CPCCOEt, a noncompetitive metabotropic glutamate receptor 1 antagonist, inhibits receptor signaling without affecting glutamate binding. Mol Pharmacol 1999; 55:453-461.
34. Carroll FY, Stolle A, Beart PM, et al. BAY36-7620: A potent non-competitive mGlu1 receptor antagonist with inverse agonist activity. Mol Pharmacol 2001; 59:965-973.

35. Bruno V, Copani A, Kn pfel T, et al. Activation of metabotropic glutamate receptors coupled to inositol phospholipid hydrolysis amplifies NMDA-induced neuronal degeneration in cultured cortical cells. Neuropharmacology 1995; 34:1089-1098.
36. Bruno V, Battaglia G, Copani A, et al. Activation of class II or class III metabotropic glutamate receptors protects cultured cortical neurons against excitotoxic degeneration. Eur J Neurosci 1995; 7:1906-1913.
37. Brabet I, Parmentier M-L, De Colle C, et al. Comparative effect of L-CCG-I, DCG-IV and _-carboxyl-L-glutamate on all cloned metabotropic glutamate receptor subtypes. Neuropharmacology 1998; 37:1043-1051.
38. Wroblewska B, Wroblewski JT, Pschenichkin S, et al. N-acetylaspartylglutamate selectively activates mGluR3 receptors in transfected cells. J Neurochem 1997; 69:174-181.
39. M ller T, De Vry, J, Schreiber R, et al. BAY 36-7620, a selective metabotropic glutamate receptor1 antagonist with anticonvulsant and neuroprotective properties. Soc Neurosci Abs 2000; 26:1650.
40. Annoura H, Fukunaga A, Uesugi M, et al. A novel class of antagonists for metabotropic glutamate receptors, 7-(hydroxyimino)cyclopropa[b]chromen-1a-carboxylates. Bioorganic Medical Chemistry Letters 1996; 6:7763-7766.
41. Monn JA, Valli MJ, Massey SM, et al. Synthesis, pharmacological characterization, and molecular modeling of heterobicyclic amino acids related to (+)-2-aminobicyclo[3.1.0] hexane-2,6-dicarboxylic acid (LY354740): Identification of two new potent, selective, and systemically active agonists for group II metabotropic glutamate receptors. J Med Chem 1999; 42:1027-1040.
42. Spooren WPJM, Gasparini F, Salt TE, et al. Novel allosteric antagonists shed light on mglu5 receptors and CNS disorders. Trends Pharmacol Sci 2001; 22:in press.
43. Walker K, Bowes M, Panesar M, et al. Metabotropic glutamate receptor subtype 5 (mGlu5) and nociceptive function, I. Selective blockade of mGlu5 receptors in models of acute, persistent and chronic pain. Neuropharmacology 2001; 40:1-9.
44. Bruno V, Battaglia G, Ksiazek I, et al. Selective activation of mGlu4 metabotropic glutamate receptors is protective against excitotoxic neuronal death. J Neurosci 2000; 20:6413-6420.
45. Henrich-Noack P, Flor PJ, Sabelhaus CF, et al. Distinct influence of the group III metabotropic glutamate receptor agonist (R,S)-4-phosphonophenylglycine [(R,S)-PPG] on different forms of neuronal damage. Neuropharmacology 2000; 39:911-917.
46. Nicoletti F, Bruno V, Catania MV, et al. Group-I metabotropic glutamate receptors: Hypotheses to explain their dual role in neurotoxicity and neuroprotection. Neuropharmacology 1999; 38:1477-1484.
47. Kawabata S, Tsutsumi R, Kohara A, et al. Control of calcium oscillations by phosphorylation of metabotropic glutamate receptors. Nature 1996; 383:89-92.
48. Pin JP, Duvoisin R. The metabotropic glutamate receptors: structure and functions. Neuropharmacology 1995; 34:1-26.
49. Chuang SC, Bianchi R, Wong RK. Group-I mGluR activation turns on a voltage-gated inward current in hippocampal pyramidal cells. J. Neurophysiol. 2000; 83:2844-2853.
50. Pisani A, Calabresi P, Centonze D, et al. Enhancement of NMDA responses by group-I metabotropic glutamate receptor activation in striatal neurones. Br J Pharmacol 1997; 120:1007-1014.
51. Jia Z, Lu Y, Henderson J, et al. Selective abolition of the NMDA component of long-term potentiation in mice lacking mGluR5. Learn. Mem. 1998; 5:331-343.
52. Ugolini A, Corsi M, Bordi F. Potentiation of NMDA and AMPA responses by the specific mGluR5 agonist CHPG in spinal cord motoneurons. Neuropharmacology 1999; 38:1569-1576.
53. Awad H, Hubert GW, Smith Y, et al. Activation of metabotropic glutamate receptor 5 has direct excitatory effects and potentiates NMDA receptor currents in neurons of the subthalamic nucleus. J Neurosci 2000; 20:7871-7879.

54. Doherty AJ, Palmer MJ, Bortolotto ZA, et al. A novel, competitive mGlu(5) receptor antagonist (LY344545) blocks DHPG-induced potentiation of NMDA responses but not the induction of LTP in rat hippocampal slices. Br J Pharmacol 2000; 131:239-244.
55. Salt TE, Binns KE. Contributions of mGlu1 and mGlu5 receptors to interactions with N-methyl-D-aspartate receptor-mediated responses and nociceptive sensory responses of rat thalamic neurons. Neuroscience 2000; 100:375-380.
56. Alagarsamy S, Marino MJ, Rouse ST, et al. Activation of NMDA receptors reverses desensitization of mGluR5 in native and recombinant systems. Nature Neurosci 1999; 2:234-240.
57. De Blasi A, Conn PJ, Pin JP, et al. Molecular determinant of metabotropic glutamate receptor signaling. Trends Pharmacol Sci 2001; 22:114-120.
58. Tu JC, Xiao B, Naisbitt S, et al. Coupling of mGluR/Homer and PSD-95 coplexes by the Shank family of postsynaptic density proteins. Neuron 1999; 23:583-592.
59. Pizzi M, Boroni F, Moraitis K, et al. Reversal of glutamate excitotoxicity by activation of PKC-associated metabotropic glutamate receptors in cerebellar granule cells relies on NR2C subunit expression. Eur J Neurosci 1999; 11:1-8.
60. Herrero I, Miras-Portugal MT, Sanchez-Prieto J. Positive feed-back of glutamate exocytosis by metabotropic presynaptic receptor stimulation. Nature 1992; 360:163-166.
61. Herrero I, Miras-Potugal MT, Sanchez-Prieto J. Rapid desensitization of the presynaptic metabotropic receptor that facilitates glutamate exocytosis. Eur J Neurosci 1994; 6:115-120.
62. Herrero I, Miras-Portugal MT, Sanchez-Prieto J. Functional switch from facilitation to inhibition in the presynapticcontrol of glutamate release by metabotropic glutamate receptors. J Biol Chem 1998; 273:1951-1958.
63. Rodriguez-Moreno A, Sistiaga A, Lerma J, et al. Switch from facilitation to inhibition of excitatory synaptic transmission by the desensitization of group-I mGluRs in the rat hippocampus. Neuron 1998; 21:1477-1486.
64. Sistiaga A, Herrero I, Conquet F, et al. The metabotropic glutamate receptor 1 is not involved in the facilitation of glutamate release in cerebro-cortical nerve terminals. Neuropharmacology 1998; 3:1485-1492.
65. Sistiaga A, Sanchez-Prieto J. Protein phosphatase 1 and 2A inhibitors prolong the switch in the control of glutamate release by group I metabotropic glutamate receptors: characterization of the inhibitory pathway. J Neurochem 2000; 75:1566-1574.
66. Bruno V, Battaglia G, Copani A, et al. An activity-dependent switch from facilitation to inhibition in the control of excitotoxicity by group-I metabotropic glutamate receptors. Eur. J. Neurosci. 2001; 13:1469-1478.
67. Ango F, Pin JP, Tu JC, et al. Dendritic and axonal targeting of type 5 metabotropic glutamate receptor is regulated by homer 1 protein and neuonal excitation. J Neurosci 2000; 20:8710-8716.
68. Lee JM, Zipfel GJ, Choi DW. The changing landscape of ischemic brain injury mechanisms. Nature 1999; 399 (Suppl.):A7-A14.
69. Allen JW, Knoblach SM, Faden AI. Activation of group I metabotropic glutamate receptors reduces neuronal apoptosis but increases necrotic cell death in vitro. Cell Death Diff 2000; 7:470-476.
70. Strasser U, Lobner D, Berhens MM, et al. Antagonists of group-I mGluRs attenuate exicitotoxic neuronal death in cortical cultures. Europ J Neurosci 1998; 10:2848-2855.
71. Battaglia G, Bruno V, Pisani A, et al. Selective blockade of type-I metabotropic glutamate receptors induces neuroprotection by enhancing GABAergic transmission. Mol Cell Neurosci 2001; in press.
72. Orlando LR, Alsdorf SA, Penney JB Jr, et al. The role of group I and group II metabotropic glutamate receptors in modulation of striatal NMDA and quinolinic acid toxicity. Exp. Neurol. 2001; 167:196-204.
73. Pellegrini-Giampietro DE, Cozzi A, Peruginelli F, et al. 1-Aminoindan-1,5-dicarboxylic acid and (S)-(+)-2-(3'-carboxyicyclo(1.1.1.)pentyl)-glycine, two mGlu1 receptor-preferring

antagonists, reduce neuronal death in in vitro and in vivo models of cerebral ischemia. Eur. J. Neurosci. 1999; 11:3637-3647.
74. Pellegrini-Giampietro DE, Peruginelli F, Meli E, et al. Protection with metabotropic glutamate 1 receptor antagonists in models of ischemic neuronal death: time-course and mechanisms. Neuropharmacol. 1999; 38:1607-1619.
75. Gereau IV RW, Conn PJ. Multiple presynaptic metabotropic glutamate receptors modulate excitatory and inhibitory synaptic transmission in hippocampal area CA1. J Neurosci 1995; 15:6879-6889.
76. Morishita W, Kirov SA, Alger BE. Evidence for metabotropic glutamate receptor activation in the induction of depolarization-induced suppression of inhibition in hippocampal CA1. J Neurosci 1998; 18:4870-4882.
77. Bushell TJ, Lee CC, Shigemoto R, et al. Modulation of synaptic transmission and differential localisation of mGluRs in cultured hippocampal autapses. Neuropharmacology 1999; 38:1553-1567.
78. Paquet M, Smith Y. Subsynaptic localization of group I metabotropic glutamate receptors at glutamatergic synapses in monkey striatum. Soc Neurosci Abs 2000; 26:740.13.
79. McNamara JO. Emerging insights into the genesis of epilepsy. Nature 1999; 399 (Suppl.):A15-A22.
80. Bruno V, Ksiazek I, Battaglia G, et al. Selective blockade of metabotropic glutamate receptor subtype 5 is neuroprotective. Neuropharmacology 2000; 39:2223-2230.
81. O Leary DM, Movsesyan V, Vicini S, et al. Selective mGluR5 antagonists MPEP and SIB-1893 decrease NMDA or glutamate-mediated neuronal toxicity through actions that reflect NMDA receptor antagonism. Br J Pharmacol 2000; 131:1429-1437.
82. Copani A, Casabona G, Bruno V, et al. The metabotropic glutamate receptor mGlu5 controls the onset of developmental apoptosis in cultured cerebellar neurons. Eur J Neurosci 1998; 10:2173-2184.
83. Allen JW, Eldadah BA, Faden AI. Beta-amyloid-induced apoptosis of cerebelalr granule cells and cortical neurons: Exacerbation by selective inhibition of group-I metabotropic glutamate receptors. Neuropharmacology 1999; 38:1243-1252.
84. Bruno V, Copani A, Battaglia G, et al. Protective effect of the metabotropic glutamate receptor agonist, DCG-IV, against excitotoxic neuronal death. Eur J Pharmacol 1994; 256:109-112.
85. Bruno V, Sureda FX, Storto M, et al. The neuroprotective activity of group-II metabotropic glutamate receptors requires new protein synthesis and involves a glial-neuronal signaling. J Neurosci 1997; 17:1891-1897.
86. Bruno V, Battaglia G, Casabona G, et al. Neuroprotection by glial metabotropic glutamate receptors is mediated by transforming growth factor-beta. J Neuroscience 1998; 18:9594-9600.
87. Buisson A, Choi DW. The inibitory mGluR agonist, S-4-carboxy-3-hydroxy-phenylglycine selectively attenuates NMDA neurotoxicity and oxygen-glucose deprivation-induced neuronal death. Neuropharmacology 1995; 34:1081-1087.
88. Buisson A, Nicole O, Docagne F, et al. Up-regulation of a serine protease inhibitor in astrocytes mediates the neuroprotective activity of transforming growth factor-β1. FASEB J 1998; 12:1683-1691.
89. D Onofrio M, Cuomo L, Battaglia G, et al. Neuroprotection mediated by group-II metabotropic glutamate receptors involves the activation of the MAP kinase pathway and requires the induction of TGF-β1. J Neurochem 2001; in press.
90. Thomsen C, Bruno V, Nicoletti F, et al. (2S,1 S,2 S,3 R)-2-(2 -Carboxy-3 -phenylcyclopropyl) glycine, a potent and selective antagonist of type 2 metabotropic glutamate receptors. Mol Pharmacol 1996; 50:6-9.
91. Battaglia G, Bruno V, Ngomba RT, et al. Selective activation of group-II metabotropic glutamate receptors is protective against excitotoxic neuronal death. Eur J Pharmacol 1998; 356:271-274.

92. Gottron F, Turetsky D, Choi D. SMI-32 antibody against non-phosphorylated neurofilaments identifies a subpopulation of cultured cortical neurons hypersensitive to kainate toxicity. Neurosci Lett 1995; 194:1-4.
93. Pizzi M, Fallacara C, Arrighi V, et al. Attenuation of excitatory amino acid toxicity by metabotropic glutamate receptor agonists and aniracetam in primary cultures of cerebellar granule cells. J Neurochem 1993; 61:683-689.
94. Ambrosini A, Bresciani L, Fracchia S, et al. Metabotropic glutamate receptors negatively coupled to adenylate cyclase inhibit N-methyl-D-aspartate receptor activity and prevent neurotoxicity in mesencephalic neurons in vitro. Mol Pharmacol 1995; 47:1057-1064.
95. Kingston AE, O Neill MJ, Bond A, et al. Neuroprotective action of novel and potent ligands of group-I and group-II metabotropic glutamate receptors. Ann NY Acad Sci 1999; 890:438-439.
96. Kingston AE, O Neill MJ, Lam A, et al. Neuroprotection by metabotropic glutamate receptor agonists: LY354740, LY379268 and LY389795. Eur J Pharmacol 1999; 377:155-165.
97. Behrens MM, Strasser U, Heidinger V, et al. Selective activation of group II mGluRs with LY354740 does not prevent neuronal excitotoxicity. Neuropharmacology 1999; 38:1621-1630.
98. Copani A, Bruno V, Battaglia G, et al. Activation of metabotropic glutamate receptors protects cultured neurons against apoptosis induced by beta-amyloid peptide. Mol Pharmacol 1995; 47:890-897.
99. Copani A, Bruno V, Barresi V, et al. Activation of metabotropic glutamate receptors prevents neuronal apoptosis in culture. J Neurochem 1995; 64:101-108.
100. Miyamoto M, Ishida M, Kwak S, et al. Agonists for metabotropic glutamate receptors in the rat delay recovery from halothane anesthesia. Eur J Pharmacol 1994; 260:99-102.
101. Henrich-Noack P, Hatton CD, Reymann KG. The mGlu receptor ligand (S)-4C3HPG protects neurons after global ischaemia in gerbils. Neuroreport 1998; 9:985-988.
102. Bond A, O Neill MJ, Hicks CA, et al. Neuroprotective effects of a systemically active group II metabotropic glutamate receptor agonist LY354740 in a gerbil model of global ischaemia. Neuroreport 1998; 9:1191-1193.
103. Bond A, Ragumoorthy N, Monn JA, et al. LY379268, a potent and selective group II metabotropic glutamate receptor agonist, is neuroprotective in gerbil global, but not focal, cerebral ischaemia. Neurosci Lett 1999; 273:191-194.
104. Lam AG, Soriano MA, Monn JA, et al. Effects of the selective metabotropic glutamate agonist LY354740 in a rat model of permanent ischaemia. Neurosci Lett 1998; 254:121-123.
105. Cai Z, Xiao F, Fratkin JD, et al. Protection of neonatal rat brain from hypoxic-ischemic injury by LY379268, a group II metabotropic glutamate receptor agonist. Neuroreport 1999; 10:3927-3931.
106. Klodzinska A, Chojnacka-Wojcik E, Pilc A. Selective group II glutamate metabotropic receptor agonist LY354740 attenuates pentetrazole- and picrotoxin-induced seizures. Pol J Pharmacol 1999; 51:543-545.
107. Klodzinska A, Bijak M, Chojnacka-Wojcik E, et al. Roles of group II metabotropic glutamate receptors in modulation of seizure activity. Naunyn Schmiedebergs Arch Pharmacol 2000; 361:283-288.
108. Allen JW, Ivanova SA, Fan L, et al. Group II metabotropic glutamate receptor activation attenuates traumatic neuronal injury and improves neurological recovery after traumatic brain injury. J Pharmacol Exp Ther 1999; 290:112-120.
109. Pizzi M, Benarese M, Boroni F, et al. Neuroprotection by metabotropic glutamate receptor agonists on kainate-induced degeneration of motor neurons in spinal cord slices from adult rat. Neuropharmacology 2000; 39:903-910.
110. Monyer H, Giffard RG, Hartley DM, et al. Oxygen or glucose deprivation-induced neuronal injury in cortical cell cultures is reduced by tetanus toxin. Neuron 1992; 8:967-973.
111. Bond A, Jones NM, Hicks CA, et al. Neuroprotective effects of LY379268, a selective mGlu2/3 receptor agonist: investigations into possible mechanism of action in vivo. J Pharmacol Exp Ther 2000; 294:800-809.

112. Ciccarelli R, Di Iorio P, Bruno V, et al. Activation of A1 adeNOSine and mGlu3 metabotropic glutamate receptors enhances the release of nerve growth factor and S-100β protein from cultured astrocytes. Glia 1999; 27:275-281.
113. Chao CC, Hu S, Kravitz FH, et al. Transforming growth factor-beta protects human neurons against beta-amyloid-induced injury. Mol Chem Neuropathol 1994; 23:159-178.
114. Prehn JHM, Bindokas VP, Marcuccilli CJ, et al. Regulation of neuronal Bcl-2 protein expression and calcium homeostasis by transforming growth factor type beta confers wide-ranging protection on rat hippocampal neurons. Proc Natl Acad Sci 1994; 91:12599-12603.
115. Prehn JH, Miller RJ. Opposite effects of TGF-beta1 on rapidly- and slowly triggered excitotoxic injury. Neuropharmacology 1996; 35:249-256.
116. Prehn JHM, Bindokas VP, Jordan J, et al. Protective effect of transforming growth factor-β1 on β-amyloid neurotoxicity in rat hippocampal neurons. Mol Pharmacol 1996; 49:319-328.
117. Copani A, Condorelli F, Caruso A, et al. Mitotic signalling by β-amyloid causes neuronal death. FASEB J 1999; 13:2225-2234.
118. Henrich-Noack P, Prehn JH, Krieglstein J. TGF-beta 1 protects hippocampal neurons against degeneration caused by transient global ischemia. Dose-response relationship and potential neuroprotective mechanisms. Stroke 1996; 27:1609-1614.
119. Meucci O, Miller RJ. Gp120-induced neurotoxicity in hippocampal pyramidal neuron cultures: protective action of TGF-beta1. J Neurosci 1996; 16:4080-4088.
120. Tomoda T, Shirasawa T, Yahagi YI, et al. Transforming growth factor-beta is a survival factor for neonate cortical neurons: coincident expression of type-1 receptors in developing cerebral cortices. Dev Biol 1996; 179:79-90.
121. Ren RF, Hawver DB, Kim RS, et al. Transforming growth factor-β protects human hNT cells from degeneration induced by β-amyloid peptide: involvement of the TGF-β type II receptor. Mol Brain Res 1997; 48:315-322.
122. Scorziello A, Florio T, Bajetto A, et al. TGF-β1 prevents gp120-induced impairment of Ca2+ homeostasis and rescues cortical neurons from apoptotic death. J Neurosci Res 1997; 49:600-607.
123. Flanders KC, Ren RF; Lippa CF. Transforming growth factor-betas in neurodegenerative disease. Prog Neurobiol 1998; 54:71-85.
124. Ruocco A, Nicole O, Docagne F, et al. A transforming growth factor-beta antagonist unmasks the neuroprotective role of this endogenous cytokine in excitotoxic and ischemic brain injury. J Cereb Blood Flow Metab 1999; 19:1345-1353.
125. Albin RL, Young AB, Penney JB. The functional anatomy of basal ganglia disorders. Trends Neurosci 1989; 12:366-375.
126. Konieczny J, Ossowska, K, Wolfarth, et al. LY354740, a group II metabotropic glutamate receptor agonist with potential antiparkinsonian properties in rats. Naunyn-Schmiedebergs Arch. Pharmacol 1998; 358:500-502.
127. Wolfarth S, Konieczny J, Lorenc-Koci E, et al. The role of metabotropic glutamate receptor (mGluR) ligands in parkinsonian muscle rigidity. Amino Acids 2000; 19:95-101.
128. Helton DR, Tizzano JP, Monn JA, et al. LY354740: A metabotropic glutamate receptor agonist which ameliorates symptoms of nicotine withdrawal in rats. Neuropharmacology 1997; 36:1511-1516.
129. Helton DR, Tizzano JP, Monn JA, et al. Anxiolytic and side-effect profile of LY354740: A potent, highly selective, orally active agonist for group II metabotropic glutamate receptors. J Pharmacol Exp Ther 1998; 284:651-660.
130. Vandergriff J, Rasmussen K. The selective mGlu2/3 receptor agonist LY354740 attenuates morphine-withdrawal-induced activation of locus coeruleus neurons and behavioral signs of morphine withdrawal. Neuropharmacology 1999; 38:217-222.
131. Popik P, Kozela E, Pilc A. A Selective agonist of group II glutamate metabotropic receptors, LY354740, inhibits tolerance to analgesic effects of morphine in mice. Br J Pharmacol 2000; 130:1425-1431.

132. Shekhar A, Keim SR. LY354740, a potent group II metabotropic glutamate receptor agonist prevents lactate-induced panic-like response in panic-prone rats. Neuropharmacology 2000; 39:1139-1146.
133. Moghaddam B, Adams BW. Reversal of phencyclidine effects by a group II metabotropic glutamate receptor agonist in rats. Science 1998; 281:1349-1352.
134. Neugebauer V, Zinebi F, Russell R, et al. Cocaine and kindling alter the sensitivity of group II and III metabotropic glutamate receptors in the central amygdala. J Neurophysiol 2000; 84:759-770.
135. Graham ME, Burgoyne RD. Activation of metabotropic glutamate receptors by L-AP4 stimulates survival of rat cerebellar granule cells in culture. Eur J Pharmacol 1994; 288:115-123.
136. Bruno V, Copani A, Bonanno L, et al. Activation of group III metabotropic glutamate receptors is neuroprotective in cortical cultures. Eur J Pharmacol 1996; 310:61-66.
137. Faden AI, Ivanova SA, Yakovlev AG, et al. Neuroprotective effects of group III mGluR in traumatic neuronal injury. J Neurotrauma 1997; 14:885-895.
138. Lafon-Cazal M, Fagni L, Guiraud MJ, et al. mGluR7-like metabotropic glutamate receptors inhibit NMDA-mediated excitotoxicity in cultured mouse cerebellar granule neurons. Eur J Neurosci 1999; 11:663-672.
139. Sabelhaus CF, Schroder UH, Breder J, et al. Neuroprotection against hypoxic/hypoglycaemic injury after the insult by the group III metabotropic glutamate receptor agonist (R, S)-4-phosphonophenylglycine. Br J Pharmacol 2000; 131:655-658.
140. DiFiglia M. Excitotoxic injury of the neostriatum: a model for Huntington s disease. TINS 1990; 13:286-289.
141. Gasparini F, Inderbitzin W, Francotte E, et al. (+)-4-Phosphonophenylglycine (PPG) a new group III selective metabotropic glutamate receptor agonist. Bioorg Med Chem Lett 2000; 10:1241-1244.
142. Pekhletski R, Gerlai R, Oversteet LS, et al. Impaired cerebellar synaptic plasticity and motor performance in mice lacking the mGluR4 subtype of metabotropic glutamate receptor. J Neurosci 1996; 16:6364-6373.
143. Bradley SR, Standaert DG, Rhodes KJ, et al. Immunohistochemical localization of subtype 4a metabotropic glutamate receptors in the rat and mouse basal ganglia. J Comp Neurol 1999; 407:33-46.
144. Kosinski CM, Risso Bradley S, Conn PJ, et al. Localization of metabotropic glutamate receptor 7 mRNA and mGluR7a protein in the rat basal ganglia. J Comp Neurol 1999; 415:266-84.
145. Thomsen C, Hampson DR. Contribution of metabotropic glutamate receptor mGluR4 to L-2-[H-3]amino-4-phosphonobutyrate binding in mouse brain. J Neurochem 1999; 71:835-840.
146. Martin G, Nie Z, Siggins GR. Metabotropic glutamate receptors regulate N-methyl-D-aspartate-mediated synaptic transmission in nucleus accumbens. J Neurophysiol 1997; 78:3028-3038.
147. Maiese K, Vincent A, Lin SH, et al. Group I and group III metabotropic glutamate receptor subtypes provide enhanced neuroprotection. J Neurosci Res 2000; 62:257-272.
148. Doble A. The role of excitotoxicity in neurodegenerative disease: Implications for therapy. Pharmacology & Therapeutics 1999; 81:163-221.
149. Wood PL. NMDA antagonists for stroke and head trauma: current status. Exp Opin Invest Drugs 1998; 7:1505-1508.
150. Abdul-Ghani AS, Attwell PJ, Singh-Kent N, et al. Anti-epileptogenic and anticonvulsant activity of L-2-amino-4-phosphonobutyrate, a presynaptic glutamate receptor agonist. Brain Res 1997; 755:202-212.
151. Chapman AG, Nanan K, Yip P, Meldrum BS. Anticonvulsant activity of a metabotropic glutamate receptor 8 preferential agonist, (R,S)-4-phosphonophenylglycine. Eur J Pharmacol 1999; 383:23-27.

152. Tang E, Yip PK, Chapman AG, et al. Prolonged anticonvulsant action of glutamate metabotropic receptor agonists in inferior colliculus of genetically epilepsy-prone rats. Eur J Pharmacol 1997; 327:109-115.

A ROLE FOR GLUTAMATE TRANSPORTERS IN NEURODEGENERATIVE DISEASES

Davide Trotti*

INTRODUCTION

In the mammalian central nervous system (CNS) L-glutamic acid acts as a chemical transmitter of excitatory signals. L-glutamate is released in the synaptic cleft and activates a multitude of highly integrated signaling pathways by binding to an array of iono- and metabotropic glutamate receptors. The termination of this chemical neurotransmission occurs via uptake of glutamate by specialized carrier proteins. The reuptake mechanism is capable of maintaining glutamate at levels below 1 μM in the synaptic cleft. Cells in the CNS possess unique Na^+- and K^+-coupled transporters with high affinity for glutamate.[1] These glutamate transporters are distinct from the recently identified vesicular glutamate transporters, which belong to the family of Na^+/phosphate and sulphate transporters.[2,3]

Both neurons and astrocytes exhibit high affinity glutamate uptake, although astroglia appears to be the primary site for glutamate clearance (uptake and biotransformation).[4] At elevated extracellular concentrations, glutamate acts as a neurotoxin and is capable of inducing degeneration of neurons.[5] This event, which is known as excitotoxicity, contributes to many neurodegenerative disease states (e.g., Huntington s disease, amyotrophic lateral sclerosis (ALS), ischemia, status epilepticus, and some encephalopathies). Downregulation or aberrant functioning of glutamate transporters may aggravate these pathological states. For example, in ischemic conditions, when decreased ATP production causes the electrochemical gradients within cells to collapse, glutamate transporters may contibute to increase the extracellular concentration of glutamate by functioning in the reversed direction,

*Department ofNeurology, Cecil B. Day Laboratory for Neuromuscular Research, Massachusetts General Hospital, Harvard Medical School, Charlestown, MA 02129.

moving glutamate from the cytoplasm to the extracellular space. On the other hand, excessive uptake of glutamate may cause glutamatergic hypofunction, one of the suspected pathophysiological mechanisms in schizophrenia.[6] The apparent involvement of glutamate carriers in different neurological disorders, as well as their capability to attenuate the glutamate neurotoxicity, has led to the identification of this transport system as a desirable therapeutic target. In this Chapter, I will review the recent insights into the functional roles, regulation and pathological implications of the high affinity glutamate transporters.

THE NA^+/K^+-DEPENDENT HIGH AFFINITY GLUTAMATE TRANSPORTERS

The identification and molecular characterization of the high affinity glutamate transporters started in 1992 when three rodent isoforms of these proteins were first cloned. Using expression cloning and *Xenopus* oocytes, Kanai and Hediger isolated a cDNA from rabbit small intestine encoding a high affinity glutamate transporter, which was named EAAC1 (excitatory amino acid carrier 1).[7] A 73 kDa glycoprotein which exhibited high affinity glutamate transport activity was purified by Kanner and co-workers to homogeneity from rat brain. An antibody was then raised against the purified protein and used to isolate a clone from a rat brain cDNA library that encoded a glutamate transporter, called GLT1 (glutamate transporter 1).[8,9] These three high affinity glutamate transporters share 50-55% amino acid sequence identity with each other. Later, human homologues and additional members of this novel transporter family were identified: Using a cloning approach based on PCR-amplification with degenerate oligonucleotides, Amara and colleagues isolated cDNAs from human motor cortex which encoded the human glutamate transporter homologues of the rodent subtypes,[10] and renamed them excitatory amino acid transporter EAAT1, EAAT2 and EAAT3 for GLAST, GLT1 and EAAC1 respectively. An additional subtype termed EAAT4 was later cloned.[11] EAAT4 exhibits 58%, 39% and 51% amino acid sequence identity to the human glutamate transporters EAAT1, EAAT2 and EAAT3, respectively. A fifth isoform of the glutamate transporter family (EAAT5) was cloned from a human retina cDNA library using a probe obtained by PCR-amplification with degenerate oligonucleotide primers of salamander retina cDNA .[12] EAAT5 exhibits 51%, 44%, 48% and 49% amino acid sequence identity to the human isoforms of the glutamate transporters EAAT1, EAAT2, EAAT3 and EAAT4, respectively.

The gene names for the human high affinity glutamate transporter are as follows: EAAT3, SLC1A1; EAAT2, SLC1A2; EAAT1, SLC1A3; EAAT4, SLC1A4; EAAT5, SLC1A5 according to the human genome database (HUGO). The acronym SLC1 refers to Solute Carrier family number 1 and A1 to family member number 1. All the mammalian glutamate transporters also have significant homology to the H^+-coupled glutamate transporters of *Escherichia coli, Bacillus stearothermophilus* and *Bacillus caldotenax*, and to the dicarboxylate transporter of *Rhizobium meliloti* with sequence identities ranging from 27 to 32%. Interestingly, there is no significant homology with other Na^+-coupled transporters such as members of the Na^+- and Cl^-dependent GABA/neurotransmitter transporter family or members of the Na^+/glucose cotransporter family.

LOCALIZATION OF GLUTAMATE TRANSPORTERS

Immunohistochemistry has revealed that GLAST and GLT1 are localized in astrocytes, while EAAC1, EAAT4 and EAAT5 are in neuronal membranes. Detailed studies have further delineated the localization of glutamate transporters to specific subcellular compartments. The neuronal transporters EAAC1 and EAAT4 appear to be localized to plasma membranes in a perisynaptic distribution. The highest density of these transporters appears to be at the edge of post-synaptic densities, rather than within the synaptic cleft. To date, most immunolocalization studies have indicated that the neuronal transporters are localized in a somatodendritic Fashion on post-synaptic spines and somas. Similarly, the astroglial glutamate transporters have a polarized distribution. Both GLAST and GLT1 are localized to astroglial membranes that immediately oppose synaptic cleft regions of the neuropil. In mammalian studies, it has been demonstrated that GLAST is highly expressed in the molecular layer of the cerebellum and less in the hippocampus, superior colliculus, and substantia gelatinosa of the spinal cord. In contrast, GLT1 expression is generally high throughout all brain regions and the spinal cord but it is largely absent from the white matter tracts. EAAC1 is selectively enriched in neurons of the hippocampus, cerebellum, and basal ganglia whereas EAAT4 is largely confined to the soma and dendrites of the Purkinje cells of the cerebellum, and EAAT5 is located in retinal ganglion cells (for review, see ref. 1). Glutamate transporters have also been localized in peripheral tissues. The glutamate transporter GLT1 was found in the liver and EAAC1 in intestine, kidney, lung and heart.[13,14]

FUNCTIONAL PROPERTIES, STOICHIOMETRY AND KINETICS OF THE GLUTAMATE TRANSPORTERS

High affinity glutamate transporters are coupled to the inwardly directed electrochemical gradients of Na^+ and H^+, and to the outwardly directed gradient of K^+. This unique coupling stoichiometry allows efficient removal of glutamate from extracellular fluids such as the synaptic cleft, the cerebrospinal fluid (CSF), the intestinal lumen and the lumen of renal proximal tubules. Knowledge of the precise coupling stoichiometry is important because it not only determines the concentrating capacity of these transporters but it also has pathological implications. The coupling stoichiometry of high affinity glutamate transporters was originally estimated using a variety of preparations. However, the initial studies provided ambiguous overall stoichiometries. Nevertheless, their combined information suggested that high affinity glutamate transporters link uphill glutamate transport to the cotransport of 2 Na^+

ions, the countertransport of 1 K^+ ion , and the cotransport of 1 H^+ ion or the countertransport of 1 OH^-.[15,16]

The question of whether the glutamate transporter-mediated uptake is coupled to the cotransport of H^+ or the countertransport of OH^- was later assessed by Zerangue and Kavanaugh. These authors used voltage clamping techniques coupled with a pH-sensitive fluorescent dye to monitor electrical currents and pH changes associated with flux of glutamate mediated by the neuronal glutamate transporter EAAT3. They demonstrated that, unlike L-glutamate and L-cysteate, transport of an equivalent amount of L-cysteine, a neutral amino acid substrate of EAAT3,[17] did not result in the marked intracellular acidification. If OH^- is countertransported, it would be expected that the transport of L-cysteine, as well as L-glutamate and L-cysteate, would acidify the cells. Therefore, it was proposed that H^+ is cotransported with amino acids as a thiolate, sulphate, or carboxylate ion pair, leading to an intracellular pH change which is dependent on the pK_a of the amino acid transported. After intracellular release, cysteine ($pK_a = 8.3$) remains predominantly protonated, whereas glutamic and cycteic acids ($pK_a < 5$) release the proton, because the intracellular pH is around ~7.3.

Zerangue and Kavanaugh also re-examined the coupling stoichiometry of the human neuronal transporter EAAT3.[17] They determined the reversal potential of the glutamate evoked uptake currents as a function of the concentration of coupling ions. The studies revealed that 3 rather than 2 Na^+ ions are co-transported with each glutamate molecule (see Fig.1). This implies that 2 positive charges instead of 1 are translocated with each glutamate molecule. Based on this stoichiometry it was calculated that the transporters can concentrate glutamate ~10^6 fold inside the cells under physiological conditions. Later, Attwell and co-wokers reported an identical coupling stoichiometry for the glial transporter GLT1 expressed in Chinese hamster ovary cells.[18,19]

The functional characterization of EAAT4 has led to the identification of a substrate gated anion conductance. This feature is also displayed by other glutamate transporter family members, and the anion permeabilities decrease in the order EAAT5-4 >>GLAST>EAAC1>>GLT1. The substrate-evoked uptake currents of EAAT4 expressed in *Xenopus* oocytes reverse at around -20 mV and consist of two components: An electrogenic transporter current reflecting translocation of substrate across the membrane, and a Cl^- current. The latter has the characteristics of a substrate-gated anion channel. Cl^- is mainly translocated in the presence of glutamate or related substrates and the anions movement is not thermodynamically coupled to the substrate transport. Chloride anions are therefore not necessary for substrate translocation. These results strongly indicate that EAAT4 itself functions as a Cl^- channel.[11] One possible explanation of the anion conductance phenomenon is that the conformation of EAAT4 when transporting glutamate or aspartate may have a loose structure in which a path for Cl^- is created. Consistent with this hypothesis is the observation that the Cl^- conductance is larger when aspartate is transported instead of glutamate, because aspartate is smaller than glutamate.[20] By using rapid applications of glutamate to outside-out patches excised from GLT1 transfected HEK 293 cells, Otis and

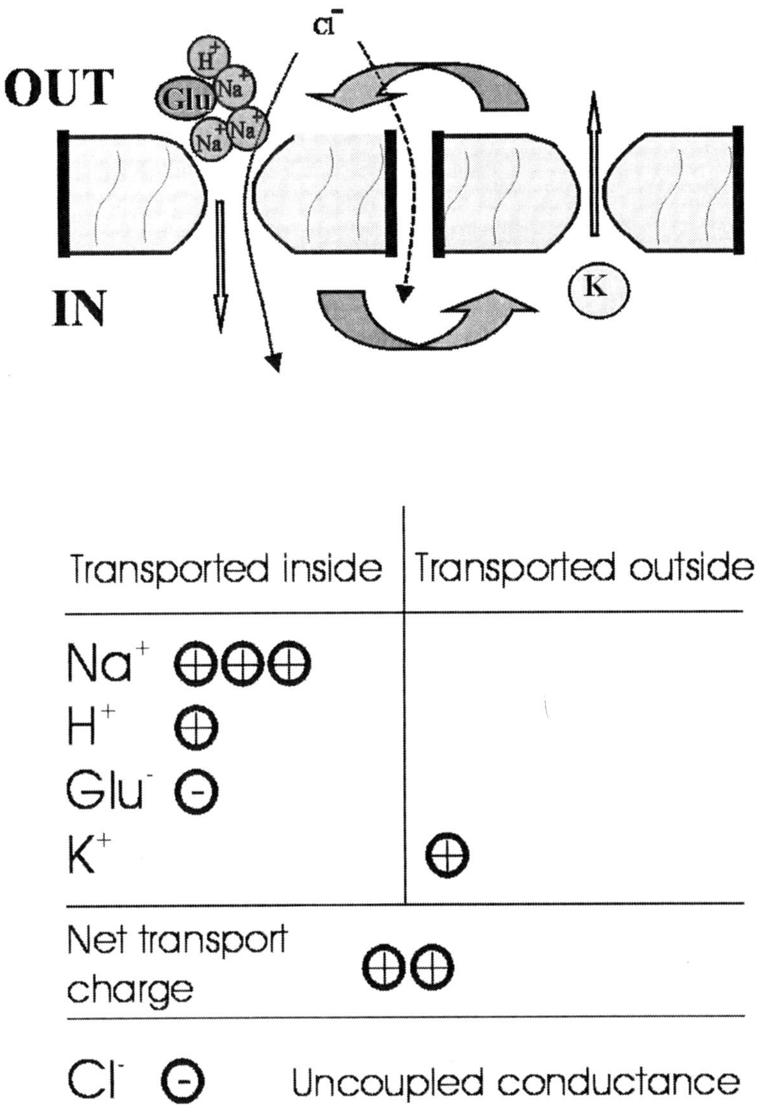

Figure 1. Stoichiometry of the glutamate transporters. Under normal circumstances, glutamate transport involves loading of the empty carrier with glutamate$^-$/H$^+$ and 3 Na$^+$, following the translocation of the fully loaded carrier across the plasma membrane (charge translocation step) and release of the cotransported ions at the intracellular space. Thereafter, 1 K$^+$ binds to the carrier inside and promotes the relocation of the empty carrier. This implies that 2 positive net charges are moved inside the cell for each glutamate molecule taken up. For net uptake of glutamate, the transporters must complete this cycle. If K$^+$ does not bind, the transporter does not enter the relocation step, binds Na$^+$ and glutamate$^-$ at the inside of the cell and translocates back in the reverse direction. In this case, the transporter behaves like an exchanger. The net uptake of glutamate is also accompanied by a chloride conductance which is not thermodynamically coupled to the transport cycle. In other words, glutamate transport can take place also in the hypothetically absence of chloride anions.

Kavanaugh demonstrated that both anion and stoichiometric currents display similar kinetics, suggesting that anion channel gating and stoichiometric charge movement are linked to early transitions in the transport cycle.[21] Thus, glutamate transporters appear to possess intrinsic structures that can function as chloride channels (Fig.2).

GLUTAMATE TRANSPORTER TOPOLOGY

Although the structure function relationship of glutamate transporters has not yet been sorted out in detail, several pieces of information are now available from site-directed mutagenesis and pharmacological studies. The glutamate transporters have a unique, highly conserved, long hydrophobic stretch near the C-terminus. Grunewald and Kanner performed cysteine-scanning mutagenesis to examine the accessibility of amino acid residues in the hydrophobic stretch of the glutamate transporter GLT1 using sulfhydryl-reactive reagents and cysteine-scanning mutagenesis.[22,23] Based on these studies, the membrane model shown in Figure 2 was constructed. A very similar topology was also deduced from studies of Amara and colleagues based on extracellular accessibilty of substituted cysteines in the glutamate transporter GLAST.[24,25]

The three dimensional structure of glutamate transporters is not available at present. However, in *Xenopus laevis* oocytes expressing EAAT3, freeze-fracture electron microscopy revealed distinct 10-nm particles, which appeared in the protoplasmic face of the oocyte only after expression of EAAT3.[26] The total number of the 10-nm particles was linearly correlated with EAAT3 maximum carrier-mediated charge movements, suggesting that the particle represented functional EAAT3 in the plasma membrane. The cross-sectional area of EAAT3 in the plasma membrane predicted the presence of 35 ± 3 transmembrane α-helices in the transporter complex. This information along with secondary structure models (6-8 transmembrane α-helices) suggested that the particles were pentagonal in which five well-defined domains could be identified. Based on the five-fold symmetry and the projections from all five domains to the extracellular space, the transporter complex was predicted to be in the appearance of a penton-based pyramid. It is speculated that, although the EAAT3 monomer can perform secondary active transport, the chloride channel mode seen in glutamate transporters is related to the oligomeric assembly as well as many multimeric ion channels whose subunits surround and contribute to the lining of the ion permeation pathway.

ROLE OF GLUTAMATE TRANSPORTERS IN THE EXCITATORY NEUROTRANSMISSION

Glutamate transporters are thought to play important roles in the termination of glutamatergic synaptic transmission. This idea is based on the evidence that excitatory synapses possess reuptake systems for glutamate clearance and that the released glutamate cannot be enzymatically inactivated in the synaptic cleft because of the

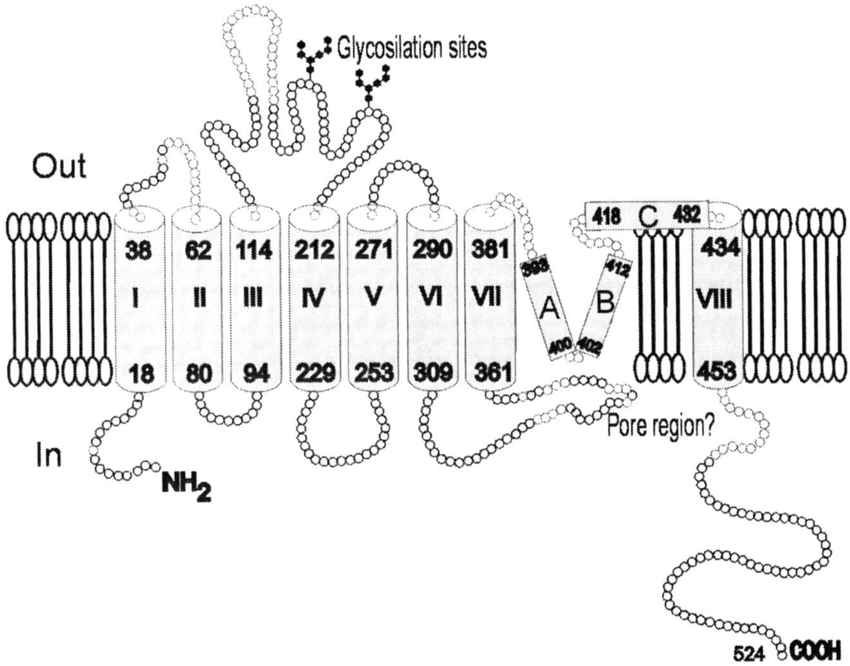

Figure 2. Membrane topology model of glutamate transporters. The topology shown is based on data obtained by cisteine scanning mutagenesis on GLT1. The model features 8 α-helical trans-membrane domains (#1-8), a large extracellular glycosylated loop between transmembrane domains 3 and 4, a reentrant loop between trans-membrane domains 7 and 8 (indicated as A and B) and a loop (indicated as C) which is predicted to extend partially into the translocation pore between trans-membrane domains 7 and 8.

lack of a specific enzyme. Mennerick and Zorumski inhibited glutamate uptake by glial depolarization rather than with uptake inhibitors[27] and they found that inhibition of glutamate uptake prolongs autoptically-induced NMDA responses known to exhibit slow or little desensitization, whereas it did not affect the time course of the excitatory post-synaptic currents (EPSCs) mediated by non-NMDA receptors, which exhibit Fast desensitization. In contrast, if desensitization of non-NMDA receptor-mediated responses was slowed down by cyclothiazide, the investigators were able to show that glutamate transporter inhibition significantly prolonged the time course of non-NMDA receptor-mediated autoptic EPSCs. Also consistent with these findings, the non-NMDA receptor-mediated EPSCs recorded in cerebellar Purkinje cells after parallel fiber stimulation was prolonged by glutamate uptake inhibition because the EPSCs of Purkinje cells normally exhibits a slower decay time course. Because of the Fast desensitization of non-NMDA receptor-mediated responses, the authors concluded that glutamate transporters do not significantly affect the decay time course of the Fast EPSC. Clements and colleagues empirically estimated the decay time course of the glutamate concentration in the synaptic cleft and concluded that it is slower than that derived from theoretical calculation based on simple diffusion out of the cleft.[28] Consequently, these investigators proposed that the diffusion of glutamate is restricted by neighboring structures. This indicated that altering the location, density, and/or the function of glutamate transporters could influence the time course of the transmitter clearance in the synaptic cleft. Kavanaugh and co-workers determined the turnover rate of the human glutamate transporter EAAT2 and EAAT1. They found that the time constant for a complete cycle of transport at -80mV and 22°C is approximately 70 msec for EAAT2[29] and 20 msec for EAAT1.[30] This is significantly slower than the estimated glutamate decay time constant in hippocampal synapses which is 1-2 ms.[28] This difference was predicted to be true also at physiological temperatures. Therefore, it can be argued as to whether glutamate transporters really constitute a major mechanism for removing released glutamate. Binding of glutamate to the transporters, however, exhibits Fast kinetics. By applying glutamate with a time resolution of 100 microseconds using the laser-pulse photolysis technique of caged glutamate, steady-state and pre-steady-state kinetics of EAAC1 were analyzed.[31] The results indicate that a rapid glutamate-binding step occurrs on a submillisecond time scale and precedes a subsequent slower electrogenic glutamate translocation across the membrane within few milliseconds after binding. The transition to an anion-conducting state is, however, delayed with respect to the onset of glutamate transport.

Thus, even though the turnover rate is low as indicated by the studies of Kavanaugh and co-workers, the glutamate binding step itself is thought to be Fast enough to participate in the removal of the released glutamate from the synaptic cleft. Consistent with this, Auger and Attwell demonstrated that transporters can remove glutamate on the same time scale that ligand-gated ion channel are open in response to released glutamate at the synapse.[32] Therefore, instead of merely maintaining a low glutamate concentration around the synaptic cleft to promote glutamate clearance by diffusion, glutamate transporters can remove glutamate with an effective time constant of 0.8 ms at 25 °C or a predicted time constant of 0.5 ms at 37 °C. This is comparable to the time needed for diffusion of glutamate away from the synaptic cleft and Faster than the synaptic current decay.[32]

As Clements and colleagues predicted, the geometry of the synapse is also an important determinant of the decay of the glutamate concentration at the synaptic cleft. For example, in the molecular layer of the cerebellar cortex, Purkinje cells and interneurons receive a common excitatory input from parallel fibers. The AMPA/kainate receptor-mediated parallel fiber EPSC recorded in Purkinje cells decays much more slowly than that recorded in interneurons. This slowness of decay does not reflect the deactivation kinetics of the postsynaptic receptors but results from the continued presence of transmitter glutamate due to retarded transmitter diffusion around the dendritic spines.[33] Thus, both diffusion of glutamate out of the synaptic cleft and binding of glutamate to glutamate transporters are the major factors affecting the decay of the glutamate concentration.

Based on the stoichiometry of glutamate transporters (cotransport of one glutamate-H^+ with three Na^+ ions and the counter-transport of one K^+ ion) and the prevailing ionic environment, it can be calculated that glutamate transporters concentrate glutamate more than 10,000-fold across cell membranes.[18] For example, the glutamate concentration in the cerebrospinal fluid (CSF) is kept at ~1 μM whereas the intracellular glutamate concentration in neurons is as high as 10 mM. Because of the high concentrating capacity of the glutamate transporters, these proteins are thought to play a major role in maintaining the extracellular glutamate concentration at low levels and to protect neurons from the excitotoxic action of glutamate. The importance of glutamate transporters in protecting neurons from glutamate excitotoxicity was experimentally demonstrated by Rothstein and coworkers who treated organotypic rat spinal cord culture with the glutamate uptake inhibitors *threo*-3-hydroxyaspartate (THA).[34] They observed motor neuron degeneration which was characterized by a slow onset and which was gradually progressing. In order to determine which glutamate transporter subtype was more important to protect neurons from glutamate excitotoxicity, Rothstein and coworkers used organotypic culture preparation and incubated the spinal cord explant with antisense oligonucleotides corresponding to the N-terminal part of cloned glutamate transporters.[35] They found that antisense oligonucleotides corresponding to the glial glutamate transporters GLT1 and GLAST induced similar motor neuron-specific degeneration as was observed when they used glutamate uptake inhibitors, whereas antisense oligonucleotides for the neuronal glutamate transporter EAAC1 did not induce neurodegeneration. The investigators also applied antisense oligonucleotides to the cerebroventricle of alert rats. Cerebroventricular administration of antisense GLT1 or GLAST oligonucleotides also resulted in the degeneration of neurons whereas EAAC1 antisense oligonucleotides did not induce neurodegeneration. Thus, primarily the glial but not the neuronal glutamate transporters protect neurons from glutamate excitotoxicity. The importance of glial glutamate transporters to protect neurons form glutamate was also demonstrated in the glutamate transporter knockout transgenic mice. Consistent with the results from antisense oligonucleotide knockout experiments, the GLT1 and GLAST knockout mice exhibited increased susceptibility to glutamate-mediated brain injury, whereas EAAC1- knockout mice did not show neurodegeneration. An important question arises why only knockouts of glial but not neuronal glutamate transporters resulted in neurodegeneration. The answer to this question is related to the different functional role glial and neuronal glutamate transporters play in the CNS. In neurons, the intracellular glutamate concentration is up to ~ 10 mM. Considering the low extracellular glutamate concentration of ~1 μM, this results in a steep glutamate concentration gradient across neuronal cell membranes which is essentially equal to the maximal concentration capacity of high affinity glutamate transporters. Consequently, under normal conditions, neuronal glutamate transporters are almost at equilibrium and have little capacity to take up glutamate. It is therefore not too surprising that the knockout of neuronal EAAC1 did not induce neurodegeneration. In glial cells, however, glutamate is taken up continuously and is then rapidly converted to glutamine by glutamine synthetase, an enzyme present in glial cells but not in neurons. The intracellular glutamate concentration in glia is therefore as low as ~50 μM. Therefore, glial glutamate transporters are not at equilibrium and keep pumping glutamate into glial cells, which generates a continuous flow of glutamate from glutamatergic synapse (source) to glial cells (sink).

REGULATION OF GLUTAMATE TRANSPORTERS

The glutamate transporter subtypes can be regulated both acutely and chronically. There is evidence for substrate-induced up-regulation of transporter activity. Long-term treatment of astrocytes with glutamate upregulates GLAST via activation of the kainate-preferring glutamate receptors.[36] Glutamate also appears to regulate transporter activity by directly interacting with the transporters. For example, treatment with glutamate transporter substrates rapidly (within minutes) causes a redistribution of the glutamate transporter subtype GLAST from intracellular compartments to the cell surface in astrocyte cultures.[37] Similarly, the activity-dependent translocation of a transporter protein to the cell surface, has been demonstrated for EAAT4 in the BT4C glioma cell line.[38] A different line of evidence also suggests that neuronal and non-neuronal factors, rather than a single neurotransmitter, are required for the expression of functionally active glutamate transporters. A search for the compounds that can mimick the action of neuron-conditioned media has revealed several soluble (poly)peptides, including hormones, growth and trophic factors, that are capable of regulating the expression of glutamate transporters.

GLT1 can be induced in astroglial cultures by pituitary adenylate cyclase-activating peptide (PACAP), a neuron-derived peptide[39] and by epidermal growth factor (EGF) receptor activation.[40] In C6 glioma, the *Wnt-1* gene product, an autocrine and paracrine soluble factor, induces GLT1 expression.[41] Growth hormone stimulates GLT1 expression in mouse placenta, whereas insulin-like growth factor II (IGF-II) downregulates EAAT4. Physiological concentrations of IGF-II ensure maintenance of GLT1, GLAST and EAAC1 at normal levels.[42] Several growth factors that are neuroprotective also increase transport activity. For example, platelet-derived growth factor (PDGF) increases cell surface expression of EAAC1 in C6 glioma cells, but has no effect on transporter expression levels.[43] Activity-dependent neurotrophic factor (ADNF) enhances basal glutamate transport in neocortical synaptosomes and attenuates oxidative stress-induced impairment of glutamate uptake, as does basic fibroblast growth factor (bFGF) and nerve growth factor (NGF).[44]

The functional activity of many membrane proteins, including various transporters and receptors, can also be rapidly regulated by changing their expression at the plasma membrane. This regulated trafficking of proteins is generally independent of protein synthesis and is not directly related to degradation, but is instead related to redistribution of receptor, transporter or other protein to or from the plasma membrane. For example, the signaling through many different types of receptors is regulated by changing the cell surface expression. Frequently agonist binding and receptor activation causes internalization of the receptor and serves to attenuate receptor signaling. With regard to trafficking of transporter proteins, one of the best studied examples is the insulin- and PKC-induced increases in cell surface expression of the GLUT4 subtype of glucose transporter that occurs in adipocytes, cardiac myocytes, and skeletal muscle (for review, see Czech and Corvera, 1999; Pessin et al, 1999). This increase in transporter activity is generally assumed to facilitate clearance of glucose from plasma after ingestion of carbohydrates. In recent years, several groups have demonstrated that the cell surface expression of most of the neurotransmitter transporters is also regulated to various stimuli and intracellular signaling pathways. Recent studies suggest that the cell surface expression of most of the subtypes of glutamate transporters can also be regulated.

Regulation of the cell surface expression of EAAC1 appears to be controlled by many of the same signaling pathways that regulate GLUT4. In earlier studies several groups provided evidence that PKC may regulate the activity of glutamate transporters. In C6 glioma, a system that endogenously expresses only EAAC1, phorbol 12-myristate 13-acetate (PMA) doubles glutamate transport activity within minutes.[45,46] This regulation, however, appear to be cell-specific as EAAC1 is downregulated by PMA in other cell types.[47]

The cell surface expression of both GLAST and EAAT4 is increased in response to substrates.[38]

In addition to growth factors and activation of PKC, glutamate transporters are regulated by their associating proteins. By means of yeast two-hybrid screening, Rothstein and co-workers have found distinct proteins interacting with EAAT4 and EAAC1. Two proteins, GTRAP41 and GTRAP48 (for glutamate transporter associated protein) specifically interact with the intracellular carboxy-terminal domain of EAAT4 and modulate its glutamate transport activity.[48] GTRAP41 is a predicted 2,388 amino acid protein and possesses 87% identity with α-spectrin III. GTRAP48 is a 1,527 amino acid protein with significant homology with p115 RhoGEF and possesses a PDZ domain and some distinctive domains characteristic of guanine nucleotide exchange factors for the Rho family of G-proteins. The expression of either GTRAP41 or GTRAP48 resulted in the increase in the V_{max} value of the transport without altering K_m value. Therefore, it is speculated that EAAT4 couples to actin cytoskeleton and to a Rho GTPase signaling via GTRAP41 and GTRAP48.

Interestingly, EAAC1 interacts with a completely different protein.[49] Rothstein and co-workers have also found a protein GTRAP3-18 that interacts with the intracellular carboxy-terminal domain of EAAC1. GTRAP3-18 is a predicted 188 amino acid hydrophobic protein with four possible transmembrane domains. GTRAP3-18 possesses a sequence identity (95%) to E18 encoded by a vitamin-A-responsive gene. In contrast to GTRAP41 and GTRAP48 that activate EAAT4-mediated transport, GTRAP3-18 reduces the EAAC1-mediated glutamate transport by decreasing the affinity to the transporter. It was also indicated that retinoic acid upregulates GTRAP3-18 expression and consequently inhibits EAAC1-mediated transport. This effect of retinoic acid is specific to EAAC1, because retinoic acid had no effect on the GLT-1 or EAAT4-mediated transport.

GLUTAMATE TRANSPORTERS IN DISEASE STATES

The availability of transporter specific reagents has stimulated studies of transporter expression in animal and human models of disease. Several of these studies indicate that the glutamate transporters are differentially regulated in various pathological conditions. Malfunctioning of glutamate transporters can obviously produce lesions of different severity. Thus, antisense knockdown of glutamate transporters in vivo reveals a major role in excitotoxicity and prevention of epileptic seizures.[35] Therefore, glutamate transporters have been suspected to play a central role in pathophysiology of neurodegenerative diseases with excitotoxic components (e.g., amyotrophic lateral sclerosis, Alzheimer s disease, epilepsy, etc.). However, in most cases it is not clear whether the downregulation of glutamate transport represents a primary defect or an event secondary to the insult, or both.

Ischemia

Reversed operation of glutamate transporters appears to contribute to the rise in extracellular glutamate and the subsequent excitotoxicity that occurs during an ischemic insult.[50] Because glutamate uptake is driven by the energy stored in the form of electrochemical gradients across the plasma membranes, the impairment of the ionic gradients caused by insufficient energy supply which occurs during brain ischemia results in a decrease in concentrating capacity of glutamate transporters and favors their reversed operation. Neurons have a much higher content of glutamate than glial cells, therefore neuronal glutamate transporters are more likely to run in reverse in ischemia and to contribute to the extracellular rise in glutamate to excitotoxic levels. Death of hippocampal neurons and a corresponding loss of EAAC1 in gerbils following transient global cerebral ischemia is preceded by a down-regulation of GLT1.[51] This suggests GLT1 dysfunction as a contributing factor for the hippocampal neuronal death following transient global cerebral ischemia.

Hypoxia-ischemia causes downregulation of neuronal EAAC1 (55%) and astroglial GLT1 (15%) in newborn piglet striatum, and some neuronal expression of GLT1 is also observed after the insult.[52] The levels of GLT1 mRNA and protein in rat hippocampal CA1 region are modestly decreased in the postischemic phase[53] but the expression of the other glial isoform GLAST and the neuronal isoform EAAC1 were not altered. In contrast, EAAC1 levels are transiently increased both in neuronal and oligodendroglial cells after transient forebrain ischaemia, and GLAST and GLT levels in CA1 region and layer V pyramidal neurons of cerebral cortex are not affected.[54] In photochemically-induced focal cortical ischemia, GLAST mRNA levels changed biphasically, with maximal decrease at 12 hours after the insult, followed by a steady increase in expression up to 72 hours after the lesion.[55,56] After neonatal hypoxic-ischemic cerebellar encephalopathy, EAAT4 levels in Purkinje neurons decrease, while GLAST expression decreases in the molecular level and becomes detectable in the Purkinje and inner granule cell layers.[57] During ischemia, glutamate transporter function can also be directly regulated by polyunsaturated fatty acids, particularly by arachidonic acid, and reactive oxygen species (ROS), both of which are produced and liberated in response to excessive glutamate receptor activation.[58-62] Arachidonic acid reversibly reduces glutamate uptake in astrocytes and synaptosomes and inhibits purified and liposome-reconstituted rat glutamate transporter GLT1 directly from the water phase and not via incorporation into the phospholipid bilayer.[63,64] The transport activities of GLT1, EAAC1, and GLAST are equally inhibited by oxidants applied extracellularly also via a direct action on the transporter protein.[65] Arachidonic acid metabolites may be a source of reactive oxygen radicals and, vice versa, these radicals may stimulate phospholipase A2, hence production of arachidonic acid. However, arachidonic acid and free oxygen radicals interfere with glutamate transporters in an additive manner, apparently via distinct mechanism.[66] The mechanism of the ROS-mediated inhibition suggests direct oxidation of cysteine sulfhydryl groups and, possibly, nitrosylation and/or nitration of tyrosine residues, although existence of a supplementary pathway cannot be excluded.[67-69]

Traumatic Injury

The effects of acute injury on transporter expression have been studied both in vitro and in vivo. Traumatic brain injury (TBI) induced by controlled cortical impact significantly decreased GLT1 and GLAST protein levels in rats after the trauma.[70] GLT1 and GLAST proteins were quantitated by immunoblotting in the ipsilateral and contralateral cortex at 2, 6, 24, 72, and 168 h after the injury. Sham-operated rats served as control. TBI resulted in a significant decrease in GLT1 (by 20-45%) and GLAST (by 30-50%) protein levels between 6 and 72 h after the injury. Decreased glial glutamate transporter function may contribute to the increased extracellular glutamate that may mediate the excitotoxic neuronal damage after TBI. In GLAST-deficient mice, the volume of edema developing after acute cerebellar injury is significantly larger than in corresponding controls animals. Thus, GLAST may play active role in reducing excitotoxic cerebellar damage after acute injury.[71]

Amyotrophic Lateral Sclerosis

Amyotrophic lateral sclerosis (ALS) is an age-related neurodegenerative disorder of motor neurons in the spinal cord, motor cortex and brain stem. Increased glutamate levels followed by excitotoxicity has been indicated as one possible pathogenic event occurring in ALS.[72,73] In particular, reduced glutamate uptake has been observed in motor cortex from post-mortem brains of ALS patients.[74] A loss of the glial glutamate transporter GLT1 has been reported in a subset of sporadic ALS patients.[75] Similarly, a reduction in levels of GLT1 protein are found in the spinal cord of transgenic $SOD1^{G85R}$ mice as well as elevation of extracellular glutamate levels,[76] suggesting that glutamate transport impairment is a common component in the pathway to motor neuron degeneration. The molecular mechanisms leading to GLT1 loss in ALS are not yet understood. Levels of GLT1 mRNA are unchanged,[77] suggesting that the reduction of GLT1 is not due to decreased transcription of mRNA, but rather to some irresolute event at the translation or post-translation level. In a subset of sporadic ALS patients, Rothstein and colleagues found abnormal mRNA editing for GLT1 and the presence of aberrant spliced forms of GLT1, which manifested dominant negative properties and caused reduced expression of the wild type transporter in transiently transfected cells.[78] However, this finding is controversial. Other investigators failed to find a correlation between the presence of aberrant GLT1 transcripts and ALS, as these transcripts were also detected in tissue specimens from control and Alzheimer s disease subjects.[79,80] The case of ALS illustrates the involvement of glutamate transporters in a neurodegenerative pathology that is also characterized by oxidative-stress. Studies by Rosen and colleagues[81] linked ALS to reactive oxygen species toxicity, since they showed that about 20% of familial ALS patients carry missense mutations in the gene encoding Cu^{2+}/Zn^{2+} superoxide dismutase (SOD1). Mice and rats made transgenic for SOD1 carrying human mutations develop selective motorneuron pathology strongly resembling ALS.[76,82,83] Signs of oxidative stress have been detected in sporadic ALS tissues as well, suggesting that the etiology of sporadic forms may involve mechanisms similar to the SOD1-dependent defects of familial ALS. Mutations in SOD1 not only reduce the capacity to detoxify superoxide, but actually catalyze the formation of more deleterious oxidants, such as OH°.

Several lines of evidence suggest a link between oxidative stress induced by SOD1 mutants and glutamate-induced cell death. SOD1 mutant-induced damage may promote excessive glutamatergic stimulation, owing to insufficient clearance of glutamate from the extracellular space by the glial glutamate transporter GLT1. A loss of GLT1 has been observed in post-mortem tissue and in the spinal cord of transgenic SOD1G85R mice. GLT1 is vulnerable to oxidative stress and is one of the proteins oxidatively damaged by 4-hydroxynonenal.[84] In addition, the mutant SOD1 proteins directly impair EEAT2 in the presence of hydrogen peroxide, but have no effect on EAAC1.[85] Therefore, it is possible that the loss of GLT1 observed in ALS may be related to oxidative damage induced by the SOD1 mutation. To further define the mechanism of this effect, the oxidant-vulnerable site of GLT1 has been mapped to the carboxyl-terminal domain. The precise site(s) of oxidant attack within that domain has not yet been defined, but it seems unlikely that oxidation of a single residue is responsible for the loss of activity. Whether the oxidation involves intra- or intermolecular reactions or both is also not known. Oxidation does not alter the mechanistic properties of GLT1 such as affinity for glutamate or the stoichiometry, suggesting that the oxidized molecules are either non-functional or form a more rigid structure that impairs the transporter dynamics. Proteins exposed to oxidative damage have altered structure, and are likely to undergo spontaneous internalization and increased proteolytic fragmentation. This process may account for the loss of GLT1 immunoreactivity detected in the transgenic model for ALS, as well as in human patients.

Using single-strand conformation polymorphism analysis of genomic DNA, Brown and colleagues reported a mutation in the GLT1 gene associated with sporadic ALS. This mutation substitutes an asparagine for a serine at position 206 (N206S)[86] and avoided the glycosylation of the carrier at that site. It was also found that the GLT1-N206S mutant had a pronounced reduction in the transport rate and a substantial dominant negative impact on the wild-type activity. The decreased rate of transport was mainly because of decrease trafficking of the mutated carrier to the plasma membrane. Moreover, the GLT1-N206S exhibited an increased reverse transport capability.[87] These combined effects significantly impair the ability of GLT1 to clear glutamate at the synaptic cleft of neurons. The evidence that a missense mutation of GLT1 is present in a patient with sporadic ALS, and that such a mutation may affect the capacity of a cell to regulate the glutamate concentration at synapses, supports the concept that this mutation may contribute to excitotoxicity that occurs in ALS.

Epilepsy

Epilepsy is a multifactorial disease. As the main excitatory neurotransmitter, glutamate plays a role in the disease. There are observations of elevated levels of gutamate reported in patients with various forms of epilepsy. It is not yet known if defective uptake played a role, although reduced expression of glutamate transporters was shown to cause or be associated with seizures in animal models. Homozygous knockout mice lacking GLT1 develop spontaneous lethal seizures,[88] while antisense-knockdown of EAAC1 leads to epileptic seizures and mild neurotoxicity.[35] However, EAAC1 deficient transgenic mice develop dicarboxylic aminoaciduria, with no apparent signs of neurodegeneration.[89] In an attempt to examine the relationship between epilepsy and glutamate transporters, the levels of transporter expression have been examined in several different animal models of epilepsy and in human tissues with varying results.

Electrical kindling in rat hippocampus CA1 area causes a 60% increase in EAAC1 4-5 weeks after initiation of kindling, while GLT1 and GLAST levels are not affected.[90] Electrical kindling also increases EAAC1 expression in piriform cortex and amygdala, but causes a decrease in GLAST expression in these brain regions.[91] GLAST and GLT1 mRNA levels are increased in thalamic and cortical structures of genetically absence epilepsy rat from Strasbourg (GAERS).[92] Iron administration to induce epilepsy causes decreased expression of GLAST mRNA and increased expression of EAAC1 mRNA for up to 30 days following the first seizure in rat cortex and hippocampus, while GLT1 mRNA level was only transiently affected.[93] In genetically epilepsy-prone (GEP) rats, audiogenic seizures cause a reduction in GABA and glutamate transporter mRNAs but no corresponding reduction in GLT1 and EAAC1 protein levels.[94] Kainic acid-induced limbic seizures cause an increase in GLAST mRNA in hippocampus, and no change is observed in animals that are injected with kainic acid but remain seizure-free, suggesting that this increase is related to the seizures and not to kainate receptor activation.[95] Kainate-induced seizures in adult, but not in young animals (younger than 21 days), causes a rapid downregulation of EAAC1 protein and mRNA and an opposite effect on GLT1 in certain hippocampal regions.[96] In temporal lobe epilepsy no differences in transporter expression are observed in patients with non-hippocampal sclerosis, but astroglial EAAT2 is decreased in CA1 stratum radiatum in patients with hippocampal sclerosis. EAAC1 and GLAST are increased in stratum granulosum/pyramidal and CA2/3 stratum radiatum, respectively.[97] It appears that increased expression of EAAC1 is one of the most consistent findings associated with epileptic seizures, but the effects on EAAC1 and the other transporters may be both dependent on the brain region and the model being studied. Further studies are necessary to firmly establish the role of glutamate transporters in epilepsy.

Alzheimer s Disease

Alzheimer s disease is a chronic degenerative neurological disorder. Hallmarks of the disease are accumulation of β-amyloid peptide, a major constituent of the amyloid plaques characteristic of the disease, as well as widespread oxidative injury. β-amyloid peptide generates free radicals and lipid peroxidation products which are able to inhibit glutamate transporters.[65,68] In contrast, the secreted form of amyloid precursor protein (APP) stimulate glutamate uptake both in cultured astrocytes and in vivo.[98] Because the glutamate transporters are the only proteins able to remove extracellular glutamate, inhibition of these proteins, either by direct action on the transporters or indirectly through alteration of the electrochemical energy gradients, may have pathogenic implications. The above reports suggest that inhibition of glutamate uptake through inactivation of the transporters might aggravate Alzheimer s disease, although this has so far not been directly shown. A different study based on antibodies reported reduced levels of GLT1 protein in frontal cortex of Alzheimer s patients but normal levels of GLT1 mRNA as well as of GLAST and EAAC1.[99] It is interesting to note that trangenic mice expressing mutant amyloid precursor protein appear to have reduced GLT1 and GLAST protein levels.[100]

Huntington s Disease

Huntington s disease (HD) is a late-onset neurodegenerative disease for which the mutation is CAG/polyglutamine repeat expansion. The R6 mouse lines expressing the HD mutation develop a movement disorder that is preceded by the formation of neuronal polyglutamine aggregates. The phenotype is likely caused by a widespread neuronal dysfunction, whereas neuronal cell death occurs late and is very selective. A decreased mRNA level of the major astroglial glutamate transporter (GLT1) in the striatum and cortex of these mice is accompanied by a concomitant decrease in glutamate uptake.[101] In contrast, the expression of the glutamate transporters GLAST and EAAC1 remained unchanged. These changes in expression occured prior to any evidence of neurodegeneration and suggest that a defect in astrocytic glutamate uptake may contribute to the phenotype and neuronal cell death in HD.

Parkinson s Disease

Some recent data suggested a possible malfunctioning of glutamate transporters, hence the contribution of glutamate neurotoxicity, in pathogenesis of Parkinson s disease. 1-methyl-4-phenyl-1,2,3,6-tetrahydropyridine (MPTP), a neurotoxin known to target dopaminergic neurons in Parkinson s disease, reversibly inhibits astroglial glutamate uptake in cultured astrocytes.[102] Inhibition of glutamate transport also occurs via oxidation products of dopamine.[103]

Hepatic Encelophathy

Hepatic encelopathy is a neurological disorder resulting from acute or chronic liver failure. Hyperammonemic condition appears to be a key factor in the pathogenesis and seems to be linked to altered glutamate uptake function and expression. Ammonia speeds up glutamate uptake by increasing cytoplasmic pH.[104] The long-term effect of ammonia, and possibly other metabolites in liver failure, is a reduction in glutamate uptake and this has been proposed as a reason for the ammonia toxicity. GLT1 mRNA and protein are decreased in frontal cortex of rats with ischemic liver failure.[105-107] GLT1 mRNA is also downregulated in hyperammonemic rats, and in rats with thioacetamide-induced acute liver failure. Portocaval anastomosis (PCA) that leads to liver atrophy, sustained hyperammonemia, and increased brain ammonia, causes biphasic changes in GLT1, GLAST, and EAAC1 levels in rat cerebellum.[109] At least some of these effects may be caused by an accumulation of ammonia that accompanies liver disease, as ammonia decreases GLAST mRNA expression in astrocytes.[110]

Glutamate Transporters and Mercury Neurotoxicity

Sensory disturbance and ataxia were described in the 1950s in Minamata (Japan) in a number of neurological cases. These neurological symptoms were attributed to the ingestion of fish contaminated with methylmercury (MeHg).[111,112] Poisoning with elementary mercury (Hg°) vapours was also shown to produce similar neurological symptoms. Several lines of evidence indicate that glutamate-mediated excitotoxicity is probably involved. For example, glutamate receptor (NMDA type) antagonists effectively blocked the neurotoxic action of mercury in cerebral neuronal cultures.[113] MeHg and Hg° would act, at least in part, through the common oxidation product mercuric mercury (Hg^{2+}). Submicromolar concentrations of Hg^{2+} were found to inhibit selectively the uptake of excitatory amino acids in cultured astrocytes. MeHg mimicked this effect but not other divalent cations.[114-116] It was also shown that mercuric chloride uncouples glutamate uptake from the co-transport of 1 H^+,[117] and that micromolar concentration of it is sufficient to induce a dramatic inhibition of the recombinant glutamate transporters GLT1, GLAST and EAAC1 functionally reconstituted in liposomes.[67]

CONCLUDING REMARKS

An apparent involvement of glutamate transporters in the pathophysiology of many neurodegenerative diseases has led to the identification of this transport system as a desirable therapeutic target. In recent years, important advances have been made in understanding basic molecular and cellular mechanisms governing the expression and regulation of high affinity glutamate transporters. It appeared that most of the transporter subtypes could be regulated both acutely and chronically. The search for regulatory mechanisms has led to the identification of several hormones, growth and trophic factors that are capable of interfering with the expression of glutamate transporters. Furthermore, several non-transportable inhibitors of glutamate uptake have been introduced and efforts to synthesize transporter subtype-specific inhibitors are currently underway. These achievements are expected to facilitate further studies on the roles of individual transporter subtypes and to develop new strategies for diagnosis and treatment of diseases associated with malfunctioning of glutamate transporters.

REFERENCES

1. Danbolt NC. Glutamate uptake. Prog Neurobiol 2001; 65:1-105.
2. Bellocchio EE, Reimer RJ, Fremeau RT, Jr., Edwards RH. Uptake of glutamate into synaptic vesicles by an inorganic phosphate transporter. Science 2000; 289:957-960.
3. Takamori S, Rhee JS, Rosenmund C, Jahn R. Identification of a vesicular glutamate transporter that defines a glutamatergic phenotype in neurons. Nature 2000; 407:189-194.
4. Danbolt NC, Storm-Mathisen J, Kanner BI. An [Na^+ + K^+]coupled L-glutamate transporter purified from rat brain is located in glial cell processes. Neuroscience 1992; 51:295-310.
5. Choi DW. Excitotoxic cell death. J Neurobiol 1992; 23:1261-1276.
6. Carlsson M, Carlsson A. Interactions between glutamatergic and monoaminergic systems within the basal ganglia Implications for schizophrenia and Parkinson s disease. Trends Neurosci 1990; 13:272-276.
7. Kanai Y, Hediger MA. Primary structure and functional characterization of a high-affinity glutamate transporter. Nature 1992; 360:467-471.
8. Pines G, Danbolt NC, Bjoras M, Zhang Y, Bendahan A, Eide L et al. Cloning and expression of a rat brain L-glutamate transporter [published erratum appears in Nature 1992 Dec 24-31;360(6406):768]. Nature 1992; 360:464-467.
9. Storck T, Schulte S, Hofmann K, Stoffel W. Structure, expression, and functional analysis of a Na(+)-dependent glutamate/aspartate transporter from rat brain. Proc Natl Acad Sci USA 1992; 89:10955-10959.
10. Arriza JL, Kavanaugh MP, Fairman WA, Wu YN, Murdoch GH, North RA et al. Cloning and expression of a human neutral amino acid transporter with structural similarity to the glutamate transporter gene family. J Biol Chem 1993; 268:15329-15332.
11. Fairman WA, Vandenberg RJ, Arriza JL, Kavanaugh MP, Amara SG. An excitatory amino-acid transporter with properties of a ligand-gated chloride channel. Nature 1995; 375:599-603.

12. Arriza JL, Eliasof S, Kavanaugh MP, Amara SG. Excitatory amino acid transporter 5, a retinal glutamate transporter coupled to a chloride conductance. Proc Natl Acad Sci USA. 1997; 94:4155-4160.
13. Hediger MA, Welbourne TC. Introduction: Glutamate transport, metabolism, and physiological responses. Am J Physiol 1999; 277:F477-480.
14. Hediger MA. Glutamate transporters in kidney and brain. Am J Physiol 1999; 277:F487-492.
15. Bouvier M, Szatkowski M, Amato A, Attwell D. The glial cell glutamate uptake carrier countertransports pH-changing anions. Nature 1992; 360:471-474.
16. Nussberger S, Foret F, Hebert SC, Karger BL, Hediger MA. Nonradioactive monitoring of organic and inorganic solute transport into single *Xenopus* oocytes by capillary zone electrophoresis. Biophys J 1996; 70:998-1005.
17. Zerangue N, Kavanaugh MP. Flux coupling in a neuronal glutamate transporter. Nature 1996; 383:634-637.
18. Levy LM, Warr O, Attwell D. Stoichiometry of the glial glutamate transporter GLT-1 expressed inducibly in a Chinese hamster ovary cell line selected for low endogenous Na^+-dependent glutamate uptake. J Neurosci 1998; 18:9620-9628.
19. Levy LM, Attwell D, Hoover F, Ash JF, Bjoras M, Danbolt NC. Inducible expression of the GLT-1 glutamate transporter in a CHO cell line selected for low endogenous glutamate uptake [published erratum appears in FEBS Lett 1998 May 1;427(1):152]. FEBS Lett 1998; 422:339-342.
20. Wadiche JI, Amara SG, Kavanaugh MP. Ion fluxes associated with excitatory amino acid transport. Neuron 1995; 15:721-728.
21. Otis TS, Kavanaugh MP. Isolation of current components and partial reaction cycles in the glial glutamate transporter EAAT2. J Neurosci 2000; 20:2749-57.
22. Grunewald M, Bendahan A, Kanner BI. Biotinylation of single cysteine mutants of the glutamate transporter GLT-1 from rat brain reveals its unusual topology. Neuron 1998; 21:623-632.
23. Grunewald M, Kanner BI. The accessibility of a novel reentrant loop of the glutamate transporter GLT-1 is restricted by its substrate. J Biol Chem 2000; 275:9684-9689.
24. Seal RP, Amara SG. A reentrant loop domain in the glutamate carrier EAAT1 participates in substrate binding and translocation. Neuron 1998; 21:1487-1498.
25. Seal RP, Leighton BH, Amara SG. A model for the topology of excitatory amino acid transporters determined by the extracellular accessibility of substituted cysteines. Neuron 2000; 25:695-706.
26. Eskandari S, Kreman M, Kavanaugh MP, Wright EM, Zampighi GA. Pentameric assembly of a neuronal glutamate transporter. Proc Natl Acad Sci USA 2000; 97:8641-8646.
27. Mennerick S, Zorumski CF. Glial contributions to excitatory neurotransmission in cultured hippocampal cells. Nature 1994; 368:59-62.
28. Clements JD, Lester RA, Tong G, Jahr CE, Westbrook GL. The time course of glutamate in the synaptic cleft. Science 1992; 258:1498-1501.
29. Wadiche JI, Arriza JL, Amara SG, Kavanaugh MP. Kinetics of a human glutamate transporter. Neuron 1995; 14:1019-1027.
30. Wadiche JI, Kavanaugh MP. Macroscopic and microscopic properties of a cloned glutamate transporter/chloride channel. J Neurosci 1998; 18:7650-7661.
31. Grewer C, Watzke N, Wiessner M, Rauen T. Glutamate translocation of the neuronal glutamate transporter EAAC1 occurs within milliseconds. Proc Natl Acad Sci USA 2000; 97:9706-9711.
32. Auger C, Attwell D. Fast removal of synaptic glutamate by postsynaptic transporters. Neuron 2000; 28:547-558.
33. Barbour B, Keller BU, Llano I, Marty A. Prolonged presence of glutamate during excitatory synaptic transmission to cerebellar Purkinje cells. Neuron 1994; 12:1331-1343.

34. Rothstein JD, Jin L, Dykes-Hoberg M, Kuncl RW. Chronic inhibition of glutamate uptake produces a model of slow neurotoxicity. Proc Natl Acad Sci USA 1993; 90:6591-6595.
35. Rothstein JD, Dykes-Hoberg M, Pardo CA, Bristol LA, Jin L, Kuncl RW et al. Knockout of glutamate transporters reveals a major role for astroglial transport in excitotoxicity and clearance of glutamate. Neuron 1996; 16:675-686.
36. Gegelashvili G, Civenni G, Racagni G, Danbolt NC, Schousboe I, Schousboe A. Glutamate receptor agonists up-regulate glutamate transporter GLAST in astrocytes. Neuroreport 1996; 8:261-265.
37. Duan S, Anderson CM, Stein BA, Swanson RA. Glutamate induces rapid upregulation of astrocyte glutamate transport and cell-surface expression of GLAST. J Neurosci 1999; 19:10193-200.
38. Gegelashvili G, Dehnes Y, Danbolt NC, Schousboe A. The high-affinity glutamate transporters GLT1, GLAST, and EAAT4 are regulated via different signalling mechanisms. Neurochem Int 2000; 37:163-170.
39. Figiel M, Engele J. Pituitary adenylate cyclase-activating polypeptide (PACAP), a neuron-derived peptide regulating glial glutamate transport and metabolism. J Neurosci 2000; 20:3596-3605.
40. Zelenaia O, Schlag BD, Gochenauer GE, Ganel R, Song W, Beesley JS et al. Epidermal growth factor receptor agonists increase expression of glutamate transporter GLT-1 in astrocytes through pathways dependent on phosphatidylinositol 3-kinase and transcription factor NF-kappaB. Mol Pharmacol 2000; 57:667-678.
41. Palos TP, Zheng S, Howard BD. Wnt signaling induces GLT-1 expression in rat C6 glioma cells. J Neurochem 1999; 73:1012-1023.
42. Matthews JC, Beveridge MJ, Dialynas E, BartkeA, Kilberg MS, Novak DA. Placental anionic and cationic amino acid transporter expression in growth hormone overexpressing and null IGF-II or null IGF-I receptor mice. Placenta 1999; 20:639-650.
43. Sims KD, Straff DJ, Robinson MB. Platelet-derived growth factor rapidly increases activity and cell surface expression of the EAAC1 subtype of glutamate transporter through activation of phosphatidylinositol 3-kinase. J Biol Chem 2000; 275:5228-5237.
44. Guo ZH, Mattson MP. Neurotrophic factors protect cortical synaptic terminals against amyloid and oxidative stress-induced impairment of glucose transport, glutamate transport and mitochondrial function. Cereb Cortex 2000; 10:50-57.
45. Davis KE, Straff DJ, Weinstein EA, Bannerman PG, Correale DM, Rothstein JD et al. Multiple signaling pathways regulate cell surface expression and activity of the excitatory amino acid carrier 1 subtype of Glu transporter in C6 glioma. J Neurosci 1998; 18:2475-2485.
46. Dowd LA, Robinson MB. Rapid stimulation of EAAC1-mediated Na^+-dependent L-glutamate transport activity in C6 glioma cells by phorbol ester. J Neurochem 1996; 67:508-516.
47. Trotti D, Peng JB, Dunlop J, Hediger MA. Inhibition of the glutamate transporter EAAC1 expressed in *Xenopus* oocytes by phorbol esters. Brain Res 2001; 914:196-203.
48. Jackson M, Song W, Liu MY, Jin L, Dykes-Hoberg M, Lin CI et al. Modulation of the neuronal glutamate transporter EAAT4 by two interacting proteins. Nature 2001; 410:89-93.
49. Lin CI, Orlov I, Ruggiero AM, Dykes-Hoberg M, Lee A, Jackson M et al. Modulation of the neuronal glutamate transporter EAAC1 by the interacting protein GTRAP3-18. Nature 2001; 410:84-88.
50. Rossi DJ, Oshima T, Attwell D. Glutamate release in severe brain ischaemia is mainly by reversed uptake. Nature 2000; 403:316-321.
51. Raghavendra Rao VL, Rao AM, Dogan A, Bowen KK, Hatcher J, Rothstein JD et al. Glial glutamate transporter GLT-1 down-regulation precedes delayed neuronal death in gerbil hippocampus following transient global cerebral ischemia. Neurochem Int 2000; 36:531-537.
52. Martin LJ, Brambrink AM, Lehmann C, Portera-Cailliau C, Koehler R, Rothstein J et al. Hypoxia-ischemia causes abnormalities in glutamate transporters and death of astroglia and neurons in newborn striatum. Ann Neurol 1997; 42:335-348.

53. Torp R, Lekieffre D, Levy LM, Haug FM, Danbolt NC, Meldrum BS et al. Reduced postischemic expression of a glial glutamate transporter, GLT1, in the rat hippocampus. Exp Brain Res 1995; 103:51-58.
54. Gottlieb M, Domercq M, Matute C. Altered expression of the glutamate transporter EAAC1 in neurons and immature oligodendrocytes after transient forebrain ischemia. J Cereb Blood Flow Metab 2000; 20:678-687.
55. Yan YP, Yin KJ, Sun FY. Effect of glutamate transporter on neuronal damage induced by photochemical thrombotic brain ischemia. Neuroreport 1998; 9:441-446.
56. Yin KJ, Yan YP, Sun FY. Altered expression of glutamate transporter GLAST mRNA in rat brain after photochemically induced focal ischemia. Anat Rec 1998; 251:9-14.
57. Inage YW, Itoh M, Wada K, Takashima S. Expression of two glutamate transporters, GLAST and EAAT4, in the human cerebellum: Their correlation in development and neonatal hypoxic-ischemic damage. J Neuropathol Exp Neurol 1998; 57:554-562.
58. Siesjo BK, Agardh CD, Bengtsson F. Free radicals and brain damage. Cerebrovasc Brain Metab Rev 1989; 1:165-211.
59. Siesjo BK. Calcium and cell death. Magnesium 1989; 8:223-237.
60. Siesjo BK, Agardh CD, Bengtsson F, Smith ML. Arachidonic acid metabolism in seizures. Ann NY Acad Sci 1989; 559:323-339.
61. Siesjo BK, Bengtsson S, Grampp W, Theander S. Calcium, excitotoxins, and neuronal death in the brain. Ann NY Acad Sci 1989; 568:234-251.
62. Siesjo BK, Bengtsson F. Calcium fluxes, calcium antagonists, and calcium-related pathology in brain ischemia, hypoglycemia, and spreading depression: a unifying hypothesis. J Cereb Blood Flow Metab 1989; 9:127-140.
63. Volterra A, Trotti D, Cassutti P, Tromba C, Salvaggio A, Melcangi RC et al. High sensitivity of glutamate uptake to extracellular free arachidonic acid levels in rat cortical synaptosomes and astrocytes. J Neurochem 1992; 59:600-606.
64. Trotti D, Volterra A, Lehre KP, Rossi D, Gjesdal O, Racagni G et al. Arachidonic acid inhibits a purified and reconstituted glutamate transporter directly from the water phase and not via the phospholipid membrane. J Biol Chem 1995; 270:9890-9895.
65. Trotti D, Rossi D, Gjesdal O, Levy LM, Racagni G, Danbolt NC et al. Peroxynitrite inhibits glutamate transporter subtypes. J Biol Chem 1996; 271:5976-5979.
66. Volterra A, Trotti D, Racagni G. Glutamate uptake is inhibited by arachidonic acid and oxygen radicals via two distinct and additive mechanisms. Mol Pharmacol 1994; 46:986-992.
67. Trotti D, NussbergerS, Volterra A, Hediger MA. Differential modulation of the uptake currents by redox interconversion of cysteine residues in the human neuronal glutamate transporter EAAC1. Eur J Neurosci 1997; 9:2207-2212.
68. Trotti D, Rizzini BL, Rossi D, Haugeto O, Racagni G, Danbolt NC et al. Neuronal and glial glutamate transporters possess an SH-based redox regulatory mechanism. Eur J Neurosci 1997; 9:1236-1243.
69. Trotti D, Danbolt NC, and Volterra A. Glutamate transporters are oxidant-vulnerable: a molecular link between oxidative and excitotoxic neurodegeneration? Trends Pharmacol Sci 1998; 19:328-334.
70. Rao VL, Baskaya MK, Dogan A, Rothstein JD, Dempsey RJ. Traumatic brain injury down-regulates glial glutamate transporter (GLT-1 and GLAST) proteins in rat brain. J Neurochem 1998; 70:2020-2027.
71. Watase K, Hashimoto K, Kano M, Yamada K, Watanabe M, InoueY, et al. Motor discoordination and increased susceptibility to cerebellar injury in GLAST mutant mice. Eur J Neurosci 1998; 10:976-988.
72. Rothstein JD. Excitotoxicity and neurodegeneration in amyotrophic lateral sclerosis. Clin Neurosci 1995; 3:348-359.
73. Rothstein JD. Excitotoxic mechanisms in the pathogenesis of amyotrophic lateral sclerosis. Adv Neurol 1995; 68:7-20.

74. Rothstein JD, Martin LJ, Kuncl RW. Decreased glutamate transport by the brain and spinal cord in amyotrophic lateral sclerosis. N Engl J Med 1992; 326:1464-1468.
75. Rothstein JD, Van Kammen M, Levey AI, Martin LJ, Kuncl RW. Selective loss of glial glutamate transporter GLT-1 in amyotrophic lateral sclerosis. Ann Neurol 1995; 38:73-84.
76. Bruijn LI, Becher MW, Lee MK, Anderson KL, Jenkins NA, Copeland NG et al. ALS-linked SOD1 mutant G85R mediates damage to astrocytes and promotes rapidly progressive disease with SOD1-containing inclusions. Neuron 1997; 18:327-338.
77. Bristol LA, Rothstein JD. Glutamate transporter gene expression in amyotrophic lateral sclerosis motor cortex. Ann Neurol 1996; 39:676-679.
78. Lin C-LG, Bristol LA, Jin L, Dykes-Hoberg M, Crawford T, Clawson L et al. Aberrant RNA processing in a neurodegenerative disease: The cause for absent EAAT2, a glutamate transporter, in amyotrophic lateral sclerosis. Neuron 1998; 20: 589-602.
79. Meyer T, Fromm A, Munch C, Schwalenstocker B, Fray AE, Ince PG et al. The RNA of the glutamate transporter EAAT2 is variably spliced in amyotrophic lateral sclerosis and normal individuals. J Neurol Sci 1999; 170:45-50.
80. Flowers JM, Powell JF, Leigh PN, Andersen P, Shaw CE. Intron 7 retention and exon 9 skipping EAAT2 mRNA variants are not associated with amyotrophic lateral sclerosis. Ann Neurol 2001; 49:643-649.
81. Rosen DR, Siddique T, Patterson D, Figlewicz DA, Sapp P, Hentati A et al. Mutations in Cu/Zn superoxide dismutase gene are associated with familial amyotrophic lateral sclerosis. Nature 1993; 362:59-62.
82. Nagai M, Aoki M, Miyoshi I, Kato M, Pasinelli P, Kasai N et al. Rats expressing human cytosolic copper-zinc superoxide dismutase transgenes with amyotrophic lateral sclerosis: Associated mutations develop motor neuron disease. J Neurosci 2001; 21:9246-9254.
83. Gurney ME, Pu J, Chiu AY, Dal Canto MC, Polchow CY, Alexander DD et al. Motor neuron degeneration in mice that express a human Cu,Zn superoxide dismutase mutation [published erratum appears in Science 1995 Jul 14;269(5221):149]. Science 1994; 264:1772-1775.
84. Pedersen WA, Fu W, Keller JN, Markesbery WR, Appel S, Smith RG et al. Protein modification by the lipid peroxidation product 4-hydroxynonenal in the spinal cords of amyotrophic lateral sclerosis patients. Ann Neurol 1998; 44:819-824.
85. Trotti D, Rolfs A, Danbolt NC, Brown RH, Jr., Hediger MA. SOD1 mutants linked to amyotrophic lateral sclerosis selectively inactivate a glial glutamate transporter [published erratum appears in Nat Neurosci 1999 Sep;2(9):848]. Nat Neurosci 1999; 2:427-433.
86. Aoki M, Lin CL, Rothstein JD, Geller BA, Hosler BA, Munsat TL et al. Mutations in the glutamate transporter EAAT2 gene do not cause abnormal EAAT2 transcripts in amyotrophic lateral sclerosis. Ann Neurol 1998; 43:645-653.
87. Trotti D, Aoki M, Pasinelli P, Berger UV, Danbolt NC, Brown RH, Jr. et al. Amyotrophic Lateral Sclerosis-linked Glutamate Transporter Mutant Has Impaired Glutamate Clearance Capacity. J Biol Chem 2000; 276:576-582.
88. Tanaka K, Watase K, Manabe T, Yamada K, Watanabe M, Takahashi K et al. Epilepsy and exacerbation of brain injury in mice lacking the glutamate transporter GLT-1. Science 1997; 276:1699-1702.
89. Peghini P, Janzen J, Stoffel W. Glutamate transporter EAAC-1-deficient mice develop dicarboxylic aminoaciduria and behavioral abnormalities but no neurodegeneration. Embo J 1997; 16:3822-3832.
90. Ghijsen, W.E., A.I. da Silva Aresta Belo, M. Zuiderwijk, and F.H. Lopez da Silva. 1999. Compensatory change in EAAC1 glutamate transporter in rat hippocampus CA1 region during kindling epileptogenesis. Neurosci Lett. 276:157-60.
91. Miller HP, Levey AI, Rothstein JD, Tzingounis AV, Conn PJ. Alterations in glutamate transporter protein levels in kindling-induced epilepsy. J Neurochem 1997; 68:1564-1570.
92. Ingram EM, Tessler S, Bowery NG, Emson PC. Glial glutamate transporter mRNAs in the genetically absence epilepsy rat from Strasbourg. Brain Res Mol Brain Res 2000; 75:96-104.

93. Doi T, Ueda Y, Tokumaru J, Mitsuyama Y, Willmore LJ. equential changes in glutamate transporter mRNA levels during Fe(3+)-induced epileptogenesis. Brain Res Mol Brain Res 2000; 75:105-112.
94. Akbar MT, Rattray M, Williams RJ, Chong NW, Meldrum BS. Reduction of GABA and glutamate transporter messenger RNAs in the severe-seizure genetically epilepsy-prone rat. Neuroscience 1998; 85:1235-1251.
95. Nonaka M, Kohmura E, Yamashita T, Shimada S, Tanaka K, Yoshimine T et al. Increased transcription of glutamate-aspartate transporter (GLAST/GluT-1) mRNA following kainic acid-induced limbic seizure [published erratum appears in Brain Res Mol Brain Res 1998 Oct 1;60(2):310]. Brain Res Mol Brain Res 1998; 55:54-60.
96. Simantov R, Crispino M, Hoe W, Broutman G, Tocco G, Rothstein JD et al. Changes in expression of neuronal and glial glutamate transporters in rat hippocampus following kainate-induced seizure activity. Brain Res Mol Brain Res 1999; 65:112-123.
97. Mathern GW, Mendoza D, Lozada A, Pretorius JK, Dehnes Y, Danbolt NC et al. Hippocampal GABA and glutamate transporter immunoreactivity in patients with temporal lobe epilepsy. Neurology 1999; 52:453-472.
98. Masliah E, Raber J, Alford M, Mallory M, Mattson MP, Yang D et al. Amyloid protein precursor stimulates excitatory amino acid transport. Implications for roles in neuroprotection and pathogenesis. J Biol Chem 1998; 273:12548-12554.
99. Li S, Mallory M, Alford M, Tanaka S, Masliah E. Glutamate transporter alterations in Alzheimer disease are possibly associated with abnormal APP expression. J Neuropathol Exp Neurol 1997; 56:901-911.
100. Masliah E, Alford M, Mallory M, Rockenstein E, Moechars D, Van Leuven F. Abnormal glutamate transport function in mutant amyloid precursor protein transgenic mice. Exp Neurol 2000; 163:381-387.
101. Lievens JC, Woodman B, Mahal A, Spasic-Boscovic O, Samuel D, Kerkerian-Le Goff L et al. Impaired glutamate uptake in the R6 Huntington s disease transgenic mice. Neurobiol Dis 2001; 8:807-821.
102. Hazell AS, Itzhak Y, Liu H, Norenberg MD. 1-Methyl-4-phenyl-1,2,3,6-tetrahydropyridine (MPTP) decreases glutamate uptake in cultured astrocytes. J Neurochem 1997; 68:2216-2219.
103. Berman SB, Hastings TG. Inhibition of glutamate transport in synaptosomes by dopamine oxidation and reactive oxygen species. J Neurochem 1997; 69:1185-1195.
104. Mort D, Marcaggi P, Grant J, Attwell D. Effect of acute exposure to ammonia on glutamate transport in glial cells isolated from the salamander retina. J Neurophysiol 2001; 86:836-844.
105. Knecht K, Michalak A, Rose C, Rothstein JD, Butterworth RF. Decreased glutamate transporter (GLT-1) expression in frontal cortex of rats with acute liver failure. Neurosci Lett 1997; 229:201-203.
106. Chan H, Butterworth RF. Evidence for an astrocytic glutamate transporter deficit in hepatic encephalopathy. Neurochem Res 1999; 24:1397-1401.
107. Chan H, Hazell AS, Desjardins P, Butterworth RF. Effects of ammonia on glutamate transporter (GLAST) protein and mRNA in cultured rat cortical astrocytes. Neurochem Int 2000; 37:243-248.
108. Norenberg MD, Huo Z, Neary JT, Roig-Cantesano A. The glial glutamate transporter in hyperammonemia and hepatic encephalopathy: Relation to energy metabolism and glutamatergic neurotransmission. Glia 1997; 21:124-133.
109. Suarez I, Bodega G, Fernandez B. Modulation of glutamate transporters (GLAST, GLT-1 and EAAC1) in the rat cerebellum following portocaval anastomosis. Brain Res 2000; 859:293-302.
110. Zhou BG, Norenberg MD. Ammonia downregulates GLAST mRNA glutamate transporter in rat astrocyte cultures. Neurosci Lett 1999; 276:145-148.
111. Malm O, Branches FJ, Akagi H, Castro MB, Pfeiffer WC, Harada M et al. Mercury and methylmercury in fish and human hair from the Tapajos river basin, Brazil. Sci Total Environ 1995; 175:141-150.

112. Harada M. Minamata disease: Methylmercury poisoning in Japan caused by environmental pollution. Crit Rev Toxicol 1995; 25:1-24.
113. Park ST, Lim KT, Chung YT, Kim SU. Methylmercury-induced neurotoxicity in cerebral neuron culture is blocked by antioxidants and NMDA receptor antagonists. Neurotoxicology 1996; 17:37-45.
114. Aschner M, Yao CP, Allen JW, Tan KH. Methylmercury alters glutamate transport in astrocytes. Neurochem Int 2000; 37:199-206.
115. Aschner M. Mercury toxicity. J Pediatr 2001; 138:450-451.
116. Aschner M. Astrocytes as modulators of mercury-induced neurotoxicity. Neurotoxicology 1996; 17:663-669.
117. Nagaraja TN, Brookes N.Mercuric chloride uncouples glutamate uptake from the countertransport of hydroxyl equivalents. Am J Physiol 1996; 271:C1487-1493

PURINES AND NEUROPROTECTION

Trevor W. Stone*

ABSTRACT

The activation of adenosine A1, A2 and A3 receptors can protect neurones against damage generated by mechanical or hypoxic/ischaemic insults as well as excitotoxins. A1 receptors are probably effective by suppressing transmitter release and producing neuronal hyperpolarisation. They are less likely to be of therapeutic importance due to the plethora of side effects resulting from A1 agonism, although the existence of receptor subtypes and recent synthetic chemistry efforts to increase ligand selectivity, may yet yield clinically viable compounds.

Activation of A_{2A} receptors can protect neurons, although there is much uncertainty as to whether agonists are acting centrally or via a peripheral mechanism such as altering blood flow or immune cell function. Selective antagonists at the A_{2A} receptor, such as 4-(2-[7-amino-2-{2-furyl}{1,2,4}triazolo{2,3-a}{1,3,5}triazin-5-yl-amino]ethyl)phenol (ZM 241385) and 7-(2-phenylethyl)-5-amino-2-(2-furyl)-pyrazolo-[4,3e]-1,2,4-triazolo[1,5-c]pyrimidine (SCH 58261), can also protect against neuronal death produced by ischaemia or excitotoxicity. In addition, A_{2A} receptor antagonists can reduce damage produced by combinations of subthreshold doses of the endogenous excitotoxin quinolinic acid and free radicals. Since the A2A receptors do not seem to be activated by normal endogenous levels of adenosine, their blockade should not generate significant side effects, so that A_{2A} receptor antagonists appear to be promising candidates as new drugs for the prevention of neuronal damage. Adenosine A3 receptors have received less attention to date, but agonists are clearly able to afford protection against damage when administered chronically.

---

*Division of Neuroscience and Biomedical Systems, West Medical Bldg, University of Glasgow, Glasgow G12 8QQ, UK.

Given the disappointing lack of success of NMDA receptor antagonists in human stroke patients, despite their early promise in animal models, it is possible that A_{2A} receptor antagonists could have a far greater clinical utility.

ADENOSINE

Adenosine is an ubiquitous compound, participating in the formation of ATP and nucleic acids. As a result of efflux from cells by diffusion or membrane transporters, and partly as a result of the catabolism of extracellular ATP, adenosine exists in the extracellular fluid at resting concentrations of 1mM or less.[1-6] This concentration can rise to 30mM or more during and soon after a period of cerebral ischaemia.[7-9] Once in the extracellular space, adenosine can mediate a variety of effects at both presynaptic and postsynaptic sites which modulate neuronal activity and, usually, have the overall effect of reducing excitability.

At presynaptic terminals, adenosine can inhibit the release of many neurotransmitters, including the major excitatory transmitter glutamate,[10-12] which has been increasingly implicated in mediating the neuronal damage which accompanies cerebral traumatic (mechanical) injuries and strokes.[13] Adenosine has also been shown to induce hyperpolarisation of neurones in the hippocampus[14,15] and striatum[16-18] and it is possible that such a direct inhibition of neuronal activity could contribute to neuroprotection. This combination of presynaptic and postsynaptic inhibitory activity made adenosine a clear candidate for suppressing neuronal activity and glutamate release resulting from a cerebral infarct, and thus affording neuroprotection. The identification of the nature of the receptors responsible for these actions thus became a prime target for the development of purines as neuroprotectants. Parallel work on the immune system has also yielded data which could explain many of the protectant effects of purines and could lead to the inclusion of several peripheral targets in the design of novel compounds.

ADENOSINE RECEPTORS

There are at least four cell membrane receptors able to respond to adenosine in the extracellular fluid : A_1, A_{2A}, A_{2B} and A_3. The first of these to receive extensive study in relation to neuroprotection were the A1 receptors, but more recent work indicates that the A_{2A} receptors may represent a more therapeutically attractive target.

A1 receptors

The A1 receptors are distributed widely around the brain, though occurring in their highest density in the hippocampus and neocortex and being found on both neurons and microglia.[19-25] They have a similar distribution in human brain[26-28] (Fig. 1). The receptors are found both presynaptically and postsynaptically, distributed on both the somata and dendrites of pyramidal neurons in all CA regions of the hippocampus.[24,25]

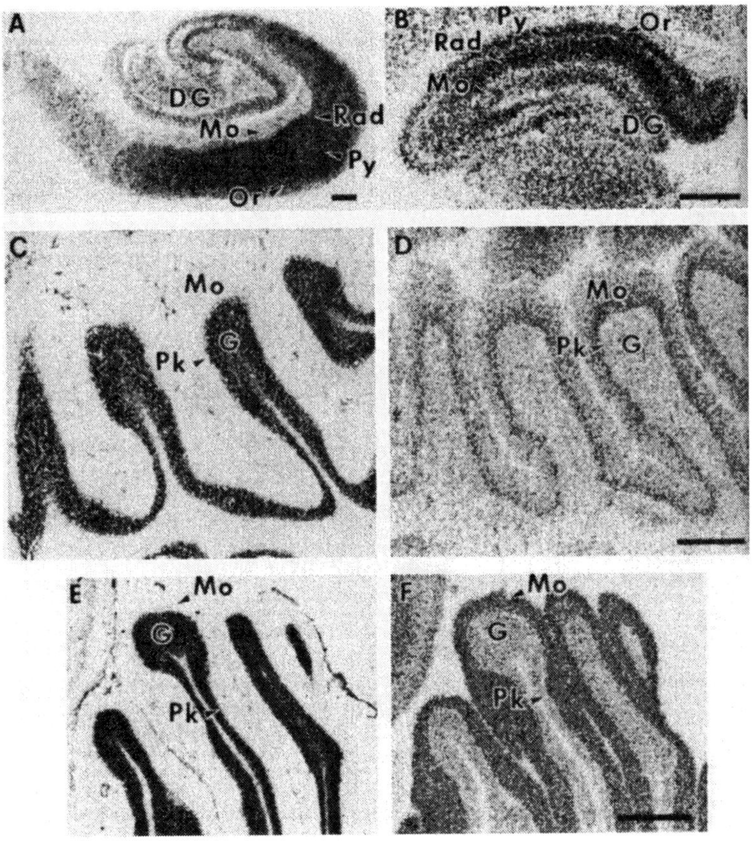

Figure 1. Autoradiograms showing the localization of binding of [3H] CHA in sections from rat brain. Above human (A) and rat (B) hippocampus. Below human (D) and rat (F) cerebellum. Tissues were incubated with 3nM [3H]CHA. To the left of the autoradiograms of the cerebellum, the corresponding tissue sections stained with cresyl violet (C and E) are shown for comparison. DG, dentate gyrus; G granule cell layer; Mo, molecular layer; Or, stratum oriens; Pk, Purkinje cell layer; Py, pyramidal layer; Rad, stratum radiatum. Bars = 1mm. (Reproduced with permission from ref. 26).

A1 receptors and Neuroprotection

The first experimental indication that A1 receptors could be neuroprotective was reported by von Lubitz et al[29] with the finding that the A1 receptor agonist N6-cyclohexyladenosine (CHA) protected neurones against damage following an ischaemic episode in gerbils. Subsequent work has adequately supported this view, with clear evidence for a neuroprotective action of a variety of selective adenosine A_1 receptor agonists against ischaemic damage.[29-37]

Protection against systemic kainate-induced neurotoxicity by the A1 receptor agonist R-phenylisopropyladenosine (PIA) (Fig. 2) was studied using conventional

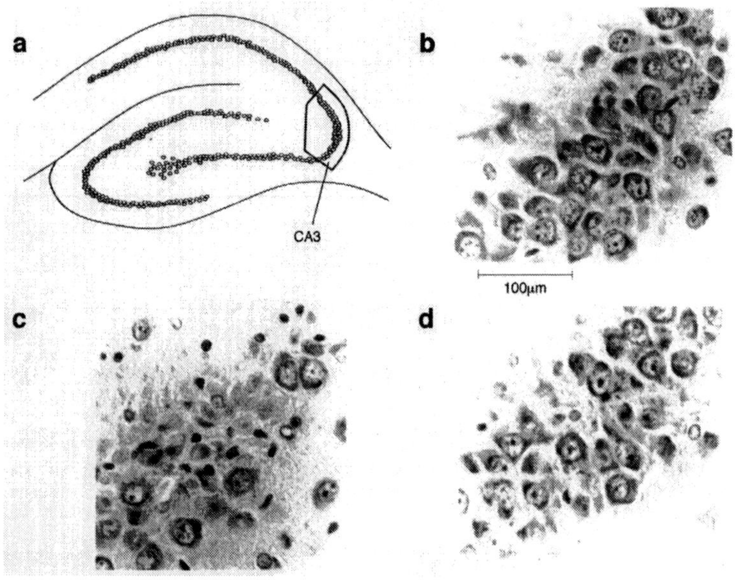

Figure 2. Kainate damage and R-PIA protection in the hippocampal CA3 region. (a) is a line diagram indicating the approximate area examined. (b) illustrates a section from a control vehicle-treated animal. (c) illustrates a section from a kainate-treated animal; there is a mixture of normal neurons and necrotic cells with contracted, darkly staining nuclei and eosinophilic cytoplasm. (d) illustrates a section from an animal treated with kainate plus R-PIA; the cells here appear normal. Calibration bar = 100μm. (Reproduced with permission from ref. 40).

histology and the use of the benzodiazepine site ligand PK11195[38,39] and extended previous work by revealing protection in a number of non-hippocampal regions.[40] This protection could be prevented by the simultaneous administration of the A1 adenosine receptor antagonist 1,3-dipropyl-8-cyclopentylxanthine (DPCPX), confirming that the protection involved adenosine A1 receptors. At a dose of 25 mg\kg R-PIA fully protected against kainate induced damage in the basolateral amygdaloid nuclei, the pyriform cortex and rhinal fissure in 8 of 11 animals in this group. In these animals the cell groups appeared normal, with no evidence of loss of structure or organisation, a normal number of neurones, and no sign of reactive gliosis. The more posteriorly located regions examined the entorhinal cortex at interaural line 3.4mm, the posteromedial cortical amygdaloid nucleus and the amygdalopyriform transition showed a degree of damage comparable with the kainate animals, with no apparent protection.

These results are consistent with earlier studies including work from other laboratories which demonstrated protection by adenosine analogues against toxin or excitotoxin induced damage.[41-43] The ability to protect may even extend to agents which do not directly activate purine receptors, but which elevate the levels of endogenous adenosine.[9,44]

As yet it remains unclear why PIA was able to reduce protection in some but not all regions of the brain. It is possible that this difference reflects an absence of adenosine receptors in the resistant areas, or that the mechanisms of excitotoxicity are different in these regions. These would be important observations to pursue as they may provide clues as to the mechanisms or neuronal damage, the role of amino acids and other agents in cell death, and the mechanisms of protection by purines.

An intriguing and very important paper by Lau and Mouradian[45] revealed that the protective activity of an A1 agonist (CHA) was also apparent in mice when damage was induced by the dopaminergic neurotoxin 1-methyl-4-phenyl-1,2,3,6-tetrahydropyridine (MPTP), and Ongini et al[46] have found reduced damage by MPTP in mice lacking A_{2A} receptors. This result clearly has implications for Parkinson s disease, since an A1 receptor agonist may be of value in this condition if the mechanisms of disease and MPTP-induced degeneration are comparable. The mechanism of protection against MPTP, however, remains unclear, although an interference with oxidative stress and mitochondrial oxidation systems may be involved. Alternatively, since antagonists at the NMDA receptors are also able to reduce toxicity caused by MPTP, it is possible that the same mechanisms are operating for A1 receptor protection against MPTP and against kainate or NMDA receptor activation.

Time Course of A1 Receptor Protection

It is very relevant to any discussion of the value of purine receptor targeting and clinical potential, to have a window of therapeutic opportunity such that a drug could be administered up to several hours after the occurrence of a cerebral insult and still retain protective activity. It has been shown that A1 agonists such as PIA can indeed be given up to 2 hours after the administration of kainic acid and remain neuroprotective (Fig. 3), implying that neuronal circuitry and/or transmitter release relevant to the neurodestructive process continue over at least this time frame.[47] Against ischaemia-induced damage, cyclohexyladenosine (CHA) remains protective when administered up to at least 30 minutes following the cerebral.[29] Similarly an adenosine-amine congener (ADAC) has been shown to have an acceptable time window of several hours after cerebral ischaemia in gerbils.[48] Since ADAC also shows fewer of the haemodynamic problems associated with other A1 agonists,[49] this compound may prove clinically useful. More recently, it has been shown that ADAC can be used chronically in low doses and still retain its neuroprotective efficacy.[30] Since most of the other purines examined produce opposite effects after acute and chronic administration, and many patients may require treating at least sub-chronically, this observation represents another possible advantage of ADAC over other known A1 receptor agonists.

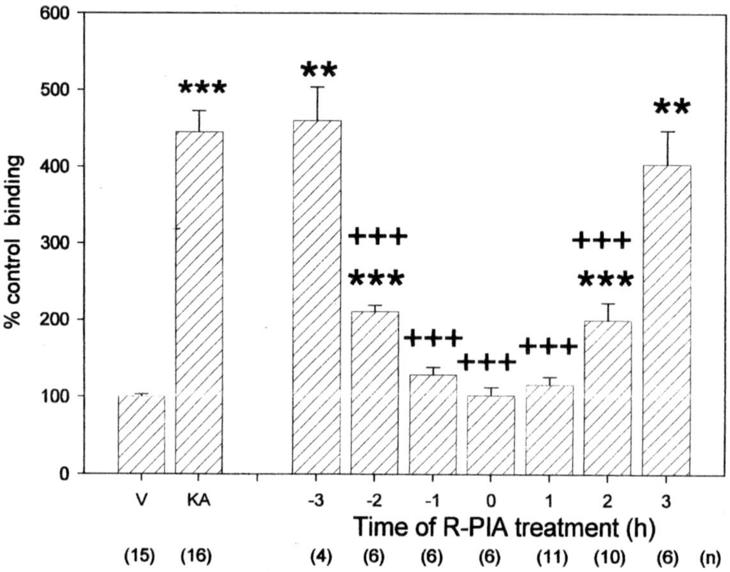

Figure 3. Time course of protection by R-phenylisopropyladenosine (R-PIA) against kainate toxicity. Rats were injected i.p. with a single dose of 25μg/kg of R-PIA at various times before or after kainate injection. Clonazepam (0.2mg/kg) was injected i.p. 10 minutes before i.p. kainate injection (t0). Binding is expresses as a percentage of the same day controls. Columns indicate mean ? s.e.m. V = saline and methanol/saline vehicle treatments (t0), KA = kainic acid 10mg/kg with methanol/saline (t0). ** $P < 0.01$; *** $P < 0.001$ significance versus vehicle control; ++ $P < 0.01$; +++ $P < 0.001$ significance versus kainate. (Reproduced with permission from ref. 39).

Mechanisms of A1 Receptor Neuroprotection

Suppression of Transmitter Release

Evidence from a variety of sources indicates the involvement of excitatory amino acids in several CNS disorders in which the most prominent feature is neuronal degeneration. The neuronal damage resulting from focal cerebral ischaemia, for example, can be prevented by agents which block the neurotoxic effects of amino acid agonists applied into the CNS 50-53.

Studies in which the destruction of afferent pathways reduces kainate induced excitation and neurodegeneration indicated that intact terminals are required for kainate excitotoxicity.[54,58] This may result from a presynaptic action of kainate to evoke the release of other compounds such as glutamate and aspartate[59-63] which are primarily responsible for the subsequent degeneration, or it may indicate the need for such presynaptic factors to exert a permissive action on the effects of kainate. Since adenosine and its analogues act at A1 nucleoside receptors to suppress the release of a variety of neurotransmitters including glutamate[8,10,64-65] acetylcholine,[66,67] dopamine[68,69] it may be this activity which could lead to the blockade of amino acid release induced by kainate and thus cause protection.

Arguing against a mechanism based on inhibiting glutamate release are studies reporting that the extracellular concentrations of glutamate do not correlate with the degree of neuronal damage in the hippocampus[70] but these studies involved ischaemic damage, and different mechanisms could be involved in ischaemic and excitotoxic protection. The systemic administration of R-PIA, CHA or an adenosine uptake inhibitor was also reported not to suppress the elevation of glutamate levels in brain which occurs during ischaemia.[11,36,71] On the other hand, cyclopentyladenosine (CPA) decreased ischaemia-induced release of glutamate and aspartate in a rat four-vessel occlusion model.[72] Theophylline increased glutamate release during ischaemia,[11] implying a degree of restraint by the high endogenous levels of adenosine which would be produced by ischaemia. A recent careful analysis of the simultaneous concentrations of purines, glutamate, neuronal damage and behavioural sequelae uncovered a significant relationship between these parameters, indicating the expected correlation between lower extracellular glutamate and decreased cell damage.[73] It is certainly relevant that other agents which can suppress the release of excitatory amino acids, such as the pyrimidine derivative BW1002C87 (5-2,3,5-trichlorophenyl)pyrimidine-2,4-diamine-1,10-ethanesulphonate) and kappa opiate receptor agonists are able to prevent neuronal damage in studies of ischaemia.[74,75]

Another amino acid which might be modulated by adenosine receptors in an important way is glycine. The A1 receptor agonist CHA depressed the rise of extracellular glycine levels observed after cerebral ischaemia in rabbits, even though there was no parallel change in the levels of glutamate.[71] An absence of change of glutamate levels was also reported by Kano et al.[76] In view of the fact that increased glycine concentrations can enhance the activation of NMDA receptors, it may be that this effect on glycine is qualitatively as well as quantitatively a significant contributor to the protection by A1 receptors.

Calcium Fluxes

Another potential mechanism is likely to be the reduction of calcium influx which has been demonstrated in neuronal and cardiac tissues.[33,77-80] This may be related to the frequently observed ability of A1 receptors to modulate potassium conductances of several types including the ATP-sensitive potassium channels in heart and hippocampal neurons.[16,81,82] There appear to be neuronal chloride conductances which are also sensitive to purines, resulting in an increase chloride influx which should contribute to neuronal inhibition in most areas of the brain.[83,84]

Adenosine as an Endogenous Neuroprotectant

One question which arises from these reports, but which still cannot be answered satisfactorily, is whether endogenous adenosine functions as an endogenous protective agent against cerebral ischaemia and excitotoxic neuronal damage. Certainly, there is a massive efflux of adenosine from cells during and following hypoxia or ischaemia, such that the extracellular concentration of adenosine may rise from the normal basal levels of around 1mM to 100mM or more.[7,85] A strong argument for a protective role of endogenous adenosine, however, is presented by the several reports that A1 adenosine receptor blockers, such as 8-cyclopentyl-1,3-dipropylxanthine (DPCPX), are able to increase the degree of neuronal damage resulting from ischaemia or excitotoxins.[86-88] On the other hand, some groups have claimed that doses of DPCPX sufficient to block the protective activity of exogenous A1 receptor agonists, are not able to increase toxicity when applied alone.[89]

These arguments are confounded by the finding that activation of NMDA receptors suppresses neuronal presynaptic sensitivity to A1 receptor activation[90,91] a phenomenon recently confirmed by the use of paired-pulse recordings.[92] The consequence of this would be that the release of glutamate, acting partly on NMDA receptors, would tend to prevent any protective effect of endogenous adenosine.

A1 Receptor Subtypes and Clinical Value

Although of scientific interest, the possible use of A1 adenosine receptor agonists as clinically useful neuroprotectants has not been received with much enthusiasm, largely due to the fact that the administration of A1 receptor agonists can induce a wide range of systemic effects which could lead to serious side effects and complications in patients. These effects include a depression of blood pressure, intestinal motility, exocrine and endocrine secretions and cardiac function, mainly due to the suppression by A1 receptors of the release of transmitters such as acetylcholine and norepinephrine and direct effects on ion channels. It is unlikely, therefore, that A1 receptor agonists will find therapeutic value unless there is a molecular or functional distinction between those receptors responsible for neuroprotection and those responsible for potential side effects. There have been occasional suggestions that A1 receptor subtypes may exist[93] but, although there are undoubted species differences in A1 receptors, there has been little support for receptor variations within a species until recently.

The synthesis of a novel series of compounds has revealed several which have clear neuroprotective activity in gerbil and mouse cerebral ischaemia models, but which exhibit a substantially lower tendency to depress blood pressure than previous A1 receptor agonists.[94] These compounds restore the possibility that A1 receptor agonists might be found with a neuroprotective selectivity sufficient to enter clinical trials.

Acute V Chronic Administration

Soon after the realisation that adenosine receptors could mediate neuroprotection, caffeine, a non-selective A1 and A2 receptor antagonist was shown to protect against ischaemic damage when it had been administered chronically.[95,96] Single, acute injections of selective A1 antagonists including DPCPX exacerbate ischaemic damage.[87,88] Conversely, the chronic administration of these same compounds induced increased damage and neuroprotection.[88] This result presumably reflects the compensatory changes of receptor density which follow the prolonged presence of any receptor ligand, although Traversa et al[97] did not detect any changes of A1 receptor binding after chronic administration of antagonists.

A_{2A} RECEPTORS

Several recent reviews have summarised the basic state of knowledge on A_{2A} receptors in the brain.[98-100] A2 receptors are subdivided into A_{2A} and A_{2B} varieties on the basis of their pharmacology, most notably the higher affinity of the former for 2-[4-(2-carboxyethyl)-phenylethylamino]-5 N-ethyl-carboxamido-adenosine (CGS 21680). The A_{2A} receptors are found in highest density in the striatum, especially on the GABA-releasing striopallidal projection neurons and on striatal cholinergic neurons,[101-104] and limbic areas such as nucleus accumbens and olfactory tubercle. They are also present within the hippocampus and cerebral cortex,[104,105] but the receptors in these latter regions appear to differ pharmacologically from the classical striatal population.[106] Within the hippocampus functional A_{2A} receptors are localised to the CA1 and CA3 regions. In human brain, A_{2A} receptors were reported to be confined largely to striatal regions,[107] but later work has claimed their presence more widely.[23]

A2 Receptor Agonists and Neuroprotection

Agonists acting at A_{2A} receptors afford protection against a variety of insults including ischaemia[87,108] and excitotoxins such as kainate.[109,110] The selective A_1 receptor antagonist DPCPX (50μg/kg), in combination with CGS 21680 (0.1mg/kg) showed no significant blockade of the protection afforded by CGS 21680 alone against kainate, suggesting that A_1 receptor activation by CGS 21680 (responsible for its anticonvulsant effects[111]) may account for only a small proportion of the protection afforded by this compound, at the doses used. However, the protection could be largely prevented by 8-(p-sulphophenyl)-theophylline (8PST), a non-selective, A1 and A2 adenosine receptor antagonist, which does not penetrate the blood-brain barrier. The experiments with 8-PST therefore, may imply a peripheral component to the mechanism of action of CGS 21680. 8-PST alone had been shown previously to increase the damage associated with kainate,[89] presumably by blocking the peripheral effects of endogenously released adenosine.

In addition, the direct injection of CGS 21680 into the hippocampus failed to afford protection. Together, these observations suggest that the A_{2A} agonist protection involves a peripheral site of action, possibly on the vascular system to facilitate the removal of kainate, or to increase the delivery of nutrients to compromised cells.

A moderately selective agonist, N6-[2-(3,5-dimethoxyphenyl)-2-(2-methylphenyl)-ethyl]adenosine (DPMA), which is 13-fold selective for the rat brain A_2 receptor versus the A_1 site ($K_i = 4.4nM$),[112,113] also significantly decreased neurone degeneration in the CA1 and CA2 areas of the hippocampus at a dose of 1.0mg/kg. CGS 21680 is 140-fold selective for the A_{2A} versus the A_1 receptor, with a slightly lower affinity than DPMA for the A_{2A} subtype (K_i=15nM)[114] and has no discernible activity at the low affinity A_{2B} receptor.[101,115]

The limited protection which has been reported with another A_{2A} receptor agonist (APEC)[116] may be due more to an affinity for adenosine A_3 receptors,[117] as both APEC administration and A_3 receptor stimulation result in a very similar pattern of protection against ischaemia: in both cases chronic, but not acute treatment increases neuronal survival.[116,118]

A2 Receptor Antagonists and Neuroprotection

The first surprising evidence for neuroprotection by an A2A receptor antagonist was reported by Gao & Phillis[119] using the relatively non-selective A2 receptor antagonist 5-amino-9-chloro-2-(2-furyl)-1,2,4-triazolo[1,5-c]quinazoline (CGS 15943). It was found that this agent could protect against cerebral ischaemia in a gerbil model. This result was subsequently confirmed using the compounds 8-(3-chlorostyryl)caffeine (CSC) and 4-amino-1-phenyl[1,2,4]-triazolo[4,3-a]quinoxaline (CP66,713)[87] and was later replicated and extended using other ischaemic models and a range of receptor ligands.[108,116, 120] Protection has also now been demonstrated against excitotoxins using the kainate and quinolinic acid models of excitotoxicity[109,110] (Fig. 4). While many of these earlier studies examined protection against global cerebral ischaemia a relatively unusual event in humans protection has also been demonstrated against focal ischaemic damage.[121]

The antagonist 4-(2-[7-amino-2-{2-furyl}{1,2,4}-triazolo{2,3-a}-(1,3,5}triazin-5-yl-amino]ethyl)phenol (ZM 241385), which is 80-fold selective for A_{2A} versus A_{2B} receptors, and 500-1000-fold selective for A_{2A} versus A_1 receptors,[122] can protect the hippocampus against damage produced by kainate to the same extent as the agonist ligand CGS 21680. When given in conjunction with CGS 21680, the combination of agonist plus antagonist produced, paradoxically, an additive effect with total neuronal conservation of all hippocampal neurones,[109,110] rather than some sort of antagonistic interaction. Since both the A_{2A} agonist and antagonist are neuroprotective in their own right, their mutual potentiation may indicate that the mechanism by which they achieve protection is different for the two compounds. It is not likely that ZM 241385 exerts its neuroprotective effects through the A_{2B} or A3 adenosine receptor subtypes, since it has low selectivity and potency at these receptors.[123]

There appears to be only a single report claiming that an A_{2A} receptor antagonist (CSC) increases, rather than decreases the neurotoxic effects of NMDA in the hippocampus.[124] However, confirmation of much of the preceding work using pharmacological antagonists of A_{2A} receptors has now been obtained using mice lacking these sites.[125,126] In these knockout animals, ischaemic brain injury is decreased compared with normal mice.

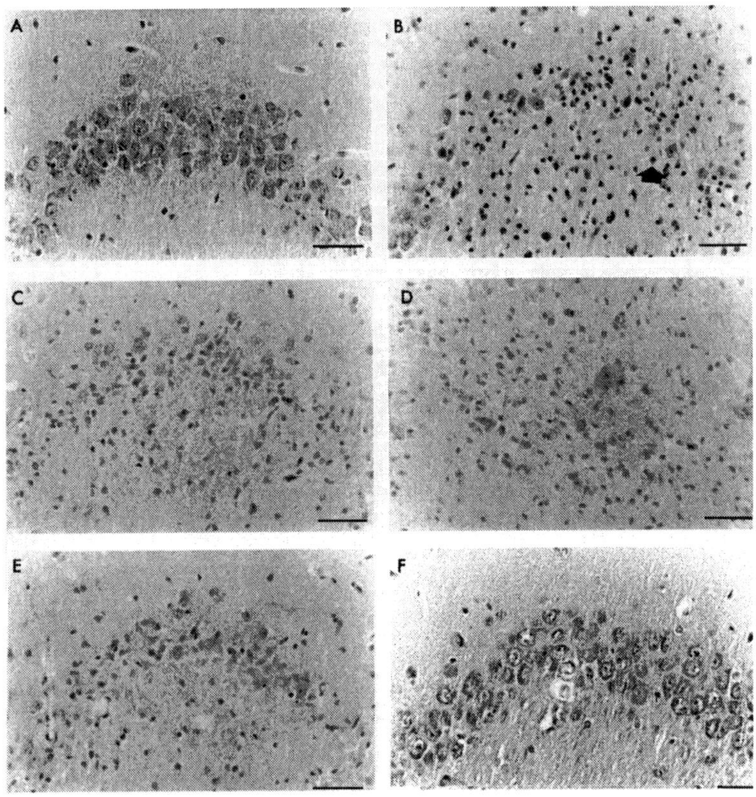

Figure 4. (A-E) Photomicrographs of the CA3 region of the hippocampus following unilateral intrahippocampal injections. The excitotoxic damage to the hippocampus is apparent when comparing the saline treated animal (A) with the same region following the administration of 0.25nmols kainate (B). While a few damaged cells are present after kainate treatment, the region shows few remaining viable neurons, while the gliotic reaction to the damage is reflected in the infiltration of the area with the small, darkly staining nuclei of glial cells. This is also the case for C-E which represent damage following CGS21680 (2.5nmol) + kainate (0.25nmol) (C), adenosine (2.5nmol) + kainate (0.25nmol) (D) and R-PIA (0.25nmol) + kainate (0.25nmol) (E), respectively. Protection of the CA3 region by ZM241385 (2.5nmol) + kainate (0.25nmol) is shown in (F), where morphology is similar to that in (A). Bar = 50μm. (Reproduced with permission from ref. 109).

Mechanisms of A_{2A} Receptor Antagonist Protection

One of the often observed actions of A_{2A} receptor stimulation is the increase of transmitter release. Although the A_{2A} receptor agonist CGS 21680 does not modify the resting levels of extracellular glutamate in brain, it has been reported to increase the efflux of this amino acid in response to ischaemia in the rat[72, 127] (Fig. 5). This is entirely consistent with a growing body of evidence showing an excitatory action of A_{2A} receptor activation on transmitter release, including glutamate,[104,128] an effect probably produced by increasing presynaptic calcium influx.[129] Consistent with this

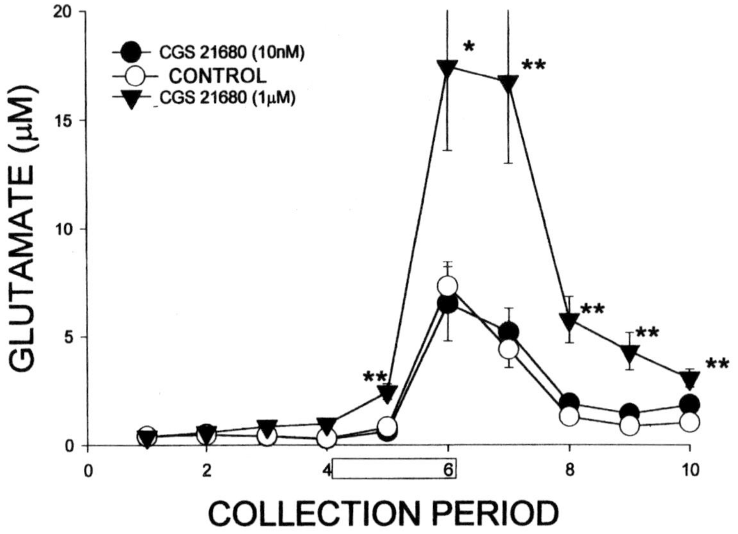

Figure 5. CGS21680 enhance the ischaemia-evoked release of glutamate from the rat cerebral cortex. Superfusate samples were collected at 10 minute intervals and analysed by HPLC. The duration of four-vessel occlusion was 20 minutes (box). Basal collection periods (1 and 2) were followed by drug administration (arrow). The two post-drug basal collections (3 and 4) were succeeded by two collection periods during the ischaemia (5 and 6) and four collections during the period of reperfusion (7-10). Significant differences in amino acid release between the control group and drug treatment groups were determined by one-way ANOVA (*$P < 0.05$; **$P < 0.01$). Values are mean ± S.E.M. (Reproduced with permission from ref. 127).

is the finding that the A_{2A} receptor antagonist CGS 15943 could depress glutamate release, possibly contributing to its neuroprotective effects by blocking the effects of endogenous adenosine at A_{2A} receptors.[72]

An observation requiring further work and explanation is that A_{2A} receptor activation can increase depolarisation-induced acetylcholine release from the hippocampus[130] but not striatum or neocortex.[131] This regional variability may, if it also applies to the release of excitatory and/or inhibitory amino acids, could have a significant influence on the sites within the brain at which protection can be mediated by A_{2A} receptor ligands.

Equally, it must be emphasised that the increase of transmitter release applied to inhibitory compounds as well as excitatory. Thus, part of the neuroprotection by A_{2A} receptor agonists may result, for example, from the enhancement of the release of inhibitory compounds such as GABA from neurons or glia.[103,132]

Receptor Interactions

One mechanism which may contribute to neuroprotection by A_{2A} receptor antagonists is the modulation of receptor-receptor interactions. Evidence from biochemical[133] and electrophysiological work [134] has supported the concept, first mooted by Cunha et al,[104] that A_{2A} receptors can interfere with activity of A1 receptors. This interaction may possibly be the result of direct, intramembrane interactions between A1 and A_{2A} receptors. The presence of the A_{2A} agonist CGS 21680 reduces neuronal sensitivity to CPA,[134] and induces a low-affinity receptor site for 2-chloro-N6-cyclopentyladenosine (CCPA),[133] the latter effect being mediated possibly by protein kinase C. A similar interaction has been described by Latini et al[135] in experiments examining the effects of purines on synaptic depression during a period of artificial ischaemia in vitro. The activation of A_{2A} receptors in hippocampal slices prevented the depression of transmission mediated by A1 receptors. If this interaction operates in vivo, then the presence of an A_{2A} receptor antagonist should release, or unmask A1 receptors, and allow the neuroprotective effect of these receptors to be seen (discussed above).

Thus, in a situation in which the extracellular level of adenosine is substantially elevated, as it is following kainate administration, adenosine may reach levels sufficient to activate A_{2A} receptors and inhibit a proportion of A_1 receptors. This may be relevant to the mechanism of neuroprotection by ZM 241385 in that its effects can be reduced by the A1 receptor antagonist DPCPX, despite the fact that ZM 241385 has no agonist activity at A1 receptors.[109,110] When ZM 241385 is administered subsequently, its blockade of A_{2A} receptors would effectively release A_1 receptors and these might then produce protection as was discussed in the opening paragraphs of this review. The A_1 and A_{2A} receptors can co-exist on the same cells and Correia-de-Sa and Ribeiro[136] first showed that at the neuromuscular junction the action of A_{2A} receptor-facilitated neurotransmitter release predominated over A_1-induced inhibition. Similarly Cunha et al[137] concluded that endogenously formed adenosine preferentially activated A_{2A} receptors in the hippocampus, enhancing synaptic transmission. Electrophysiological extracellular and intracellular studies also confirm the view that A_{2A} receptors inhibit A1 receptor activation.[134]

Interference with Reactive Oxygen Species

As a result of the compromised mitochondrial function which follows a period of oxygen deprivation, there is a greatly enhanced generation of reactive oxygen species (free radicals) such as peroxide, hydroxyl radical, superoxide and peroxynitrite. These radical species are believed to be responsible for most of the molecular and cellular disruption which accompanies ischaemia and the calcium overload which accompanies excitotoxicity.[138-141] Adenosine A_2 receptors can decrease the generation of superoxide anion by neutrophils.[142,143] In view of the apparent neuroprotective activity of A2A receptor antagonists it was, therefore, of interest to know whether A_{2A} receptor antagonists could also modify damage which involved the presence of reactive oxygen species.

Table 1 summarises results in which hippocampal damage was induced by a combination of quinolinic acid and reactive oxygen species. Quinolinic acid is a selective agonist at N-methyl-D-aspartate (NMDA) receptors.[144-146] It is also an

Table 1. Neuronal survival in animals treated with intrahippocampal injections of quinolinic acid alone or in combinations with a mixture of xanthine and xanthine oxidase, or S-nitroso-N-acetyl-penicillamine (SNAP)

Treatment	Neuronal survival (% controls)
Quinolinic acid	95.5 ± 4.2
Xanthine / xanthine oxidase (X / XO)	102.2 ± 14.5
Quinolinic acid + X / XO	34.4 ± 5.2*
SNAP	114.4 ± 22.0
Quinolinic acid + SNAP	54.4 ± 5.0*

* $p < 0.05$ compared with quinolinic acid alone.

endogenous compound, produced as a major metabolite of tryptophan. At low dose (50nmols) quinolinic acid did not produce any significant neuronal loss when injected alone into the hippocampus. This dose is approximately 40% of the dose normally needed to produce damage. Similarly, a low dose of a mixture of xanthine and xanthine oxidase, a well-recognised superoxide and hydroxyl-radical generating system, produced no damage when injected alone. When a combination of both these compounds was administered, substantial cell loss was recorded (Table 1).

Since quinolinic acid is an agonist at NMDA receptors, neuronal damage can normally be prevented completely by analogues of the glutamate receptor antagonist kynurenic acid (another tryptophan metabolite described by Perkins & Stone,[147] see refs. 145,148). The damage produced by the combined use of quinolinic acid and the free radical generating system was only partly decreased by 5,7-dichlorokynurenic acid, suggesting that neuronal loss was not caused by a free radical-mediated enhancement of NMDA receptor toxicity. It would seem more likely that the interaction is the result of an NMDA receptor-mediated enhancement of free radical toxicity, or to a completely different mechanism of one or both of the agents.

Interestingly, toxicity produced by the quinolinate/free radical combination could be suppressed by three different A_{2A} receptor antagonists CSC, ZM 241385 and SCH 58261. This result emphasises that A_{2A} receptor antagonists may have a potentially wide use as neuroprotectants against cell injury in many situations where damage is produced by either glutamate receptor activation, or oxygen radical generation, or their combination. It is presumably the latter which account for neuroprotection during ischaemia, when glutamate and free radical levels are known to rise.

The functional importance of this interaction is that the kynurenine pathway, responsible for the generation of endogenous quinolinic acid, is upregulated in glia and macrophages in response to inflammation or immune activation.[146] The cellular insult imposed by an ischaemic stress or other injury will thus lead to an increased generation of quinolinic acid together with increased levels of oxygen radicals from these cells. Their mutual potentiation could therefore account for a significant fraction of the overall cell damage which results.

Anti-Inflammatory Effects

The ability to reduce the toxic effects of oxygen radicals is probably only one aspect of the anti-inflammatory activity of adenosine which has received much attention in peripheral organ systems. Adenosine itself suppresses phagocytosis, free radical generation and cell adherence by blood white cells activated by immune stimulation[142,149-151] and a substantial accumulation of evidence indicates that it is the A_{2A} receptor which is largely responsible for these effects,[152,153] probably via the unusual mechanism of activating a serine/threonine protein phosphatase.[154] Certainly, human neutrophils possess abundant numbers of A_{2A} receptors.[155] However, some cells such as neutrophils also possess A1 receptors, and Cronstein et al[156] have shown that these promote chemotaxis and other features of the immune response. The opposite effects of A1 and A_{2A} receptors may form a feedback loop in which the normally low concentrations of adenosine, via A1 receptors, are able to enhance the sensitivity of white cells to immune triggers, whereas at high concentrations, once the immune response has been activated, A_{2A} receptors act to limit the degree of activation.[156]

Adenosine also antagonises the pro-inflammatory activity of TNFα[157] and inhibits the release of this, other cytokines and complement from a range of cells.[158-160] Activation of A2A receptors specifically inhibits the production of IL-12 by human monocytes but increases the generation of IL-10.[161] This, and other modifications in the relative release of a range of cytokines could hold significant implications for the effects of adenosine receptor activation on immune and inflammatory processes, including those concerned with the production of neuronal damage.

Adenosine Receptors and Glial Cells

A number of purines are able to increase astrocyte proliferation[162,163] and promote the astrogliosis which follows brain injury by activating A_{2A} receptors.[164] Whether this effect contributes to neuroprotection, however, remains to be seen, since an increased gliosis might be expected to lead to increased local concentrations not only of anti-inflammatory cytokines, but also of injurious oxygen radicals and damaging products of the kynurenine pathway such as quinolinic acid.

Clues from the Heart?

A strong argument that purine neuroprotection may involve the anti-inflammatory actions rather than suppression of glutamate release or direct effects on neurons arises from the repeated demonstration that purines can also protect the heart against damage following a period of ischaemia and reperfusion.[165,166] Most of the purine anti-inflammatory actions can be demonstrated in cardiac tissue, including the suppression of TNFα expression,[167-170] and include modulation of neutrophil adherence to myocytes.[171] In the latter case, the A_{2A} receptors are again inhibitory while A1 receptors promote cell adherence. However, while the latter action should in theory increase post-ischaemic damage, Lozza et al[172] have claimed that both A1 and A_{2A} receptor agonists protect against myocardial ischaemic/reperfusion injury. This is certainly a confused field of research, since Casati et al[173] reported that A1 agonists but not A_{2A} receptors, mediated cardiac protection whereas Cargnoni et al[174] claimed precisely the opposite. Both groups agreed, however, that changes of coronary blood flow were not responsible for the protection they observed. Consistent with a protective effect is the discovery that overexpression of A1 receptors is associated with resistance to cardiac ischaemic damage.[166]

Modulation of Apoptosis

Although necrosis plays a large part in the cell death which occurs with excitotoxic damage apoptosis, characterised by cell shrinkage, organelle relocalisation and compaction, chromatin condensation and formation of apoptotic bodies,[175] may account for a substantial proportion of neuronal degeneration.[176-178] The excitotoxin kainic acid induces morphological and biochemical features also suggesting the occurrence of apoptosis.[177,179-181] Kainic acid has been used extensively in neurobiological research as it preferentially damages neurones in the limbic system, particularly the hippocampus. Its administration, systemically or centrally is considered to provide a valuable experimental model for temporal lobe epilepsy and global ischaemia. Observation of the electrophoresis laddering pattern, indicative of DNA fragmentation, has been noted in rats in the hippocampus, entorhinal and sensory cortex at 18 and 72 hours following intraperitoneal injections of 10 mg/kg kainate.[182] Using a higher dose of 15 mg/ kg, Weiss et. al[177] observed TdT-mediated dUTP-biotin nick end labelling (TUNEL) detectable DNA fragmentation in the thalamus, amygdala, cortex and the CA1 region of the hippocampus. Further TUNEL-positive staining in the CA3 region was apparent only following seizures. The staining remained at similar levels for at least 7 days, suggesting a long-lasting, dynamic process. Kainate has also induced the expression of immediate-early genes such as *c-fos*,[183] believed by many to be necessary for apoptosis,[184] as well as the expression of target effector genes .[185]

When animals were injected with kainate intraperitoneally, TUNEL staining of sections from these animals revealed apoptotic cells only in the CA1 region.[186] TUNEL positive cells were observed only in three animals in which previous haematoxylin and eosin staining of serial sections had revealed moderate to high levels of damage in the CA1 region (50-90%) and not in an animal with only 25% damage. Within the CA1 area, TUNEL-positive cells were the same cells observed as damaged by haematoxylin and eosin staining. No apoptotic cells were apparent in the CA3a region (or any other pyramidal region) in any of these animals. A linear correlation was found in the CA3a region between the percentage of apoptosis and neuronal damage, with the amount of apoptosis increasing proportionally with increased damage. This increase in TUNEL positive staining in the CA1 region was not observed in animals treated with saline only. The simultaneous treatment of animals with CGS 21680 or ZM 241385 prevented any occurrence of apoptosis.

After kainate injections directly into the hippocampus, TUNEL-positive cells were noted only in areas of damage, including the injection tract. Damage in the CA3a region after kainate injection was near maximal in all animals studied. Following TUNEL staining many apoptotic cells were also apparent in the CA3a region. In the CA1 region there was both a lack of damage and TUNEL staining in all sections. As for kainate alone, animals co-injected with CGS 21680 and kainate exhibited both 100 % damage in the CA3a region and a high percentage of apoptotic neurones. Some variation in this pattern of high damage, high TUNEL-positive staining, was observed in animals treated with ZM 241385 and kainate.

This study also showed that the mechanism of protection observed with the systemically administered adenosine compounds, did not involve the inhibition of either necrotic or apoptotic cell death. No increase or decrease in TUNEL staining was evident. In the CA1 region of animals given an intraperitoneal injection of kainate, the degree of apoptosis correlated with damage such that there was always a high proportion of TUNEL staining in areas where damage was observed and low or no staining in less damaged regions. In conjunction with the experiments with kainate alone, no co-administered drug induced TUNEL staining within the CA3a region. This suggests that protection is mediated at a point much earlier along the neurodegenerative pathway, before cells have undergone necrosis. Ankarcrona et. al[187] suggested that apoptosis only occurs after a population of cells, which failed to recover mitochondrial integrity, underwent necrosis. If the mechanisms of protection with any of the above compounds involved reducing this delayed apoptosis, a high degree of damage would be expected without the presence of apoptotic cells.

This increase in apoptosis is not observed after the injection of ZM 241385 in the intrahippocampal injection model. This may be due to the lack of involvement of the inflammatory response in this regimen. That there is a complete absence of apoptosis in the CA3a region after co-injection of ZM 241385 with kainate, is suggestive of a role for the A_{2A} receptor in the promotion of apoptosis. The A_{2A} receptor may, therefore, play an important role in determining the pathway by which certain neuronal populations die during kainate-induced excitotoxicity, although this role may be dependent on experimental conditions.

Protection in Neurodegenerative Disorders

Although much of the interest in the therapeutic value of purine receptor ligands has centred on protection following strokes, there remains the possibility that overactivation of glutamate receptors may contribute to neurodegenerative disorders such as Alzheimer s disease and Huntington s disease. This possibility is the rationale for studying the protective effects of agents against excitotoxins. It has already been noted that A1 agonists and A_{2A} receptor agonists and antagonists can protect against kainic acid induced damage.[38,109,110] One of the excitotoxins of greatest relevance, however, is quinolinic acid, a tryptophan metabolite for which the evidence for a role in some degenerative disorders is substantial (see refs. 145,146 for reviews). Reggio et al[188] have reported that an A2 receptor antagonist, DMPX, can protect against neuronal loss induced by quinolinic acid injected into the striatum a frequently used model of Huntington s disease.

A2 Receptors and Aging

An interesting observation has recently been reported by Corsi et al[13,189] that the agonist CGS 21680 only increased the spontaneous efflux of glutamate and GABA in young, but not old rats, although it enhanced potassium-evoked release similarly in both groups of animals. This may have implications for the utility of A_{2A} receptor ligands in treating older patients after cerebral ischaemia, since chronic treatment might show fewer side effects attributable to increased basal release of glutamate, while retaining neuroprotective activity against the depolarisation-induced release occurring during and immediately after cerebral ischaemia or trauma.

Acute versus Chronic Treatment

There seems to be a less marked difference, compared with A1 ligands, in the neuroprotective effects of A2 receptors ligands following acute and chronic treatments. Evidence to date indicates that the qualitative effects are similar for both agonists and antagonists after acute or chronic administration, correlating with the failure to detect changes of receptor properties in vivo[116] or in vitro[190] in the maintained presence of these compounds. von Lubitz et al[116] found that the A2 receptor agonist APEC, applied acutely, improved post-ischaemic blood flow but failed to exhibit any protection of neurons, while the antagonist CSC did afford protection despite having no effect on blood flow. The latter result supports the evidence from other groups who have used acute administration of drugs almost exclusively. However, the chronic administration of CSC proved to have opposite effects to the acute, leading to an increased post-insult blood flow but a lack of neuroprotection.

Therapeutic Implications

These various findings have aroused great interest in the search for new drugs which could be used to slow or prevent the neuronal damage which characterises neurodegenerative conditions such as Alzheimer s disease, Parkinson s disease, and Huntington s disease. That interest is attributable not only to the efficacy of the compounds available, but also to the fact that they should be relatively free of major side effects. Whereas the use of A1 receptor agonists would lead to a suppression of transmitter release at many sites within the central and peripheral nervous system, A_{2A} receptor antagonists should not be limited in this way. The A_{2A} receptors are relatively low affinity sites for adenosine, and the levels of adenosine present normally in most biological fluids is probably insufficient to activate them. Activation would only be expected under conditions in which there is a large increase in the generation and release of adenosine.

In practice this means that A_{2A} antagonists have little effect on heart rate, blood pressure, or other vital signs under normal conditions. During ischaemia in the brain, however, the levels of adenosine may rise to levels at which A_{2A} receptors are activated. Stimulation of A_{2A} receptors is known to increase the release of the excitotoxic amino acid glutamate,[127] which would tend to cause or facilitate the occurrence of damage. Under these circumstances, A_{2A} receptor antagonists should reduce the enhanced release of glutamate and thus decrease the extent of neuronal damage. Their beneficial activity would therefore be restricted to those areas of brain experiencing ischaemia, with little or no effect on other areas of the body.

A particularly exciting aspect of A_{2A} receptor protection is that it may contribute to the long-term benefits of treating patients with Parkinson s disease with A_{2A} receptor antagonists. It is clear that A_{2A} receptors potently modulate cell sensitivity to dopamine receptors, accounting for the beneficial effects of adenosine antagonists in this disease.[191-193] This phenomenon has led to clinical trials with A_{2A} receptor antagonists in Parkinson s disease with promising, though as yet unpublished, results. In lower primates, A_{2A} antagonists are certainly effective against toxin-induced models of the disorder.[194,195]

A_{2A} Receptor Sub-Populations

In any drug research programme, the identification of receptor subtypes always raises hopes that novel compounds might be generated able to have more functionally, or anatomically, selective effects than earlier generation compounds. In the case of A_{2A} receptors, it now appears highly likely that there are indeed at least two sub-populations of receptor, one characteristic of the striatum and exhibiting high affinity for CGS21680, and a second site found in several regions of brain including cerebral cortex and hippocampus with a lower affinity.[196,197] These sites can best be distinguished by SCH 58261.[198] Recent work suggests that the presence of A_{2A} receptor sub-populations may even be reflected in overt measures of motor behaviour [199] and consistent with the suggestion that the striatal site shows a high sensitivity to SCH 58261, with the extrastriatal receptor being more sensitive to ZM 241385.

A3 receptors

The A3 population of purinoreceptors has only been known since the cloning work of Zhou et al,[200] and they remain the least well studied. They do occur on neuronal and glial cells membranes, even in human tissue,[201] although with much lower density in the brain than in peripheral tissues such as lungs and heart. A number of relatively selective agonist and antagonist ligands have now become available for affecting A3 sites, and it is apparent that suitable manipulation can lead to neuroprotection. A note of caution has been introduced into the study of A3 receptors by Rivkees et al[202] who have recently failed to detect A3 receptors or their corresponding mRNA in brain tissue, and have also claimed that proposed A3 selective pharmacological ligands show high affinity for A1 receptors. This report, therefore, implies that results obtained with A3 receptor ligands should be viewed with caution until any involvement of A1 receptors has been fully excluded.

A3 Receptors and Neuroprotection

The evidence to date does suggest that the acute administration of A3 receptor agonists can induce neuronal death and can exacerbate the neuronal loss produced by cerebral ischaemia.[118] Agonists at A3 receptors also promote neuronal death both directly and by enhancing glutamate-induced damage, in neuronal cultures.[203] As with the A1 receptor studies, however, chronic administration of A3 receptor agonists can induce protection. The acute administration of the A3 receptor agonist iodobenzyl-5 N-methyl-carboxamidoadenosine (IB-MECA) reduced post-ischaemic blood flow and produced an exacerbation of ischaemic damage in the gerbil brain, whereas chronic administration of the same ligand resulted in increased restoration of blood flow and marked neuroprotection.[118] The protective effect of IB-MECA has now been shown to be associated with a preservation of cytoskeletal proteins such as microtubule-associated protein and an increased presence of glial fibrillary acidic protein in the damaged areas[204] suggesting that both neurons and glia are involved in the protective activity.

In addition to increasing ischaemic damage, the acute administration of A3 agonists can reduce the protection afforded by an A1 agonist.[30] What is not clear is whether there is any contribution to this effect of A3 ligands acting directly on other receptors. For example, A3 receptors can induce a loss of sensitivity of A1 receptors in the hippocampus[205] and, since A1 receptors are known to be neuroprotective and A1 antagonists increase damage (as discussed above), this effect might lead to, or contribute to, the observed enhancement of neuronal injury by A3 receptor agonists.

Mechanisms of A3 Receptor-Mediated Protection

Agonists at A3 receptors reduce the production of several cytokines from cells, such as interleukin-10 (IL-10), IL-12, interferon-γ and tumour necrosis factor-α (TNFα)[206] when the generation of these has been elevated in mice suffering infections. This action seems to represent a powerful anti-inflammatory consequence of stimulating A3 receptors and, as discussed earlier for A_{2A} receptors, may contribute significantly to the protection against neuronal damage when this involves the development of an inflammation-like process. However, the role of A3 receptors remains controversial. Hannon et al[153] could find no contribution of A3 receptors to adenosine inhibition of the oxidative burst in human neutrophils, whereas Ezeamuzie and Phillips[207] reported that A3 receptors were responsible for suppressing superoxide generation in human eosinophils. It remains unclear whether this phenomenon is related to the report that A3 receptors increase the levels of several antioxidant enzymes including superoxide dismutase in endothelial cells,[208] leading to a reduction of tissue lipid peroxidation. Sajjadi et al[209] reported on pharmacological evidence that A3 receptor activation was largely responsible for the suppression by adenosine of TNFα secretion by a human macrophage cell line (U937). Interestingly, the production of other cytokines such as IL-1β, IL-6 an IL-8 were unchanged. Similar results, indicating the A3 receptor is capable of suppressing TNFα secretion were obtained using murine macrophages.[210] The inhibition of TNFα secretion was recently used as part of the screening tests for developing new chemical ligands for A3 receptors.[211]

Certainly, the fact that A3 receptor ligands can protect the heart against ischaemic/reperfusion injury[212] indicates that mechanisms for protection operate which must be independent of glutamate release, and presumably involve calcium movements, potassium conductances or actions on white cells. A3 receptor agonists depress neutrophil functions and open ATP-dependent potassium channels, for example, both mechanisms being potential contributors to protection against ischaemia-reperfusion injury.[213,214]

Abbracchio et al[215] have reported that the A3 receptor agonist chloro-IB-MECA induced a number of cytoskeletal changes together with alterations in overall cell morphology and the expression of the GTP-binding protein Rho, in astroglial cell cultures. These observations indicate that A3 receptor activation can induce intracellular modifications which could result in more subtle responses of cells to injurious or protectant ligands than simply a change of intracellular calcium and the classical transduction systems.

Modulation of Adenosine Concentrations

Adenosine itself has been used only rarely in studies of neuroprotection, largely because it is so efficiently removed by uptake processes and the destructive enzymes adenosine deaminase and adenosine kinase. However, post-ischaemic injury to the spinal cord has been prevented by the local perfusion of solutions containing adenosine, the continuous presence of the purine negating the effects of metabolism and removal.[216]

Although the emphasis in the preceding discussion has been on agonists and antagonists acting directly at receptor sites, there is also evidence that modulating the levels of adenosine in the extracellular fluid may also afford neuroprotection, presumably by raising the local concentration of adenosine to one at which it is itself able to activate A1 or A3 receptors. Several inhibitors of adenosine kinase or adenosine deaminase have been shown to prevent ischaemia-induced neuronal loss.[217-219]

In summary, it is clear that purines are potent modulators of neuronal viability and protection against a range of insults. In the continuing search for new therapeutic agents to manage the scourge of strokes and neurodegenerative damage in modern society, purines will remain as potentially valuable agents, especially those, such as A_{2A} receptor antagonists, which should bring few side effects.

ABBREVIATIONS

ADAC	Adenosine amine congener
APEC	2-[(2-aminoethylamino)-carbonyl-ethylphenyl-ethylamino]-5-N- ethylcarboxamido-adenosine
BDNF	brain-derived neurotrophic factor
BW1002C87	(5-2,3,5-trichlorophenyl)pyrimidine-2,4-diamine-1,10-ethane-sulphonate
CCPA	2-chloro-N6-cyclopentyladenosine
CGS 21680	2-[4-(2-carboxyethyl)-phenylethylamino]-5 N-ethyl-carboxamido- adenosine
CGS 15943	5-amino-9-chloro-2-(2-furyl)-1,2,4-triazolo[1,5-c]quinazoline
CHA	N6-cyclohexyladenosine
CSC,	8-(3-chlorostyryl)caffeine
CP 66,9713	4-amino-1-phenyl[1,2,4]-triazolo[4,3-a]quinoxaline
CPX	8-cyclopentyl-1,3-dipropylxanthine (=DPCPX)
DMPX	3,7-dimethyl-1-propargylxanthine
DPCPX	8-cyclopentyl-1,3-dipropylxanthine (=CPX)
DPMA	(N6-[2-(3,5-dimethoxyphenyl)-2-(2-methylphenyl)-ethyl]-adenosine
IB-MECA	2-chloro-N6-(3-iodobenzyl)adenosine-5 N-methyluronamide
KF17837	(E/Z)-7-methyl-8-(3,4-dimethoxystyryl)-xanthine
NECA	5 N-ethylcarboxamido-adenosine hydrochloride
NMDA	N-methyl-D-aspartate
PIA	((R)-(N6-phenylisopropyl)adenosine
8-PST	8-(p-sulphophenyltheophylline)
SCH 58261	7-(2-phenylethyl)-5-amino-2-(2-furyl)-pyrazolo-[4,3e]-1,2,4-triazolo[1,5-c]-pyrimidine
TdT,	terminal deoxynucleotidyl-transferase
TUNEL	TdT-mediated dUTP-biotin nick end labelling
ZM 241385,	4-(2-[7-amino-2-{2-furyl}{1,2,4}-triazolo{2,3-a}-(1,3,5}triazin-5-yl-amino]ethyl)phenol

REFERENCES

1. Zetterstrom T, Vernet L, Ungerstedt U et al. Purine levels in the intact brain, studied with an implanted perfused hollow fibre. Neurosci Lett 1982; 29:111-115.
2. Ballarin M, Herrera-Marschitz M, Casas M et al. Striatal adenosine levels measured in vivo by microdialysis in rats with unilateral dopamine denervation. Neurosci Lett 1987; 83:338-344.
3. Chen Y, Graham DI, Stone TW. Release of endogenous adenosine and its metabolites by the activation of NMDA receptors in the rat hippocampus in vivo. Brit J Pharmacol 1992; 106:632-638.
4. Pazzagli M, Corsi C, Latini S et al. In vivo regulation of extracellular adenosine levels in the cerebral cortex by NMDA and muscarinic receptors. Europ J Pharmacol 1994; 254:277-282.

5. Carswell H V, Graham D I and Stone T W. Kainate-evoked release of adenosine from the hippocampus of the anaesthetised rat: Possible involvement of free radicals. J Neurochem 1997; 68:240-247.
6. Latini S, Bordoni F, Pedata F et al. Extracellular adenosine concentrations during in vitro ischaemia in rat hippocampal slices. Brit J Pharmacol 1999; 127:729-739.
7. Hagberg H, Andersson P, Lacarewicz J et al. Extracellular adenosine, inosine, hypoxanthine, and xanthine in relation to tissue nucleotides and purines in rat striatum during transient ischaemia. J Neurorchem 1987; 49:227-231.
8. Andine P, Rudolphi KA, Fredholm BB et al. Effect of propentofylline (HWA285) in extracellular purines and excitatory amino acids in CA1 of rat hippocampus during transient ischemia. Brit J Pharmacol 1990; 100:814-818.
9. Dux E, Fastbom J, Ungerstedt U et al. Protective effect of adenosine and a novel xanthine derivative propentofylline on the cell damage after bilateral carotid occlusion in the gerbil hippocampus. Brain Res 1990; 516:248-256.
10. Corradetti, R, Conte, G, Moroni et al. Adenosine decreases aspartate and glutamate release from rat hippocampal slices. Europ J Pharmacol 1984; 140:19-26.
11. Heron A, Lasbennes F, Seylez J. Adenosine modulation of amino acid release in rat hippocampus during ischaemia and veratridine depolarisation. Brain Res 1993; 608:27-32.
12. Poli A, Lucchi R, Vibio M, Barnabei O. Adenosine and glutamate modulate each other s release from rat hippocampal synaptosomes. J Neurochem 1991; 57:298-306.
13. Corsi C, Melani A, Bianchi L et al. Effect of adenosine A_{2A} stimulation on GABA release from the striatum of young and aged rats in vivo. NeuroReport 1999; 10:3933-3937.
14. Gerber U, Greene RW, Haas HL et al. Characterisation of inhibition mediated by adenosine in the hippocampus of the rat in vitro. J Physiol 1989; 417:567-578.
15. Ameri A, Jurna I. Adenosine A1 and non-A1 receptors: Intracellular analysis of the actions of adenosine agonists and antagonists in rat hippocampal neurones. Brain Res 1991; 546:69-78.
16. Trussel LD, Jackson MB. Adenosine activated potassium conductance in cultured striatal neurones. Proc Nat Acad Sci USA 1985; 82:4857-4861.
17. Thompson SM, Haas HL, Gahwiler BH. Comparison of the actions of adenosine at presynaptic and postsynaptic receptors in the rat hippocampus in vitro. J Physiol 1992; 451:347-363.
18. Stone TW, ed. Adenosine in the Nervous System. London: Academic Press, 1991.
19. Goodman RR, Snyder SH. Autoradiographic localisation of adenosine receptors in rat brain using [3H]cyclohexyladenosine. J Neurosci 1982; 2:1230-1241.
20. Lee KS, Reddington M. Autoradiographic evidence for multiple CNS binding sites for adenosine derivative. Neuroscience 1986; 19:535-549.
21. Fiebich BL, Biber K, Lieb K et al . Cyclooxygenase-2 expression in rat microglia is induced by adenosine A(2a)-receptors. Glia 1996; 18:152-160.
22. Rivkees SA, Price SL, Zhou FC. Immunohistochemical detection of A1 adenosine receptors in rat brain with emphasis on localisation in the hippocampal formation,cerebral cortex, cerebellum and basal ganglia. Brain Res 1995; 677:193-203.
23. Svenningsson P, Hall H, Sedvall G, Fredholm BB. Distribution of adenosine receptors in the postmortem human brain: an extended autoradiographic study. Synapse 1997; 27:322-335.
24. Ochiishi T, Chen L, Yukawa A et al. Cellular localisation of adenosine A1 receptors in rat forebrain: Immunohistochemical analysis using adenosine A1 receptor-specific monoclonal antibody. J Comp Neurol 1999; 411:301-316.
25. Ochiishi T, Saitoh Y, Yukawa A et al. High levels of adenosine A1 receptor-like immunoreactivity in the CA2/CA3a region of the adult rat hippocampus. Neuroscience 1999; 93:955-967.
26. Fastbom J, Pazos A, Probst A et al. Adenosine A1-receptors in human brain: Characterisation and autoradiographic visualisation. Neurosci Lett 1986; 65:127-132.

27. Fastbom J, Pazos A, Palacios JM. The distribution of adenosine A1 receptors and 5 nucleotidase in the brain of some commonly used experimental animals. Neuroscience 1987; 22:813-826.
28. Fastbom J, Pazos A, Probst A, et al. Adenosine A1 receptors in the human brain: A quantitative autoradiographic study. Neuroscience 1987; 22:827-839
29. von Lubitz DKEJ, Dambrosia JM, Redmond DJ. Protective effect of cyclohexyladenosine in treatment of cerebral ischaemia in gerbils. Neuroscience 1989; 30:451-462.
30. von Lubitz DJKE, Lin RCS, Bischofberger N et al. Protection against ischemic damage by adenosine amine congener, a potent and selective adenosine A1 receptor agonist. Europ J Pharmacol 1999; 369:313-317.
31. Roucher P. Meric P. Correze JL et al. Metabolic effects of R-phenylisopropyladenosine during reversible forebrain ischemia studied by in vivo 31P nuclear magnetic resonance spectroscopy. J Cereb Blood Flow Metab 1991; 11:453-458.
32. Miller LP, Hsu C. Therapeutic potential for adenosine receptor activation in ischemic brain injury. J Neurotrauma 1992; 9:S563-S577.
33. Rudolphi KA, Schubert P, Parkinson FE et al. Adenosine and brain ischemia. Cerebrovasc Brain Metab Revs 1992; 4:346-369
34. Tominaga K, Shibata S, Watanabe S. A neuroprotective effect of adenosine A1-receptor agonists on ischemia-induced decrease in 2-deoxyglucose uptake in rat hippocampal slices. Neurosci Lett 1992; 145:67-70.
35. Daval J-L, Von Lubitz DJKE, Deckert J et al. Protective effect of cyclohexyladenosine A_1-receptors, guanine nucleotide and forskolin binding sites following transient brain ischaemia: a quantitative autoradiographic study. Brain Res 1989; 491:212-226.
36. Heron A, Lekiettre D, Le Peillet E et al. Effects of an A_1 adenosine receptor agonist on the neurochemical, behavioural and histological consequences of ischaemia. Brain Res 1994; 641:217-224.
37. Domenici MR, Scotti de Carolis A, Sagratella A. Block by N6-L-phenylisopropyladenosine of the electrophysiological and morphological correlates of hippocampal ischaemic injury in the gerbil. Brit J Pharmacol 1996; 118:1551-1557.
38. MacGregor DG, Stone TW. Inhibition by the adenosine analogue, (R-)-N6-phenylisopropyladenosine, of kainic acid neurotoxicity in rat hippocampus after systemic administration. Brit J Pharmacol 1993; 109:316-321.
39. MacGregor DG, Miller WJ, Stone TW. Mediation of the neuroprotective action of R-phenylisopropyladenosine through a centrally located adenosine A1 receptor. Brit J Pharmacol 1993; 110:470-476.
40. MacGregor DG, Jones PA, Maxwell WL, Graham DI, Stone TW. Prevention by a purine analogue of kainate-induced neuropathology in rat hippocampus. Brain Res 1996; 725:115-120.
41. Arvin B, Neville LF, Pan J, Roberts PJ. 2-Chloroadenosine attenuates kainic acid-induced toxicity within the rat striatum: relationship to release of glutamate and Ca2+ influx. Brit J Pharmacol 1989; 98:225-235.
42. Connick JH, Stone TW. Quinolinic acid neurotoxicity: protection by intracerebral phenylisopropyl adenosine (PIA) and potentiation by hypotension. Neurosci Lett 1989; 101:191-196.
43. Finn SF, Swartz KJ, Beal MF. 2-Chloroadenosine attenuates NMDA, kainate and quisqualate toxicity. Neurosci Lett 1991; 126:191-194.
44. Parkinson FE, Rulolphi KA, Fredholm BB. Propentofylline: a nucleoside transport inhibitor with neuroprotective effects in cerebral ischemia. Gen Pharmacol 1994; 25:1053-1058.
45. Lau Y-S, Mouradian MM. Protection against acute MPT-induced dopamine depletion in mice by adenosine. J Neurochem 1993; 60:768-771.
46. Ongini E, Monopoli A, Impagnatiello F et al. Dual actions of A2A adenosine receptor antagonists on motor dysfunction and neurodegenerative processes. Drug Develop Res 2001; 52:379-386.

47. Miller WJ, MacGregor DG, Stone TW. Time course of purine protection against kainate-induced increase in hippocampal [3H]-PK11195 binding. Brain Res Bull 1994; 34:133-136.
48. Von Lubitz DKJE, Lin RC-S, Paul IA et al. Postischemic administration of adenosine amine congener (ADAC): Analysis of recovery in gerbils. Europ J Pharmacol 1996; 316:171-176.
49. Bischofberger N, Jacobson KA, von Lubitz DKJE. Adenosine A1 receptor agonists as clinically viable agents for treatment of ischemic brain disorders Ann NY Acad Sci 1997; 825:23-29.
50. Simon RP. Swan JH, Griffiths T et al. Blockade of NMDA receptors may protect against ischaemic damage in the brain. Science 1984; 226:850-852.
51. Gill R, Foster AC, Woodruff GN. Systemic administration of MK-801 protects against ischaemia induced hippocampal neurodegerneration in the gerbil. J Neurosci 1987; 7:3343-3349.
52. Park CK, Nehls DG, Graham DI et al. Focal cerebral ischaemia in the cat: treatment with the glutamate antagonist MK-801 after induction of ischaemia. J Cereb Blood Flow Metab 1988; 8:757-762.
53. Park CK, Nehls G, Graham DI et al. The glutamate antagonist MK-801 reduces focal ischemic brain damage in the rat. Ann. Neurol. 1988; 24:543-551.
54. Kohler C, Schwarcz R, Fuxe K. Perforant path transections protect hippocampal granule cells from kainate lesion. Neurosci Lett 1978; 10:241-246.
55. McGeer EG, McGeer PL, Singh K. Kainate-induced degeneration of neostriatal neurones: dependency upon cortical spinal tract. Brain Res 1978; 139:381-383.
56. Nadler JV, Cuthbertson GJ. Kainic acid neurotoxicity toward hippocampal formation: dependence on specific excitatory pathways. Brain Res 1980; 195:47-56.
57. Heggli DA, Malthe-Sorensen D. Systemic injection of kainic acid: Effect on neurotransmitter markers in pyriform cortex, amygdaloid complex and hippocampus and protection by cortical lesioning and anticonvulsants. Neuroscience 1982; 7:1257-1264.
58. Okazaki MM, Nadler JV. Protective effects of mossy fibre lesions against kainic acid induced seizures and neuronal degeneration. Neuroscience 1988; 26:763-781.
59. Ferkany JW, Zaczek R, Coyle, JT. Kainic acid stimulates excitatory amino acid neurotransmitter release at presynaptic receptors. Nature 1982; 298:757-759.
60. Lehmann A, Isacsson H, Hamberger A. Effects of in vivo administration of kainic acid on the extracellular amino acid pool in the rabbit hippocampus. J Neurochem 1983; 40:1314-1320.
61. Jacobson I, Hamberger A. Kainic acid-induced changes of extracellular amino acid levels, evoked potentials and EEG activity in the rabbit olfactory bulb. Brain Res 1985; 348:289-296.
62. Connick JH, Stone TW. The effects of kainate, beta-kainate and quinolinic acids on the release of endogenous amino acids from rat brain slices. Biochem Pharmacol 1986; 35:3631-3635.
63. Virgili M, Poll A, Contestabile A et al. Synaptosomal release of newly synthesised or recently accumulated amino acids; differential effects of kainic acid on naturally occurring excitatory amino acids and on [D-3H]-aspartate. Neurochem Intern 1986; 9:29-33.
64. Fastbom J, Fredholm BB. Inhibition of (3H)glutamate release from rat hippocampal slices by L-PIA. Acta Physiol Scand 1985; 125:121-123.
65. Butcher SP, Bullock R, Graham DI et al. Correlation between amino acid release and neuropathologic outcome in rat brain following middle cerebral artery occlusion. Stroke 1990; 21:1727-1733.
66. Spignoli G, Pedata F, Pepeu G. A1 and A2 adenosine receptors modulate acetylcholine release from brain slices. Europ J Pharmacol 1984; 97:341-342.
67. Brown SJ, James S, Reddington M et al. Both A1 and A_{2A} purine receptors regulate striatal acetylcholine release. J Neurochem 1990; 55:31-38.
68. Michaelis ML, Michaelis EK, Myers SL. Adenosine modulation of synaptosomal dopamine release. Life Sci 1979; 24:2083-2092.

69. Chowdhury M, Fillenz M. Presynaptic adenosine A2 and NMDA receptors regulate dopamine synthesis in rat striatal synaptosomes. J Neurochem 1991; 56:1783-1788.
70. Mitani A, Andou Y, Kataoka K. Selective vulnerability of hippocampal CA1 neurons cannot be explained in terms of an increase in glutamate concentration during ischemia in the gerbil: brain microdialysis study. Neuroscience 1992; 48:307-313.
71. Cantor SL, Zornow MH, Miller LP et al. The effect of cyclohexyladenosine on the periischemic increases of hippocampal glutamate and glycine in the rabbit. J Neurochem 1992; 59:1884-1892.
72. Simpson RE, O Regan MH, Perkins LM et al. Excitatory transmitter amino acid release from the ischaemic rat cerebral cortex: Effects of adenosine recptor agonists and antagonists. J Neurochem 1992; 58:1683-1690.
73. Melani A, Pantoni L, Corsi C et al. Striatal outflow of adenosine, excitatory amino acids, GABA, and taurine in awake, freely moving rats after middle cerebral artery occlusion Correlation with neurological deficit and histopathological damage. Stroke 1999; 30:2448-2454.
74. Ochoa MC, Jackson TA, Aaron CS et al. Antagonism of kainic acid lesions in the mouse hippocampus by U-54494A and U-50488H. Life Sci 1992; 51:1135-1143.
75. Graham SH, Chen J, Sharp FR et al. Limiting ischaemic injury by inhibition of excitatory amino acid release. J Cereb Blood Flow Metab 1993; 13:88-97.
76. Kano T, Katayama Y, Kawamata T et al. Propentofylline administered by microdialysis attenuates ischemia-induced hippocampal damage but not excitatory amino acid release in gerbils. Brain Res 1994; 641:149-154.
77. Fredholm BB, Dunwiddie TV. How does adenosine inhibit transmitter release? Trends Pharmacol Sci 1988; 9:130-134.
78. Dolphin AC, Prestwich SA. Pertussis toxin reverses adenosine inhibition of neuronal glutamate release. Nature 1985; 316:148-150.
79. Scholz KP, Miller RJ. Inhibition of quantal transmitter release in the absence of calcium influx by a G protein-linked adenosine receptor at hippocampal synapses. Neuron 1991; 8:1139-1150.
80. Greene RW, Haas HL The electrophysiology of adenosine in the mammalian CNS. Progr Neurobiol 1991; 36:329-341.
81. Regenold JT, Illes P. Inhibitory adenosine A1-receptors on rat locus coeruleus neurones. An intracellular electrophysiological study. 1990; Naunyn-Schmied Arch Pharmacol 341:225-231.
82. Hosseinzadeh H, Stone TW. Tolbutamide blocks postsynaptic but not presynaptic effects of adenosine on hippocampal CA1 neurons. J Neural Transm 1998; 105:161-172.
83. Mager R, Ferroni S, Schubert P Adenosine modulates a voltage-dependent chloride conductance in cultured hippocampal neurons. Brain Res 1990; 532:58-62.
84. Schubert P, Ferroni S, Mager R Pharmacological blockade of chloride pumps on chloride channels reduces the adenosine-mediated depression of stimulus train-evoked calcium fluxes in rat hippocampal slices. Neurosci Lett 1991; 124:174-177.
85. Globus M Y T, Busto R, Dietrich W D et al. Effect of ischaemia on the in vivo release of striatal dopamine, glutamate, and gamma-aminobutyric acid studied by intracerebral microdialysis. J Neurochem 1988; 51:1455-1464.
86. Rudolphi KA, Keil M, Hinze HJ. Effect of theophylline on ischemically induced hippocampal damage in Mongolian gerbils: a behavioural and histopathological study. J Cereb Blood Flow Metab 1987; 7:74-81.
87. Phillis JW. The effects of selective A1 and A_{2A} adenosine receptor antagonists on cerebral ischemic injury in the gerbil. Brain Res 1995; 705:79-84.
88. von Lubitz DJKE, Lin RC-S, Melman N et al. Chronic administration of selective adenosine A_1 receptor agonist or antagonist in cerebral ischaemia. Europ J Pharmacol 1994; 256:161-167.
89. MacGregor DG, Stone TW. Blockade by 1,3-dipropyl-8-cyclopentylxanthine (CPX) of purine protection against kainate neurotoxicity. Brain Res 1994; 644:339-342.
90. Bartrup JT, Stone TW. Activation of NMDA receptor-coupled channels suppresses the inhibitory action of adenosine on hippocampal slices. Brain Res 1990; 530:330-334.

91. Bartrup JT, Addae JI, Stone TW. Interaction between adenosine and excitatory agonists in rat hippocampal slices. Brain Res 1991; 564:323-327.
92. Nikbakht M-R, Stone TW. Activation of NMDA receptors suppresses the presynaptic effects of adenosine. Brit J Pharmacol 2001; 133:155P
93. Gustaffson LE, Wiklund CU, Wiklund NP et al. Identification of subclasses of adenosine A1 receptors. Suggestion of A1a and A1b subclasses and evidence from antagonist data. In: Ribeiro JA, ed. Adenosine Receptors in the Nervous System. London: Taylor & Francis, 1989:194.
94. Knutsen LJS, Lau J, Petersen H et al. N-substituted adenosines as novel neuroprotective A1 agonists with diminished hypotensive effects. J Med Chem 1999; 42:3463-3477.
95. Rudolphi KA, Keil M, Fastbom J et al. Ischemic damage in gerbil hippocampus is reduced following upregulation of adenosine A1 receptors by caffeine treatment. Neurosci Lett 1989; 103:275-280.
96. Sutherland GR, Peeling J, Lesiuk J et al. The effects of caffeine on ischemic neuronal injury as determined by magnetic resonance imaging and histopathology. Neuroscience 1991; 42:171-182.
97. Traversa U, Rosati AM, Florio C et al. Effects of chronic administration of adenosine antagonists on adenosine A1 and A_{2A} receptors in mouse brain. In vivo 1994; 8:1073-1978.
98. Latini S, Pazzagli M, Pepeu G et al. A2 adenosine receptors: Their presence and neuromodulatory role in the CNS. Gen Pharmacol 1996; 27:925-933.
99. Ongini E, Fredholm BB. Pharmacology of adenosine A_{2A} receptors. Trends Pharmacol Sci 1996; 17:364-372.
100. Moreau JL, Huber G. Central adenosine A_{2A} receptors: An overview. Brain Res Revs 1999; 31:65-82.
101. Jarvis MF, Williams M. Direct autoradiographic localisation of adenosine A_2 receptors in the brain using the A_2-selective agonist, [3H]CGS21680. Europ J Pharmacol 1989; 168:243-246.
102. Schiffmann SN, Jacobs O, Vanderhaegen JJ. Striatal restricted adenosine A2 receptor (RDC8) is expressed by enkephalin but not by substance P neurons: An in situ hybridisation histochemistry study. J Neurochem 1991; 57:1062-1067.
103. Kurokawa M, Kirk IP, Kirkpatrick KA et al. Inhibition by KF17837 of adenosine A_{2A} receptor-mediated modulation of striatal GABA and acetylcholine release. Brit J Pharmacol 1994; 113:43-48.
104. Cunha RA, Johansson B, Van der Ploeg I et al. Evidence for functionally important adenosine A_{2A} receptors in the rat hippocampus. Brain Res 1994; 649:208-216.
105. Dixon AK, Gubitz AK, Sirinathsinghji DJS et al. Tissue distribution of adenosine receptor mRNAs in the rat. Brit J Pharmacol 1996; 118:1461-1468.
106. Cunha RA, Johansson B, Constantino MD et al. Evidence for high affinity binding sites for the adenosine A_{2A} receptor agonist [3H]CGS21680 in the rat hippocampus and cerebral cortex that are different from striatal A_{2A} receptors. Naunyn-Schmiedeberg s Arch Pharmacol 1996; 353:261-271.
107. Martinez-Mir MI, Probst A, Palacios JM. Adenosine A2 receptors: selective localisation in the human basal ganglia and alterations with disease. Neuroscience 1991; 42:697-706.
108. Sheardown MJ, Knutsen LJS. Unexpected neuroprotection observed with the adenosine A_{2A} receptor agonist CGS21680. Drug Develop Res 1996; 39:108-114.
109. Jones PA, Smith RA, Stone TW. Protection against intrahippocampal kainate excitotoxicity by intracerebral administration of an adenosine A_{2A} receptor antagonist. Brain Res 1998; 800:328-335.
110. Jones PA, Smith RA, Stone TW. Protection against kainate-induced excitotoxicity by adenosine A_{2A} receptor agonists and antagonists. Neuroscience 1998; 85:229-237.
111. Sperk G. Kainic acid seizures in the rat. Prog Neurobiol 1994; 42:1-32.
112. Bridges A J, Bruns R F, Ortwine D F et al. N6-[2-(3,5-dimethoxyphenyl)-2-(2-methylphenyl)-ethyl]adenosine and its uronamide derivatives. Novel adenosine agonists and antagonists with

both high affinity and high selectivity for the adenosine A_2 receptor. J Med Chem 1988; 31:1282-1285.

113. Merkel LA, Lappe RW, Rivera LM et al. Demonstration of vasorelaxant activity with an A_1-selective adenosine agonist in porcine coronary artery: involvement of potassium channels. J Pharmacol Exp Ther 1992; 260:437-443.
114. Hutchison AJ, Webb RL, Oei HH et al. CGS21680, an A_2 selective adenosine receptor agonist with preferential hypotensive activity. J Pharmacol Exp Ther 1989; 251:47-55.
115. Lupica CR, Cass WA, Zahniser NR et al. Effects of the selective adenosine agonist CGS21680 on in vitro electrophysiology, cAMP formation and dopamine release in rat hippocampus and striatum. J Pharmacol Exp Ther 1990; 252:1134-1141.
116. von Lubitz DKJE, Lin RC-S and Jacobson KA. Cerebral ischaemia in gerbils: Effects of acute and chronic treatment with adenosine A_{2A} receptor agonist and antagonist. Europ J Pharmacol 1995; 287:295-302.
117. Kim HO, Ji X-D, Sidiqqi SM et al. 2-Substitution of N6-benzyladenosine-5 -uronamides enhances selectivity for A_3 adenosine receptors. J Med Chem 1994; 21:3614-3621.
118. von Lubitz DJKE, Lin RC-S, Popik P et al. Adenosine A_3 receptor stimulation and cerebral ischaemia. Europ J Pharmacol 1994; 26:59-67.
119. Gao Y, Phillis JW. CGS 15943, an adenosine A2 receptor antagonist, reduces cerebral ischemic injury in the mongolian gerbil. Life Sci 1994; 55:PL61-PL65.
120. Monopoli A, Lozza G, Forloni A et al. Blockade of adenosine A_{2A} receptors by SCH 58261 results in neuroprotective effects in cerebral ischaemia in rats. NeuroReport 1998; 9:3955-395.
121. Ongini E, Adami M, Ferri C et al. Adenosine A_{2A} receptors and neuroprotection. Ann NY Acad Sci 1997; 825:30-48
122. Palmer TM, Poucher SM, Jacobson KA et al. ^{125}I-4-(2-[7-amino-2-{2-furyl}{1,2,4}triazolo{2,3-a}{1,3,5}triazin-5-yl-amino]ethyl)phenol, a high affinity antagonist radioligand selective for the A_{2A} adenosine receptor. Mol Pharmacol 1995; 48:970-974.
123. Poucher SM, Keddie JR, Singh P et al. The in vitro pharmacology of ZM241385, a potent non-xanthine, A_{2A} selective adenosine receptor antagonist. Brit J Pharmacol 1995; 115:1096-1102.
124. Robledo P, Ursu G, Mahy N. Effects if adenosine and GABA-A receptor antagonists on NMDA-induced neurotoxicity in the rat hippocampus. Hippocampus 1999; 9:527-533.
125. Bona E, Aden U, Gilland E et al. Neonatal cerebral hypoxia-ischemia: The effect of adenosine receptor antagonists. Neuropharmacology 1997; 9:1327-1338
126. Chen JF, Huang Z, Ma J et al. A_{2A} adenosine receptor deficiency attenuates brain injury induced by transient focal ischemia in mice. J Neurosci 1999; 19:9192-9200
127. O Regan MH, Simpson RE, Perkins LM et al. The selective A_2 adenosine receptor agonist CGS21680 enhances excitatory amino acid release from the ischaemic rat cerebral cortex. Neurosci Lett 1992; 138:169-172.
128. Sebastiao AM, Ribeiro JA. Evidence for the presence of excitatory A2 adenosine receptors in the rat hippocampus. Neurosci Lett 1992; 138:41-44.
129. Goncalves ML, Cunha RA, Ribeiro JA. Adenosiné A_{2A} receptors facilitate 45Ca2+ uptake through class A calcium channels in rat hippocampal CA3 but not CA1 synaptosomes. Neurosci Lett 1997; 238:73-77.
130. Jin S, Fredholm BB. Adenosine A_{2A} receptor stimulation increases release of acetylcholine from rat hippocampus but not striatum, and does not affect catecholamine release. Arch Pharmacol 1997; 355:48-56.
131. Broad RM, Fredholm BB. A1, but not A_{2A} adenosine receptors modulate electrically stimulated [14C]acetylcholine release from rat cortex. J Pharmacol Exp Therap 1996; 277:193-197.
132. Mayfield RD, Suzuki F, Zahniser NR. Adenosine A_{2A} receptor modulation of electrically evoked endogenous GABA release from rat globus pallidus. J Neurochem 1993; 60:2334-2337.

133. Dixon AK, Widdowson L, Richardson PJ. Desensitisation of the adenosine A1 receptor by the A_{2A} receptor in the rat striatum. J Neurochem 1997; 69:315-321.
134. O Kane EM, Stone TW. Interactions between A1 and A2 adenosine receptor-mediated responses in the rat hippocampus in vitro. Europ J Pharmacol 1998; 362:17-25.
135. Latini S, Bordoni F, Corradetti R et al. Effect of A_{2A} adenosine receptor stimulation and antagonism on synaptic depression induced by in vitro ischaemia in rat hippocampal slices. Brit J Pharmacol 1999; 128:1035-1044.
136. Correia-de-Sa P, Ribeiro JA. Tonic adenosine A_{2A} receptor activation modulates nicotinic autoreceptor function at the rat neuromuscular junction. Europ J Pharmacol 1994; 271:349-355.
137. Cunha RA, Correia-de-Sa P, Sebasti o AM et al. Preferential activation of excitatory adenosine receptors at rat hippocampal and neuromuscular synapses by adenosine formed from released adenine nucleotides. Brit J Pharmacol 1996; 119:253-260.
138. Cheng Y, Sun AY. Oxidative mechanisms involved in kainate-induced cytotoxicity in cortical neurones. Neurochem Res 1994; 19:1557-1564.
139. Kitagawa K, Matsumoto M, Oda T et al. Free radical generation during brief period of cerebral ischaemia may trigger delayed neuronal death. Neuroscience 1990; 35:551-558.
140. MacGregor DG, Higgins MJ, Jones PA et al. Ascorbate attenuates the systemic kainate-induced neurotoxicity in the rat hippocampus. Brain Res 1996; 727:133-144.
141. O Neill M, Canney M, Feeney C et al. Protective effects of exifone, a free radical scavenger, against ischaemia-induced hippocampal neurodegeneration in the gerbil. Med Sci Res 1994; 22:589-590.
142. Cronstein BN, Rosenstein ED, Kramer SB. Adenosine: A physiologic modulator of superoxide anion generation by human neutrophils. Adenosine acts via an A2 receptor on human neutrophils. J Immunol 1985; 135:1366-1371.
143. Cronstein BN, Daguma L, Nichols D et al. The adenosine/neutrophil paradox resolved: human neutrophils possess both A1 and A2 receptors that both promote chemotaxis and inhibit O_2 generation respectively. J Clin Invest 1990; 85:1150-1157.
144. Stone TW, Perkins MN. Quinolinic acid: A potent endogenous excitant at amino acid receptors in the CNS. Europ J Pharmacol 1981; 72:411-412.
145. Stone TW. The Neuropharmacology of quinolinic acid and kynurenic acid. Pharmacol Revs 1993; 45:309-379.
146. Stone TW. Kynurenines in the CNS: from endogenous obscurity to clinical relevance. Progr Neurobiol 2001; 64:185-218
147. Perkins MN, Stone TW. An iontophoretic investigation of the action of convulsant kynurenines and their interaction with the endogenous excitant quinolinic acid. Brain Res 1982; 247:184-187.
148. Stone TW. The development and therapeutic potential of kynurenic acid and kynurenine derivatives for CNS neuroprotection. Trends Pharmacol Sci 2000; 21:149-154
149. Cronstein BN, Kubersky SM, Weissman G et al. Engagement of adenosine receptors inhibits hydrogen peroxide-release by activated human neutrophils. Clin Immunol Immunopathol 1987; 42:76-85
150. Cronstein BN. Adenosine, an endogenous anti-inflammatory agent. J Appl Physiol 1994; 76:5-13.
151. Burkey TH, Webster RD. Adenosine inhibits fMLP-stimulated adherence and superoxide anion generation by human neutrophils at an early step in signal transduction. Biochem Biophys Acta 1993; 1175:312-318.
152. Dianzani C, Brunelleschi S, Viano I et al. Adenosine modulation of primed human neutrophils. Europ J Pharmacol 1994; 263:223-226.
153. Hannon JP, Bray-French KM, Phillips RM et al. Further pharmacological characterization of the adenosine receptor subtype mediating inhibition of oxidative burst in human isolated neutrophils. Drug Development Res 1998; 43:214-224

154. Revan S, Montesinos MC, Naime D et al. Adenosine A2 receptor occupancy regulates stimulated neutrophil function via activation of a serine/threonine protein phosphatase. J Biol Chem 1996; 271:17114-17118.
155. Varani K, Gessi S, Dionisotti S et al. [3H]-SCH58261 labelling of functional A_{2A} receptors in human neutrophil membranes. Brit J Pharmacol 1998; 123:1723-1731.
156. Cronstein BN, Levin RI, Phillips MR et al. Neutrophil adherence to endothelium is enhanced via adenosine A_1 receptors and inhibited via adenosine A_2 receptors. J Immunol 1992; 148:2201-2206.
157. Barnes CR, Mandell GL, Carper HT et al. Adenosine modulation of TNFα-induced neutrophil activation. Biochem Pharmacol 1995; 50:1851-1857.
158. Lappin D, Whaley K. Adenosine A2 receptors on human monocytes modulate C2 production. Clin Exp Immunol 1984; 57:454-460.
159. LeVraux V, Chen YL, Masson I et al. Inhibition of human monocyte TNF production by adenosine receptor agonists. Life Sci 1993; 52:1917-1924.
160. Ritchie PK, Spangelo BL, Krzymoski DK et al. Adenosine increases interleukin-6 release and decreases TNF release from rat adrenal zona glomerulosa cells, ovarian cells, anterior pituitary cells and peritoneal macrophages. Cytokine 1997; 9:187-198.
161. Link AA, Kino T, Worth JA et al. Ligand activation of the adenosine A2A receptors inhibits IL-12 production by human monocytes. J Immunol 2000; 164:436-442.
162. Rathbone MP, Middlemiss PJ, DeLuca B et al. Extracellular guanosine increases astrocyte cAMP: inhibition by adenosine A2 antagonists. NeuroReport 1991; 2:661-664.
163. Rathbone MP, Middlemiss PJ, Gysbers JW et al. Trophic effects of purines in neurons and glial cells. Progr Neurobiol 1999; 59:663-690.
164. Hindley S, Herman MAR, Rathbone MP. Stimulation of astrogliosis in vivo by extracellular ADP or an adenosine A2 receptor agonist. J Neurosci Res 1994; 38:399-406.
165. Zhao ZQ, McGee DS, Nakanishi K et al. Receptor-mediated cardioprotective effects of endogenous adenosine are exerted primarily during reperfusion after coronary occlusion in the rabbit. Circulation 1993; 88:709-719.
166. Matherne GP, Linden J, Byford AM et al. Transgenic A1 adenosine receptor overexpression increases myocardial resistance to ischemia. Proc Nat Acad Sci USA 1997; 94:6541-6546.
167. Meldrum DR, Cain BS, Cleveland JC et al. Adenosine decreases post-ischemic cardiac TNFα production: Anti-inflammatory implications for preconditioning and transplantation. Immunology 1997; 92:472-477.
168. Wagner DR, Combes A, McTiernan C et al. Adenosine inhibits lipopolysaccharide-induced cardiac expression of TNFα. Circ Res 1998; 82:47-56.
169. Wagner DR, McTiernan C, Sanders VJ et al. Adenosine inhibits lipopolysaccharide-induced secretion of TNFα in the failing human heart. Circulation 1998; 97:521-524.
170. Cain BS, Meldrum DR, Dinarello CA et al. adenosine reduces cardiac TNFα production and human myocardial injury following ischemia-reperfusion. J Surg Res 1998; 76:117-123.
171. Bullough DA, Magill MJ, Firestein GS et al. Adenosine activates A2 receptors to inhibit neutrophil adhesion and injury to isolated cardiac myocytes. J Immunol 1995; 155:2579-2586.
172. Lozza G, Conti A, Ongini E et al. Cardioprotective effects of adenosine A1 and A_{2A} receptor agonists in the isolated heart. Pharmacol Res 1997; 35:57-64.
173. Casati C, Forlani A, Lozza G et al. Hemodynamic changes do not mediate the cardioprotection induced by the A1 adenosine receptor agonist CCPA in the rabbit. Pharmacol Res 1997; 35:51-55.
174. Cargnoni A, Ceconi C, Boraso A et al. Role of A_{2A} receptors in the modulation of myocardial reperfusion damage. J Cardiovasc Pharmacol 1999; 33:883-893.
175. Kerr JFR. Shrinkage necrosis: A distinct mode of cellular death. J Pathol 1971; 105:13-20.
176. Honkaniemi J, Massa SN, Breckinridge M et al. Global ischaemia induces apoptosis-associated genes in hippocampus. Mol Brain Res 1996; 42:79-88.
177. Weiss S, Cataltepe O, Cole AJ. Anatomical studies of DNA fragmentation in rat brain after systemic kainate administration. Neuroscience 1996; 74:541-551.
178. Sastry PS, Rao KS. Apoptosis and the Nervous System. J Neurochem 2000; 74:1-20.

179. Nishiyama K, Kwak S, Takekoshi S et al. In situ nick end-labelling detects necrosis of hippocampal pyramidal cells induced by kainic acid. Neurosci Lett 1996; 212:139-142.
180. Pollard H, Charriaut-Marlangue C, Cantagrel S et al. Kainate induced apoptotic cell death in hippocampal neurones. Neuroscience 1994; 63:7-18.
181. Simonian NA, Getz RL, Leveque JC et al. Kainate induces apoptosis in neurones. Neuroscience 1996; 74:675-683.
182. Filipkowski RK, Hetman M, Kaminska B et al. DNA fragmentation in rat brain after intraperitoneal administrtion of kainate. Neuroreport 1994; 5:1538-1540.
183. Kasof GM, Mandelzys A, Maika SD et al. Kainic acid-induced neuronal death is associated with DNA damage and a unique immediate-early gene response in c-fos-lacZ transgenic rats. J Neuroscience 1995; 15:4238-4249.
184. Smeyne RJ, Vendrell M, Hayward M et al. Continuous c-fos expression precedes programmed cell death in vivo. Nature 1993; 363:166-169.
185. Konopka D, Norwicka D, Filipkowski R K et al. Kainate-evoked secondary gene expression in the rat hippocampus. Neurosci Lett 1995; 185:167-170.
186. Stone TW, Jones PA, Smith RA. Neuroprotection by A2A receptor antagonists. Drug Develop Res 2001; 52:323-330.
187. Ankarcrona M, Dypbukt JM, Bonfoco E et al. Glutamate-induced neuronal death: A succession of necrosis or apoptosis depending on mitochondrial function. Neuron 1995; 15:961-973.
188. Reggio R, Pezzola A, Popoli P. The intrastriatal injection of an adenosine A_{2A} receptor antagonist prevents frontal cortex EEG abnormalities in a rat model of Huntington s disease. Brain Res 1999; 831:315-318.
189. Corsi C, Melani A, Bianchi L et al. Striatal A_{2A} adenosine receptors differentially regulate spontaneous and K+-evoked glutamate release in vivo in young and aged rats. NeuroReport 1999; 10:687-691.
190. Abbracchio MP, Fogliatto G, Paoletti AM et al. Prolonged in vitro exposure of rat brain slices to adenosine analogues: selective desensitsation of A1 but not A2 receptors. Europ J Pharmacol 1992; 227: 17-324.
191. Mally J, Stone TW. The effect of theophylline on parkinsonian symptoms. J Pharm Pharmacol 1994; 46:515-517
192. Mally J, Stone TW. Potential role of adenosine antagonist therapy in the treatment of pathological tremor disorders. Pharmacol Therap 1996; 72:243-250.
193. Mally J, Stone TW. Potential of adenosine A_{2A} receptor antagonists in the treatment of movement disorders. CNS Drugs 1998; 10:311-320.
194. Grondin R, Bedard PJ, Tahar AH et al. Antiparkinsonian effect of a new selective adenosine A_{2A} receptor antagonist in MPTP-treated monkeys. Neurology 1999; 52:1673-1677.
195. Kanda T, Jackson MJ, Smith LA et al. Adenosine A_{2A} antagonist: A novel antiparkinsonian agent that does not provoke dyskinesia in parkinsonian monkeys. Ann Neurol 1998; 43:507-513.
196. Johannson B, Fredhom BB. Further characterisation of the binding of the adenosine receptor agonist [3H]CGS21680 to rat brain using autoradiography. Neuropharmacology 1995; 34:393-403.
197. Johansson B, Georgiev V, Parkinson FE et al. The binding of the adenosine A2-selective agonist [3H]CGS 21680 to rat cortex differs from its binding to rat striatum. Europ J Pharmacol 1993; 247:103-110.
198. Lindstrom K, Ongini E, Fredholm BB. The selective A_{2A} receptor antagonist SCH 58261 discriminates between two different binding sites for [3H]CGS 21680 in the rat brain. Arch Pharmacol 1996; 354:539-541.
199. El-Yacoubi M, Ledent C, Parmentier M et al. SCH 58261 and ZM241395 differentially prevent the motor effects of CGS 21680 in mice: Evidence for a functional atypical adenosine A_{2A} receptor. Europ J Pharmacol 2000; 401:63-77.

200. Zhou QY, Li C, Olah ME et al. Molecular cloning and characterisation of an adenosine receptor: the A3 receptor. Proc Nat Acad Sci USA 1992; 89:7432-7436.
201. Jacobson KA. Adenosine A3 receptors: novel ligands and paradoxical effects. Trends Pharmacol Sci 1998; 19:184-191.
202. Rivkees SA, Thevananther S, Hao H. Are A3 adenosine receptors expressed in the brain? NeuroReport 2000; 11:1025-1030.
203. Sei Y, Lubitz DKJE, Abbracchio MP et al. Adenosine A3 receptor agonist-induced neurotoxicity in rat cerebellar granule neurons. Drug Develop Res. 1997; 40:267-273.
204. von Lubitz DKJ, Lin RCS, Boyd M et al. Chronic administration of adenosine A3 receptor agonist and cerebral ischemia: Neuronal and glial effects. Europ J Pharmacol 1999; 367:157-163.
205. Dunwiddie TV, Diao L, Kim HO et al. Activation of hippocampal adenosine A3 receptors produces a desensitisation of A1 receptor-mediated responses in rat hippocampus. J Neurosci 1997; 17:607-614
206. Hasko G, Nemeth ZH, Vizi ES et al. An agonist of adenosine A3 receptors decreases interleukin-12 and interferon-γ production and prevents lethality in endotoxemic mice. Europ J Pharmacol 1998; 358:261-268.
207. Ezeamuzie CI, Phillips E. Adenosine A3 receptors on human eosinophils mediate inhibition of degranulation and superoxide anion release. Brit J Pharmacol 1999; 127:188-194.
208. Maggirwar SB, Dhanraj DN, Somani SM et al. Adenosine acts as an endogenous activator of the cellular antioxidant defense system. Biochem Biophys Res Comm 1994; 201:508-515.
209. Sajjadi FG, Takabayashi K, Foster AC et al. Inhibition of TNF-α expression by adenosine: Role of A3 adenosine receptors. J Immunol 1996; 156:3435-3442.
210. McWhinney CD, Dudley MW, Bowlin TL et al. Activation of adenosine A3 receptors on macrophages inhibits TNFa. Europ J Pharmacol 1996; 310:209-216.
211. Knutsen LJS, Sheardown MJ, Roberts SM et al. Adenosine A1 and A3 selective N-alkoxypurines as novel cytokine modulators and neuroprotectants. Drug Develop Res 1998; 45:214-221.
212. Liang BT, Jacobson KA. A physiological role of the adenosine A3 receptor: Sustained cardioprotection Proc Nat Acad Sci 1998; 95:6995-6999.
213. Jordan JE, Thourani VH, Auchampach JA et al. A3 adenosine receptor activation attenuates neutrophil function and neutrophil-mediated reperfusion injury. Amer J Physiol 1999; 277:H1895-H1905.
214. Thourani VH, Nakamura M, Ronson RS et al. Adenosine A3 receptor stimulation attenuates postischemic dysfunction through K(ATP) channels. Amer J Physiol 1999; 277:H228-H235.
215. Abbracchio MP, Rainaldi G, Giammarioli AM et al. The A3 adenosine receptor mediates cell spreading, reorganisation of actin cytoskeleton and distribution of Bcl-x(L): Studies in human astroglioma cells. Biochem Biophys Res Comm 1997; 241:297-304.
216. Herold JA, Kron IL, Langenburg SE et al. Complete prevention of postischemic spinal cord injury by means of regional infusion with hypothermic saline and adenosine. J Thorac Cardiovasc Surg 1994; 107:536-542.
217. Phillis JW, O Regan MH. Prevention of ischemic brain injury by adenosine receptor activation. Drug Develop Res 1993; 28:390-394.
218. Jiang N, Kowaluk EA, Lee CH et al. Adenosine kinase inhibition protects against transient focal ischemia in rats. Europ J Pharmacol 1997; 320:131-137.
219. Lin Y, Phillis JW. Deoxycoformycin and oxypurinol: Protection against focal ischemic brain injury in the rat. Brain Research 1992; 571:272-280.

HEAT SHOCK PROTEINS AND NEUROPROTECTION

Midori A. Yenari*

ABSTRACT

In response to many metabolic disturbances and injuries including stroke, neurodegenerative disease, epilepsy and trauma, the cell mounts a stress response with induction of a variety of proteins, most notably the 70 kD heat shock protein (Hsp70). The possibility that stress proteins might be neuroprotective was suspected because Hsp70, in particular, was induced to high levels in brain regions that were relatively resistant to injury. Hsp70 expression was also correlated with the phenomenon of induced tolerance. With the availability of transgenic animals and gene transfer, has it become increasingly clear that such heat shock proteins do indeed protect cells from injury. Several reports have now shown that selective overexpression of Hsp70 leads to protection in several different models of nervous system injury. This review will cover these studies, along with potential mechanisms by which Hsp70 might mediate cellular protection.

INTRODUCTION

It is well known that cells respond to external stress in a highly conserved, stereotypical Fashion. The stress response results in gene expression following such environmental challenges such as high temperatures, ischemia, excitotoxin exposure and other stresses which result in protein denaturation (see recent reviews refs. 1,2). Heat shock proteins (Hsps) are induced by stressful stimuli and are thought to assist in the maintenance of cellular integrity and viability. Hsps consist of both stress

*Department of Neurosurgery, Stanford University, 1201 Welch Road, MSLS Building P304, Stanford, CA 94305-5487.

inducible and constitutive family members, as well as members that are associated with specific organelles. Constitutively synthesized Hsps perform housekeeping functions. They function as molecular chaperones by helping nascent polypeptides assume their proper conformation by binding to nascent proteins via their C terminal domain. Hsps are also involved in antigen presentation, steroid receptor function, intracellular trafficking, nuclear receptor binding, and apoptosis.[1,3] However, many are also upregulated by stress. Inducible Hsps prevent protein denaturation and incorrect polypeptide aggregation during exposure to physiochemical insults. Hsps may prevent protein unfolding or aggregation and enhance cell survival.

The 70 kD family of stress proteins is one of the most extensively studied. Included in this family are Hsc70 (heat shock cognate, the constitutive form), Hsp70 (the inducible form is also referred to as Hsp72), Grp75 (a constitutively expressed mitochondrial glucose regulated protein), Grp78 (a constitutively expressed glucose regulated protein found in the endoplasmic reticulum) and heme oxygenase-1 (HO-1, a stress protein involved in bilirubin metabolism). Other members of this family include Hsp40, Hsp90, Hsp27 and Hsp chaperone proteins such as the DNAJ homologs (HDJ-1 and HDJ-2 in humans), the latter being proteins which assist other Hsps in protein folding. Following a variety of central nervous system insults, Hsp70 is expressed at especially high levels and is present in the cytosol, nucleus and endoplasmic reticulum. Because inducible Hsp70 is not usually detectable under normal conditions and its expression is the most robust of the Hsps, it is often regarded as a diagnostic marker for stress. The stress proteins of the Hsp70 family function as chaperones, interacting transiently with many proteins in an ATP dependent manner. Denatured proteins are thought to serve as the stimulus for stress protein induction. These proteins activate heat shock factors (HSFs) within the cytosol by dissociating other Hsps that are normally bound to HSFs. Once liberated, HSFs are phosphorylated and form trimers. The trimers then enter the nucleus and bind to heat shock elements (HSEs) within the promoters of different heat shock genes leading to transcription and Hsp translation.[3,4] Once expressed, Hsp70 binds to then quickly releases denatured proteins in an ATP dependent Fashion. (Fig. 1) There appear to be two main functional domains within Hsp70. The N terminal end contains an ATP binding domain whereas the C terminal part contains a substrate binding domain. Substrate binding cannot occur in the absence of ATP binding which is regulated by an EEVD motif within the C terminus[5]. Therefore, various stresses resulting in protein denaturation stimulate Hsp70 expression, which presumably acts to restore protein structure and function.

In the nervous system, the heat shock proteins are induced in a variety of pathologic states including cerebral ischemia, neurodegenerative diseases, epilepsy and trauma. Expression has been detected in a variety of cell populations within the nervous system including neurons, glia, and endothelial cells.[6] Although Hsp70 has been long thought to protect cells by preserving tertiary protein structure and preventing protein aggregation, direct evidence has been lacking. Gene transfer techniques and transgenic animal strain have now made it possible to selectively overexpress Hsps to better understand the precise role they play in cellular injury.

This review will focus on recent results that help to elucidate the role of Hsp70 in neuroprotection.

WHERE AND WHEN IS HSP70 EXPRESSED?

Several groups using various models of experimental nervous system stress and injury have studied the anatomic and temporal expression of Hsp70. In a model of global cerebral ischemia (a model which results in cerebral injury similar to that following cardiac arrest), Hsp70 mRNA was detected in the hippocampus within hours of ischemia onset and decreased when neurons were lost.[4] Protein expression followed a few hours later[7]. Brief periods of ischemia (3-8 minutes) resulted in Hsp70 protein expression within neurons and some glia after 24 hours, but longer durations of ischemia (10-20 minutes)[4,7] showed decreased Hsp70 protein expression. Following 60 minutes of transient focal cerebral ischemia (a model of stroke), Hsp70 protein expression was similar patterns were observed with persistent expression as far out as 7 days.[4,8] Hsp70 was observed within neurons and astrocytes at the infarct periphery, but only endothelial cells expressed it within brain regions where ischemia was the most severe (striatum)[8]. With increasing duration of middle cerebral artery (MCA) occlusion, graded levels of Hsp70 expression are observed within neurons, microglia and endothelial cells that decreased after the most severe ischemic insults.[4,9,10] It has previously been suggested that glia can transfer Hsp70 to neurons; therefore, neuronal Hsp70 expression may be linked to a potential protective mechanism from glia.[1]

In models of excitotoxicity and seizures, kainic acid (KA) administration resulted in widespread Hsp70 induction, particularly within cortical and hippocampal neurons.[11,12] Hsp70 mRNA was present in dentate granule cells within 6 hours of KA application and protein was observed by 12 hours, then decreased after 24 hours.[11] Protein expression has been observed to be especially prominent within brain regions known to be resistant to injury in this model, whereas less expression was seen in degenerating neurons.[13] On the other hand, Hsp70 was observed within hippocampal neuron populations such as CA1, CA3 and dentate granule cells,[12] and appeared to precede and correlate with the extent of neuronal damage assessed a few days later.[14] Furthermore, MK801, an antagonist of the N-methyl-D-aspartate receptor, reduced both Hsp70 expression and KA induced neuronal injury.[12]

There appear to be similar patterns of Hsp70 expression in different in vivo models of cerebral stress, with graded expression depending on the severity of the insult and the cell population. In general, Hsp70 is observed in the brain within several hours of the insult, and persists for a few days.

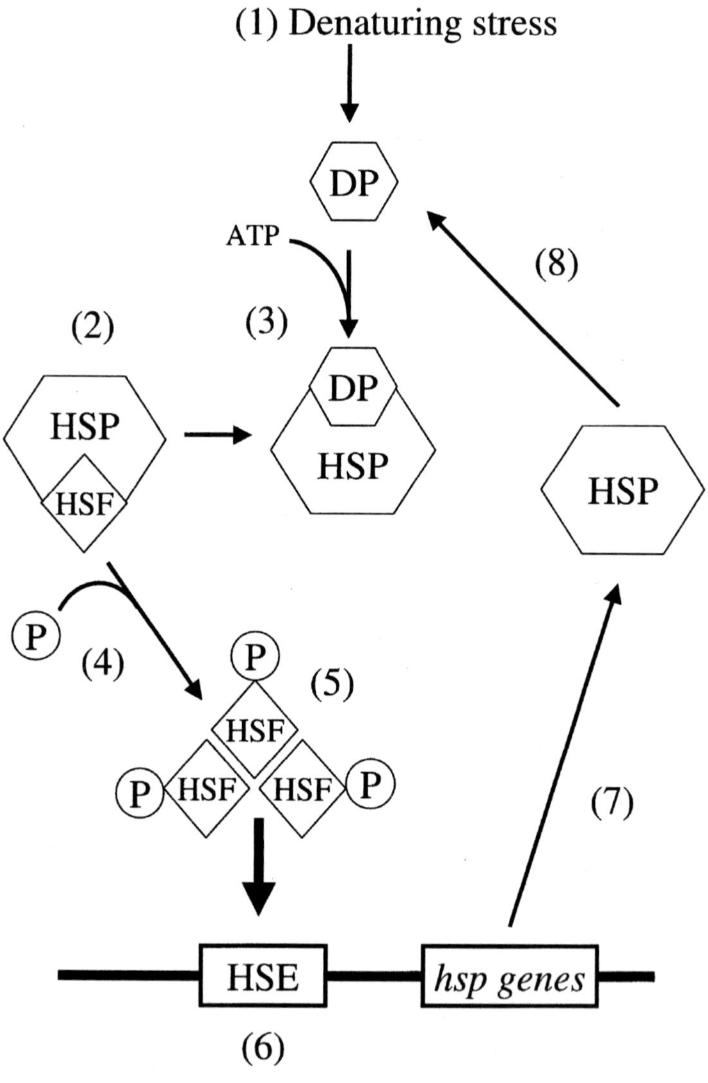

Figure 1. Mechanism of heat shock protein activation. Various stresses result in protein denaturation (1). Heat shock proteins (Hsp) are normally bound to heat shock factors (HSF) (2), but dissociate in the presence of denatured proteins (DP). Once dissociated, Hsps are free to bind DPs, followed by rapid release. These are ATP requiring steps (3). More Hsps are generated when HSFs phosphorylate (P) (4) and trimerize (5). These trimers can bind heat shock elements (HSE) (6) that are contained within the promoters of the Hsps and generate more protein (7). Newly generated Hsps can then bind more denatured proteins (8).

CORRELATIVE EVIDENCE FOR A NEUROPROTECTIVE ROLE

Stress proteins are increased in resistant cell populations following injury: Whether Hsp70 serves a protective role against various cerebral insults has been debated. It has long been known that certain cell populations are less vulnerable than others to various cerebral insults such as global cerebral ischemia and KA toxicity. Following ischemia, several groups have reported that Hsp70 protein is expressed only in cell populations that will ultimately recover from cerebral ischemia[10] such as infarct borders (focal cerebral ischemia) and dentate granules (global ischemia). With infarction, expression is absent within brain tissue and is restricted to vascular endothelial cells.[10,13] Similar results were observed in the KA model.[4,13] These studies imply that Hsp70 is expressed in cell populations known to be resistant to injury; therefore, Hsp70 may be responsible for the observed resistance.

In contrast, other data suggest that Hsp70 is expressed in vulnerable cell populations or is expressed regardless of the fate of the cell.[4,15] These contrasting observations could be limited by the severity of the insults, and whether or not protein translation has occurred. Dysjunction of Hsp70 mRNA and protein expression following focal cerebral ischemia has been observed. While Hsp70 mRNA was expressed extensively, protein was observed only within potentially salvageable peri-infarct regions.[1,4] but not in damaged regions. Therefore, Hsp70 may be capable of rescuing injured cells provided the protein has been translated.

Hsp expression correlates with protection in tolerance paradigms: Whether Hsp70 is playing a protective role against various insults could not be addressed in the above studies. However, Hsp70 s increased expression led investigators to study its potential role in the protective effect of induced tolerance , a phenomenon whereby a prior sublethal insult leads to protection against a subsequent severe insult. For instance, thermal or chemical stress protects against excitotoxic insults such as glutamate exposure in cultured neuronal cells and in whole animal models.[16] Prior heat shock, sublethal ischemia or chemical toxins protect against subsequent ischemia.[1,4] Induction of Hsp70 and protection from subsequent injury has been demonstrated in neuron cultures and in whole animal models of thermal stress, global and focal cerebral ischemia (Reviewed by Massa et al[4] and Parsell et al[17]). The expression of Hsp70 also correlated with the period within which tolerance was observed, leading some to believe that Hsp70 may explain the observed phenomenon.[18] Interestingly, the interval of tolerance was associated increases in neuronal Hsp70 and activated astrocytic Hsp27.[19] However, these sublethal insults cause a host of changes in protein expression and metabolism rather than selectively increasing Hsp70 synthesis, and the role of Hsp70 in mediating the protective effects has been far from decisive. In fact, protection in tolerance experiments has been noted even when HSP70 and other new protein synthesis is blocked,[20] suggesting that induced tolerance does not always require protein production.[17]

NEUROPROTECTION WITH HSP70 OVEREXPRESSION

To directly test a protective role of Hsp70, cells can be made to selectively overexpress the protein, or expression can even be blocked. This can now be accomplished using transgenic animal models, gene transfer or antisense oligonucleotides. A number of studies have been published where Hsp70 has been overexpressed in various non-neuronal cell lines, and found protection against numerous stresses including heat shock,[21] oxidative stress,[22] apoptotic stimuli[23,24] and ischemia like conditions.[25,26] In the nervous system, Hsp70 overexpression in cultured hippocampal[27,28] and peripheral[29-32] neurons and glia[29,30] similarly protected against insults such as heat shock and metabolic stresses. Hsp70 expression can also be suppressed with antisense oligonucleotides that inhibit transcription.[33] Using this approach, Sato et al[33] found that the protection from induced tolerance was reversed with Hsp70 blockade using antisense oligonucleotides. At the in vivo level, a few laboratories have generated transgenic mice capable of overexpressing *Hsp70* from a constitutive promoter.[34-36] In models of myocardial ischemia, these mice were found to have reduced infarct volume,[36] improved recovery of ATP stores[35] and better contractile recovery.[34]

However, Hsp70 overexpression is not protective in all instances. Using a defective herpes simplex virus (HSV) vector, Fink et al[28] showed that Hsp70 overexpression protected cultured hippocampal neurons from severe heat shock, but failed to protect against direct application of glutamate or 3-nitropropionic acid (3-NP), a mitochondrial toxin. Wagstaff et al[37] showed that Hsp70 overexpression protected cultured peripheral neurons from thermal and simulated ischemia, but not apoptotic stimuli. Observations from our laboratory have shown that glial cell cultures from brains of transgenic mice that overexpress Hsp70 are resistant to hydrogen peroxide injury, but are less resistant to other injury paradigms.[38] Interestingly, hippocampal neurons were more resistant to certain stresses compared to cortical neuron cultures. These results suggest that Hsp70 protects against some but not all kinds of central nervous system injury, and that the protective effects may be related to the nature and severity of the insults. It has previously been shown that the response to heat shock and Hsp70 expression can vary depending on the type and age of the cell;[39] therefore, it is likely that the protective effects may similarly be dependent on such factors.

Hsp protection following ischemia and ischemia like insults: Nervous system ischemia can be modeled in vitro by exposing cultures to ischemia like conditions such as substrate and/or oxygen deprivation,[38,40] excitotoxin exposure[28] or incubation in simulated ischemic buffers[32]. Recent work from our group showed that cultured hippocampal, but not cortical neurons from transgenic mice overexpressing Hsp70 were protected from excitotoxin exposure and oxygen and glucose deprivation.[38] Glial cultures isolated from this same mouse strain were also resistant to substrate deprivation. Higher levels of transgene expression could be attained by using retroviral vectors which increased Hsp70 protein greater than 10 fold (R. G. Giffard, personal communication). Xu et al[41] and Papadopoulos et al[40] used such retroviral vectors to overexpress Hsp70 in astrocyte cultures. Exposure to isolated glucose or combined oxygen and glucose deprivation led to robust glial survival following either insult. Conversely, suppression of Hsp70 in hippocampal neuron cultures with an antisense oligonucleotide worsened injury following heat shock.[33]

At the whole animal model level, cerebral ischemia models using Hsp70 overexpressing transgenic mice have been reported. Whereas various strains of Hsp70 transgenic mice are protected against myocardial ischemia,[34-36] this has not always been the case for brain ischemia. The reasons for these discrepancies are not clear, but could be due to strain differences in transgene expression within different organ systems, or because brain regions have varying susceptibilities depending on the model studied. A few groups have studied permanent MCA occlusion in various Hsp70 transgenic mice strains with conflicting results. Constitutive overexpression of Hsp70 in one strain using a cytomegalovirus (CMV) enhancer combined with a β-actin promoter resulted in near complete protection determined by overall reduction in infarct size.[42] Another group using a similar model but a different mouse strain found only improved hippocampal neuron survival by 24 h; however, overall infarct size was not affected.[43] In this latter group, Hsp70 expression was under the control of a β-actin (constitutive) promoter. In yet a third study, our group[38] examined a third strain of Hsp70 transgenic mice and failed to find any differences between infarct size or hippocampal neuron survival. These differing results could be due to background strain differences and the limitations of transgenic animals, such as developmental alterations in other biochemical systems caused by transgene overexpression. The extent of Hsp70 expression might also be different between the transgenic strains, since different promoters were used. The transgenic model containing the CMV enhancer increased transgene expression by 5-10 fold.[42] In contrast, the model using the β-actin promoter alone resulted in only 2 fold increases.[38] Therefore, higher transgene levels could account for the marked protection in the first study,[42] and the less robust or lack of protection in the latter two studies.[38,43]

Whether lifelong Hsp70 expressing animals have alterations in other systems is unclear, though the gross phenotype appears normal. Therefore, viral vector mediated transfer of Hsp70 can be used to study neuroprotection in wildtype animals. Our laboratories have recently taken advantage of the neurotropic properties of herpes simplex virus (HSV) to preferentially overexpress Hsp70 in neurons.[28] HSV vectors are capable of transfecting approximately 70% or more cortical neurons following direct injection into the brain.[44] One of the major constraints in overexpressing genes in vivo has been the relatively low efficacy and efficiency of viral vector mediated gene transfer. To circumvent this problem, bipromoter vectors were developed which contained both the genes for *Hsp70* and a reporter gene, *lacZ*.[45-47] With this approach, it becomes possible to identify the subset of transfected cells that also express the transgene of interest. Following direct intracerebral injection, vector mediated Hsp70 protein and reporter gene (β-galactosidase or β-gal) expression occurred about 9-12 hours later. The number of targeted Hsp70 overexpressing neurons were differentiated from cells that expressed endogenous Hsp70 by identifying the β-gal positive cells (Fig. 2A). By counting the number of β-gal positive cells in the ischemic side of the

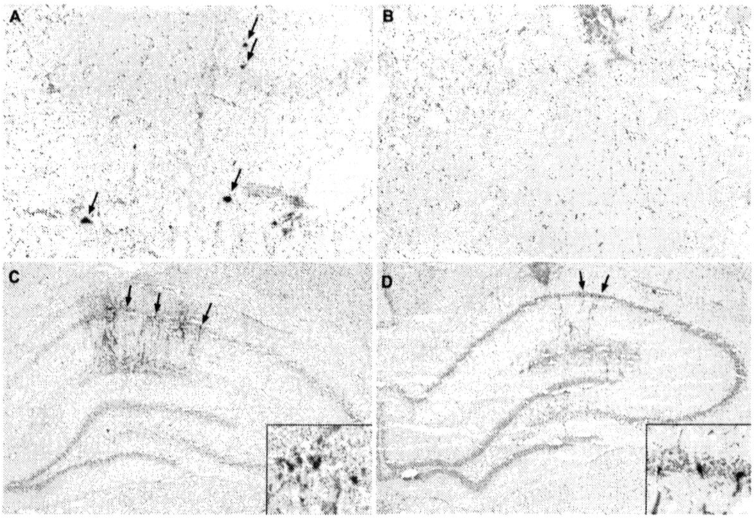

Figure 2. Hsp70 protects neurons from a variety of central nervous system insults. Several surviving, beta-galactosidase (β-gal) positive cells (arrows) which received a Hsp70 overexpressing vector are evident within an ischemic striatum 48h after 1 hour of middle cerebral artery occlusion (stroke) (A) compared to one receiving control vector where virtually no surviving cells are seen (B). Following 8 minutes of forebrain ischemia, more transfected, β-gal positive CA1 neurons are present within a hippocampus given Hsp70 vector (C) versus control vector (D). Insets show detail of transfected neurons within the hippocampal CA1 subsector. Brain sections are counterstained with cresyl violet. (C & D are reprinted with permission from: Yenari MA et al. In: J Krieglstein & S Klumpp (eds), Pharmacology of Cerebral Ischemia, 457-472 © 2000 Medpharm Scientific Publishers; Yenari MA et al. In Slikker W & Trembly B (eds), Neuroprotective Agents: 5th International Conference, 939:340-357 2001 NY Acad Sci; Yenari MA et al. Neurol Res 23:543-552 2001 Forefront Publishing Group.)

brain and comparing it to β-gal counts from the non ischemic contralateral side, we found greater numbers of striatal neurons 48 hours after 1 hour of MCA occlusion when vector was injected 12 hours prior to ischemia onset[48] (Figs. 2A-B and 3). However, overall infarct size was not affected given the limited number of cells the vectors can target. Similar results were observed against KA induced excitotoxicity where viral vectors were injected into the hippocampi of animals 15 hours before systemic KA administration (Fig. 3). Hippocampal dentate neuron survival at 24 hours was improved with Hsp70 overexpression. Hsp70 overexpression also protected against global cerebral ischemia. Using the HSV vectors, our group overexpressed Hsp70 within the vulnerable CA1 neurons of the hippocampus (Fig. 2C-D).[49] Animals were then subjected to 8 minutes of forebrain ischemia, and the proportion of surviving, vector targeted cells compared to sham injured transfected controls were determined 3 days later. Like the results from the focal cerebral ischemia and KA models, more Hsp70 transfected CA1 hippocampal neurons remained compared to control vector transfected (Fig. 3[49]). Another study by Kelly et al[50] used adenoviral vectors to transfer Hsp70 to striata of mice. Following 20 minutes forebrain ischemia

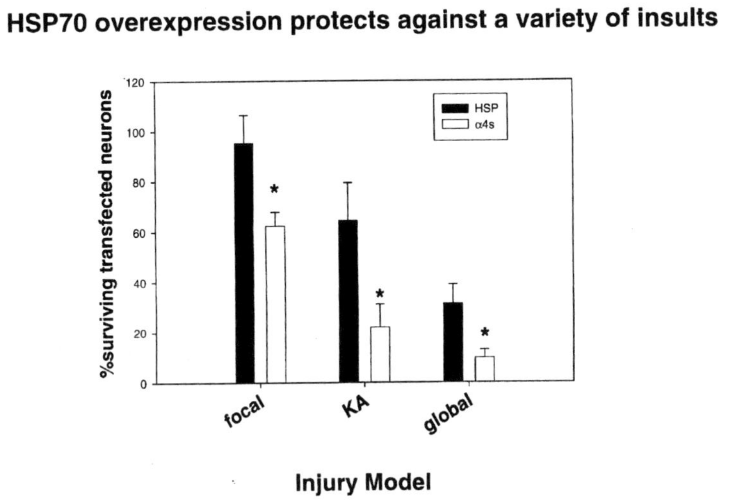

Figure 3. Hsp70 overexpression protects against a variety of insults in vivo. Following focal cerebral ischemia, normalizing counts of β-galactosidase positive neurons within the ischemic striatum to the contralateral non-ischemic side showed significantly better survival among the group where Hsp70 was overexpressed (focal). Similar protection was observed when Hsp70 was overexpressed in hippocampal neurons in a model of excitotoxin exposure due to kainic acid (KA) and in a model of forebrain ischemia (global) (HSP=Hsp70 vector treated, α4s=control vector treated, *$p<0.05$).

in their model, they, too, found neuroprotection following intrastriatal injection of Hsp70 expressing vectors compared to β-galactosidase expressing control vectors. These data indicate that Hsp70 is protective at the in vivo level in adult wildtype animals

Hsp protection in neurodegenerative diseases: Hsps also appear to protect against neurodegeneration in models of polyglutamine expansion disease. Polyglutamine diseases constitute a heterogeneous set of diseases characterized by the cellular accumulation of glutamine aggregates. These diseases contain a genetic defect characterized by repeating trinucleotide CAG motifs. CAG repeats result in expanses of glutamine which result in misfolded, aggregated protein, leading to toxicity and cell death. Human trinucleotide repeat diseases include Huntington s disease, Kennedy s disease (spinobulbar muscular atrophy), certain spinocerebellar ataxic disorders and others. Overexpressing various Hsps in cells transformed with polyglutamine sequences has led to reduction in protein aggregates and less toxicity. By transfecting cells with genes that express proteins associated with spinocerebellar degeneration, such as mutant ataxin-1[51] or ataxin-3,[52] protein aggregates could be observed. Many of these aggregates were associated with Hsps, particularly Hsp70, indicating that these expanded proteins can themselves elicit a stress response. After transfecting such cells containing these mutant ataxins, aggregates could be suppressed by overexpression of the Hsp40 chaperones HDJ-1 and HDJ-2.[51, 52] In similar models of Kennedy s disease, cell lines were transfected with mutant androgen

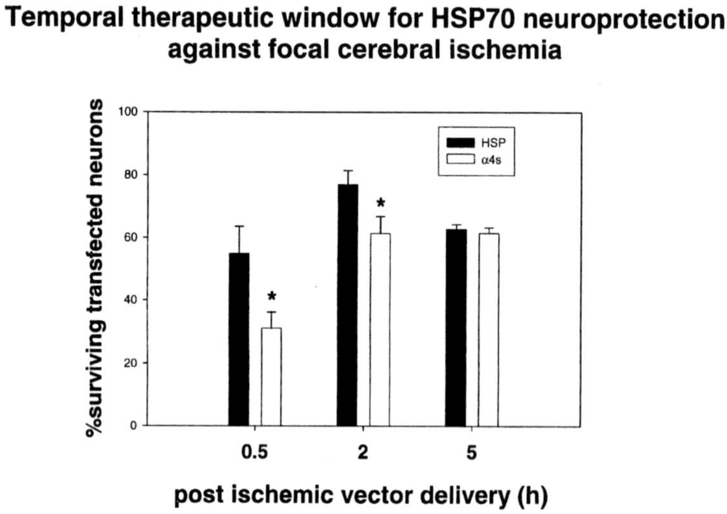

Figure 4. Hsp70 protects striatal neurons when delivered after stroke onset. The proportion of surviving vector targeted neurons is improved even when Hsp70 vector is delivered 0.5 and 2 h post ischemia, but is no longer protective when administered at 5 h (HSP=Hsp70 vector treated, α4s=control vector treated, *$p<0.05$). Percentages represent the number of remaining β-galactosidase positive neurons within the ischemic striatum compared to counts in the contralateral, nonischemic striatum for each animal.

receptor. Overexpression of Hsp40 and Hsp70[53] or HDJ-1[54] could also suppress protein aggregates. In a model of Huntington s disease, these same Hsps could inhibit the assembly of abnormal polyglutamine containing huntingtin protein into amyloid fibrils.[55] Recently, a murine model of spinocerebellar ataxia-1 (SCA-1) was crossed with a transgenic mouse model which overexpressed HSP70.[56] SCA-1 mice overexpressing Hsp70 performed better on behavioral tests, and Pukinje cell neuropathology was improved by Hsp70 overexpression. However, characteristic nuclear inclusions were not altered by HSP70 expression. Nevertheless, such data are consistent with the notion that Hsps protect cells from injury through their chaperone functions, and possibly by preventing protein aggregation.

Hsps and clinical relevance: Whether Hsp70 can protect cells when delivered after an insult would have obvious clinical implications. Our group recently found that overexpressing Hsp70 up to 2 hours following middle cerebral artery occlusion resulted in improved striatal neuron survival.[57] When Hsp70 expressing vector was delivered 5 hours post insult, neuroprotection was lost (Fig. 4). Since the HSV vectors do not begin to express transgene until 4-6 hours post transfection,[45-47] this would suggest a temporal therapeutic window of 6-8 hours post stroke.

Another potentially clinically applicable approach for utilizing the neuroprotective properties of the Hsps might be to pharmacologically induce them in the brain. Geldanamycin, a benzoquinoid ansamycin, is one such compound currently being tested for cancer therapy.[58] It binds to Hsp90, and possibly induces other Hsps by releasing HSFs to bind HSEs and increase Hsp70 expression.[59] In a recent presentation by Lu et al,[60] geldanamycin induced expression of several Hsps including Hsp70 and HO-1. Intrathecal and intraperitoneal administration lead to neuroprotection within hippocampal CA1 when given 6 hours prior to 5 minutes of forebrain ischemia in gerbils. Whether geldanamycin might protect the brain when given after the onset of cerebral ischemia is not yet known. However, Xiao et al[61] showed that it can prevent glutamate induced oxidative toxicity in hippocampal cell cultures, even if given 4 hours after glutamate exposure.

POTENTIAL MECHANISMS OF PROTECTION

The mechanism of protection with Hsp70 is believed to be related to its chaperone functions leading to prevention of protein malfolding and aggregation. Polyglutamine diseases are known to have accumulations of abnormal protein chains. Following both focal[62] and global[63] cerebral ischemia, protein aggregates have been described within vulnerable cell populations. Hsps appear to interfere with the formation of such aggregates for some neurodegenerative disease, but has yet to be demonstrated in ischemia models.

Nevertheless, certain observations suggest that Hsp s chaperone functions may play a role in protection from injury through improving function of several different proteins. For instance, there may be a relationship between stress protein protection and oxidative injury. Work by Polla and colleagues[64] demonstrated that Hsp70 induction correlated best with protection of a cell line from hydrogen peroxide induced oxidative injury. In cultured glial cells, Xu and Giffard[41] noted that protection with Hsp70 overexpression against glucose deprivation or hydrogen peroxide exposure was associated with increased glutathione levels, suggesting that Hsp70 may protect cells through an antioxidant mechanism, perhaps by improving protein stability of endogenous antioxidants. Prolonged expression of *Hsp70* mRNA among Cu-Zn superoxide dismutase (SOD1) transgenic mice was observed following focal[65] and global[66] cerebral ischemia and kainic acid induced injury.[67] The reasons for these observations are not clear, but might be due to an overall reduction in oxidative stress, which permits extended expression of Hsp70. Prolonged Hsp70 expression could then enhance SOD1 s protective effect through its chaperone functions. Another possibility is that increased levels of SOD1 without compensatory increases in glutathione peroxidase could lead to an accumulation of hydrogen peroxide, another reactive species. Consequently, hydrogen peroxide could lead to Hsp70 induction due to the increased oxidative stress.

Hsps may protect by mechanisms unrelated to their chaperone function as well. Insults such as ischemia, hypoxia and heat shock are notable for toxic increases in intracellular calcium that lead to activation of various proteases and endonucleases and ultimately, cell death. In non-neuronal cell lines, overexpression of Hsp70 attenuates this influx of calcium by desensitizing other ion regulating system such as the $Na+/Ca++$ exchanger (reviewed by Kiang & Tsokos[3]). Hsp70 may also alter other proteins or genes known to be involved in ischemic and excitotoxic injury and

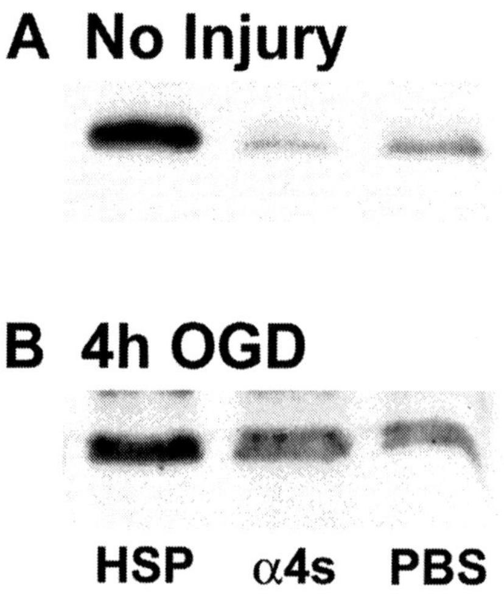

Figure 5. Hsp70 increases Bcl-2 expression: Cultured hippocampal and cortical neurons transfected with defective herpes simplex viral vectors express higher levels of Bcl-2 protein. Western blots show increased Bcl-2 protein in cell extracts of neurons 24 hours after transfection with Hsp70 (HSP) compared to control vector (α4s) and mock (PBS) transfection (A). Following 4 hours oxygen-glucose deprivation (OGD), Hsp70 transfected neuronal cultures still express higher levels of Bcl-2 (B). Courtesy of R.G. Giffard and L.J. Xu, Dept. of Anesthesiology, Stanford University.

inflammation by inhibiting nitric oxide synthase generation and nuclear translocation of the transcription factor, NF-κB.[68,69] Others have shown that prior thermal stress leads to inhibition of the inflammatory response, and this inhibition was associated with increased levels of Hsp70 induction and decreased nuclear NF-κB translocation.[69,70] It has been speculated that Hsp70 could interact with NF-κB s inhibitor protein, IκB, and prevent IκB phosphorylation and NF-κB dissociation.[68] Anther possibility is that Hsp70, which can also translocate to the nucleus, could also compete for the same nuclear pore as NF-κB.[68,71]

Recent studies in non-neuronal cell lines have shown that Hsp70 s protective effect may also be due to anti-apoptotic mechanisms. Apoptosis, or programmed cell death is known to occur in pathological states either by activation of specific death receptors (Reviewed by Ashkenazi & Dixit[72]), or internally, via mitochondrial release of cytochrome C (Reviewed by Green & Reed[73]). Central to mitochondrial based apoptosis is the assembly of the so-called apoptosome. This occurs when procaspase-9 binds to Apaf-1 in the cytosol, and becomes activated when cytochrome C is release from the mitochondria to the cytosol. This release of cytochrome C is blocked by the anti-apoptotic protein, Bcl-2. Activated caspase-9 then leads to activation of various effector caspases including caspase-3.

Several papers have now established that Hsps can interfere with apoptosis in various systems (Fig. 6). Overexpression of Hsp70 in lymphoid tumor cell lines

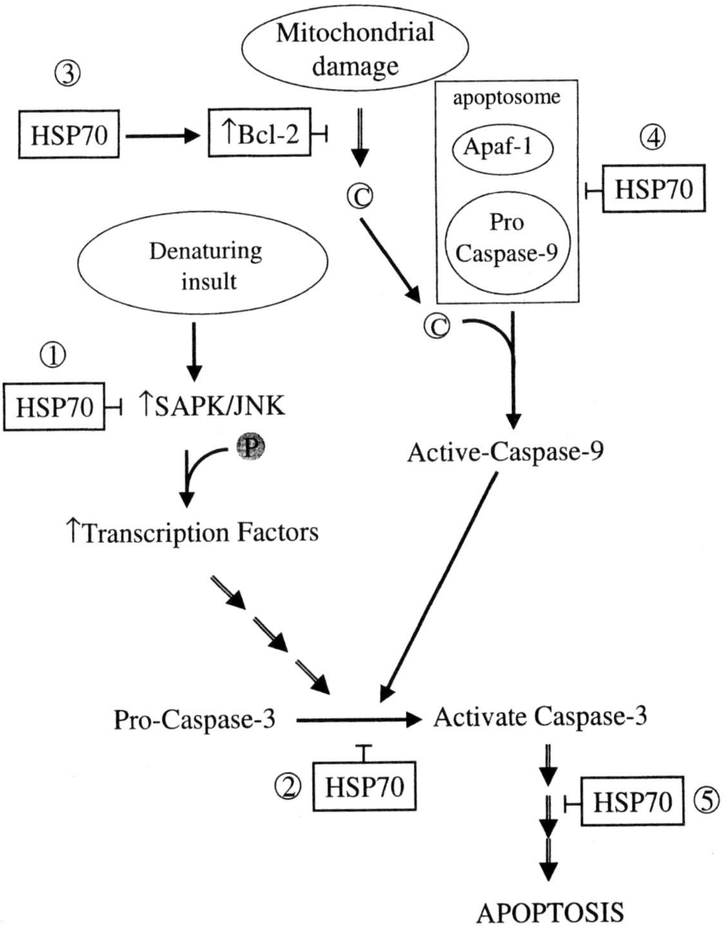

Figure 6. Protection from apoptosis by Hsp70 may have multiple sites of action. In non-neuronal cell cultures, Hsp70 appears to interfere with several different steps in apoptosis. Following an appropriate apoptotic insult (e.g., thermal stress, tumor necrosis factor, etc.), the stress activated protein kinase (SAPK/JNK) pathway is activated. Activated SAPK/JNK phosphorylates (P) various transcription factors. Transcription factor phosphorylation leads to increases in various enzymes (including other caspases) which cleave inactive (pro) forms of various caspases to their active form. Apoptosis can also be activated via mitochondrial damage leading to cytosolic release of cytochrome C (C). Cytochrome C then activates complexes of pro-caspase-9 and Apaf-1 (which constitute the so-called apoptosome), leading to activation of caspase-9. Caspase-9 then activates caspase-3, an effector caspase. Hsp70 appears to interfere with these pathways: 1) before SAPK/JNK activation, but only when Hsp70 expression is induced,[21] 2) after SAPK/JNK activation and before caspase activation,[21,74,75] 3) by increasing Bcl-2 expression which in turn inhibits cytochrome C release,[49,83] 4) interferes with Apaf-1 and preventing recruitment of pro-caspase-9 into the apoptosome,[76,77] and 5) at an as yet unknown point downstream of caspase activation.[23]

appears to inhibit apoptosis by blocking caspase activation and activity.[21,74,75] It has recently been shown that Hsp70 could inhibit caspase activation by interfering with Apaf-1, and prevent the recruitment of procaspase-9 to the apoptosome.[76,77] How this occurs is not yet clear, but has been hypothesized to be due a direct competition between Hsp70 and procaspase-9 for Apaf-1 binding.[77,78] However, Mosser et al[21] demonstrated that Hsp70 did not appear to interfere with caspase-3 processing, and other data suggest that Hsp70 interferes with apoptosis downstream of caspase activation.[23] Yet other studies in tumor cells lines have shown that Hsp70 may block stress kinase (SAPK/JNK) activation when Hsp70 overexpression is induced,[21] but not when constitutively expressed.[21,23,24] However, Hsps do not appear to block Fas mediated, receptor activated apoptosis.[78,79] Together, these studies suggest that Hsp70 probably acts at multiple sites to confer protection in models of apoptosis.

Whether Hsp70 prevents apoptosis from occurring in the brain has yet to be definitively shown; however, cells with DNA fragmentation (detected by DNA nick end labeling) following focal cerebral ischemia rarely expressed Hsp70 protein.[80] Other related stress proteins have been shown to inhibit apoptosis in the brain. For instance, Bcl-2, an anti-apoptotic protein, is increased in neurons of mice which overexpress heme oxygenase (HO-1),[81] and GRP78 appears to inhibit caspase activation in cultured hippocampal rat neurons.[82] Our group also found that Hsp70 overexpression also increased levels of Bcl-2 protein. Following viral vector transfection, Bcl-2 was present in uninjured Hsp70 expressing neurons compared to control vector transfected cultures (Fig. 5A). Following oxygen glucose deprivation, Hsp70 overexpression still resulted in increased Bcl-2 levels (Fig. 5B). We observed similar patterns in vivo where higher proportions of CA1 hippocampal neurons transfected with Hsp70 also expressed Bcl-2 following 8 minutes of forebrain ischemia.[49] These findings would be consistent with those of others who reported that Hsps decrease cytochrome C release following hydrogen peroxide exposure.[83] In this scenario, it is conceivable that Hsps, by an as yet unknown mechanism, could increase Bcl-2 expression, which in turn block cytochrome C release and subsequent effector caspase activation. However, a few reports using in vitro models showed that Hsp70 and Hsp90 do not protect against apoptotic stress in peripheral neurons,[84] but that Hsp27 does.[32] These latter observations might suggest that stress proteins, though not necessarily Hsp70, do play a role in blocking neuronal apoptosis.

Other mechanisms of protection might also be involved. Hsp70 s protein binding properties are dependent on ATP;[5] however, insults such as ischemia are especially notable for reduction in metabolic stores despite the observed neuroprotection from a few different laboratories. Interestingly, Hsp70 variants containing mutations in the ATP binding domain still protect tumor cells from heat shock.[85] Therefore, ATP independent mechanisms underlying the observed neuroprotection is plausible, but have yet to be tested in appropriate models.

CONCLUSION

Recent studies from different laboratories have now established that the stress response provides the organism with a cellular process for self preservation, and that stress proteins themselves can directly protect cells from death. The specific mechanisms underlying this protection are not well understood, but are likely multifactorial, encompassing a wide range of cellular chaperone functions from the prevention of protein aggregation to interfering with various death cascades. Capitalizing on the cell s natural response to stress is an attractive therapeutic target for a variety of nervous system diseases.

ACKNOWLEDGMENTS

This work was funded in part by National Institutes of Health grants # P01 NS37520 and R01 NS40516 and American Heart Association Grant #0060091Y. The author would like to thank the many lab members in the Departments of Neurosurgery, Biology and Anesthesiology at Stanford University who contributed to some of the work presented here, and Ms. Beth Houle for preparation of the figures.

REFERENCES

1. Sharp FR, Massa SM, Swanson RA. Heat-shock protein protection. Trends Neurosci 1999; 22(3):97-99.
2. Yenari MA, Giffard RG, Sapolsky RM et al. The neuroprotective potential of heat shock protein 70 (HSP70). Mol Med Today 1999; 5(12):525-531.
3. Kiang JG, Tsokos GC. Heat shock protein 70 kDa: Molecular biology, biochemistry, and physiology. Pharmacol Ther 1998; 80(2):183-201.
4. Massa SM, Swanson RA, Sharp FR. The stress gene response in brain. Cerebrovasc Brain Metab Rev 1996; 8(2):95-158.
5. Freeman BC, Myers MP, Schumacher R et al. Identification of a regulatory motif in Hsp70 that affects ATPase activity, substrate binding and interaction with HDJ-1. Embo J 1995; 14(10):2281-2292.
6. Foster JA, Brown IR. Differential induction of heat shock mRNA in oligodendrocytes, microglia, and astrocytes following hyperthermia. Brain Res Mol Brain Res 1997; 45(2):207-128.
7. Vass K, Welch WJ, Nowak TS, Jr. Localization of 70-kDa stress protein induction in gerbil brain after ischemia. Acta Neuropathol (Berl) 1988; 77(2):128-135.
8. Kinouchi H, Sharp FR, Hill MP et al. Induction of 70-kDa heat shock protein and Hsp70 mRNA following transient focal cerebral ischemia in the rat. J Cereb Blood Flow Metab 1993; 13(1):105-115.
9. Soriano MA, Planas AM, Rodriguez-Farre E et al. Early 72-kDa heat shock protein induction in microglial cells following focal ischemia in the rat brain. Neurosci Lett 1994; 182(2):205-207.
10. Nowak TS, Jr., Jacewicz M. The heat shock/stress response in focal cerebral ischemia. Brain Pathol 1994; 4(1):67-76.

11. Armstrong JN, Plumier JCL, Robertson HA et al. The inducible 70,000 molecular/weight heat shock protein is expressed in the degenerating dentate hilus and piriform cortex after systemic administration of kainic acid in the rat. Neuroscience 1996; 74(3):685-93.
12. Zhang X, Boulton AA, Yu PH. Expression of heat shock protein-70 and limbic seizure-induced neuronal death in the rat brain. Eur J Neurosci 1996; 8:1432-1440.
13. Sharp FR. Stress proteins are sensitive indicators of injury in the brain produced by ischemia and toxins. J Toxicol Sci 1995; 20(4):450-453.
14. Gass P, Prior P, Kiessling M. Correlation between seizure intensity and stress protein expression after limbic epilepsy in the rat brain. Neuroscience 1995; 65(1):27-36.
15. Chopp M, Li Y, Dereski MO et al. Neuronal injury and expression of 72-kDa heat-shock protein after forebrain ischemia in the rat. Acta Neuropathol (Berl) 1991; 83(1):66-71.
16. Yang YL, Lin MT. Heat shock protein expression protects against cerebral ischemia and monoamine overload in rat heatstroke. Am J Physiol 1999; 276(6 Pt 2):H1961-H1967.
17. Parsell DA, Lindquist S. Heat shock proteins and stress tolerance. In: Morimoto RI, Tissieres A, Georgopoulos C, eds. The Biology of Heat shock proteins and Molecular Chaperones. New York: Cold Spring Harbor Laboratory Press, 1994:457-494.
18. Yang YL, Lu KT, Tsay HJ et al. Heat shock protein expression protects against death following exposure to heatstroke in rats. Neurosci Lett 1998; 252(1):9-12.
19. Currie RW, Ellison JA, White RF et al. Benign focal ischemic preconditioning induces neuronal Hsp70 and prolonged astrogliosis with expression of Hsp27. Brain Res 2000; 863(1-2):169-181.
20. Bader SB, Price BD, Mannheim-Rodman LA et al. Inhibition of heat shock gene expression does not block the development of thermotolerance. J Cell Phys 1992; 151:56-62.
21. Mosser DD, Caron AW, Bourget L et al. Role of the human heat shock protein Hsp70 in protection against stress-induced apoptosis. Molec Cellular Biol 1997; 17(9):5317-5327.
22. Bellmann K, Jaattela M, Wissing D et al. Heat shock protein Hsp70 overexpression confers resistance against nitric oxide. FEBS Lett 1996; 391(1-2):185-188.
23. Jaattela M, Wissing D, Kokholm K et al. Hsp70 exerts its anti-apoptotic function downstream of caspase-3-like proteases. Embo J 1998; 17(21):6124-6134.
24. Buzzard KA, Giaccia AJ, Killender M et al. Heat shock protein 72 modulates pathways of stress-induced apoptosis. J Biol Chem 1998; 273(27):17147-17153.
25. Mestril R, Chi SH, Sayen MR et al. Expression of inducible stress protein 70 in rat heart myogenic cells confers protection against simulated ischemia-induced injury. J Clin Invest 1994; 93(2):759-767.
26. Williams RS, Thomas JA, Fina M et al. Human heat shock protein 70 (Hsp70) protects murine cells from injury during metabolic stress. J Clin Invest 1993; 92(1):503-508.
27. Beaucamp N, Harding TC, Geddes BJ et al. Overexpression of Hsp70i facilitates reactivation of intracellular proteins in neurones and protects them from denaturing stress. FEBS Lett 1998; 441(2):215-219.
28. Fink SL, Chang LK, Ho DY et al. Defective herpes simplex virus vectors expressing the rat brain stress-inducible heat shock protein 72 protect cultured neurons from severe heat shock. J Neurochem (1997); 68(3):961-969.
29. Uney JB, Kew JN, Staley K et al. Transfection-mediated expression of human Hsp70i protects rat dorsal root ganglian neurones and glia from severe heat stress. FEBS Lett 1993; 334(3):313-316.
30. Uney JB, Staley K, Tyers P et al. Transfection with Hsp70i protects rat dorsal root ganglia neurones and glia from heat stress. Gene Ther 1994; 1 Suppl 1:S65.
31. Mailhos C, Howard MK, Latchman DS. Heat shock proteins hsp90 and Hsp70 protect neuronal cells from thermal stress but not from programmed cell death. J Neurochem 1994; 63(5):1787-1795.
32. Wagstaff MJ, Smith J, Collaco-Moraes Y et al. Delivery of a constitutively active form of the heat shock factor using a virus vector protects neuronal cells from thermal or ischaemic stress but not from apoptosis. Eur J Neurosci 1998; 10(11):3343-3350.

33. Sato K, Saito H, Matsuki N. HSP70 is essential to the neuroprotective effect of heat-shock. Brain Res 1996; 740(1-2):117-123.
34. Plumier JC, Ross BM, Currie RW et al. Transgenic mice expressing the human heat shock protein 70 have improved post-ischemic myocardial recovery. J Clin Invest 1995; 95(4):1854-1860.
35. Radford NB, Fina M, Benjamin IJ et al. Cardioprotective effects of 70-kDa heat shock protein in transgenic mice. Proc Natl Acad Sci USA 1996; 93(6):2339-2342.
36. Marber MS, Mestril R, Chi SH et al. Overexpression of the rat inducible 70-kD heat stress protein in a transgenic mouse increases the resistance of the heart to ischemic injury. J Clin Invest 1995; 95(4):1446-1456.
37. Wagstaff MJ, Collaco-Moraes Y, Smith J et al. Protection of neuronal cells from apoptosis by Hsp27 delivered with a herpes simplex virus-based vector. J Biol Chem 1999; 274(8):5061-5069.
38. Lee J, Yenari M, Sun G et al. Differential neuroprotection from human heat shock protein 70 overexpression in in vitro and in vivo models of ischemia and ischemia-like conditions. Exp Neurol 2001; 170(1):129-139.
39. Nishimura RN, Dwyer BE. Evidence for different mechanisms of induction of HSP70i: A comparison of cultured rat cortical neurons with astrocytes. Brain Res Mol Brain Res 1996; 36(2):227-239.
40. Papadopoulos MC, Sun XY, Cao J et al. Over-expression of HSP-70 protects against combined oxygen-glucose deprivation. Neuroreport 1996; 7(2):429-432.
41. Xu L, Giffard RG. HSP70 protects murine astrocytes from glucose deprivation injury. Neuroscience Letters 1997; 224(1):9-12.
42. Rajdev S, Hara K, Kokubo Y et al. Mice overexpressing rat heat shock protein 70 are protected against cerebral infarction. Ann Neurol 2000; 47(6):782-791.
43. Plumier JC, Krueger AM, Currie RW et al. Transgenic mice expressing the human inducible Hsp70 have hippocampal neurons resistant to ischemic injury. Cell Stress Chaperones 1997; 2(3):162-167.
44. Yenari MA, Minami M, Sun GH et al. Calbindin d28k overexpression protects striatal neurons from transient focal cerebral ischemia. Stroke 2001; 32(4):1028-1035.
45. Yenari M, Dumas T, Sapolsky R et al. Gene therapy for treatment of cerebral ischemia using defective herpes simplex viral vectors. Neurol Res 2001; 23:543-552.
46. Yenari M, Dumas T, Sapolsky R et al. Gene therapy for treatment of cerebral ischemia using defective herpes simplex viral vectors. In: Slikker W, Trembly B, eds. Neuroprotective Agents: 5th International Conference, Ann NY Acad Sci. Vol. 939. New York: New York Academy of Sciences, 2001:340-357.
47. Yenari MA, Dumas T, Sapolsky RM et al. The role of gene therapy in acute cerebral ischemia. In: Krieglstein K, S. Klumpp, eds. Pharmacology of Cerebral ischemia. Stuttgart: Medpharm Scientific Publishers, 2000:457-472.
48. Yenari MA, Fink SL, Sun GH et al. Gene therapy with HSP72 is neuroprotective in rat models of stroke and epilepsy. Ann Neurol 1998; 44(4):584-591.
49. Zhang Z, Kelly S, Sun GH et al. HSP72 gene transfer protects CA1 neurons from global cerebral ischemia by improving Bcl-2 expression. Soc Neurosci Abs 2001; 27:20.11.
50. Kelly S, Uney JB, McCulloch J. Adenovirus HSP70 gene transfer ameliorate damage following global ischaemia. J Cereb Blood Flow Metab 2001; 21(1):S23.
51. Cummings CJ, Mancini MA, Antalffy B et al. Chaperone suppression of aggregation and altered subcellular proteasome localization imply protein misfolding in SCA1. Nat Genet 1998; 19(2):148-154.
52. Chai Y, Koppenhafer SL, Bonini NM et al. Analysis of the role of heat shock protein (Hsp) molecular chaperones in polyglutamine disease. J Neurosci 1999; 19(23):10338-10347.
53. Kobayashi Y, Kume A, Li M et al. Chaperones Hsp70 and Hsp40 suppress aggregate formation and apoptosis in cultured neuronal cells expressing truncated androgen receptor protein with expanded polyglutamine tract. J Biol Chem 2000; 275(12):8772-8778.

54. Stenoien DL, Cummings CJ, Adams HP et al. Polyglutamine-expanded androgen receptors form aggregates that sequester heat shock proteins, proteasome components and SRC-1, and are suppressed by the HDJ-2 chaperone. Hum Mol Genet 1999; 8(5):731-741.
55. Muchowski PJ, Schaffar G, Sittler A et al. Hsp70 and hsp40 chaperones can inhibit self-assembly of polyglutamine proteins into amyloid-like fibrils. Proc Natl Acad Sci USA 2000; 97(14):7841-7846.
56. Cummings CJ, Sun Y, Opal P et al. Over-expression of inducible HSP70 chaperone suppresses neuropathology and improves motor function in SCA1 mice. Hum Mol Genet 2001; 10(14):1511-1518.
57. Hoehn B, Ringer T, Fink S et al. Postischemic overexpression of HSP72 by gene therapy protects striatal neurons after experimental stroke. J Cereb Blood Flow Metab 2001; 21(suppl. 1):S410.
58. Neckers L, Schulte TW, Mimnaugh E. Geldanamycin as a potential anti-cancer agent: Its molecular target and biochemical activity. Invest New Drugs 1999; 17(4):361-373.
59. Zou J, Guo Y, Guettouche T et al. Repression of heat shock transcription factor HSF1 activation by HSP90 (HSP90 complex) that forms a stress-sensitive complex with HSF1. Cell 1998; 94(4):471-480.
60. Lu A, Nee A, Parmentier S et al. Geldanamycin, a HSP90 binding drug, induces heat shock proteins in vivo and in vitro, and protects against in vivo cerebral ischemia. J Cereb Blood Flow Metab 2001; 21(1):S231.
61. Xiao N, Callaway CW, Lipinski CA et al. Geldanamycin provides posttreatment protection against glutamate- induced oxidative toxicity in a mouse hippocampal cell line. J Neurochem 1999; 72(1):95-101.
62. Hu BR, Janelidze S, Ginsberg MD et al. Protein aggregation after focal brain ischemia and reperfusion. J Cereb Blood Flow Metab 2001; 21(7):865-875.
63. Hu BR, Martone ME, Jones YZ et al. Protein aggregation after transient cerebral ischemia. J Neurosci 2000; 20(9):3191-3199.
64. Polla BS, Kantengwa S, Francois D et al. Mitochondria are selective targets for the protective effects of heat shock against oxidative injury. Proc Natl Acad Sci USA 1996; 93(13):6458-6463.
65. Kamii H, Kinouchi H, Sharp FR et al. Prolonged expression of Hsp70 mRNA following transient focal cerebral ischemia in transgenic mice overexpressing CuZn-superoxide dismutase. J Cereb Blood Flow Metab 1994; 14(3):478-486.
66. Kondo T, Murakami K, Honkaniemi J et al. Expression of Hsp70 mRNA is induced in the brain of transgenic mice overexpressing human CuZn-superoxide dismutase following transient global cerebral ischemia. Brain Res 1996; 737(1-2):321-326.
67. Kondo T, Sharp FR, Honkaniemi J et al. DNA fragmentation and Prolonged expression of c-fos, c-jun, and Hsp70 in kainic acid-induced neuronal cell death in transgenic mice overexpressing human CuZn-superoxide dismutase. J Cereb Blood Flow Metab 1997; 17(3):241-256.
68. Feinstein DL, Galea E, Reis DJ. Suppression of glial nitric oxide synthase induction by heat shock: effects on proteolytic degradation of IkappaB-alpha. Nitric oxide 1997; 1(2):167-176.
69. Guzhova IV, Daricva ZA, Melo AR et al. Major stress protein Hsp70 interacts with NF-κB regulatory complex in human T-lymphoma cells. Cell Stress Chaperones 1997; 2(2):132-139.
70. Heneka MT, Sharp A, Klockgether T et al. The heat shock response inhibits NF-kappaB activation, nitric oxide synthase type 2 expression, and macrophage/microglial activation in brain. J Cereb Blood Flow Metab 2000; 20(5):800-811.
71. Silver PA. How proteins enter the nucleus. Cell 1991; 64(3):489-497.
72. Ashkenazi A, Dixit VM. Death receptors: Signaling and modulation. Science 1998; 281(5381):1305-1308.
73. Green DR, Reed JC. Mitochondria and apoptosis. Science 1998; 281(5381):1309-1312.
74. Gabai VL, Meriin AB, Mosser DD et al. Hsp70 prevents activation of stress kinases. A novel pathway of cellular thermotolerance. J Biol Chem 1997; 272(29):18033-18037.

75. Gabai VL, Meriin AB, Yaglom JA et al. Role of Hsp70 in regulation of stress-kinase JNK: implications in apoptosis and aging. FEBS Lett 1998; 438(1-2):1-4.
76. Saleh A, Srinivasula SM, Balkir L et al. Negative regulation of the Apaf-1 apoptosome by Hsp70. Nat Cell Biol 2000; 2(8):476-483.
77. Beere HM, Wolf BB, Cain K et al. Heat-shock protein 70 inhibits apoptosis by preventing recruitment of procaspase-9 to the Apaf-1 apoptosome. Nat Cell Biol 2000; 2(8):469-475.
78. Beere HM, Green DR. Stress management Heat shock protein-70 and the regulation of apoptosis. Trends Cell Biol 2001; 11(1):6-10.
79. Creagh EM, Cotter TG. Selective protection by hsp 70 against cytotoxic drug-, but not Fas-induced T-cell apoptosis. Immunology 1999; 97(1):36-44.
80. States BA, Honkaniemi J, Weinstein PR et al. DNA fragmentation and HSP70 protein induction in hippocampus and cortex occurs in separate neurons following permanent middle cerebral artery occlusions. J Cereb Blood Flow Metab 1996; 16(6):1165-1175.
81. Panahian N, Yoshiura M, Maines MD. Overexpression of heme oxygenase-1 is neuroprotective in a model of permanent middle cerebral artery occlusion in transgenic mice. J Neurochem 1999; 72(3):1187-1203.
82. Yu Z, Luo H, Fu W et al. The endoplasmic reticulum stress-responsive protein GRP78 protects neurons against excitotoxicity and apoptosis: suppression of oxidative stress and stabilization of calcium homeostasis. Exp Neurol 1999; 155(2):302-314.
83. Creagh EM, Carmody RJ, Cotter TG. Heat shock protein 70 inhibits caspase-dependent and -independent apoptosis in Jurkat T cells. Exp Cell Res 2000; 257(1):58-66.
84. Wyatt S, Mailhos C, Latchman DS. Trigeminal ganglion neurons are protected by the heat shock proteins Hsp70 and hsp90 from thermal stress but not from programmed cell death following nerve growth factor withdrawal. Brain Res Mol Brain Res 1996; 39(1-2):52-56.
85. Li GC, Li L, Liu RY et al. Heat shock protein Hsp70 protects cells from thermal stress even after deletion of its ATP-binding domain. Proc Natl Acad Sci USA 1992; 89(6):2036-2040.

III. NEUROTROPHIC FACTORS AS NEUROPROTECTIVE AGENTS

NEUROTROPHINS

Georg Dechant[1] and Harald Neumann[2,3]

SUMMARY

Nerve growth factor was the first identified protein with anti-apoptotic activity on neurons. This prototypic neurotrophic factor, together with the three structurally and functionally related growth factors brain-derived neurotrophic factor (BDNF), neurotrophin-3 (NT3) and neurotrophin-4/5 (NT4/5), forms the neurotrophin protein family. TargeT cells for neurotrophins include many neurons affected by neurodegenerative diseases such as Alzheimer s disease, Parkinson s disease, amyotrophic lateral sclerosis and peripheral polyneuropathies. In addition, the neurotrophins act on neurons affected by other neurological and psychiatric pathologies including ischemia, epilepsy, depression and eating disorders. Work with cell cultures and animal models provided solid support for the hypothesis that neurotrophins prevent neuronal death. While no evidence exists that a lack of neurotrophins underlies the etiology of any neurodegenerative disease, these studies have spurred on hopes that neurotrophins might be useful symptomatic-therapeutic agents. However first clinical trials led to variable results and severe side effects were observed. For future therapeutic use of the neurotrophins it is therefore crucial to expand our knowledge about their physiological functions as well as their pharmacokinetic properties. A major challenge is to develop methods for their application in effective doses and in a precisely timed and localized fashion.

FROM NEUROTROPHIN PHYSIOLOGY TO THERAPY

The human brain undergoes continuous structural remodeling in response to signals originating from inside and outside of the body. At a molecular level these changes can be very subtle and involve minor modifications of synaptic proteins. At

[1]Neurobiochemistry and [2]Neuroimmunology, Max-Planck-Institute of Neurobiology, Am Klopferspitz 18a, 82152 Martinsried,Germany and [3]Neuroimmunology, European Neuroscience Institute G ttingen, Waldweg 33, 37073 G ttingen, Germany.

a cellular level, dendritic spines or nerve cell arborizations are reshaped. Finally, entire nerve cells are newly generated or removed, even in the mature brain (reviewed in ref. 1). In the developing nervous system these changes are particularly drastic and hence are more easily observed and investigated. However, similar or identical molecular mechanisms are still at work in the adult mature brain.

An intricate molecular machinery comprised of many signaling proteins including protein growth factors and their receptors implements these changes in the nervous system. The neurotrophins, a family of protein growth factors, are prime candidates for molecular mediators of such neuronal plasticity. During embryonic and postnatal development these proteins regulate survival, differentiation and specification of neurons (reviewed by ref. 2). In the adult nervous system they modulate or trigger fast synaptic responses and cause changes in the function of synapses and in the morphology of neurites. Consequently they influence higher systemic functions such as behavior, learning, memory formation and cognition (reviewed in ref. 3).

In view of the important physiological function that the neurotrophins exert in the brain and in the peripheral nervous system, they are candidate agents for the treatment of neurological and psychiatric diseases. Hopes in these proteins were furthered by a large number of pre-clinical studies showing that most of the neuronal populations affected by neurodegenerative diseases respond to neurotrophins and express neurotrophin receptors, although there is no direct proof that a lack of neurotrophins is causally involved in the manifestation of any neurodegenerative pathology. Neurotrophins could inhibit or delay degenerative processes in neurons under numerous experimental conditions. Furthermore, the expression levels of neurotrophins and their receptors are strongly regulated in pathophysiological situations arguing for their involvement in the cellular responses to pathological processes.

Unfortunately the high therapeutic expectations in neurotrophins resulted in insufficient pre-clinical preparations of clinical trials (reviewed in ref. 4). In particular the need for precisely controlled, timed and localized delivery as well as the potential side effects have apparently been underestimated. In the following, we first describe the neurotrophin molecules, their receptors and the signaling pathways triggered by them. We then briefly summarize the evidence for neuroprotective effects of the neurotrophins with an emphasis on in vivo studies. In this context we give reference to the initial clinical trials performed with neurotrophins as well as to the problems encountered therein. Finally we attempt to point to possible solutions for these problems as an outlook.

STRUCTURE AND PHYSIOLOGICAL FUNCTIONS OF NEUROTROPHINS AND THEIR RECEPTORS

During the last decade, the knowledge of how cell survival and death is controlled has expanded at an ever increasing rate. In each cell, an intrinsic cell suicide program exists that is under constant control by cell-intrinsic and extrinsic stimuli.[5] Also, the survival and death of neurons depends strongly on extrinsic signaling by neighboring or interacting cell types. Many of these signals are soluble growth factors such as the neurotrophins nerve growth factor (NGF), brain-derived neurotrophic factor (BDNF), neurotrophin-3 (NT-3), and neurotrophin-4/5 (NT-4/5) (Fig. 1). These neurotrophic factors signal by binding to two different types of transmembrane receptor proteins on responsive neurons. One class of neurotrophin receptors are the tropomyosin receptor kinase (Trk) proteins that form a small subfamily of the large family of receptor tyrosine kinases (Fig. 2). The second molecular entity that mediates neurotrophin signaling is the p75 neurotrophin receptor, a protein that binds all neurotrophins and is a member of the tumor necrosis factor receptor (TNF-R) family (Fig. 3).

Our view of the precise interplay between the neurotrophins and their receptors as well as of the receptor mediated cellular signals in neurons is still fragmentary. In order to gain a better understanding of how neurotrophins protect neurons under pathophysiological conditions it is therefore of central importance to study the neurotrophin receptors and their intracellular signaling first under normal conditions. One important and largely enigmatic observation is that the response of neurons to neurotrophins can be very diverse depending on input from other signaling pathways and changes during development. Neurotrophin signaling is therefore often referred to as dependent on the "cellular context". Consequently, results obtained with one type of neuron cannot reliably (and should not) be extrapolated onto other cells. As it will be discussed below, such an extrapolation of the effects of NGF on sensory neurons and of BDNF on spinal motoneurons had a negative impact on the first clinical studies with BDNF in amyotrophic lateral sclerosis (ALS).

The Neurotrophins

The neurotrophins are four distinct small basic proteins (Fig. 1a). In each case the active growth factor protein is a homodimer that is formed by identical peptide chains of about 120 amino acids each. The two monomers cohere by non-covalent chemical bonds. Highly conserved pairs of cystein residues form a so-called cystein knot motif within each monomer that stabilizes the conformation of the proteins.[6] Whereas the overall structure of the four neurotrophin-molecules is very similar, differences between them become apparent upon closer inspection (Fig. 1b). Each molecule is uniquely characterized by a specific pattern of charged basic or acidic residues exposed on its surface. Differences between the individual neurotrophins are particularly evident within peptide loops protruding from the core of the molecules that is itself formed by beta sheets (reviewed in ref. 7). These loops mediate the main contacts between the neurotrophins and their specific receptors on the surface of responsive cells. This blend of structural similarities and differences in the structure of the molecules is reflected by the biological properties of the neurotrophins (reviewed in ref. 8). The specific molecular differences between the neurotrophins not only determine their receptor interactions, but also profoundly influence their

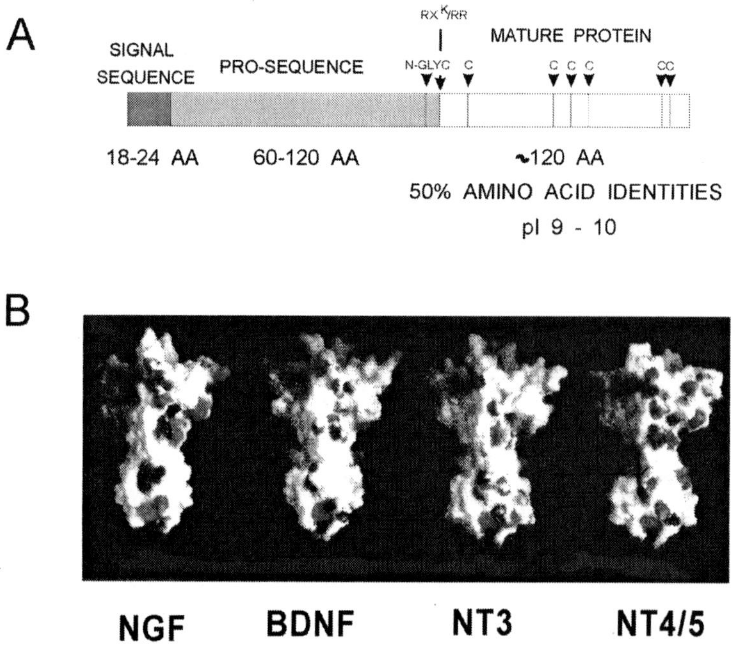

Figure 1. Structure of the neurotrophins.

A) The neurotrophins are a family of protein growth factors with neurotrophic (nerve nourishing) biological activity. Other growth factor families with similar neurotrophic functions include fibroblast growth factors, transforming growth factor beta homologues, insulin like growth factors and neuropoietic cytokines (not shown). The neurotrophin growth factor family consists of four proteins: nerve growth factor (NGF), brain-derived neurotrophic factor (BDNF), neurotrophin-3 (NT3) and neurotrophin-4 (also known as neurotrophin 4/5, NT4/5). The neurotrophins have a characteristic protein structure that distinguishes them from other growth factor families. The carboxyterminal- biologically active part of the proteins is about 50% identical in all 4 neurotrophins (depicted in yellow). Identities include a characteristic pattern of 6 cystein residues (C) that form three conserved cystein bridges. The aminoterminal half of the proteins is functional in the biosynthesis of the proteins and includes a signal peptide (red) mediating membrane transport and a pro-peptide (orange) that appears important for protein folding. The mature monomer is released from the pro-sequence by proteolytic cleavage at a conserved Lysine (K)/Arginine (R) rich sequence.

B) In their mature form the neurotrophins form homodimers with a characteristic elongated protein fold of β-sheets and protruding loops (McDonald and Chao 1995). Ligand-receptor interactions are determined by binding of the region of the ligands that are specific for each of the four neurotrophins. Colours depict the charged residues exposed on the surface of the molecules (basic residues in blue, acidic residues in red). This pattern of negative and positive charges is characteristic for each neurotrophin and determines interactions with specific transmembrane receptors.

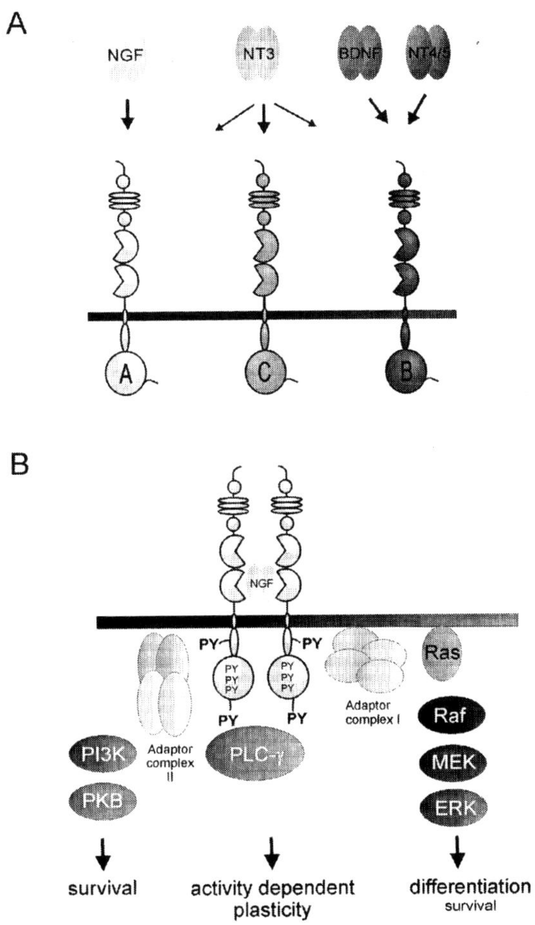

Figure 2. The Trk family of neurotrophins transmembrane receptors.
A) All neurotrophins bind to one or more receptor of the Trk protein family. (tropomyosine receptor kinase, TrkA, B, C) Trk receptors are a small subfamily consisting of three members out of the large superfamily of several hundred proteins with tyrosine kinase enzymatic activity that also includes receptors for EGF, insulin, FGF, PDGF and numerous other growth factors. The Trk receptor tyrosine kinases are characterized by a specific combination of structural motives: In their extracellular domains: three tandem repeat leucine-rich motives are flanked by two cystein clusters. The main contacts between the Trk receptors and their ligands occur within two Ig-like C2 type domains. The protein sequences of the intracellular, enzymatically active, tyrosine kinase domains are highly conserved in the three receptors. The neurotrophins bind to the Trk receptors with strong (thick arrows) and weak interactions (thin arrows). Both strong and weak (also sometimes called preferred and non-preferred) interactions are relevant in vivo.
B) Simplified sketch of Trk receptor signaling. Upon dimerization several tyrosine residues become phosphorylated (PY) in the intracellular domains of the Trk receptors. This enables association of signaling molecules, in particular various adaptor proteins to the receptors that in turn trigger different cellular responses, including differentiation and neurite outgrowth, promotion of survival through blockade of apoptotic pathways and regulation of activity dependent plasticity. Duration and extent of signaling pathway activation depends on the type of cell and on the Trk receptor activated. The pathways cross talk to each other (not shown).

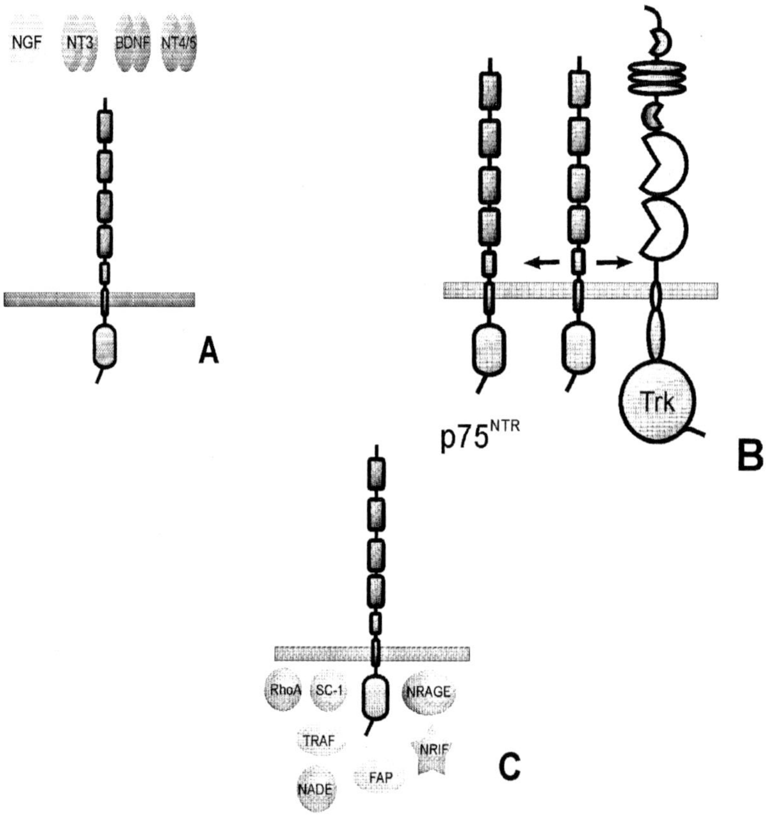

Figure 3. Protein interaction of the p75 neurotrophin receptor
The second type of neurotrophin receptor is a molecule called p75. This protein, previously also known under the names NGF-receptor or low-affinity neurotrophin receptor, binds all four neurotrophins. Its protein structure defines it as a member of the TNF-R/Fas family of transmembrane receptors.
A) Binding of its neurotrophin ligands occurs within the four cystein-rich domains in its extracellular part. The kinetics of p75 binding is slightly different for all four neurotrophins and is strongly influenced by the cellular environment. Both binding and functional experiments show that p75 discriminates between its ligands and triggers intracellular signaling pathways differently dependent on which ligand is associated.
B) p75 forms complexes with itself and with all three Trk receptors. While the functional significance of the formation of homomeric p75 complexes is currently not understood, biochemical experiments with the Trk receptors have demonstrated, that p75 association can modulate the interactions between the Trk receptor and their ligands and Trk receptor auto-phosphorylation.
C) The intracellular domain of p75 remotely resembles the death domain found in several members of the TNF-R/Fas family. In contrast to the enzymatically active Trk receptor, p75 signaling appears to be mediated merely by association of intracellular proteins (for a review of the individual molecules see ref. 3). The subsequent signaling events, which are only partly understood, can trigger cell death, cell cycle arrest and local changes in neurite formation or retraction.

biophysical and pharmacokinetic behavior. The neurotrophins have a tendency to stick unspecifically to surfaces, for example to cellular membranes. Their stickiness depends on a mixture of hydrophobic and hydrophilic interactions and is an important parameter for their use as therapeutics, since it is a limiting factor for their diffusion through organs and tissues such as the brain parenchyma. The diffusion behavior is different for each of the neurotrophins, for example NGF[9] shows a higher diffusion rate compared to BDNF.[10,11]

Heterodimers of two different neurotrophin monomers can be formed in test tubes but were not detected under physiological conditions.[12,13] The generation and application of synthetic heterodimers is an interesting pharmacological approach that might broaden the spectrum of responding neurons. For example, for the treatment of sensory neurons in the peripheral nervous system heterodimers between NGF and NT3 might act simultaneously on NGF and NT3 responsive subpopulations (see below). Another interesting aspect of neurotrophin molecules is that their biosynthesis and secretion in the mature nervous system is controlled by neuronal activity. Frequently the neurotrophins seem to be made available to the surrounding tissue on demand. This mechanism enables the neurotrophins to adjust size and function of nerve cell populations to the precise requirement of the structures they innervate (reviewed by ref. 14). As discussed in the following sections, it will be important for the future therapeutic use of neurotrophins to devise strategies for a similarly controlled exposure of diseased neurons to these factors in order to minimize adverse side effects, as they have been observed following systemic treatment.

The Neurotrophin Receptors

The membrane receptors for neurotrophins comprise three receptor tyrosine kinases belonging to the tropomyosin receptor kinases (Trk) family (reviewed by ref. 15) and p75, a molecule that groups within the large family of TNF-receptor homologs (reviewed by ref. 16). The molecular interplay between these receptors is complex and incompletely understood,[17] but as a net result it triggers cellular effects, which are highly ligand and cell type specific. The interactions of neurotrophins with their receptors are frequently analyzed on cell lines overexpressing neurotrophin receptors in a recombinant form. In these highly defined in vitro approaches strong binding interactions and strong activation of intracellular signaling pathways were observed which reflect in principle the mechanism occuring in primary nerve cells (reviewed in refs. 15,18). However, it should be kept in mind that the response of primary neurons to neurotrophins is under the control of fine-tuned cell specific parameters. In particular the antagonistic view that Trk receptors stimulate survival, whereas p75 signaling triggers apoptosis is certainly an over-simplified perspective.[17] Another context-dependent and variable parameter is the affinity of neuronal receptors for their neurotrophin ligands, which can vary by several magnitudes between 10^{-12} M and 10^{-8}M. The frequently expressed statement that Trk receptors form binding sites of high affinity for their neurotrophin ligands, whereas p75-binding sites are of low affinity is not correct. It derives from biochemical in vitro studies and is not thoroughly backed up by studies with primary neurons. As a matter of fact, data on the binding properties of defined receptor molecules on neurons in situ are sparse especially in the CNS. Therefore a precise dose response curve has to be established for each cell type as a prerequisite for the establishment of treatment paradigms.

The spectra of responsive neurons are distinct and characteristic for each neurotrophin ligand, but they are also partly overlapping. For example, cholinergic forebrain neurons respond strongly to NGF and weakly to BDNF, whereas spinal motoneurons respond strongly to BDNF but are negatively affected by NGF (see below). Cellular responses to activation of different neurotrophin receptors by the same ligand can be very diverse: for example NGF can stimulate the survival of some cell types while it kills others (reviewed by ref. 19). Compared to other growth factor families, which typically have a much broader range of targeT cells including numerous non-neuronal cell types, neurotrophins act predominantly on a restricted number of neuronal populations. This should be an advantage for their therapeutic use.

The interactions between the neurotrophins and their receptors are best analyzed on the sensory neurons on the dorsal root ganglia (DRG). A large number of biochemical, cell culture and genetic studies were performed with these cells. The expression pattern of the Trk-receptors was analyzed in the various sub-populations of sensory DRG neurons and was compared to the biological effects of their ligand and the results revealed an excellent correlation between the Trk receptor expression and responsiveness of subpopulations of DRG neurons. For example, the large sized, proprioceptive neurons express TrkC and respond to NT3 while small sized nociceptitive neurons express TrkA and depend on NGF. These findings were most convincingly corroborated in vivo by the analysis of mice carrying targeted mutations in receptor or ligand genes (reviewed by ref. 20).

In the peripheral nervous system (PNS), for almost all neuronal subpopulations expressing Trk receptors, biological functions could be defined for the corresponding ligands. By contrast, the correlation between Trk receptor expression and responsiveness is less obvious in the brain. One clearly defined case is that of the cholinergic forebrain neurons, which, like a small number of non-cholinergic nuclei express the NGF receptor TrkA.[21] Consistently, cholinergic forebrain neurons respond strongly to NGF.[22] By contrast, the BDNF and NT4/5 receptor TrkB is widely expressed in the mature CNS and is highly developmentally regulated.[23] The finding that the TrkB gene is expressed in various protein isoforms further complicates a direct correlation between ligand function and TrkB receptor expression. The individual TrkB variants, such as TrkB.T1, with a truncatation in the intracellular domain, are generated by alternative splicing of the primary gene transcript and are regulated independently of each other.[24,25] The situation for TrkC is complicated too. It is expressed in a variety of CNS structures preferentially during early ontogenetic development and down-regulated in the healthy mature brain.[26]

The second type of neurotrophin receptor is the p75 neurotrophin receptor, a 75 kDa glycoprotein. Despite having been the subject of a large number of recent studies, its function has only begun to be understood (reviewed by refs. 27,28). Studying this receptor is rendered difficult by the lack of a simple and reliable experimental read-out system for its activity. This is in contrast to the enzymatically active receptor of the Trk family, whose activity can easily be assessed by analysis of their tyrosine phosphorylation. The p75 neurotrophin receptor triggers signaling by non-covalent binding to intracellular molecules (Fig. 3C). The cellular effects of its signaling cascade are diverse. In certain biological structures, including neuronal precursor populations in the retina and spinal cord, p75 can cause the activation of apoptotic cell death (reviewed in ref. 29). Furthermore, it exerts local regressive effects on neuritic growth.[30] Besides its role as an independently signaling receptor, p75 also binds to all three Trk receptors and this physical interaction can modify the tyrosine kinase activity as well as the ligand specificity of the Trk receptors.[31] In a reciprocal Fashion the Trk receptors have a blocking effect on the downstream signaling of p75.[32] The diverse biochemical and functional interactions between the Trk and p75 neurotrophin receptors render them a demanding object for future studies. Yet evidence is accumulating that the fine-tuning of these ligand-receptor interactions critically determines the distinct cellular responses. Hence the neurotrophin receptors should not merely be considered as simple mechanical triggers of intracellular signaling pathways. Consequently we have to realize, that based on our current fragmentary knowledge, the response of a cell to neurotrophin treatment cannot always be predicted, even when the expressed set of receptors has been determined.

Evidently, the biochemical properties of the cellular neurotrophin receptors are important for neurotrophin therapy. In vivo work has revealed that the effects of neurotrophins are very dose sensitive and strongly depend on the appropriate concentration.[33] Treatment regimes should be adjusted to these findings, for example by developing spaced versus chronic infusion protocols.[33,34] The underlying molecular mechanisms remain to be determined. One potential explanation could be down-regulation of the receptors. In particular the biological effects of BDNF at supraoptimal doses might be hampered by a specific molecular property of its receptor TrkB. Exposure to excess BDNF causes rapid internalization of TrkB followed by subsequent proteolytic degradation of the receptors.[35] Consequently the neuron is desensitized for BDNF after prolonged periods of treatment.[36] Desensitization is observed at different levels for the neurotrophin receptors and depends on so far unresolved molecular properties.[37,38] In contrast to TrkB, TrkA is not downregulated. Consequently, TrkA phosphorylation is detected for a much longer time period after NGF injection into the brain and the treatment leads to a prolonged stimulation of receptor expression.[21,39] The difference in cell membrane cycling between the TrkA and TrkB receptors was attributed to a short peptide stretch in their intracellular domains.[35] Within the brain the truncated T1 isoform of the TrkB receptor appears to regulate the liquor concentrations of BDNF. It is strongly expressed in the ependymal cells and mediates internalization of BDNF.[40] TrkB.T1 might form a ligand-specific removal or sequestering system for BDNF and NT4/5, but not for NGF potentially explaining the differences in diffusion of neurotrophins in the brain.

All neurotrophins interact with more than one functional receptor with different kinetics and affinities (Figs. 2a and 3a). At low, presumably physiological, concentrations neurotrophins bind to their preferred receptor and exert very specific effects. At elevated doses, their receptor spectrum is changed and consequently the biological response becomes less defined. Excessively applied neurotrophins can lead to adverse toxic effects through various receptor dependent mechanisms such as receptor competition or free radical formation.[41-43] The degree of receptor promiscuity is different for each of the neurotrophins and involves both Trk and p75 neurotrophin receptors. In fact NT3, the most promiscuous ligand, binds to all known neurotrophin receptors. Yet it would be wrong to call these interactions unspecific since evidence is accumulating that such non-preferred interactions are indeed physiological.[44,45] Obviously the variety of cellular responses at elevated neurotrophin concentration emphasizes the need for a precise control of an exposure regime.

Neurotrophin Signaling Mechanisms

Of all trk receptors the signaling cascades originating at the NGF receptor TrkA are best understood (reviewed by ref. 46), whereas our knowledge about the signaling of TrkB and TrkC is much less detailed.[47] In many aspects Trk receptors are typical tyrosine kinases. Receptor dimerization upon ligand binding leads to autophosphorylation of the dimeric receptor complex at several defined tyrosine kinase residues.[48] Three major intracellular signaling pathways originating at TrkA have been identified in biochemical and more recently in genetic experiments (Fig. 2b). The first pathway leads to the activation of the ERK MAP-kinase signaling module through a cascade that includes Ras. This pathway is primarily triggered by the association of Grb2/SOS complexes to phosphorylated adaptor proteins, most prominently Shc or FRS2. The activation of this pathway initiates predominantly differentiation and neuritogenesis, but also neuronal survival (reviewed by ref. 49). The second pathway stimulates predominantly neuronal survival by activation of protein kinase B (PKB)/AKT, which subsequently leads to inactivation of the pro-apoptotic protein Bad and to phosphorylation of the forkhead transcription factor.[50] This anti-apoptotic pathway is activated by TrkA through its association with the large multi-adaptor proteins insulin receptor substrates 1 (IRS1) and IRS2, but also Gab-1.[51,52] The IRS proteins then mediate PI3-kinase activation. The third pathway involves binding of PLC-γ to the carboxyterminal tyrosine residue 785 of TrkA. Activated PLC-γ stimulates production of the lipid second messenger IP3. The function of this neurotrophin-induced pathway is less defined compared to the other two. It appears to be important for synaptic plasticity and neurotrophin mediated neurotrophin release.[53] Furthermore it also provides a potential link between the signaling of neurotrophins and electrical activity in the CNS through the control of intracellular Ca^{2+} levels. Physiological neuronal activity stimulates survival responses, while lack of activity promotes neuronal cell death. For example, blockade of preganglionic transmission onto sympathetic neurons increases the number of dying neurons during development.[54] In the CNS, reduced neuronal survival has been observed in the developing and adult brain following blockade of electric activity.[55] Neurotrophin mediated changes in cortical dendritic morphology also require neuronal activity.[56]

It is important to note that the cellular responses to signaling of the three trk receptors can be surprisingly diverse although these proteins share extensive primary sequence identities in their intracellular enzymatic domains.[57] Hence, the same concept of diversity and similarity that applies to the structure and function of the ligand also is applicable to the receptors. The molecular explanation for the observed differences of Trk receptor signaling might reside in the differential use of intracellular adaptor proteins homologous to IRS, Shc or Gab-1. Distinct association kinetics of these molecules and binding to different consensus sites in the receptors may explain the observed specificity of Trk receptor signaling.

Our knowledge about the intracellular signaling mechanisms of the neurotrophin receptor p75 remain largely elusive. Recent data indicate that p75 mediates local effects on neurite outgrowth via interference with the activity of the small GTPase RhoA.[30] Another p75-triggered pathway leads, via ceramide production by sphingomyelin hydrolysis,[58,59] to the activation of the transcription factor NFκb.[60] This pathway might serve as an anti-apoptotic signal balancing the activation of cell death cascades.[61,62] That p75 activation can also actively kill cells is an important consideration for potential therapeutic use of neurotrophins as neuroprotective agent.[63,64] The precise composition of the cell death pathway triggered by p75 remains to be established. Biochemical and genetic evidence involves several of the recently identified protein interactors (reviewed by ref. 3). Downstream effects may include caspase 3 activation,[65] activation of the JNK MAPK- module[66] as well as activation of p53 and p73.[67,68]

Cellular Neuroprotective Mechanisms

It may be expected that neurotrophins would have two beneficial effects in neurodegenerative diseases. First, these proteins might enhance long term survival of damaged neurons and second, they might maintain or re-induce the physiological functional status of affected neurons.

A core problem for the treatment of neurological diseases is the incomplete knowledge about the underlying pathophysiological processes. It appears that neurons are killed by both necrotic and apoptotic mechanisms (see reviews in refs. 69,70). Interestingly, the same neurotrophic extracellular stimuli that regulate ontogenetic cell death and survival appear to be re-activated under pathophysiological situations in the mature nervous system. Therefore neurotrophins are valuable experimental tools for the study of neuroprotective mechanisms. They not only exert their activity on specific neuronal populations, they also prevent apoptotic or excitotoxic neuronal death by activating multiple signaling pathways. In various models of disease neurotrophins activate the tyrosine kinase receptors TrkA, TrkB and TrkC in order to prevent degeneration of injured mature CNS neurons in models of chronic degenerative diseases as well as in lesion paradigms. However, in many experimental conditions it is currently not known whether neurotrophin treatment interferes with apoptosis, necrosis or even directly with specific neuropathological mechanisms such as the generation of reactive oxygen species or excitatory cell death.[71,72] For neurodegenerative diseases it remains a matter of debate to which degree the bcl2/caspase3 mechanism of apoptosis that regulates ontogenetic cell death is also causing pathophysiological neuronal death.[73,74] The neuroprotective action of neurotrophins on neurons damaged by excitotoxins, toxic chemical agents or injury indicates that these factors also interfere with necrotic mechanisms (reviewed by ref. 75). Hence it appears that neurotrophins activate different cellular mechanisms to protect neurons in a cell specific manner.

Neurotrophins have broader biological effects on neurons than just regulation of cell death. They also regulate subcellular processes underlying neuronal plasticity and maintenance of structural neuronal integrity. Since neurodegenerative diseases are accompanied by substantial structural damage and reactive plasticity, neurotrophic therapy might be advantageous also in this regard. As outlined above, the effects of neurotrophins on survival and neuronal morphology are mediated by distinct intracellular signaling pathways and consequently these effects can be separated in animal models of neuronal cell death in vivo.[76,77] In the cerebral cortex neurotrophins stimulate both growth of and retraction of dendritic branches indicating that the effect of neurotrophins on neuronal morphology is not invariably growth promoting.[78] Again this observation emphasizes the necessity to further study the role of neurotrophins in maintenance and refinement of neuronal structures and at the same time it argues against systemic neurotrophin treatment, since the potential of interference of exogenously applied neurotrophins with functional cortical circuitries has to be a major concern (see below). In this context it is of interest to note that neurotrophins are themselves expressed in many neuronal subtypes and their production and secretion is itself regulated by exogenous neurotrophin supply.[79-81] Transneuronal induction of neurotrophin gene expression is well documented in the autonomous nervous system and fulfills physiological functions there.[82] Similar phenomena are likely to exist also inside the CNS, for example in the corticospinal tract.[83] The possibility of transneuronal signaling also argues against systemic neurotrophin treatment.

NEUROTROPHINS IN ANIMAL MODELS OF PATHOLOGICAL SITUATIONS AND CLINICAL TRIALS

Much of the hope that neurotrophins might be useful therapeutics is based on the dynamically regulated expression of these factors and their receptors in animal models of neurological diseases. Neurotrophins strongly inhibit or delay degenerative processes in a variety of in vivo and in vitro models of human neurodegenerative disease such as Alzheimer s disease, Parkinson s disease, amyotrophic lateral sclerosis and peripheral sensory neuropathy. Consequently neurotrophins have been used for clinical trials with the rationale that these proteins would inhibit degeneration of nerve cells and lead to re-establishment of lost synaptic connections. However, despite the convincing success in animal models of disease, clinical trails with neurotrophins were initially disappointing.

Cortical and Hippocampal Neurons

Strong neuroprotective effects have been observed with neurotrophins in models of acute neuronal injury. BDNF and TrkB are up-regulated as a part of the endogenous response to cerebral ischemia adjacent to the lesion[84,85] and the infarct size is reduced by application of BDNF or NT4/5.[86-88] Neurotrophins and their receptors are strongly and differentially regulated after induction of seizures[89-91] and soluble TrkB fusion proteins inhibit kindling development.[92] In lesion models, BDNF protects neurons of the corticospinal tract[93] and NT3 potentiates the axonal growth promoting effect of the IN1 antibody after corticospinal tract lesion.[94]

Cholinergic Forebrain Neurons

All major neuronal subpopulations affected by Alzheimer s disease, in particular cholinergic cells in the basal forebrain, but also neurons of the hippocampus, cerebral cortex and the locus coeruleus, respond to at least one neurotrophin. The cholinergic deficit in the CNS might represent an ideal target for neurotrophin therapy, since most of the cholinergic neurons of the Nucleus Meynert complex respond very selective to NGF.[95] The NGF receptor TrkA is expressed in most cholinergic neurons in the forebrain and a limited number of neurons in other non-cholinergic nuclei.[96,97] In experiments with rodents and subsequently with primates, a protective effect of NGF against atrophy was demonstrated on cholinergic neurons after fimbria-fornix lesion.[98] NGF prevents cholinergic atrophy not only if experimentally induced by a lesion, but also in aged animals. These cellular effects are paralleled by behavioral improvements of aged animals, whereas in normal younger animals NGF surprisingly had an adverse effect.[99-101] A comparison with other growth factors shows that NGF is the most efficacious neurotrophin acting on cholinergic neurons.[102-104] However BDNF and NT4/5 also exerts significant neuroprotective effects on cholinergic forebrain neurons.[105,106] Interestingly, BDNF gene expression levels are reduced in Alzheimer patients, concomitant with a loss of cholinergic basal forebrain neurons, although the causal relation to neurodegeneration remains open.[107] NGF levels are increased in Alzheimer patients. Most likely it accumulates in the cortex as a consequence of the reduction in retrograde transport and removal by cholinergic fibers. There has been an initial report that mice with a partial p75 gene deletion scientific have a higher number of cholinergic neurons in the medial septum. This report caused a scientific dispute and later on had to be retracted.[108,109] Recent results obtained with a novel mutant p75 mouse line generated in our laboratory indicate that p75 signaling can cause the death or atrophy of developing cholinergic neurons in the medial septum. A careful anatomical and histochemical analysis of ChAT-immunoreactive neurons revealed that mice in which the p75 gene is completely inactivated have a highly significant increase in cell number at postnatal day 15.

Many of the surplus neurons are maintained for at least 3 months (T. Naumann, E. Casademunt, personal communication). It will be interesting to see, whether this increase in ChAT positive neurons leads to detectable behavioral changes in these animals.

Two studies report on the effect of NGF applied by intraventricular injection to patients suffering from Alzheimer s disease. In both reports some beneficial effects were observed, which were opposed, however, by adverse side effects such as weight loss and induction of painful sensations.[110,111] Hence it appears that this form of NGF application is not well suited for treatment of Alzheimer s disease.

Mesencephalic Dopaminergic Neurons

BDNF and NT4/5 support the survival of dopaminergic neurons isolated from the substantia nigra in cell cultures and protect them against 6-OHDA and MPP^+ toxicity.[112,113] However, in lesion experiments in adult animals no protection by neurotrophins against the loss of dopamine was observed following MPTP application.[114] On adult axotomized cells BDNF and NT3 promote survival and differentiation of dopaminergic neurons.[115] In unlesioned animals chronic injection of BDNF into the substantia nigra lead to conflicting results with either hypo- or hyperfunction of the dopaminergic system indicating complex interactions and a potential for significant side-effects.[116,117] Furthermore, the expression of the BDNF/ NT4/5 receptor TrkB expression is low under physiological conditions in the cells of the substantia nigra.[118,119]

Spinal and Facial Motoneurons

Spinal motoneurons express the BDNF receptors TrkB[120] and mice with a targeted deletion of the TrkB gene have less motoneurons.[121] These results are in conflict with the analysis of mice carrying a targeted deletion in the TrkB ligands. Mice carrying a targeted mutation in the genes encoding BDNF and NT4/5 show no reduction in motoneuron number.[122,123] Motoneuron axotomy increases BDNF and TrkB gene expression in motoneurons.[124] In neuronal cell cultures, motoneuron survival is mediated by both NT4/5 and BDNF.[125] In vivo pharmacological administration of BDNF rescues embryonic motoneurons from naturally occurring death and axotomized neonatal motoneurons are also rescued by BDNF in vivo.[126-128] Most notably, adult rat motoneurons in the facial nucleus can be rescued with BDNF after axotomy, indicating that mature neurons maintain their responsiveness to this neurotrophin.[129] The common neurotrophin receptor p75 is expressed on embryonic motoneurons and on mature cells after injury.[130,131] During early development p75 triggers cell death of some immature neurons in the spinal cord.[132] Therefore, it should be considered that neurotrophins promote not only cell survival but also could trigger regressive phenomena on motoneurons as described above for cholinergic neurons in the basal forebrain. In newborn rats increased cell death is observed after application of NGF.[126] This NGF mediated cell death is dependent on p75 and likely to be caused by ligand competition for the binding to functional Trk receptors.[42] The effect of the absence of p75 receptor on the regeneration of adult axotomized facial motoneurons has recently been tested in our complete knock out animals. No differences were observed in neuronal survival and speed of axonal regeneration (A. Gschwendtner, G. Raivich, personal communication), which is in contrast to a previous report that describes improved motoneuron regeneration in the absence of full-length p75.[133]

No indication exists that the degeneration of motoneurons in the anterior horn and in the brainstem in amyotrophic lateral sclerosis is linked to a lack of trophic support. Nevertheless the strong response during development and in lesion paradigms of spinal motoneurons to growth factors such as insulin-like growth factor 1 (IGF1), glial cell line derived neurotrophic factor (GDNF), ciliary neuronotrophic factor (CNTF) and in particular BDNF renders ALS a good candidate disease for neurotrophic therapy.[134] However, a clinical phase III trial of systemically applied BDNF in ALS was recently abandoned. It is questionable whether therapeutically effective concentrations of BDNF have reached the soma of spinal motoneurons in this clinical trial. Motoneurons of all central neurons were considered attractive targeT cells since their axons in the periphery were thought to have free access to growth factors applied systemically. Subsequently those factors could be retrogradly transported and exert their activity on the cell soma in the spinal cord. The hypothesis of such a retrograde activity of BDNF is mainly based on the findings that sensory and sympathetic neurons effectively take up and transport the neurotrophin NGF.[135] However, the retrograde transport of NGF might be strongly facilitated by some of the specific properties of peripheral NGF-responsive neurons, which are characterized by unmyelinated fibers and free nerve endings. By contrast the BDNF-sensitive motoneurons have thickly myelinated fibers and synapses deeply buried in the muscle. While BDNF is effectively retrogradely transported by motoneuron axons upon injection into the sciatic nerve,[136] motoneuron endplates are very poor entry points for systemically applied BDNF. Intraventricular injection of BDNF provides an alternative application route and work with rodents has demonstrated that adult facial motoneurons were equally rescued by BDNF administered intraventricularly and systemically.[129] However the pre-clinical studies performed in rodents are difficult to compare with the situation in humans, where the distances between peripherally applied factor and the cell soma are considerably longer. Therefore, large animals, such as sheep, which have diffusion distances comparable to human beings, have been employed in pre-clinical studies.[11] Unfortunately the planned clinical trials based on these experiments are now in jeopardy after the disappointing result with systemically applied BDNF.[137]

Peripheral Neurons

Peripheral sensory neurons represent the best-investigated targeT cells for neurotrophins. Basically all embryonic and most adult sensory neurons express Trk receptors often in combination with p75. Of special interest are cells in the dorsal root ganglia.[138,139] Surprisingly, adult, but not embryonic sensory neurons survive in culture in the absence of neurotrophins, while they maintain their neurotrophin responsiveness.[140,141] These findings indicate a role of neurotrophins in the regulation of specific functional properties of adult sensory neurons. This is corroborated by in vivo results with application of neurotrophins or blocking antibodies directed against them.[142] In axotomized cells the expression of genes encoding neurotrophins and their receptor are downregulated and the expression is restored by neurotrophin treatment.[143] In patients with diabetes neurotrophin levels are increased in skin under neuropathological conditions.[144] NGF primarily acts on adult nociceptors characterized by small fiber diameter and substance P expression (reviewed by ref. 145). NT3 acts preferentially on large-fiber neurons, prevents demyelination of Fast proprioceptive fibers under pathological conditions and reverses the characteristic electrophysiological slowing down of conduction velocity.[146] Furthermore it protects rat sensory neurons from cisplatin-induced toxicity.[147] Another potential peripheral target for neurotrophin treatment are the structures of the auditory system. Interestingly not only the auditory and vestibular neurons, but also the hair cells depend on neurotrophin support during development.[148] Both BDNF and NT3 have strong protective effects against ototoxic treatment.[149,150] Finally NT3 acts potently on the neurons of developing enteric nervous system[151] and TrkC is expressed on mature enteric neurons[152] suggesting this factor as a potential treatment for gut diseases.

Recombinant human nerve growth factor has been tested in phase II and phase III clinical trials for the treatment of patients with small fiber neuropathies. In a phase II trial with AIDS patients NGF-treatment caused significant relief of deafferentiation pain.[153] The trials with patients suffering from diabetic polyneuropathy gave rise to variable results. A first phase II trial initially indicated beneficial effects on sensory neurological scores,[154] a result that could not be corroborated in a subsequent phase III trial.[155] In all studies NGF caused mild myalgia and injection pain as an adverse effect.[156] The underlying cellular mechanism for these painful sensations is not fully elucidated. But it is well documented that NGF treatment up-regulates Substance P expression in sensory neurons and induces mast cell degranulation.[157] Recently it was shown that NGF activates the capsaicin receptor through TrkA mediated activation of phospho-lipase C.[158]

Topical application of NGF has substantial beneficial effects on ulcer healing, potentially in an indirect manner as a consequence of its activity on sensory innervation. In a study with patients suffering from corneal keratitis local NGF application restored corneal integrity in all cases[159] and wound healing was induced in rheumatoid arthritis patients with foot ulcers.[160]

NON-NEUROPROTECTIVE AND SIDE EFFECTS OF NEUROTROPHINS

Besides their cellular neuroprotective function, the neurotrophins influence also higher brain function and modulate the functions of non-neuronal organs, in particular the immune and the vascular system. While these effects on non-neuronal cells might well be exploitable for therapies such as autoimmune diseases, they could give rise to adverse side effects during treatment of neurodegenerative diseases.

Changes in Monoaminergic Brain Functions and Behavior

In addition to their neuroprotective role, neurotrophins are also potent regulators of neuronal gene transcription. In the serotonergic system BDNF regulates the expression of serotonergic markers including tryptophan hydroxylase and the serotonine transporter SERT, but not neuronal survival.[161,162] Similarly, neurotrophin induced changes of neurotransmitter function might occur on many nerve cells, including noradrenergic neurons,[163] dopaminergic and cholinergic neurons. BDNF infusion alters serotonin levels and serotonergic axonal sprouting in the brain[164,165] and modifies firing patterns of serotonergic and dopaminergic neurons.[166,167] Therefore neurotrophin treatment might result in psychogenic or behavioral effects. BDNF intraventricular infusion indeed induces behavioral changes in mice.[168] Heterozygote BDNF mutant mice are hyperaggressive and hyperphagic.[169,170] Conversely, infusion of BDNF into normal animals causes abrupt weight loss.[164] Taken together these findings indicate that satiety and locomotor nuclei in the hypothalamus are controlled by BDNF.

Effects on Memory Formation and Mental State

Neurotrophins change the shape and functional properties of central neurons in numerous ways. For example, already milliseconds after applications of BDNF to slice cultures, ion channel properties are modified and action potentials are triggered.[171] After seconds and minutes the release of neurotransmitter is facilitated and biochemical modifications of synaptic components are observed, such as phosphorylation of synapsins.[172,173] After days to weeks axons and dendrites change their shape.[78,174,175] Experiments with knockout mice have demonstrated that in hippocampal neurons the activity of TrkB and BDNF are required for long term potentiation (LTP) of synapses, a molecular mechanism associated with memory formation.[176,177] Furthermore, hippocampal mechanisms of pathological depression might be controlled by BDNF.[178,179] In animal models of depression BDNF infusion had antidepressant effects.[180] For cortical neurons it is predicted that their exposure to excess neurotrophins leads to more and stronger synapses, for example in GABA-ergic inhibitory neurons responding to BDNF (reviewed by ref. 181). These changes may result in unbalanced local neuronal circuitries in several brain areas including in particular the neocortex. Indeed, neurotrophins effectively change the induction phase of the kindling model of epilepsy.[182-184]

Stimulation of Non-Neuronal Cells

While neurotrophins exert their activity pre-dominantly on neurons, there are also effects described on non-neuronal cells. Schwann cells express large amounts of the p75 receptor and are known targets for neurotrophins. In particular, NGF activates the transcription factor NFκB in Schwann cells[60] and leads to changes in Schwann cell migration.[185] Following NGF treatment in vivo Schwann cell hyperplasia has been observed.[186] Another glial target for neurotrophin actions are oligodendrocytes. NT3 is a mitogen for oligodendrocyte precursor[187] and both BDNF and NT3 stimulate oligodendrocyte proliferation during axonal regeneration.[188] In contrast, NGF kills cultured mature oligodendrocytes via a p75-mediated mechanism.[63] A similar pro-apoptotic effect of NGF was also recently described for Schwann cells.[189]

In the immune system NGF is an autocrine survival factor for memory B cells[190] and stimulates IgG4 production of human B cells.[191] NGF also mediates profound anti-inflammatory effects on T cells and antigen presenting cells such as macrophages and microglia. Recently, we demonstrated that NGF suppresses the inducibility of MHC class II on microglial cells.[192] Furthermore, NGF directly acts on monocytes and inhibits their trans-endothelium migration into the cerebrum.[193] The role and the effect of NGF in experimental autoimmune encephalomyelitis were analyzed in several studies. Intraventricular application of NGF protected marmosets against EAE.[194] T cells overexpressing NGF inhibited the clinical symptoms and reduced the infiltration of monocytes in EAE[193] (Fig. 4).

Another tissue responsive to neurotrophins is the cardiovascular system. Recently, BDNF was identified as a potent survival regulator of endothelial cells in the heart.[195] Therefore its systemic application might change the properties of mature large vessels. We have recently analyzed the vascular system of p75 null mutant mice and found that the walls of the large elastic vessels in embryos lacking p75 are thinner compared to wild type animals with a corresponding strong tendency to form hemorrhages (von Schack et al, unpublished observation). NT3 mutant mouse embryos display severe heart malformations.[196] Neurotrophins and their receptors are expressed under pathophysiological conditions in adult blood vessels.[197] Taken together these results strongly indicate a role for the neurotrophins in the development and maintenance of large blood vessels. Hence cardiovascular problems after systemic application of neurotrophins need to be considered as potential side effects.

POTENTIAL IMPROVEMENTS OF NEUROTROPHIN THERAPY

The physiology of neurotrophins and their receptors is complex and, as outlined above, the list of observed and potential side effects is steadily growing. It appears

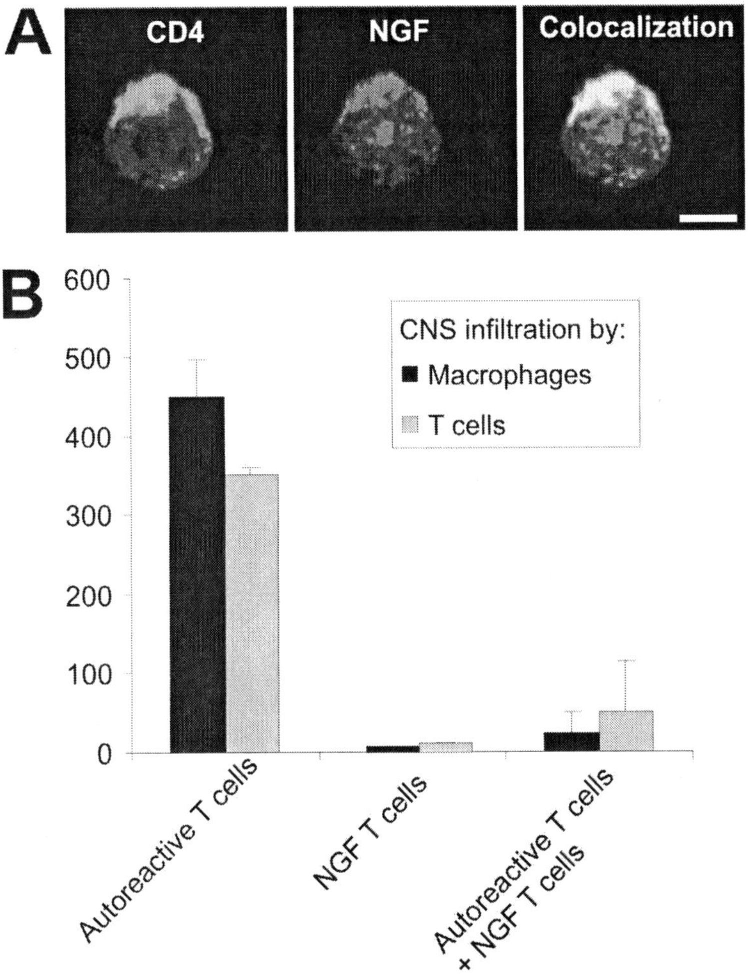

Figure 4. T lymphocytes as a therapeutical vehicle of neurotrophins into the CNS
A) Cultured T cells were retrovirally transduced with the NGF gene. The NGF-transduced T cells produce substantial amounts of NGF, which can be detected in T cells by confocal laser scanning microscopy after immunolabeling with an antibody directed against NGF. Double labeling with the antibody directed against CD4 identified the T cells. Scale bar: 5mm.
B) NGF-transduced T cells were used for therapy of experimental autoimmune encephalomyelitis, an animal model of multiple sclerosis. The disease was experimentally induced in rats by autoreactive (MBP-specific) T cells that stimulate brain inflammation. The NGF-transduced T cells migrate into the brain and release their therapeutic load, the NGF molecule. The NGF-transduced T cells improved the clinical signs (data not shown) and reduced the number of T cells and macrophages invading the spinal cord. Data are presented as mean +/- SD.

that this complexity has been underestimated during the first clinical trials. At the same time it has become apparent that systemic neurotrophin treatment is not very effective. New methods of local application and means to control adverse side effects should be developed.

Modification of Neurotrophins and their Pharmacology

The interactions between neurotrophins and their receptors can be modulated by the generation of mutant growth factors by recombinant DNA technology. In one approach mutated neurotrophin proteins have been developed that bind to multiple Trk receptors and to p75.[198-200] Mutant growth factors with enhanced receptor specificity compared to the naturally occurring proteins have also been generated.[201] In order to solve the problem of the short half-lives of neurotrophins in vivo, new agents are required which are more resistant to proteolytic degradation.[202] These could, for example, consist of synthetic peptides or antibody agonists for neurotrophin receptors.[203,204] The potential use of heteromeric neurotrophins combined from two different monomers has already been mentioned above.[205] In addition, conjugation of neurotrophins with other proteins such as the transferrin receptor might enable their passage through the blood-brain barrier.[206] However, the existing conjugates cross the blood-brain barrier only very inefficiently.

Stimulation of local endogenous neurotrophin synthesis or release with small molecular weight compounds has to be considered as an alternative to the more invasive application of exogenous growth factors (reviewed by ref. 207). Endogenous neurotrophin production and release is part of the response of the nervous system to injuries. In the brain, the mechanism that links neuronal activity to the expression of the BDNF gene has begun to be understood.[208,209] However the neurotrophin genes have multiple promoters and are expressed in tissue specific transcripts indicating a complex network of gene regulatory mechanisms.[210] Correspondingly multiplex regulatory mechanisms appear to determine the expression of the Trk receptor genes.[211] One advantage of activating the endogenous neurotrophin receptor expression or activity might be the stimulation of autocrine loops. Improved responsiveness to neurotrophic factors can be expected to strengthen existing connections or to generate additional ones. This might lead to increased overall electric activity, which in turn stimulates neurons to produce and secrete endogenous neurotrophins by increasing intracellular Ca^{2+} levels.[53] Consequently such an autocrine or paracrine cellular system, within limits, might stabilize itself.

Alternative Application Methods and Growth Factor Cocktails

Neurotrophins, like other large polypeptides, can only poorly cross the blood-brain barrier unless it breaks down in diseases such as ischemia.[212] Alternative methods of neurotrophin application in the CNS must therefore be developed. Current approaches to solve this problem include gene therapy using different viral vectors such as adenoviruses, adeno-associated virus, and lentiviruses.[213-215] Activated and genetically engineered T-cell lines, which secrete recombinant neurotrophins after transmigration into the brain parenchyma, have been experimentally used as therapeutic vehicles.[216] For example, T cells retrovirally transduced with the NGF gene migrate into the brain of a rat with experimental autoimmune encephalomyelitis improving the disease and inhibiting immune cells infiltration into the brain tissue (Fig. 4). Other strategies include implantation of gelfoam pledgets from which neurotrophins are slowly released[217,218] or insertion of small capsules containing neurotrophin secreting cells.[219,220]

An important therapeutic concept for future studies will be to exploit synergistic mechanisms between the signaling cascades of neurotrophins and other growth factors. This appears to be particularly important for CNS neurons, which typically depend on more than one growth factor for their survival. In long term it seems likely that combinations of neurotrophins and other growth factor families with neurotrophic activity, such as IGFs, GDNF-like molecules, FGFs or neuropoietic cytokines, in particular CNTF, will be more effective than a single growth factor. Cocktails of growth factors can be designed in which each growth factor is present at suboptimal concentrations, which would limit receptor availability and consequently side effects. Such cocktails have already been shown to act synergistically on motoneurons.[221-223] It might also turn out to be desirable to block adverse effects of endogenous cytokines, which act antagonistically to neurotrophins.[224] Thus, blocking action of TNF or CD95 with neutralizing antibodies might be a useful therapeutic approach to support neurotrophin effects.

Neurotrophins and Cell Replacement Therapy

Apart from the immediate use of the neurotrophins as therapeutic agents, it is also likely that one or more of the neurotrophins will be an integral part of treatment protocols for stem cells with the aim to differentiate them into specific neuronal cell types for cell replacement therapy.[225-227]

From the rather disappointing results of the clinical trials performed so far the conclusion can be drawn, that systemic and probably also intraventricular infusion of neurotrophins are rather ineffective and prone to give rise to severe side effects. One alternative approach could be to combine cell replacement therapy with neurotrophin gene transfer into stem cells prior to transplantation.[228,229] This approach appears particularly attractive since neurotrophins have the potential to not only prevent or slow down neurodegeneration, but also to foster the establishment of synaptic connections of stem cell derived neurons. Stem cells can be transplanted to defined brain structures and even more importantly they can be engineered to produce neurotrophin transgenes under exogenous control dependent on the absence or presence of small molecular weight substances such as doxycycline, ecdysone or tamoxifen.[230] Such an approach would warrant localized expression and at the same time allow to interrupt the therapy as soon as side effects are observed. However, in light of the current excitement about the great potential of cell replacement therapies for neurodegenerative diseases, it is recommended to take a look back on the clinical trials with neurotrophins. They also have been prematurely advertised as miracle drugs. As desirable as it would be, it cannot reasonably be expected that cell replacement therapies will escape the phases of experimental problems and frustration that the neurotrophins went through.

ACKNOWLEDGMENTS

The authors are grateful to Dr. Alexander Fl gel, Dr . Hans Lassmann and Ingeborg Haarmann for help in the experiments with GFP transduced T cells. We thank Dr. Hartmut Wekerle and Dr. Yves-Alain Barde for continuous support and Dr. Hans Thoenen and Dr. Michael Sendtner for their helpful comments on the manuscript.

REFERENCES

1. Gage FH. Mammalian neural stem cells. SCIENCE 2000; 287:1433-1438.
2. Barde Y-A. The Nerve Growth Factor Family. Prog Growth Factor Res 1990; 2:237-348.
3. Bibel M, Barde Y-A. Neurotrophins: Key regulators of cell fate and cell shape in the vertebrate nervous system. Genes Dev 2000; 14:2919-2937.
4. Thoenen H. Treatment of degenerative disorders of the nervous system: From helpless descriptive categorization to rational therapeutic approaches. In: Ingoglia NA, Murray N, eds. Axonal Regeneration in the Central Nervous System. New York: Dekker M, 2001:675-697
5. Goyal L. Cell death inhibition: keeping caspases in check. Cell 2001; 104:805-808.
6. McDonald NQ, Lapatto R, Murray-Rust J et al. New protein fold revealed by a 2.3-resolution crystal structure of nerve growth factor. Nature 1991; 354:411-414.
7. McDonald NQ, Chao MV. Structural determinants of neurotrophin action. J Biol Chem 1995; 270:19669-19672.
8. Lewin GR, Barde YA. Physiology of the Neurotrophins. Ann Rev Neurosci 1996; 19:289-317.
9. Tria MA, Fusco M, Vantini G et al. Pharmacokinetics of nerve growth factor (NGF) following different routes of administration to adult rats. Exp Neurol 1994; 127:178-183.
10. Yan Q, Matheson C, Sun J et al. Distribution of intracerebral ventricularly administered neurotrophins in rat brain and its correlation with trk receptor expression. Exp Neurol 1994; 127:23-36.
11. Dittrich F, Ochs G, Grosse-Wilde A et al. Pharmacokinetics of intrathecally applied BDNF and effects on spinal motoneurons. Exp Neurol 1996; 141:225-239.
12. Radziejewski C, Robinson RC. Heterodimers of the neurotrophic factors: Formation, isolation, and differential stability. Biochemistry 1993; 32:13350-13356.
13. Arakawa T, Haniu M, Narhi LO et al. Formation of heterodimers from three neurotrophins, nerve growth factor, neurotrophin-3, and brain-derived neurotrophic factor. J Biol Chem 1994; 269:27833-27839.
14. Thoenen H. Neurotrophins and activity-dependent plasticity. Prog Brain Res 2000; 128:183-191.
15. Barbacid M. The Trk family of neurotrophin receptors. J Neurobiol 1994; 25:1386-1403.
16. Grell M, Clauss M. TNF and TNF receptor superfamily. In: Bona CA, Revillard J-P, eds. Cytokines and Cytokine Receptors. New York: Harwood Academic Press, 2000:118-148.
17. Dechant G. Molecular interactions between neurotrophin receptors. Cell Tissue Res 2001; (online version http://dx.doi.org/10.1007/s004410100378)
18. Dechant G, Rodr guez-T bar A, Barde Y-A. Neurotrophin Receptors. Prog Neurobiol 1994; 42:347-352.
19. Davies AM. Neurotrophins: The yin and yang of nerve growth factor. Curr Biol 1997; 7:R38-R40.

20. Snider WD. Functions of the neurotrophins during nervous system development: What the knockouts are teaching us. Cell 1994; 77:627-638.
21. Holtzman DM, Li Y, Parada LF et al. p140trk mRNA marks NGF-responsive forebrain neurons: evidence that trk gene expression is induced by NGF. Neuron 1992; 9:465-478.
22. Hefti F. Nerve growth factor promotes survival of septal cholinergic neurons after fimbrial transections. J Neurosci 1986; 6 :2155-2162.
23. Klein R, Martin-Zanca D, Barbacid M et al. Expression of the tyrosine kinase receptor gene TrkB is confined to the murine embryonic and adult nervous system. Development 1990; 109:845-850.
24. Klein R, Conway D, Parada LF et al. The TrkB tyrosine protein kinase gene codes for a second neurogenic receptor that lacks the catalytic kinase domain. Cell 1990; 61:647-656.
25. Middlemas DS, Lindberg RA, Hunter T. trk B, a neural receptor protein-tyrosine kinase: evidence for a full-length and two truncated receptors. Mol Cell Biol 1991; 11:143-153.
26. Tessarollo L, TsoulFas P, Martin-Zanca D et al. *TrkC*, a receptor for neurotrophin-3, is widely expressed in the developing nervous system and in non-neuronal tissues. Development 1993; 118:463-475.
27. Barker PA. p75NTR: A study in contrasts [see comments]. Cell Death Differ 1998; 5:346-356.
28. Dobrowsky RT, Carter BD. p75 neurotrophin receptor signaling: Mechanisms for neurotrophic modulation of cell stress? J Neurosci Res 2000; 61:237-243.
29. Frade JM, Barde YA. Nerve growth factor: Two receptors, multiple functions. BioEssays 1998; 20:137-145.
30. Yamashita T, Tucker KL, Barde YA. Neurotrophin binding to the p75 receptor modulates Rho activity and axonal outgrowth. Neuron 1999; 24:585-593.
31. Bibel M, Hoppe E, Barde YA. Biochemical and functional interactions between the neurotrophin receptors *trk* and $p75^{NTR}$. EMBO J 1998; 616-622.
32. Majdan M, Miller FD. Neuronal life and death decisions functional antagonism between the Trk and p75 neurotrophin receptors. Int J Dev Neurosci 1999; 17:153-161.
33. Vejsada R, Sagot Y, Kato AC. BDNF-mediated rescue of axotomized motor neurones decreases with increasing dose. NeuroReport 1994; 5:1889-1892.
34. Vejsada R, Sagot Y, Kato AC. Quantitative comparison of the transient rescue effects of neurotrophic factors on axotomized motoneurons in vivo. Eur J Neurosci 1995; 7:108-115.
35. Sommerfeld MT, Schweigreiter R, Barde YA et al. Down-regulation of the neurotrophin receptor TrkB following ligand binding. Evidence for an involvement of the proteasome and differential regulation of TrkA and TrkB. J Biol Chem 2000; 275:8982-8990.
36. Carter BD, Zirrgiebel U, Barde Y-A. Differential regulation of $p21^{ras}$ activation in neurons by nerve growth factor and brain-derived neurotrophic factor. J Biol Chem 1995; 270:21751-21757.
37. Frank L, Ventimiglia R, Anderson K et al. BDNF down-regulates neurotrophin responsiveness, TrkB protein and TrkB mRNA levels in cultured rat hippocampal neurons. Eur J Neurosci 1996; 8:1220-1230.
38. Widmer HR, Ohsawa F, Knusel B et al. Down-regulation of phosphatidylinositol response to BDNF and NT-3 in cultures of cortical neurons. Brain Res 1993; 614:325-334.
39. Knusel B, Kaplan DR, Hefti F. Intraparenchymal NGF injections in adult and aged rats induce long- lasting Trk tyrosine phosphorylation. Exp Neurol 1996; 139:121-130.
40. Biffo S, Offenh user N, Carter BD et al. Selective binding and internalisation by truncated receptors restrict the availability of BDNF during development. Development 1995; 121:2461-2470.
41. Von Bartheld CS, Kinoshita Y, Prevette D et al. Positive and negative effects of neurotrophins on the isthmo-optic nucleus in chick embryos. Neuron 1994; 12:639-654.
42. Wiese S, Metzger F, Holtmann B et al. The role of $p75^{NTR}$ in modulating neurotrophin survival effects in developing motoneurons. Eur J Neurosci 1999; 11:1668-1676.
43. Klocker N, Cellerino A, Bahr M. Free radical scavenging and inhibition of nitric oxide synthase potentiates the neurotrophic effects of brain-derived neurotrophic factor on axotomized retinal ganglion cells In vivo. J Neurosci 1998; 18:1038-1046.

44. Davies AM, Minichiello L, Klein R. Developmental changes in NT3 signalling via TrkA and TrkB in embryonic neurons. EMBO J 1995; 14:4482-4489.
45. Wyatt S, Pi n LGP, Ernfors P et al. Sympathetic neuron survival and TrkA expression in NT3-deficient mouse embryos. EMBO J 1997; 16:3115-3123.
46. Friedman WJ, Greene LA. Neurotrophin signaling via Trks and p75. Exp Cell Res 1999; 253:131-142.
47. Yuen EC, Mobley WC. Early BDNF, NT-3, and NT-4 signaling events. Exp Neurol 1999; 159:297-308.
48. Cunningham ME, Greene LA. A function-structure model for NGF-activated TRK. EMBO Journal 1998; 17:7282-7293.
49. Kaplan DR, Miller FD. Signal transduction by the neurotrophin receptors. Curr Op Cell Biol 1997; 9:213-221.
50. Datta SR, Brunet A, Greenberg ME. Cellular survival: a play in three Akts. Genes Dev 1999; 13:2905-2927.
51. Holgado-Madruga M, Moscatello DK, Emlet DR et al. Grb2-associated binder-1 mediates phosphatidylinositol 3-kinase activation and the promotion of cell survival by nerve growth factor. Proc Natl Acad Sci USA 1997; 94:12419-12424.
52. Yamada M, Ohnishi H, Sano S et al. Insulin receptor substrate (IRS)-1 and IRS-2 are tyrosine-phosphorylated and associated with phosphatidylinositol 3-kinase in response to brain-derived neurotrophic factor in cultured cerebral cortical neurons. J Biol Chem 1997; 272:30334-30339.
53. Canossa M, Gartner A, Campana G et al. Regulated secretion of neurotrophins by metabotropic glutamate group I (mGluRI) and Trk receptor activation is mediated via phospholipase C signalling pathways. EMBO J 2001; 20:1640-1650.
54. Maderdrut JL, Oppenheim RW, Prevette D. Enhancement of naturally occurring cell death in the sympathetic and parasympathetic ganglia of the chicken embryo following blockade of ganglionic transmission. Brain Res 1988; 444:189-194.
55. Ghosh A, Carnahan J, Greenberg ME. Requirement for BDNF in activity-dependent survival of cortical neurons. SCIENCE 1994; 263:1618-1623.
56. McAllister AK, Katz LC, Lo DC. Neurotrophin regulation of cortical dendritic growth requires activity. Neuron 1996; 17:1057-1064.
57. Brodski C, Schn rch H, Dechant G. Neurotrophin-3 promotes the cholinergic differentiation of sympathetic neurons. Proc Natl Acad of Sci USA 2000; 97:9683-9688.
58. Dobrowsky RT, Jenkins GM, Hannun YA. Neurotrophins induce sphingomyelin hydrolysis Modulation by Co- expression of $p75^{NTR}$ with Trk receptors. J Biol Chem 1995; 270:22135-22142.
59. Dobrowsky RT, Carter BD. Coupling of the p75 neurotrophin receptor to sphingolipid signaling. Ann NY Acad Sci 1998; 845:32-45.
60. Carter BD, Kaltschmidt C, Kaltschmidt B et al. Selective activation of NF-kappaB by nerve growth factor through the neurotrophin receptor p75. Science 1996; 272:542-545.
61. Hamanoue M, Middleton G, Wyatt S et al. p75-mediated NF-kappaB activation enhances the survival response of developing sensory neurons to nerve growth factor. Mol Cell Neurosci 1999; 14:28-40.
62. Goldhawk DE, Meakin SO, Verdi JM. Subpopulations of rat $B2^+$ neuroblasts exhibit differential neurotrophin responsiveness during sympathetic development. Dev Biol 2000; 218:367-377.
63. Casaccia-Bonnefil P, Carter BD, Dobrowsky RT et al. Death of oligodendrocytes mediated by the interaction of nerve growth factor with its receptor p75. Nature 1996; 383:716-719.
64. Frade JM, Rodr guez-T bar A, Barde YA. Induction of cell death by endogenous nerve growth factor through its p75 receptor. Nature 1996; 383:166-168.
65. Agerman K, Baudet C, Fundin B et al. Attenuation of a caspase-3 dependent cell death in NT4- and p75-deficient embryonic sensory neurons. Mol Cell Neurosci 2000; 16:258-268.
66. Yoon SO, Casaccia-Bonnefil P, Carter B et al. Competitive signaling between TrkA and p75 nerve growth factor receptors determines cell survival. J Neurosci 1998; 18:3273-3281.

67. Aloyz RS, Bamji SX, Pozniak CD et al. P53 is essential for developmental neuron death as regulated by the TrkA and p75 neurotrophin receptors. J Cell Biol 1998; 143:1691-1703.
68. Pozniak CD, Radinovic S, Yang A et al. An anti-apoptotic role for the p53 family member, p73, during developmental neuron death. SCIENCE 2000; 289:304-306.
69. Budd SL, Nicholls DG. Mitochondria in the life and death of neurons. Essays Biochem 1998; 33:43-52.
70. Martin LJ. Neuronal cell death in nervous system development, disease, and injury. Int J Mol Med 2001; 7:455-478.
71. Dugan LL, Creedon DJ, Johnson EM, Jr. et al. Rapid suppression of free radical formation by nerve growth factor involves the mitogen-activated protein kinase pathway. Proc Natl Acad Sci USA 1997; 94:4086-4091.
72. Koh JY, Gwag BJ, Lobner D et al. Potentiated necrosis of cultured cortical neurons by neurotrophins. Science 1995; 268:573-575.
73. Davies AM. The Bcl-2 family of proteins, and the regulation of neuronal survival. Trends Neurosci 1995; 18:355-358.
74. Martin LJ. Neuronal death in amyotrophic lateral sclerosis is apoptosis: Possible contribution of a programmed cell death mechanism. J Neuropathol Exp Neurol 1999; 58:459-471.
75. Lindholm D. Role of neurotrophins in preventing glutamate induced neuronal cell death. J Neurol 1994; 242:S16-S18.
76. Sagot Y, Dubois-Dauphin M, Tan SA et al. Bcl-2 overexpression prevents motoneuron cell body loss but not axonal degeneration in a mouse model of a neurodegenerative disease. J Neurosci 1995; 15:7727-7733.
77. Patel TD, Jackman A, Rice FL et al. Development of sensory neurons in the absence of NGF/TrkA signaling in vivo. Neuron 2000; 25:345-357.
78. McAllister AK, Katz LC, Lo DC. Opposing roles for endogenous BDNF and NT-3 in regulating cortical dendritic growth. Neuron 1997; 18:767-778.
79. Leing rtner A, Heisenberg C-P, Kolbeck R et al. Brain-derived neurotrophic factor increases neurotrophin- 3 expression in cerebellar granule neurons. J Biol Chem 1994; 269:828-830.
80. Canossa M, Griesbeck O, Berninger B et al. Neurotrophin release by neurotrophins: Implications for activity- dependent neuronal plasticity. Proc Natl Acad Sci USA 1997; 94:13279-13286.
81. Kr ttgen A, M ller JC, Heymach JV, Jr. et al. Neurotrophins induce release of neurotrophins by the regulated secretory pathway. Proc Natl Acad Sci USA 1998; 95:9614-9619.
82. Causing CG, Gloster A, Aloyz R et al. Synaptic innervation density is regulated by neuron-derived BDNF. Neuron 1997; 18:257-267.
83. Giehl KM. Trophic dependencies of rodent corticospinal neurons. Rev Neurosci 2001; 12:79-94.
84. Merlio JP, Ernfors P, Kokaia Z et al. Increased production of the TrkB protein tyrosine kinase receptor after brain insults. Neuron 1993; 10:151-164.
85. Kokaia Z, Andsberg G, Yan Q et al. Rapid alterations of BDNF protein levels in the rat brain after focal ischemia: Evidence for increased synthesis and anterograde axonal transport. Exp Neurol 1998; 154:289-301.
86. Beck T, Lindholm D, Castren E et al. Brain-derived neurotrophic factor protects against ischemic cell damage in rat hippocampus. J Cereb Blood Flow Metab 1994; 14:689-692.
87. Tsukahara T, Yonekawa Y, Tanaka K et al. The role of brain-derived neurotrophic factor in transient forebrain ischemia in the rat brain. Neurosurgery 1994; 34:323-331.
88. Chan KM, Lam DT, Pong K et al. Neurotrophin-4/5 treatment reduces infarct size in rats with middle cerebral artery occlusion. Neurochem Res 1996; 21:763-767.
89. Gall C, Lauterborn J, Bundman M et al. Seizures and the regulation of neurotrophic factor and neuropeptide gene expression in brain. Epilepsy Res Suppl 1991; 4:225-245.
90. Ernfors P, Bengzon J, Kokaia Z et al. Increased levels of messenger RNAs for neurotrophic factors in the brain during kindling epileptogenesis. Neuron 1991; 7:165-176.
91. Mudo G, Jiang XH, timmusk t et al. Change in neurotrophins and their receptor mRNAs in the rat forebrain after status epilepticus induced by pilocarpine. Epilepsia 1996; 37:198-207.

92. Binder DK, Routbort MJ, Ryan TE et al. Selective inhibition of kindling development by intraventricular administration of TrkB receptor body. J Neurosci 1999; 19:1424-1436.
93. Giehl KM, Tetzlaff W. BDNF and NT-3, but not NGF, prevent axotomy-induced death of rat corticospinal neurons in vivo. Eur J Neurosci 1996; 8:1167-1175.
94. Schnell L, Schneider R, Kolbeck R et al. Neurotrophin-3 enhances sprouting of corticospinal tract during development and after adult spinal cord lesion. Nature 1994; 367:170-173.
95. Hefti F, Weiner WJ. Nerve growth factor and Alzheimer s disease. Ann Neurol 1986; 20:275-281.
96. Vazquez ME, Ebendal T. Messenger RNAs for trk and the low-affinity NGF receptor in rat basal forebrain. NeuroReport 1991; 2:593-596.
97. Venero JL, Hefti F. TrkA NGF receptor expression by non-cholinergic thalamic neurons. NeuroReport 1993; 4:959-962.
98. Hefti F, Lapchak PA. Pharmacology of nerve growth factor in the brain. Adv Pharmacol 1993; 24:239-273.
99. Will B, Hefti F. Behavioural and neurochemical effects of chronic intraventricular injections of nerve growth factor in adult rats with fimbria lesions. Behav Brain Res 1985; 17:17-24.
100. Fischer W, Wictorin K, Bjoerklund A et al. Amelioration of cholinergic neuron atrophy and spatial memory impairment in aged rats by nerve growth factor. Nature 1987; 329:65-68.
101. Markowska AL, Koliatsos VE, Breckler SJ et al. Human nerve growth factor improves spatial memory in aged but not in young rats. J Neurosci 1994; 14:4815-4824.
102. Widmer HR, Knusel B, Hefti F. BDNF protection of basal forebrain cholinergic neurons after axotomy: complete protection of p75NGFR-positive cells. NeuroReport 1993; 4:363-366.
103. Koliatsos VE, Price DL, Gouras GK et al. Highly selective effects of nerve growth factor, brain-derived neurotrophic factor, and neurotrophin-3 on intact and injured basal forebrain magnocellular neurons. J Comp Neurol 1994; 343:247-262.
104. Fischer W, Sirevaag A, Wiegand SJ et al. Reversal of spatial memory impairments in aged rats by nerve growth factor and neurotrophins 3 and 4/5 but not by brain-derived neurotrophic factor. Proc Natl Acad Sci USA 1994; 91:8607-8611.
105. Kn sel B, Winslow JW, Rosenthal A et al. Promotion of central cholinergic and dopaminergic neuron differentiation by brain-derived neurotrophic factor but not neurotrophin 3. Proc Natl Acad Sci USA 1991; 88:961-965.
106. Alderson RF, Wiegand SJ, Anderson KD et al. Neurotrophin-4/5 maintains the cholinergic phenotype of axotomized septal neurons. Eur J Neurosci 1996; 8:282-290.
107. Phillips HS, Hains JM, Armanini M et al. BDNF mRNA is decreased in the hippocampus of individuals with Alzheimer s disease. Neuron 1991; 7:695-702.
108. Peterson DA, Leppert JT, Lee KF et al. Basal forebrain neuronal loss in mice lacking neurotrophin receptor p75. Science 1997; 277:837-838.
109. Hagg T, Van der Zee CEEM, Ross GM et al. Basal forebrain neuronal loss in mice lacking neurotrophin receptor p75 Response. Science 1997; 277:838-839.
110. Olson L, Nordberg A, von Holst H et al. Nerve growth factor affects 11C-nicotine binding, blood flow, EEG, and verbal episodic memory in an Alzheimer patient (case report). J Neural Transm Park Dis Dement Sect 1992; 4:79-95.
111. Eriksdotter JM, Nordberg A, Amberla K et al. Intracerebroventricular infusion of nerve growth factor in three patients with Alzheimer s disease. Dement Geriatr Cogn Disord 1998; 9:246-257.
112. Hyman C, Hofer M, Barde Y-A et al. BDNF is a neurotrophic factor for dopaminergic neurons of the substantia nigra. Nature 1991; 350:230-232.
113. Spina MB, Squinto SP, Miller J et al. Brain-derived neurotrophic factor protects dopamine neurons against 6-hydroxydopamine and N-methyl-4-phenylpyridinium ion toxicity: Involvement of the glutathione system. J Neurochem 1992; 59:99-106.
114. Knusel B, Beck KD, Winslow JW et al. Brain-derived neurotrophic factor administration protects basal forebrain cholinergic but not nigral dopaminergic neurons from degenerative changes after axotomy in the adult rat brain. J Neurosci 1992; 12:4391-4402.

115. Hagg T. Neurotrophins prevent death and differentially affect tyrosine hydroxylase of adult rat nigrostriatal neurons in vivo. Exp Neurol 1998; 149:183-192.
116. Altar CA, Boylan CB, Jackson C et al. Brain-derived neurotrophic factor augments rotational behavior and nigrostriatal dopamine turnover *in vivo*. Proc Natl Acad Sci USA 1992; 89:11347-11351.
117. Lapchak PA, Beck KD, Araujo DM et al. Chronic intranigral administration of brain-derived neurotrophic factor produces striatal dopaminergic hypofunction in unlesioned adult rats and fails to attenuate the decline of striatal dopaminergic function following medial forebrain bundle transection. Neuroscience 1993; 53:639-650.
118. Ringstedt T, Lagercrantz H, Persson H. Expression of members of the *trk* family in the developing postnatal rat brain. Dev Brain Res 1993; 72:119-131.
119. Numan S, Seroogy KB. Expression of *Trk*B and *Trk*C mRNAs by adult midbrain dopamine neurons: A double-label in situ hybridization study. J Comp Neurol 1999; 403:295-308.
120. Koliatsos VE, Clatterbuck RE, Winslow JW et al. Evidence that brain-derived neurotrophic factor is a trophic factor for motor neurons in vivo. Neuron 1993; 10:359-367.
121. Klein R, Smeyne RJ, Wurst W et al. Targeted disruption of the *TrkB* neurotrophin receptor gene results in nervous system lesions and neonatal death. Cell 1993; 75:113-122.
122. Liu X, Ernfors P, Wu H et al. Sensory but not motor neuron deficits in mice lacking NT4 and BDNF. Nature 1995; 375:238-241.
123. Conover JC, Erickson JT, Katz DM et al. Neuronal deficits, not involving motor neurons, in mice lacking BDNF and/or NT4. Nature 1995; 375:235-238.
124. Kobayashi NR, Bedard AM, Hincke MT et al. Increased expression of BDNF and TrkB mRNA in rat facial motoneurons after axotomy. Eur J Neurosci 1996; 8:1018-1029.
125. Sendtner M, Holtmann B, Hughes RA. The response of motoneurons to neurotrophins. Neurochem Res 1996; 21:831-841.
126. Sendtner M, Holtmann B, Kolbeck R et al. Brain-derived neurotrophic factor prevents the death of motoneurons in newborn rats after nerve section. Nature 1992; 360:757-759.
127. Yan Q, Elliott J, Snider WD. Brain-derived neurotrophic factor rescues spinal motor neurons from axotomy-induced cell death. Nature 1992; 360:753-755.
128. Johnson JE, Qin-Wei Y, Prevette D et al. Brain-derived proteins that rescue spinal motoneurons from cell death in the chick embryo: Comparisons with target- derived and recombinant factors. J Neurobiol 1995; 27:573-589.
129. Yan Q, Matheson C, Lopez OT et al. The biological responses of axotomized adult motoneurons to brain- derived neurotrophic factor. J Neurosci 1994; 14:5281-5291.
130. Ernfors P, Henschen A, Olson L et al. Expression of nerve growth factor receptor mRNA is developmentally regulated and increased after axotomy in rat spinal cord motoneurons. Neuron 1989; 2:1605-1613.
131. Koliatsos VE, Crawford TO, Price DL. Axotomy induces nerve growth factor receptor immunoreactivity in spinal motor neurons. Brain Res 1991; 549:297-304.
132. Frade JM, Barde YA. Genetic evidence for cell death mediated by nerve growth factor and the neurotrophin receptor p75 in the developing mouse retina and spinal cord. Development 1999; 126:683-690.
133. Ferri CC, Moore FA, Bisby MA. Effects of facial nerve injury on mouse motoneurons lacking the p75 low-affinity neurotrophin receptor. Journal of Neurobiology 1998; 34:1-9.
134. Yan Q, Miller JA. The use of trophic factors in degenerative motoneuron diseases. Exp Neurol 1993; 124:60-63.
135. Stoeckel K, Guroff G, Schwab M et al. The significance of retrograde axonal transport for the accumulation of systemically administered nerve growth factor (NGF) in the rat superior cervical ganglion. Brain Res 1976; 109:271-284.
136. Curtis R, Tonra JR, Stark JL et al. Neuronal injury increases retrograde axonal transport of the neurotrophins to spinal sensory neurons and motor neurons via multiple receptor mechanisms. Mol Cell Neurosci 1998; 12:105-118.

137. Ochs G, Penn RD, Giess R et al. A trial of recombinant methionyl human Brain-Derived Neurotrophic Factor (BDNF) given by intrathecal (IT) infusion to patients with Amyotrophic Lateral Sclerosis (ALS). Neurology 1998; A246.
138. Phillips HS, Armanini MP. Expression of the trk family of neurotrophin receptors in developing and adult dorsal root ganglion neurons. Philos Trans R Soc Lond B Biol Sci 1996; 351:413-416.
139. Karchewski LA, Kim FA, Johnston J et al. Anatomical evidence supporting the potential for modulation by multiple neurotrophins in the majority of adult lumbar sensory neurons. J Comp Neurol 1999; 413:327-341.
140. Lindsay RM. Nerve growth factors (NGF, BDNF) enhance axonal regeneration but are not required for survival of adult sensory neurons. J Neurosci 1988; 8:2394-2405.
141. Lindsay RM, Harmar AJ. Nerve growth factor regulates expression of neuropeptide genes in adult sensory neurons. Nature 1989; 337:362-364.
142. Mendell LM. Neurotrophin action on sensory neurons in adults: an extension of the neurotrophic hypothesis. Pain 1999; Suppl 6:S127-S132.
143. Verge VM, Gratto KA, Karchewski LA et al. Neurotrophins and nerve injury in the adult. Philos Trans R Soc Lond B Biol Sci 1996; 351:423-430.
144. Anand P. Neurotrophins and peripheral neuropathy. Philos Trans R Soc Lond B Biol Sci 1996; 351:449-454.
145. Lewin GR. Neurotrophins and the specification o neuronal phenotype. Phil Trans R Soc London 1996;405-411.
146. Gao WQ, Dybdal N, Shinsky N et al. Neurotrophin-3 reverses experimental cisplatin-induced peripheral sensory neuropathy. Ann Neurol 1995; 38:30-37.
147. Apfel SC, Arezzo JC, Lipson L et al. Nerve growth factor prevents experimental cisplatin neuropathy. Ann Neurol 1992; 31:76-80.
148. Ernfors P, Van De WT, Loring J et al. Complementary roles of BDNF and NT-3 in vestibular and auditory development. Neuron 1995; 14:1153-1164.
149. Ernfors P, Duan ML, Elshamy WM et al. Protection of auditory neurons from aminoglycoside toxicity by neurotrophin-3. Nat Med 1996; 2:463-467.
150. Staecker H, Kopke R, Malgrange B et al. NT-3 and/or BDNF therapy prevents loss of auditory neurons following loss of hair cells. NeuroReport 1996; 7:889-894.
151. Chalazonitis A, Rothman TP, Chen J et al. Neurotrophin-3 induces neural crest-derived cells from fetal rat gut to develop *in vitro* as neurons or glia. J Neurosci 1994; 14:6571-6584.
152. Sternini C, Su D, Arakawa J et al. Cellular localization of Pan-trk immunoreactivity and TrkC mRNA in the enteric nervous system. J Comp Neurol 1996; 368:597-607.
153. McArthur JC, Yiannoutsos C, Simpson DM et al. A phase II trial of nerve growth factor for sensory neuropathy associated with HIV infection. AIDS Clinical Trials Group Team 291. Neurology 2000; 54:1080-1088.
154. Apfel SC, Kessler JA, Adornato BT et al. Recombinant human nerve growth factor in the treatment of diabetic polyneuropathy. NGF Study Group. Neurology 1998; 51:695-702.
155. Apfel SC, Schwartz S, Adornato BT et al. Efficacy and safety of recombinant human nerve growth factor in patients with diabetic polyneuropathy: A randomized controlled trial. rhNGF Clinical Investigator Group. JAMA 2000; 284:2215-2221.
156. Petty BG, Cornblath DR, Adornato BT et al. The effect of systemically administered recombinant human nerve growth factor in healthy human subjects. Ann Neurol 1994; 36:244-246.
157. Mazurek N, Weskamp G, Erne P et al. Nerve growth factor induces mast cell degranulation without changing intracellular calcium levels. FEBS Lett 1986; 198:315-320.
158. Chuang H, Prescott ED, Kong H et al. Bradykinin and nerve growth factor release the capsaicin receptor from PtdIns(4,5)P2-mediated inhibition. Nature 2001; 411:957-962.
159. Lambiase A, Rama P, Bonini S et al. Topical treatment with nerve growth factor for corneal neurotrophic ulcers. N Engl J Med 1998; 338:1174-1180.

160. Tuveri M, Generini S, Matucci-Cerinic M et al. NGF, a useful tool in the treatment of chronic vasculitic ulcers in rheumatoid arthritis. Lancet 2000; 356:1739-1740.
161. Siuciak JA, Clark MS, Rind HB et al. BDNF induction of tryptophan hydroxylase mRNA levels in the rat brain. J Neurosci Res 1998; 52:149-158.
162. Galter D, Unsicker K. Sequential activation of the 5-HT1$_A$ serotonin receptor and TrkB induces the serotonergic neuronal phenotype. Mol Cell Neurosci 2000; 15:446-455.
163. Arenas E, Persson H. Neurotrophin-3 prevents the death of adult central noradrenergic neurons *in vivo*. Nature 1994; 367:368-371.
164. Pelleymounter MA, Cullen MJ, Wellman CL. Characteristics of BDNF-induced weight loss. Exp Neurol 1995; 131:229-238.
165. Mamounas LA, Altar CA, Blue ME et al. BDNF promotes the regenerative sprouting, but not survival, of injured serotonergic axons in the adult rat brain. J Neurosci 2000; 20:771-782.
166. Shen RY, Altar CA, Chiodo LA. Brain-derived neurotrophic factor increases the electrical activity of pars compacta dopamine neurons in vivo. Proc Natl Acad Sci USA 1994; 91:8920-8924.
167. Celada P, Siuciak JA, Tran TM et al. Local infusion of brain-derived neurotrophic factor modifies the firing pattern of dorsal raphe serotonergic neurons. Brain Res 1996; 712:293-298.
168. Martin-Iverson MT, Todd KG, Altar CA. Brain-derived neurotrophic factor and neurotrophin-3 activate striatal dopamine and serotonin metabolism and related behaviors: Interactions with amphetamine. J Neurosci 1994; 14:1262-1270.
169. Lyons WE, Mamounas LA, Ricaurte GA et al. Brain-derived neurotrophic factor-deficient mice develop aggressiveness and hyperphagia in conjunction with brain serotonergic abnormalities. Proc Natl Acad Sci USA 1999; 96:15239-15244.
170. Kernie SG, Liebl DJ, Parada LF. BDNF regulates eating behavior and locomotor activity in mice. EMBO J 2000; 19:1290-1300.
171. Kafitz KW, Rose CR, Thoenen H et al. Neurotrophin-evoked rapid excitation through TrkB receptors. Nature 1999; 401:918-921.
172. Lohof AM, Ip NY, Poo M. Potentiation of developing neuromuscular synapses by the neurotrophins NT-3 and BDNF. Nature 1993; 363:350-353.
173. Jovanovic JN, Czernik AJ, Fienberg AA et al. Synapsins as mediators of BDNF-enhanced neurotransmitter release. Nat Neurosci 2000; 3:323-329.
174. Ruit KG, Osborne PA, Schmidt RE et al. Nerve growth factor regulates sympathetic ganglion cell morphology and survival in the adult mouse. J Neurosci 1990; 10:2412-2419.
175. McAllister AK, Lo DC, Katz LC. Neurotrophins regulate dendritic growth in developing visual cortex. Neuron 1995; 15:791-803.
176. Korte M, Carroll P, Wolf E et al. Hippocampal long-term potentiation is impaired in mice lacking brain-derived neurotrophic factor. Proc Natl Acad Sci USA 1995; 92:8856-8860.
177. Minichiello L, Korte M, Wolfer D et al. Essential role for TrkB receptors in hippocampus-mediated learning. Neuron 1999; 24 :401-414.
178. Duman RS, Vaidya VA, Nibuya M et al. Stress, antidepressant treatments, and neurotrophic factors: Molecular and cellular mechanisms. The Neuroscientist 1995; 6:351-360.
179. Duman RS, Heninger GR, Nestler EJ. A molecular and cellular theory of depression. Arch Gen Psych 1997; 54:597-606.
180. Siuciak JA, Lewis DR, Wiegand SJ et al. Antidepressant-like effect of brain-derived neurotrophic factor (BDNF). Pharmacol Biochem Behav 1997; 56:131-137.
181. Marty S, Berzaghi MD, berninger b. Neurotrophins and activity-dependent plasticity of cortical interneurons. Trends Neurosci 1997; 20:198-202.
182. Adams B, Sazgar M, Osehobo P et al. Nerve growth factor accelerates seizure development, enhances mossy fiber sprouting, and attenuates seizure-induced decreases in neuronal density in the kindling model of epilepsy. J Neurosci 1997; 17:5288-5296.
183. Van der Ze CE, Rashid K, Le K et al. Intraventricular administration of antibodies to nerve growth factor retards kindling and blocks mossy fiber sprouting in adult rats. J Neurosci 1995; 15: 5316-5323.

184. Reibel S, Larmet Y, Le BT et al. Brain-derived neurotrophic factor delays hippocampal kindling in the rat. Neuroscience 2000; 100:777-788.
185. Anton ES, Weskamp G, Reichardt LF et al. Nerve growth factor and its low-affinity receptor promote Schwann cell migration. Proc Natl Acad Sci USA 1994; 91:2795-2799.
186. Winkler J, Ramirez GA, Kuhn HG et al. Reversible Schwann cell hyperplasia and sprouting of sensory and sympathetic neurites after intraventricular administration of nerve growth factor. Ann Neurol 1997; 41:82-93.
187. Barres BA, Raff MC, Gaese F et al. A crucial role for neurotrophin-3 in oligodendrocyte development. Nature 1994; 367:371-375.
188. McTigue DM, Horner PJ, Stokes BT et al. Neurotrophin-3 and brain-derived neurotrophic factor induce oligodendrocyte proliferation and myelination of regenerating axons in the contused adult rat spinal cord. J Neurosci 1998; 18:5354-5365.
189. Syroid DE, Maycox PJ, Soilu-Hanninen M et al. Induction of postnatal schwann cell death by the low-affinity neurotrophin receptor in vitro and after axotomy. J Neurosci 2000; 20:5741-5747.
190. Torcia M, Bracci-Laudiero L, Lucibello M et al. Nerve growth factor is an autocrine survival factor for memory B lymphocytes. Cell 1996; 85:345-356.
191. Kimata H, Yoshida A, Ishioka C et al. Nerve growth factor specifically induces human IgG4 production. Eur J Immunol 1991; 21:137-141.
192. Neumann H, Misgeld T, Matsumuro K et al. Neurotrophins inhibit major histocompatibility class II inducibility of microglia: involvement of the p75 neurotrophin receptor. Proc Natl Acad Sci USA 1998; 95 :5779-5784.
193. Flugel A, Matsumuro K, Neumann H et al. Anti-inflammatory activity of nerve growth factor in experimental autoimmune encephalomyelitis: inhibition of monocyte transendothelial migration. Eur J Immunol 2001; 31:11-22.
194. Villoslada P, Hauser SL, Bartke I et al. Human nerve growth factor protects common marmosets against autoimmune encephalomyelitis by switching the balance of T helper cell type 1 and 2 cytokines within the central nervous system. J Exp Med 2000; 191:1799-1806.
195. Donovan MJ, Lin MI, Wiegn P et al. Brain derived neurotrophic factor is an endothelial cell survival factor required for intramyocardial vessel stabilization. Development 2000; 127:4531-4540.
196. Donovan MJ, Hahn R, Tessarollo L et al. Identification of an essential nonneuronal function of neurotrophin 3 in mammalian cardiac development. Nat Genetics 1996; 14:210-213.
197. Donovan MJ, Miranda RC, Kraemer R et al. Neurotrophin and neurotrophin receptors in vascular smooth muscle cells. Regulation of expression in response to injury. Am J Pathol 1995; 147:309-324.
198. Ilag LL, Curtis R, Glass D et al. Pan-neurotrophin 1: a genetically engineered neurotrophic factor displaying multiple specificities in peripheral neurons in vitro and in vivo. Proc Natl Acad Sci USA 1995; 92:607-611.
199. Urfer R, TsoulFas P, Soppet D et al. The binding epitopes of neurotrophin-3 to its receptors TrkC and gp75 and the design of a multifunctional human neurotrophin. EMBO J 1994; 13:5896-5909.
200. Friedman WJ, Black IB, Persson H et al. Synergistic trophic actions on rat basal forebrain neurons revealed by a synthetic NGF/BDNF chimaeric molecule. Eur J Neurosci 1995; 7:656-662.
201. Ryden M, Murray-Rust J, Glass D et al. Functional analysis of mutant neurotrophins deficient in low-affinity binding reveals a role for p75LNGFR in NT-4 signalling. EMBO J 1995; 14:1979-1990.
202. Saragovi HU, Gehring K. Development of pharmacological agents for targeting neurotrophins and their receptors. Trends Pharmacol Sci 2000; 21:93-98.
203. LeSauteur L, Maliartchouk S, Le Jeune H et al. Potent human p140-TrkA agonists derived from an anti-receptor monoclonal antibody. J Neurosci 1996; 16:1308-1316.
204. Maliartchouk S, Feng Y, Ivanisevic L et al. A designed peptidomimetic agonistic ligand of TrkA nerve growth factor receptors. Mol Pharmacol 2000; 57:385-391.

205. Treanor JJS, Schmelzer C, Knusel B et al. Heterodimeric neurotrophins induce phosphorylation of Trk receptors and promote neuronal differentiation in PC12 cells. J Biol Chem 1995; 270:23104-23110.
206. Friden PM, Walus LR, Watson P et al. Blood-brain barrier penetration and in vivo activity of an NGF conjugate. Science 1993; 259:373-377.
207. Hefti F. Pharmacology of neurotrophic factors. Annu Rev Pharmacol Toxicol 1997; 37:239-267.
208. Saudou F, Finkbeiner S, Devys D et al. Huntingtin acts in the nucleus to induce apoptosis but death does not correlate with the formation of intranuclear inclusions. Cell 1998; 95:55-66.
209. Shieh PB, Ghosh A. Molecular mechanisms underlying activity-dependent regulation of BDNF expression. J Neurobiol 1999; 41:127-134.
210. timmusk t, Palm K, Metsis M et al. Multiple promoters direct tissue-specific expression of the rat BDNF gene. Neuron 1993; 10:475-489.
211. Ma L, Merenmies J, Parada LF. Molecular characterization of the TrkA/NGF receptor minimal enhancer reveals regulation by multiple cis elements to drive embryonic neuron expression. Development 2000; 127:3777-3788.
212. Pan W, Banks WA, Kastin AJ. Permeability of the blood-brain barrier to neurotrophins. Brain Res 1998; 788:87-94.
213. Gravel C, Gotz R, Lorrain A et al. Adenoviral gene transfer of ciliary neurotrophic factor and brain- derived neurotrophic factor leads to long-term survival of axotomized motor neurons. Nat Med 1997; 3:765-770.
214. Mandel RJ, Gage FH, Clevenger DG et al. Nerve growth factor expressed in the medial septum following in vivo gene delivery using a recombinant adeno-associated viral vector protects cholinergic neurons from fimbria-fornix lesion-induced degeneration. Exp Neurol 1999; 155:59-64.
215. Kordower JH, Emborg ME, Bloch J et al. Neurodegeneration prevented by lentiviral vector delivery of GDNF in primate models of Parkinson s disease. Science 2000; 290:767-773.
216. Flugel A, Willem M, Berkowicz T et al. Gene transfer into CD4+ T lymphocytes: Green fluorescent protein-engineered, encephalitogenic T cells illuminate brain autoimmune responses. Nat Med 1999; 5:843-847.
217. Hoffman D, Wahlberg L, Aebischer P. NGF released from a polymer matrix prevents loss of ChAT expression in basal forebrain neurons following a fimbria-fornix lesion. Exp Neurol 1990; 110:39-44.
218. Shibayama M, Hattori S, Himes BT et al. Neurotrophin-3 prevents death of axotomized Clarke s nucleus neurons in adult rat. J Comp Neurol 1998; 390:102-111.
219. Hoffman D, Breakefield XO, Short MP et al. Transplantation of a polymer-encapsulated cell line genetically engineered to release NGF. Exp Neurol 1993; 122:100-106.
220. Winn SR, Hammang JP, Emerich DF et al. Polymer-encapsulated cells genetically modified to secrete human nerve growth factor promote the survival of axotomized septal cholinergic neurons. Proc Natl Acad Sci USA 1994; 91:2324-2328.
221. Haase G, Kennel P, Pettmann B et al. Gene therapy of murine motor neuron disease using adenoviral vectors for neurotrophic factors. Nat Med 1997; 3:429-436.
222. Sendtner M. Gene therapy for motor neuron disease. Nat Med 1997; 3:380-381.
223. Sagot Y, Rosse T, Vejsada R et al. Differential effects of neurotrophic factors on motoneuron retrograde labeling in a murine model of motoneuron disease. J Neurosci 1998; 18:1132-1141.
224. Nagatsu T, Mogi M, Ichinose H et al. Changes in cytokines and neurotrophins in Parkinson s disease. J Neural Transm Suppl 2000;277-290.
225. Vicario-Abejon C, Johe KK, Hazel TG et al. Functions of basic fibroblast growth factor and neurotrophins in the differentiation of hippocampal neurons. Neuron 1995; 15:105-114.
226. Vicario-Abejon C, Collin C, TsoulFas P et al. Hippocampal stem cells differentiate into excitatory and inhibitory neurons. Eur J Neurosci 2000; 12:677-688.
227. Leventhal C, Rafii S, Rafii D et al. Endothelial trophic support of neuronal production and recruitment from the adult mammalian subependyma. Mol Cell Neurosci 1999; 13:450-464.

228. Rubio F, Kokaia Z, Arco A et al. BDNF gene transfer to the mammalian brain using CNS-derived neural precursors. Gene Ther 1999; 6:1851-1866.
229. Martinez-Serrano A, Lundberg C, Horellou P et al. CNS-derived neural progenitor cells for gene transfer of nerve growth factor to the adult rat brain: Complete rescue of axotomized cholinergic neurons after transplantation into the septum. J Neurosci 1995; 15:5668-5680.
230. Moutsatsos IK, Turgeman G, Zhou S et al. Exogenously regulated stem cell-mediated gene therapy for bone regeneration. Mol Ther 2001; 3:449-461.

FIBROBLAST GROWTH FACTORS AND NEUROPROTECTION

Christian Alzheimer[1] and Sabine Werner[2]

INTRODUCTION FIBROBLAST GROWTH FACTORS (FGFS) AND FGF RECEPTORS

The FGFs comprise a large group of structurally similar polypeptide mitogens which currently includes 22 different members. The first members of this family, FGF1 (acidic FGF, aFGF) and FGF2 (basic FGF, bFGF), were described in 1986 by Jaye et al[1] and Abraham et al.[2] In the meantime, 20 other FGFs have been discovered which were designated FGF3 FGF23. FGF19 has so far only been described in humans and FGF15 only in mice, and it has been suggested that FGF19 is the human ortholog of mouse FGF15. All members of the FGF family range in molecular weight from 17 to 34 kDa in vertebrates, and some of them are glycosylated (for review see ref. 3).

Most of the FGFs have a classical aminoterminal signal sequence and are, therefore, efficiently secreted via the endoplasmic reticulum-Golgi secretory pathway. FGF9, 16 and 20 lack an obvious amino-terminal signal peptide, but are nevertheless secreted. FGFs11-14 are thought to remain intracellular, and secretion of these FGFs has not yet been observed. Although FGF1 and FGF2 lack a signal sequence and are normally located in the cytoplasm or in the nucleus, they can be found on the cell surface and within the extracellular matrix. They are either released from damaged cells or secreted by an alternative exocytotic mechanism that bypasses the endoplasmic reticulum-Golgi pathway (for review see refs. 3,4).

FGFs have been shown to interact with three different types of binding partners: heparan sulphate proteoglycans,[3-5] a cysteine-rich transmembrane FGF-binding

[1]Institute of Physiology, University of Munich, Pettenkoferstr. 12, D-80336 Munich, Germany; Present address: Institute of Physiology, University of Kiel, 24098 Kiel, Germany and [2]Institute of Cell Biology, ETH-Z rich, H nggerber g, CH-8093 Zurich, Switzerland.

protein which appears to be involved in the regulation of intracellular FGF trafficking,[6] and four high-affinity transmembrane FGF receptors of the tyrosine kinase family which are responsible for signal transduction.[7] The FGF receptors, FGFR1 FGFR4, are transmembrane protein tyrosine kinases with either two or three immunoglobulin-like domains and a heparin binding sequence in the extracellular part of the receptor. Alternative mRNA splicing generates receptors with different carboxyterminal halves of the third immunoglobulin-like domain, designated IIIb or IIIc isoforms. This alternative splicing event is regulated in a cell- and tissue-specific manner, and has been shown to dramatically affect ligand-receptor binding specificity.[7] The different members of the FGF family bind to the receptor splice variants with different affinities. Most FGFs bind to a specific subset of FGF receptors. FGF1, however, binds to all known receptors, and FGF7 specifically interacts with the IIIb isoform of FGFR2.[8]

A characteristic feature of FGFs is their interaction with heparin or heparan sulphate proteoglycans. These interactions stabilize FGFs and may limit their diffusion and release into interstitial spaces. Most importantly, the interaction of FGFs with heparin or heparan sulphate proteoglycans is essential for the activation of the signaling receptor.[3,5]

Most members of the FGF family have a broad mitogenic spectrum. They stimulate proliferation of a variety of cells of mesodermal, ectodermal and also of endodermal origin.[3,4] The only known exception is FGF7, which seems to act only on epithelial cells, at least in the adult organism.[9] FGFs are not only mitogenic, but they also have the capacity to regulate migration and differentiation in vitro and in vivo. Finally, some of the FGFs have been shown to be cytoprotective (see below).[3,9]

FGFs and their receptors are expressed at multiple sites in the developing and adult animal, suggesting important roles of FGFs in development and tissue homeostasis. This hypothesis was confirmed by the wide variety of phenotypic abnormalities observed in FGF and FGFR knockout animals.[3] Finally, abnormalities in FGF/FGF receptor signaling are associated with human disease, including cancer and various genetic disorders.

DISTRIBUTION OF FGFS IN ADULT BRAIN

The brain is a rich source of various FGFs. While the expression pattern of FGF2 has been studied in detail (for review, see refs. 10,11), data on the mRNA expression and the immunoreactivity of the other FGFs is relatively sparse. Both FGF1 and FGF2 have been localized to glial cells and neurons of the CNS. Notably, high levels of FGF1 were found in neurons of brain regions that are at high risk for neurodegenerative diseases such as Alzheimer«s disease (magnocellular forebrain cholinergic neurons), Parkinson«s disease (substantia nigra neurons) and amyotrophic lateral sclerosis (motor neurons).[11] Most, if not all astrocytes of the brain express FGF2, whereas a prominent neuronal localization appears to be confined mainly to neurons in the CA2 region of the hippocampus.[12] In both astrocytes and CA2 neurons, FGF2 immunoreactivity is localized primarily in the nucleus and to a lesser extent in the cytoplasm and the processes of stained cells.[13] *Fgf5* mRNA is widely distributed at low levels in the brain, with several loci of *Fgf5* expression found in the cerebral cortex, hippocampus and thalamus. At least some of the *Fgf5* expressing cells appear to be neurons.[14] *Fgf7* mRNA could not be detected in any of the postnatal brain regions examined so far.[15,16] *Fgf9* mRNA is moderately to weakly expressed in widespread regions of the brain, with some preference to brain regions involved in motor control (red nucleus, parts of cerebellum, oculomotor nucleus). The cellular localization of *Fgf9* mRNA indicated that this FGF family member is mainly produced by neurons.[17,18] Expression of the *Fgf10* gene is spatially restricted to some regions of the brain, including hippocampus, thalamus, and several nuclei in midbrain and brainstem, with a preferential expression in neurons rather than in glial cells.[15] Interestingly, with its predominant expression in the CA1 and CA3, but not in the CA2 region of the hippocampus, the spatial distribution of *Fgf10* mRNA is just opposite to that of *Fgf2* mRNA, which is largely confined to the CA2 region, suggesting that the two FGFs have distinct functions in the hippocampus. Finally, *Fgf14* mRNA was recently detected in the brain, with FGF14-1b as the predominant isoform. The developmental expression pattern suggests that FGF14 plays a significant role in cerebellar development.[19]

All FGF receptors are expressed in the brain. The *Fgfr1* gene is significantly expressed in select neuronal populations, but also in astrocytes and oligodendrocytes.[20,21] FGFR2 and FGFR3 are predominantly found in astrocytes and oligodendrocytes,[22,23] whereas FGFR4 is almost exclusively expressed in neurons of the medial habenular nucleus.[24]

FGFS AND NEURONAL DEVELOPMENT

FGFs and their receptors (FGFRs) have been implicated in many aspects of neuronal development, where they promote proliferation, differentiation and axonal branching. For example, FGF8 regulates growth and patterning of the midbrain and the anterior forebrain,[25,26] FGF3 plays an essential role in the development of the inner ear,[27] and FGF8, FGF14, FGF15, FGF17, and FGF18 are expressed in the developing cerebellum.[19,28] Demonstrating the significance of FGFs for appropriate cerebellar development, a phenotype displaying gait defects was observed in $FGF17^{-/-}$ and $FGF8^{\Delta 2,3n/+}$ mice, albeit with low penetrance.[28] FGF2 appears to be intimately involved in cortical development (for review see refs. 29,30). It supports the survival of neurons from many regions of the fetal rat brain, including hippocampus, cortex, thalamus and striatum.[31,32] In addition to its neurotrophic effects, FGF2 acts as a mitogenic factor, stimulating the proliferation of neuronal progenitor cells. Most interestingly in the context of this chapter, FGF2 is capable of activating a latent neurogenic program in neural stem cells from diverse regions of the adult brain and of stimulating neurogenesis in the mature CNS.[33-35] FGF2 is present in the telencephalon as early as E9.5 and high levels are found in the cerebral cortex throughout neurogenesis and into adulthood.[10] FGF2 appears to exert a dual function on early phases of cortical neuroectoderm cell proliferation and later phases of differentiation. Since astrocytes appear to be the predominant source of FGF2, with only few neuronal populations displaying FGF2 immunoreactivity (see above), glial cells might provide trophic support to neuronal cells.[12] Despite accumulating evidence from in vitro studies, which strongly implicate FGF2 as a neurotrophic factor, it was not before the generation of mice lacking FGF2 that the issue of whether endogenous FGF2 is essential for survival of cortical neurons was settled. Neurohistological analysis of FGF2 knockout mice revealed abnormalities in the cytoarchitecture of the neocortex, combined with a significant reduction in neuronal density, most pronounced in layer V.[36,37] It was concluded that one important role of FGF2 during cortical neurogenesis may be to amplify the progenitor pool for pyramidal projection neurons.[38] Data from several developmental studies indicate that FGF2 also acts as a target-derived factor that promotes axon branching of cortical neurons. Specifically, FGF2 may induce the formation of collateral axon branches by its effects on the morphology and the behavior of the primary growth cone.[39] The strong effect of FGF2 on neurite morphogenesis has been linked to the increase by FGF2 of L-type Ca^{2+} channels in fetal neurons.[40]

UPREGULATION OF FGFS AFTER BRAIN INJURY

The expression of many neurotrophic factors in CNS neurons is regulated by physiological stimuli, such as afferent synaptic activity, suggesting that these factors participate in functional and/or morphological changes associated with neuronal plasticity. More dramatic alterations in the expression pattern of many neurotrophic factors are induced by several forms of acute brain injury.[41] Since the late 1980«s, numerous publications have accumulated evidence for a pronounced upregulation of FGFs in the cellular response to acute brain injury. Among the FGF family members, FGF2 clearly emerged as the central player in acute CNS damage, and we will hence focus the following section on this factor. FGF2 upregulation was demonstrated after mechanical brain injury,[42-48] after ischemic insults,[49,50] and after convulsive seizures.[51-54] Three days after mechanical cortical injury, macrophages are the predominant FGF2 immunoreactive cells at the lesion site. A second phase of increased FGF2 expression, which peaks at about one week post-lesion, is mediated by activated astrocytes and microglia, particularly at the border between the neuronal and scar tissue. Although multiple cells within the lesioned CNS, including astrocytes, microglia, neurons and vascular endothelial cells, might all be able to express FGF2, glial cells appear to be the primary source of the newly synthesized FGF2. In support of this notion, the gradual fall in FGF2 levels one week after injury parallels the concomitant decrease in reactive glia.[45] Interestingly, an increase in the intensity of FGF2 immunoreactivity was also detected within the extracellular matrix surrounding the lesion, suggesting that FGF2 was released from astrocytes and becomes available to neurons.[46] A similar time course of induction of glial FGF2 synthesis was observed after transient forebrain ischemia.[49] In contrast to the slow, but sustained upregulation of FGF2 in traumatic or ischemic lesion models, a faster and transient response was induced by epileptic convulsions not associated with neuronal loss. For example, seizures induced by microinjection of the $GABA_A$ receptor antagonist, bicuculline, in the deep prepiriform cortex lead to a significant increase in *Fgf2* mRNA in the entorhinal cortex, the hippocampus and the olfactory bulb within 5 hours of epileptic activity.[51] However, if seizure activity was severe enough to produce neuronal loss and reactive astrocytosis, a long-term induction of *Fgf2* gene expression for up to 2 weeks was observed in the damaged region.[53] Although astrocytes appear to represent the main source of FGF2 in seizure models, elevated *Fgf2* mRNA signals were also observed in select neuronal populations.[54,55]

Upregulation of FGF2 also has been implicated in the protective effect of cortical spreading depression (SD) against subsequent ischemic damage.[56,57] Cortical SD is a rapid and nearly complete, but reversible depolarization of a large population of neurons, which propagates in a slow wave-like fashion through the gray matter (for review see ref. 58 and Chapter 5 of this book). The protective effect of SD against ischemic insults might last for several days suggesting that some kind of sustained downmodulation of neuronal vulnerability is initiated. Right now, it is not clear whether the upregulation by SD of FGF2 is a pivotal mechanism affording enhanced protection against subsequent stroke. However, with the generation of FGF2 deficient mice (see above), this intriguing issue should be resolved in the near future.

NEUROPROTECTIVE EFFECTS OF FGF2

Among the various growth factors and cytokines studied so far, the neuroprotective and neurotrophic profile of FGF2 is best documented (reviewed in ref. 59). If administered during or within hours of acute injury, systemic or intracerebroventricular (icv.) FGF2 reduces infarct size after stroke,[60-66] diminishes histopathologic damage associated with fluid percussion injury,[67] affords neuroprotection against N-methyl-D-aspartate (NMDA) and kainate receptor-mediated excitotoxicity,[60,68,69] and prevents the death of axotomized CNS neurons.[70-74] It is worth noting that FGF2, which is highly neuroprotective against seizure-induced long-term behavioral deficits, might also act as a convulsant at higher concentrations.[75] If administered 1 day after focal cerebral infarction, intracisternal FGF2, although no longer reducing infarct size, is still capable of enhancing behavioral recovery.[76] Correspondingly, blockade of FGF2 by neutralizing antibodies retarded recovery of forelimb manipulatory abilities after unilateral suction lesion of the motor cortex.[77] Further support for a beneficial role of FGF2 in acute brain injury comes from a study in which mice expressing a bovine FGF2 transgene in the brain showed increased resistance to hypoxemic-ischemic cerebral damage.[78] The finding that intravenous FGF2 rapidly crosses the blood brain barrier in adult brain, with the levels in blood plasma and cerebrospinal fluid rising in parallel, bears particular significance for the application of this factor in a clinical setting.[34,78A] As a result of the outstanding benefits of FGF2 in various animal lesion models and its high cerebral bioavailability after intravenous injection, clinical trials of intravenous FGF2 (Fiblast®) in acute stroke were conducted. In a North American phase II/III trial, 302 patients suffering from acute ischemic stroke were enrolled between August

1997 and May 1998, when enrollment was halted by the Data Safety Monitoring Committee following an interim analysis, which revealed an unfavorable risk-to-benefit ratio in stroke patients treated with FGF2 versus those treated with placebo (information according to the Internet Stroke Center at www.strokecenter.org).

Data on the possible neuroprotective potential of other FGFs is sparse. Several studies found beneficial effects of FGF1 (acidic FGF) in animal models of acute ischemic or excitotoxic brain damage.[79-83] A recent study demonstrated that FGF7 (keratinocyte growth factor) prevents ischemia-induced delayed neuronal death in the hippocampal CA1 region of the gerbil brain.[84] Finally, FGF8 was shown to protect cultured hippocampal neurons from oxidative insult.[85]

FGFS AND GLIA

Given that FGF2 is mitogenic for oligodendrocytes and astrocytes, and stimulates migration and functional differentiation of astrocytes, one might wonder whether, in addition to its neuroprotective effects, FGF2 may also have the undesirable side effect of promoting scar formation (for review see ref. 86 and Chapter 4 of this book). Although the involvement of FGF2 in scar formation appears minor compared to that of other potent fibrotic agents such as TGF-β_1, several reports lend credence to the notion that FGF2 is not an innocent player in the glial response to acute brain injury. For example, the reactive gliosis following mechanical or electrolytic lesions in the neonatal and adult brain was significantly augmented and accelerated by FGF2 injected into the lesion site just after the lesion was performed.[87-89] Along the same lines, FGF2, if injected into various regions of the noninjured adult rat brain, produced a glial reaction that resembled the reactive gliosis seen after brain injury.[89] Supporting its involvement in extracellular matrix remodeling after injury, FGF2 was found to increase the production of tenascin-C mRNA and protein in cultured hippocampal astrocytes.[90] It should be pointed out, however, that the effects of FGF2 on glial cells might also indirectly promote survival of select neuronal populations, as FGF2 (and FGF1) were found to stimulate nerve growth factor (NGF) synthesis and secretion by astrocytes.[91] Since intercellular communication between astrocytes through gap junctions is essential to many of their functions, the finding that FGF2, FGF5 and FGF9 downregulate astroglial gap junctions and functional coupling in a brain region-specific fashion is of particular interest.[92] It is well documented that acute brain injury is associated with downregulation of connexin 43, the predominant component of astroglial gap junctions. One might thus speculate whether the lesion-induced upregulation of FGFs is causally linked to the uncoupling of astrocytes.

NEUROPROTECTIVE MECHANISMS OF FGF2

Ionotropic Glutamate Receptors

The seminal study by Mattson et al[93] in cultured hippocampal neurons provided compelling evidence that FGF2 raises the threshold for glutamate neurotoxicity and reduces the rise in intracellular Ca^{2+} associated with glutamate receptor activation. Since then, a number of studies by Mattson and coworkers and other groups have elaborated on the mechanisms underlying the protective effect of FGF2 against excitotoxic damage. For example, FGF2 was found to exert differential effects on glutamate receptor subtype expression: Whereas it selectively increases the AMPA-receptor subunit GluR1 in hippocampal neurons,[94] a functional 71kDA NMDA receptor that mediates Ca^{2+} influx and neurotoxicity is downregulated by FGF2 in the same preparation.[95] In functional terms, this differential effect of FGF2 on glutamate receptor expression translates into an increase in AMPA receptor-mediated, but a decrease in NMDA receptor-mediated Ca^{2+} elevations, with the latter serving as a significant excitoprotective mechanism.[94] Providing further insights into the mechanisms underlying the inhibitory effect of FGF2 on NMDA receptor-mediated Ca^{2+} influx, a recent study demonstrated that chronic treatment (hours to days) of cultured hippocampal neurons by FGF2 potentiates Ca^{2+}-dependent inactivation of NMDA receptor currents through a calcineurin-dependent mechanism.[96] In marked contrast, short-term exposure of acutely dissociated hippocampal neurons to FGF2 produced a selective enhancement of NMDA receptor-mediated increases in cytosolic Ca^{2+}.[97]

Ca^{2+} Homeostasis, Mitochondrial Dysfunction and Reactive Oxygen Species

In addition to reducing Ca^{2+} influx through NMDA receptors, FGF2 also operates on a second line of defense against the loss of Ca^{2+} homeostasis and the concurrent mitochondrial dysfunction. For example, FGF2 increases the synthesis of calbindin D28k,[98] a Ca^{2+}-binding protein thought to exert an excitoprotective role in CNS neurons.[99] Perhaps even more importantly, FGF2 increases the activity of antioxidant enzymes, such as superoxide dismutase and glutathione reductase.[100,101] Extending the beneficial effects of FGF2 into the range of chronic neurodegenerative diseases such as Alzheimer«s disease, Mattson and coworkers recently demonstrated that FGF2 also attenuates oxidative stress and mitochondrial dysfunction induced by amyloid peptide Aβ [102] and mitigates the enhanced neuronal vulnerability to excitotoxicity in cultured hippocampal neurons from presenilin1 mutant knock-in mice.[103]

Apoptosis and Neurogenesis

Two recent studies, one performed in vitro and one in vivo, implicated FGF2 in anti-apoptotic pathways. In cultured hippocampal neurons, FGF2 prevented apoptosis induced by NO donors. Whereas NO donor-induced apoptosis was typically associated with downregulation of Bcl-2, upregulation of Bax and subsequent caspase-3-like activation, pretreatment with FGF2 abrogated the changes in Bcl-2 and Bax protein levels as well as the caspase-3-like activation.[104] Somewhat similar findings were found in an animal stroke model in which permanent occlusion of the right middle cerebral artery (MCAO) causes focal cerebral infarction. In animals receiving intravenous infusion of FGF2 for 3 h, beginning at 30 min after MCAO, FGF2 prevented the reduction of immunoreactivity of the anti-apoptotic protein Bcl-2, which is typically observed in untreated stroke animals. In contrast to the in vitro study mentioned above, FGF2 did not alter immunoreactivity to the pro-apoptotic proteins Bax, caspase-1, and caspase-3. Nevertheless, FGF2 produced a substantial decrease in apoptotic neurons, especially in the border (penumbra) of the infarct, which is exactly the zone predominantly spared by FGF2 treatment.[105] With respect to the neuroprotective effect of FGF2 in stroke models, it is worth noting that upregulation by FGF2 of endothelial NO synthase (eNOS)[106] and subsequent increase in cerebral blood flow (CBF)[107] is not the predominant protective mechanism because FGF2 retains its infarct-reducing efficacy in eNOS deficient mice, in which CBF is not increased by FGF2.[108]

In addition to preventing lesion-induced apoptosis, FGF2 was also reported to promote neurogenesis in the adult dentate gyrus in response to injury.[109] The dentate gyrus belongs to the hippocampal formation and contains into adulthood neuroprogenitor cells that are able to divide and differentiate into granule cells.[110,111] Whereas kainate injection or MCAO induced appreciable neurogenesis in the dentate gyrus of adult control mice, the number of newly generated neurons, as indicated by bromodeoxyuridine (BrdU) incorporation into nuclei of dentate granule cells, was significantly decreased in FGF2 deficient mice, but could be restored to control levels after icv. injection of a herpes simplex virus-1 amplicon vector carrying the FGF2 gene.[109] These data suggest that the lesion-associated upregulation of FGF2 not only promotes survival of neurons exposed to various forms of acute damage, but plays also a critical role in neuronal repair.

Interaction Between FGF2 and Activin A

In a collaborative study between our laboratories, we made the intriguing observation that induction of activin A, a member of the transforming growth factor-β (TGF-β) superfamily (see next chapter), is essential for the neuroprotective action of FGF2 in vivo.[112] Like the other members of the TGF-β superfamily, activins are dimeric proteins, consisting of two β_A subunits (activin A), two β_B subunits (activin B) or a β_A and a β_B subunit (activin AB). In addition, β_C, β_D, and β_E subunits have been identified, although the corresponding proteins have not been characterized in detail. The β_A and β_B subunits can also dimerize with a homologous α subunit, leading to the formation of inhibin A ($\alpha\beta_A$) or inhibin B ($\alpha\beta_B$). Previous findings from our and other laboratories already pointed to a possible involvement of activin A in the early neuronal response to injury.[113-117] Since activin A promotes survival of midbrain and hippocampal neurons in vitro,[118,119] reduces ischemic brain injury in infant rats,[120] and protects striatal and midbrain neurons against neurotoxic damage,[118,121] we speculated that the lesion-induced upregulation of activin A might serve an excitoprotective function. A first indication for a possible interaction between FGF2 and activin A came from a comparison between the activin β_A mRNA expression pattern in lesioned and in FGF2-protected hippocampi. In our lesion model, exogenous FGF2 abolished the neuronal damage in the CA3 region of the ipsilateral hippocampus which is typically seen after intracerebroventricular (icv.) application of this excitotoxin (Fig. 1A). Unexpectedly, FGF2 also produced a strikingly stronger upregulation of β_A mRNA than KA injection alone (Fig. 1B). FGF2 substantially augmented β_A mRNA expression on the ipsilateral side at 6h and 24h post-lesion, without appreciably influencing the time course of signal elevation. Two independent sets of experiments corroborated the hypothesis that activin A is crucially involved in the neuroprotective effects ascribed to FGF2. First, recombinant activin A was as effective as exogenous FGF2 in preventing excitotoxic neuronal loss (Fig. 1C). Second, the activin-binding protein follistatin, which neutralizes activin in vitro and in vivo,[122,123] consistently abrogated the beneficial action of FGF2 in the KA lesion model (Fig. 1D). Control experiments dispelled concerns that follistatin caused neuronal damage by a mechanism other than its activin-neutralizing action. A different type of interaction between FGF2 and the TGF-β superfamily was reported from cultured midbrain dopaminergic neurons, in which the neurotrophic effect of FGF2 was mediated by TGF-β1-3.[124] In contrast to activin A, which is of neuronal origin in lesioned hippocampus, the effect of the TGF-βs was mediated by co-cultured glial cells.

The main neuroprotective mechanisms of FGF2 are summarized in Fig. 2. As it is becoming evident that induction of activin A is an essential step in the signaling cascade affording neuroprotection after application of FGF2, it remains to be determined which of the effects, originally attributed to FGF2, are directly mediated by this growth factor, and which require upregulation of activin A.

SUMMARY AND CONCLUSIONS

Several members of the FGF family, in particular FGF2, are intimately involved in neuronal protection and repair after ischemic, metabolic or traumatic brain injury. Expression of *Fgf2* mRNA and protein is strongly upregulated after neuronal damage, with glial cells as the predominant source. Given its survival-promoting effects on cultured neurons, exogenous FGF2 was tested in several animal models of stroke and excitotoxic damage, in which it consistently proved protective against neuronal loss. FGF2 affords neuroprotection by interfering with a number of signaling pathways, including expression and gating of NMDA receptors, maintenance of

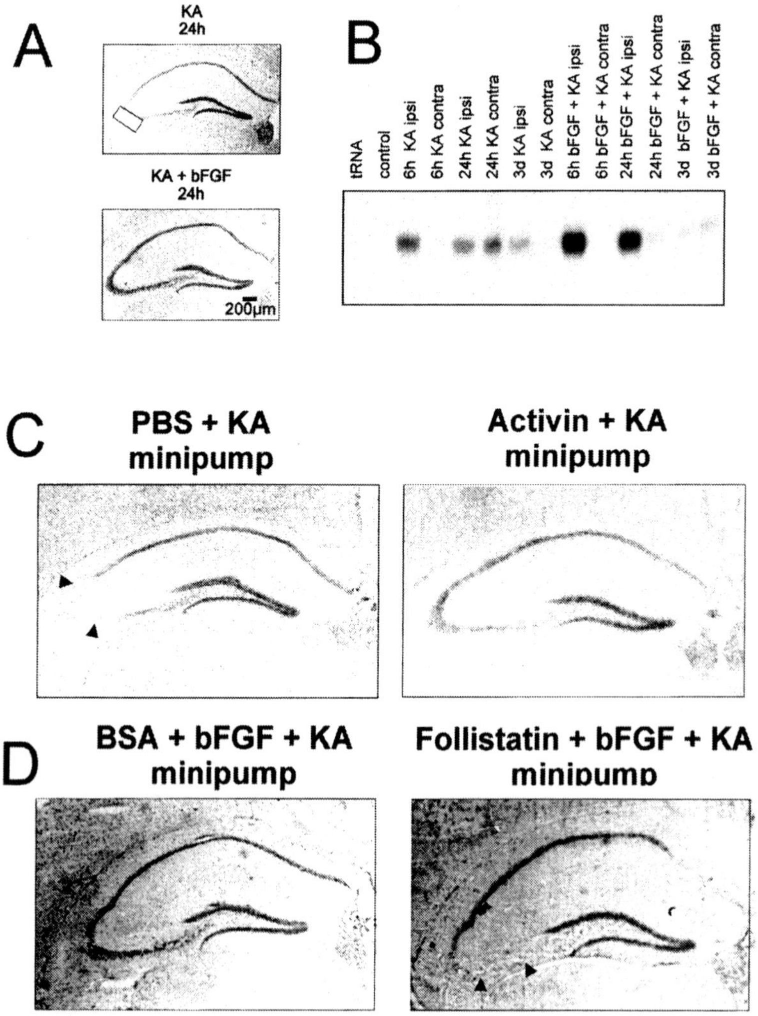

Figure 1. Induction of activin A is an essential step for the neuroprotective action of FGF2 (bFGF) in an animal model of select excitotoxic neuronal loss. A: Intracerebroventricular (icv.) injection of FGF2 prevents the loss of hippocampal CA3 neurons (lower micrograph), which is typically seen in the ipsilateral CA3 region after icv. kainic acid (KA). Box in upper microphotograph indicates site of lesion in this Nissl-stained hippocampal section. B: RNase protection assays were performed using ipsilateral (ipsi) and contralateral (contra) hippoccampi from mice receiving an icv. injection of the excitotoxin kainic acid (KA) or from mice co-injected with KA and FGF2. FGF2 strongly enhanced the lesion-associated induction of activin β_A mRNA. C: Continuous icv. application of exogenous recombinant activin A by means of an osmotic minipump protected hippocampal CA3 neurons against excitotoxic KA lesion (right micrograph). Arrowheads in left micrograph (control with solvent) indicate area of neuronal loss. D: The protective action of icv. FGF2 (left micrograph) was abrogated in the presence of the activin-neutralizing protein, follistatin (arrowheads in right micrograph indicate area of neuronal loss). Reprinted with permission from Tretter et al, Nature Med. 2000; 6:812-815.

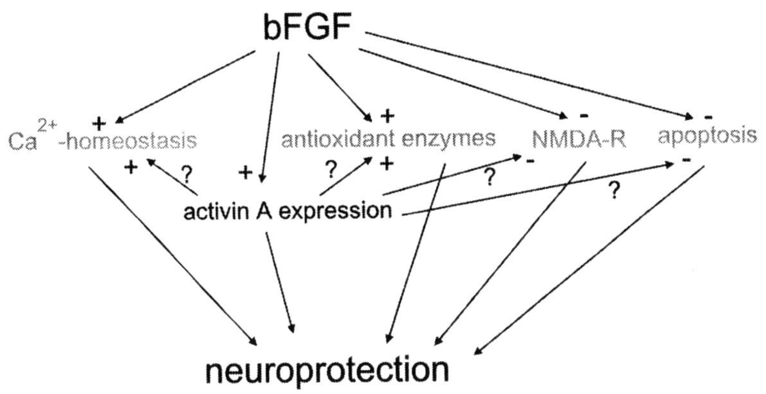

Figure 2. Synopsis of neuroprotective signaling pathways influenced by FGF2. (-) and (+) indicate attenuation and enhancement, respectively, of respective target by FGF2. Induction by FGF2 of activin A plays a central role, but it remains to determined which of the signaling pathways are directly, and which are indirectly, i.e., through activin A, modulated by FGF2.

Ca^{2+} homeostasis and regulation of ROS detoxifying enzymes. FGF2 prevents apoptosis by strengthening anti-apoptotic pathways and promotes neurogenesis in adult hippocampus after injury. The protective action of FGF2 has been linked to its augmenting effect on the lesion-induced upregulation of activin A, a member of the TGF-β superfamily. Despite the well-documented benefits of FGF2 in animal models of stroke, there is currently no clinical development in stroke, after a phase II/III trial with FGF2 in acute stroke patients was discontinued because of an unfavorable risk-to-benefit ratio. As the molecular targets of FGF2 are going to be unraveled over the next years, new therapeutic strategies will hopefully emerge that enable us to influence the various protective mechanisms of FGF2 in a more specific fashion.

ACKNOWLEDGEMENTS

The authors thank David Ornitz for comments on the manuscript. The work of the authors was supported by the Deutsche Forschungsgemeinschaft (DFG), the Bundesministerium f r Bildung und Forschung (Bmb+f), and the Human Frontier Science Program.

REFERENCES

1. Jaye M, Howk R, Burgess W et al. Human endothelial cell growth factor: Cloning, nucleotide sequence, and chromosome localization. Science 1986; 233:541-545.
2. Abraham JA, Mergia A, Whang JL et al. Nucleotide sequence of a bovine clone encoding the angiogenic protein, basic fibroblast growth factor. Science 1986; 233:545-548.
3. Ornitz DM, Itoh N. Fibroblast growth factors. Genome Biol 2001; 2:REVIEWS3005.

4. Basilico C, Moscatelli D. The FGF family of growth factors and oncogenes. Adv Cancer Res 1992; 59:115-165.
5. Ornitz DM. FGFs, heparan sulfate and FGFRs: complex interactions essential for development. BioEssays 2000; 22:108-112.
6. Zuber ME, Zhou Z, Burrus LW et al. Cysteine-rich FGF receptor regulates intracellular FGF-1 and FGF-2 levels. J Cell Physiol 1997; 170:217-227.
7. Johnson DE, Williams LT. Structural and functional diversity in the FGF receptor multigene family. Adv Cancer Res 1993; 60:1-41.
8. Ornitz DM, Xu J, Colvin JS et al. Receptor specificity of the fibroblast growth factor family. J Biol Chem 1996; 271:15292-15297.
9. Werner S. Keratinocyte growth factor: A unique player in epithelial repair processes. Cytokine Growth Factor Rev 1998; 9:153-165.
10. Baird A. Fibroblast growth factors: Activities and significance of non-neurotrophin neurotrophic growth factors. Curr Opin Neurobiol 1994; 4:78-86.
11. Eckenstein FP, Andersson C, Kuzis K et al. Distribution of acidic and basic fibroblast growth factors in the mature, injured and developing rat nervous system. Prog Brain Res 1994; 103:55-64.
12. Woodward WR, Nishi R, Meshul CK et al. Nuclear and cytoplasmic localization of basic fibroblast growth factor in astrocytes and CA2 hippocampal neurons. J Neurosci 1992; 12:142-152.
13. Williams TE, Meshul CK, Cherry NJ et al. Characterization and distribution of basic fibroblast growth factor-containing cells in the rat hippocampus. J Comp Neurol 1996; 370:147-158.
14. Haub O, Drucker B, Goldfarb M. Expression of the murine fibroblast growth factor 5 gene in the adult central nervous system. Proc Natl Acad Sci USA 1990; 87:8022-8026.
15. Hattori Y, Yamasaki M, Konishi M et al. Spatially restricted expression of fibroblast growth factor-10 mRNA in the rat brain. Brain Res Mol Brain Res 1997; 47:139-146.
16. Mason IJ, Fuller-Pace F, Smith R et al. FGF-7 (keratinocyte growth factor) expression during mouse development suggests roles in myogenesis, forebrain regionalisation and epithelial-mesenchymal interactions. Mech Dev 1994; 45:15-30.
17. Tagashira S, Ozaki K, Ohta M et al. Localization of fibroblast growth factor-9 mRNA in the rat brain. Brain Res Mol Brain Res 1995; 30:233-241.
18. Colvin JS, Feldman B, Nadeau JH et al. Genomic organization and embryonic expression of the mouse fibroblast growth factor 9 gene. Dev Dyn 1999; 216:72-88.
19. Wang Q, McEwen DG, Ornitz DM. Subcellular and developmental expression of alternatively spliced forms of fibroblast growth factor 14. Mech Dev 2000; 90:283-287.
20. Gonzalez AM, Berry M, Maher PA et al. A comprehensive analysis of the distribution of FGF-2 and FGFR1 in the rat brain. Brain Res 1995; 701:201-226.
21. Beer HD, Vindevoghel L, Gait MJ et al. Fibroblast growth factor (FGF) receptor 1-IIIb is a naturally occurring functional receptor for FGFs that is preferentially expressed in the skin and the brain. J Biol Chem 2000; 275:16091-16097.
22. Peters K, Ornitz D, Werner S et al. Unique expression pattern of the FGF receptor 3 gene during mouse organogenesis. Dev Biol 1993; 155:423-430.
23. Miyake A, Hattori Y, Ohta M et al. Rat oligodendrocytes and astrocytes preferentially express fibroblast growth factor receptor-2 and -3 mRNAs. J Neurosci Res 1996; 45:534-541.
24. Itoh N, Yazaki N, Tagashira S et al. Rat FGF receptor-4 mRNA in the brain is expressed preferentially in the medial habenular nucleus. Brain Res Mol Brain Res 1994; 21:344-348.
25. Crossley PH, Martinez S, Martin GR. Midbrain development induced by FGF8 in the chick embryo. Nature 1996; 380:66-68.
26. Martinez S, Crossley PH, Cobos I et al. FGF8 induces formation of an ectopic isthmic organizer and isthmocerebellar development via a repressive effect on Otx2 expression. Development 1999; 126:1189-1200.
27. Represa J, Leon Y, Miner C et al. The int-2 proto-oncogene is responsible for induction of the inner ear. Nature 1991; 353:561-563.

28. Xu J, Liu Z, Ornitz DM. Temporal and spatial gradients of FGF8 and FGF17 regulate proliferation and differentiation of midline cerebellar structures. Development 2000; 127:1833-1843.
29. Temple S, Qian X. bFGF, neurotrophins, and the control or cortical neurogenesis. Neuron 1995; 15:249-252.
30. Vaccarino FM, Schwartz ML, Raballo R et al. Fibroblast growth factor signaling regulates growth and morphogenesis at multiple steps during brain development. Curr Top Dev Biol 1999; 46:179-200.
31. Walicke P, Cowan WM, Ueno N et al. Fibroblast growth factor promotes survival of dissociated hippocampal neurons and enhances neurite extension. Proc Natl Acad Sci USA 1986; 83:3012-3016.
32. Walicke PA, Baird A. Trophic effects of fibroblast growth factor on neural tissue. Prog Brain Res 1988; 78:333-338.
33. Palmer TD, Ray J, Gage FH. FGF-2-responsive neuronal progenitors reside in proliferative and quiescent regions of the adult rodent brain. Mol Cell Neurosci 1995; 6:474-486.
34. Wagner JP, Black IB, DiCicco-Bloom E. Stimulation of neonatal and adult brain neurogenesis by subcutaneous injection of basic fibroblast growth factor. J Neurosci 1999; 19:6006-6016.
35. Palmer TD, Markakis EA, Willhoite AR et al. Fibroblast growth factor-2 activates a latent neurogenic program in neural stem cells from diverse regions of the adult CNS. J Neurosci 1999; 19:8487-8497.
36. Ortega S, Ittmann M, Tsang SH et al. Neuronal defects and delayed wound healing in mice lacking fibroblast growth factor 2. Proc Natl Acad Sci USA 1998; 95:5672-5677.
37. Dono R, Texido G, Dussel R et al. Impaired cerebral cortex development and blood pressure regulation in FGF-2-deficient mice. EMBO J 1998; 17:4213-4225.
38. Raballo R, Rhee J, Lyn-Cook R et al. Basic fibroblast growth factor (FGF2) is necessary for cell proliferation and neurogenesis in the developing cerebral cortex. J Neurosci 2000; 20:5012-5023.
39. Szebenyi G, Dent EW, Callaway JL et al. Fibroblast growth factor-2 promotes axon branching of cortical neurons by influencing morphology and behavior of the primary growth cone. J Neurosci 2001; 21:3932-3941.
40. Shitaka Y, Matsuki N, Saito H et al. Basic fibroblast growth factor increases functional L-type Ca2+ channels in fetal rat hippocampal neurons: implications for neurite morphogenesis in vitro. J Neurosci 1996; 16:6476-6489.
41. Isackson PJ. Trophic factor response to neuronal stimuli or injury. Curr Opin Neurobiol 1995; 5:350-357.
42. Finklestein SP, Apostolides PJ, Caday CG et al. Increased basic fibroblast growth factor (bFGF) immunoreactivity at the site of focal brain wounds. Brain Res 1988; 460:253-259.
43. Frautschy SA, Walicke PA, Baird A. Localization of basic fibroblast growth factor and its mRNA after CNS injury. Brain Res 1991; 553:291-299.
44. Logan A, Frautschy SA, Baird A. Basic fibroblast growth factor and central nervous system injury. Ann N Y Acad Sci 1991; 638:474-476.
45. Logan A, Frautschy SA, Gonzalez AM et al. A time course for the focal elevation of synthesis of basic fibroblast growth factor and one of its high-affinity receptors (flg) following a localized cortical brain injury. J Neurosci 1992; 12:3828-3837.
46. Gomez-Pinilla F, Lee JW, Cotman CW. Basic FGF in adult rat brain: cellular distribution and response to entorhinal lesion and fimbria-fornix transection. J Neurosci 1992; 12:345-355.
47. Gomez-Pinilla F, Cotman CW. Transient lesion-induced increase of basic fibroblast growth factor and its receptor in layer VIb (subplate cells) of the adult rat cerebral cortex. Neuroscience 1992; 49:771-780.
48. Reilly JF, Kumari VG. Alterations in fibroblast growth factor receptor expression following brain injury. Exp Neurol 1996; 140:139-150.

49. Takami K, Iwane M, Kiyota Y et al. Increase of basic fibroblast growth factor immunoreactivity and its mRNA level in rat brain following transient forebrain ischemia. Exp Brain Res 1992; 90:1-10.
50. Lippoldt A, Andbjer B, Rosen L et al. Photochemically induced focal cerebral ischemia in rat: time dependent and global increase in expression of basic fibroblast growth factor mRNA. Brain Res 1993; 625:45-56.
51. Riva MA, Gale K, Mocchetti I. Basic fibroblast growth factor mRNA increases in specific brain regions following convulsive seizures. Brain Res Mol Brain Res 1992; 15:311-318.
52. van der Wal EA, Gomez-Pinilla F, Cotman CW. Seizure-associated induction of basic fibroblast growth factor and its receptor in the rat brain. Neuroscience 1994; 60:311-323.
53. Riva MA, Donati E, Tascedda F et al. Short- and long-term induction of basic fibroblast growth factor gene expression in rat central nervous system following kainate injection. Neuroscience 1994; 59:55-65.
54. Gall CM, Berschauer R, Isackson PJ. Seizures increase basic fibroblast growth factor mRNA in adult rat forebrain neurons and glia. Brain Res Mol Brain Res 1994; 21:190-205.
55. Bugra K, Pollard H, Charton G et al. aFGF, bFGF and flg mRNAs show distinct patterns of induction in the hippocampus following kainate-induced seizures. Eur J Neurosci 1994; 6:58-66.
56. Matsushima K, Schmidt-Kastner R, Hogan MJ et al. Cortical spreading depression activates trophic factor expression in neurons and astrocytes and protects against subsequent focal brain ischemia. Brain Res 1998; 807:47-60.
57. Kawahara N, Ruetzler CA, Mies G et al. Cortical spreading depression increases protein synthesis and upregulates basic fibroblast growth factor. Exp Neurol 1999; 158:27-36.
58. Somjen GG. Mechanisms of spreading depression and hypoxic spreading depression- like depolarization. Physiol Rev 2001; 81:1065-1096.
59. Ay H, Ay I, Koroshetz WJ et al. Potential usefulness of basic fibroblast growth factor as a treatment for stroke. Cerebrovasc Dis 1999; 9:131-135.
60. Nozaki K, Finklestein SP, Beal MF. Basic fibroblast growth factor protects against hypoxia-ischemia and NMDA neurotoxicity in neonatal rats. J Cereb Blood Flow Metab 1993; 13:221-228.
61. Koketsu N, Berlove DJ, Moskowitz MA et al. Pretreatment with intraventricular basic fibroblast growth factor decreases infarct size following focal cerebral ischemia in rats. Ann Neurol 1994; 35:451-457.
62. Fisher M, Meadows ME, Do T et al. Delayed treatment with intravenous basic fibroblast growth factor reduces infarct size following permanent focal cerebral ischemia in rats. J Cereb Blood Flow Metab 1995; 15:953-959.
63. Tanaka R, Miyasaka Y, Yada K et al. Basic fibroblast growth factor increases regional cerebral blood flow and reduces infarct size after experimental ischemia in a rat model. Stroke 1995; 26:2154-2158.
64. Jiang N, Finklestein SP, Do T et al. Delayed intravenous administration of basic fibroblast growth factor (bFGF) reduces infarct volume in a model of focal cerebral ischemia/reperfusion in the rat. J Neurol Sci 1996; 139:173-179.
65. Bethel A, Kirsch JR, Koehler RC et al. Intravenous basic fibroblast growth factor decreases brain injury resulting from focal ischemia in cats. Stroke 1997; 28:609-615.
66. Nakata N, Kato H, Kogure K. Protective effects of basic fibroblast growth factor against hippocampal neuronal damage following cerebral ischemia in the gerbil. Brain Res 1993; 605:354-356.
67. Dietrich WD, Alonso O, Busto R et al. Posttreatment with intravenous basic fibroblast growth factor reduces histopathological damage following fluid-percussion brain injury in rats. J Neurotrauma 1996; 13:309-316.
68. Kirschner PB, Henshaw R, Weise J et al. Basic fibroblast growth factor protects against excitotoxicity and chemical hypoxia in both neonatal and adult rats. J Cereb Blood Flow Metab 1995; 15:619-623.

69. Liu Z, Holmes GL. Basic fibroblast growth factor is highly neuroprotective against seizure-induced long-term behavioural deficits. Neuroscience 1997; 76:1129-1138.
70. Anderson KJ, Dam D, Lee S et al. Basic fibroblast growth factor prevents death of lesioned cholinergic neurons in vivo. Nature 1988; 332:360-361.
71. Peterson DA, Lucidi-Phillipi CA, Murphy DP et al. Fibroblast growth factor-2 protects entorhinal layer II glutamatergic neurons from axotomy-induced death. J Neurosci 1996; 16:886-898.
72. Otto D, Frotscher M, Unsicker K. Basic fibroblast growth factor and nerve growth factor administered in gel foam rescue medial septal neurons after fimbria fornix transection. J Neurosci Res 1989; 22:83-91.
73. Cummings BJ, Yee GJ, Cotman CW. bFGF promotes the survival of entorhinal layer II neurons after perforant path axotomy. Brain Res 1992; 591:271-276.
74. Himmelseher S, Pfenninger E, Georgieff M. Effects of basic fibroblast growth factor on hippocampal neurons after axonal injury. J Trauma 1997; 42:659-664.
75. Liu Z, Holmes GL. Basic fibroblast growth factor-induced seizures in rats. Neurosci Lett 1997; 233:85-88.
76. Kawamata T, Alexis NE, Dietrich WD et al. Intracisternal basic fibroblast growth factor (bFGF) enhances behavioral recovery following focal cerebral infarction in the rat. J Cereb Blood Flow Metab 1996; 16:542-547.
77. Rowntree S, Kolb B. Blockade of basic fibroblast growth factor retards recovery from motor cortex injury in rats. Eur J Neurosci 1997; 9:2432-2441.
78. MacMillan V, Judge D, Wiseman A et al. Mice expressing a bovine basic fibroblast growth factor transgene in the brain show increased resistance to hypoxemic-ischemic cerebral damage. Stroke 1993; 24:1735-1739.
78A. Cuevas P, Carceller F, Munoz-Willery I et al. Intravenous fibroblast growth factor penetrates the blood-brain barrier and protects hippocampal neurons against ischemia-reperfusion injury. Surg Neurol 1998; 49:77-83.
79. Sasaki K, Oomura Y, Suzuki K et al. Acidic fibroblast growth factor prevents death of hippocampal CA1 pyramidal cells following ischemia. Neurochem Int 1992; 21:397-402.
80. MacMillan V, Walton-Roche K, Davis J. Acidic fibroblast growth factor infusion reduces ischemic CA1 hippocampal damage in the gerbil. Can J Neurol Sci 1993; 20:37-40.
81. Cuevas P, Carceller F, Reimers D et al. Acidic fibroblast growth factor rescues gerbil hippocampal neurons from ischemic apoptotic death. Neurol Res 1998; 20:271-274.
82. Hossain MA, Fielding KE, Trescher WH et al. Human FGF-1 gene delivery protects against quinolinate-induced striatal and hippocampal injury in neonatal rats. Eur J Neurosci 1998; 10:2490-2499.
83. Li AJ, Oomura Y, Sasaki K et al. Protective effect of acidic fibroblast growth factor against ischemia- induced learning and memory deficits in two tasks in gerbils. Physiol Behav 1999; 66:577-583.
84. Sadohara T, Sugahara K, Urashima Y et al. Keratinocyte growth factor prevents ischemia-induced delayed neuronal death in the hippocampal CA1 field of the gerbil brain. NeuroReport 2001; 12:71-76.
85. Mark RJ, Fuson KS, Keane-Lazar K et al. Fibroblast growth factor-8 protects cultured rat hippocampal neurons from oxidative insult. Brain Res 1999; 830:88-93.
86. Logan A, Berry M. Transforming growth factor-beta 1 and basic fibroblast growth factor in the injured CNS. Trends Pharmacol Sci 1993; 14:337-342.
87. Eclancher F, Perraud F, Faltin J et al. Reactive astrogliosis after basic fibroblast growth factor (bFGF) injection in injured neonatal rat brain. Glia 1990; 3:502-509.
88. Menon VK, Landerholm TE. Intralesion injection of basic fibroblast growth factor alters glial reactivity to neural trauma. Exp Neurol 1994; 129:142-154.
89. Eclancher F, Kehrli P, Labourdette G et al. Basic fibroblast growth factor (bFGF) injection activates the glial reaction in the injured adult rat brain. Brain Res 1996; 737:201-214.

90. Mahler M, Ferhat L, Gillian A et al. Tenascin-C mRNA and tenascin-C protein immunoreactivity increase in astrocytes after activation by bFGF. Cell Adhes Commun 1996; 4:175-186.
91. Yoshida K, Gage FH. Fibroblast growth factors stimulate nerve growth factor synthesis and secretion by astrocytes. Brain Res 1991; 538:118-126.
92. Reuss B, Hertel M, Werner S et al. Fibroblast growth factors-5 and -9 distinctly regulate expression and function of the gap junction protein connexin43 in cultured astroglial cells from different brain regions. Glia 2000; 30:231-241.
93. Mattson MP, Murrain M, Guthrie PB et al. Fibroblast growth factor and glutamate: opposing roles in the generation and degeneration of hippocampal neuroarchitecture. J Neurosci 1989; 9:3728-3740.
94. Cheng B, Furukawa K, O Keefe JA et al. Basic fibroblast growth factor selectively increases AMPA-receptor subunit GluR1 protein level and differentially modulates Ca^{2+} responses to AMPA and NMDA in hippocampal neurons. J Neurochem 1995; 65:2525-2536.
95. Mattson MP, Kumar KN, Wang H et al. Basic FGF regulates the expression of a functional 71 kDa NMDA receptor protein that mediates calcium influx and neurotoxicity in hippocampal neurons. J Neurosci 1993; 13:4575-4588.
96. Boxer AL, Moreno H, Rudy B et al. FGF-2 potentiates Ca^{2+}-dependent inactivation of NMDA receptor currents in hippocampal neurons. J Neurophysiol 1999; 82:3367-3377.
97. Abe K, Saito H. Selective enhancement by basic fibroblast growth factor of NMDA receptor-mediated increase of intracellular Ca^{2+} concentration in hippocampal neurons. Brain Res 1992; 595:128-132.
98. Collazo D, Takahashi H, McKay RD. Cellular targets and trophic functions of neurotrophin-3 in the developing rat hippocampus. Neuron 1992; 9:643-656.
99. Mattson MP, Rychlik B, Chu C et al. Evidence for calcium-reducing and excito-protective roles for the calcium-binding protein calbindin-D28k in cultured hippocampal neurons. Neuron 1991; 6:41-51.
100. Mattson MP, Lovell MA, Furukawa K et al. Neurotrophic factors attenuate glutamate-induced accumulation of peroxides, elevation of intracellular Ca^{2+} concentration, and neurotoxicity and increase antioxidant enzyme activities in hippocampal neurons. J Neurochem 1995; 65:1740-1751.
101. Hou JG, Cohen G, Mytilineou C. Basic fibroblast growth factor stimulation of glial cells protects dopamine neurons from 6-hydroxydopamine toxicity: involvement of the glutathione system. J Neurochem 1997; 69:76-83.
102. Mark RJ, Keller JN, Kruman I et al. Basic FGF attenuates amyloid beta-peptide-induced oxidative stress, mitochondrial dysfunction, and impairment of Na^+/K^+-ATPase activity in hippocampal neurons. Brain Res 1997; 756:205-214.
103. Guo Q, Sebastian L, Sopher BL et al. Neurotrophic factors [activity-dependent neurotrophic factor (ADNF) and basic fibroblast growth factor (bFGF)] interrupt excitotoxic neurodegenerative cascades promoted by a PS1 mutation. Proc Natl Acad Sci USA 1999; 96:4125-4130.
104. Tamatani M, Ogawa S, Nunez G et al. Growth factors prevent changes in Bcl-2 and Bax expression and neuronal apoptosis induced by nitric oxide. Cell Death Differ 1998; 5:911-919.
105. Ay I, Sugimori H, Finklestein SP. Intravenous basic fibroblast growth factor (bFGF) decreases DNA fragmentation and prevents downregulation of Bcl-2 expression in the ischemic brain following middle cerebral artery occlusion in rats. Brain Res Mol Brain Res 2001; 87:71-80.
106. Rosenblatt S, Irikura K, Caday CG et al. Basic fibroblast growth factor dilates rat pial arterioles. J Cereb Blood Flow Metab 1994; 14:70-74.
107. Tanaka R, Miyasaka Y, Yada K et al. Basic fibroblast growth factor increases regional cerebral blood flow and reduces infarct size after experimental ischemia in a rat model. Stroke 1995; 26:2154-2158.
108. Huang Z, Chen K, Huang PL et al. bFGF ameliorates focal ischemic injury by blood flow-independent mechanisms in eNOS mutant mice. Am J Physiol 1997; 272:H1401-H1405.

109. Yoshimura S, Takagi Y, Harada J et al. FGF-2 regulation of neurogenesis in adult hippocampus after brain injury. Proc Natl Acad Sci USA 2001; 98:5874-5879.
110. Altman J, Das GD. Autoradiographic and histological evidence of postnatal hippocampal neurogenesis in rats. J Comp Neurol 1965; 124:319-335.
111. Kaplan MS, Hinds JW. Neurogenesis in the adult rat: electron microscopic analysis of light radioautographs. Science 1977; 197:1092-1094.
112. Tretter YP, Hertel M, Munz B et al. Induction of activin A is essential for the neuroprotective action of basic fibroblast growth factor in vivo. Nature Med 2000; 6:812-815.
113. Tretter YP, Munz B, H bner G et al. Strong induction of activin expression after hippocampal lesion. NeuroReport 1996; 7:1819-1823.
114. Lai M, Sirimanne E, Williams CE et al. Sequential patterns of inhibin subunit gene expression following hypoxic-ischemic injury in the rat brain. Neuroscience 1996; 70:1013-1024.
115. Lai M, Gluckman P, Dragunow M et al. Focal brain injury increases activin βA mRNA expression in hippocampal neurons. NeuroReport 1997; 8:2691-2694.
116. Hubner G, Alzheimer C, Werner S. Activin: a novel player in tissue repair processes. Histol Histopathol 1999; 14:295-304.
117. Munz B, Hubner G, Tretter Y et al. A novel role of activin in inflammation and repair. J Endocrinol 1999; 161:187-193.
118. Krieglstein K, Suter-Crazzolara C, Fischer WH et al. TGF-beta superfamily members promote survival of midbrain dopaminergic neurons and protect them against MPP+ toxicity. EMBO J 1995; 14:736-742.
119. Iwahori Y, Saito H, Torii K et al. Activin exerts a neurotrophic effect on cultured hippocampal neurons. Brain Res 1997; 760:52-58.
120. Wu DD, Lai M, Hughes PE et al. Expression of the activin axis and neuronal rescue effects of recombinant activin A following hypoxic-ischemic brain injury in the infant rat. Brain Res 1999; 835:369-378.
121. Hughes PE, Alexi T, Williams CE et al. Administration of recombinant human Activin-A has powerful neurotrophic effects on select striatal phenotypes in the quinolinic acid lesion model of Huntington s disease. Neuroscience 1999; 92:197-209.
122. Nakamura T, Takio K, Eto Y et al. Activin-binding protein from rat ovary is follistatin. Science 1990; 247:836-838.
123. De Winter JP, Ten Dijke P, De Vries CJM et al. Follistatins neutralize activin bioactivity by inhibition of activin binding to its type II receptors. Mol Cell Endocrinol 1996; 116:105-114.
124. Krieglstein K, Reuss B, Maysinger D et al. Transforming growth factor-beta mediates the neurotrophic effect of fibroblast growth factor-2 on midbrain dopaminergic neurons. Eur J Neurosci 1998; 10:2746-2750.

TGF-βS AND THEIR ROLES IN THE REGULATION OF NEURON SURVIVAL

Klaus Unsicker[1] and Kerstin Krieglstein[2]

ABSTRACT

Transforming growth factor-βs (TGF-βs) are a still growing superfamily of cytokines with widespread distribution and diverse biological functions. They fall into several subfamilies including the TGF-βs 1, 2, and 3, the bone morphogenetic proteins (BMPs), the growth/differentiation factors (GDFs), activins and inhibins, and the members of the glial cell line-derived neurotrophic factor family. Following a brief description of their general roles and signaling in development, maintenance of homeostasis, and disease, we shall focus on their distribution in the CNS and their involvement in regulating neuron survival and death.

INTRODUCTION

TGF-βs are now widely recognized as a prototype of multifunctional and contextually acting growth factors in the regulation of key events of development, disease, and repair.[1-3] Although they were discovered more than 20 years ago, their functions in the CNS have only more recently started to be elucidated.[4-6] The term transforming growth factor (TGF) alludes to the capacity to transform , i.e., to induce anchorage-independent growth of normal rat kidney cells and fibroblasT cells lines.[7,8] Soon, it became apparent that the transforming capacity of the factor was in fact the result of the combined actions of at least two molecules that were named TGF-α and TGF-β. A host of additional functions of TGF-βs were discovered over the past years including important roles in cell cycle control, regulation of

[1]Neuroanatomy and Interdisciplinary Center for Neurosciences (IZN), University of Heidelberg, Im Neuenheimer Feld 307, 2. OG, D-69120 Heidelberg, Germany and [2]Department of Anatomy and Neuroanatomy, University of G ttingen, Kreuzber gring 36, D-37075 G ttingen, Germany.

early development, differentiation, extracellular matrix formation, hematopoesis, angiogenesis, chemotaxis, and immune functions. TGF-β is the prototypic molecule of a growth factor superfamily presently comprising more than 50 members. Other prominent members in the TGF-β superfamily include activins, bone morphogenetic proteins (BMPs), and the glial cell line-derived neurotrophic factor (GDNF) subfamily. A common structural feature of these molecules is the cystine knot motif,[9] which is also shared by other cytokines including the neurotrophins and platelet-derived growth factor (PDGF).

To recognize TGF-β as an important factor for CNS functions took almost a decade. We know now that TGF-β holds key roles in regulation of neuron survival[10,11] and orchestration of repair processes in the nervous system.[12] A number of excellent reviews exist illuminating TGF-βs and TGF-β subfamilies in general,[13-16] TGF-β signal transduction,[17] and TGF-βs, particularly TGF-β1, in neurodegenerative diseases and ischemic injury.[12,18-20] This review intends to provide information on the role of the TGF-βs proper (i.e., TGF-β1, -β2, and -β3) in the developing and adult nervous system with regard to the regulation of neuron survival and death, but will also briefly touch on the regulation of neuron survival by other members of the TGF-β superfamily.

TGF-βS: A BRIEF OVERVIEW OF THEIR MOLECULAR BIOLOGY, BIOCHEMISTRY AND SIGNALING

According to sequence similarities TGF-β proteins have been grouped into four subfamilies including the glial cell line-derived neurotrophic factor (GDNF)-family, the TGF-β family, activins, and the Dpp and Vg1-related (DVR) group, which is often referred to as the BMP family. Furthermore, the TGF-β superfamily harbours several distantly related members, e.g., the inhibin α-chain and Mullerian inhibiting substance (MIS).

The TGF-β Family

Until now, five TGF-β isoforms have been isolated. TGF-β1, -β2, and -β3 are mammalian isoforms;[21-23] TGF-β4 and TGF-β5 represent the chick- and *Xenopus*-homologues, respectively, of TGF-β1.[24,25] TGF-βs are expressed in numerous tissues, such as mesenchyme, connective tissues, endothelium, platelets, immune and bone cells.[1] Effects of distinct isoforms largely depend on the type and differentiation state of targeT cells and on the presence of other cytokines.[6,26]

Activins and Inhibins

Activins are formed by homo- or heterodimerization of two inhibin-β subunits. Inhibins arise from dimerization of one β-subunit with the inhibin-α subunit. Activins and Inhibins were termed based on the antagonistic activities observed in several biological systems. While activins stimulated synthesis of follicle-stimulating hormone (FSH) in pituitary gonadotrops,[27] inhibins blocked FSH production.[28] Activins and inhibins also regulate steroid hormone synthesis by granulosa and placental cells[29] and are important in erythroid differentiation[30] and induction of mesoderm in embryonic *Xenopus* explants.[31]

DVR Group

The DVR group (also referred to as the family of bone morphogenetic proteins, BMPs) is the largest family within the TGF-β superfamily, with about 30 members identified in vertebrates and invertebrates. Although initially named by virtue of their osteoinductive capacity,[32] BMPs exhibit a much wider spectrum of functions ranging from sculpting the body plan to regulating development of many organs including the nervous system.[33] In addition to numerous mammalian growth/differentiation factors (GDFs), the subfamily includes the Drosophila proteins 60A, Decapentaplegic (Dpp) and Screw, and several molecules isolated from *Xenopus*. Dpp organizes dorso-ventral patterning in Drosophila.[34] Vegetal 1 (Vg1) is encoded by a maternal mRNA localized to vegetal blastomeres in *Xenopus* embryos.[35] Local post-transcriptional activation of the Vg1-precursor protein leads subsequently to formation of axial mesoderm.[36] In mouse embryos the gene nodal is important for mesoderm formation and organization of axial structures.[37]

GDNF Family

GDNF is the founder member of a distant TGF-β subfamily, which comprises GDNF, Neurturin, Persephin, and Artemin/Neublastin.[38-42] GDNF was first isolated from a glial cell line (B49) conditioned medium as a survival factor for cultured midbrain dopaminergic neurons.[38] In the presence of TGF-β the molecule is a potent trophic factor for many populations of peripheral and CNS neurons.[10,11,16] Although GDNF and its congeners exhibit the typical structural features of TGF-βs, they do not signal through TGF-β serine/threonine kinase receptors, but employ a receptor complex, which consists of the tyrosine kinase c-Ret and a glycosyl-phosphatidylinositol (GPI)-anchored α-receptor.[16,43]

Divergent Genes

These include, amongst others, Mullerian inhibiting substance (MIS), a product of embryonic Sertoli cells that induces regression of the Mullerian duct in male embryos.[44] Two TGF-β related factors have been identified in *C. elegans*: UNC-129 is required to guide pioneer motor axons along the dorsoventral axis of the nematode.[45] DAF-7 controls *C. elegans* larval development by transducing environmental cues promoting recovery from dauer state.[46]

Molecular Structure of TGF-βs

TGF-βs are typical secreted molecules consisting of (i) an aminoterminal signal peptide and (ii) a pro-domain of variable length involved in correct folding, dimerization of subunits, and regulation of factor activity.[3] Dibasic cleavage sites (RXXR-motifs) serve the cleavage of the pro-domain from the mature part of the protein generating a carboxyterminal fragment of 110 to 140 amino acids. Homo- or heterodimerization of these C-terminal domains generates the biologically active molecules. Based on crystallography studies and NMR analysis, respectively, of TGF-β2,[47,48] BMP-7,[49] and TGF-β1,[50,51] the common structural motif of the TGF-β superfamily comprises seven characteristic cysteine residues within the mature domains of the monomeres. Six cysteines form three intramolecular disulfide bonds, which form the so-called cystine knot. Two of these disulfide bridges form an eight-membered ring, which is traversed by the third cystine bond. The seventh cysteine makes the intermolecular cystine bridge that links the two monomers into a functional dimer. This bond provides additional stabilization of hydrophobic interactions that are the main driving force for dimerization. The cystine knot motif is the common feature of a structural superfamily of growth factors including neurotrophins, members of the PDGF family and glycoprotein hormones.[52] Both, the six cysteines forming the cystine knot, and the four antiparallel β-strands of NGF, PDGF, and TGF-β are located at topologically equivalent positions suggesting that these molecules arose from a common monomeric ancestor. However, similarity of the proteins is restricted to the architecture of the protomeres. In contrast, although all types of cystine knot growth factors are dimeric, the active molecules differ in orientation of the monomeric subunits and in the dimerizing interfaces. Consequently, the polypeptide families employ different types of kinase receptors resulting in diverse biological capacities of the ligands.

TGF-β Receptors

TGF-β binds to several cell surface proteins termed type I (53 kDa), type II (70-100 kDa), and type III receptors (200-400 kDa) according to their approximate sizes.[53] The type III receptors, betaglycan and endoglin, supposedly modulate ligand access to the signalling receptors,[54,55] while receptors I and II (TβR-I and TβR-II) are responsible for signal transduction. TβR-I and TβR-II represent serine/threonine kinase receptors; they serve signal transduction of all TGF-β family members, except for GDNF and related factors. Although both types of receptors have functional extracellular ligand-binding and cytoplasmatic kinase regions, physical and functional interaction between the two receptors is required for signal transduction. Receptor I requires receptor II for ligand binding, and receptor I and the kinase activity of receptor II are required for signalling.[56] By first binding to the type II receptor, TGF-β triggers recruitment of the type I receptor into the complex, which provides the basis for downstream signaling. This mode of sequential ligand binding is characteristic for TβR-II, ActR-II, ActR-IIb, and AMR-II, which are able to bind ligand on their own.[17] The alternative cooperative binding mode is typical of BMP receptors, which only bind ligand with high affinity when expressed together.[57,58] In view of the dimeric structure of the ligands, the receptor complexes are probably tetrameric structures of two receptor I and two receptor II molecules.[59]

The mechanism of receptor activation is well established for TGF-β receptors and has been extended to activin receptors.[60,61] Phosphorylation of serine and threonine residues in the GS box, a conserved glutathione- and serine-rich sequence in the juxtamembrane region of TβR-I by the constitutively active TβR-II is an essential event in TGF-β signaling.[62,63] This results in activation of the TβR-I kinase and allows receptor I to propagate the signal to downstream elements.

The GDNF family occupies a special position among TGF-βs with regard to signaling. Its members employ a heteromeric receptor complex consisting of the receptor tyrosine kinase Ret,[64,65] and a GPI-linked α receptor. Until now, four different α-receptors have been identified in mammals.[43,66,67] Individual GFR-αs differ in their affinities to different GDNF family members and act as adaptor molecules for presenting ligands to the receptor kinase.[16]

TGF-β-Mediated Signal Transduction

The discovery of Smad proteins has been a breakthrough in the understanding of TGF-β-mediated signal transduction. The Drosophila gene Mothers against decapentaplegic (Mad), which constitutes a central component in Dpp signal transduction,[68,69] has led to the discovery of three homologous proteins in *C. elegans*, called Sma-2, -3, and -4.[70] After the founder members Sma and Mad, vertebrate homologues were named Smads. The Smad family can be subdivided into three subgroups. The receptor-regulated Smads (R-Smads) are direct substrates of specific type I receptor kinases and act in a pathway-restricted fashion. While Smad-1, and possibly Smad-5, and -8 specifically interact with BMP type I receptors,[71,72] Smad-2 and -3 associate with type I receptors of activin and TGF-β.[73-75] The second subfamily contains the common mediator Smads (Co-Smads); they associate with activated R-Smads forming heteromeric complexes that translocate to the nucleus and activate transcriptional responses. Smad-4, originally identified as the product of the tumor suppressor gene Deleted in Pancreatic Carcinoma Locus 4 (DPC-4)[76] is the only representative of this class in vertebrates and binds to different pathway-specific Smads.[74,77] The third class comprises the inhibitory Smads (anti-Smads) that counteract the activity of Co-Smad/R-Smad complexes. Smad-6 and -7 are the most divergent mammalian Smad proteins antagonizing TGF-β and BMP receptor signalling at three different levels.[78] Inhibition is achieved by specific, stable association with activated type I receptors, preventing binding and activation of R-Smads,[79,80] by competition for binding to Smad-4, [17] or by repressive functions of putative target genes prior to signal-induced transcriptional activation.[81]

The molecular pathway of TGF-β signaling from cell membrane to the nucleus can be illustrated by Smad-2. Upon activation of the heteromeric receptor complex and phosphorylation of serine residues within the GS-box of TβR-I or Act-RI, Smad-

2 is transiently recruited into the activated receptor complex. The process is mediated by chaperones like SARA [82] and terminated upon Smad phosphorylation by the type I receptor.[83] Phosphorylation is a key step in Smad activation and involves serines in conserved carboxy-terminal SSXS-motifs that are restricted to the group of R-Smads.[84] Mutation of these carboxy-terminal residues prevents downstream signalling.[85] Released Smad-2 associates within the cytosol with Smad-4, and the heteromeric protein complex translocates into the nucleus, where it associates with DNA-binding subunits such as FAST-1 and modulates transcription of specific target genes.[86] Likely, the cell type specific repertoire of interacting transcription factors as well as recruitment of co-activators and co-repressors decides which genes are activated upon stimulation with a given ligand.[87,88] The signalling function of Smad-2 is terminated by conjugation with ubiquitin and degradation by the proteasome.[89]

EXPRESSION OF TGF-βS AND TβRS IN THE NERVOUS SYSTEM

TGF-β2 and -β3 immunoreactivities and sites of their synthesis are widely distributed in the developing and adult CNS and peripheral nervous system (PNS) of mouse, rat, and chick.[4,90-92] Three aspects are of special importance, (i) the virtual ubiquity of TGF-β in all areas of the CNS as well as in the the PNS, (ii) consistent co-expression of TGF-β2 and TGF-β3 in neurons, astroglial, and Schwann cells, and (iii) the almost complete lack or low levels, respectively, of TGF-β1 in the unlesioned nervous system.

Development

A comprehensive immunocytochemical study has revealed the earliest, prominent stainings in mouse embryos at E12.5 to E13 along peripheral nerves, in radial glial cells, and along CNS axon tracts.[4] At these early stages, nerve cell bodies do not show detectable levels of TGF-β immunoreactivity suggesting that cells accompanying axon bundles are the earliest sites of TGF-β storage. At E15 TGF-β2/β3 become detectable in brain, spinal cord, and peripheral neurons. Although TGF-β immunoreactivity is prominent in radial glia from very early stages of development onwards, it is not detectable in the ventricular zones throughout the CNS suggesting that TGF-β may not be involved in the control of cell divisions and fate of neural progenitors in this area. Following the onset of migration of neuronal cells, the subventricular zone, subplate, and lamina I of the telencephalic cortex stain intensely for TGF-β. As they arise astroglial cells also start to be immunoreactive for TGF-β2/β3.[4,91,92] Comprehensive in situ hybridization studies on TGF-β and TβR mRNA expressions in the developing mammalian brain have not been performed. TβRII mRNA is apparently widely distributed in the CNS;[93] localization by in situ hybridization has revealed TβRII mRNA in the embryonic rat hindbrain at E14 along radially extending structures, which may represent radial glial cells.[94]

The distribution of TGF-β mRNA and protein in the embryonic chick PNS and CNS largely matches that seen in the mammalian nervous system.[95] Thus, TGF-β3 mRNA as well as TGF-β2 and -β3 immunoreactivities are not detectable in the E3 neural tube, although both TGF-β2 and TGF-β3 are highly expressed in notochord, dermomyotome and wall of the dorsal aorta. At E5 and later, TGF-β3 synthesis is clearly visible in the presumptive white matter of spinal cord and brain. Again, both immunocytochemistry and in situ hybridization suggest expression of TGF-β2 and -β3 in radial glia at this period of development, particularly in midline radial glia. Together, these data suggest that the possibility for transfer of TGF-β from perineuronal glia (radial glia, astroglia and Schwann cells) to neurons, as documented in epithelial/mesenchymal systems, should be investigated.

Adult Nervous System

TGF-β immunoreactivity in the adult rat nervous system is widespread.[90] TGF-β1 is restricted to meninges, while TGF-β2 and -β3 occur in both neurons and in glial cells (astrocytes and Schwann cells, not oligodendrocytes). With regard to neurons, levels of immunoreactive TGF-β2 and -β3 vary greatly depending on neuron type and brain regions. Highest proportions (<80%) of TGF-β-immunoreactive neuronal cell bodies are in cortical layers 2/3 and 5, in the hippocampus, piriform cortex, retina, and ventral spinal cord.[90] Other regions, as most thalamic and hindbrain areas, except for motoneuron and aminergic nuclei, are largely devoid of TGF-β-immunoreactive neuronal cell bodies. TGF-β2 mRNA has been demonstrated for the postnatal and adult mouse hippocampal, dentate gyrus, and Purkinje neurons.[96] TGF-β3 mRNA is also widespread in the adult CNS, but at much lower levels than TGF-β2 mRNA.

Several reports suggest that TGF-β immunoreactivity in the human brain is less pronounced than in adult rat brain.[97,98] There is faint neuronal staining for TGF-β2 in human neocortex, hippocampus, and entorhinal cortex. TGF-β3 also occurs in neurons in several brain regions. Glial cells exhibit only weak (TGF-β2) or no staining (TGF-β3) in the human brain.

The chromaffin cell, a model neuron, is probably the most thoroughly studied cell in the nervous system with regard to TGF-β synthesis, storage, and release. Chromaffin cells are derivatives of the neural crest and its sympathoadrenal lineage, closely related to sympathetic neurons, and concentrated in the adrenal medulla, from where they can be easily isolated in large amounts for biochemical studies.[10,99,100] Adrenal medullary chromaffin cells synthesize, store, and release all three isoforms of TGF-β. Carbachol, a cholinergic agonist, induces release of TGF-β (assayable by the mink lung epithelial cell assay) that can be partially blocked with verapamil.[10] Together with the visualization of TGF-β immunoreactivity in chromaffin cells both *in situ* and in culture this suggests that TGF-β can be liberated from chromaffin cells, at least in part, by the regulated pathway of secretion. Whether the regulated mode of release also applies to other types of neuronal cells is currently being investigated.

TGF-βS AND THE REGULATION OF PROLIFERATION, SURVIVAL AND DIFFERENTIATION OF NEURONS

Consistent with its wide distribution in the nervous system, TGF-β has been shown to exert multiple effects on neurons and glial cells both in vitro and in vivo. Since this review focuses on the implications of TGF-β in the regulation of neuron survival and death, readers interested in TGF-β functions in glial cells are referred to a recent review.[101]

TGF-β Controls Proliferation of Neuronal and Neuroendocrine Cells

Although TGF-β has not been detected in the ventricular zones of the developing brain and may therefore not be implicated in the regulation of neural stem cell proliferation, there are several examples showing that TGF-β may act as an inhibitor of neural cell proliferation in certain populations of neural cells. Cerebellar granule cells have an unusual history in that they undergo two subsequent waves of proliferation and migration. They express TGF-β2 at high levels which peak at postnatal day 10. In culture, they release TGF-β into the medium. Granule cell proliferation is significantly decreased by TGF-β consistent with an antimitogenic role of TGF-β in the generation of this cerebellar neuron population.[96]

Chromaffin cells, neuroendocrine cells of neural crest origin, proliferate throughout their whole lifespan, in contrast to sympathetic neurons, to which they are closely related. Several factors have been identified that suppress chromaffin cell divisions in vitro, including glucocorticoids and TGF-β.[102] While glucocorticoids do not seem to have an in vivo relevance, as shown in glucocorticoid receptor-deficient mice,[103] neutralization of TGF-β using systemically applied pan-TGF-β antibodies markedly increases numbers of adrenal chromaffin cells and BrdU incorporation in chick embryos suggesting an in vivo role of TGF-β in the regulation of proliferation of this neuroendocrine cell type.[104] Interestingly, numbers of sympathoadrenal progenitor cells, the common progenitors for sympathetic neurons and chromaffin cells, do not seem to be affected by neutralisation of endogenous TGF-β, as shown by antibody treatments of quail embryos between E2.5 and E4, i.e., prior to migration of SA cells into the adrenal anlagen.[105]

TGF-β in Neuronal Survival and Death

TGF-β Has Neurotrophic Effects in vitro Under Very Complex Culture Conditions

Martinou and co-workers[106] and Chalazonitis et al[107] were the first to describe survival promoting effects on cultured motoneurons and sensory neurons, respectively. However, culture systems employed were very complex. Thus, motoneurons were grown on a feeder layer of astrocytes and sensory neurons in the presence of non-neuronal cells. In another highly complex culture system of mid-brain dopaminergic neurons all three mammalian TGF-β isoforms promoted the survival of dopaminergic neurons and prevented MPP^+ toxicity.[108-110] TGF-β also acts on cultured serotonergic neurons, but promotes the transmitter phenotype rather than survival.[111,112] Together, these data suggested that TGF-β can mimic neurotrophic factors in respective in vitro assays.

Reducing the Complexity of Culture Conditions Abolishes Neurotrophic Effects of TGF-β for Most Neuron Populations Studied

In order to characterize potential neurotrophic roles of TGF-βs in more detail, we and others used highly purified neuron populations, low-density cultures, and fully defined serum-free conditions. In the absence of non-neuronal cells and serum, TGF-β does not promote survival in assays of embryonic chick ciliary, dorsal root, and sympathetic ganglionic neurons.[10,112,113] This also applies to embryonic rat dorsal root ganglionic neurons. The only neuron population discovered so far, which responds to TGF-β with increased survival, are spinal cord motoneurons.[10,114] The failure of TGF-β to elicit survival is not due to a lack of TGF-β receptors; expression of TβRII has been documented for ciliary, sensory, and motoneurons.[10,113]

TGF-β Is an Important Co-Trophin for Established Neurotrophic Factors

Although TGF-β certainly does not fall into the category of neurotrophic factors, it is crucially involved in the regulation of neuron survival in vitro and in vivo. When neurons are grown in the presence of low and subthreshold concentrations of established trophic factors, i.e., neurotrophins, neurokines, and fibroblast growth factors,[112,113] TGF-β clearly shifts the dose-response curve to lower neurotrophic factor concentrations.[113] Under such conditions, even subthreshold concentrations of a neurotrophic molecule can elicit a prominent survival-promoting effect. The biological significance of TGF-β in modulating neurotrophic factor efficacy becomes even more striking when the endogenous TGF-β, which is synthesized by neurons, is neutralized by TGF-β antibodies.[112,113] Neutralizing endogenous TGF-β may deprive established neurotrophic factors, as, e.g., NGF, of 50% or more of their survival promoting capacity. The biological relevance of the synergy of TGF-β and neurotrophic factors is also highlighted by the fact that neurotrophic factors as, e.g., NGF, CNTF and FGF-2, upregulate TGF-β synthesis and release from neurons.[112,113]

A particularly striking example of TGF-β/neurotrophic factor synergisms is the co-operativity of TGF-β and GDNF. When grown in serum-free media and in the absence of exogenous TGF-β, GDNF, a molecule widely acknowledged as a neurotrophic factor, failed to promote survival of a large number of peripheral and CNS neuron populations.[10] At a molecular level, some aspects of the TGF-β/GDNF crosstalk have begun to be clarified. When treated with phosphatidyl-inositol phospholipase C, which liberates GPI-linked α receptors within the receptor complexes of, e.g., CNTF and GDNF, ciliary neurons lose responsiveness to CNTF and GDNF, respectively. In the presence of TGF-β, however, both CNTF and GDNF promote survival suggesting that TGF-β may protect or restore the respective α receptors. The biological relevance of the TGF-β/GDNF synergy is underscored by two types of observations. First, both molecules are co-stored in and -liberated from secretory granules of adrenal medullary chromaffin cells.[10] Second, neutralization of endogenous TGF-β in a lesion paradigm of preganglionic sympathetic neurons abolishes the rescuing effect of GDNF on this neuron population.[11]

TGF-β in Neuronal Differentiation

TGF-β has been implicated in neuronal differentiation, i.e., regulation of neurite growth, transmitter synthesis and in synapse formation. TGF-β1 and -β2 have been reported to cause neurite sprouting and elongation of hippocampal neurons in vitro.[115] TGF-β1 has also been shown to promote re-elongation of injured axons of cultured hippocampal neurons.[112] These in vitro data were expanded by showing that TGF-β can also increase the number of neurites as well as neurite length in an explant of dorsal root ganglia.[95]

The regulation of Ca^{2+} activated K^{+}-channels in developing chick ciliary ganglion neurons has been shown to be regulated by two different target-derived TGF-β isoforms, which produce opposing effects on the electrophysiological differentiation of these neurons. TGF-β1 causes the developmental functional expression and stimulation of Ca^{2+} activated K^{+}-channels.[117] Opposingly to TGF-β1, TGF-β3 inhibited the functional expression of these Ca^{2+} activated K^{+}-channels leaving other voltage activated Ca^{2+} currents unchanged.[118]

TGF-β has been shown to have a prominent role in long-term synaptic facilitation (24-48 h) in isolated Aplysia ganglia.[119] TGF-β did not affect short- (5-15 min) or intermediate-term (2-4 h) facilitation. Neutralization of TGF-β abolished the well-established facilitation mediated by serotonin suggesting that TGF-β is an important component within the synaptic signalling cascade for adult neuronal plasticity.

A potential role of TGF-β in neuromuscular signal transduction has been suggested based on the localization of all three TGF-β isoforms at the postsynaptic area of neuromuscular junctions identified by acetylcholinesterase reaction and acetylcholine receptor antibodies.[120,121]

Protection of Lesioned Neurons by Administration of TGF-β

In vitro Studies

Several lines of evidence suggest that TGF-β can prevent neuron death in a variety of lesion paradigms in vitro and in vivo. Paradoxically, TGF-β can also aggravate neuronal injury, and mechanisms underlying this ambiguity of TGF-β in the regulation of neuronal survival are still enigmatic. Generally, it is unclear, however, whether protection of (usually chemically) lesioned neurons or induction of apoptosis, respectively, by TGF-β requires co-signaling by other cytokines and neurotrophic molecules, as demonstrated on cultured unlesioned neurons (see above).

Several studies have addressed potential roles of TGF-β in beta-amyloid induced neuron death. For example, pretreatment of cultured hippocampal neurons lesioned with the beta-amyloid peptide fragment 25-35 with TGF-β resulted in a 60% increase in cell viability as determined by MTT incorporation.[122] A similar, yet less pronounced protective effect was seen in two mouse and human neuroblastoma cell lines. Protection was shown to be mediated through the TGF-β type II receptor.[123] Treatment with TGF-β following exposure to beta amyloid peptide also partially prevented neuron death. These results corroborate other studies, in which TGF-β was shown to protect hippocampal neurons from beta-amyloid neurotoxicity.[124]

TGF-β also protects hippocampal neurons in vitro against short-term exposure to glutamate or NMDA and the resulting intracellular calcium overload.[125] TGF-β pretreatment reduces calcium increases,[126] not only in the excitotoxicity paradigm, but also, when intracellular calcium homeostasis is perturbed in response to treatment with the HIV-1 coat protein gp 120.[127] Paradoxically, TGF-β potentiates rather than ameliorates slowly triggered excitotoxic death induced by long-term treatment with glutamate or kainate[125] and aggravates NMDA-induced neuron death in mixed neuron-glial cultures.[128,129]

TGF-β has also proven its efficacy in other in vitro lesion models, as, e.g., intoxication of midbrain dopaminergic neurons by MPP+.[108]

In vivo Studies

Administration of TGF-β in a variety of CNS lesion paradigms in vivo,[12,19] ranging from target deprivation to ischemia and models of Parkinsonism, has not conclusively revealed an overall and robust potency of TGF-β to protect lesioned neurons.

In a model of neuronal target deprivation in the adult rat, Blottner and collaborators[130] have demonstrated that substitution of the target tissue by TGF-β prevents death of neurons that die following destruction of the target. The model chosen was the neuronal axis that connects preganglionic sympathetic neurons in the spinal cord with the adrenal medulla. This model permits to selectively destroy the adrenal medulla and replace it by gelfoam soaked with growth factors to be tested.[131]

In a 6-hydroxydopamine model of Parkinsonism TGF-β, in contrast to GDNF, failed to prevent death of dopaminergic nigrostriatal neurons,[132] despite its proven capacity to protect intoxicated dopaminergic neurons in vitro.[109,110] A general problem with most studies in which TGF-β had failed to show a neuroprotective potential or had minor effects only is the failure to demonstrate biological activity of the particular charge of TGF-β applied.

Protection of Lesioned Neurons by TGF-β Deprivation

Lesion experiments in chick embryos (limb bud ablation, i.e., axotomy of motor, sensory, and autonomic axons) have clearly shown that neutralisation of endogenous TGF-β using pan-TGF-β antibodies that recognize all three TGF-β isoforms results in the protection of 50% of the total population of spinal cord motoneuron population that otherwise dies in the presence of TGF-β. This experiment therefore unravels TGF-β as a neuron survival antagonizing principle by demonstrating an essential requirement of TGF-β in vivo to execute apoptosis of axotomized motoneurons and, possibly other neuron populations, too. The ambiguity of TGF-β in terms of affecting conditions suitable for neuronal regeneration is highlighted by the observation that TGF-β1 apparently plays a key role in structuring the glial scar, a major chemical barrier for regenerating axons following mechanical lesioning: neutralization of TGF-β1 decreases astroglial scarring, without, however, clearly promoting neurite regeneration.[133] Consistent with a role of TGF-β in scarring, administration of TGF-β1 into the subarachnoidal space after spinal cord injury produces pronounced scar formation. Although the TGF-β treatment initially facilitates functional recovery of hindlimb motor control, the final outcome is not better.[134]

Together, the pivotal role of TGF-β in promoting neuron survival in one context and preventing it in another one is far from being understood, as long as context is insufficiently defined in molecular terms.

REGULATION OF NEURON SURVIVAL AND MAINTENANCE BY MEMBERS OF THE TGF-β SUPERFAMILY OTHER THAN TGF-βS PROPER

Glial Cell Line-Derived Neurotrophic Factor

GDNF, first characterized by its neurotrophic activity for mesencephalic dopaminergic neurons,[38] has received extensive attention mostly because of its clinical potential in the therapy of Parkinson«s disease, but also because of its mode of signaling through a receptor tyrosine kinase, c-ret, and a GPI-linked alpha receptor, GFR-α. GDNF was the first representative of a new TGF-β subfamily, which comprises at present four members, GDNF, neurturin, persephin, and artemin.[38-42] GDNF and other members of the GDNF family have been extensively reviewed with respect to their neurotrophic actions and molecular bases of signaling.[16,135-137] We will therefore only briefly summarize important features of GDNF activities in vivo in relation to the lesioned nigrostriatal system, impaired motoneurons, and ischemia.

Multiple studies have proven the prominent survival-promoting effects of GDNF on lesioned nigrostriatal dopamine neurons.[16] There are two particularly convincing features of GDNF protectivity, its protective effect following application after lesioning[132] and its protective effects in non-human primates.[138-141] These findings have been more recently supplemented by optimizing sites of administration of recombinant GDNF[142,143] and GDNF encoded by adenoviral and lentiviral vectors.[144-147] The GDNF-related molecule neurturin has been shown to protect not only the dopaminergic innervation of the striatum, but also striatal projection neurons in a rat model of Huntington«s disease.[148]

Viral vector-mediated GDNF gene transfer has also been shown to be effective on lesioned adult motoneurons.[149-152] In addition to GDNF, neurturin and persephin have been successfully applied to lesioned postnatal motoneurons.[39-41,153]

Potential roles of endogenous GDNF family ligands and their receptors in coping with brain insults have been proposed based on their regulation in vivo.[154]

GDNF has also proven its neurotrophic potency in several models of brain ischemia by various routes of application including adenoviral gene transfer.[155-157] An important mode of action of GDNF concerns the reduction of NMDA-induced calcium influx by the mitogen-activated protein kinase pathway.[158] GDNF also seems to be an important mediator of neuroprotective actions of other growth factors that are beneficial in glutamate toxicity, including fibroblast growth factor-2 and brain-derived neurotrophic factor.[159]

Activin

Activin has well-documented neurotrophic and protective effects in the lesioned nervous system, which have been reviewed by Alzheimer and Werner in the preceding Chapter of this book.

Bone Morphogenetic Proteins and Growth/Differentiation Factors

In contrast to the vast literature that exists of BMPs and their role in gliogenesis (for a review see Mehler et al, ref. 14), there are only a few reports on neurotrophic actions of BMPs. In vitro BMPs can promote survival of dopaminergic midbrain neurons[160] and induce the serotonergic phenotype in cells cultured from the embryonic rat raphe.[161] BMP-2 has been reported to promote survival and differentiation of striatal GABAergic neurons;[162] OP-1 enhances dendritic growth from cerebral cortical neurons in vitro.[163] BMP-4/7 have crucial roles in the induction of specific neuronal phenotypic features, most notably catecholaminergic properties of peripheral and central neurons in vitro and in vivo (Rohrer, personal communication).[164-166] Similarly, BMP-2 induces neurotrophin-3 responsiveness in developing rat sympathetic neurons[167] and cholinergic properties in CNS neurons.[168]

Among the GDFs, GDF-5 and GDF-15 have been shown to have neurotrophic effects on midbrain dopaminergic neurons in vitro and in vivo.[109,169-170]

CONCLUSIONS

The available data on the distribution and the wide range of functions of TGF-βs in the nervous system support the notion that TGF-β2 and -β3 are probably the most important cytokines with modulatory actions in this system. Their principal roles seem to be related to regulating precursor cell divisions, migration of neural precursors, neuron and glial cell survival, differentiation, and synaptic plasticity. Mice (conditionally) deficient for TGF-β2 and -β3 or TβRII may be expected to shed more light on neural TGF-β functions in the future. Among the other subfamilies of the TGF-β superfamily, the GDNF family occupies a prominent position regarding neurotrophic functions. Neurotrophic roles of BMPs and GDFs have been less extensively investigated.

ACKNOWLEDGEMENTS

We thank the Deutsche Forschungsgemeinschaft for continuing support of our research on TGF-βs.

REFERENCES

1. Roberts AB, Sporn MB. The transforming growth factor-βs. In: Sporn MB, Roberts AB, Eds. Handbook of Experimental Pharmacology. Heidelberg: Springer Verlag, 1990;95:419-472.
2. Wahl SM. Transforming growth factor beta (TGF-β) in inflammation: A cause and a cure. J Clin Immunol 1992; 12:66-74.
3. Kingsley DM. The TGF-β superfamily: New members, new receptors, and new genetic tests of function in different organisms. Genes Dev 1994; 8:133-146.
4. Flanders KC, L decke G, Engels S, Cissel DS, Roberts AB, Kondaiah P et al. Localization and actions of transforming growth factor-βs in the embryonic nervous system. Development 1991; 113:183-191.
5. Finch CE, Laping NJ, Morgan TE, Nicholson NR, Pasinetti GM. TGF-β1 is an organizer of responses to neurodegeneration. J Cell Biochem 1993; 53:314-322.
6. Krieglstein K, Rufer M, Suter-Crazzolara C, Unsicker K. Neural functions of the transforming growth factors β. Int J Dev Neurosci 1995; 13:301-315.
7. Moses HL, Branum EL, Proper JA, Robinson RA. Transforming growth factor production by chemically transformed cells. Cancer Res 1981; 41:2842-2848.
8. Roberts AB, Anzano MA, Lamb LC, Smith JM, Sporn MB. New class of transforming growth factors potentiated by epidermal growth factor: isolation from non-neoplastic tissues. Proc Natl Acad Sci USA 1981; 78:5339-5343.
9. McDonald NQ, Hendrickson WA. A structural superfamily of growth factors containing a cystine knot motif. Cell 1993; 73:421-424.
10. Krieglstein K, Henheik P, Farkas L, Jaszai J, Galter D, Krohn K et al. GDNF requires TGF-β for establishing its neurotrophic activity. J Neurosci 1998; 18:9822-9834.
11. Schober A, Hertel R, Arumae U, Farkas L, Jaszai J, Krieglstein K et al. Glial cell line-derived neurotrophic factor rescues target-deprived sympathetic spinal cord neurons but requires transforming growth factor-beta as cofactor in vivo. J Neurosci 1999; 19:2008-2015.
12. Flanders KC, Fen RF, Lippa CF. Transforming growth factor-βs in neurodegenerative disease. Prog Neurobiol 1998; 54:71-85.

13. Lawrence DA. Transforming growth factor-beta: A general review. Eur Cytokine Netw 1996; 7:363-374.
14. Mehler MF, Mabie PC, Zhang D, Kessler JA. Bone morphogenetic proteins in the nervous system. Trends Neurosci 1997; 20:309-317.
15. Ebendal T, Bengtsson H, S derstr m S. Bone morphogenetic proteins and their receptors: Potential functions in the brain. J Neurosci Res 1998; 51:139-146.
16. Unsicker K, Suter-Crazzolara C, Krieglstein K. Neurotrophic roles of GDNF and related factors. In: Hefti F, ed. Handbook of Experimental Pharmacology. Heidelberg: Springer Verlag, 1999; 134:189-224.
17. Massague J. TGF-beta signal transduction. Annu Rev Biochem 1998; 67:753-791.
18. Kiefer R, Streit WJ, Toykka KV, Kreutzberg GW, Hartung HP. Transforming growth factor-beta 1: A lesion-associated cytokine of the nervous system. Int J Dev Neurosci 1995; 13:331-339.
19. Pratt BM, McPherson JM. TGF-β in the central nervous system: Potential roles in ischemic injury and neurodegenerative diseases. Cytokine Growth Factor Rev 1997; 8:267-292.
20. Raivich G, Jones JJ, Werner A, Bluthmann H, Doetschman T, Kreutzberg GW. Molecular signals for glial activation: pro- and anti-inflammatory cytokines in the injured brain. Acta Neurochir Suppl 1999; 73,21-30.
21. Ohta M, Greenberger JS, Anklesaria P, Bassols A, Massague J. Two forms of transforming growth factor-beta distinguished by multipotential haematopoietic progenitor cells. Nature 1987; 329:539-541.
22. Madisen L, Webb NR, Rose TM, Marquardt H, Ikeda T, Twardzik D et al. Transforming growth factor-beta 2: cDNA cloning and sequence analysis. DNA 1988; 7:1-8.
23. Derynck R, Lindquist PB, Lee A, Wen D, Tamm J, Graycar JL et al. A new type of transforming growth factor-beta, TGF-beta 3. EMBO J 1988; 7:3737-3743.
24. Burt DW, Paton IR. Evolutionary origins of the transforming growth factor-beta gene family. DNA Cell Biol 1992 ;11:497-510.
25. Burt DW, Law AS. Evolution of the transforming growth factor-beta superfamily. Prog Growth Factor Res 1994; 5:99-118.
26. Nathan C, Sporn M. Cytokines in context. J Cell Biol 1991;113:981-986.
27. Vale W, Rivier J, Vaughan J, McClintock R, Corrigan A, Woo W et al. Purification and characterization of an FSH releasing protein from porcine ovarian follicular fluid. Nature 1986; 321:776-779.
28. Ling N, Ying SY, Ueno N, Esch F, Denoroy L, Guillemin R. Isolation and partial characterization of a Mr 32,000 protein with inhibin activity from porcine follicular fluid. Proc Natl Acad Sci USA 1985; 82:7217-7221.
29. Petraglia F, Vaughan J, Vale W. Inhibin and activin modulate the release of gonadotropin-releasing hormone, human chorionic gonadotropin, and progesterone from cultured human placenta cells. Proc Natl Acad Sci USA 1989; 86:5114-5117.
30. Eto Y, Tsuji T, Takezawa M, Takano S, Yokogawa X, Shibai H. Purification and characterization of erythroid differentiation factor (EDF) isolated from human leukemia cell line THP-1. Biochem Biophys Res Commun 1987; 142:1095-1103.
31. Smith JC, Price BM, Van Nimmen K, Huylebroeck D. Identification of a potent *Xenopus* mesodem-inducing factor as a homologue of activin A. Nature 1990; 345:729-731.
32. Wozney JM, Rosen V, Celeste AJ, Mitsock LM, Whitters MJ, Kriz RW et al. Novel regulators of bone formation: molecular clones and activities. Science 1988; 242:1528-1534.
33. Hogan BL. Bone morphogenetic proteins: multifunctional regulators of vertebrate development. Genes Dev 1996; 10:1580-1594.
34. Ferguson EL, Anderson KV. Decapentaplegic acts as a morphogen to organize dorsal-ventral pattern in the *Drosophila* embryo. Cell 1992; 71:451-461.
35. Weeks DL, Melton DA. A maternal mRNA localized to the vegetal hemisphere in *Xenopus* eggs codes for a growth factor related to TGF-beta. Cell 1987; 51:861-867.

36. Thomsen GH, Melton DA. Processed Vg1 protein is an axial mesoderm inducer in *Xenopus*. Cell 1993; 74:433-441.
37. Zhou X, Sasaki H, Lowe L, Hogan BL, Kuehn MR. Nodal is a novel TGF-beta-like gene expressed in the mouse node during gastrulation. Nature 1993; 361:543-547.
38. Lin LF, Doherty DH, Lile JD, Bektesh S, Collins F. GDNF: A glial cell line-derived neurotrophic factor for midbrain dopaminergic neurons. Science 1993; 260:1130-1132.
39. Kotzbauer PT, Lampe PA, Heuckeroth RO, Golden JP, Creedon DJ, Johnson EM Jr. et al. Neurturin, a relative of glial-cell-line-derived neurotrophic factor. Nature 1996; 384:467-470.
40. Milbrandt J, de Sauvage FJ, Fahrner TJ, Baloh RH, Leitner ML, Tansey MG et al. Persephin, a novel neurotrophic factor related to GDNF and neurturin. Neuron 1998; 20:1-20.
41. Baloh RH, Tansey MG, Lampe PA, Fahrner TJ, Enomoto H, Simburger KS et al. Artemin, a novel member of the GDNF ligand family, supports peripheral and central neurons and signals through the GFRalpha3-RET receptor complex. Neuron 1998; 21:1291-1302.
42. Rosenblad C, Gronborg M, Hansen C, Blom N, Meyer M, Johansen J et al. In vivo protection of nigral dopamine neurons by lentiviral gene transfer of the novel GDNF-family member neublastin/artemin. Mol Cell Neurosci 2000; 15:199-214.
43. Treanor JJ, Goodman L, de Sauvage F, Stone DM, Poulsen KT, Beck CD et al. Characterization of a multicomponent receptor for GDNF. Nature 1996; 382:80-83.
44. Blanchard MG, Josso N. Source of the anti-Mullerian hormone synthesized by the fetal testis: Mullerian-inhibiting activity of fetal bovine Sertoli cells in tissue culture. Pediatr Res 1974; 8:968-971.
45. Colavita A, Krishna S, Zheng H, Padgett RW, Culotti JG. Pioneer axon guidance by UNC-129, a *C. elegans* TGF-beta. Science 1998; 281:706-709.
46. Ren P, Lim CS, Johnsen R, Albert PS, Pilgrim D, Riddle DL. Control of *C. elegans* larval development by neuronal expression of a TGF-beta homolog. Science 1996; 274:1389-1391.
47. Daopin S, Piez KA, Ogawa Y, Davies DR. Crystal structure of transforming growth factor-beta 2: an unusual fold for the superfamily. Science 1992; 257:369-373.
48. Schlunegger MP, Gr tter MG. An unusual feature revealed by the crystal structure at 2.2 A resolution of human transforming growth factor-β2. Nature 1992; 358:430-434.
49. Griffith DL, Keck PC, Sampath TK, Rueger DC, Carlson WD. Three-dimensional structure of recombinant human osteogenic protein 1: Structural paradigm for the transforming growth factor β superfamily. Proc Natl Acad Sci USA 1996;93:878-883.
50. Archer SJ, Bax A, Roberts AB, Sporn MB, Ogawa Y, Piez KA et al. Transforming growth factor beta 1: Secondary structure as determined by heteronuclear magnetic resonance spectroscopy. Biochemistry 1993; 32:1164-1171.
51. Hinck AP, Archer SJ, Qian SW, Roberts AB, Sporn MB, Weatherbee JA et al. Transforming growth factor beta 1: Three-dimensional structure in solution and comparison with the X-ray structure of transforming growth factor beta 2. Biochemistry 1996; 35:8517-8534.
52. Sun PD, Davis DR. The cystine-knot growth-factor superfamily. Annu Rev Biophys Biomol Struct 1995; 24:269-291.
53. Massague J. The transforming growth factor-beta family. Annu Rev Cell Biol 1990; 6:597-641.
54. Cheifetz S, Bellon T, Cales C, Vera S, Bernabeu C, Massague J et al. Endoglin is a component of the transforming growth factor-beta receptor system in human endothelial cells. J Biol Chem 1992; 267:19027-19030.
55. Lopez-Casillas F, Wrana JL, Massague J. Betaglycan presents ligand to the TGF beta signaling receptor. Cell 1993; 73:1435-1444.
56. Wrana JL, Attisano L, Carcamo J, Zentella A, Doody J, Laiho M et al. TGF beta signals through a heteromeric protein kinase receptor complex. Cell 1992 ;71:1003-1014.
57. Liu F, Ventura F, Doody J, Massague J. Human type II receptor for bone morphogenic proteins (BMPs): Extension of the two-kinase receptor model to the BMPs. Mol Cell Biol 1995; 15:3479-3486.

58. Rosenzweig BL, Imamura T, Okadome T, Cox GN, Yamashita H, ten Dijke P et al. Cloning and characterization of a human type II receptor for bone morphogenetic proteins. Proc Natl Acad Sci USA 1995; 92:7632-7636.
59. Yamashita H, ten Dijke P, Franzen P, Miyazono K, Heldin CH. Formation of hetero-oligomeric complexes of type I and type II receptors for transforming growth factor-beta. J Biol Chem 1994; 269:20172-20178.
60. Attisano L, Wrana JL, Montalvo E, Massague J Activation of signalling by the activin receptor complex. Mol Cell Biol 1996; 16:1066-1073.
61. Willis SA, Zimmerman CM, Li LI, Mathews LS. Formation and activation by phosphorylation of activin receptor complexes. Mol Endocrinol 1996; 10:367-379.
62. Franzen P, Heldin CH, Miyazono K. The GS domain of the transforming growth factor-beta type I receptor is important in signal transduction. Biochem Biophys Res Commun 1995; 207:682-689.
63. Wieser R, Wrana JL, Massague J. GS domain mutations that constitutively activate T beta R-I, the downstream signaling component in the TGF-beta receptor complex. EMBO J 1995; 14:2199-2208.
64. Durbec P, Marcos-Gutierrez CV, Kilkenny C, Grigoriou M, Wartiowaara K, Suvanto P et al. GDNF signalling through the Ret receptor tyrosine kinase. Nature 1996; 381:789-793.
65. Trupp M, Arenas E, Fainzilber M, Nilsson AS, Sieber BA, Grigoriou M et al. Functional receptor for GDNF encoded by the c-ret proto-oncogene. Nature 1996; 381:785-789.
66. Thompson J, Doxakis E, Pinon LG, Strachan P, Buj-Bello A, Wyatt S et al. GFRalpha-4, a new GDNF family receptor. Mol Cell Neurosci 1998; 11:117-126.
67. Jing S, Yu Y, Fang M, Hu Z, Holst PL, Boone T et al. GFRalpha-2 and GFRalpha-3 are two new receptors for ligands of the GDNF family. J Biol Chem 1997; 272:33111-33117.
68. Raftery LA, Twombly V, Wharton K, Gelbart WM. Genetic screens to identify elements of the decapentaplegic signaling pathway in *Drosophila*. Genetics 1995; 139:241-254.
69. Sekelsky JJ, Newfeld SJ, Raftery LA, Chartoff EH, Gelbart WM. Genetic characterization and cloning of mothers against dpp, a gene required for decapentaplegic function in *Drosophila melanogaster*. Genetics 1995; 139:1347-1358.
70. Savage C, Das P, Finelli AL, Townsend SR, Sun CY, Baird SE et al. *Caenorhabditis elegans* genes sma-2, sma-3, and sma-4 define a conserved family of transforming growth factor beta pathway components. Proc Natl Acad Sci USA 1996; 93:790-794.
71. Hoodless PA, Haerry T, Abdollah S, Stapleton M, O Connor MB, Attisano L et al. MADR1, a MAD-related protein that functions in BMP2 signaling pathways. Cell 1996; 85:489-500.
72. Liu F, Hata A, Baker JC, Doody J, Carcamo J, Harland RM, Massague J. A human Mad protein acting as a BMP-regulated transcriptional activator. Nature 1996; 381:620-623.
73. Chen Y, Lebrun JJ, Vale W. Regulation of transforming growth factor beta- and activin-induced transcription by mammalian Mad proteins. Proc Natl Acad Sci USA 1996; 93:12992-12997.
74. Lagna G, Hata A, Hemmati-Brivanlou A, Massague J. Partnership between DPC4 and SMAD proteins in TGF-beta signalling pathways. Nature 1996; 383:832-836.
75. Zhang Y, Feng X, We R, Derynck R. Receptor-associated Mad homologues synergize as effectors of the TGF-beta response. Nature 1996; 383:168-172.
76. Hahn SA, Schutte M, Hoque AT, Moskaluk CA, da Costa LT, Rozenblum E et al. DPC4, a candidate tumor suppressor gene at human chromosome 18q21.1. Science 1996; 271:350-353.
77. Heldin CH, Miyazono K, ten Dijke P. TGF-beta signalling from cell membrane to nucleus through SMAD proteins. Nature 1997; 390:465-471.
78. Hill CS. The Smads. Int J Biochem Cell Biol 1999; 31:1249-1254.
79. Nakao A, Afrakhte M, Moren A, Nakayama T, Christian JL, Heuchel R et al. Identification of Smad7, a TGFbeta-inducible antagonist of TGF-beta signalling. Nature 1997; 389:631-635.
80. Imamura T, Takase M, Nishihara A, Oeda E, Hanai J, Kawabata M et al. Smad6 inhibits signalling by the TGF-beta superfamily. Nature 1997; 389:622-626.

81. Itoh S, Landstrom M, Hermansson A, Itoh F, Heldin CH, Heldin NE et al. Transforming growth factor beta1 induces nuclear export of inhibitory Smad7. J Biol Chem 1998; 273:29195-29201.
82. Tsukazaki T, Chiang TA, Davison AF, Attisano L, Wrana JL. SARA, a FYVE domain protein that recruits Smad2 to the TGFbeta receptor. Cell 1998; 95:779-791.
83. Macias-Silva M, Abdollah S, Hoodless PA, Pirone R, Attisano L, Wrana JL. MADR2 is a substrate of the TGFbeta receptor and its phosphorylation is required for nuclear accumulation and signaling. Cell 1996; 87:1215-1224.
84. Kretzschmar M, Liu F, Hata A, Doody J, Massague J. The TGF-beta family mediator Smad1 is phosphorylated directly and activated functionally by the BMP receptor kinase. Genes Dev 1997; 11:984-995.
85. Kretzschmar M, Massague J. Smads: mediators and regulators of TGF-beta signaling. Curr Opin Genet Dev 1998; 8:103-111.
86. Chen X, Weisberg E, Fridmacher V, Watanabe M, Naco G, Whitman M. Smad4 and FAST-1 in the assembly of activin-responsive factor. Nature 1997; 389:85-89.
87. Feng XH, Zhang Y, Wu RY, Derynck R. The tumor suppressor Smad4/DPC4 and transcriptional adaptor CBP/p300 are coactivators for Smad3 in TGF-beta-induced transcriptional activation. Genes Dev 1998; 12:2153-2163.
88. Wotton D, Lo RS, Lee S, Massague J. A Smad transcriptional corepressor. Cell 1999; 97:29-39.
89. Lo RS, Massague J. Ubiquitin-dependent degradation of TGF-beta-activated Smad2. Nat Cell Biol 1999; 1:472-478.
90. Unsicker K, Flanders KC, Cissel DS, Lafyatis R, Sporn MB. Transforming growth factor beta isoforms in the adult rat central and peripheral nervous system. Neuroscience 1991; 44:613-625.
91. Pelton RW, Saxena B, Jones M, Moses HL, Gold LI. Immunhistochemical localization of TGF-β1, TGF-β2 and TGFβ3 in mouse embryo: Expression patterns suggest multiple roles during embryonic development. J Cell Biol 1991; 115:1091-1105.
92. Pelton RW, Dickinson ME, Moses HL, Hogan BLM. In-situ hybridization analysis of TGF-β3 RNA expression during mouse development: comparative studies with TGF-β1 and β2. Development 1991; 110:609-620.
93. B ttner M, Unsicker K, Suter-Crazzolara C. Expression of TGF- β type II receptor mRNA in the CNS. NeuroReport 1996; 7:2903-2907.
94. Galter D, Bottner M, Unsicker K. Developmental regulation of the serotonergic transmitter phenotype in rostral and caudal raphe neurons by transforming growthfactor-betas. J Neurosci Res 1999; 56:531-538.
95. Unsicker K, Meier C, Krieglstein K, Sartor BM, Flanders KC. Expression, localization, and function of transforming growth factor-βs in embryonic chick spinal cord, hindbrain, and dorsal root ganglia. J Neurobiol 1996; 29:262-276.
96. Constam DB, Schmid P, Aguzzi A, Schachner M, Fontana A. Transient production of TGF-β_2 by postnatal cerebellar neurons and its effect on neuroblast proliferation. Eur J Neurosci 1994; 6:766-778.
97. Peress NS, Perillo E. Differential expression of TGF-beta 1, 2 and 3 isotypes in Alzheimer s disease: a comparative immunohistochemical study with cerebral infarction, aged human and mouse control brains. J Neuropathol Exp Neurol 1995; 54:802-811.
98. Flanders KC, Lippa CF, Smith TW, Pollen DA, Sporn MB. Altered expression of transforming growth factor-beta in Alzheimer s disease. Neurology. 1995; 45:1561-1569.
99. Unsicker K. The chromaffin cell: paradigm in cell development and growth factor biology. J Anat 1993; 183:207-221.
100. Unsicker K, Krieglstein K. Growth factors in chromaffin cells. Prog Neurobiol 1996; 48:307-324.

101. B ttner M, Krieglstein K, Unsicker K. The transforming growth factor-betas: Structure, signaling, and roles in nervous system development and functions. J Neurochem 2000; 75:2227-2240.
102. Wolf N, Krohn K, Bieger S, Frodin M, Gammeltoft S, Krieglstein K et al. Transforming growth factor-beta, but not ciliary neurotrophic factor, inhibits DNA synthesis of adrenal medullary cells in vitro. Neuroscience 1999; 90:629-41.
103. Finotto S, Krieglstein K, Schober A, Deimling F, Lindner K, Bruhl B et al. Analysis of mice carrying targeted mutations of the glucocorticoid receptor gene argues against an essential role of glucocorticoid signalling for generating adrenal chromaffin cells. Development 1999; 126:2935-2944.
104. Combs SE, Krieglstein K, Unsicker K. Reduction of endogenous TGF-beta increases proliferation of developing adrenal chromaffin cells in vivo. J Neurosci Res 2000; 59:379-383.
105. Combs SE, Krieglstein K, Ernsberger U, Unsicker K. Immuno-neutralization of endogenous TGF-β does not affect phenotypid development of sympathoadrenal progenitors into adrenal chromaffin cells. Mech Dev 2001; 109:295-302.
106. Martinou JC, le van Thai A, Valette A, Weber MJ Transforming growth factor β_1 is a potent survival factor for rat embryo motoneurons in culture. Dev Brain Res 1990; 52:175-181.
107. Chalazonitis A, Kalberg J, Twardzik DR, Morrison RS, Kessler JA. Transforming growth factor beta has neurotrophic actions on sensory neurons in vitro and is synergistic with nerve growth factor. Dev Biol 1992; 152:121-132.
108. Krieglstein K, Unsicker K. Transforming growth factor-β promotes survival of midrain dopaminergic neurons and protects them against N-methyl-4-phenylpyridinium ion toxicity. Neuroscience 1994; 63:1189-1196.
109. Krieglstein K, Suter-Crazzolara C, Fischer WH, Unsicker K. TGF-β superfamily members promote survival of midbrain dopaminergic neurons and protect them against MPP+ toxicity. EMBO J 1995; 14:736-742.
110. Poulsen KT, Armanini MP, Klein RD, Hynes MA, Phillips HS, Rosenthal A. TGF beta 2 and TGF beta 3 are potent survival factors for midbrain dopaminergic neurons. Neuron 1994; 13:1245-1252.
111. Galter D, Unsicker K. Regulation of the transmitter phenotype of rostral and caudal groups of cultured serotonergic raphe neurons. Neuroscience 1999; 88:549-559.
112. Krieglstein K, Unsicker K. Distinct modulatory actions of TGF-β and LIF on neurotrophin-mediated survival of developing sensory neurons. Neurochem Res 1996; 21:849-856.
113. Krieglstein K, Farkas L, Unsicker K. TGF-beta regulates the survival of ciliary ganglionic neurons synergistically with ciliary neurotrophic factor and neurotrophins. J Neurobiol 1998; 37:563-572.
114. Gouin A, Bloch-Gallego E, Tanaka H, Rosenthal A, Henderson CE. Transforming growth factor-beta 3, glial cell line-derived neurotrophic factor, and fibroblast growth factor-2, act in different manners to promote motoneuron survival in vitro. J Neurosci Res 1996; 43:454-464.
115. Ishihara A, Saito H, Abe K. Transforming growth factor-beta 1 and -beta 2 promote neurite sprouting and elongation of cultured rat hippocampal neurons. Brain Res 1994; 639:21-25.
116. Abe K, Chu PJ, Ishihara A, Saito H. Transforming growth factor-beta 1 promotes re-elongation of injured axons of cultured rat hippocampal neurons. Brain Res 1996; 723:206-209.
117. Cameron JS, Lhuillier L, Subramony P, Dryer SE. Developmental regulation of neuronal K+ channels by target-derived TGF beta in vivo and in vitro. Neuron 1998; 21:1045-1053.
118. Cameron JS, Dryer L, Dryer SE. Regulation of neuronal K(+) currents by target-derived factors: opposing actions of two different isoforms of TGFbeta. Development 1999; 126:4157-4164
119. Zhang F, Endo S, Cleary LJ, Eskin A, Byrne JH. Role of transforming growth factor-beta in long-term synaptic facilitation in Aplysia. Science 1997; 275:1318-1320.

120. Toepfer M, Fischer P, Abicht A, Lochmuller H, Pongratz D, Muller-Felber W. Localization of transforming growth factor beta in association with neuromuscular junctions in adult human muscle. Cell Mol Neurobiol 1999; 19:297-300.
121. McLennan IS, Koishi K, Zhang M, Murakami N. The non-synaptic expression of transforming growth factor-beta 2 is neurally regulated and varies between skeletal muscle fibre types. Neuroscience 1998; 87:845-853.
122. Ren RF, Flanders KC. Transforming growth factors-beta protect primary rat hippocampal neuronal cultures from degradation induced by beta-amyloid peptide. Brain Res 1996; 732:16-24.
123. Ren RF, Hawver DB, Kim RS, Flanders KC. Transforming growth factor beta protects human hNT cells from degradation induced by beta-amyloid peptide:involvement of the TGF-beta type II receptor. Mol Brain Res 1997; 48:315-322.
124. Prehn JH, Bindokas VP, Jordan J, Galindo MF, Ghadge GD, Roos RP et al. Protective effect of transforming growth factor-beta 1 on beta-amyloid neurotoxicity in rat hippocampal neurons. Mol Pharmacol 1996; 49:319-328.
125. Prehn JHM, Miller RJ. Opposite effects of TGF-β1 on rapidly- and slowly-triggered excitotoxic injury. Neuropharmacology 1996; 35:249-256.
126. Prehn JHM, Bindokas VP, Marcuccilli CJ, Krajewski S, Reed JC, Miller RJ. Regulation of Bcl2 protein expression and calicum homeostasis by transforming growth factor β confers wide-ranging protection of hippocampal neurons. Proc natl Acad Sci USA 1994; 91:12599-12603.
127. Meucci O, Miller RJ. pg120-induced neurotoxicity in hippocampal pyramidal neuron cultures: protective action by TGF-beta1. J Neurosci 1996; 16:4080-4088.
128. Chao CC, Hu S, Tsang M, Weatherbee JM, Molitor TW, Anderson WR et al. Effects of transforming growth factor-β on murine astrocyte glutamine synthase activity. J Clin Invest 1992; 90:1786-1793.
129. Kane CJM, Brown GJ, Phelan KD. Transforming growth factor-β2 increases NMDA recepptor-mediated excitotoxicity in rat cerebral cortical neurons independently of glia. Neurosci Lett 1996; 204:93-96.
130. Blottner D, Wolf N, Lachmund A, Flanders KC, Unsicker K. TGF-β rescues target-derived preganglionic sympathetic neurons in the spinal cord. Eur J Neurosci 1996; 8:202-210.
131. Schober A, Unsicker K. Growth and neurotrophic factors regulating development and maintenance of sympathetic preganglionic neurons. Intl Rev Cytology 2001; 205:37-76
132. Sauer H, Rosenblad C, Bj rklund. A. Glial cell line-derived neurotrophic factor but not transforming growth factor beta 3 prevents delayed degeneration of nigral dopaminergic neurons following striatal 6-hydroxydopamine lesion. Proc natl Acad Sci USA 1995; 92:8935-8939.
133. Logan A, Berry M, Gonzalez AM, Frautschy SA, Sporn MB, Baird A. Effects of transforming growth factor β1 on scar production in the injured central nervous system of the rat. Eur J Neurosci 1994; 3:355-363.
134. Hamada Y, Ikata T, Katoh S, Katoh K, Niwa M, Tsutsumishita Y et al. Effects of exogenous transforming growth factor-beta 1 on spinal cord injury in rats. Neurosci Lett 1996; 203:97-100.
135. Baloh RH, Enomoto H, Johnson EM Jr., Milbrandt J. The GDNF family ligands and receptors Implications for neural development. Curr Op Neurobiol 2000; 10:103-110.
136. Airaksinen MS, Titievsky A, Saarma M. GDNF family neurotrophic factor signaling: four masters, one servant? Mol Cell Neurosci 1999;13:313-325.
137. Saarma M. GDNF recruits the signaling crew into lipid rafts. Trends in Neurosci 2001; 24:427-429.
138. Gash DM, Gerhardt GA, Hoffer BJ. Effects of glial cell line-derived neurotrophic factor on the nigrostriatal dopamine system in rodents and nonhuman primates.Adv Pharmacol 1998; 42:911-5

139. Kordower JH, Emborg ME, Bloch J, Ma SY, Chu Y, Leventhal L et al. Neurodegeneration prevented by lentiviral vector delivery of GDNF in primate models of Parkinson s disease. Science 2000; 290:767-773.
140. Kozlowski DA, Bremer E, Redmond DE Jr., George D, Larson B, Bohn MC. Quantitative analysis of transgene protein, mRNA, and vector DNA following injection of an adenoviral vector harboring glial cell line-derived neurotrophic factor into the primate caudate nucleus. Mol Ther 2001; 3:256-261.
141. Iravani MM, Costa S, Jackson MJ, Tel BC, Cannizzaro C, Pearce RK et al. GDNF reverses priming for dyskinesia in MPTP-treated, L-DOPA-primed common marmorsets. Eur J Neurosci 2001; 13:597-608.
142. Kirik D, Rosenblad C, Bj rklund A. Preservation of a functional nigrostriatal dopmaine pathway by GDNF in the intrastriatal 6-OHDA lesion model depends on the site of administration of the trophic factor. Eur J Neurosci 2000; 12:3871-3882.
143. Rosenblad C,, Kirik D, Bj rklund A. Sequential administration of GDNF into the substantia nigra and striatum promotes dopamine neuron survival and axonal sprouting but not striatal reinnervation of functional recovery in the partial 6-OHDA lesion model. Exp Neurol 2000; 161:503-516.
144. Choi-Lundberg DL, Lin Q, Schallert T, Crippens D, Davidson BL, Chang YN et al. Behavioral and cellular protection of rat dopmainergic neurons by an adenoviral vector encoding glial cell line-derived neurotrophic factor. Exp Neurol 1998; 154:261-275.
145. Kirik D, Rosenblad C, Bj rklund A, Mandel RJ. Long-term rAAV-mediated gene transfer of GDNF in the rat Parkinson s model: intrastriatal but not intranigral transduction promotes functional regeneration in the lesioned nigrostriatal system. J Neurosci 2000; 20:4686-4700.
146. Bensadoun JC, Deglon N, Tseng JL, Zurn AD, Aebischer P. Lentiviral ectors as a gene delivery system in the mouse midbrain: cellular and behavioral improvements in a 6_OHDA model of Parkinson s disease using GDNF. Exp Neurol 2000; 164:15-24.
147. Bj rklund A, Kirik D, Rosenblad C, Georgievskka B, Lundberg C, Mandel RJ. Towards a neuroprotective gene therapy for Parkinson s disease: use of adenovirus, AVV and lentivirus vectors for gene transfer of GDNF to the nigrostriatal system in the rat Parkinson model. Brain Res 2000; 886:82-98.
148. Perez-Navarro E, Akerud P, Marco S, Canals JM, Tolosa E, Arenas E et al. Neurturin protects striatal neurons but not interneurons in a rat model of Huntington s disease. Neuroscience 2000; 98:89-96.
149. Hottinger AF, Azzouz M, Deglon N, Aebischer P, Zurn AD. Complete and long-term rescue of lesioned adult motoneurons by lentiviral-mediated expression of glial cell line-derived neurotrophic factor in the facial nucleus. J Neurosci 2000; 20:5587-5593.
150. Sakamoto T, Watabe K, Ohashi T, Kawazoe Y, Oyanagi K, Inoue K et al. Adenoviral vector-mediated GDNF gene transfer prevents death of adult facial motoneurons. Neuroreport 2000; 11:1857-1860.
151. Sakurai M, Abe K, Hayashi T, Setoguchi Y, Yaginuma G, Meguro T et al. Adenovirus-mediated glial cell line-derived neurotrophic factor gene delivery reduces motor neuron injury after transient spinal cord ischemia in rabbits. J Thorac Cardiovasc Surg 2000; 120:1148-1157.
152. Watabe K, Sakamoto T, Ohashi T, Kawazoe Y, Oyanagi K, Takeshima T et al. Adenoviral gene transfer of glial cell line-derived neurotrophic factor to injured adult motoneurons. Hum Cell 2001; 14:7-15.
153. Bilak MM, Shifrin DA, Corse AM, Bilak SR, Kuncl RW. Neuroprotective utility and neurotrophic action of neurturin in postnatal motor neurons:comparison with GDNF and persephin. Mol Cell Neurosci 1999; 13:326-336.
154. Kokaia Z, Airaksinen MS, Nanobashvili A, arsson E, Kujamaki E, Lindvall O et al. GDNF family ligands and receptors are differentially regulated after brain insults in the rat. Eur J Neurosci 1999; 11:1202-1216.

155. Miyazaki H, Okuma Y, Fujii Y, Magasima K, Nomura Y. Gleil cell line-derived neurotrophic factor protects against delayed neuronal death after transient forebrain ischemia in rats. Neuroscience 1999; 89:643-647.
156. Tsai TH, Chen SL, Chiang YH, Lin SZ, Ma HI, Kuo SW et al. Recombinant adeno-associated virus vestor expressing glial cell line-derived neurotrophic factor reduces ischemia-induced damage. Exp Neurol 2000; 166:266-275.
157. Yagi T, Jikihara I, Fukumura M, Watanabe K, Ohashi T, Eto Y et al. Rescue of ischemic brain injury by adenoviral gene transfer of glial cell line-derived neurotrophic factor after transient global ischemia in gerbils. Brain Res 2000; 885;273-283.
158. Nicole O, Ali C, Docagne F, Plawinski L, MacKenzie ET, Vivien D et al. Neuroprotection mediated by glial cell line-derived neurotrophic factor: involvement of a reduction of NMDA-induced calcium influx by the mitogen-activated protein kinase pathway. J Neurosci 2001; 21:3024-3033.
159. Lenhard, T, Schober, A., Suter-Crazzolara C, Unsicker K. Fibroblast growth factor-2 requiresglial cell line-erived neurotrophic factor for exerting its neuroprotective actions on glutamate-lesioned hippocampal neurons. Mol Cell Neurosci 2001; under revision.
160. Jordan J, B ttner M, Schluesener H, Unsicker K, Krieglstein K. Bone morphogenetic proteins: Neurotrophic roles for midbrain dopaminergic neurons and implications of astroglial cells. Eur J Neurosci 1997; 9:1699-1710.
161. Galter D, B ttner M, Krieglstein K, Sch mig E, Unsicker K. Differential regulation of distinct phenotypic features of serotonergic neurons by bone morphogenetic proteins. Eur J Neurosci 1999; 11:2444-2452.
162. Hattori A, Katayama M, Iwasaki S, Ishii K, Tsujimoto M, Kohno M. Bone morphogenetic protein-2 promotes survival and differentiation of striatal GABAergic neurons in the absence of glial cell proliferation. J Neurochem 1999; 72:2264-2271.
163. Le Roux P, Behar S, Higgins D, Charette M. OP-1 enhamces dendritic growth from cerebral cortical neurons in vitro. Exp Neurol 1999; 160:151-163.
164. Reissmann E, Ernsberger U, Francis-West PH, Rueger D, Brickell PM, Rohrer H. Involvement of bone morphogenetic protein-4 and bone morphogenetic protein-7 in the differentiation of the adrenergic phenotype in developing sympathetic neurons. Development 1996; 122:2079-2088.
165. Shah NM, Groves AK, Anderson DJ. Alternative neural crest cell fates are instructively promoted by TGFbeta superfamily members. Cell 1996; 85:331-343.
166. Schneider C, Wicht H, Enderich J, Wegner M, Rohrer H. Bone morphogenetic proteins are required in vivo for the generation of sympathetic neurons. Neuron 1999; 24:861-870.
167. Kobayashi M. Fujii M, Kurihara K, Matsuoka I. Bone morphogenetic protein-2 and retinoic acid induce neurotrophin-3 responsiveness in develing rat sympathetic neurons. Mol Brain Res 1998; 53:206-217.
168. Lopez-Coviella I, Berse B, Krauss R, Thies RS, Blusztajn JK. Induction and maintenance of the neuronal cholinergic phenotype in the central nervous system by BMP-9. Science 2000; 289:313-316.
169. Sullivan AM, Pohl J, Blunt SB. Growth/differentiation factor 5 and glial cell line-deried neurotrophic factor enhance survival and function of dopaminergic grafts in a rat model of Parkinson s disease. Eur J Neurosci 1998; 10:3681-3688.
170. Strelau J, Sullivan A, B ttner M, Lingor P, Falkenstein E, Suter-Crazzolara C et al. GDF-15/MIC-1 is a novel trophic factor for midbrain dopaminergic neurons in vivo. J Neurosci 2000; 20:8597-8603.

VASCULAR ENDOTHELIAL GROWTH FACTOR

Hugo H. Marti*

The striking structural and anatomical parallels between the vasculature and the nervous system is reflected by the fact that these two organ systems appear to use related mechanisms during their development.[1] Thus, it is not surprising that an increasing number of vascular biologists and researchers in the neuroscience field are fascinated by molecules that play an important role in both systems.

Vascular endothelial growth factor (VEGF) is a major regulator of new blood vessel growth and an important inducer of vascular permeability. However, during the last few years, it has become apparent that VEGF has additional non-vascular functions. In particular the identification of the neuropilins, receptors for the semaphorin family of proteins that mediate neuronal axon pathfinding, as coreceptors for various members of the VEGF family, as well as the detection of VEGF receptors on neurons and astrocytes, suggests that VEGF can act as neurotrophic and neuroprotective factor in the central as well as in the peripheral nervous system. Thus, VEGF is a pleiotrophic factor with many important functions in the brain. This Chapter summarizes the current knowledge on VEGF expression, regulation and function in the adult nervous system.

THE MEMBERS OF THE VEGF FAMILY

VEGF (also designated as VEGF-A) is the founding member of a family of homodimeric glycoproteins that are structurally related to the platelet-derived growth factors (PDGF); this family also includes placenta growth factor (PlGF), VEGF-B, VEGF-C, VEGF-D and VEGF-E. This VEGF family of proteins bind selectively with different affinities to at least five distinct receptors (Fig. 1). Three of them belong to the superfamily of receptor tyrosine kinases and are termed VEGF receptor-1 (VEGFR-1), also called Flt-1 (fms-like tyrosine kinase 1), VEGFR-2, also called

*Max-Planck-Institute for Physiological and Clinical Research, Parkstrasse 1, D-61231 Bad Nauheim, Germany; Present address: Institute of Physiology, University of Z rich, Winterthurerstrasse 190, CH-8057 Z rich, Switzerland.

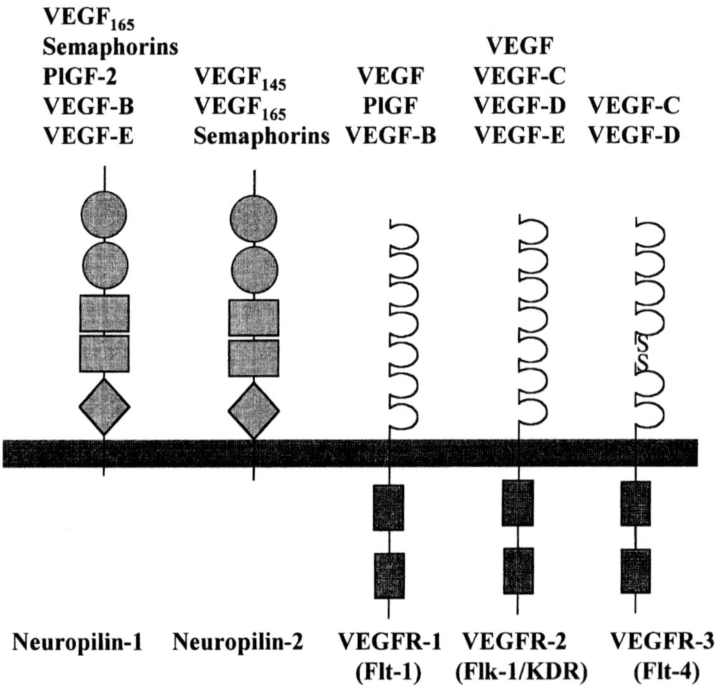

Figure 1. The receptors for VEGF. Neuropilin receptors have five extracellular domains, two each with homologies to complement components C1r and C1s (circles) and coagulation factors V and VIII (rectangles), as well as a MAM (meprin, A5, receptor tyrosine phosphatase μ) domain (diamond). In addition to the semaphorins, neuropilins bind selective members of the VEGF family of growth factors. VEGF receptors possess seven immunoglobulin-like domains in the extracellular part and a split intracellular tyrosine kinase domain, and bind different members of the VEGF family. VEGFR-3 is proteolytically cleaved in the fifth immunglobulin-like domain, but the resulting chains remain linked by a disulfide bond. (Adapted from refs. 3-5).

KDR (kinase insert-domain containing receptor) in humans, and Flk-1 (fetal liver kinase 1) in rodents, respectively, as well as VEGFR-3, also called Flt-4. The fourth and fifth receptors are neuropilin-1 and neuropilin-2.[2] Most of the research on the VEGF family so far, especially with respect to the expression pattern and function in the nervous system, has focused on VEGF-A.

The secreted PlGF shows a strong structural homology to VEGF[6] and exists as three isoforms, PlGF-1, PlGF-2 and PlGF-3, which result from alternative splicing from a single gene locus.[5] PlGF is predominantly expressed in placenta and binds exclusively to VEGFR-1.[7] Although no PlGF expression has been detected in human cortical brain tissue, expression was demonstrated in various tumors of the CNS.[8,9] PlGF induces both proliferation and chemotaxis of endothelial cells in vitro, and is angiogenic in vivo.[6,10,11] These results support the proposal that PlGF is involved in tumor angiogenesis and that it is not required for vascular development in the brain.

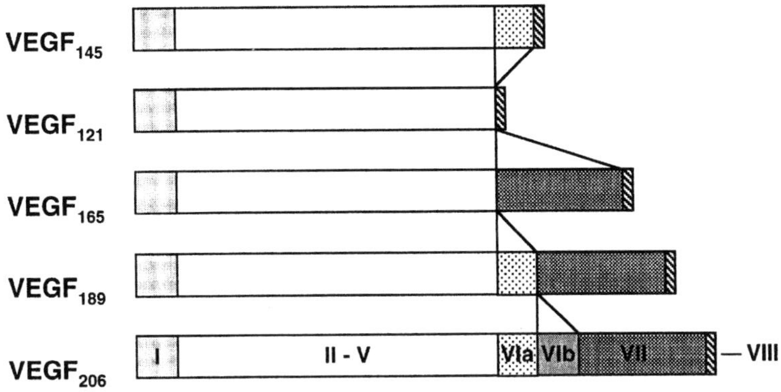

Figure 2. The human isoforms of VEGF-A. At least five different isoforms are generated by alternative splicing from a single gene. Exon I, encoding for a signal peptide, exons II V and exon VIII are present in all isoforms. Exons III and IV are required for binding to VEGFR-1 and VEGFR-2, respectively. The isoforms differ by the presence or absence of exon VI, that mainly mediates heparin binding and exon VII, which seems to be responsible for interaction with neuropilin-1. (Adapted from refs. 5,31).

A recent report further demonstrates that embryonic angiogenesis in mice is not affected by PlGF deficiency, but that PlGF contributes to angiogenesis under pathological conditions.[12]

VEGF-B binds to both VEGFR-1 and neuropilin-1. VEGF-B is widely expressed, most prominently in heart and skeletal muscle, but also in mouse and human brain.[13,14] The highest levels of VEGF-B in the brain are detected in neuronal-like cells of the hippocampus and the cerebral cortex.[15] VEGF-B is implicated in angiogenesis by its role in the regulation of extracellular matrix degradation, cell adhesion and migration of endothelial cells.[14]

VEGF-C is a ligand for both VEGFR-2 and VEGFR-3.[16] VEGF-C is synthesized as a prepropeptide and subsequently undergoes proteolytic maturation.[17] VEGF-C mRNA is found in several tissues including heart, placenta, ovary, and small intestine,[18] although it is undetectable in the normal brain.[15] As VEGFR-3 is the main receptor for VEGF-C and is predominantly expressed on lymphatic endothelium, VEGF-C was therefore considered to be the prototypic lymphangiogenic factor.[19] However, VEGF-C has also been shown to act on vascular endothelial cells both in vitro and in vivo.[20-22] Indeed, mice which lacked a functional *vegfr-3* gene showed defective blood vessel development in early stage mouse embryos.[23] Thus, the VEGF-C/VEGFR-3 system has an essential role not only for lymphatic vessel formation but also for angiogenesis.

VEGF-D is structurally very similar to VEGF-C and it also binds to VEGFR-2 and VEGFR-3.[24] VEGF-D is mitogenic for endothelial cells and thus may play a role in endothelial cell regulation. The expression of VEGF-D is prominent in heart and skeletal muscle, as well as in mesenchymal cells of the lung and skin, but shows only very weak expression in the brain.[24,25]

VEGF-E is the collective term for a group of proteins with homology to VEGF-A that are encoded by certain strains of the *orf* parapoxvirus, which affects goats, sheep and occasionally humans.[5] It possesses about 25% amino acid identity to mammalian VEGF.[26] VEGF-E binds with high affinity to VEGFR-2 and neuropilin-1, but neither to VEGFR-1 nor to VEGFR-3, inducing vascular permeability and potent angiogenic activity both in vitro and in vivo.[27-29]

Finally, VEGF-A is ubiquitously expressed at low levels within the CNS by astrocytes and neurons.[30] VEGF-A exists as several isoforms, derived from a single gene by alternative splicing (Fig. 2). The smallest form, $VEGF_{121}$, is 121 amino acids long and does not bind to heparin, while the four larger forms, $VEGF_{145}$, $VEGF_{165}$, $VEGF_{189}$ and $VEGF_{206}$, all bind heparin with increasing affinity. $VEGF_{189}$ and $VEGF_{206}$ remain cell-attached while the smaller isoforms are secreted.[5,10,31] VEGF was first isolated in 1983 as a factor leading to increased vascular permeability in tumors and was thus called vascular permeability factor (VPF).[32] In 1989, VPF was independently isolated and cloned as an endothelial cell-specific mitogen.[33] VEGF is expressed in virtually all cells in the body, however; expression in endothelial cells has mainly been found in vitro in cell culture, while it is undetectable in vivo.[34] These findings indicate that VEGF acts as a paracrine factor on neighboring endothelial cells carrying the VEGF receptors rather than in an autocrine fashion. VEGF is such a potent regulator of vascular development that its dosage must be tightly regulated. Disruption of even a single allele of the VEGF gene in mice results in embryonic lethality due to severe vascular defects.[35,36] VEGF-A binds to receptors, VEGFR-1 and VEGFR-2. Recently, the neuropilins have been identified as coreceptors for specific VEGF isoforms. While neuropilin-1 is a receptor only for the $VEGF_{165}$ isoform,[37] neuropilin-2 binds both $VEGF_{165}$ and $VEGF_{145}$.[38] Targeted inactivation of VEGFR-1 and VEGFR-2 as well as neuropilin-1 in mice resulted in defects of blood vessel formation and embryonic lethality, demonstrating further the importance of VEGF-A for appropriate vascular development.[39-41]

In summary, VEGF-A and VEGF-B appear to be the only VEGF members expressed at significant levels in the CNS, whereas PlGF may be activated during pathological situations. While VEGF-A fulfills an important role in vascular development of the brain,[42] it remains to be established whether VEGF-B has any physiological function in the CNS. I will, therefore, for the remaining part of this Chapter focus on the regulation and function of VEGF-A in the nervous system.

REGULATION OF VEGF AND VEGF RECEPTOR EXPRESSION

A variety of physiological and pathological processes are associated with upregulation of components of the VEGF/VEGFR-system, including embryogenesis, the female reproductive cycle, pregnancy, wound healing, tumor growth, diabetic retinopathy and ischemic diseases (for a review see refs. 3,43). In the search for mechanisms and factors capable of influencing VEGF expression during these processes, many cytokines and growth factors have been shown to modulate VEGF gene expression.[5] For example, TNFα and bFGF are able to induce VEGF gene expression in vitro;[44,45] also glucose deficiency has been shown to increase VEGF expression.[46] The most important and intensively studied inducer of VEGF gene expression, however, is hypoxia, which was demonstrated for the first time in the perinecrotic areas of glioblastomas.[47,48] Under hypoxic conditions VEGF expression is mediated through the activation of specific hypoxia-inducible transcription factors, HIF-1 and HIF-2.[49,50] In addition, VEGF upregulation during hypoxia is also achieved by an increase in the stability of its mRNA[51] and by the efficient hypoxic translation of the VEGF mRNA which is mediated by an internal ribosomal entry site.[52]

HIF-1 is an ubiquitously expressed master regulator of oxygen homeostasis. It is a heterodimeric transcription factor that is composed of HIF-1α and HIF-1β (also known as the arylhydrocarbon receptor nuclear translocator ARNT) protein subunits. Whereas HIF-1β/ARNT is constitutively expressed, HIF-1α expression is induced in hypoxic cells. HIF-1 binds to the promoter/enhancer elements of hitherto more than 20 known hypoxia-inducible genes and stimulates their transcription. These genes play essential roles in the physiologic adaptation to hypoxia, including glycolysis, erythropoiesis, angiogenesis and vascular remodeling.[53] Furthermore, it became evident that HIF-1 plays a central role in many human diseases, including myocardial ischemia, stroke and tumor growth.[54] Recently, a close homologue of HIF-1α was identified and termed HIF-2α.[50,55-57] In addition, it has been shown that ARNT2, a conserved ARNT homologue that is highly expressed in neurons, forms functional HIF complexes in vivo, leading to the hypothesis that HIF-1α/ARNT2 heterodimers may specifically mediate transcriptional responses in the nervous system.[58]

HIF-1α mRNA is constitutively expressed in virtually all organs and cells[59] implicating a regulation of HIF-1 activity at a posttranscriptional level.[60] HIF-1α protein levels are downregulated in normoxic cells by ubiquitination and rapid proteasomal degradation, a process that is mediated by the von Hippel-Lindau tumor suppressor protein (pVHL).[53,61] The binding of pVHL to the α subunit requires a specific, oxygen-dependent, prolyl hydroxylation within the degradation domain of HIF-1α (Fig. 3).[62,63] The involved prolyl-hydroxylase might therefore act as an oxygen sensor. As a result HIF-1α and also HIF-2α proteins are undetectable in normoxic tissues.[64] However, expression of HIF-1α increases dramatically in hypoxic tissues, including the brain (e.g., during chronic hypoxia or ischemia).[65-67]

As a consequence of HIF-1 activation, VEGF transcription is increased during hypoxia.[49] VEGF expression is upregulated mainly in astrocytes and neurons during hypoxic and ischemic events,[34,67,68] but also in injured brain or spinal cord tissue.[69-71] In addition, gene expression of both VEGFR-1 and VEGFR-2 is activated in hypoxic tissues.[68,72,73] Whereas VEGFR-1 is a target gene for HIF-1 and is therefore directly upregulated by hypoxia,[34,74] VEGFR-2 is not induced after short term hypoxic exposure[34] but rather after prolonged periods of hypoxia, such as those occurring during ischemic events.[68,75] VEGFR-2 has no HIF-1 consensus site in its promoter region[74] but may be activated by HIF-2.[76] Increased VEGF levels may activate

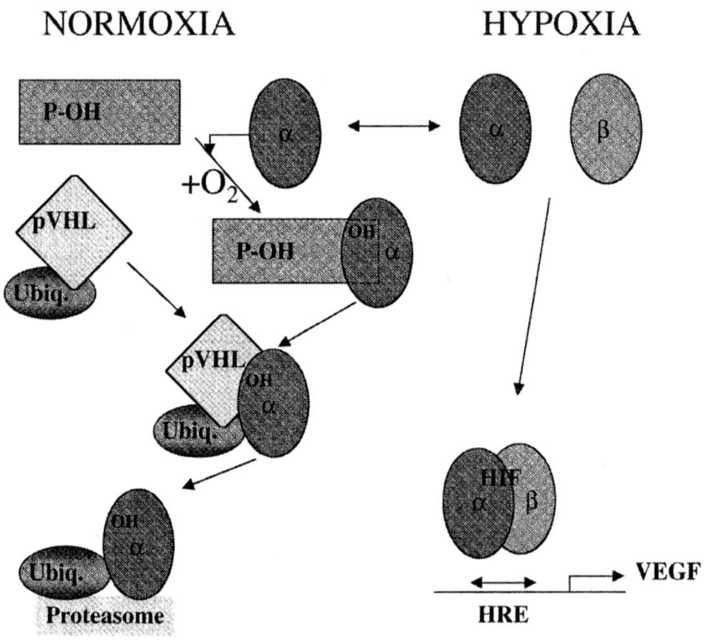

Figure 3. Model of oxygen sensing, signaling and gene regulation. Under hypoxic conditions, HIF-1α (α) is stable and can translocate to the nucleus where it heterodimerizes with HIF-1β/ARNT (β) and after binding to hypoxia-response elements (HRE), activates gene transcription, of e.g., VEGF. When, however, the intracellular oxygen tension is high, a prolyl hydroxylase enzyme (P-OH) modifies the HIF-1α subunits resulting in the hydroxylation of a specific proline residue (OH). This modification is sufficient for binding of HIF-1α to the von Hippel-Lindau tumor suppressor protein (pVHL) which then targets the α subunit for proteolytic degradation in the proteasome via ubiquitination (ubiq.). (Adapted from refs. 53,82).

VEGFR-2 expression through a positive feedback loop indicating a further regulatory pathway.[77,78] The third receptor for VEGF, neuropilin-1, is upregulated in endothelial cells of cerebral blood vessels from hypoxic tissue after focal cerebral ischemia.[79] For the other members of the VEGF family, as well as for VEGFR-3, upregulation by hypoxia has not been demonstrated conclusively, and it is so far unknown whether hypoxia-response elements allowing transcriptional regulation by either HIF-1 or HIF-2 are present in these genes.[80,81]

PLEIOTROPIC ACTION OF VEGF IN THE CNS

VEGF has long been considered a selective endothelial cell mitogen that promotes vasculogenesis and angiogenesis. In addition, it initiates endothelial procoagulant activity and induces vascular permeability. These effects are compatible with the expression of VEGF receptors specifically on vascular endothelial cells.

However, other cell types, including hematopoietic progenitor cells, monocytes and trophoblasts also express functional VEGF receptors (for a review see ref. 2). More recently, additional cellular targets of VEGF were discovered in the CNS. Astrocytes and neurons have been reported to express either or both VEGFR-1 and VEGFR-2 during cerebral ischemia.[83] VEGFR-2 expression has also been described in neural progenitor cells of the retina,[84] and expression of VEGFR-1 was found on reactive astrocytes after VEGF stimulation,[85] and in other models involving CNS trauma.[86] These findings led to the hypothesis that VEGF might act as a direct neurotrophic or even neuroprotective factor. This concept is supported by the finding that other pro-angiogenic growth factors such as bFGF and PDGF have been shown to be neurotrophic as well.[87] Furthermore, the identification of the neuropilins as coreceptors for different isoforms of VEGF[37,38] implicates VEGF in the processes of neuronal growth and repair. Thus, VEGF may affect neuronal survival in a variety of pathological conditions *indirectly* by inducing angiogenesis and vascular permeability but also by a *direct* action on neurons and astrocytes.

Direct Neurotrophic/Neuroprotective Actions of VEGF

The first evidence for a neurotrophic function for VEGF came from a series of experiments performed in the laboratory of Jeffrey Rosenstein, using murine brain explant cultures in vitro. The study was initially designed to investigate angiogenic reactions in the CNS after VEGF administration. This group found an increase in enlarged, dilated and branching vessels after VEGF stimulation in rat cortical slice explant cultures but also expression of VEGFR-1 on reactive astrocytes.[85] Subsequently, in mesencephalic explant cultures they detected that VEGF administration led to the proliferation of astrocytes and also prevented dopaminergic neurons from cell death.[88] The significance of VEGF-mediated astrocyte proliferation is unknown. Recent studies support the assertion that astrocytes are actively involved in the generation and maintenance of neuronal synapses.[89] VEGF may sustain these effects on neurons by providing the necessary glial bed and thus lead to an indirect stabilization of neuronal function. This idea is supported by the finding that expression of both VEGFR-1 and VEGFR-2 in M ller cells of the developing retina is required for normal retinal development, from which it is surmised that the development of blood vessels and neural tissue takes place in a coordinated fashion, guided by the expression of VEGF receptors in glial cells.[90] In addition, direct effects of VEGF on neuronal cell lines and primary neurons were also reported, including the role of VEGF in promoting the development and survival of rat photoreceptor cells.[91] Also, addition of VEGF to the immortalized hippocampal neuronal cell line HN33 resulted in reduced cell death that was associated with an in vitro model of cerebral ischemia by doubling the number of surviving cells after 24 hours of combined oxygen and glucose deprivation.[92] Furthermore, VEGF protects rat primary embryonic hippocampal neurons against glutamate-induced neurotoxicity. [93] Six hours of pretreatment with VEGF before exposure to glutamate resulted in increased cell survival, inhibited upregulation of caspase-3-like activity and also attenuated DNA laddering. These results imply that VEGF is able to block neuronal apoptosis in specific cell lineages. Further evidence in support of this hypothesis comes from a recent in vivo study, where topical application of VEGF to the surface of an ischemic brain led to a significant reduction of infarct volume and neuronal damage some 24 hours after onset of experimentally induced stroke.[94] The positive effect of VEGF administration after 24 hours is supportive of a direct protective effect of VEGF on the neurons itself. An indirect neuroprotective effect mediated by inducing angiogenesis (see below) would need more time, as new vessel growth during cerebral ischemia is initiated only 24 to 48 hours after occlusion of the artery.[68] Alternatively, VEGF may improve survival of hypoxic endothelial cells in the penumbra surrounding the infarcted area, thus preserving a residual blood flow which then contributes to better neuronal survival.

In the peripheral nervous system, VEGF has also been implicated in neurotrophic and neuroprotective effects. VEGF is expressed by neurons of the superior cervical ganglia (SCG) and dorsal root ganglia (DRG), and VEGFR-2 is present on neurons within SCG and DRG, and also on Schwann cells.[87] In these different cell types, VEGF and VEGFR-2 expression demonstrates temporal changes during early postnatal life, from which it may be concluded that there is developmental regulation of VEGF activity in peripheral ganglia.[95] Addition of VEGF to explanted ganglia promoted cell survival, axonal outgrowth, and proliferation of Schwann cells.[87] Topical application of VEGF to injured peripheral nerves, therefore, could have a beneficial effect on nerve regeneration by promoting the invasion of Schwann cells and new vessel growth.[96,97] In addition, it has been shown that VEGF gene transfer in a rabbit ischemic hindlimb model, as well as in diabetic rats, had favorable effects on peripheral nerve function.[98,99] The role of VEGF, and especially its inducibility by hypoxia, on neuronal survival was very recently underlined by a fascinating study in the laboratory of Peter Carmeliet. Transgenic mice with a deletion of the hypoxia-response element in the promoter of the VEGF gene demonstrated a reduction of baseline levels of VEGF specifically in neural tissue, and hypoxia-induced VEGF expression in the spinal cord and the brain was blunted. At the age of five months, the transgenic animals developed a progressive motor neuron degeneration, which was reminiscent of human amyotrophic lateral sclerosis.[100]

In the search for mechanisms that mediate the neuronal functions of VEGF, the interaction between VEGF and the neuropilins deserves attention. Neuropilins were initially identified according to their ability to bind secreted semaphorins of subclass 3.[1] However, this semaphorin-neuropilin complex is unable to transmit a signal to the cell because neuropilins lack a signaling-competent cytoplasmic domain. Therefore, the binding of an additional class of receptors, the plexins, is required to form a fully functional semaphorin receptor complex. Alternatively, neuropilins can use VEGFR-2 as a coreceptor by interacting with specific isoforms of VEGF. The presence of neuropilin-1 on endothelial cells enhances the affinity of $VEGF_{165}$ for its signaling receptor VEGFR-2, leading to increased endothelial cell chemotaxis and mitogenesis.[37] Furthermore, neuropilin-1 may also use the tyrosine kinase activity of VEGFR-1 for signaling after semaphorin binding.[101] In nervous tissue, secreted

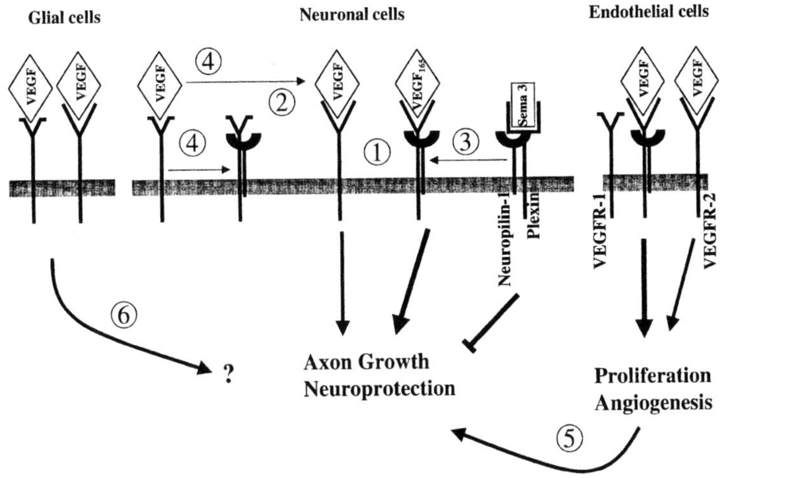

Figure 4. Hypothetical model of VEGF-mediated neuroprotection. Various VEGF receptors are expressed on glial, neuronal and endothelial cells. 1) VEGF signals through binding to VEGFR-2 on neurons. Additional complex formation with neuropilin-1 enhances this signal. 2) Neuropilin-1 binds the sink receptor VEGFR-1 thereby releasing VEGF, which then can bind to and signal through VEGFR-2. 3) Complex formation between VEGF and neuropilin-1 inhibits formation of the semaphorin-plexin-neuropilin complex, thereby counteracting the repellent and apoptotic function of semaphorin 3A. This in turn leads to axonal outgrowth and neuroprotection. 4) Regulation of VEGFR-1 expression which is able to bind both VEGF and neuropilin-1, in turn determines the amount of available VEGF for binding to VEGFR-2 as well as of neuropilin-1 for the formation of the semaphorin-plexin-neuropilin complex. 5) VEGF-mediated endothelial cell proliferation leads to new vessel growth, thereby allowing enhanced oxygen and nutrient delivery to neuronal cells which in turn increases cell survival. 6) VEGF-mediated proliferation of astrocytes might help to stabilize neuronal function.

semaphorins control cell migration by forming tissue gradients that are sensed by receptor complexes consisting of plexins and neuropilins.[4] Semaphorins act as repelling cues for the guidance of axons in developing nervous tissue. In the adult organism, semaphorins seem to inhibit axon growth after injury and maintain established neural pathways. Secreted semaphorins and VEGF compete with each other for binding to their natural receptors neuropilin-1 and VEGFR-2, respectively. To further complicate the situation, it has been demonstrated that neuropilin-1 binds with high affinity to VEGFR-1, thereby inhibiting the binding of neuropilin-1 to $VEGF_{165}$.[102] VEGFR-1 lacks a significant signaling activity and has therefore been considered to negatively regulate the VEGF-dependent activity by functioning as a sink receptor.[2] Based on these results, a number of scenarios emerge, as to how VEGF action on neuronal cells could be mediated (Fig. 4). 1) VEGF can bind to VEGFR-2 present on neurons and signal via the kinase domain of this receptor. An additional complex formation with neuropilin-1 could modulate or enhance this signal. 2) As VEGFR-1 can act as a sink receptor, preventing VEGF binding to the signaling VEGFR-2, the binding of neuropilin-1 to VEGFR-1 could block this sink function and thereby increase the availability of VEGF for VEGFR-2 leading to an increased activity via this receptor. 3) The neurotrophic action of VEGF could be an indirect one. As semaphorins inhibit axonal growth, VEGF could compete with semaphorins for neuropilin binding, thereby blocking the repellent function of semaphorins on neurons. This would result in an increased axonal outgrowth and better survival of neurons as previously demonstrated.[87,97,101] 4) Regulated expression of VEGFR-1 could modulate the level of available neuropilin-1 and thus influence the affinity of VEGFR-2 for VEGF as well as signaling via the semaphorin-plexin complexes. The balance between VEGF and semaphorin availability at their common receptors VEGFR-1, VEGFR-2 and neuropilin-1 may therefore modulate the migration, apoptosis, survival and proliferation of neural cells.[101] The situation may be even more complex, as it is currently unknown whether neuropilin-1 and VEGFR-2 are able to directly interact. Furthermore, neuropilin-2 and the plexins may interfere with the other receptors as well. On the other hand, it is evident that semaphorins by competing with VEGF for binding to neuropilin-1 could influence the angiogenesis reaction mediated by VEGF.[37]

Angiogenesis

VEGF can also influence neuronal survival indirectly by inducing new blood vessel growth, thereby allowing increased transportation of oxygen and nutrients to hypoxic tissue. Formation and remodeling of new blood vessels is governed by three processes: vasculogenesis, angiogenesis and arteriogenesis. Vasculogenesis is defined as the differentiation of mesodermal progenitor cells (angioblasts) into endothelial cells in situ where they subsequently aggregate and form a primary vascular plexus.[103] Vasculogenesis occurs primarily during embryonic development. However, recent studies suggest that endothelial precursor cells may exist in the bone marrow and contribute to new vessel formation or vascular remodeling by vasculogenesis also in the adult organism.[104] Angiogenesis is defined as the formation of new blood vessels by sprouting of endothelial cells from pre-existing vessels or by intravascular subdivision (intussusception).[105] Angiogenesis further refines the primitive embryonic vascular plexus and includes remodeling, a process that transforms the relatively uniformly sized vasculature into the network of small and large vessels, that finally undergoes maturation by recruiting perivascular cells, such as smooth muscle cells and pericytes. Angiogenesis is an important process during embryogenesis but also occurs in the adult in response to altered metabolic requirements, e.g., it can be triggered by hypoxia. The growth of new blood vessels is a complex process that requires the coordinated interaction of endothelial cells with the tissue environment.[3] Transient phases of angiogenesis occur in various physiological processes in the adult organism, for example during the female reproductive cycle in the ovary and in the uterus, as well as during pregnancy and in wound healing. Furthermore, pathological neovascularization occurs in a variety of diseases, such as ophthalmic and rheumatic diseases, psoriasis, hemangioblastoma, solid tumor growth and also during ischemic diseases.[106] Finally, arteriogenesis is the rapid proliferation of pre-existing collateral vessels that occurs in ischemic tissue (for a review see ref. 107).

The brain, with its high rate of oxidative metabolism, must rely on a steady supply of oxygen. A prolonged hypoxic period will lead to a reduction of tissue oxygenation with detrimental effects on proper brain function. In order to maintain oxygen delivery to the brain, the organism increases the vascular density in this organ, resulting in smaller intercapillary distances, which will finally restore tissue oxygenation.[108,109] The brain is thus capable of structural and functional plasticity to balance energy supply and demand.[110] Prolonged hypoxia increases VEGF expression in adult brain,[34,111] thus implicating VEGF in the observed increase in vascular density. VEGF-mediated new vessel growth, can thus be considered as an physiologically adaptive response to tissue hypoxia. In addition, various pathophysiological events are associated with tissue hypoxia, such as brain injury, cerebral ischemia and also a variety of neurodegenerative diseases. Thus, although tissue hypoxia can lead to neuronal damage it is also the major stimulus capable of activating endogenous protective strategies. For example, activation of HIF-1 can promote cell survival in hypoxic and ischemic tissues via the upregulation of angiogenic and neuroprotective factors such as VEGF or erythropoietin.[68,112] Indeed, increased expression of HIF-1 and HIF-2 has been reported during stroke.[66,67] Induction of angiogenic processes through exogenous administration of VEGF (therapeutic angiogenesis) may therefore be important for the treatment of stroke, as has been shown for other ischemic disease such as hindlimb ischemia or myocardial infarction.[3,43,113] However, a major concern precluding the use of VEGF in the CNS is the fact that this growth factor induces significant vascular permeability in vivo.

Vascular Permeability

Brain edema formation as a consequence of the disruption of the blood-brain barrier is a major problem in a variety of diseases of the CNS, including brain trauma, tumor growth and stroke. The upregulation of VEGF which occurs during these pathological processes might be responsible for the observed increase in vascular permeability and subsequent brain edema formation. Chronic hypoxia leads not only to an upregulation of VEGF mRNA in the brain, but also to an increased permeability in cortical vessels.[114] Similarly, high altitude brain edema formation is associated with an increased VEGF production in the brain that can be blocked by a neutralizing anti-VEGF antibody (H.H.M., manuscript submitted). These results support the proposal that a strategy aimed at blunting brain edema formation by blocking VEGF action might be useful to enhance neuronal survival during brain trauma or stroke. Indeed, antagonism of VEGF action by using a VEGFR-1-IgG fusion protein, which sequesters VEGF, reduced edema formation and infarct size in a mouse model of cerebral ischemia.[115] Thus, the usage of VEGF as therapeutic agent is context-dependent and therefore needs to be tightly controlled with respect to both time and dosage. This is evident from a recent study using a fibrin clot model of middle cerebral artery occlusion (MCAO) in rats. Late administration of VEGF (48 hours

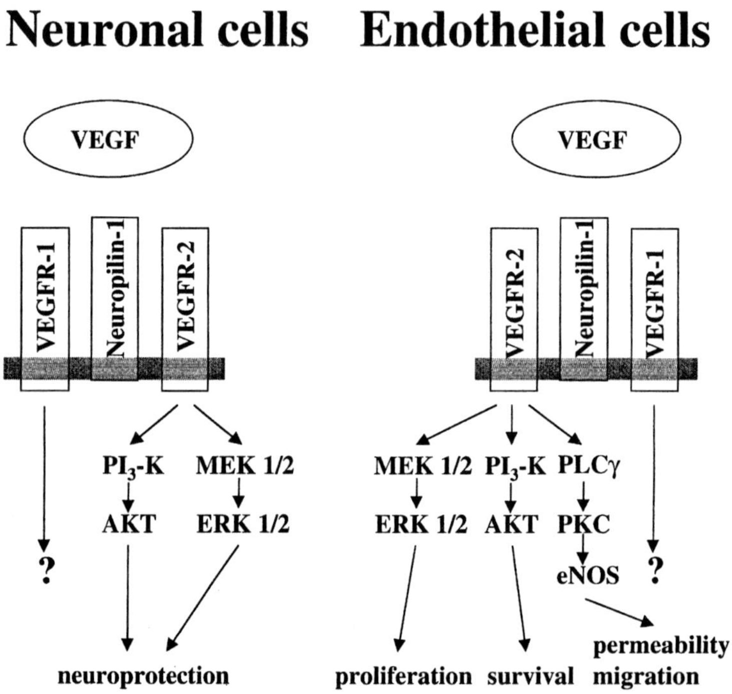

Figure 5. Major signal transduction pathways induced by VEGF. In endothelial cells and neurons VEGF signaling through activated VEGFR-2 involves the MAP-kinase pathway (MEK ERK) as well as the PI_3 kinase AKT pathway. In endothelial cells, activation of the MAP-kinase pathway mediates cell proliferation, while signaling through the PI_3 kinase-AKT pathway leads to cell survival. Recruitment of phospholipase C gamma (PLC-γ) followed by activation of protein kinase C (PKC) and endothelial nitric oxide synthetase (eNOS), finally, is involved in induction of vascular permeability. In neurons, both PI_3 kinase-AKT and MAP-kinase pathways seem to be involved in VEGF-mediated neuroprotection. All these signaling events are mediated through activation of VEGFR-2, while confirmed data are not available about the role of VEGFR-1 and neuropilin-1. (Adapted from ref. 2).

after MCAO) to the ischemic rats enhanced angiogenesis in the ischemic penumbra and significantly improved neurological recovery. However, early postischemic (1 hour after MCAO) administration of VEGF increased blood-brain barrier leakage and the size of the subsequent ischemic lesion.[116] Thus, a therapeutic regimen for the treatment of stroke evolving from these results could potentially consist of the early inhibition of VEGF function thereby minimizing brain edema formation, followed by a later active VEGF treatment leading to angiogenesis and neuroprotection.

INTRACELLULAR SIGNALING EVENTS

An important question in the field of neuroprotective aspects of VEGF is; through which intracellular signaling cascades are the various VEGF effects mediated? With regard to the possible therapeutic use of VEGF in the CNS, it would be helpful to know whether these several different VEGF-mediated effects use different signaling pathways which can be specifically activated (e.g., by low molecular compounds interfering with one pathway but not affecting the other). For example, it has been shown that distinct members of the Src kinase family (Src, Yes, Fyn) are required for different VEGF-mediated processes.[117] Mice lacking Src or Yes, but not Fyn, showed no vascular permeability in response to VEGF administration, whereas induction of angiogenesis and endothelial cell survival required Src family kinase activity in general. In accordance with these results, blockage of Src, but not Fyn activity in mice provided cerebral protection following stroke, probably via a reduction in brain edema formation.[118]

In endothelial cells, VEGFR-2 is considered to be the major signaling receptor, while VEGFR-1 acts as a sink to trap an excess of VEGF. Endothelial proliferation is mediated via the Ras-Raf-MAP(mitogen-activated protein)-kinase pathway, while protein kinase C (PKC) activation is involved in endothelial migration and vascular permeability (for a review see ref. 2). The role of VEGF as a survival factor for endothelial cells is mediated by the phosphoinositol 3 (PI_3) kinase-AKT signaling pathway.[119] This situation may be similar in neuronal cells. A recent study, using antisense oligonucleotides for VEGFR-1 or VEGFR-2, supported the hypothesis that VEGFR-2, but not VEGFR-1, mediate neuroprotective effects. However, involvement of neuropilins was not investigated in this study. Neuroprotection was associated with the activation of PI_3 kinase-AKT pathway as well as MEK/ERK signaling pathway.[93] In the setting of cerebral ischemia *in vivo* and in various *in vitro* model of neuronal death, the neuroprotective function of VEGF is associated with an activation of VEGFR-2 expression and increased levels of phosphorylated AKT, implicating the PI_3 kinase-AKT pathway.[73,92,120] Figure 5 summarizes the major signal transduction pathways that are induced by activated VEGFR-2 in endothelial cells and neurons.

CONCLUSION

The VEGF/VEGFR receptor system is activated in a variety of pathological conditions in the CNS, most prominent during cerebral ischemia. But also in brain injury and neurodegenerative diseases, such as Alzheimer s disease, involvement of VEGF is discussed. VEGF expression is markedly enhanced in Alzheimer s disease implicating compensatory mechanisms to counter insufficient vascularity or reduced perfusion apparent in this condition.[121] VEGF might therefore be an important factor in the pathophysiology of diseases that are associated with tissue hypoxia, and may has also an important role for their treatment. VEGF can exert its neuroprotective function by at least two mechanisms. VEGF acts directly on neurons expressing various VEGF receptor types, thereby mediating its neuroprotective effect. Alternatively, VEGF induces angiogenesis, thereby allowing increased oxygen and nutrient transportation to ischemic or hypoxic tissues, which ultimately leads to protection of stressed neurons and possibly other cell types in the area. VEGF or VEGF-like agents may therefore be useful for the treatment of neurodegenerative as well as ischemic disorders of the CNS. A major risk of this form of therapy is that VEGF may also damage neurons by its effect of increasing permeability leading to brain edema formation. This precludes VEGF therapy as an efficient treatment for ischemic conditions such as stroke because serious adverse side effects may occur. As indicated above, a potential stroke therapy could therefore consist of early VEGF inhibition followed by a period of active VEGF treatment. Alternatively, one might search for additional factors capable of inducing angiogenesis and increasing neuronal survival, but devoid of the harmful effects associated with edema formation. The hematopoietic growth factor erythropoietin might be such a factor. Erythropoietin is expressed in the CNS, is induced by hypoxia like VEGF and has a very similar pattern of expression and functional role during cerebral ischemia. Indeed, erythropoietin has been shown to be neuroprotective as well as angiogenic in the brain,[112] but does not seem to induce vascular permeability. Thus, therapeutic application of erythropoietin, which has been widely used as human therapeutic in the clinic, might be a more promising therapeutic approach. An alternative to the treatment of ischemic brain injury by application of VEGF alone, might be the combination of different angiogenic and neuroprotective growth factors. The combined application of VEGF with angiopoietins has been shown to protect vessels from VEGF-induced vascular leakage.[122] However, the use of VEGF to treat patients with diseases of the CNS may lead to the induced growth of dormant tumors or facilitate atherosclerotic plaque progression.[123] Thus, much more research is needed to establish the role of VEGF as neuroprotective agent, and whether the concept of therapeutic angiogenesis using VEGF is applicable as therapy for the treatment of neurodegenerative or ischemic diseases such as Alzheimer s disease or stroke. Furthermore, the potential damage to nervous tissue which may be caused by blocking an essential survival signal for neuronal cells mediated through the VEGF pathway should be avoided under any circumstances. The rapidly increasing number of reports demonstrating neurotrophic or neuroprotective functions of VEGF therefore warrants a careful re-evaluation of all antiangiogenic strategies (e.g., for treatment of solid tumor growth).

ACKNOWLEDGMENTS

I gratefully acknowledge Sabine Raab and Marcel Groot for helpful discussion and Christopher Mitchell and Friedemann Kiefer for critically reading the manuscript. I was supported by a fellowship from the Max Planck Society.

REFERENCES

1. Shima DT, Mailhos C. Vascular developmental biology: Getting nervous. Curr Opin Genet Dev 2000; 10:536-542.
2. Clauss M. Molecular biology of the VEGF and the VEGF receptor family. Semin Thromb Hemost 2000; 26:561-569.
3. Breier G. Functions of the VEGF/VEGF receptor system in the vascular system. Semin Thromb Hemost 2000; 26:553-559.
4. Tamagnone L, Comoglio PM. Signalling by semaphorin receptors: Cell guidance and beyond. Trends Cell Biol 2000; 10:377-383.
5. Robinson CJ, Stringer SE. The splice variants of vascular endothelial growth factor (VEGF) and their receptors. J Cell Sci 2001; 114:853-865.
6. Maglione D, Guerriero V, Viglietto G et al. Isolation of a human placenta cDNA coding for a protein related to the vascular permeability factor. Proc Natl Acad Sci USA 1991; 88:9267-9271.
7. Park JE, Chen HH, Winer J et al. Placenta growth factor. Potentiation of vascular endothelial growth factor bioactivity, *in vitro* and *in vivo*, and high affinity binding to Flt-1 but not to Flk-1/KDR. J Biol Chem 1994; 269:25646-25654.
8. Hatva E, Kaipainen A, Mentula P et al. Expression of endothelial cell-specific receptor tyrosine kinases and growth factors in human brain tumors. Am J Pathol 1995; 146:368-378.
9. Nomura M, Yamagishi S, Harada S et al. Placenta growth factor (PIGF) mRNA expression in brain tumors. J Neuro-Oncol 1998; 40:123-130.
10. Ferrara N, Davis-Smyth T. The biology of vascular endothelial growth factor. Endocr Rev 1997; 18:4-25.
11. Clauss M. Functions of the VEGF receptor-1 (Flt-1) in the vasculature. Trends Cardiovasc Med 1998; 8:241-245.
12. Carmeliet P, Moons L, Luttun A et al. Synergism between vascular endothelial growth factor and placental growth factor contributes to angiogenesis and plasma extravasation in pathological conditions. Nature Med 2001; 7:575-583.
13. Olofsson B, Pajusola K, Kaipainen A et al. Vascular endothelial growth factor B, a novel growth factor for endothelial cells. Proc Natl Acad Sci USA 1996; 93:2576-2581.
14. Olofsson B, Korpelainen E, Pepper MS et al. Vascular endothelial growth-factor B (VEGF-B) binds to VEGF receptor-1 and regulates plasminogen activator activity in endothelial cells. Proc Natl Acad Sci USA 1998; 95:11709-11714.
15. Lagercrantz J, Farnebo F, Larsson C et al. A comparative study of the expression patterns for *vegf*, *vegf-b* / *vrf* and *vegf-c* in the developing and adult mouse. BiochimBiophys Acta 1998; 1398:157-163.
16. Joukov V, Pajusola K, Kaipainen A et al. A novel vascular endothelial growth factor, VEGF-C, is a ligand for the flt4 (VEGFR-3) and KDR (VEGFR-2) receptor tyrosine kinases. EMBO J 1996; 15:290-298.
17. Joukov V, Sorsa T, Kumar V et al. Proteolytic processing regulates receptor specificity and activity of VEGF-C. EMBO J 1997; 16:3898-3911.
18. Lee J, Gray A, Yuan J et al. Vascular endothelial growth factor-related protein: A ligand and specific activator of the tyrosine kinase receptor Flt4. Proc Natl Acad Sci USA 1996; 93:1988-1992.
19. Jeltsch M, Kaipainen A, Joukov V et al. Hyperplasia of lymphatic vessels in VEGF-C transgenic mice. Science 1997; 276:1423-1425.
20. Witzenbichler B, Asahara T, Murohara T et al. Vascular endothelial growth factor-C (VEGF-C/VEGF-2) promotes angiogenesis in the setting of tissue ischemia. Am J Pathol 1998; 153:381-394.

21. Enholm B, Jussila L, Karkkainen M et al. Vascular endothelial growth factor-C: A growth factor for lymphatic and blood vascular endothelial cells. Trends Cardiovasc Med 1998; 8:292-297.
22. Cao Y, Linden P, Farnebo J et al. Vascular endothelial growth factor C induces angiogenesis *in vivo*. Proc Natl Acad Sci USA 1998; 95:14389-14394.
23. Dumont DJ, Jussila L, Taipale J et al. Cardiovascular failure in mouse embryos deficient in VEGF receptor-3. Science 1998; 282:946-949.
24. Achen MG, Jeltsch M, Kukk E et al. Vascular endothelial growth factor D (VEGF-D) is a ligand for the tyrosine kinases VEGF receptor 2 (Flk1) and VEGF receptor 3 (Flt4). Proc Natl Acad Sci USA 1998; 95:548-553.
25. Yamada Y, Nezu J, Shimane M et al. Molecular cloning of a novel vascular endothelial growth factor, VEGF-D. Genomics 1997; 42:483-488.
26. Lyttle DJ, Fraser KM, Fleming SB et al. Homologs of vascular endothelial growth factor are encoded by the poxvirus orf virus. J Virol 1994; 68:84-92.
27. Ogawa S, Oku A, Sawano A et al. A novel type of vascular endothelial growth factor, VEGF-E (NZ-7 VEGF), preferentially utilizes KDR/Flk-1 receptor and carries a potent mitotic activity without heparin-binding domain. J Biol Chem 1998; 273:31273-31282.
28. Meyer M, Clauss M, Lepple-Wienhues A et al. A novel vascular endothelial growth factor encoded by Orf virus, VEGF-E, mediates angiogenesis via signalling through VEGFR-2 (KDR) but not VEGFR-1 (Flt-1) receptor tyrosine kinases. EMBO J 1999; 18:363-374.
29. Wise LM, Veikkola T, Mercer AA et al. Vascular endothelial growth factor (VEGF)-like protein from orf virus NZ2 binds to VEGFR2 and neuropilin-1. Proc Natl Acad Sci USA 1999; 96:3071-3076.
30. Ogunshola OO, Stewart WB, Mihalcik V et al. Neuronal VEGF expression correlates with angiogenesis in postnatal developing rat brain. Develop Brain Res 2000; 119:139-153.
31. Neufeld G, Cohen T, Gengrinovitch S et al. Vascular endothelial growth factor (VEGF) and its receptors. FASEB J 1999; 13:9-22.
32. Senger DR, Galli SJ, Dvorak AM et al. Tumor cells secrete a vascular permeability factor that promotes accumulation of ascites fluid. Science 1983; 219:983-985.
33. Keck PJ, Hauser SD, Krivi G et al. Vascular permeability factor, an endothelial cell mitogen related to PDGF. Science 1989; 246:1309-1312.
34. Marti HH, Risau W. Systemic hypoxia changes the organ-specific distribution of vascular endothelial growth factor and its receptors. Proc Natl Acad Sci USA 1998; 95:15809-15814.
35. Carmeliet P, Ferreira V, Breier G et al. Abnormal blood vessel development and lethality in embryos lacking a single VEGF allele. Nature 1996; 380:435-439.
36. Ferrara N, Carver-Moore K, Chen H et al. Heterozygous embryonic lethality induced by targeted inactivation of the VEGF gene. Nature 1996; 380:439-442.
37. Soker S, Takashima S, Miao HQ et al. Neuropilin-1 is expressed by endothelial and tumor cells as an isoform-specific receptor for vascular endothelial growth factor. Cell 1998; 92:735-745.
38. Gluzman-Poltorak Z, Cohen T, Herzog Y et al. Neuropilin-2 is a receptor for the vascular endothelial growth factor (VEGF) forms VEGF-145 and VEGF-165. J Biol Chem 2000; 275:18040-18045.
39. Fong G-H, Rossant J, Gertsenstein M et al. Role of the Flt-1 receptor tyrosine kinase in regulating the assembly of vascular endothelium. Nature 1995; 376:66-70.
40. Shalaby F, Rossant J, Yamaguchi TP et al. Failure of blood-island formation and vasculogenesis in flk-1- deficient mice. Nature 1995; 376:62-66.
41. Kawasaki T, Kitsukawa T, Bekku Y et al. A requirement for neuropilin-1 in embryonic vessel formation. Development 1999; 126:4895-4902.
42. Breier G, Albrecht U, Sterrer S et al. Expression of vascular endothelial growth factor during embryonic angiogenesis and endothelial cell differentiation. Development 1992; 114:521-532.
43. Marti HH, Risau W. Angiogenesis in ischemic disease. Thromb Haemost 1999; 82 (Suppl.):44-52.

44. Ryuto M, Ono M, Izumi H et al. Induction of vascular endothelial growth factor by tumor necrosis factor α in human glioma cells. Possible roles of SP-1. J Biol Chem 1996; 271:28220-28228.
45. Stavri GT, Zachary IC, Baskerville PA et al. Basic fibroblast growth factor upregulates the expression of vascular endothelial growth factor in vascular smooth muscle cells. Synergistic action with hypoxia. Circulation 1995; 92:11-14.
46. Shweiki D, Neeman M, Itin A et al. Induction of vascular endothelial growth factor expression by hypoxia and by glucose deficiency in multicell spheroids: Implications for tumor angiogenesis. Proc Natl Acad Sci USA 1995; 92:768-772.
47. Shweiki D, Itin A, Soffer D et al. Vascular endothelial growth factor induced by hypoxia may mediate hypoxia-initiated angiogenesis. Nature 1992; 359:843-845.
48. Plate KH, Breier G, Weich HA et al. Vascular endothelial growth factor is a potential tumour angiogenesis factor in human gliomas in vivo. Nature 1992; 359:845-848.
49. Forsythe JA, Jiang B-H, Iyer NV et al. Activation of vascular endothelial growth factor gene transcription by hypoxia-inducible factor 1. Mol Cell Biol 1996; 16:4604-4613.
50. Ema M, Taya S, Yokotani N et al. A novel bHLH-PAS factor with close sequence similarity to hypoxia-inducible factor 1α regulates the VEGF expression and is potentially involved in lung and vascular development. Proc Natl Acad Sci USA 1997; 94:4273-4278.
51. Ikeda E, Achen MG, Breier G et al. Hypoxia-induced transcriptional activation and increased mRNA stability of vascular endothelial growth factor (VEGF) in C6 glioma cells. J Biol Chem 1995; 270:19761-19766.
52. Stein I, Itin A, Einat P et al. Translation of vascular endothelial growth factor mRNA by internal ribosome entry: Implications for translation under hypoxia. Mol Cell Biol 1998; 18:3112-3119.
53. Wenger RH. Mammalian oxygen sensing, signalling and gene regulation. J Exp Biol 2000; 203:1253-1263.
54. Semenza GL. HIF-1 and human disease: One highly involved factor. Genes Dev 2000; 14:1983-1991.
55. Tian H, McKnight SL, Russell DW. Endothelial pas domain protein 1 (EPAS1), a transcription factor selectively expressed in endothelial cells. Genes Dev 1997; 11:72-82.
56. Flamme I, Fr hlich T, von Reutern M et al. HRF, a putative basic helix-loop-helix-PAS-domain transcription factor is closely related to hypoxia-inducible factor-1α and developmentally expressed in blood vessels. Mech Develop 1997; 63:51-60.
57. Hogenesch JB, Chan WK, Jackiw VH et al. Characterization of a subset of the basic-helix-loop-helix-PAS superfamily that interacts with components of the dioxin signaling pathway. J Biol Chem 1997; 272:8581-8593.
58. Maltepe E, Keith B, Arsham AM et al. The role of ARNT2 in tumor angiogenesis and the neural response to hypoxia. Biochem Biophys Res Commun 2000; 273:231-238.
59. Wenger RH, Rolfs A, Marti HH et al. Nucleotide sequence, chromosomal assignment and mRNA expression of mouse hypoxia-inducible factor-1α. Biochem Biophys Res Commun 1996; 223:54-59.
60. Wenger RH, Kvietikova I, Rolfs A et al. Hypoxia-inducible factor-1α is regulated at the post-mRNA level. Kidney Int 1997; 51:560-563.
61. Maxwell PH, Wiesener MS, Chang GW et al. The tumour suppressor protein VHL targets hypoxia-inducible factors for oxygen-dependent proteolysis. Nature 1999; 399:271-275.
62. Ivan M, Kondo K, Yang HF et al. HIFa targeted for VHL-mediated destruction by proline hydroxylation: Implications for O_2 sensing. Science 2001; 292:464-468.
63. Jaakkola P, Mole DR, Tian YM et al. Targeting of HIF a to the von Hippel-Lindau ubiquitylation complex by O_2-regulated prolyl hydroxylation. Science 2001; 292:468-472.
64. Talks KL, Turley H, Gatter KC et al. The expression and distribution of the hypoxia-inducible factors HIF-1α and HIF-2α in normal human tissues, cancers, and tumor-associated macrophages. Am J Pathol 2000; 157:411-421.

65. Chavez JC, Agani F, Pichiule P et al. Expression of hypoxia-inducible factor-1α in the brain of rats during chronic hypoxia. J Appl Physiol 2000; 89:1937-1942.
66. Bergeron M, Yu AY, Solway KE et al. Induction of hypoxia-inducible factor-1 (HIF-1) and its target genes following focal ischaemia in rat brain. Eur J Neurosci 1999; 11:4159-4170.
67. Jin KL, Mao XO, Nagayama T et al. Induction of vascular endothelial growth factor and hypoxia-inducible factor-1α by global ischemia in rat brain. Neuroscience 2000; 99:577-585.
68. Marti HJH, Bernaudin M, Bellail A et al. Hypoxia-induced vascular endothelial growth factor expression precedes neovascularization after cerebral ischemia. Am J Pathol 2000; 156:965-976.
69. Papavassiliou E, Gogate N, Proescholdt M et al. Vascular endothelial growth factor (vascular permeability factor) expression in injured rat brain. J Neurosci Res 1997; 49:451-460.
70. Bartholdi D, Rubin BP, Schwab ME. VEGF mRNA induction correlates with changes in the vascular architecture upon spinal cord damage in the rat. Eur J Neurosci 1997; 9:2549-2560.
71. Sk ld M, Cullheim S, Hammarberg H et al. Induction of VEGF and VEGF receptors in the spinal cord after mechanical spinal injury and prostaglandin administration. Eur J Neurosci 2000; 12:3675-3686.
72. Tuder RM, Flook BE, Voelkel NF. Increased gene expression for VEGF and the VEGF receptors KDR/flk and flt in lungs exposed to acute or chronic hypoxia, Modulation of gene expression by nitric oxide. J Clin Invest 1995; 95:1798-1807.
73. Jin KL, Mao XO, Nagayama T et al. Induction of vascular endothelial growth factor receptors and phosphatidylinositol 3 -kinase/Akt signaling by global cerebral ischemia in the rat. Neuroscience 2000; 100:713-717.
74. Gerber H-P, Condorelli F, Park J et al. Differential transcriptional regulation of the two vascular endothelial growth factor receptor genes; flt-1, but not flk- 1/KDR, is up-regulated by hypoxia. J Biol Chem 1997; 272:23659-23667.
75. Li J, Brown LF, Hibberd MG et al. VEGF, *flk-1*, and *flt-1* expression in a rat myocardial infarction model of angiogenesis. Am J Physiol 1996; 270:H1803-H1811.
76. Kappel A, R nicke V, Damert A et al. Identification of vascular endothelial growth factor (VEGF) receptor-2 (*Flk-1*) promoter/enhancer sequences sufficient for angioblast and endothelial cell-specific transcription in transgenic mice. Blood 1999; 93:4284-4292.
77. Kremer C, Breier G, Risau W et al. Up-regulation of flk-1/vascular endothelial growth factor receptor 2 by its ligand in a cerebral slice culture system. Cancer Res 1997; 57:3852-3859.
78. Shen B-Q, Lee DY, Gerber H-P et al. Homologous up-regulation of KDR/Flk-1 receptor expression by vascular endothelial growth factor *in vitro*. J Biol Chem 1998; 273:29979-29985.
79. Zhang ZG, Tsang W, Zhang L et al. Up-regulation of neuropilin-1 in neovasculature after focal cerebral ischemia in the adult rat. J Cereb Blood Flow Metab 2001; 21:541-549.
80. Gleadle JM, Ebert BL, Firth JD et al. Regulation of angiogenic growth factor expression by hypoxia, transition metals, and chelating agents. Am J Physiol 1995; 268:C1362-C1368.
81. Enholm B, Paavonen K, Ristimaki A et al. Comparison of VEGF, VEGF-B, VEGF-C and Ang-1 m-RNA regulation by serum, growth factors, oncoproteins and hypoxia. Oncogene 1997; 14:2475-2483.
82. Zhu H, Bunn HF. How do cells sense oxygen? Science 2001; 292:449-451.
83. Lennmyr F, Ata KA, Funa K et al. Expression of vascular endothelial growth factor (VEGF) and its receptors (Flt-1 and Flk-1) following permanent and transient occlusion of the middle cerebral artery in the rat. J Neuropathol Exp Neurol 1998; 57:874-882.
84. Yang XJ, Cepko CL. Flk-1, a receptor for vascular endothelial growth-factor (VEGF), is expressed by retinal progenitor cells. J Neurosci 1996; 16:6089-6099.
85. Rosenstein JM, Mani N, Silverman WF et al. Patterns of brain angiogenesis after vascular endothelial growth factor administration *in vitro* and *in vivo*. Proc Natl Acad Sci USA 1998; 95:7086-7091.

86. Krum JM, Rosenstein JM. VEGF mRNA and its receptor *flt-1* are expressed in reactive astrocytes following neural grafting and tumor cell implantation in the adult CNS. Exp Neurol 1998; 154:57-65.
87. Sondell M, Lundborg G, Kanje M. Vascular endothelial growth factor has neurotrophic activity and stimulates axonal outgrowth, enhancing cell survival and Schwann cell proliferation in the peripheral nervous system. J Neurosci 1999; 19:5731-5740.
88. Silverman WF, Krum JM, Mani N et al. Vascular, glial and neuronal effects of vascular endothelial growth factor in mesencephalic explant cultures. Neuroscience 1999; 90:1529-1541.
89. Temburni MK, Jacob MH. New functions for glia in the brain. Proc Natl Acad Sci USA 2001; 98:3631-3632.
90. Robinson GS, Ju M, Shih S-C, Xu X, Mcmahon G, Caldwell R, Smith LEH. Nonvascular role for VEGF:VEGFR-1, 2 activity is critical for neural retinal development. FASEB J (March 20, 2001) 10.1096/fj.00-0598fje.
91. Yourey PA, Gohari S, Su JL et al. Vascular endothelial cell growth factors promote the in vitro development of rat photoreceptor cells. J Neurosci 2000; 20:6781-6788.
92. Jin KL, Mao XO, Greenberg DA. Vascular endothelial growth factor: direct neuroprotective effect in *in vitro* ischemia. Proc Natl Acad Sci USA 2000; 97:10242-10247.
93. Matsuzaki H, Tamatani M, Yamaguchi A, Namikawa K, Kiyama H, Vitek MP et al. Vascular endothelial growth factor rescues hippocampal neurons from glutamate-induced toxicity: Signal transduction cascades. FASEB J (March 12, 2001) 10.1096/fj.00-0495fje.
94. Hayashi T, Abe K, Itoyama Y. Reduction of ischemic damage by application of vascular endothelial growth factor in rat brain after transient ischemia. J Cereb Blood Flow Metab 1998; 18:887-895.
95. Sondell M, Kanje M. Postnatal expression of VEGF and its receptor flk-1 in peripheral ganglia. Neuroreport 2001; 12:105-108.
96. Sondell M, Lundborg G, Kanje M. Vascular endothelial growth factor stimulates Schwann cell invasion and neovascularization of acellular nerve grafts. Brain Res 1999; 846:219-228.
97. Sondell M, Sundler F, Kanje M. Vascular endothelial growth factor is a neurotrophic factor which stimulates axonal outgrowth through the flk-1 receptor. Eur J Neurosci 2000; 12:4243-4254.
98. Schratzberger P, Schratzberger G, Silver M et al. Favorable effect of VEGF gene transfer on ischemic peripheral neuropathy. Nature Med 2000; 6:405-413.
99. Schratzberger P, Walter DH, Rittig K et al. Reversal of experimental diabetic neuropathy by VEGF gene transfer. J Clin Invest 2001; 107:1083-1092.
100. Oosthuyse B, Moons L, Storkebaum E et al. Deletion of the hypoxia-response element in the vascular endothelial growth factor promoter causes motor neuron degeneration. Nat Genet 2001; 28:131-138.
101. Bagnard D, Vaillant C, Khuth S-T et al. Semaphorin 3A-vascular endothelial growth factor-165 balance mediates migration and apoptosis of neural progenitor cells by the recruitment of shared receptor. J Neurosci 2001; 21:3332-3341.
102. Fuh G, Garcia KC, de Vos AM. The interaction of neuropilin-1 with vascular endothelial growth factor and its receptor Flt-1. J Biol Chem 2000; 275:26690-26695.
103. Risau W, Flamme I. Vasculogenesis. Annu Rev Cell Dev Biol 1995; 11:73-91.
104. Asahara T, Takahashi T, Masuda H et al. VEGF contributes to postnatal neovascularization by mobilizing bone marrow-derived endothelial progenitor cells. EMBO J 1999; 18:3964-3972.
105. Risau W. Mechanisms of angiogenesis. Nature 1997; 386:671-674.
106. Folkman J. Angiogenesis in cancer, vascular, rheumatoid and other disease. Nat Med 1995; 1:27-31.
107. Buschmann I, Schaper W. Arteriogenesis versus angiogenesis: Two mechanisms of vessel growth. News Physiol Sci 1999; 14:121-125.
108. Harik SI, Hritz MA, LaManna JC. Hypoxia-induced brain angiogenesis in the adult rat. J Physiol (London) 1995; 485:525-530.

109. Boero JA, Ascher J, Arregui A et al. Increased brain capillaries in chronic hypoxia. J Appl Physiol 1999; 86:1211-1219.
110. LaManna JC, Harik SI. Brain metabolic and vascular adaptations to hypoxia in the rat. Adv Exp Med Biol 1997; 428:163-167.
111. Kuo N-T, Benhayon D, Przybylski RJ et al. Prolonged hypoxia increases vascular endothelial growth factor mRNA and protein in adult mouse brain. J Appl Physiol 1999; 86:260-264.
112. Marti HH, Bernaudin M, Petit E et al. Neuroprotection and angiogenesis: A dual role of erythropoietin in brain ischemia. News Physiol Sci 2000; 15:225-229.
113. Greenberg DA. Angiogenesis and stroke. Drug News Perspect 1998; 11:265-270.
114. Ment LR, Stewart WB, Fronc R et al. Vascular endothelial growth factor mediated reactive angiogenesis in the postnatal developing brain. Develop Brain Res 1997; 100:52-61.
115. van Bruggen N, Thibodeaux H, Palmer JT et al. VEGF antagonism reduces edema formation and tissue damage after ischemia/reperfusion injury in the mouse brain. J Clin Invest 1999; 104:1613-1620.
116. Zhang ZG, Zhang L, Jiang Q et al. VEGF enhances angiogenesis and promotes blood-brain barrier leakage in the ischemic brain. J Clin Invest 2000; 106:829-838.
117. Eliceiri BP, Paul R, Schwartzberg PL et al. Selective requirement for Src kinases during VEGF-induced angiogenesis and vascular permeability. Mol Cell 1999; 4:915-924.
118. Paul R, Zhang ZG, Eliceiri BP et al. Src deficiency or blockade of Src activity in mice provides cerebral protection following stroke. Nature Med 2001; 7:222-227.
119. Gerber HP, McMurtrey A, Kowalski J et al. Vascular endothelial growth factor regulates endothelial cell survival through the phosphatidylinositol 3'-kinase/Akt signal transduction pathway. Requirement for Flk-1/KDR activation. J Biol Chem 1998; 273:30336-30343.
120. Jin KL, Mao XO, Greenberg DA. Vascular endothelial growth factor rescues HN33 neural cells from death induced by serum withdrawal. J Mol Neurosci 2000; 14:197-203.
121. Kalaria RN, Cohen DL, Premkumar DRD et al. Vascular endothelial growth factor in Alzheimer s disease and experimental cerebral ischemia. Mol Brain Res 1998; 62:101-105.
122. Thurston G, Rudge JS, Ioffe E et al. Angiopoietin-1 protects the adult vasculature against plasma leakage. Nature Med 2000; 6:460-463.
123. Celletti FL, Waugh JM, Amabile PG et al. Vascular endothelial growth factor enhances atherosclerotic plaque progression. Nature Med 2001; 7:425-429.

IV. ADVANCES IN DRUG DELIVERY TO CNS NEURONS

BLOOD-BRAIN BARRIER DRUG TARGETING ENABLES NEUROPROTECTION IN BRAIN ISCHEMIA FOLLOWING DELAYED INTRAVENOUS ADMINISTRATION OF NEUROTROPHINS

William M. Pardridge*

ABSTRACT

The blood-brain barrier (BBB) is the rate-limiting step in the translation of neurotrophin neuroscience into clinically effective neurotherapeutics. Since neurotrophins do not cross the BBB, these proteins cannot be used for neuroprotection following intravenous administration, and it is not feasible to administer these molecules by intra-cerebral injection in human stroke. The present studies describe the development of the chimeric peptide brain drug targeting technology and the use of brain-derived neurotrophic factor (BDNF) chimeric peptides in either global or regional brain ischemia. The BDNF chimeric peptide is formed by conjugation of BDNF to a monoclonal antibody (MAb) to the BBB transferrin receptor, and the MAb acts as a molecular Trojan Horse to ferry the BDNF across the BBB via transport on the endogenous BBB transferrin receptor. High degrees of neuroprotection in transient forebrain ischemia, permanent middle cerebral artery occlusion, or reversible middle cerebral artery occlusion are achieved with the delayed intravenous administration of BDNF chimeric peptides. In contrast, no neuroprotection is observed following the intravenous administration of unconjugated BDNF, because the neurotrophin does not cross the BBB in vivo.

*Department of Medicine, UCLA School of Medicine, Los Angeles, CA 90095-1682.

BLOOD-BRAIN BARRIER, NEUROTROPHINS AND NEUROLOGICAL DISEASE

The blood-brain barrier (BBB) is the rate-limiting step in the translation of molecular neurosciences into clinically effective neurotherapeutics (Fig. 1). If neurotherapeutics are developed in the absence of any consideration to BBB transport, then the central nervous system (CNS) drug development pathway invariably leads to program termination.[1] Termination arises because essentially 100% of large molecule neuropharmaceuticals do not cross the BBB and > 98% of small molecule neuropharmaceuticals do not cross the BBB.[2] Despite these facts, the translation of neuroscience into neurotherapeutics invariably takes place in the absence of any consideration of BBB transport properties, and the neurotrophins are a case study of this problem. Not surprisingly, and despite the fact that the neurotrophin genes were cloned more than 10 years ago, there is not a single CNS disease that is currently being treated with neurotrophic factors. Indeed, virtually all clinical trials with neurotrophins have been halted by large pharmaceutical companies.

Neurotrophin Drug Development

The history of neurotrophin drug development is outlined in Figure 2. There are more than 30 neurotrophic factors that are powerful neuroprotective agents should these molecules be delivered to the target sites within the brain or spinal cord. These target sites all lie behind the BBB. Therefore, it is not possible to develop neurotrophins as clinically effective neurotherapeutics, because (a) these molecules do not cross the BBB, and (b) neurotrophin drug development took place in the absence of any consideration to BBB transport. The pitfalls of developing drugs for the brain that do not cross the BBB are illustrated in the case of amytrophic lateral sclerosis (ALS), which was treated in the 1990s with 3 different neurotrophins, including ciliary neurotrophic factor (CNTF), insulin-like growth factor (IGF)-1, and brain-derived neurotrophic factor (BDNF). The CNTF clinical trials were based only on in vitro tissue culture data and some in vivo data in developing animals.[3] The IGF-1 clinical trials were based on tissue culture data and in vivo data on the sciatic nerve,[3] which is outside the CNS and the BBB. An example of a tissue culture finding with nerve growth factor (NGF) is shown in Figure 2, which demonstrates neuronal differentiation in a dorsal root ganglia following exposure to NGF.[4] On the basis of such experiments, patients with ALS were treated with CNTF, IGF-1, or BDNF in separate costly phase III clinical trials. In all 3 trials, the neurotrophic factor was administered by subcutaneous injection, even though the target neurons to be treated in ALS resided behind the BBB or blood-spinal cord barrier. None of these neurotrophic factors crossed the BBB and not surprisingly, all 3 phase III clinical trials failed in the treatment of ALS. Subsequently, neurodegenerative conditions such as Parkinson s disease (PD) were treated with neurotrophic factors

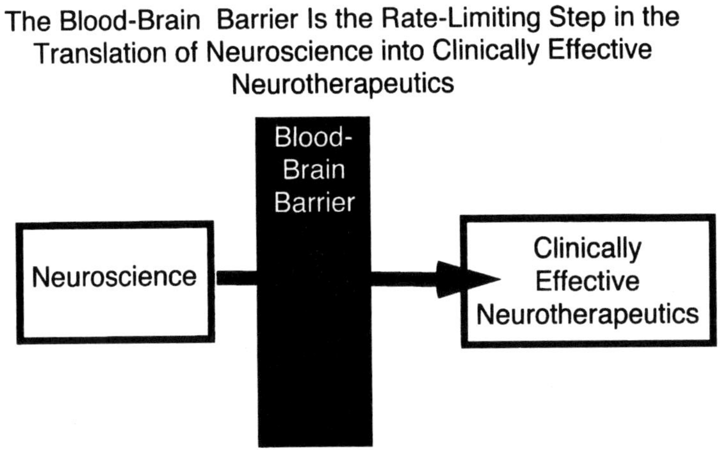

Figure 1. Neuroscience, blood-brain barrier, and clinically effective neurotherapeutics.

such as glial-derived neurotrophic factor (GDNF) by intracerebral ventricular (ICV) infusion.[5]

Limitations of ICV Infusion as a Brain Drug Delivery System

Figure 2 shows an autoradiogram of rat brain obtained 24 hours after the single injection of radiolabeled BDNF into the lateral ventricle.[6] The autoradiography demonstrates that the neurotrophic factor does not distribute beyond the ependymal epithelial layer lining the ventricular surface. This study illustrates that the ICV infusion of drug is a poor means of drug delivery to brain parenchyma. ICV infusion of drug allows for drug distribution to the surface of the brain. However, owing to the limitations of diffusion within the brain and to the rapid rate of bulk flow of CSF through the ventricular flow tracks, the majority of drug that is injected into the ventricule is rapidly distributed into the peripheral circulation rather than penetrate into brain parenchyma by diffusion.[7] The second phenomenon demonstrated by the autoradiography in Figure 2 is that CSF flow takes place in a unidirectional manner. Molecules injected into the lateral ventricule quickly move to the third ventricle, then to the fourth ventricle, then over the convexities of the brain and are absorbed into the superior sagittal sinus across the arachnoid villae without significant penetration to the contralateral brain. The third property of ICV drug infusion demonstrated by the autoradiography in Figure 2 is that whereas the brain parenchyma is exposed to very little neurotrophic factor, the ependymal surface of the brain is exposed to very high concentrations of neurotrophic factor. This causes significant sub-ependymal gliosis following the ICV infusion of neurotrophic factors in animals.[8,9] The ICV infusion of GDNF resulted in such toxicity without efficacy in the treatment of Parkinson s disease that the clinical trials were halted.[5]

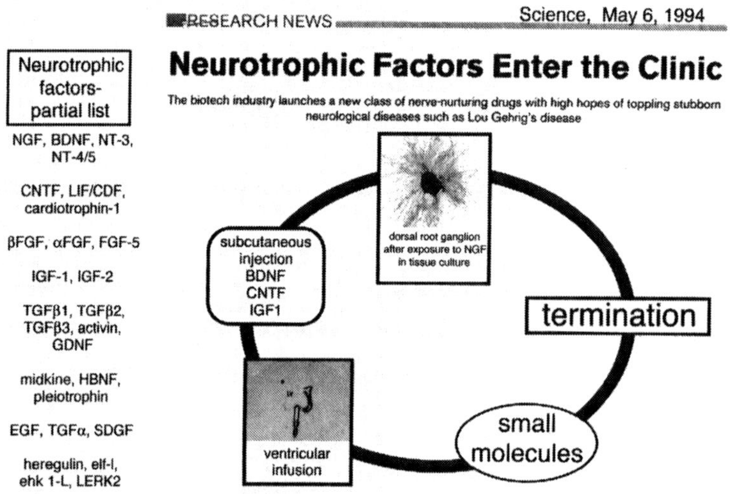

Figure 2. There are >30 different neurotrophic factors known to date and many have neuroprotective and restorative effects when injected directly into the brain.[3] However, these molecules do not cross the BBB. Preclinical research and brain drug development of the neurotrophins was largely restricted to cell culture,[3] and the effects of nerve growth factor (NGF) on the dorsal root ganglion in cell culture is shown.[4] Clinical trials for neurotrophic factors such as brain derived neurotrophic factor (BDNF), ciliary neurotrophic factor (CNTF), or insulin like growth factor (IGF)-1, were pefomed in the 1990s, and the neurotrophic factor was administered by subcutaneous injection in these trials. None of these neurotrophic factors cross the BBB and the phase III clinical trials failed. Subsequently, pharmaceutical companies attempted to administer neurotrophic factors to the brain by ventricular infusion. However, as shown in the autoradiogram, the distribution of a neurotrophic factor into brain following administration into a lateral ventricle is restricted to the ipsilateral ependymal surface of the brain at 24 hours after administration.[6] Given the poor penetration into the brain of neurotrophic factors following either peripheral or ICV administration, these drug development programs were halted. It was believed that neurotrophic factor small molecules would be discovered and that these small molecules would cross the BBB. However, peptidomimetic small molecules, should they be discovered, would still need brain drug targeting systems. Therefore, the fate of the neurotrophic factor drug development program ultimately is termination, because no BBB drug targeting strategy is available. Abbreviations used: NT, neurotrophin; FGF, fibroblastic growth factor; TGF, transforming growth factor; GDNF, glial derived neurotrophic factor; HBNF, heparin binding growth factor; EGF, epidermal growth factor; SDGF, Schwannoma-derived growth factor; LIF, leukemia inhibitory factor; CDF, cholinergic neuron differentiation factor. The partial list of neurotrophic factors is adapted from Hefti.[3]

Small Molecules

The administration of neurotrophic factors by either subcutaneous administration or ICV infusion resulted in either no therapeutic effects or enhanced toxicity. Neurotrophin drug development then evolved into small molecules and in this approach, cloned neurotrophic factor receptors are expressed and analyzed by high throughput screening (HTS) methodologies, which involves the screening of hundreds of thousands of small molecule drug candidates. This approach has not led to the discovery of neurotrophic factor small molecule peptidomimetics, because small molecule peptidomimetics tend to be antagonists, not agonists.[3] Even if a small molecule neurotrophic factor peptidomimetic agonist was discovered, it is unlikely this molecule would undergo transport across the BBB in pharmacologically significant amounts. While it is often stated that small molecules freely cross the blood-brain barrier, in fact, only a certain type of small molecule crosses the BBB. Small molecules that do cross the BBB have the dual molecular characteristics of (a) molecular weight under a 400-500 Dalton threshold and (b) lipid solubility.[2] The kinds of small molecules selected in receptor-based HTS programs have molecular weights above 400-500 Daltons and/or high degrees of hydrogen bonding and low levels of lipid solubility. The end result is program termination of the neurotrophin CNS drug development program, once the small molecule candidate is found to not cross the BBB. The termination of neurotrophin CNS drug development is most unfortunate considering the years of effort that have been devoted to neurotrophin molecular neuroscience and also considering the millions of individuals suffering from chronic neurodegenerative conditions of the brain. The individuals suffering from Alzheimer s disease (AD), Parkinson s disease, Huntington s disease, stroke, and other neurodegenerative conditions could benefit from neurotrophin neuropharmaceuticals. The problem is that the neurotrophin molecule must undergo a molecular reformulation, within the context of a BBB drug targeting technology, so that the neurotrophin is able to traverse the BBB and distribute into brain parenchyma following intravenous or subcutaneous administration.

This chapter will review noninvasive neurotrophin delivery to the brain using the chimeric peptide BBB drug targeting technology. The work demonstrates that profound degrees of neuroprotection can be achieved in either global or permanent or reversible regional brain ischemia following delayed intravenous administration of a neurotrophic factor, providing the neurotrophin is conjugated to a BBB drug targeting system.

CHIMERIC PEPTIDE TECHNOLOGY

BBB Transport

The capillary endothelial cells of the brain of all vertebrates have unique anatomic specializations characterized by (a) high resistance epithelial-like tight junctions that eliminate any paracellular pathway, and (b) a paucity of pinocytosis across the endothelial cell, which eliminates the transcellular pathway of solute movement from blood to brain interstitium.[2] The BBB is laid down within the first trimester of human fetal life and restricts the movement of essentially all molecules between blood and brain. There may be molecular movement across the BBB via one of two generalized pathways: (a) free diffusion via lipid solubility and (b) catalyzed transport. Free diffusion occurs for molecules that have a molecular weight under a 400-500 Dalton threshold and have high degrees of lipid solubility and low hydrogen bonding. Catalyzed transport occurs via one of several endogenous transport systems within the BBB and this may take place on either carrier-mediated transport (CMT) systems or receptor-mediated transport (RMT) systems within the brain capillary endothelium. The CMT systems transport small molecule nutrients in milliseconds, and include the Glut1 glucose transporter, the LAT1 large neutral amino acid transporter, the MCT1 monocarboxylic acid transporter, and the CNT2 adenosine transporter.[1] Some of the RMT systems expressed at the BBB are shown in Figure 3 and include the insulin receptor (IR), the transferrin receptor (TfR), the IGF2 or IGF1 receptor, or the leptin receptor.[10-13] Despite the presence of an IGF1 or an IGF2 receptor at the BBB, there is minimal transport of these molecules into brain from blood following intravenous administration, because the IGFs are strongly bound > 99.9% by IGF specific binding proteins in the blood.[14]

Chimeric peptides

Based on the observation that peptide receptors are present on the brain capillary endothelium and that some of these mediate the transcytosis of the circulating peptide through the BBB, the chimeric peptide technology was developed.[7] A chimeric peptide is formed when a nontransportable drug is conjugated to a brain drug delivery vector. The latter may be an endogenous peptide, a modified protein, or a peptidomimetic monoclonal antibody (MAb) that undergoes receptor-mediated or absorptive-mediated transcytosis through the BBB in vivo on one of the endogenous transport systems shown in Figure 3. Insulin undergoes RMT through the BBB in vivo on the BBB insulin receptor.[15] However, the intravenous administration of a drug/insulin conjugate could lead to hypoglycemia by triggering insulin receptors in peripheral tissues. The conjugation of drugs to transferrin (Tf) is problematical because the Tf/drug conjugate would compete with the very high concentration (25 μM) of endogenous Tf in the blood. These problems are obviated with the use of peptidomimetic MAbs, which bind exofacial epitopes on the BBB receptor that are removed from the binding site of the endogenous ligand. This binding of the MAb to the BBB receptor allows the MAb to piggy-back across the BBB on the endogenous RMT system. Any drug attached to the peptidomimetic MAb may also undergo transport across the BBB on the endogenous RMT system.[16]

RMT on the BBB Transferrin Receptor

The outline of the receptor-mediated transcytosis of circulating transferrin through the BBB via the TfR is shown in Figure 4D. Transcytosis of transferrin through the BBB in vivo was demonstrated with both a capillary depletion technique and thaw mount autoradiography.[17] The transcytosis of a transferrin receptor peptidomimetic MAb, OX26, was demonstrated with electron microscopy.[16] In these experiments, a 5 nm gold conjugate of the OX26 MAb was infused in the carotid artery of anesthetized rats for 10 minutes and the infusion was followed by saline clearing of the cerebral microvasculature and by perfusion fixation with glutaralde-

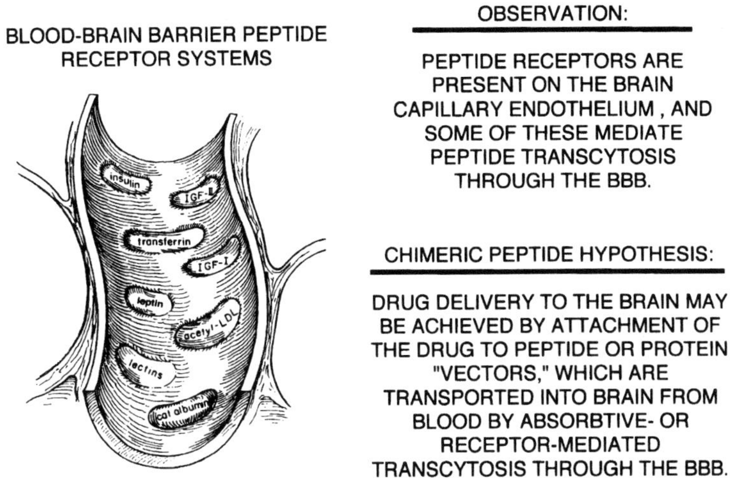

Figure 3. Blood-barrier peptide receptors and chimeric peptide hypothesis. Reproduced from ref. 1.

hyde. Immunogold silver enhancement at the light microscopic level is shown in Figure 4A and demonstrates localization of the OX26 MAb/gold conjugate within the capillary endothelium of brain. The tissue was processed for electron microscopy and OX26 MAb/gold conjugate was visible on the luminal surface of the brain capillary endothelium, as shown in Figure 4B. Whenever the OX26/gold conjugate was found in the intra-endothelial compartment of the capillary endothelium, these molecules were always found clustered in 100 nm endosomal structures, as shown in low magnification in Figure 4B and high magnification in Figure 4C. The exocytosis of the OX26/gold conjugate into the brain interstitium is demonstrated in Figure 4C.

Transferrin Transcytosis at the BBB is Bi-directional

The transcytosis of transferrin in the blood to brain direction was initially demonstrated with internal carotid artery perfusion of radiolabeled transferrin.[17-18] The receptor-mediated transcytosis of transferrin in the brain to blood direction was demonstrated recently with the brain efflux index (BEI) method.[19] These studies showed that both holo-transferrin and apo-transferrin were rapidly exported from brain to blood via reverse transcytosis on the BBB transferrin receptor and that the apo-transferrin was transcytosed approximately 2.5-fold faster than the holo-transferrin. There are other examples of macromolecule secretion from brain to blood via reverse transcytosis across the BBB. IgG molecules injected into the brain are rapidly exported to blood via reverse transcytosis on the BBB Fc receptor.[20] Whereas the endothelial TfR mediates bidirectional transcytosis of Tf across the BBB, the brain endothelial FcR mediates the unidirectional transcytosis of IgG molecules in the brain to blood direction only.[19-20]

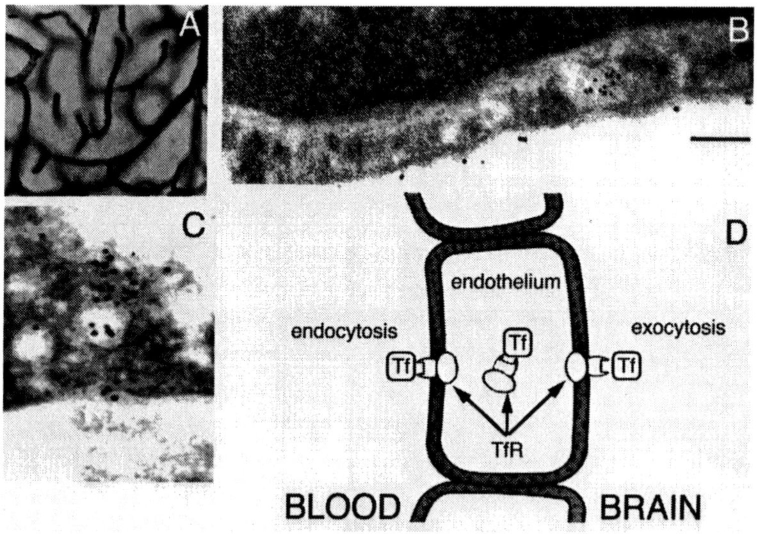

Figure 4. (A) Silver-enhanced vibratome section of rat brain perfused for 10 min with OX26/ 5 nm gold conjugate in vivo. The coronal section is at the level of the frontal cortex. The magnification bar = 10 μm. (B) Electron micrograph of a brain capillary endothelial cell after perfusion of rat brain in vivo with the OX26/ gold conjugate. The study shows dispersed OX26/gold attached to the luminal plasma membrane and clustered in endosome-like vesicles. Magnification bar = 100 nm. (C) High magnification showing clustering of OX26/ gold conjugate within endosomal structures as well as exocytosis of the conjugate into brain interstitial fluid, via transport across the abluminal membrane. (D) Receptor-mediated transcytosis (RMT) of transferrin (Tf) through the brain capillary endothelium via the transferrin receptor (TfR). Reproduced from ref. 16.

Species Specific BBB Transport Vectors

The OX26 antibody is a mouse MAb to the rat TfR and this antibody is active only in rats.[21] The OX26 MAb has no biologic activity in mice as it does not bind the mouse TfR (Fig. 5). Recently, 2 different rat MAbs to the mouse TfR, the 8D3 MAb or the RI7-217 MAb, were demonstrated to undergo receptor-mediated transcytosis across the mouse BBB via the endogenous TfR in this species.[21] Brain drug delivery studies in Old World primates such as the Rhesus monkey are possible using the 83-14 murine MAb to the human insulin receptor (HIR).[22] This MAb is highly active at the human BBB and is also active at the BBB of Old World primates such as the Rhesus monkey. The HIR MAb is not active in New World primates such as Squirrel monkey.[22] The murine 83-14 MAb cannot be used for brain drug targeting in humans owing to the high immunogenecity of a mouse protein in humans. However, the HIR MAb was recently genetically engineered to enable human applications.[23] As discussed in later sections of this chapter, the chimeric 83-14 MAb retained 100% of the binding activity for the human insulin receptor, and is a highly efficacious brain drug delivery vector in Rhesus monkeys. Second and third

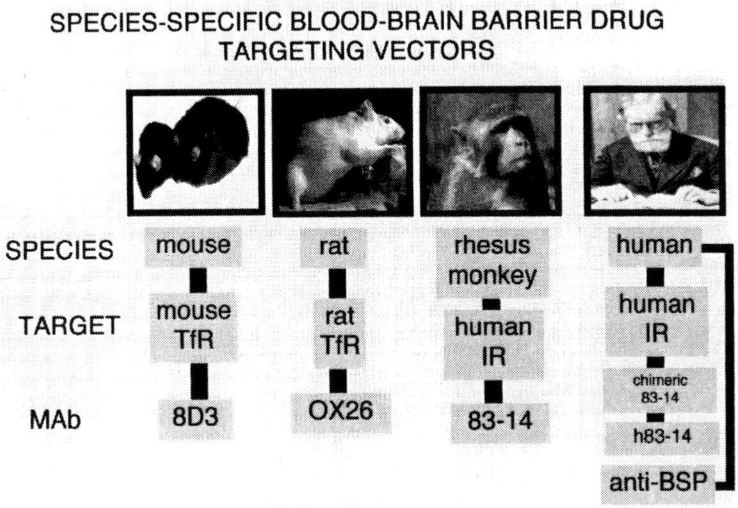

Figure 5. Monoclonal antibodies (MAb) are species-specific with respect to the reactivity to either the transferrin receptor (TfR) or the insulin receptor (IR). BSP = brain capillary specific protein. Reproduced from ref. 1.

generation BBB drug delivery systems in humans will be humanized forms of the 83-14 MAb and MAbs to brain capillary specific proteins (BSP).[1]

Avidin-biotin Technology in Brain Drug Delivery

Subsequent to the discovery of brain drug targeting vectors for a given species, it is necessary to devise suitable technologies for high efficiency coupling or conjugation of the drug to the transport vector.[1] These goals are achieved with the use of avidin-biotin technology in brain drug targeting. In this approach, the drug is mono-biotinylated in parallel with the production of a vector/avidin or vector/streptavidin (SA) fusion protein. Owing to the extremely high affinity binding of biotin by avidin, there is instantaneous capture of the biotinylated therapeutic by the avidin/vector fusion protein. Given these properties, a two vial format for drug administration was developed. In this approach, the BBB vector, fused to the avidin or SA, is prepared in one vial. The monobiotinylated therapeutic is prepared in another vial. Just prior to intravenous injection of the conjugate, the two vials are mixed to allow for rapid formation of the entire drug/MAb conjugate, which is joined through the avidin/biotin linkage, as depicted in Figure 6. Because avidin or SA has 4 biotin binding sites, there would be the formation of high molecular weight aggregates if the drug had degrees of biotinylation higher than a single biotin residue per drug molecule. Therefore, it is essential that the drug be mono-biotinylated. Genetically engineered vector/avidin or vector/SA fusion genes and fusion proteins

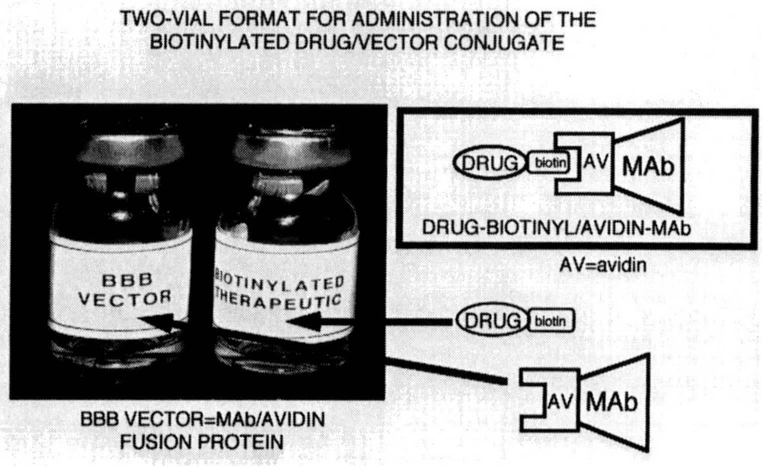

Figure 6. Two vial format for administration of drug/vector conjugates using avidin-biotin technology. The BBB transport vector vial contains a genetically engineered fusion protein of avidin (AV) and a peptidomimetic monoclonal antibody (MAb). Separately, the drug is mono-biotinylated. The entire conjugate, which is shown in the inset, is formed instantaneously upon mixture of the two vials. Owing to the very high affinity of avidin binding of biotin, there is instantaneous capture of the biotinylated drug by the MAb/AV fusion protein. Reproduced from ref. 1.

have been produced and purified and the biologic activity of these proteins has been demonstrated with in vivo brain drug targeting studies.[24,25]

TARGETING CHIMERIC NEUROTROPHIC FACTORS TO THE BRAIN

Vasoactive Intestinal Peptide (VIP)

VIP Neurotherapeutics

VIP is the principle endogenous vasodilator in the CNS,[26] and VIP is also a neuroprotective agent.[27] The topical application of VIP to brain blood vessels results in vasodilatation.[28] However, there is no enhancement of cerebral blood flow following the intravenous or intracarotid administration of VIP in multiple species,[29-31] because this neuropeptide does not cross the BBB.[32] In order to develop a VIP chimeric peptide that would be pharmacologically active in brain following intravenous administration, a series of experiments were performed to engineer a monobiotinylated form of VIP (bioVIP). This VIP chimeric peptide was then used to augment cerebral blood flow (CBF) in conscious rats following intravenous administration.[33]

Monobiotinylation of VIP Analogue

The biotin residue can be placed on a variety of functional groups including primary amines on lysine residues, carboxyl moieties on glutamate or aspartate residues, or sulfhydryl moieties on cysteine residues. VIP has 3 internal lysine residues and Lys^{15} can be biotinylated without any loss of biologic activity.[34] Therefore, the lysine residues at positions 20 and 21 were converted to arginine residues to prevent biotinylation at these sites (Fig. 7). The amino terminus was acetylated to prevent biotinylation and also to render the peptide resistant to aminopeptidase activity. The Ile at position 26 was converted to alanine to prolong the duration of action and the methionine at position 17 was converted to norleucine to enable iodination of the peptide (Fig. 7). The structure of the VIP analogue was confirmed by fast atom bombardment mass spectrometry.[32] The VIPa was biotinylated with NHS-XX-biotin, where NHS = N-hydroxysuccinimide and XX = bisaminohexanoyl. The XX linker is 14 atoms long and is positioned between the e-amino moiety of Lys^{15} on the VIP and the biotin group. The VIPa-XX-biotin was bound by a conjugate of streptavidin (SA) and the OX26 MAb to form the chimeric peptide shown in Figure 8A. The biologic activity of the VIP analogue was demonstrated with a mammalian VIP radioreceptor assay.[33] Although conjugation of the VIP to the OX26/SA BBB delivery system did diminish the affinity of the neuropeptide for the VIP receptor, there was still retention of biologic activity in the pharmacologically active range following conjugation of the neuropeptide to the MAb vector (Fig. 8B).

Cerebral Blood Flow Enhancement by VIP Chimeric Peptides

Conscious rats with pre-implanted indwelling femoral artery and femoral vein catheters were administered 1 of 4 different formulations intravenously: saline, OX26 MAb alone, the mono-biotinylated VIP analogue (bio-XX-VIPa) alone, or the VIP chimeric peptide.[33] These drugs were administered to conscious rats by intravenous administration at low systemic doses (20 μg/kg or 5 μg/rat) of the VIP. Administration of the unconjugated bio-XX-VIPa, without conjugation to the OX26/SA delivery system, increased blood flow in salivary gland by 350%, but had no effect on brain blood flow (Fig. 9). Although the VIP analogue did not cross the BBB in vivo, this neuropeptide was freely transported across the porous capillaries perfusing salivary gland. Salivary gland capillaries are richly innervated by VIPergic nerve endings which result in vasodilatation in that exocrine organ.[35] In contrast, the administration of the VIP chimeric peptide caused no increase in salivary gland blood flow, but resulted in a 60% increase in CBF in conscious rats following intravenous injection (Fig. 9). The increase in CBF following intravenous administration of the VIP chimeric peptide was observed because the VIP was able to undergo receptor-mediated transcytosis across the BBB on the endogenous transferrin

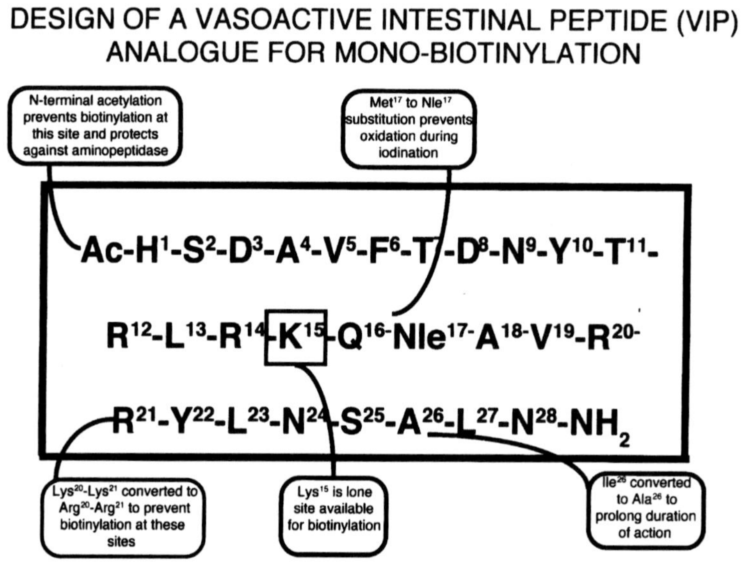

Figure 6. Two vial format for administration of drug/vector conjugates using avidin-biotin technology. The BBB transport vector vial contains a genetically engineered fusion protein of avidin (AV) and a peptidomimetic monoclonal antibody (MAb). Separately, the drug is mono-biotinylated. The entire conjugate, which is shown in the inset, is formed instantaneously upon mixture of the two vials. Owing to the very high affinity of avidin binding of biotin, there is instantaneous capture of the biotinylated drug by the MAb/AV fusion protein. Reproduced from ref. 1.

receptor.[33] In contrast, conjugation of the VIP neuropeptide to the OX26/SA delivery system actually impeded transport across salivary gland capillaries. The molecular weight of the bio-XX-VIP, approximately 4000 Daltons, was effectively increased to 204,000 Daltons, following conjugation to the 200,000 Dalton OX26/SA drug targeting system. The size of the VIP chimeric peptide was too large to enable molecular diffusion across the small pore system of capillaries in exocrine glands such as salivary glands. Therefore, the ratio of CBF to salivary gland blood flow (SBF) was increased 10-fold following conjugation of the VIP analogue to the BBB drug targeting system (Fig. 9). The CBF/SBF ratio is a measure of the therapeutic index of the VIP chimeric peptide. The 10-fold increase in this therapeutic index illustrates the targeting capabilities of the chimeric peptide technology. Conjugation of the neuropeptide to the BBB drug targeting system not only causes selective uptake into brain, but also causes decreased uptake in peripheral tissues.[33]

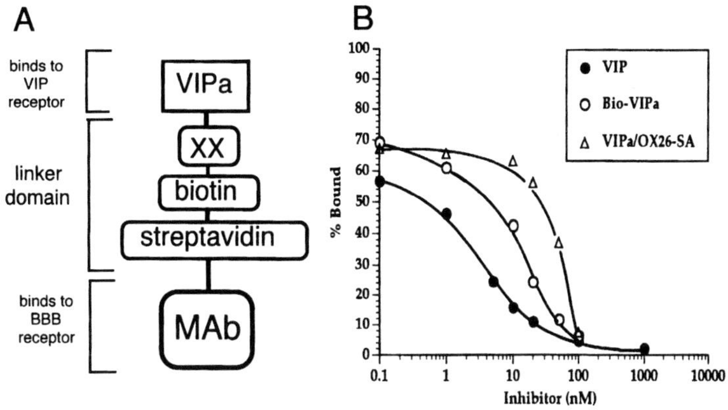

Figure 8 (A) The structure of the VIP chimeric peptide is comprised of three domains. The first domain is the VIP analogue (VIPa) that binds the VIP receptor. The second domain is the linker domain, which is comprised of streptavidin, which binds the biotin, which is conjugated to an internal lysine residue on the VIPa via a 14-atom bis-aminohexanoyl (-XX-) linker. The third domain is the targeting domain comprised of a monoclonal antibody (MAb) that targets an endogenous peptide receptor on the BBB such as the transferrin receptor or the insulin receptor. (B) Competition curves in a radioreceptor assay using rat lung membranes and [^{125}I] mammalian VIP as the tracer. The ED_{50} values for the VIP, the biotin-XX-VIPa (bioVIPa), and the biotin-XX-VIPa conjugated to OX26/SA (VIPa/OX26-SA) are 3, 12, and 45 nM, respectively. A 100 nM concentration of the OX26/SA conjugate, without VIPa attached, had no effect on the binding of [^{125}I]-VIP. Reproduced from ref. 33

Brain-Derived Neurotrophic Factor (BDNF)

Neurotrophin Surface Charge and Plasma Pharmacokinetics

The members of the NGF-like family of neurotrophins, which includes NGF, BDNF, neurotrophin (NT)-3, and NT-4/5, all have high degrees of structural homology and are highly cationic neuropeptides. An electrostasis model of NGF is shown in Figure 10 and indicates there is a cationic groove comprised of lysine and arginine residues down the center of the molecule with segregation of the anionic charges attached to glutamate and aspartate residues on the periphery of the cationic groove.[36] The cationic groove is responsible for neurotrophin interaction with one of the specific trk receptors.[37] The cationic surface charge of the NGF-like neurotrophin also causes rapid uptake by liver, which results in rapid systemic clearance of the neurotrophins from blood, which is characterized by a $t_{1/2} < 5$ minutes.[38] In pharmacokinetic terms, the plasma area under the concentration curve (AUC) is reduced and the NGF-like neurotrophins all have very poor pharmacokinetic profiles. The plasma pharmacokinetics of the neurotrophins was considered only in the late stages of neurotrophin CNS drug development.[3] However, the pharmacokinetic profile of a neurotrophin, or any drug, should be considered early in the CNS

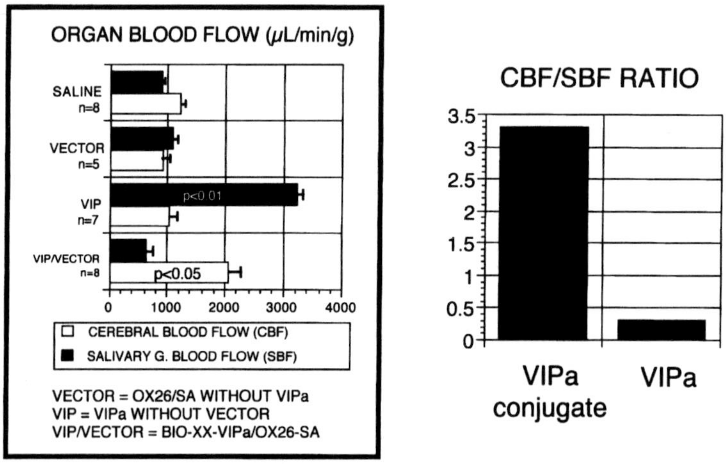

Figure 9. (Left) Organ blood flow in brain and salivary gland. (Right) Ratio of cerebral blood flow (CBF) to salivary gland blood flow (SBF) in rats administered VIP alone or VIP chimeric peptide. Reproduced from ref. 65.

drug development process. This is because the actual uptake of a drug by brain following intravenous administration, expressed as % of injected dose (ID)/g brain, is a dual function of the BBB permeability-surface area (PS) product, and the plasma AUC,[1]

$$\%ID/g = PS \times AUC$$

The BBB PS product is a function of the BBB drug targeting system, and the plasma AUC is a function of the pharmacokinetic profile. If the drug is rapidly removed from plasma, the AUC is reduced, and the brain uptake (%ID/g) is reduced proportionately. Therefore, in neurotrophin drug development, it is advisable to reformulate the protein to both (a) enable BBB transport, and (b) optimize the plasma AUC. The rapid removal of proteins from blood can be reversed by protein pegylation.

Protein Carboxyl-directed Pegylation

Polymers of polyethylene glycol (PEG) of either 2000 or 5000 Daltons molecular weight, designated PEG^{2000} or PEG^{5000}, respectively, can be conjugated to the surface of NGF-like neurotrophins to prevent the rapid uptake of these cationic proteins from blood.[39] PEG conjugation is called pegylation. Prior work on protein pegylation involved attachment of the PEG moieties to e-amino groups on internal lysine residues. However, the NGF-like neurotrophins lose biologic activity when the surface lysine residues are modified,[40] principally because the lysine and arginine residues comprise the cationic groove that interacts with the specific trk receptors.[37] An alternative to amino-directed protein pegylation is carboxyl-directed protein

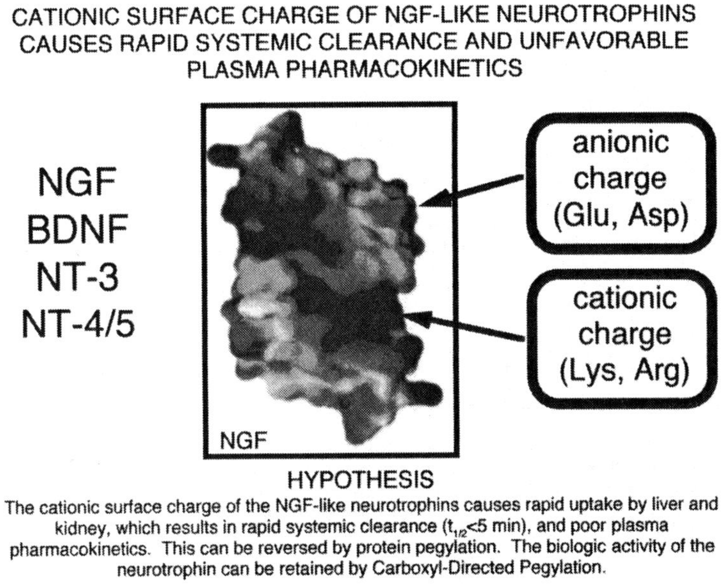

Figure 10. Electrostatic charge model of NGF showing the cationic groove as represented by the black coloration in the central part of the molecule. This is comprised of the cationic charges of lysine (Lys) and arginine (Arg) residues. Conversely, the anionic charges shown in gray on the periphery of the molecule are comprised of glutamate (Glu) and aspartate (Asp) residues. Reproduced from ref. 36.

pegylation.[41] In this approach, PEG hydrazide derivatives are used to attach the PEG polymers to the carboxyl moieties of surface glutamate or aspartate residues, which are segregated from the cationic groove (Fig. 10). Sodium dodecylsulfate polyacrylamide gel electrophoresis (SDS-PAGE) demonstrated the increased molecular weight of the BDNF following conjugation with either PEG^{2000} or PEG^{5000} on surface carboxyl residues (Fig. 11A). The BDNF-PEG^{2000} had an average molecular weight of 28,000 Daltons and the BDNF-PEG^{5000} had an average molecular weight of 50,000 Daltons. These observations indicated approximately 5-7 PEG monomers were attached per BDNF monomer, indicating that approximately half of the surface glutamate or aspartate residues were conjugated. The pharmacokinetic profile of unconjugated BDNF, BDNF-PEG^{2000}, or BDNF-PEG^{5000} was measured in rats following iodination of the BDNF analogues. The plasma clearance in ml/min/kg of the 3 different forms of BDNF are shown in Figure 11B and indicate the plasma clearance is reduced 73% and 94% by carboxyl-directed pegylation of BDNF by PEG^{2000} and $PEG5^{5000}$, respectively.[41]

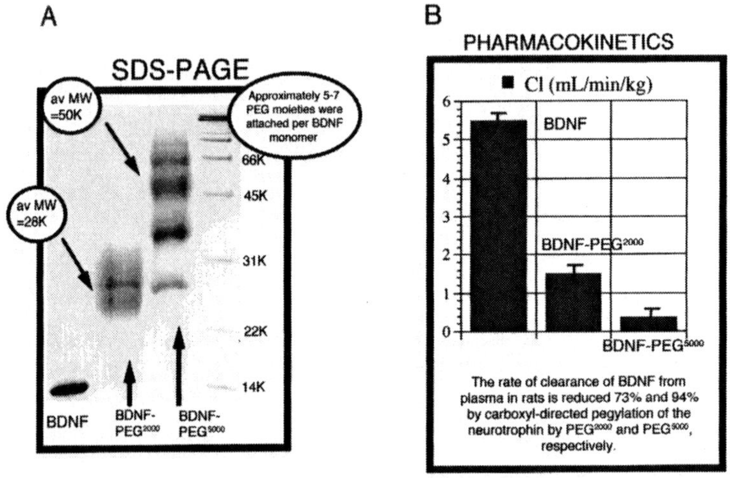

Figure 11. (Left) Sodium dodecylsulfate polyacrylamide gel electrophoresis (SDS-PAGE) of BDNF, BDNF-PEG^{2000}, and BDNF-PEG^{5000}. (Right) Plasma pharmacokinetics showing the rate of clearance (Cl) of [^{125}I]-BDNF, [^{125}I]-BDNF-PEG^{2000}, and [^{125}I]-BDNF-PEG^{5000}. Reproduced from ref. 41.

BDNF Chimeric peptides: Combined Use of Carboxyl-Directed Protein Pegylation Technology and Avidin-Biotin Technology

To enable monobiotinylation of BDNF-PEG^{2000}, the neurotrophin was pegylated with both hydrazide-PEG^{2000} and hydrazide-PEG^{2000}-biotin in a 7:1 molar ratio.[42] Following conjugation of the BDNF-PEG-biotin to the OX26/SA and purification by gel filtration chromatography, a molecular analysis of the BDNF chimeric peptide was performed with SDS-PAGE, film autoradiography, and Western blotting, as shown in Figure 12B. The average molecular weight of the BDNF-PEG^{2000}-biotin was 45,000 Daltons based on either Coomassie blue staining or film autoradiography (Fig. 12). The incorporation of the biotin residue into the BDNF-PEG^{2000}-biotin was demonstrated by Western blotting, as shown in lane 5 of Figure 12B.

The brain uptake of the BDNF chimeric peptide was measured at 60 minutes following intravenous injection in anesthetized rats and the level of brain uptake was 0.07% of injected dose (ID) per gram brain (Fig. 12C). This level of brain uptake is comparable to the brain uptake of morphine, a neuroactive small molecule, following intravenous injection.[43] In contrast, the brain uptake of the unconjugated BDNF was negligible, indicating BDNF does not cross the BBB (Fig. 12C).

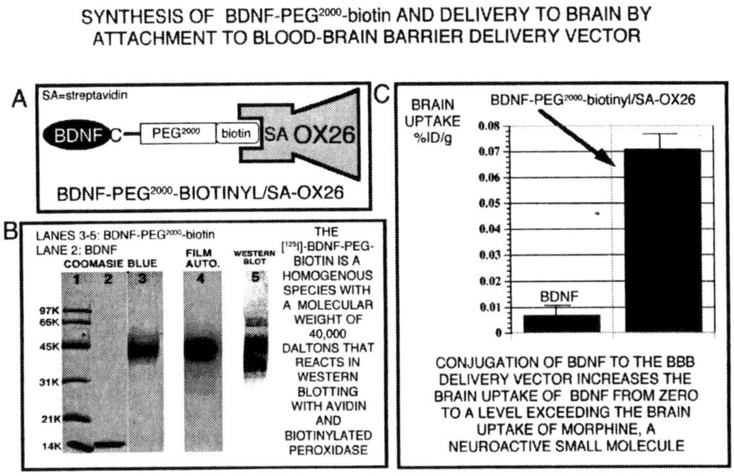

Figure 12. (A) Structure of BDNF chimeric peptide. (B) SDS-PAGE, film autoradiography (auto.), and Western blotting of [^{125}I]-BFNF-PEG^{2000}-biotin. (C) Brain uptake of either the [^{125}I]-BDNF or the [^{125}I]-BFNF-PEG^{2000}-biotin conjugated to SA-OX26 in anesthetized rats. Reproduced from 14.

Biologic Activity of BDNF Chimeric peptides: TrkB Autophosphorylation

3T3 cells permanently transfected with the TrkB receptor were serum starved, and then exposed to 1-100 ng/ml of either BDNF, BDNF-PEG^{2000}-biotin, BDNF-PEG^{2000}-biotinyl/SA-OX26, or SA/OX26 without BDNF, as shown in Figure 13. The cells were then lysed, immunoprecipitated with a trk B antiserum, separated on SDS-PAGE, followed by Western blotting of phosphotyrosine residues using an anti-phosphotyrosine antibody.[42] The phosphotyrosine Western blots were scanned and the quantitative results are shown in Figure 13. These results show there is complete retention of biologic activity of the BDNF following either pegylation and formation of the BDNF-PEG^{2000}-biotin or conjugation of the BDNF-PEG^{2000}-biotin to SA-OX26, which is a conjugate of streptavidin and the OX26 MAb. In contrast, there is no autophosphorylation of the TrkB receptor following exposure of the cells to the SA-OX26 targeting system without BDNF attached (Fig. 13).

BDNF Chimeric Peptides and Transient Forebrain Ischemia

The transient forebrain ischemia (TFI) model[44] for evaluating CNS pharmacological effects of intravenously administered neurotrophins is shown in Figure 14. The animal is maintained on a respirator and is anesthetized with 70% nitrous oxide, 30% oxygen and 0.5% halothane.[45] Body temperature was maintained with a

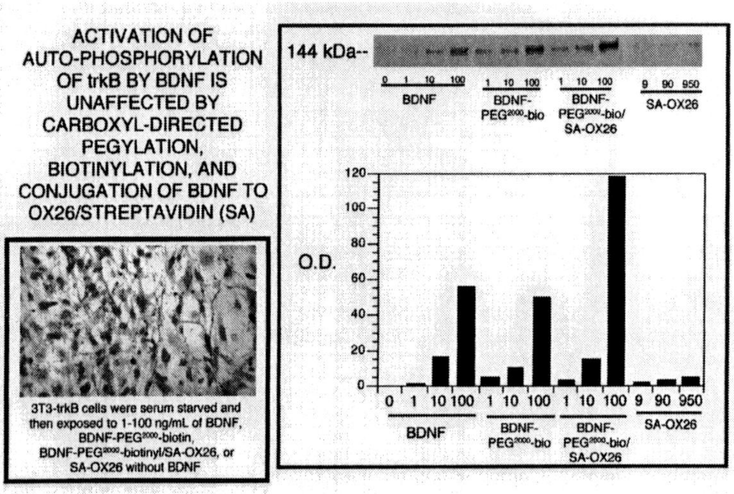

Figure 13. (Left) 3T3 cells that were permanently transfected with a gene encoding the TrkB BDNF receptor are immunstained using an antiserum to TrkB, and this shows abundant expression of the TrkB in the majority of these cells. (Right) The top panel shows the antiphosphotyrosine Western blot. The bottom panel is the densitometric scan of the Western blot and shows retention of the biologic activity of the BDNF following pegylation and conjugation to the SA-OX26 delivery system. Reproduced from ref. 42.

thermal blanket. The systolic blood pressure was measured with a rat tail amplifier. Prior to induction of ischemia, the halothane component of the anesthesia was discontinued. Blood was collected via the femoral artery catheter and arterial blood gases were measured. Muscle paralysis was induced with intravenous suxamethonium (0.5 mg/kg) and the electroencephalogram (EEG) was recorded via 2 anchored metal screws with an oscillograph and preamplifier. After the blood gas values were stabilized, forebrain ischemia was induced by clamping both common carotid arteries, and by lowering the blood pressure to a level < 50 mm Hg with a single intravenous injection of 5 mg/kg of trimethaphan and by phlebotomy via the femoral artery catheter. Cerebral ischemia was confirmed by isoelectric EEG.[45] After 12 minutes of the ischemic insult, the carotid clamps were released and the shed blood was slowly reinfused at a rate of 2 ml/min. The animal was allowed to recover under a heating lamp for 4 hours and was returned to the vivarium. The animals were sacrificed 7 days later and coronal sections were Nissl stained. The pyramidal neurons of the hippocampal CA1, CA3, and CA4 regions were counted with a measuring grid under light microscopy at 100X magnification. Only neurons with a visible nucleolus were scored. In this TFI model, there is a selective loss of the hippocampal CA1 pyramidal neurons at 7 days following the TFI event, as shown in Figure 14.

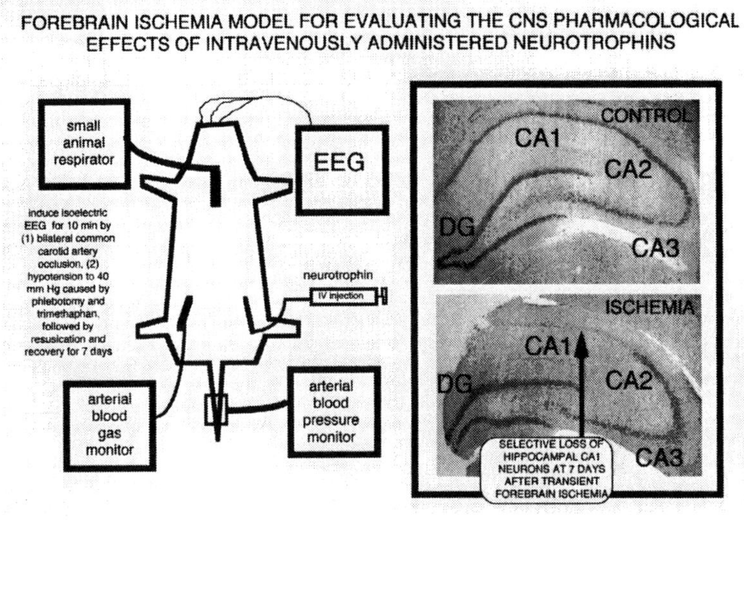

Figure 14. Transient forebrain ischemia model.

The animals subjected to the TFI and isoelectric EEG were treated with either saline, unconjugated BDNF, unconjugated OX26 MAb, or the BDNF/OX26 chimeric peptide. The BDNF chimeric peptide is designated BDNF-PEG^{2000}-biotin/SA-OX26.[45] The bi-functionality of the BDNF chimeric peptide is shown in Figure 15. Despite the protein pegylation and attachment to the BBB drug targeting system, the BDNF is still biologically active and binds actively to the neuronal TrkB receptor to mediate neuroprotection.[42] Moreover, the BDNF chimeric peptide also binds to the transferrin receptor on the BBB, and this enables transport of the BDNF chimeric peptide across the BBB from blood to brain following intravenous administration. The extended bridge comprised of the PEG^{2000} and the biotin/SA releases any steric hindrance between the 2 components of the chimeric peptide, and this enables retention of the bi-functional nature of the conjugate, and binding to both the neuronal TrkB and BBB Tf receptors (Fig. 15).

During the 10 minute ischemic period, the blood pressure ranged from 22±2 to 27±6 mm Hg in all 4 treatment groups.[45] There were no significant differences in the physiologic parameters, including body temperature or blood gases in the 4 groups. Ischemia was confirmed in all rats by demonstration of isoelectric EEG, which is illustrated in Figure 16. The animals were treated intravenously immediately after recovery from the ischemic period. Seven days after the ischemic episode, the animals were sacrificed for Nissl staining and neuron counting of the hippocampal CA1 sector. These data show a 68±10% decrease in the hippocampal CA1 neurons

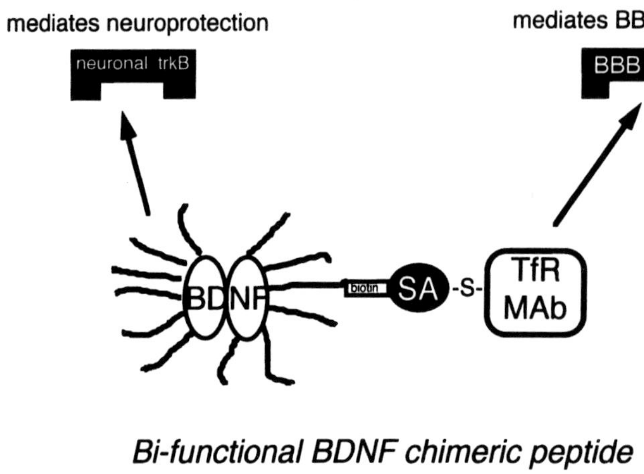

Figure 15. BDNF chimeric peptide binds to both the BBB transferrin receptor (TfR) to mediate uptake into brain from blood, and to the neuronal TrkB receptor, to mediate pharmacologic action within the brain.

in the rats subjected to TFI (Fig. 16) and there was no therapeutic benefit observed with the intravenous administration of either unconjugated BDNF or unconjugated OX26 MAb. In contrast, there was complete normalization of the CA1 pyramidal neuron density following the intravenous administration of the BDNF chimeric peptide (Fig. 16). Representative Nissl stains are shown in Figure 16. The neurons were reduced in number and size in the CA1 sector of the hippocampus in the rats subjected to TFI and treated with unconjugated BDNF. These brains appeared no different from the animals treated with the saline control,[45] which is expected considering that BDNF does not cross the BBB.[41] Moreover, the BBB is not disrupted following a 10 minute transient forebrain ischemia until very late stages when chances for neuroprotection are no longer present. Since the BBB is intact in the period immediately following global ischemia,[46] and since BDNF does not cross the BBB,[41] neuroprotection in brain would not be expected following intravenous administration of BDNF unless this neurotrophin was conjugated to a BBB drug targeting system. When this is done, there is complete normalization of the neuron density in the CA1 sector of the hippocampus in the animals treated with the BDNF chimeric peptide (Fig. 16).

BDNF Chimeric Peptides and Permanent Middle Cerebral Artery Occlusion

The transient forebrain ischemia model is representative of global brain ischemia such as following cardiac arrest. Regional brain ischemia caused by thrombotic or embolic stroke is modeled by the middle cerebral artery occlusion (MCAO) method.[47]

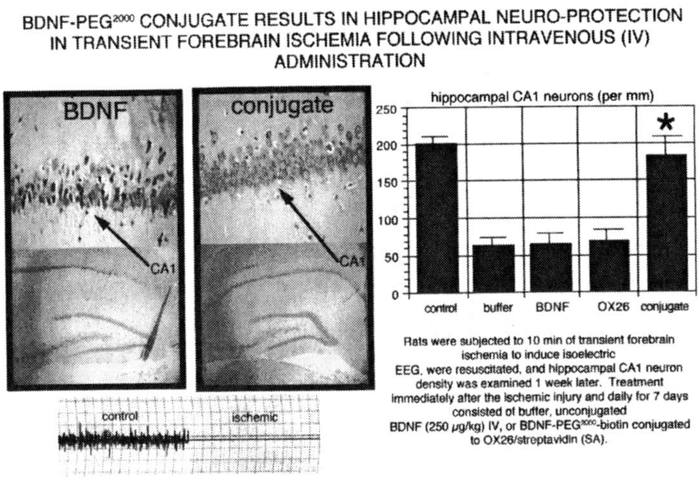

Figure 16. (Left) Nissl staining at high and low magnification ofbrain taken from rats subjected to transient forebrain ischemia and treated with either unconjugated BDNF or BDNF conjugate. (Right) Density of hippocampal CA1 neurons in control rats and four different groups of treated rats following 10 minutes of transient forebrain ischemia. Reproduced from 45.

Both permanent and reversible forms of MCAO have been examined. Initial studies were performed with the permanent MCAO model.[48] Adult Sprague-Dawley rats were anesthetized with 70% nitrous oxide, 30% oxygen, and 0.5% halothane, and systolic blood pressure and arterial blood gases were normalized. Body temperature and blood glucose were also measured. The right common carotid artery and the right external carotid artery were exposed and the occipital artery and superior thyroidal artery were electrocoagulated. The right pterygopalatine artery was ligated and the right common carotid artery was clamped and a 4-0 nylon suture was inserted retrogradely via arteriectomy of the external carotid artery into the internal carotid artery. The suture was slowly advanced until resistance was felt. The external carotid artery was ligated and the common carotid artery clamp was released and the skin incision was sutured. After surgery, rats were allowed to recover spontaneous breathing and were kept for 24 hours in their cages with free access to food and water. Animals were treated intravenously with 1 of 4 different formulations: the BDNF chimeric peptide (BDNF-MAb conjugate), the MAb alone, the BDNF alone, or saline.[48] The dose of BDNF administered intravenously was 1, 5, or 50 μg/rat. The BDNF chimeric peptide was administered at either 0, 1, or 2 hours following permanent occlusion of the middle cerebral artery. Following sacrifice, the brain was removed, chilled, and 6x2 mm coronal slices were prepared with a brain matrix.

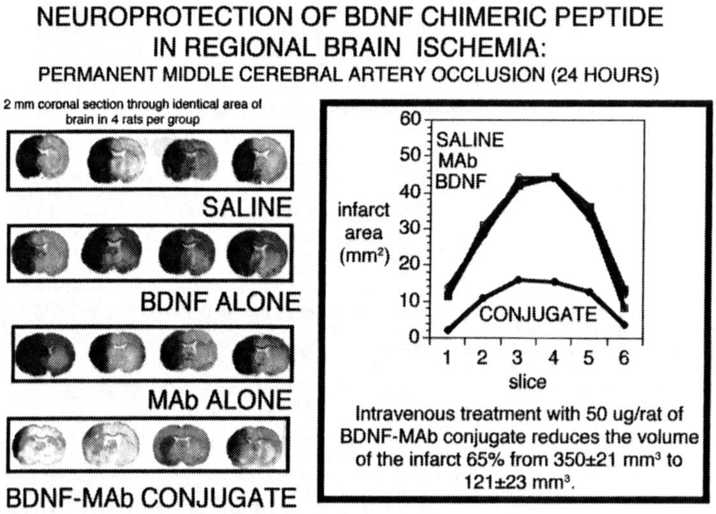

Figure 17. (Left) TTC stains of coronal sections of rat brain following 24 hour permanent MCAO. There were four treatment groups. (Right) The infarct area for each of six different coronal slices is shown, and these values were used to compute the volume of the infarct. Reproduced from 48.

Slice 1 is the most rostral and slice 6 is the most caudal section. These sections were stained with 2% 2,3,5 triphenyltetrazolium chloride (TTC). The TTC stained viable brain tissue dark red while infarcted tissue is unstained. After staining, the sections were fixed in 10% formalin and then scanned on a 1200 dpi Umax scanner. The images were transferred to Adobe Photoshop 5.5 on a G4 Power Macintosh and quantified using NIH image software. The border between the infarcted and noninfarcted tissue was outlined with the image analysis system and the area of infarction was measured by subtracting the area of the noninfarcted ipsilateral hemisphere from that of the contralateral hemisphere for each of the 6 coronal slices. In addition, the area of brain tissue not stained by TTC was computed and the area of brain edema was calculated for each coronal slice. The infarct areas or brain edema areas for each slice were averaged and multiplied by a total thickness of brain (12 mm) to give the infarct and edema volumes. The statistical significance at the $p < 0.05$ level was determined by analysis of variance (ANOVA) with Bonferroni correction.[48]

The TTC stains for 16 rats are shown in Figure 17. The scanned images of each coronal section were inverted in Photoshop to generate the images shown in Figure 17. The infarcted area, which is not stained by TTC, appears black in the inverted image, and the healthy brain, which stains red by TTC, appears white in the inverted image. The data show that unconjugated BDNF, unconjugated OX26, and saline,

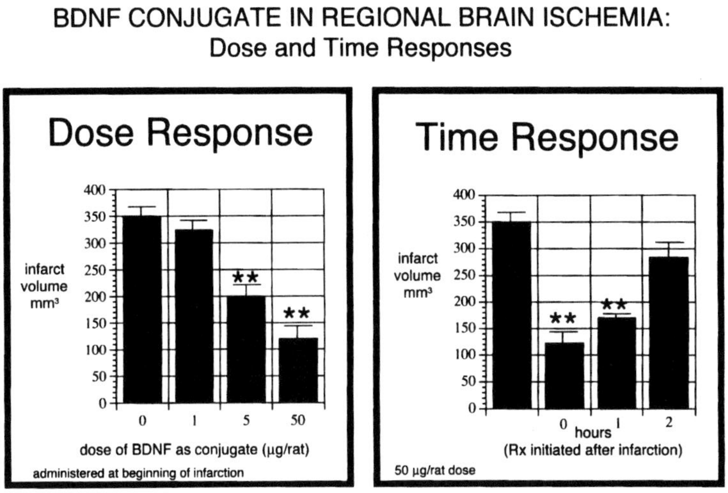

Figure 18. Dose response (Left) and time response (Right) of BDNF chimeric peptides in permanent middle cerebral artery occlusion. Reproduced from 48.

have no effect on the stroke volume in the permanent MCAO model. However, intravenous administration of the BDNF chimeric peptide (BDNF-MAb conjugate) at a dose of 50 μg/rat of BDNF, reduced the infarct volume by 65% and reduced the edema volume by 66%.[48]

A dose response relationship was examined by reducing the BDNF administered in the form of the conjugate from 50 μg/rat to 5 μg/rat or 1 μg/rat and the infarct volumes of these animals are shown in Figure 18. The intermediate dose of BDNF chimeric peptide, 5 μg/rat, resulted in a 43% reduction in infarct volume in the total hemisphere and a 71% reduction in hemispheric edema volume.[48] There was no significant effect on the infarct or edema volume with a very low dose of BDNF chimeric peptide, 1 μg/rat. A time response study was also performed at the higher dose of BDNF chimeric peptide (50 μg/rat). In this study, the BDNF chimeric peptide was delayed and not administered until either 1 or 2 hours after permanent occlusion of the middle cerebral artery. The BDNF chimeric peptide reduced the infarct and the edema volume by 52% and 59%, respectively, following delayed treatment given 1 hour after permanent occlusion of the middle cerebral artery. There was no significant reduction in infarct volume when the administration of the BDNF chimeric peptide was delayed until 2 hours after permanent occlusion of the middle cerebral artery (Fig. 18).

The permanent MCAO study shows that the BDNF chimeric peptide, but not the native BDNF, enables neuroprotection in regional brain ischemia following delayed noninvasive (intravenous) administration of the neurotrophin. Unconjugated or native BDNF has no neuroprotective effects in brain because (a) BDNF does not cross the BBB[41] and (b) the BBB is not disrupted during the treatment period following occlusion of the middle cerebral artery.[49] In the absence of hyperglycemia, the BBB is not disrupted for at least 6 hours after occlusion of the middle cerebral artery.[49] In contrast, neuronal loss in regional ischemia occurs in the immediate period following occlusion of the artery when there is no BBB disruption.[50]

The infarct volume is progressively decreased as the dose of BDNF chimeric peptide is increased from 1 to 5 to 50 μg/rat. The brain uptake of the BDNF chimeric peptide in the rat is approximately 0.1% ID/g (Fig. 12). Therefore, the brain concentration of BDNF after the 5-50 μg/rat doses is increased 5-50 ng/g brain.[48] The effect of the 5 μg/rat dose of intravenous BDNF chimeric peptide is comparable to the pharmacologic effects of ICV infusion of 2 μg/day of BDNF administered directly into the lateral ventricle.[51] However, in these studies, it was necessary to begin the ICV infusion of the BDNF 24 hours before occlusion of the middle cerebral artery. The BDNF reaches brain parenchyma slowly by diffusion when the drug is administered by ICV infusion (Fig. 2). The efficacy of diffusion decreases with the square of the diffusion distance and this accounts for the poor penetration of BDNF into brain parenchyma following ICV infusion (Fig. 2). The diffusion distance in the human brain is 1000-fold greater than the diffusion distance in the rat brain, and the ICV infusion of neurotrophic factors has had not beneficial effect in human neuropathology.[5] Moreover, it is not feasible to administer neuroprotective agents by craniotomy in human stroke, much less prior to the cerebral insult.

BDNF Chimeric peptides and Reversible Middle Cerebral Artery Occlusion

The reperfusion associated with reversible brain ischemia is representative of human stroke and reperfusion can aggravate the development of brain edema in focal ischemia and accentuate neuronal loss.[52] Therefore, a series of experiments was performed to determine the neuroprotection of intravenous BDNF chimeric peptides in a model using one hour reversible occlusion of the middle cerebral artery.[53] In addition, these studies examine the long term effects of neuroprotection with the BDNF chimeric peptide and infarct volumes are measured at both 24 hours and 7 days after reversible MCAO. In these studies, the middle cerebral artery was occluded for 60 minutes. In these treatment schedules, the BDNF chimeric peptide was administered intravenously in the femoral vein via a 30 gauge needle at 1 hour after middle cerebral artery occlusion or at the beginning of reperfusion. Alternatively, the administration of the BDNF chimeric peptide was delayed 2 hours after arterial occlusion, which is 1 hour after reperfusion. The hemispheric infarct zones were subdivided into cortical and subcortical infarct areas. None of the treatments resulted in a decrease in subcortical infarct volume (Fig. 19). Although unconjugated BDNF had no effect on the cortical infarct volume, the single intravenous injection of 50 μg/rat of the BDNF chimeric peptide administered at 60 minutes after MCAO resulted in a 68% reduction in cortical stroke volume (Fig. 19). If the intravenous adminis-

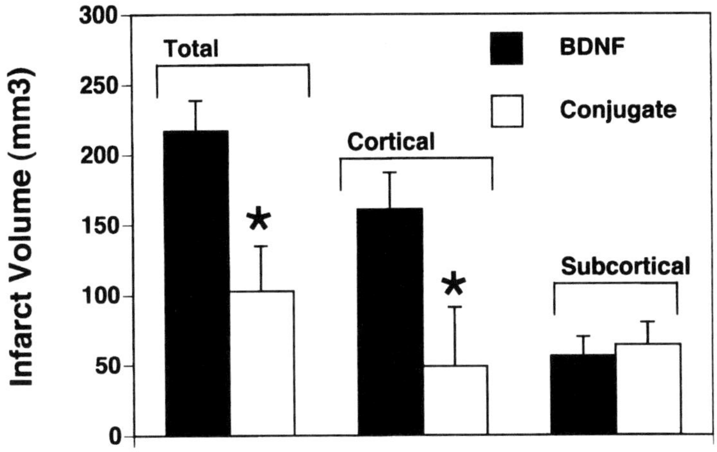

Figure 19. Total (hemispheric), cortical, and subcortical infarct volume at 7 days after a 60 minute MCAO in rats treated with either a single intravenous injection of BDNF or BDNF chimeric peptide at a dose of 50 μg/rat at 1 hour after occlusion of the middle cerebral artery. Mean ± S.D. (n = 4 rats per group). *p < 0.01 difference between BDNF and the conjugate. Reproduced from 53.

tration of the BDNF chimeric peptide was delayed 2 hours after insertion of the catheter, there was a 31% reduction ($p < 0.05$) in the cortical infarct volume.[53]

In the seven day study, additional groups of rats were subjected to 1 hour of MCAO and treated with 50 μg/rat of BDNF or BDNF chimeric peptide. The animals were sacrificed 7 days later and the hemispheric cortical and subcortical infarct volumes were measured. Intravenous administration of the unconjugated BDNF caused no decrease in either total hemispheric infarct volume or cortical infarct volume and these results were comparable to the saline treated animals (Fig. 20). Conversely, the total hemispheric infarct volume at 7 days was reduced 53% ($p < 0.01$) by the BDNF chimeric peptide, compared to the total hemispheric infarct volume following treatment with the unconjugated BDNF.[53] The cortical infarct volume was reduced 70% ($p < 0.01$) with the BDNF chimeric peptide compared to the unconjugated BDNF (Fig. 20). The TTC stains of 6 different coronal sections obtained 7 days after treatment with either the unconjugated BDNF or the BDNF chimeric peptide are shown in Figure 20.

Neuroprotection with unconjugated BDNF following intravenous administration has been reported in a 2 hour reversible MCAO model and in this study, BDNF was infused over a 3 hour period and the infusion was started 30 minutes after occlusion of the artery.[54] There was a 55% reduction of cortical stroke volume following the intravenous infusion of 300 μg/rat of unconjugated BDNF. This dose is 6-fold higher than the dose of BDNF chimeric peptide administered in the studies

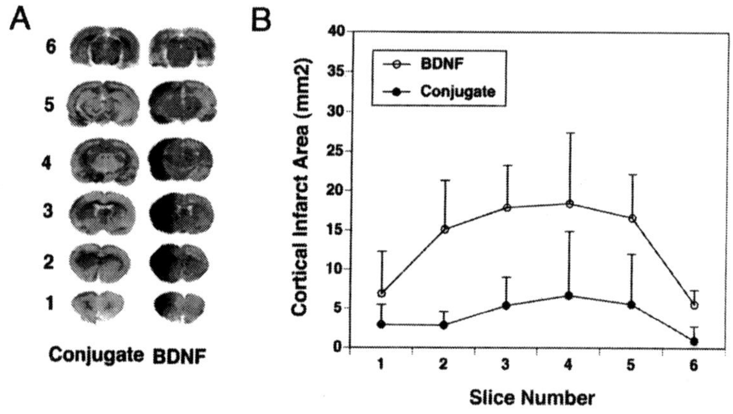

Figure 20. (A) TTC stains are shown for 2 different rats for all 6 coronal slabs. A representative rat from the group treated with the BDNF chimeric peptide (conjugate) and a representative rat from the group treated with unconjugated BDNF are shown. (B) The area of the cortical infarct region in 6 coronal slabs is shown for the animals treated with either unconjugated BDNF or BDNF chimeric peptide at a dose of 50 μg/rat at 1 hour after insertion of the middle cerebral artery suture. At 60 minutes after insertion of the suture, the suture was removed and the animals were treated. The animals recovered from anesthesia and were sacrificed 7 days later for measurement of infarct volumes. Data are mean ± S.D. (n = 4 rats/group). Reproduced from 53.

described in Figures 19-20. Even though very large doses of unconjugated BDNF were administered intravenously, it is unexpected that neuroprotection is achieved with intravenous unconjugated BDNF. This is because the BBB transport of BDNF is negligible when there is no BBB disruption and when the BDNF is not conjugated to a BBB drug delivery system (Fig. 12). Intravenous administration of unconjugated BDNF was neuroprotective in a reversible MCAO model that was performed with chloral hydrate anesthesia that was associated with a significant level of hyperglycemia and a plasma glucose concentration of 220±47 mg %.[54] This level of hyperglycemia, in parallel with a 2 hour reversible MCAO, causes vasculopathy and premature disruption of the BBB.[55] This level of modest hyperglycemia may accelerate opening of the BBB in focal brain ischemia and enable high doses of unconjugated neurotrophic factor to enter the brain following intravenous administration. In the absence of hyperglycemia and vasculopathy, it is unlikely that unconjugated BDNF is neuroprotective following intravenous administration, because this protein does not cross the BBB.

One study suggests that BDNF can cross the BBB following intravenous administration, because radioactivity can be recovered in brain following the intravenous administration of [^{125}I] BDNF.[56] However, this is an artifact arising from peripheral metabolism of this highly cationic neurotrophin that is rapidly removed from blood by peripheral tissues, particularly the liver.[38, 41] This rapid uptake by peripheral tissues and metabolic degradation is followed by the release of radiolabeled low molecular weight metabolites such as iodotyrosine back to the bloodstream.[1] The [^{125}I]-tyrosine may then cross the BBB on the large neutral amino acid transporter, and account for the radioactivity in brain following intravenous injection of [^{125}I]-BDNF. This interpretation is supported by prior work using 2 different methodologies. In the first approach, there was no measurable uptake of radioactivity by brain following the intravenous injection of [^{125}I]-BDNF in rats,[41] when the peripheral metabolism of the neurotrophic factor was completely suppressed by pegylation of the neurotrophin (Fig. 11B). Second, the brain uptake of radioactivity following the intravenous administration of radiolabeled neuropeptide is suppressed 10-fold when the neuropeptide is labeled with 111-Indium, as opposed to 125-Iodine.[57] Peptide degradation products labeled with 111-Indium, which are formed by metabolism in peripheral tissues, are not re-exported back to blood and are not taken up across the BBB in the form of low molecular weight radiolabeled degradation products.[57] The interpretation of artifacts in relation to brain uptake of radiolabeled neuropeptides has been recently reviewed.[1]

In summary, intravenous doses of unconjugated BDNF at a level of 5-50 μg/rat results in no neuroprotection in regional brain ischemia.[48,53] In contrast, these doses of BDNF chimeric peptide are highly neuroprotective in regional brain ischemia owing to transport of the BDNF chimeric peptide across the BBB in vivo on the endogenous BBB transferrin receptor (Fig. 15). Doses of BDNF chimeric peptide as low as 5 μg/rat result in substantial neuroprotection following intravenous administration.[48] Moreover, the neuroprotection is observed following delayed intravenous administration.[48,53] Neuroprotection is possible only during the first 1-3 hours after focal brain ischemia,[50] when the BBB is not usually disrupted.[49] Therefore, neurotrophin neuropharmaceuticals must be enabled to undergo transport across the BBB following intravenous administration if these agents are to be neuroprotective in regional brain ischemia.

TARGETING GENE THERAPEUTICS TO THE BRAIN

The chronic neurodegenerative diseases of the brain may also be amenable to brain gene therapy. Genes encoding neurotrophic factors and inserted in viral vectors cause neuroprotection following the intracerebral implantation of the viral vector in brain.[58,59] However, the intracerebral implantation of viral vectors may not be feasible for human brain gene therapy for 2 reasons. First, the viral vectors, either adenovirus or herpes simplex virus (HSV), are highly immunogenic and all individuals have a pre-existing immunity to both of these common viruses. A single intracerebral injection of either adenovirus or HSV results in dose-dependent inflammatory reactions in the brain leading to demyelination.[60,61] The viral induced demyelination has been observed in rodent, primate, and human brain. The second limitation is that the intracerebral implantation of a viral vector will only allow for expression of the gene in a very circumscribed treatment volume[1] < 1 mm^3. In contrast, the areas requiring neuroprotection in a variety of CNS diseases, including AD, PD, or

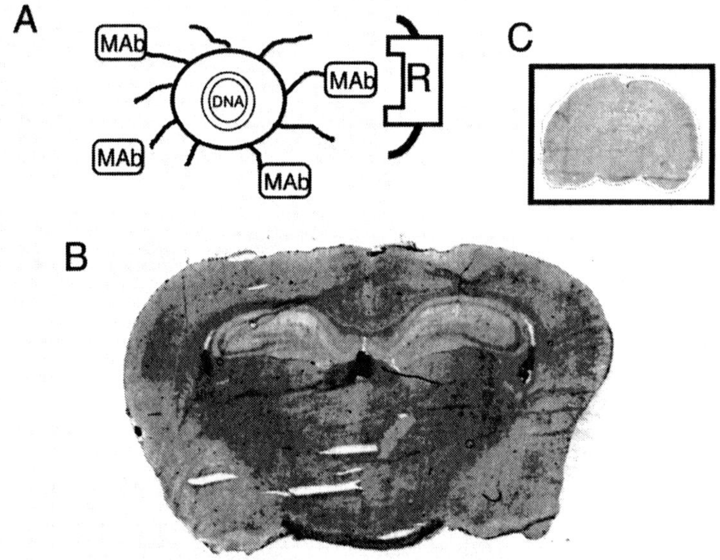

Figure 21. (A) Nonviral plasmid DNA is encapsulated in pegylated immunoliposomes constructed from neutral lipids. There are approximately 3000 strands of polyethylene glycol of 2000 Daltons molecular weight, designated PEG^{2000}, attached to the liposome surface, and about 1% of the PEG strands is conjugated with a monoclonal antibody (MAb) to a receptor (R) to mediate both transcytosis across the BBB and endocytosis into target neurons (B) β-galactosidase histochemistry of rat brain at 48 hours after intravenous injection of the β-galactosidase gene packaged inside the OX26 pegylated immunoliposome. The control brain from rats receiving no gene administration is shown in Panel C. Panels B and C were not counter-stained. Reproduced from 62.

Huntington s disease, have treatment volumes that are log orders greater than 1 mm^3. What is needed is a targeting technology for delivering therapeutic genes across the BBB following intravenous administration of a nonviral vector.

Noninvasive, nonviral gene targeting to the brain is possible with the application of the chimeric peptide technology.[62] In this approach, a nonviral plasmid carrying a therapeutic or exogenous gene is encapsulated in the interior of neutral 85 nm pegylated immunoliposomes, as depicted in Figure 21A. Packaging the plasmid in the interior of the liposomes protects the exogenous gene from the ubiquitous endonucleases in the body following intravenous administration. The liposome or other nanocontainer is nonimmunogenic and is formed by either natural lipids or other nonimmunogenic polymeric substances. The injection of a plasmid encapsulated in a simple liposome into the bloodstream would result in rapid uptake of the complex by cells lining the reticuloendothelial system. The nanocontainer carrying the exogenous gene may be stabilized in the bloodstream with the use of pegylation technology.[62] In this approach, several thousand strands of PEG^{2000} are attached to the surface of the liposome using lipid-PEG^{2000} conjugates. However, the injection of a pegylated liposome into the bloodstream would not allow for targeting to the brain. Brain gene targeting is accomplished by tethering targeting MAbs to the tips of approximately 1% of the PEG strands as depicted in Figure 21A. These MAbs trigger transcytosis of the liposome carrying the gene across the BBB and also trigger endocytosis of the therapeutic gene into the target neurons of brain.[62] Targeting across the second barrier comprised of the neuronal or glial cell membrane in brain is achieved owing to expression of the TfR on the plasma membrane of brain cells.[63]

Two different reporter genes were used in initial evaluation of this brain gene targeting technology, luciferase and b-galactosidase.[62] The β-galactosidase histochemistry of rat brain obtained 48 hours after a single intravenous injection of the pegylated immunoliposome carrying the gene is shown in Figure 21B. The brain expresses the β-galactosidase gene widely as seen at low magnification (Fig. 21B). No β-galactosidase activity is observed in either control brain or in brain of animals injected with pegylated immunoliposomes wherein the OX26 MAb was replaced with the mouse IgG_{2A} isotype control (Fig. 21C). Pyramidal neurons of the CA1-CA3 sectors of the hippocampus are clearly visualized as are the choroid plexi in both lateral ventricles and in the dorsal horn and the mamillary recess of the third ventricle. The paired supraoptic nuclei of the hypothalamus at the base of the brain are viewed at low magnification. At higher magnification, the microvasculature, the choroid plexus epithelial cells, and the thalamic nuclei showed β-galactosidase gene expression. Gene expression in neurons was observed throughout the brain in a region specific manner.

These studies demonstrate it is possible to target therapeutic genes widely throughout the brain in a noninvasive way following a simple intravenous injection of a nonviral therapeutic gene packaged within the interior of pegylated immunoliposomes. The development of noninvasive, nonviral gene therapy of the brain requires a molecular formulation derived from the use of advanced drug targeting technology that brings together multiple disciplines including liposome technology, polymer technology, monoclonal antibody targeting technology, genetic engineering, and the molecular biology of therapeutic gene discovery. There are many diseases where it will be useful to have widespread expression of an exogenous gene throughout the brain including inborn errors of metabolism such as lysosomal storage disorders, fragile X syndrome, Rett s syndrome, Canavan s disease, and the inherited epilepsies. Other conditions that would be amenable to noninvasive, nonviral gene therapy of the brain includes cerebral acquired immune deficiency syndrome (AIDS), brain tumors, and neurodegenerative conditions such as Parkinson s disease or Alzheimer s disease. In the case of Parkinson s disease, it would be desirable to augment dopamine production only in the basal ganglia, and not in the cerebral cortex. Region specific expression of an exogenous gene using targeting technology is possible in the future with the use of cell-specific promoters at the 5 -end of the gene and cell-specific mRNA stabilizers at the 3 -end of the gene.[62] The persistence of plasmid based gene formulations in brain cells in vivo can also be optimized as

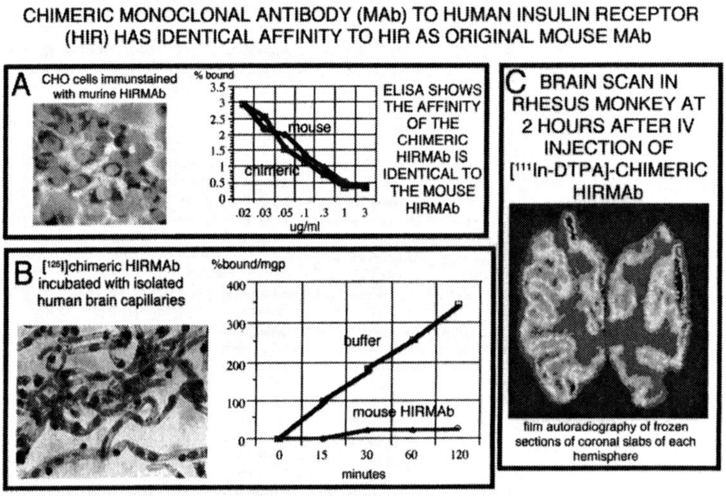

Figure 22. (A) ELISA shows equal affinity of the mouse HIR MAb and chimeric HIR MAb. The antigen in these studies was affinity purified soluble extracellular domain of the HIR that was produced from CHO cells. Immunocytochemistry was performed with the CHO cells using the 83-14 HIR MAb and the micrograph is shown in the inset. This demonstrates the CHO cell line secretes functional and soluble HIR that reacts with the HIRMAb. (B) Isolated human brain capillaries were used as an in vitro model system of the human BBB and the binding of the [^{125}I] chimeric HIR MAb to these capillaries was measured over 120 minutes in the presence of either buffer alone or buffer plus 10 μg/ml mouse HIR MAb. (C) Film autoradiogram of Rhesus monkey brain at 2 hours after intravenous injection of chimeric HIR MAb radiolabeled with 111-Indium bound to a diethyelentriaminepentaacetic acid (DTPA) chelator moiety conjugated to the chimeric HIR MAb. Reproduced from 23.

described recently.[1] The availability of the complete sequence of the human genome and the emerging applications of the genomics technologies will further augment in the future the need for the development of nonviral, noninvasive gene therapy of the brain.

DRUG TARGETING TO THE HUMAN BRAIN

The chimeric peptide technology has been reduced to practice in both rodents and primates and both neurodiagnostic and neurotherapeutic molecules have been delivered across the BBB in these animals.[1] The extension of the chimeric peptide technology to the treatment of human brain disorders requires the genetic engineering of the murine antibodies that are used to target drugs and genes via the endogenous transport systems on the BBB. Human/mouse chimeric MAbs are genetically engineered antibodies wherein approximately 85% of the sequence is of human origin. The murine sequences encoding the variable region of the heavy chain (VH) or the variable region of the light chain (VL) may be spliced into gene fragments containing the constant regions of the heavy chain (HC) and the light chain (LC) of human immunoglobulins. Recently, the murine 83-14 MAb to the HIR was genetically engineered and the human/mouse chimeric HIR MAb was produced.[23] CHO cells that have been permanently transfected with a gene encoding the soluble extracellular domain of the HIR[64] were used to establish an ELISA.[23] These studies demonstrated a complete retention of the affinity of the chimeric MAb for the HIR following the genetic engineering, relative to the affinity of the original murine HIR MAb for the human insulin receptor (Fig. 22A). The genetically engineered chimeric HIR MAb was radiolabeled with [^{125}I] and added to isolated human brain capillaries and there was rapid uptake of the chimeric HIR MAb as shown in Figure 22B. This uptake at the human BBB was suppressed by high concentrations of murine HIR MAb (Fig. 22B). The chimeric HIR MAb was conjugated with diethylenetriamine pentaacetic acid (DTPA) dianhydride and radiolabeled with 111-Indium. The radiolabeled chimeric HIR MAb was injected into anesthetized Rhesus monkeys and brain scans were obtained 2 hours after intravenous injection in the living monkey.[23] As shown by the brain scan in Figure 22C, there was avid uptake of the chimeric HIR MAb by the primate brain and a comparable high uptake in the human brain is anticipated based on the studies with human brain capillaries shown in Figure 22B.

In summary, genetically engineered chimeric HIR MAb has been produced and the chimeric antibody has identical reactivity to the human and primate HIR as the original murine antibody. This chimeric HIR MAb may be used in humans for drug targeting to the brain of neurodiagnostic and neurotherapeutic drugs that normally do not cross the BBB. Based on these preclinical studies, it is anticipated that a fusion protein of the chimeric HIRMAb and BDNF would be highly neuroprotective in human stroke and other neurodegenerative disorders following intravenous administration.

ACKNOWLEDGMENTS

Daniel Jeong skillfully prepared the manuscript. This work was supported by NIH grant NS-34698.

REFERENCES

1. Pardridge WM. Brain Drug Targeting: The Future of Brain Drug Development. Cambridge: Cambridge University Press, 2001:1-370.
2. Pardridge WM. CNS drug design based on principles of blood-brain barrier transport. J Neurochem 1998; 70:1781-1792.
3. Hefti F. Pharmacology of neurotrophic factors. Annu Rev Pharmacol Toxicol 1997;37:239-267.
4. Apfel S. Clinical Applications of Neurotrophic Factors. New York: Lippincott-Raven, 1997:5.
5. Kordower JH, Palfi S, Chen E-Y et al. Clinicopathological findings following intraventricular glial-derived neurotrophic factor treatment in a patient with Parksinson s disease. Annal Neurol 1999; 46:419-424.

6. Yan Q, Matheson C, Sun J et al. Exp Neurol 1994; 127:23-36.
7. Pardridge WM. Peptide Drug Delivery to the Brain. New York: Raven Press, 1991:1-357.
8. Winkler J, Ramirez GA, Kuhn HG et al. Reversible schwann cell hyperplasia and sprouting of sensory and sympathetic neurites after intraventricular administration of nerve growth factor. Ann Neurol 1997; 41:82-93.
9. Yamada K, Kinoshita A, Kohmura E et al. Basic fibroblast growth factor prevents thalamic degeneration after cortical infarction. J Cereb Blood Flow Metabol 1991; 11:472-478.
10. Pardridge WM, Eisenberg J, Yang J. Human blood-brain barrier insulin receptor. J Neurochem 1985; 44:1771-1778.
11. Pardridge WM, Eisenberg J, Yang J. Human blood-brain barrier transferrin receptor. Metabol 1987; 36:892-895.
12. Duffy KR, Pardridge WM, Rosenfeld RG. Human blood-brain barrier insulin-like growth factor receptor. Metabol 1988; 37:136-140.
13. Golden PL, Maccagnan TJ, Pardridge WM. Human blood-brain barrier leptin receptor. Binding and endocytosis in isolated human brain microvessels. J Clin Invest 1997; 99:14-18.
14. Clemmons DR. Insulinlike growth factor binding proteins. Trends Endocrinol Metabol 1990; 1:412-417.
15. Duffy KR, Pardridge WM. Blood-brain barrier transcytosis of insulin in developing rabbits. Brain Res 1987; 420:32-38.
16. Bickel U, Kang Y-S, Yoshikawa T et al. In vivo demonstration of subcellular localization of anti-transferrin receptor monoclonal antibody-colloidal gold conjugate within brain capillary endothelium. J Histochem Cytochem 1994; 42:1493-1497.
17. Skarlatos S, Toshikawa T, Pardridge WM. Transport of [^{125}I] transferrin through the rat blood-brain barrier in vivo. Brain Res 1995; 683:164-171.
18. Fishman JB, Rubin JB, Handrahan JV et al. Receptor-mediated transcytosis of transferrin across the blood-brain barrier. J Neurosci Res 1987; 18:299-304.
19. Zhang Y, Pardridge WM. Rapid transferrin efflux from brain to blood across the blood-brain barrier. J Neurochem 2001; 76:1597-1600.
20. Zhang Y, Pardridge WM. Mediated efflux of IgG molecules from brain to blood across the blood-brain barrier. J Neuroimmunol. 2001; 114:168-172.
21. Lee HJ, Engelhardt B, Lesley J et al. Targeting rat anti-mouse transferrin receptor monoclonal antibodies through the blood-brain barrier in the mouse. J Pharmacol Exp Ther 2000; 292:1048-1052.
22. Pardridge WM, Kang Y-S, Buciak JL et al. Human insulin receptor monoclonal antibody undergoes high affinity binding to human brain capillaries in vivo and rapid transcytosis through the blood-brain barrier in vivo in the primate. Pharm Res 1995; 12:807-816.
23. Coloma MJ, Lee HJ, Kurihara A et al. Transport across the primate blood-brain barrier of a genetically engineered chimeric monoclonal antibody to the human insulin receptor. Pharm Res 2000; 17:266-274.
24. Penichet ML, Kang Y-S, Pardridge WM et al. An anti-transferrin receptor antibody-avidin fusion protein serves as a delivery vehicle for effective brain targeting in an animal model. Initial applications in antisense drug delivery to the brain. J Immunol 1999; 163:4421-4426.
25. Li JY, Sugimura K, Boado RJ et al. Genetically engineered brain drug delivery vectors Cloning, expression, and in vivo application of an anti-transferrin receptor single chain antibody-streptavidin fusion gene and protein. Protein Engineer 1999; 12:787-796.
26. Lee TJ-F, Saito A. Vasoactive intestinal polypeptide-like substance: the potential transmitter for cerebral vasodilation. Science 1984; 224:898-901.
27. Said SI. Molecules that protect: the defense of neurons and other cells. J Clin Invest 1996; 97:2163-2164.
28. Lindvall M, Owan C. Autonomic nerves in the mammalian choroid plexus and their influence on the formation of cerebrospinal fluid. J Cereb Blood Flow Metabol 1981; 1:245-266.
29. Huffman L, Hedge GA. Effects of vasoactive intestinal peptide on thyroid blood flow and circulating thyroid hormone levels in the rat. Endocrinol 1986; 118:550-557.

30. Wilson DA, O Neill JT, Said SI et al. Vasoactive intestinal polypeptide and the canine cerebral circulation. Circ Res 1981; 48:138-148.
31. McCulloch J, Edvinsson L. Cerebral circulatory and metabolic effects of vasoactive intestinal polypeptide. Am J Physiol 1980; 238:H449-H456.
32. Bickel U, Yoshikawa T, Landaw EM et al. Pharmacologic effects in vivo in brain by vector-mediated peptide drug delivery. Proc Natl Acad Sci 1993; 90:2618-2622.
33. Wu D, Pardridge WM. CNS pharmacologic effect in conscious rats after intravenous injection of a biotinylated vasoactive intestinal peptide analogue coupled to a blood-brain barrier drug delivery system. J Pharmacol Exp Ther 1996; 279:77-83.
34. Andersson M, Marie J-C, Carlquist M et al. The preparation of biotinyl-e-aminocaproylated forms of the vasoactive intestinal peptide (VIP) as probes for the VIP receptor. FEBS 1991; 282:35-40.
35. Huffman LJ, Connors JM, Hedge GA. VIP and its homologues increase vascular conductance in certain endocrine and exocrine glands. Am J Physiol 1988; 254:435-E442.
36. Honig B, Nicholls A. Classical electrostatics in biology and chemistry. Science 1995; 268:1144-1149.
37. Ibanez CF, Ebendal T, Barbany G et al. Disruption of the low affinity receptor-binding site in NGF allows neuronal survival and differentiation by biding to the trk gene product. Cell 1992; 69:329-341.
38. Pardridge WM, Kang Y-S, Buciak JL. Transport of human recombinant brain-derived neurotrophic factor (BDNF) through the rat blood-brain barrier in vivo using vector-mediated peptide drug delivery. Pharm Res 1994; 11:738-746.
39. Abuchowski A, McCoy JR, Palczuk NC et al. Effect of covalent attachment of polyethylene glycol on immunogenicity and circulating life of bovine liver catalase. J Biol Chem 1977; 252:3582-3586.
40. Rosenberg MB, Hawrot E, Breakefield XO. Receptor binding activities of biotinylated derivatives of b-nerve growth factor. J Neurochem 1986; 46:641-648.
41. Sakane T, Pardridge WM. Carboxyl-directed pegylation of brain-derived neurotrophic factor markedly reduces systemic clearance with minimal loss of biologic activity. Pharm Res 1997; 14:1085-1091.
42. Pardridge WM, Wu D, Sakane T. Combined use of carboxyl-directed protein pegylation and vector-mediated blood-brain barrier drug delivery system optimizes brain uptake of brain-derived neurotrophic factor following intravenous administration. Pharm Res 1998; 15:576-582.
43. Wu D, Kang Y-S, Bickel U et al. Blood-brain barrier permeability to morphine-G-glucoronide is markedly reduced compared to morphine. Drug Metabol Disp 1997; 25:768-771.
44. Smith M-L, Bendek G, Dahlgren N et al. Models for studying long-term recovery following forebrain ischemia in the rat. 2. A 2-vessel occlusion model. Acta Neurol Scand 1984; 69:385-401.
45. Wu D, Pardridge WM. Neuroprotection with noninvasive neurotrophin delivery to brain. Proc Natl Acad Sci USA 1999; 96:254-259.
46. Preston E, Foster DO. Evidence for pore-like opening of the blood-brain barrier following forebrain ischemia in rats. Brain Res 1997; 4-10.
47. Longa EZ, Weinstein PR, Carlson S et al. Reversible middle cerebral artery occlusion without craniectomy in rats. Stroke 1989; 20:84-91.
48. Zhang Y, Pardridge WM. Conjugation of brain-derived neurotrophic factor to a blood-brain barrier drug targeting system enables neuroprotection in regional brain ischemia following intravenous injection of the neurotrophin. Brain Res 2001; 889:49-56.
49. Menzies SA, Betz AL, Hoff JT. Contributions of ions and albumin to the formation and resolution of ischemic brain edema. J Neurosurg 1993; 78:257-266.
50. Albayrak S, Zhao Q, Siesjo BK et al. Effect of transient focal ischemia on blood-brain barrier permeability in the rat: correlation to cell injury. Acta Neuropathol 1997; 94:158-163.
51. Schabitz W-R, Schwab S, Spranger M et al. Intraventricular brain-derived neurotrophic factor reduces infarct size after focal cerebral ischemia in rats. J Cerb Blood Flow Metabol 1997; 17:500-506.

52. Kaplan B, Brint S, Tanabe J et al. Temporal thresholds for neocortical infarction in rats subjected to reversible focal cerebral ischemia. Stroke 1991; 22:1032-1039.
53. Zhang Y, Pardridge WM. Neuroprotection in transient focal brain ischemia following delayed, intravenous administration of BDNF conjugated to a blood-brain barrier drug targeting system. Stroke 2001; 32:1378-1384.
54. Schabitz WR, Sommer C, Zoder W et al. Intravenous brain-derived neurotrophic factor reduces infarct size and counterregulates Bax and Bcl-2 expression after temporary focal cerebral ischemia. Stroke 2000; 31:2212-2217.
55. Kawai N, Keep RF, Betz AL. Hyperglycemia and the vascular effects of cerebral ischemia. Stroke 1997; 28:149-154.
56. Pan W, Banks WA, Fasold MB et al. Transport of brain-derived neurotrophic factor across the blood-brain barrier. Neuropharmacol 1998; 37:1553-1561.
57. Kurihara A, Deguchi Y, Pardridge WM. Epidermal growth factor radiopharmaceuticals: ^{111}In chelation, conjugation to a blood-brain barrier delivery vector via a biotin-polyethylene linker, pharmacokinetics, and in vivo imaging of experimental brain tumors. Bioconj Chem 1999; 10:502-511.
58. Bemelmans A-P, Horellou P, Pradier L et al. Brain-derived neurotrophic factor-mediated protection of striatal neurons in an excitotoxic rat model of Huntington s disease, as demonstrated by adenoviral gene transfer. Human Gene Ther 1999; 10:2987-2997.
59. Kitagawa H, Sasaki C, Sakai K et al. Adenovirus-mediated gene transfer of glial cell line-derived neurotrophic factor prevents ischemic brain injury after transient middle cerebral artery occlusion in rats. J Cereb Blood Flow Metabol 1999; 19:1336-1344.
60. McMenamin MM et al. A g34.5 mutant of herpes simplex I causes severe inflammation in the brain. Neurosci 1998; 83:1225-1237.
61. Lawrence MS et al. Inflammatory responses and their impact on b-galactosidase transgene expression following adenovirus vector delivery to the primate caudate nucleus. Gene Ther 1999; 6:1368-1379.
62. Shi N, Pardridge WM. Noninvasive gene targeting to the brain. Proc Natl Acad Sci USA 2000; 97:7567-7572.
63. Mash DC, Pablo J, Flynn DD et al. Characterization and distribution of transferrin receptors in the rat brain. J Neurochem 1990; 55:1972-1979.
64. Cosgrove L, Lovrecz GO, Verkuylen A et al. Purification and properties of insulin receptor ectodomain from large-scale mammalian cell culture. Protein Expression 1995; 6:789-798.
65. Pardridge WM. Vector-mediated peptide drug targeting to the brain. In: Frokjaer S, Christrup L, Krogsgaard-Larsen P, eds. Peptide and Protein Drug Research. Munksgaard, Copenhagen: Alfred Benzon Symposium 43, 1998:381-396

INVASIVE DRUG DELIVERY

Ulrike Blömer[1,2], Arnold Ganser[1] and Michaela Scherr[1]

ABSTRACT

The central nervous system is a very attractive target for new therapeutic strategies since many genes involved in neurological diseases are known and often only local low level gene expression is required. However, as the blood brain barrier on one hand prevents some therapeutic agents given systematically from exerting their activity in the CNS, it also provides an immune privileged environment. Neurosurgical technology meanwhile allows the access of nearly every single centre of the CNS and provides the surgical tool for direct gene delivery via minimal invasive surgical approaches to the brain.

Successful therapy of the central nervous system requires new tools for delivery of therapeutics in vitro and in vivo (Fig. 1). The application of therapeutic proteins via pumps into the CSF was shown to be only of limited value since the protein mostly is not sufficiently transported within the tissue and the half life of proteins limits the therapeutic success.

Direct gene delivery into the host cell has been a main strategy for years, and in the beginning the direct DNA delivery or encapsulation in liposomes or other artificial encapsulation have been applied with different success. For several years the most promising tools have been vectors based on viruses. Viruses are able to use the host cell machinery for protein synthesis, and some of them are able to stably insert into the host cell genome and provide long term transgene expression as long as the cell is alive. The increasing knowledge of viruses and their live cycle promoted the development of viral vectors that function like a shuttle to the cell, with a single round of infection either integrating or transiently expressing the transgene. Viral vectors have proven to be one of the most efficient and stable transgene shuttle into

[1]Department of Hematology and Oncology, Medical School Hannover, Carl-Neuber-Str. 1, 30625 Hannover, Germany and [2]Department of Neurosurgery, University of Kiel, Weimarer Strasse 8, D-24105 Kiel, Germany.

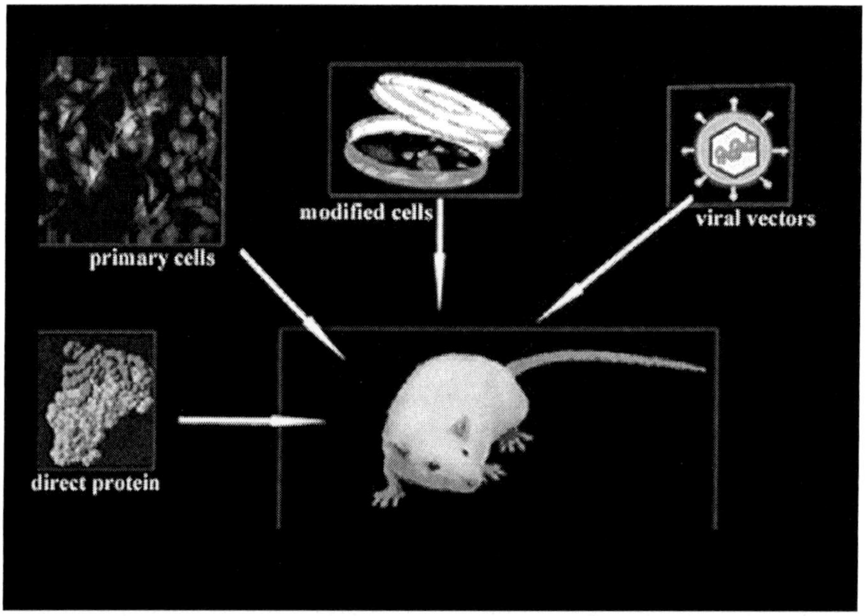

Figure 1. Therapeutical approaches in CNS disorders. Direct delivery of proteins via pumps or single injections, transplantation of fetal or embryonic cells, re-implantation of cells ex vivo genetically modified, or direct gene delivery in vivo via viral vectors.

the cell and have gained increasing importance. The limitations of some viral vectors like the adenoviral vector and adeno-associated viral vector have been improved by new constructs like HIV-1 based lentiviral vectors. The immune response caused by expression of viral proteins, or the inability of some viral vectors like the retroviral vector to infect only dividing cells have been overcome by these new constructs. Lentiviral vectors allow an efficient and stable transgene expression over years in vivo without effecting transgene expression or immune response. In this Chapter we will describe synthetic vectors, give an overview of the most common viral vectors and focus our attention on lentiviral vectors, since we consider them to be the most efficient tool for gene delivery in the CNS.

VECTORS IN GENE THERAPY

The most important goal of gene therapy should be efficient delivery of therapeutic genes and a high level of gene expression in the target area without any toxicity or immune response, leaving the target area intact. Some diseases will require long term stable transgene expression, some have to be expressed only for a short time and others will need to be regulated. Some disorders will need a systemic application, correcting a large number of cells in their genome or pathological product; others will only need transgene expression in a limited amount of cells in a specific area. The main approaches target the substitution of absent enzymes or proteins (growth factors, dopamine), the inactivation of pathological proteins (antisense oligonucleotides or ribozymes) or the delivery of pro-drugs or the induction of immune responses for cancer therapy. Pathologies of the CNS may require tight regulation of gene expression up to a complete shut off after the therapeutic approach. In the Parkinsonian brain, for example, the level of transgene expression and resulting dopamine level needs to be adjusted to the individual patient, according to symptoms and the percentage of destroyed dopaminergic cells. These regulatable vector systems are available and based on specific promoters, that can be regulated for example with antibiotics.[1-3] Other approaches may require the re-exposure and immunity as well as local toxicity need to be overcome. Because the immune response, humoral as well as cellular, is mostly due to the expression of viral proteins. Deletion of viral protein expression must be as complete as possible. Acute toxicity of viral vector preparations by foreign materials or coatings needs to be overcome, and the potential of immune responses against the transgene needs to be a further object of investigation. As most vectors like the retroviral based vectors integrate randomly, a mutagenic effect with the possibility of disruption and transcriptional activation of cellular genes, like oncogenes, cannot be excluded. Studies until now did not show mutagenesis in vitro or in vivo; however increasing efficacy and the results of the first clinical trials with viral vectors the consideration be given targeting and specific integration in pre-determined sites. Viruses, like adeno-associated viruses, already integrate specifically and evaluation of the properties of these vectors will hopefully allow for transfer into other viral vector constructs, for instance retroviral vectors. These mechanisms need to be further investigated and taken into consideration in future constructs. For specific therapy selective targeting of cells and tissues is required, either by different receptor mechanisms or by specific promoter activities. However these promoters proved to be less efficient in transgene expression.

DIRECT DNA DELIVERY AND SYNTHETIC NONVIRAL VECTORS

Nonviral gene delivery allows for gene transfer in vitro and in a limited amount in vivo. Plasmids encoding for transgenes are charged negatively and have in average 200nm hydrodynamic diameter. Numerous nonviral gene transfer tools like liposomes,[4] direct DNA precipitation, polymers, cationic lipids and endosomal lytic peptides were investigated in vitro and in vivo.[5] In vitro approaches mostly use the DNA precipitation technique with calcium phosphate (CA_3PO_4) and lysosomal fusion technology. DNA precipitation allows plasmid transfer into the cell in monolayer cell culture[6] and is based on the nonspecific pick up of foreign DNA by the host cell, depending on the precipitation reaction and precipitate size.[7] Ideally the DNA transfer rate in vitro is 10-50%; however numerous cell lines and primary cells are resistant to this method. Transfection is limited in time, because of the lack of integration. Random integration is seen in 10^{-5} to 10^{-4} cases and depends on the particle number and is multi locally possible.[8] Complement inactiviation and low integration is responsible for the lack of efficient gene transfer in in vivo approaches. The same problem occurs by direct DNA delivery, even with enhancement by convection or electrical fields. The rate of complement binding of DNA and the low rate of entry into the host cell does not result in efficient gene expression in vivo. Transgene expression has been shown in rats, mice, rabbits, and primates using the calcium phosphate technique,[9] but in direct DNA injection in animal models of Duchennes, for example, there was no therapeutic effect.[10-13]

Cationic polymers, like polyethylenimine, can optimize gene transfer in vivo, because the charge of the complex formation with the DNA favours interaction with the cell membrane, however also with complement and extracellular matrix. The best in vivo level has been achieved with polymers having a medium weight of 22 or 25 kD. Studies in the CNS with double staining revealed neurons as well as glial cells expressing the transgene.[14,15] Intraventricular delivery of polymer encapsulated DNA succeeded in an expression of single cells throughout the brain, however mosT cells did not reach a significant level of expression.[15] Liposomes, on the other hand, are spherical particles in which the membrane is composed of a lipid bilayer, encapsulating parts of the solvent and DNA in the interior. The sizes of the constructs range between 0.02 and 10 μm, they can be charged differently and they contain polymer coats and various ligands on their surface. Usually constructs resemble cell membranes and contain lecithin, cephalin, sphingomyelins and sterols. Interaction of liposomes with cells involve adsorption, lipid exchange, endocytosis and fusion. It has been speculated that some liposome interaction can include reversible cell membrane poration. Instability of the liposome has required coatings, limiting the fusion rate. Generally, the intravenously infusion of liposomes results in only a small number of expressing cells, since undirected fusion to circulating cells occurred. Local injection into the CNS revealed high toxicity and low transgene expression.

VIRAL VECTORS AND GENE DELIVERY

Viruses use the gene expression machinery of the hosT cells. In some transfected cells viral proteins are expressed episomally, in others by stably integration into the host cell genome. Some viruses force the host into lytic virus production, destroying it; others stay latent for centuries in the host system, being reactivated eventually, but keeping the host cell intact. The abilities of viruses to deliver foreign proteins to hosT cells have been the basis of the development of viral vectors. The biology of viral vectors is based on the first phase of the viral life cycle, the infection of the host cell. However, the structural phase of viral protein expression and virus particle assembly has to be prevented. At the beginning the application of still replicating viral vectors has caused diffuse immune responses as well as uncontrolled spread. Actual constructs allow for a single round of host cell infection by viral vectors (non-replicating viral vectors) reducing or completely abolishing viral protein expression which is substituted by the expression of the gene of interest or therapeutic gene. The transduction of a host cell by viral vectors therefore is an abortive, nonexplicative infection, introducing the therapeutic gene expressed from a recombinant vector. Viral vector production based on producer cell lines, either stably or transiently, substitutes for vector particle assembly, but is required to produce replication deficient particles in high titer. Viral life cycle and interaction with the producer or host cell machinery has proven to be highly complex. Even in in vitro settings this complex mechanism has not been reproducible by synthetic components.

HERPES SIMPLEX VIRAL VECTORS

Herpes viruses (HSV) are large double stranded DNA viruses, with the potential to accommodate large cassettes containing transgenes of at least 30 kb.[16] The viral capsid is encoated by a double layer lipid envelope, with glycoproteins embedded in the viral envelope mediating the host cell infection. Infection takes place by attachment at the cell surface, fusion and virus penetration.[17-19]

HSV have generated mechanisms that allow for long term persistence in an episomal state without causing disease or host cell death in an immune competent host. The application of HSV-based vectors has been especially attractive for the CNS, since HSV naturally infects and stays in neuronal cells.[20] Persistence of HSV in neurons causes no lytic cycle and leaves the neuronal function completely intact because HSV contains a specific neuronal promoter staying active during the latency phase (latency active promoter).[21] The HSV genome can be divided into two viral genes that are essential and accessory for the replication in the host cell. Accessory functions can be deleted without reducing virus growth; however complementary cells are required for substitution of essential sequences and cell propagation.

Herpes simplex virus type-1 vectors can be constructed by deletion of approximately half of the viral genome replacing it by a foreign gene. The development of such recombinant vectors reduced for their lytic cycle genes are required for long term in vivo expression. The deletion of the immediate early (IE) genes reduce the lethal effects significantly and allow high-titer HSV vector production.[22] These IE HSV-deleted vectors allow for latency persistence in neurons, and expression in non-neuronal tissues.[23] The major advantage of HSV is the efficacy of infection. However, the lack of long term expression, the still evoked immune response and the question of recombination of the herpes virus in patients caused many groups to switch to other gene transfer tools.

Amplicon vectors are defective, interfering viral genomes from high passage viral stocks and a real alternative to replication-defective recombinant genomic vectors.[24] The amplicon vector needs helper viruses like the HSV for replication and packaging. Further developments use even more deleted helper viruses, reduced for packaging signals for example and propagated in bacteria.[25] Theoretically these vector productions are helper free, since helper virus DNA does not overlap with the amplicon DNA, but contamination with recombinations in the stock are possible and may be replication competent. But until now the production of replication-competent, virus-free genomic vectors has been significantly higher in infectious vector particle yield.

Herpes viral vectors have been used in a broad range of in vitro and in vivo settings. They have been successful in treating animal models for Parkinson, cancer, PNS diseases and pain.[26-29]

ADENOVIRAL VECTORS

Adenoviruses are linear double stranded DNA viruses containing approximately 36000 bp encapsulated in a protein coat. There are over 50 different human adenoviral serotypes; however, the adenoviral vectors are from strains 2 and 5 to which most adults were exposed during their lifetime. Early attention was focused on gene transfer into the respiratory epithelium, but soon it was recognized that many different tissues are accessible for adenoviruses.[30,31] To reduce the immune response in vivo adenoviral vectors based on non-human serotypes have been pursued.[32-34] The adenoviral life cycle has been analyzed, and their genome encodes for over 50 polypeptides. After cell entry, proteins allow for efficient penetration of the nucleus. The early region 1 genome E1 quickly initiates transcription, starts viral gene expression and results in genome replication. E1, E2 and E4 are required for replication, and later on genes initiate the expression of viral structural proteins. Enormous amounts of viruses are generated per single host cell leading to cell lysis and virus spread so that titers as high as 10^{13} particles per ml can be achieved. E3 is important in the life cycle of the AD virus since it controls the immune response in infected hosts.[35] Removal of different parts of the viral genome, like the E3 region allows the insertion of a transgene of around 8 kb. The first generation of AD vectors were deleted for E1, requiring stable expression of E1 in helper cell lines for vector generation. In vivo application revealed soon that low level viral protein expression resulted in cytokine response, antigen-dependent immune response and viral mediated destruction of the host cell, limiting the transgene expression in vivo.[36-38]

Further modifications with deletion of E1, E2, and /or E4 reduced the immune response significantly; however, the AD vector expression still was not persistent in immune competent animals.[26,39-42] Although these vectors seemed to be less toxic, the infusion of a AD vector into a young human patient with a genetic disorder (OTC) was the first gene therapy with lethal outcome.

Recently developed AD viral vectors are produced with a helper dependent system[43] delivering the deleted sequences and the transgene (28-32kb) containing vector with the LTRs. The vector and its helper are separated and further purified, but still 0.1% contaminants of helper virus remain. Further clinical investigations will be necessary to study the consequences of this contamination and viral protein expression. Adenoviral vector direct injection into pancreas, liver, spine, CNS have shown transient effects; however, most settings used immuno-deficient animals. In particular, in vivo striatal injections in immune competent rats revealed a strong immune response, even weeks after the transgene expression ceased.[44] Adenoviral vectors have been applied early in clinical trials for cystic fibrosis without success and meanwhile in cancer therapy, using the strong immune response in order to enhance the anti-tumor strategy.[45]

ADENO-ASSOCIATED VIRAL VECTOR (AAV)

Adeno-associated viral vectors are single stranded DNA parvoviruses, non-pathogenic for mammals. They normally need helper virus in order to replicate and therefore are dependent on ADV for productive infection. Six different human seroptypes have been identified and viral vectors are based on the AAV-2. The AAV genome consists of the sequences encoding multiple polypeptides for viral genome replication and those encoding viral structural proteins. The packaging capacity of the AAV is around 5 kb, limiting the size for foreign genes significantly. Wild-type virus integrates directly into the chromosome 19 if the area is available,[46] however this ability is lost in AAV vectors. AAV vector production is dependent on adenoviral function or at least plasmids that deliver those sequences.[47] The helper virus is separated from vector preparations by heat-inactivation and high-titer vector preparations can be obtained.[48] Transgene expression in vivo shows a slow increase over weeks, reaching a level of steady-state. This is thought to be due to the generation of DNA itself or second strand synthesis forming of concatemers.[49-52] However, in vivo transgene expression is mostly episomal and only random integration occurs[53,54] with a very low efficiency, but long term persistency.[50]

Since the adeno-associated viral vector does not express any viral proteins, there is no immune response associated with transgene expression in vivo except the possibility for antibodies preventing re-exposure expression. It has been shown in many animal studies that the AAV vector is able to infect nondividing cells[56] however, with limited efficacy in vivo compared to other viral vectors.[44] The application in animal models for Parkinson disease, even primates, revealed biological effects.[57,58] Clinical trials are on the way for cystic fibrosis, hemophilia and muscle dystrophy.

RETROVIRUS

Retroviruses are the basis for the most efficient gene transfer system available today. Not only the transfer into a broad variety of cell types, primary as well as established cell lines, but also the stable integration of the transgene into the host cell genome have led to an extensive knowledge about these vectors. Although replicating retroviral vectors are able to transfer genes into targeT cells, their restriction in insert size limits their use. The deletion of most of the viral protein coding regions allows a larger insert (approx. 7 kb), and since the development of packaging cell lines for the generation of retroviral vectors, the lack of helper virus contamination and unwanted viral spread allow a broad range of applications in vitro and in vivo. In addition, the improvement in the packaging and vector constructs allow the generation of retroviral high-titer stocks. The major limitation of retroviral vectors until now was their inability to introduce transgenes into non-dividing cells, like terminally differentiated neurons for example. Recently the development of new lentiviral vectors, a subgroup of retroviruses, with their unique nuclear import machinery led to long-term efficient transduction and increased the application of viral vectors in vivo significantly.

BIOLOGY OF RETROVIRUSES

Retroviruses are single stranded RNA viruses, initially identified as RNA tumor viruses by Rous in spontaneous chicken sarcomas in 1960 (RSV). Later on a broad number of transforming viruses was discovered by expression of viral oncogenic sequences from host cell genomes or by integration close to cellular oncogenes.[59] Viral integration may cause the inactivation of tumor suppressing genes, as shown for the p53.[60] The discovery of human retroviruses causing T-cell leukemias/lymphomas resulted in the theory of transcriptional activation of host cell genes by virally encoded proteins.

The majority of retroviral vectors are based on the Moloney murine leukemia virus (MLV), containing 9 kb genome flanked by the two long terminal repeats (LTRs). The integrated provirus is flanked by the LTR elements that are functioning as the binding site for cellular transcription factors and initiation. The untranslated packaging signal follows the 5'LTR and is responsible for the packaging of retroviral RNA into virions.[61] The *gag* region encodes viral internal structural proteins and the *pol* region encodes reverse transcription and integrase. The envelope protein is encoded by the *env* gene region.[62] The promoter in the 5'LTR initiates the transcription of the full-length viral RNA and singly spliced viral RNA. The integrated full-length DNA is the source for the RNA packaged into virions and the spliced transcript is used to produce the envelope peptide. Proteases encoded in the *pol* region cleave the precursors and generate the final viral proteins, among which there are the viral proteases, reverse transcriptases, integrase, matrix, capsid, nucleocapsid, and surface proteins.[59] For selection of transduced cells many retroviral constructs contain a drug resistance gene, like *neo* driven by the LTR or other promoters. These genes allow the selection in vitro as well as in vivo. In the host cell cytoplasm the viral proteins assemble and build an icosahedral virion core, accommodating two single stranded RNAs. The viral envelope protein binds to specific host cell-surface receptors and is the primary determinant of the host cell range.[63] Retroviral strains differ in their ability to infect distinct host cells. Some retroviruses infect a broad cell range but no mouse cells (xenotropic); some infect only mouse cells (ecotropic); or they infect both murine and non-murine cells (amphotropic).[64] After infection of the host cell the viral RNA is reverse transcribed into a double-stranded DNA, transported to the nucleus and integrated as the provirus into the cell DNA. For integration the host cell needs to progress in the cell cycle, because the breakdown of the nuclear membrane is required for the infection with simple retroviruses.[65] Integration of the retroviral genome requires the replication of the host cell genome during or shortly after the S phase of the cell cycle.[66] The major limitation of retroviral viruses to

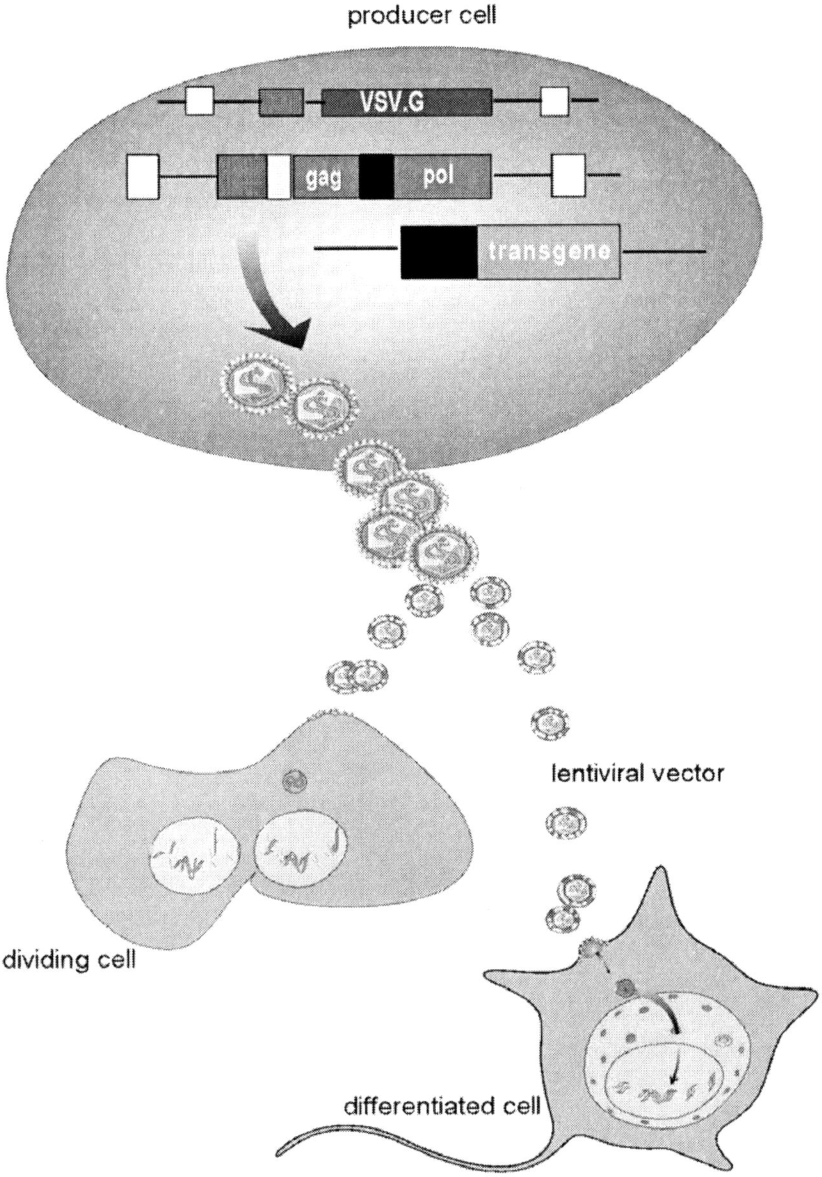

Figure 2. Generation of lentiviral vectors in triple-quadruple transfection in produces cells. Viral particels are harvested and either transferred on cells or further concentrated and injected in vivo. Lentiviral vectors are stably transducing dividing and non-dividing cells.

infect only dividing cells has been overcome by lentiviruses, a member of the retroviral family, which are able to translocate the viral DNA to the nucleus with the help of accessory viral proteinVpr and the viral gag matrix (MA),[67] although the cell is not undergoing the cell cycle.[68-72] The intergrase allows the insertion of the lentiviral provirus and infection of dividing as well as non-dividing cells in vitro and in vivo.[73,74] The integration site of the proviral genome into the host cell genome is not restricted to a single integration and is not completely random; open chromatin domains are preferred.[75,76]

Retroviruses are integrated, viral proteins are expressed and virions assembled without detremental effects on the host cell, and although retroviruses were discovered as pathogenic agents, the majority does not cause any damage to the host organism.

MOLONEY MURINE LEUKEMIA VIRUS AND LENITVIRUS BASED VECTORS

Non-replicating retroviral vector systems consist of two elements, the retroviral vector itself and the packaging cell line (Fig. 2).[77] The retroviral vector is a modified derivative of a bacterial plasmid, which does not encode any viral proteins but is determined as a shuttle for the transgene into the host cell. The packaging cell line provides the viral proteins that are necessary to generate virions containing the vector RNA, allowing the budding to the host cell, reverse transcription, and integration into the host cell DNA. The improvement of the packaging cell lines with separation of *gag* and *pol* expressed on one plasmid and the envelope on a different has led to helper free viral vectors.[78,79]

The generation of retroviral vectors that have incorporated the envelope of the vesiculo-stomatitis virus (VSV.G protein) broadened the host cell range significantly.[80] In addition, it transforms these pseudotyped viral vectors into stable and highly concentrated stocks, without the loss of transduction efficiency.[81-83] Meanwhile the improvement of packaging allows the generation of some retroviruses in stable packaging cell lines, which express the necessary retroviral proteins and provide for large amounts of VSV-G proteins by inducible expression.[84] The majority of retroviral protocols still require the production of pseudotyped retroviral vectors in transient transfection protocols in human kidney 293T cells. The titers obtained in the producer cell supernatant are around 10^5 transducing units/ml (TU/ml) and, after ultracentrifugation, 10^8 TU/ml and so studies of direct gene delivery in vivo are possible.

The most important goal in the delivery of transgenes in vitro as well as in vivo is to achieve stable and long term transgene expression at therapeutic levels. This is mainly influenced by the transcription and translation efficiency and RNA and protein stability. The promoter used to express the transgene is the main limiting factor to influence the protein level. Viral promoters like the LTR , CMV, SV40 and HSV-TK are generally very active in dividing cell.[85,86,87] but are less efficient in terminally differentiated quiescent cells and cells after transplantation.[88] Promoters derived from housekeeping genes like β actin, PGK, and DHFR[89,90] express at low levels in all cell types and cell cycle stages. Tissue specific promoters like the NSE, GFAP and preproenkephalin promoter allow however the expression of transgenes in certain cell types, especially interesting in the CNS.[91-93] Different promoter and enhancer elements can be combined to reach an efficient and cell-specific transgene expression. The combination of bacterial and eukaryotic regulatory elements meanwhile allows the regulation of transgene expression, as shown for the tetracycline repressor system.[2,3,94,95]

RETROVIRAL GENE TRANSFER EX VIVO

The inability of regular retroviral vectors to transduce terminally differentiated cells, like neurons, muscle cells and hepatocytes led to a broad number of experiments using ex vivo approaches to deliver gene products in vivo. Primary cells as well as cell lines are stably transduced in vitro, expanded in cell culture and implanted into a broad variety of species and locations. In this Chapter we will focus on the main applications of retroviral vectors in the field of gene therapy to the central nervous system.

Although fetal tissue grafts are useful as a cell replacement source, genetically modified cells for intracerebral transplantation promise far greater benefits. For example, genetically engineered cells can be autologous and therefore minimize the problems of cellular rejection. In addition, molecular biological methods allow the genetic modification of cells to produce a more controlled and broader range of desired factors than can be obtained with non-engineered cells. Following neuronal trauma, toxic compounds can be reduced or eliminated by cells engineered to produce free radical scavengers.[96] In neurodegenerative disorders, cells that lose their vital source of trophic factors, e.g., nerve growth factor (NGF), brain-derived neurotrophic factor (BDNF), and neurotrophins (NT-3 and NT4-5), can be supported by transplantation of cells modified to produce these factors.[97,98-102] The delivery of neurotransmitters and neuromodulators in models of neuronal degeneration was found to restore neuronal function in brain, although grafted cells are not able to mimic the normal dynamic functions of intercellular contact.[103]

Engineered cells may also serve as a drug delivery system in cancer therapy, delivering suicide genes or toxic compounds to rapidly dividing tumor cells. Preferential incorporation of drug sensitive genes into tumor cells enables transduced cells to produce enzymes metabolising drugs into toxic derivatives. These toxic derivatives result in the destruction of tumor cells following systematic administration of the appropriate drug, whereas the majority of healthy brain cells remain intact because they are quiescent.[104-107]

Immortalized neuronal and non-neuronal cell lines (C6, neuroblastoma, AT20) have been used for gene therapy. However, the persistent growth leads to tumor formation and limits therapeutic applications.

Non-neuronal primary fibroblasts have been studied extensively because they are easily obtained, can be maintained in cell culture for weeks and can be retrovirally modified. Contact inhibition in high density cultures leads to decreased cell division and also prevents tumor-like growth[108,109] in the CNS as well as in peripheral tissue.[88] The morphology of fibroblast grafts is similar to that of fibroblasts normally found

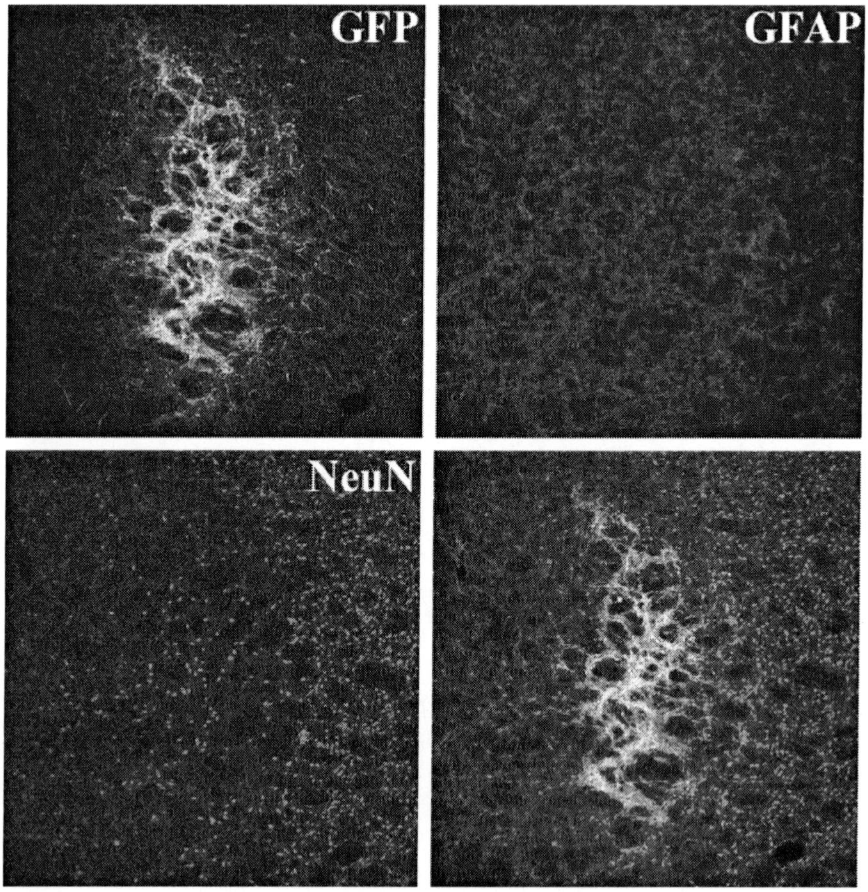

Figure 3. Confocal microscopy of coronal brain section of adult rats injected with lentiviral vector encoding for green fluorescent protein. 3 months after injection cells expressing the transgene (GFP-green) are further characterized as neurons (NeuN-red) and astrocytes (GFAP-blue).

in the skin, and viability has been demonstrated by collagen staining and abundant fibronectin production within the graft border.[101] Genetically engineered fibroblasts producing neurotrophic factors have been successfully implanted into various rodent models, diminishing the neuronal loss following surgical and toxic lesion.[110-112]

Increased survival of chromaffin cells transplanted together with peripheral nerve fragments is the result of NGF supplementation. Co-grafting of NGF-producing fibroblasts with chromaffin cells enhances both survival and transdifferentiation.[113-115]

Immature neuronal progenitor cells isolated from the adult and fetal brain have been successfully cultured and characterized.[116-119] Retrovirally immortalized rodent progenitor cell lines have been successfully transplanted in various regions of the brain[120] because they grow quickly ex vivo and therefore allow further retroviral vector modifications. The ease of culturing and manipulation of neuronal progenitor cells, their integration into the host system without uncontrolled proliferation and their potential to differentiate into mature neurons makes these cells a promising tool for ex vivo gene therapy.[121]

Retrovirally transferred oncogenes (e.g., v-myc, ras) have been used for immortalization of slowly dividing cells. These genes maintain cells in a highly mitotic undifferentiated state for as long as 22 months.[122] However transplantation of oncogene-expressing cells has revealed chromosomal damage and various cell morphologies.[123] Oncogene-expressing cells can also exhibit uncontrolled growth with resulting tumor formation.[124,125]

To obtain regulatable expression of transgenes, the temperature-sensitive mutant of SV40 large T antigen (TsA58) has been used.[126,127] SV40 regulates the expression of oncogenes at 25°C and leaves cells in an undifferentiated, rapidly dividing state. Down regulation of oncogene expression and differentiation of these cells into neurons occur at 37 °C. To externally trigger transgene expression, a regulatable retroviral vector in which the oncogene v-myc is driven by a tetracycline-controlled transactivator has also been used for conditional immortalisation of adult progenitor cells.[94] The suppression of the v-myc oncogene expression was sufficient to urge proliferating cells leave the cell cycle and induce terminal differentiation.

However, transgene expression was not stable with retroviral vectors, because of the silencing of the expression.[88,90,128]

RETROVIRAL GENE TRANSFER IN VIVO

Retroviral vectors have a broad host cell range, but their use is limited to dividing cells. As cell division is limited in the CNS, the application of this system is restricted mostly to ex vivo experiments.

To target non-dividing terminally differentiated cells, especially neurons of the CNS, the vector based on the human immune deficiency virus (HIV) has been used for direct in vivo approaches (Fig. 3).[73,74] Like other lentiviruses, HIV is able to infect dividing as well as quiescent cells, such as monocyte-derived macrophages and growth-arrested cells. This viral vector is also pseudotyped with the VSV.G protein envelope and generated in the three plasmid cotransfection system used for generation of MLV-based retroviral vectors. The transgene is expressed from the hCMV promoter. Gene delivery using this vector has been tested by intracerebral injection of highly concentrated vector (10^8 TU/ml) into the striatum, the hippocampus, the septum, the retina and the ventricle of adult rats. Six months after injection the reporter gene (β-galactosidase or green fluorescent protein) was still detectable in every injection site and terminally differentiated neurons, as well as other CNS cells were transduced. The evaluation of the transduction efficiency revealed a stable number of cells and levels of transgene expression over the whole period investigated.[129] Obvious pathological changes or signs of immune response were not detected in the rat brain tissue. In comparison control animals injected with a MLV-based retroviral vector did not express the β-galactosidase reporter gene after 6 weeks, supporting earlier findings of the shut off of transgenes in vivo. The application of lentiviral vectors have been broadened rapidly. The CNS in rodents and primates have been investigated for functional proteins and biological models for Parkinson's, Huntington's, retinal photoreceptor degeneration type VII mucopolysaccharidoses, and metachromatic leucodystrophy.[130-134] Limitations are mostly due to specific conditions like mucus in the bronchial epithelium or cell cycle specifities.[135,136] A very interesting possibility offers lentiviral gene transfer into hematopoietic stem cells.[137] promising also capacities for progenitor transduction of the CNS. Improvement in vector preparations as well as titer determination will allow for highly efficient production and large scale vector particle production.[138,139] Long-term transgene expression, stable integration and lack of expression of viral proteins associated with immune responses make this lentiviral vector a promising tool in direct in vivo gene transfer. Recently, recombinant lentiviral vectors derived from non-human lentiviruses have been developed. These vectors focus on the non-pathogenicity of these viruses in humans, probably allowing easier access to clinical applications.[140,141] However the recent generation of lentiviral vectors (HIV based) are improved in safety, deleted for more viral proteins and reducing the risk from generation to generation of vectors.

RETROVIRAL VECTOR SAFETY

In the evaluation of retroviral vector gene therapy, two major issues need to be addressed. The cotransfection in transient transfection protocols as well as stable packaging cell lines always bear the risk of generating replication competent helper viruses, however the new packaging cell lines are significantly modified.[142] Occasionally the generation of helper viruses by multiple ping-pong infection is possible.[143,144] The overlap of sequences between the specific viral vector and the precise packaging sequence present in the packaging cell line may give rise to helper viruses.[145,146] Even the third generation of packaging cells still allows the observation of transmission of viral functions to recipient cells. The design of packaging cells with precise viral sequences necessary to encode *gag-pol* and an expression vector that uses totally non-retroviral sequences will increase the safety significantly. In addition the use of non-retroviral envelope proteins decreases the risk of recombination.[84]

The second concern is the insertion site of the proviral DNA into the host cell genome; 20-50% of the proviral integration takes place in the Cpg-rich , nuclease-sensitive , transcriptionally active, euchromatic regions of the cellular DNA. Theoretically the insertion can lead to a single hit mutagenesis. Although it is very unlikely that a single mutation will result in tumorgenic cell-transformation. The remaining 50-80% integration occurs randomly. Based on the fact that only 20% of the host cell genome is actively transcribed, 40-50% of all insertion events take place in inactive sites, leading to poor viral gene expression.[147,148]

The risk of further viral infection and spread in the organism in addition is severely diminished by the effects of human serum and cerebrospinal fluid on wild-type retroviruses[149] and retroviral vectors. Russel et al report that a broad range of MLV based viral vectors with ecotropic and amphotropic envelope proteins were all inactivated by human sera, as well as NIH-3T3 cells and packaging cells derived from them.[150]

CONCLUSION

Viral vectors are the most efficient tool in gene delivery, in vitro and in vivo. Although our knowledge is growing daily, the improvement of targeting, regulation and large scale safe production of vectors is a challenging goal. First steps like production of vectors not causing any immune response have been accomplished. Some viral vectors are ready for larger clinical trials, but recent lethal outcomes of approaches with vectors like the adenoviral vector should cause us to reconsider. It is necessary to use the most recent vector generation and most recent technology to avoid these experiences, and gene therapy will be successful in the long term run.

REFERENCES

1. Clackson T. Regulated gene expression systems. Gene Ther 2000; 7:120-125.
2. Gossen M, Bujard H. Tight control of gene expression in mammalian cells by tetracycline-responsive promoters. Proc Natl Acad Sci USA 1992; 89:5547-5551.
3. Gossen M, Bonin AL, Bujard H. Control of gene activity in higher eukaryotic cells by prokaryotic regulatory elements. Trends Biochem Sci 1993; 18:471-475.
4. Yang K, Clifton GL, Hayes RL. Gene therapy for central nervous system injury: the use of cationic liposomes: an invited review. J Neurotrauma 1997; 14:281-297.
5. Ledley FD. Nonviral Gene therapy: The promise of genes as pharmaceutical products. Human Gene therapy 1995; 6:1129-1144.
6. Loyter A, Beigel M. Sendai virus envelope as a carrier for transferring membrane proteins: A quantitative estimation of virus-cell adsorption and fusion. Prog Clin Biol Res 1982; 102 pt A:337-345.
7. Loyter A, Scangos GA, Ruddle FH. Mechanisms of DNA uptake by mammalian cells: Fate of exogenously added DNA monitored by the use of fluorescent dyes. Proc Natl Acad Sci USA 1982; 79:422-426.
8. Ruddle FH. A new era in mammalian gene mapping: Somatic cell genetics and recombinant DNA methodologies. Nature 1981; 294:115-120.
9. Jiao S, Williams P, Berg RK et al. Direct gene transfer into nonhuman primate myofibers in vivo. Hum Gene Ther 1992; 3:21-33.
10. Cao L, Zheng ZC, Zhao YC et al. Gene therapy of Parkinson disease model rat by direct injection of plasmid DNA-lipofectin complex. Hum Gene Ther 1995; 6:1497-1501.
11. Danko I, Fritz JD, Latendresse JS et al. Dystrophin expression improves myofiber survival in mdx muscle following intramuscular plasmid DNA injection. Hum Mol Genet 1993; 2:2055-2061.
12. Danko I, Wolff JA: Direct gene transfer into muscle. Vaccine 1994; 12:1499-1502.
13. Danko I, Williams P, Herweijer H et al. High expression of naked plasmid DNA in muscles of young rodents. Hum Mol Genet 1997; 6:1435-1443.

14. Abdallah B, Hassan A, Benoist C et al. A powerful nonviral vector for in vivo gene transfer into the adult mammalian brain: polyethylenimine. Hum Gene Ther 1996; 7:1947-1954.
15. Goula D, Benoist C, Mantero S et al. Polyethylenimine-based intravenous delivery of transgenes to mouse lung. Gene Ther 1998; 5:1291-1295.
16. Krisky DM, Marconi PC, Oligino T et al. Rapid method for construction of recombinant HSV gene transfer vectors. Gene Ther 1997; 4:1120-1125.
17. Shieh MT, Spear PG. Herpes virus-induced cell fusion that is dependent on cell surface heparan sulfate or soluble heparin. J Virol 1994; 68:1224-1228.
18. Mettenleiter TC, Spear PG. Glycoprotein gB (gII) of pseudorabies virus can functionally substitute for glycoprotein gB in herpes simplex virus type 1. J Virol 1994; 68:500-504.
19. Steven AC, Trus BL, Booy FP et al. The making and breaking of symmetry in virus capsid assembly: glimpses of capsid biology from cryoelectron microscopy. Faseb J 1997; 11:733-742.
20. Goins WF, Krisky D, Marconi P et al. Herpes simplex virus vectors for gene transfer to the nervous system. J Neurovirol 1997; 3 Suppl 1:S80-S88.
21. Goins WF, Sternberg LR, Croen KD et al. A novel latency-active promoter is contained within the herpes simplex virus type 1 UL flanking repeats. J Virol 1994; 68:2239-2252
22. Samaniego LA, Neiderhiser L, DeLuca NA. Persistence and expression of the herpes simplex virus genome in the absence of immediate-early proteins. J Virol 1998; 72:3307-3320.
23. Glorioso JC, DeLuca NA, Fink DJ. Development and application of herpes simplex virus vectors for human gene therapy. Annu Rev Microbiol 1995; 49:675-710.
24. Spaete RR, Frenkel N. The herpes simplex virus amplicon: A new eucaryotic defective-virus cloning-amplifying vector. Cell 1982; 30:295-304.
25. Saeki Y, Ichikawa T, Saeki A et al. Herpes simplex virus type 1 DNA amplified as bacterial artificial chromosome in *Escherichia coli*: Rescue of replication-competent virus progeny and packaging of amplicon vectors. Hum Gene Ther 1998; 9:2787-2794.
26. Glorioso JC, Naldini L, Kay MA. Viral vectors for gene therapy: The art of turning infectious agents into vehicles of therapeutics. Nat Med 2001; 7:33-40.
27. Zhu J, Kang W, Wolfe JH et al. Significantly increased expression of beta-glucuronidase in the central nervous system of mucopolysaccharidosis type VII mice from the latency-associated transcript promoter in a nonpathogenic herpes simplex virus type 1 vector. Mol Ther 2000; 2:82-94.
28. Martino G, Poliani PL, Furlan R et al. Cytokine therapy in immune-mediated demyelinating diseases of the central nervous system: A novel gene therapy approach. J Neuroimmunol 2000; 107:184-190.
29. Goins WF, Lee KA, Cavalcoli JD et al. Herpes simplex virus type 1 vector-mediated expression of nerve growth factor protects dorsal root ganglion neurons from peroxide toxicity. J Virol 1999; 73:519-532.
30. Mallet J, Le Gal La Salle G, Robert JJ et al. Adenovirus mediated gene transfer to the central nervous system. Gene Ther 1994; 1 Suppl 1:S52.
31. Barkats M, Bilang-Bleuel A, Buc-Caron MH et al. Adenovirus in the brain: Recent advances of gene therapy for neurodegenerative diseases. Prog Neurobiol 1998; 55:333-341.
32. Parks RJ. Improvements in adenoviral vector technology: overcoming barriers for gene therapy. Clin Genet 2000; 58:1-11.
33. Morral N, O'Neal W, Rice K et al. Administration of helper-dependent adenoviral vectors and sequential delivery of different vector serotype for long-term liver-directed gene transfer in baboons. Proc Natl Acad Sci USA 1999; 96:12816-12821.
34. Loser P, Hillgenberg M, Arnold W et al. Ovine adenovirus vectors mediate efficient gene transfer to skeletal muscle. Gene Ther 2000; 7:1491-1498.
35. Chirmule N, Propert K, Magosin S et al. Immune responses to adenovirus and adeno-associated virus in humans. Gene Ther 1999; 6:1574-1583.

36. Brody SL, Metzger M, Danel C et al. Acute responses of non-human primates to airway delivery of an adenovirus vector containing the human cystic fibrosis transmembrane conductance regulator cDNA. Hum Gene Ther 1994; 5:821-836.
37. Gooding LR. Regulation of TNF-mediated cell death and inflammation by human adenoviruses. Infect Agents Dis 1994; 3:106-115.
38. Gooding LR. Virus proteins that counteract host immune defenses. Cell 1992; 71:5-7
39. Lusky M, Christ M, Rittner K et al. In vitro and in vivo biology of recombinant adenovirus vectors with E1, E1/E2A, or E1/E4 deleted. J Virol 1998; 72:2022-2032.
40. Armentano D, Smith MP, Sookdeo CC et al. E4ORF3 requirement for achieving long-term transgene expression from the cytomegalovirus promoter in adenovirus vectors. J Virol 1999; 73:7031-7034.
41. Christ M, Louis B, Stoeckel F et al. Modulation of the inflammatory properties and hepatotoxicity of recombinant adenovirus vectors by the viral E4 gene products. Hum Gene Ther 2000; 11:415-427.
42. Gorziglia MI, Lapcevich C, Roy S et al. Generation of an adenovirus vector lacking E1, e2a, E3, and all of E4 except open reading frame 3. J Virol 1999; 73:6048-6055.
43. Morsy MA, Caskey CT. Expanded-capacity adenoviral vectors—The helper-dependent vectors. Mol Med Today 1999; 5:18-24.
44. Blömer U, Naldini L, Kafri T et al. Highly efficient and sustained gene transfer in adult neurons with a lentivirus vector. J Virol 1997; 71:6641-6649.
45. Rosengart TK, Patel SR, Crystal RG. Therapeutic angiogenesis: protein and gene therapy delivery strategies. J Cardiovasc Risk 1999; 6:29-40.
46. Samulski RJ, Zhu X, Xiao X et al. Targeted integration of adeno-associated virus (AAV) into human chromosome 19 [published erratum appears in EMBO J 1992 March; 11(3):1228]. Embo J 1991; 10:3941-3950.
47. Samulski RJ, Chang LS, Shenk T. A recombinant plasmid from which an infectious adeno-associated virus genome can be excised in vitro and its use to study viral replication. J Virol 1987; 61:3096-3101.
48. Samulski RJ, Chang LS, Shenk T. Helper-free stocks of recombinant adeno-associated viruses: Normal integration does not require viral gene expression. J Virol 1989; 63:3822-3828.
49. Duan D, Sharma P, Yang J et al. Circular intermediates of recombinant adeno-associated virus have defined structural characteristics responsible for long-term episomal persistence in muscle tissue. J Virol 1998; 72:8568-8577.
50. Vincent-Lacaze N, Snyder RO, Gluzman R et al. Structure of adeno-associated virus vector DNA following transduction of the skeletal muscle. J Virol 1999; 73:1949-1955.
51. Yang J, Zhou W, Zhang Y et al. Concatamerization of adeno-associated virus circular genomes occurs through intermolecular recombination. J Virol 1999; 73:9468-9477.
52. Ferrari FK, Samulski T, Shenk T et al. Second-strand synthesis is a rate-limiting step for efficient transduction by recombinant adeno-associated virus vectors. J Virol 1996; 70:3227-3234.
53. Nakai H, Herzog RW, Hagstrom JN et al. Adeno-associated viral vector-mediated gene transfer of human blood coagulation factor IX into mouse liver. Blood 1998; 91:4600-4607.
54. Duan D, Fisher KJ, Burda JF et al. Structural and functional heterogeneity of integrated recombinant AAV genomes. Virus Res 1997; 48:41-56.
55. Nakai H, Iwaki Y, Kay MA et al. Isolation of recombinant adeno-associated virus vector-cellular DNA junctions from mouse liver. J Virol 1999; 73:5438-5447.
56. Monahan PE, Samulski RJ. Adeno-associated virus vectors for gene therapy: More pros than cons? Mol Med Today 2000; 6:433-440.
57. Kaplitt MG, Leone P, Samulski RJ et al. Long-term gene expression and phenotypic correction using adeno-associated virus vectors in the mammalian brain. Nat Genet 1994; 8:148-154.
58. Kaplitt MG, Xiao X, Samulski RJ et al. Long-term gene transfer in porcine myocardium after coronary infusion of an adeno-associated virus vector. Ann Thorac Surg 1996; 62:1669-1676.

59. Bishop JM, Varmus H. Functions and origins of retroviral transforming genes: RNA tumor viruses. In: Cold Spring Harbor: Cold Spring Harbor Laboraories, 1984.
60. Mowat M, Cheng A, Kimura N et al. Rearrangements of the cellular p53 gene in erythroleukaemic cells transformed by Friend virus. Nature 1985; 314:633-636.
61. Mann R, Mulligan RC, Baltimore D. Construction of a retrovirus packaging mutant and its use to produce helper-free defective retrovirus. Cell 1983; 33:153-159.
62. Coffin J. Structure of the retroviral genome. In: Furmanski P, ed. RNA Tumor Viruses. Cold Spring Harbor: Cold Spring Harbor Laboratory, 1984.
63. Miller AD, Chen F. Retrovirus packaging cells based on 10A1 murine leukemia virus for production of vectors that use multiple receptors for cell entry. J Virol 1996; 70:5564-5571.
64. Weiss R. Experimental biology and assay of RNA tumor viruses. In: Furmanski P, ed. RNA Tumor Viruses. Cold Spring Harbor: Cold Spring Harbor Laboratory, 1984.
65. Varmus HE, Padgett T, Heasley S et al. Cellular functions are required for the synthesis and integration of avian sarcoma virus-specific DNA. Cell 1977; 11:307-319.
66. Hajihosseini M, Iavachev L, Price J. Evidence that retroviruses integrate into post-replication host DNA. Embo J 1993; 12:4969-4974.
67. Heinzinger N, Bukrinsky M, Haggerty S. The Vpr protein of human immunodeficiency virus type 1 influences nuclear licalization of teh viral nucleic acids in nondividing hosT cells. Proc Natl Acad Sci USA 1994; 91:7311-7315.
68. Bukrinsky MI, Sharova N, Dempsey MP et al. Active nuclear import of human immunodeficiency virus type 1 preintegration complexes. Proc Natl Acad Sci USA 1992; 89:6580-6584.
69. Gallay P, Swingler S, Aiken C et al. HIV-1 infection of nondividing cells: C-terminal tyrosine phosphorylation of the viral matrix protein is a key regulator. Cell 1995; 80:379-388.
70. Gallay P, Swingler S, Song J et al. HIV nuclear import is governed by the phosphotyrosine-mediated binding of matrix to the core domain of integrase. Cell 1995; 83:569-576.
71. Lewis PF, Emerman M. Passage through mitosis is required for oncoretroviruses but not for the human immunodeficiency virus. J Virol 1994; 68:510-516.
72. von Schwedler U, Kornbluth RS, Trono D. The nuclear localization signal of the matrix protein of human immunodeficiency virus type 1 allows the establishment of infection in macrophages and quiescent T lymphocytes. Proc Natl Acad Sci USA 1994; 91:6992-6996.
73. Naldini L, Blomer U, Gage FH et al. Efficient transfer, integration, and sustained long-term expression of the transgene in adult rat brains injected with a lentiviral vector. Proc Natl Acad Sci USA 1996; 93:11382-11388.
74. Naldini L, Blomer U, Gallay P et al. In vivo gene delivery and stable transduction of nondividing cells by a lentiviral vector. Science 1996; 272:263-267.
75. Coffin JM. Virology. In: Fields BN, Knipe DM, eds. New York: Raven, 1990:1437-1500.
76. Rohdewohld H, Weiher H, Reik W et al. Retrovirus integration and chromatin structure: Moloney murine leukemia proviral integration sites map near DNase I-hypersensitive sites. J Virol 1987; 61:336-343.
77. Miller AD. Retroviral vectors. Curr Top Microbiol Immunol 1992; 158:1-24.
78. Danos O, Mulligan RC. Safe and efficient generation of recombinant retroviruses with amphotropic and ecotropic host ranges. Proc Natl Acad Sci USA 1988; 85:6460-6464.
79. Markowitz D, Goff S, Bank A. A safe packaging line for gene transfer: Separating viral genes on two different plasmids. J Virol 1988; 62:1120-1124.
80. Zavada J. Pseudotypes of vesicular stomatitis virus with the coat of murine leukaemia and of avian myeloblastosis viruses. J Gen Virol 1972; 15:183-191.
81. Pear WS, Nolan GP, Scott ML et al. Production of high-titer helper-free retroviruses by transient transfection. Proc Natl Acad Sci USA 1993; 90:8392-8396.
82. Burns JC, Friedmann T, Driever W et al. Vesicular stomatitis virus G glycoprotein pseudotyped retroviral vectors: Concentration to very high titer and efficient gene transfer into mammalian and nonmammalian cells [see comments]. Proc Natl Acad Sci USA 1993; 90:8033-8037.

83. Emi N, Friedmann T, Yee JK. Pseudotype formation of murine leukemia virus with the G protein of vesiculo stomatitis virus. J Virology 1991; 65:1202-1207.
84. Ory DS, Neugeboren BA, Mulligan RC. A stable human-derived packaging cell line for production of high titer retrovirus/vesicular stomatitis virus G pseudotypes. Proc Natl Acad Sci USA 1996; 93:11400-11406.
85. Hock RA, Miller AD, Osborne WR. Expression of human adenosine deaminase from various strong promoters after gene transfer into human hematopoietic cell lines. Blood 1989; 74:876-881.
86. Dai Y, Roman M, Naviaux RK et al. Gene therapy via primary myoblasts: long-term expression of factor IX protein following transplantation in vivo. Proc Natl Acad Sci USA 1992; 89:10892-10895.
87. Li M, Hantzopoulos PA, Banerjee D et al. Comparison of the expression of a mutant dihydrofolate reductase under control of different internal promoters in retroviral vectors. Hum Gene Ther 1992; 3:381-390.
88. Palmer TD, Rosman GJ, Osborne WR et al. Genetically modified skin fibroblasts persist long term after transplantation but gradually inactivate introduced genes. Proc Natl Acad Sci 1991; 88:1330-1334.
89. Apperley J, Luskey B, Williams D. Retroviral genetransfer of human adenosine deaminase in murine hematopoetic cells: Effect of selectable marker sequences on long term expression. Blood 1991; 78:307-310.
90. Scharfmann R, Axelrodt JH, Verma IM. Long-term in vivo expression of retrovirus-mediated gene transfer in mouse fibroblast implants. Proc Natl Acad Sci USA 1991; 88:4626-4630.
91. Brenner M, Kisseberth W, Su Y et al. GFAP promoter directs astrocyte-specific expression in transgenic mice. J Neuroscience 1994; 14:1030-1037.
92. Forss-Petter S, Danielson P, Catsica S et al. Trangenic mice expressing β-galactosidase in mature neurons under neuron-specific enolse promoter control. Neuron 1990; 5:187-197.
93. Kaplitt MG, Kwong AD, Kleopoulos SP et al. Preproenkephalin promoter yields region-specific and long-term expression in adult brain after direct in vivo gene transfer via a defective herpes simplex viral vector. Proc Natl Acad Sci USA 1994; 91:8979-8983.
94. Hoshimaru M, Ray J, Sah DW et al. Differentiation of the immortalized adult neuronal progenitor cell line HC2S2 into neurons by regulatable suppression of the v-myc oncogene. Proc Natl Acad Sci USA 1996; 93:1518-1523.
95. Gossen M, Bujard H. Efficacy of tetracycline-controlled gene expression is influenced by cell type: commentary. Biotechniques 1995; 19:213-216; discussion 216-217.
96. Nakao N, Kokaia Z, Odin P et al. Protective effects of BDNF and NT-3 but not PDGF against hypoglycemic injury to cultured striatal neurons. Exp-Neurol 1995; 131:1-10.
97. Frim DM, Uhler TA, Short MP et al. Effects of biologically delivered NGF, BDNF and bFGF on striatal excitotoxic lesions. Neuroreport 1993; 4:367-370.
98. Minger SL, Fisher LJ, Ray J et al. Long-term survival of transplanted basal forebrain cells following in vitro propagation with fibroblast growth factor-2. Exp Neurol 1996; 141:12-24.
99. Takayama H, Ray J, Raymon HK et al. Basic fibroblast growth factor increases dopaminergic graft survival and function in a rat model of Parkinson's disease. Nat Med 1995; 1:53-58.
100. Tuszynski MH, Roberts J, Senut MC et al. Gene therapy in the adult primate brain: Intraparenchymal grafts of cells genetically modified to produce nerve growth factor prevent cholinergic neuronal degeneration. Gene Ther 1996; 3:305-314.
101. Lucidi Phillipi CA, Gage FH, Shults CW et al. Brain-derived neurotrophic factor-transduced fibroblasts: production of BDNF and effects of grafting to the adult rat brain. J Comp Neurol 1995; 354:361-376.
102. Lucidi-Phillipi CA, Gage FH. The neurotrophic hypothesis and the cholinergic basal forebrain projection. Prog Brain Res 1993; 98:241-249.
103. Fisher LJ, Ray J. In vivo and ex vivo gene transfer to the brain. Curr Opin Neurobiol 1994; 4:735-741.
104. Culver KW, Blaese RM. Gene therapy for cancer. Trends Genet 1994; 10:174-178.

105. Ram Z, Walbridge S, Shawker T et al. The effect of thymidine kinase transduction and ganciclovir therapy on tumor vasculature and growth of 9L gliomas in rats. J Neurosurg 1994; 81:256-260.
106. Barba D, Hardin J, Ray J et al. Thymidine kinase-mediated killing of rat brain tumors. J Neurosurg 1993; 79:729-735.
107. Barba D, Hardin J, Sadelain M et al. Development of anti-tumor immunity following thymidine kinase-mediated killing of experimental brain tumors. Proc Natl Acad Sci USA 1994; 91:4348-4352.
108. Kawaja MD, Rosenberg MB, Yoshida K et al. Somatic gene transfer of nerve growth factor promotes the survival of axotomized septal neurons and the regeneration of their axons in adult rats. J Neurosci 1992; 12:2849-2864.
109. Kawaja MD, Gage FH. Morphological and neurochemical features of cultured primary skin fibroblasts of Fischer 344 rats following striatal implantation. J Comp Neurol 1992; 317:102-116.
110. Rosenberg MB, Friedmann T, Robertson RC et al. Grafting genetically modified cells to the damaged brain: restorative effects of NGF expression. Science 1988; 242:1575-1578.
111. Wolff JA, Fisher LJ, Xu L et al. Grafting fibroblasts genetically modified to produce L-dopa in a rat model of Parkinson disease. Proc Natl Acad Sci USA 1989; 86:9011-9014.
112. Frim DM, Uhler TA, Galpern WR et al. Implanted fibroblasts genetically engineered to produce brain-derived neurotrophic factor prevent 1-methyl-4-phenylpyridinium toxicity to dopaminergic neurons in the rat. Proc Natl Acad Sci USA 1994; 91:5104-5108.
113. Niijima K, Chalmers GR, Peterson DA et al. Enhanced survival and neuronal differentiation of adrenal chromaffin cells cografted into the striatum with NGF-producing fibroblasts. J Neurosci 1995; 15:1180-1194.
114. Cunningham LA, Short MP, Breakefield XO et al. Nerve growth factor released by transgenic astrocytes enhances the function of adrenal chromaffin cell grafts in a rat model of Parkinson's disease. Brain Res 1994; 658:219-231.
115. Chalmers GR, Fisher LJ, Niijima K et al. Adrenal chromaffin cells transdifferentiate in response to basic fibroblast growth factor and show directed outgrowth to a nerve growth factor source in vivo. Exp Neurol 1995; 133:32-42.
116. Gage FH, Coates PW, Palmer TD et al. Survival and differentiation of adult neuronal progenitor cells transplanted to the adult brain. Proc Natl Acad Sci USA 1995; 92:11879-11883.
117. Gage FH. Neuronal stem cells: their characterization and utilization. Neurobiol Aging 1994; 15 Suppl 2:S191.
118. Luskin MB. Restricted proliferation and migration of postnatally generated neurons derived from the forebrain subventricular zone. Neuron 1993; 11:173-189.
119. Ray J, Raymon HK, Gage FH. Generation and culturing of precursor cells and neuroblasts from embryonic and adult central nervous system. Methods Enzymol 1995; 254:20-37.
120. Gage FH, Coates PW, Palmer TD et al. Survival and differentiation of adult neuronal progenitor cells. Proc Natl Acad Sci USA 1995; 92:11879-11883.
121. Lacorazza HD, Flax JD, Snyder EY et al. Expression of human beta-hexosaminidase alpha-subunit gene (the gene defect of Tay-Sachs disease) in mouse brains upon engraftment of transduced progenitor cells. Nat Med 1996; 2:424-429.
122. Snyder EY, Deitcher DL, Walsh C et al. Multipotent neural cell lines can engraft and participate in development of mouse cerebellum. Cell 1992; 68:33-51.
123. Cepko CL. Immortalization of neural cells via retrovirus-mediated oncogene transduction. Annu Rev Neurosci 1989; 12:47-65.
124. Kiess M, Scharm B, Aguzzi A et al. Expression of ril, a novel LIM domain gene, is down-regulated in Hras-transformed cells and restored in phenotypic revertants. Oncogene 1995; 10:61-68.
125. Brustle O, Aguzzi A, Talarico D et al. Angiogenic activity of the K-fgf/hst oncogene in neural transplants. Oncogene 1992; 7:1177-1183.

126. Greenwood J, Pryce G, Devine L et al. SV40 large T immortalised cell lines of the rat blood-brain and blood-retinal barriers retain their phenotypic and immunological characteristics. J Neuroimmunol 1996; 71:51-63.
127. Jung M, Crang AJ, Blakemore WF et al. In vitro and in vivo characterisation of glial cells immortalised with a temperature sensitive SV40 T antigen-containing retrovirus. J Neurosci Res 1994; 37:182-196.
128. St. Louis D, Verma IM. An alternative approach to somatic cell gene therapy. Proc Natl Acad Sci USA 1988; 85:3150-3154.
129. Blömer U, Naldini L, Kafri T et al. Highly efficient and sustained gene transfer in adult neurons with a lentiviral vector. J Virol 1997; 71:6642-6649.
130. Blömer U, Kafri T, Randolph-Moore L et al. Bcl-xL protects adult septal cholinergic neurons from axotomized cell death. Proc Natl Acad Sci USA 1997; 95:2603-2608.
131. Blömer U, Horner PJ, Samii M et al. Genetransfer in the Adult Rat Spinal Cord with Lentiviral Vectors. J Neurotrauma 2000; 17:948.
132. Kordower JH, Bloch J, Ma SY et al. Lentiviral gene transfer to the nonhuman primate brain. Exp Neurol 1999; 160:1-16.
133. Takahashi M, Miyoshi H, Verma IM et al. Rescue from photoreceptor degeneration in the rd mouse by human immunodeficiency virus vector-mediated gene transfer. J Virol 1999; 73:7812-7816.
134. Bosch A, Perret E, Desmaris N et al. Reversal of pathology in the entire brain of mucopolysaccharidosis typeVII mice after lentivira-mediated-gene transfer. Hum Gene Ther 2000; 11:1139-1150.
135. Johnson LG, Olsen JC, Naldini L et al. Pseudotyped human lentiviral vector-mediated gene transfer to airway epithelia in vivo. Gene Ther 2000; 7:568-574.
136. Park F, Ohashi K, Chiu W et al. Efficient lentiviral transduction of liver requires cell cycling in vivo. Nat Genet 2000; 24:49-52.
137. Miyoshi H, Smith KA, Mosier DE et al. Transduction of human CD34+ cells that mediate long-term engraftment of NOD/SCID mice by HIV vectors. Science 1999; 283:682-687.
138. Scherr M, Battmer K, Blömer U et al. A Rapid and reliable Method to determine lentiviral vector titers using quantitative real time PCR. Biotechnique 2001. In press.
139. Scherr M, Battmer K, Eder M et al. Efficient gene transfer into the CNS by anion exchange chromotography purified lentiviral vectors. J Neurosur 2001; Submitted.
140. Mitrophanous K, Yoon S, Rohll J et al. Stable gene transfer to the nervous system using a non-primate lentiviral vector. Gene Ther 1999; 6:1808-1818.
141. Poeschla EM, Wong-Staal F, Looney DJ. Efficient transduction of nondividing human cells by feline immunodeficiency virus lentiviral vectors. Nat Med 1998; 4:354-357.
142. Zufferey R, Dull T, Mandel RJ et al. Self-inactivating lentivirus vector for safe and efficient in vivo gene delivery. J Virol 1998; 72:9873-9880.
143. Otto E, Jones-Trower A, Vanin EF et al. Characterization of a replication-competent retrovirus resulting from recombination of packaging and vector sequences. Hum Gene Ther 1994; 5:567-575.
144. Donahue RE, Kessler SW, Bodine D et al. Helper virus induced T cell lymphoma in nonhuman primates after retroviral mediated gene transfer. J Exp Med 1992; 176:1125-1135.
145. Miller AD, Rosman GJ. Improved retroviral vectors for gene transfer and expression. Biotechniques 1989; 7:980-982, 984-986, 989-990.
146. Adam MA, Miller AD. Identification of a signal in a murine retrovirus that is sufficient for packaging of nonretroviral RNA into virions. J Virol 1988; 62:3802-3806.
147. Shih CC, Stoye JP, Coffin JM. Highly preferred targets for retrovirus integration. Cell 1988; 53:531-537.
148. Steffen M, Scherdin U, Vertes I et al. Karyotype instability and altered differentiation of rat sarcoma cells after retroviral infection. Genes Chromosomes Cancer 1992; 4:46-57.
149. Bartholomew RM, Esser AF. Mechanism of antibody-independent activation of the first component of complement (Cl) on retrovirus membranes. Biochemistry 1980; 19:2847-2853.
150. Russell DW, Berger MS, Miller AD. The effects of human serum and cerebrospinal fluid on retroviral vectors and packaging cell lines. Hum Gene Ther 1995; 6:635-641.

V. NEUROPROTECTIVE STRATEGIES IN ANIMAL AND IN VITRO MODELS OF NEURONAL DAMAGE

ISCHEMIA AND STROKE

Matthias Endres* and Ulrich Dirnagl*

ABSTRACT

Cell death following cerebral ischemia is mediated by a complex pathophysiologic interaction of different mechanisms. In this Chapter we will outline the basic principles as well as introduce in vitro and in vivo models of cerebral ischemia. Mechanistically, excitotoxicity, peri-infarct depolarization, inflammation and apoptosis seem to be the most relevant mediators of damage and are promising targets for neuroprotective strategies.

EPIDEMIOLOGICAL DATA

The incidence of stroke in the Federal Republic of Germany (a population of 81 million people) is approximately 250-400/100.000. Hence, an estimated total of 200.000 strokes occur per year. Stroke mortality amounts irrespective of all therapeutic efforts still to 25 to 30%. Stroke is the third leading cause of death in the industrialized world. For example, in Germany there are two million stroke victims alive (for comparison, four million in the United States of America). Stroke is the leading cause of disability; its direct and indirect costs amount to approximately 30 to 40 billion US$ per year.

INTRODUCTION

The metabolism of the brain depends exclusively on oxygen and glucose and adds up to the consumption of 75 l molecular oxygen and 120 g glucose per day. In comparison to other organs this situation is unique. Albeit the weight of the brain

*Experimental Neurology, Charit , Humboldt-University of Berlin, D-10098 Berlin, Germany.

accounts for only 2% of the total body weight, the brain claims 20% of total body perfusion and 20% of total oxygen consumption. Even short durations of reduced brain perfusion leading to lack of oxygen and energy metabolites may lead to irreversible structural damage. Variable ischemia thresholds that are region and cell-type specific determine whether a respective cell will survive the insult or die. (see also Table 1)

GLOBAL VS. FOCAL ISCHEMIA

In general, there are two mechanistically distinct modes of cerebral ischemia, i.e., global and focal ischemia, respectively. *Global ischemia* in man develops after transient circulatory arrest with resuscitation or after near-drowning (Table 2). Consequently, absolute cerebral blood flow falls off from 0.8 ml/g/min to zero within seconds. Loss of consciousness follows after approximately 10 seconds. EEG activity ceases after 30 to 40 secs and a few minutes of global ischemia lead to irreversible cellular damage that evolves over days. The typical histological picture following global ischemic insults is described by delayed neuronal death sparing glial cells (sometimes even associated with astrogliosis). As a general rule, under normothermic conditions, 10 min of global ischemia are lethal in man. In the United States approximately 500 000 people/year die because of circulatory arrest leading to global ischemia.

Focal ischemia follows transient or permanent flow reduction in the territory of a cerebral artery. Typically, flow reduction is due to embolic or thrombotic vessel occlusion. In contrast to the situation after global ischemia, focal ischemia is characterized by the formation of a so-called *ischemic penumbra* . The penumbra is defined as the ischemia border-zone which is (still) metabolically active but functionally silent. While absolute regional blood flow in the ischemic core is diminished to levels < 0.1 ml/g/min, blood flow in the penumbra typically remains at 0.2-0.4 ml/g/min. The typical histological picture following focal ischemia is a pan-necrosis that includes all cell types in the brain (neurons, astrocytes, oligodendrocytes, endothelial cells).

ANIMAL MODELS OF CEREBRAL ISCHEMIA

Although stroke has been studied in many species (for example rabbits, dogs, cats, and baboons) rats and mice are the most widely investigated.[1] Mice are especially useful because of the availability of unique strains that can be genetically engineered to over- or underexpress targeted genes.[2-4] Several well-established models are available to study *global ischemia.* In the so-called four -vessel occlusion model (4VO), flow in both carotid arteries and vertebral arteries is blocked for a specified time period (Pulsinelli-Brierley -model). [6] In the two-vessel occlusion model (2VO) which is also referred to as severe forebrain ischemia , only the carotid arteries are temporarily occluded, sometimes along with mild hypotension.[7] In these models,

Table 1. Perfusion thresholds during cerebral ischemia

The following perfusion thresholds (ml/g/min) will lead to...

Loss of protein synthesis	0.55 ml/g/min
anaerobic glycolysis	0.35 ml/g/min
synaptic release of transmitters (e.g. glutamate)	0.20 ml/g/min
failure of energy metabolism	0.20 ml/g/min
anoxic depolarisation	0.15 ml/g/min

injury develops selectively in cells most vulnerable to ischemic damage such as in the CA1 sector in hippocampus, medium-sized neurons in the striatum, and Purkinje cells in cerebellum. Neurons are more susceptible than glial cells, and die over hours to days after the insult; hence the term delayed neuronal death . Experimental *focal ischemia* is most commonly studied during permanent or transient occlusion of a middle cerebral artery (MCA).[8-10] Proximal MCA occlusion can be induced by an intraluminal suture (so-called filament model) or with a vascular clip and causes injury to cortex and deep structures (striatum). Distal MCA occlusion (the so-called Brint-model) is usually produced by placing a vascular clip on a pial vessel or by cautery.[11] The occlusion typically spares striatum and primarily involves the neocortex. Pan-necrosis develops in the territory supplied by the respective artery with glial and endothelial cell death. If recirculation is established early (2 hrs or less) outcome is better (transient MCA occlusion).[12] In some ways, the reperfused brain imitates restoration of blood flow after spontaneous lysis of a thrombo-embolic clot in humans- even though reperfusion after clot lysis is certainly more complex than an on/off phenomenon as modelled by placement and retraction of an intravascular filament. During reperfusion, free radical production and NO generation are especially pronounced and contribute to reperfusion injury . [13] After longer times of ischemia, reperfusion is incomplete due to microvascular occlusion which has been termed the no-reflow phenomenon some thirty years ago. [14] Because oxygen free radicals and NO promote apoptotic cell death, transient ischemia models have become especially useful to investigate cell death in vivo[15,16] which may particularly apply for models of mild ischemia . [17-19] In these models apoptosis is prominent after 30 min MCA occlusion followed by longer reperfusion times (several days).[19,20] The pattern of cell death is reminiscent of global ischemia in that it is both selective for neurons and delayed. Mild ischemia models may be similar to transient ischemic attacks in man. In fact, changes in T1/T2-weighted MR imaging 7 days after 15 min MCAo occlusion in rats resemble those 7 to 10 days after transient ischemic attacks (TIA s) in patients with known cardiogenic embolism. However, selective neuronal death has not yet been convincingly documented following TIA s in humans.

Table 2.

Animal Models of Cerebral ischemia	Possible Correlate in Man	Histological Findings
-global ischemia		
a. permanent (circulatory arrest)	hanging, circulatory arrest, drowning	brain swelling, cell death
b. transient (e.g. 4-vessel occlusion)	circulatory arrest with resuscitation, near-drowning	delayed neuronal death astrogliosis
-focal ischemia		
a. permanent (e.g. MCAo)	hemispheral stroke (thrombotic or embolic occlusion without recanalisation)	pannecrosis, hemispheral swelling
b. transient (e.g. MCAo/reperfusion)	stroke with spontaneous recanalization or thrombolytic therapy (reperfusion injury)	pannecrosis, inflammation
c. mild (e.g. brief MCAo/reperf.)	transient ischemic attack (?)	delayed neuronal death astrogliosis

IN VITRO MODELS OF CEREBRAL ISCHEMIA

To study the effects of cerebral ischemia in post-mitotic neurons in vitro, the so-called *oxygen-glucose deprivation* (OGD) model is commonly used. Primary cultures of post-mitotic neurons from different regions of the brain (such as cortex, striatum, septum, hippocampus, etc) can be established from rat or mouse embryos (day 16 to 18). After several days in vitro (10 to 14 days) these post-mitotic cells can be exposed to a combined deprivation of oxygen and glucose. Depending on the length and severity of the insult cell death develops and can be quantified on a morphological, biochemical, or molecular basis.[21,22] Other possibilities to study ischemic damage in vitro include the studies of *brain slices*, particularly the hippocampal slice. Usually, the bathing solution is changed from a mixture of oxygen/carbon dioxide to nitrogen/carbon dioxide in the absence (hence ischemia) or presence (hence anoxia) of glucose. Generally , 5 —7 min of ischemia lead to profound cell loss in the CA1 region. Shorter insults lead to a more slowly evolving damage, requiring approximately 12 h to be manifested. This type of cell death in many ways resembles delayed neuronal death seen in vivo.

IMPORTANCE OF PHYSIOLOGIC PARAMETERS FOR STROKE OUTCOME

Both animal and clinical studies have proven that changes in *physiologic parameters* early after stroke onset influence ischemia outcome. Of great importance are changes in brain and body temperature, partial pressures of oxygen and carbon dioxide, mean arterial blood pressure and glucose metabolism which may have direct impact on the treatment of stroke patients.

Temperature: Fever is a frequent complication following stroke and increases in body and brain temperature are associated with poor stroke outcome. Experimental studies unequivocally demonstrate that hypothermia reduces lesion volume while hyperthermia increases cell death following global ischemia (Table 3) and lesion volume after focal ischemia. Reducing temperature to 30ßC for 1 h after 2 h transient occlusion of the middle cerebral artery reduced lesion volume by 50%. This has direct implications for the treatment of acute stroke victims. Even slight elevations of body temperature (>37.5ßC) should be normalized by physical andpharmacological means. Current studies investigate acute treatment protocols using hypothermia.

Oxygen and carbon dioxide: Both hypoxia as well as hypercapnia have adverse effect of stroke outcome in animal models of cerebral ischemia. Theoretically, hypercapnia has advantageous effects by inducing vasodilation. Perfusion in the penumbra, however, is decreased during hypercapnia possibly due to a steal phenomenon.

Mean arterial blood pressure: Cerebral autoregulation is severely impeded during cerebral ischemia. Hence, brain perfusion is directly dependent on mean arterial blood pressure. Recent experimental studies have indeed demonstrated that blood pressure lowering increases infarct size in spontaneously hypertensive rats.[23] As a consequence, blood pressure in patients should be maintained at relatively high levels to improve brain perfusion and thus outcome.

Glucose: Albeit loss of glucose is one of the initiators of cell death after cerebral ischemia, elevated levels of blood glucose are nevertheless deleterious following focal ischemia. Accumulation of lactate by anaerobic glycolysis leads to acidosis with adverse effects on ischemic tissue.

PATHOPHYSIOLOGICAL CASCADES FOLLOWING CEREBRAL ISCHEMIA

Loss of oxygen and energy depletion initiates a self-promoting cascade of pathophysiologic events that evolve over minutes, hours, days, and even weeks. In this Chapter we will follow a putative sequence of events that can be differentiated in (1) *excitotoxicity*, (2) *peri-infarct depolarisation*, (3) *inflammation*, and (4) *apoptosis*. Obviously, this is an oversimplification and both mechanistically and in terms of temporal evolution there is substantial overlap.[24]

Excitotoxicity and Energy Depletion

As mentioned above brain tissue has an high energy demand. Moreover, to generate energy metabolites the brain is exclusively dependent on oxidative phosphorylation. Following focal ischemia there is profound loss of both oxygen

Table 3. Effects of temperature on damage following global cerebral ischemia

Protocol	Damage in CA1, %		
	37...C	35...C	33...C
15 min of 2-vessel occlusion in rat[108]	95	70	20
20 min of 4-vessel occlusion in rat[109]	95	70	20
5 min of 2-vessel occlusion in gerbil[110]	80	60	0
20 min of 2-vessel occlusion in rat	27	95	45

and glucose. Within minutes energy-dependent ion channels become activated which leads to the loss of the membrane potential.[25] Both neuronal and non-neuronal cells become depolarized and voltage-dependent Ca^{++}-channels are activated. Depolarization also induces release of neurotransmitters such as excitatory amino acids (glutamate) from presynaptic axon terminals into the synaptic cleft.[26] Since re-uptake mechanisms have failed, these excitatory amino acids subsequently accumulate in the synaptic cleft and induce activation of ligand-gated Ca^{++}-permeable channels (such as of the NMDA and AMPA subtype) as well as metabotropic Ca^{++}-channels.

Subsequently, there is influx of other extracellular ions (Na^+, Cl^-) that accumulate intracellulary and lead to cell swelling and intracellular (cytotoxic) *edema.* This intracellular edema can be visualized by magnetic resonance tomography using diffusion weighted imaging (DWI) techniques, providing the most sensitive non-invasive modality for detecting ischemia-induced tissue injury. Using perfusion weighted imaging (PWI) it is possible to compare regions of compromised blood flow (PWI) with already lesioned tissue (DWI). The difference between both volumes (mismatch) might correspond to salvageable penumbra tissue. The latter hypothesis is currently being tested in a number of clinical trials.

The most important trigger for all subsequent events leading to cellular disruption and cell death is the intracellular *increase of calcium-ions*: Ca^{++} acts as a universal second and third messenger and triggers via enzyme induction multiple cytoplasmic and nuclear cascades: Ca^{++} activates a number of important enzymes: (1) proteolytic enzymes (such as calpain, gelsolin etc.) with subsequent degradation of the cytoskeleton and structural proteins (e.g., actin, laminin, spectrin, microtubuli-associated proteins);[27-29] (2) xanthine oxidase and phospholipases (such as phospholipase A2) which leads to membrane degradation, (3) cyclooxygenase leading to free radical generation (4) neuronal type nitric oxide synthase (nNOS) leading to the production of nitric oxide (NO) and NO-derived radical peroxynitrite by the reaction with superoxide.[30,31]

Nitric oxide plays a critical role during cerebral ischemia. Importantly, NO may exert beneficial as well a deleterious effects depending on the time-point and compartment of its production. Hence, analysis of the role of NO during cerebral ischemia has led to much confusion in the literature and was described as a "double-edged sword". NO is synthetized from L-arginine and oxygen by NO synthases (NOS). There are two constitutive isoforms, neuronal (type I) and endothelial (type III), which are calcium/calmodulin-dependent, and one inducible isoform (iNOS, type II). Small quanta of NO synthetized by constitutive NOS regulate a wide variety of physiological functions such as blood pressure, vascular tone, permeability and neurotransmission.[32] iNOS can be induced in microglia, astrocytes, endothelium, and vascular smooth muscle. Once expressed, it is continuously active, irrespective of intracellular calcium levels and leads to high output NO synthesis leading to cytotoxicity and inflammatory actions. Like iNOS, nNOS can also generate high amounts of NO and cause cytotoxicity under pathophysiological conditions, due to the above mentioned intracellular rise of Ca^{++}.[21] Although nNOS positive neurons comprise only 1-2% of all neurons they possess extensive branching. Of note, the nNOS positive neurons themselves are surprisingly resistant against injury. Cortical levels of NO increase strikingly from approximately 10 nM to 2 μM within a few minutes after MCA occlusion. Animals lacking nNOS expression (nNOS knockout animals) have 38% smaller cerebral infarcts following permanent MCAo than control mice, unequivocally demonstrating the deleterious role of nNOS-derived NO during cerebral ischemia. NO neurotoxicity is mediated most likely by peroxynitrite formed by the reaction of NO with superoxide, a complex that rapidly decomposes into hydroxyl radicals, which are a highly reactive species.

An attractive downstream candidate for mediating NO-induced neurotoxicity is activation of the nuclear enyzme *poly(ADP-ribose)polymerase* (PARP). Formation of single-stranded DNA nicks (for example by peroxynitrite) is an obligatory stimulus of PARP leading to the formation of poly-ADP-ribose chains and depletion of its intracellular substrate NAD. [33,34] PARP is activated within minutes following cerebral ischemia and reperfusion[35] and inhibition of PARP activation or deletion of the PARP gene confers protection after ischemia.[35,36] Consistent with the notion that PARP is a perpetrator in NO-mediated neurotoxicity nNOS knockout animals had strikingly reduced levels of PARP activation within ischemic tissue.[37]

On the other hand, NO produced by the *endothelial NOS isoform* may exert protective effects via augmentation of cerebral blood flow, inhibition of platelet aggregation and leukocyte activation. Animals lacking expression of the endothelial NOS subtype (Type III) have been generated (eNOS knockout mice).[38] Expectedly, these animals have elevated arterial blood pressure. Moreover, eNOS knockout animals develop enlarged cerebral infarcts following middle cerebral artery occlusion, demonstrating a protective role of type III NOS during cerebral ischemia. The susceptibility of eNOS mutants to ischemic injury may be due to their diminished capacity to adapt to reduced perfusion pressure (i.e.dilate) at the margins of an ischemic lesion. This coupled to enhanced platelet and neutrophil adhesion, renders eNOS mutants susceptible to injury. Consistent with this notion, blocking NOS activity by nitro-L-arginine administration increased infarct size in nNOS knockout animals, presumably due to inhibition of the constitutively expressed eNOS isoform.[39] In conclusion, genetic evidence suggests that two NOS isoforms, i.e., nNOS and iNOS, contribute to ischemic injury, most likely due to the generation of NO-derived radicals, while NO generated by another NOS isoform, i.e., eNOS, is protective by its effect on blood flow and platelet aggregation. Pharmacologic approaches for acute treatment should be directed at selective inhibition of nNOS and iNOS isoforms while eNOS activity should be augmented. Recently, *statins* (HMG-CoA reductase inhibitors) were shown to selectively upregulate eNOS activity and NO production after chronic administration. Mice treated for 14 days with statins had increased eNOS message and activity, augmented cerebral blood flow and decreased markers of platelet activation, and had significantly smaller stroke sizes after MCAo.[40,41]

Another important source for the generation of reactive oxygen species (ROS) are *mitochondria,* which may themselves be damaged by ROS initiating a vicious cycle.[42,43] Specifically, the inner mitochondrial membrane is severely damaged by the generation of radicals and also by formation of a so-called mitochondrial permeability transition pore (MPT), which leads to or gan swelling, generation of additional ROS and finally cessation of ATP production. This may be coupled to the specific or unspecific release of cytochrome c into the cytosol which has been recognized as a specific upstream trigger for the initiation of the proteolytic caspase cascade leading to apoptotic cell death.[44]

Hemodynamic, metabolic, and ionic phenomena are not homogeneous in different ischemic brain regions. In the center or core of the ischemic region, cerebral blood flow is less than 20% of normal levels. In this region, permanent and anoxic depolarizations develop minutes after onset of ischemia. Subsequently, cells are killed rapidly by lipolysis, proteolysis and disruption of the cytoskeleton leading to loss of energy and ion homeostasis.[45] Between this core region and normal, non-ischemic tissue lies the so-called penumbra, an ischemic borderzone which is functionally silent but metabolically (still) active. In theory, the whole penumbra region can be salvaged by timely therapeutic intervention. Over time, however, and without treatment the core grows on the cost of the penumbra , i.e., the penumbra can progress to infarction by mechanisms described in this Chapter.[46-48] Hence, the goal of all therapeutic interventions is to protect the ischemic penumbra region. Albeit there is good evidence for the existence of a penumbra during human stroke,[49,50] its size and pathophysiologic relevance remains unclear.[51]

GLUTAMATE RECEPTORS AND EXCITOTOXICITY

Activation of *glutamate receptors* and subsequent calcium influx leading to increased intra-cellular calcium levels ($[Ca^{++}]i$) may be the most important early trigger of subsequent cell death after cerebral ischemia. Consequently, pharmacological inhibition of these receptors is an attractive treatment approach for stroke. Three different ligand-gated glutamate receptors can be differentiated based on their respective pharmacological affinities:(1) N-methyl-D-aspartate (NMDA) receptor, (2) α-amino-3 hydroxy-5 methyl-4- isoxazolone-proprionic acid (AMPA)

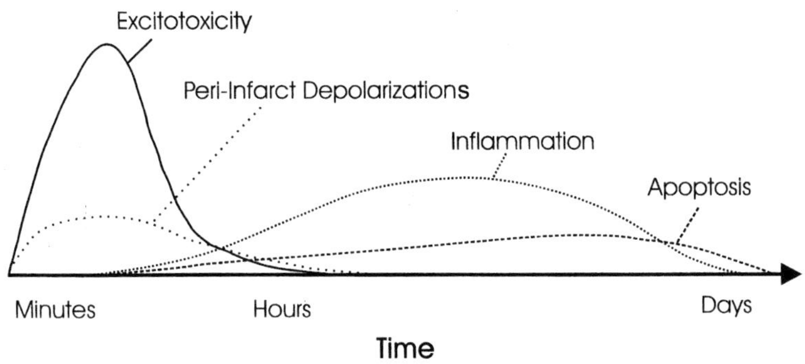

Figure 1. Putative cascade of damaging events in focal cerebral ischemia.
The x-axis reflects the evolution of the cascade over time, while the y-axis illustrates the impact of each element of the cascade on final outcomce. (adapted from Dirnagl et al, TINS 22:391-397)

receptor, and (3) kainate receptor. In addition, there are metabotropic G-protein coupled receptors.

NMDA receptors are permeable for Ca^{++}, Na^{+} and K^{+} ions. Following NMDA-receptor activation three stages inducing neurotoxicity can be differentiated: (1) changes in the intracellular milieu (induction), (2) exponential increase in $[Ca^{++}]i$ (amplification), (3) neuronal degeneration (expression). There are several pharmacologically distinct receptor antagonists available: NMDA receptor antagonists (such as MK-801) inhibit almost completely intracellular Ca^{++}-influx in primary neurons following "*oxygen-glucose-deprivation*" *in vitro*. NMDA receptor blockage using competitive and non-competitive receptor antagonists confers robust neuroprotection in animal model of focal (but not global) cerebral ischemia. Administration of NMDA receptor antagonists is neuroprotective only within a very limited window of opportunity (minutes to a few hours), however . Moreover, severe adverse effects of NMDA receptor blocker were demonstrated including exogenous psychosis and specific cell loss in the limbic system.[52,53] Currently, site-specific antagonists (for example for the glutathione site) are being tested as a possible alternative.

AMPA receptors mediate Na^{+} and K^{+} currents. Na^{+}-influx via AMPA as well as kainate receptors leads in turn to secondary influx of Ca^{++}-ions. Pharmacological blockade of these receptors via specific inhibitors confers significant reduction of infarct volume in animal stroke models. Of note, the treatment window is somewhat better than with NMDA receptor antagonists.[54] In contrast to the above listed ligand-gated glutamate receptors, the role of metabotropic glutamate receptors during cerebral ischemia has so far remained unclear. Some studies suggest a protective role for some classes of metabotropic receptors.[55]

TISSUE ACIDOSIS

Following oxygen deprivation glucose metabolism is changed to anaerobic glycolysis leading to lactate accumulation and *tissue acidosis.* For a long time acidosis was recognized as a central pathomechanism of tissue injury following stroke. For example, neurons in culture die rapidly when exposed to low pH. This has been explained by increased radical production and other mechanisms at low pH which may mediate damage. Recent observations, however, have shed new light on the dogma that acidosis is always dismal after cerebral ischemia. For example, NMDA receptors become inactivated at low pH levels protecting neurons from further Ca^{++}-influx.

PROTEIN SYNTHESIS AND EARLY GENE EXPRESSION

Even slight reductions of regional cerebral blood flow compromise total protein synthesis. In spite of these inhibitory effects on general protein synthesis there are some genes and gene products that are upregulated both on a transcriptional and translational level. Among these are the so-called *immediate early genes* whose gene products act as transcription factors. Via expression of these early genes both protective and destructive cascades are initiated.

Peri-Infarct Depolarization

Loss of energy and glutamate release induce depolarization of neurons and non-neuronal cells. In the core of the ischemic lesion this may induce so-called anoxic depolarization: Due to lack of energy equivalent the cells are unable to repolarize and are destined to die. In the border-zone (penumbra) of the lesion, however , cells are able to re-polarize, but at the expense of further energy depletion. Waves of de- and re-polarization may establish which are called *peri-infarct depolarizations.*[56,57]Evidence of this phenomenon has been provided in several animal models and a number of observations argue that it may have pathophysiological relevance. (1) Peri-infarct depolarizations (PIDs) may occur several times per hour and can be detected up to 6 —8 hours after ischemia onset. (2) Lesion volume and cell loss correspond to the number of PIDs.[58] (3) Reduction of PIDs by therapeutic intervention reduces infarct volume.[59] (4) Induction of additional PIDs by topical KCl administration further increases lesion volume. Albeit experimental evidence for PIDs as a relevant pathomechanism after stroke is convincing, so far evidence for the occurrence of PIDs after stroke in man is lacking.[60]

Inflammation

Expression of *proinflammatory genes*, such as NF-κB, hypoxia-inducible factor and interferon 1b is triggered early after onset of ischemia.[61-63] These

Table 4. Apoptosis vs. necrosis

	Apoptosis	Necrosis
DNA	oligonucleosomal laddering	degradation
nucleus	chromatin margination	pyknosis
membrane integrity	conserved	destroyed
inflammation	no	yes
cell volume (oncosis)	reduced	increased
cell fragmentation	yes (apoptotic bodies)	no (lysis)

Adapted from Ann Neurol 38:839-851 (1995)

proinflammatory genes may initiate the production of several mediators of inflammation including platelet activation factor and tumor necrosis factor. In turn, these factors themselves initiate a cascade of events that includes the expression of adhesion molecules of endothelial cells (intercellular and vascular adhesion molecules; ICAM and VCAM as well as different selectins). *Adhesion molecules* mediate rolling, adhesion (sticking) and finally migration of leukocytes through the vascular wall.[64-66] In a first wave neutrophils invade the brain which is followed by macrophages and monocytes. The latter express typical chemokines (e.g., interleukin-8 and monocyte chemoattractant protein-1) leading to a vicious cycle.

However, not only blood-born cells mediate inflammatory responses after ischemia but also brain-derived immunocompetent cells such as *microglia.* Microglial cells constitute the primary immunoeffector cells of the brain and amount to up to 20% of total brain cell number. Within hours after onset of an insult these cells become activated microglia. The question, however, whether inflammation after cerebral ischemia is good or bad has not unequivocally been resolved although the literature is in favor of a destructive effect. For example, (1) induction of neutropenia, (2) inhibition of several inflammatory mediators (such as IL-1b) or (3) adhesion molecules significantly protects from cell death and improves outcome after stroke.[67,68]

There are a plethora of molecular and biological mechanisms that contribute to inflammation-mediated cellular damage: The cerebral microcirculation becomes severely compromised by leukocyte plugging of small vessels. Neurons and macrophages may induce toxic enzymes such as inducible NO synthase (iNOS) or cyclooxygenase (COX). *iNOS* is produced by invading neutrophils which may lead to increased NO production. With the use of pharmacological inhibition or genetic deletion (iNOS knockout mouse) it has been unequivocally demonstrated that iNOS exerts neurotoxic effects during cerebral ischemia. Of note, the pathomechanism of iNOS induced cytotoxicity is a delayed one: iNOS becomes expressed only 24 h after the onset of the insult and administration of iNOS inhibitors is neuroprotective even when administered one day after stroke onset.[63,69,70] *COX-2* on the other hand is produced by neurons. It amplifies ischemic damage by production of superoxide anions and toxic prostanoids.[71] The role of yet another inflammatory mediator, *TNF-alpha*, during cerebral ischemia has so far remained controversial.[72,73]

Apoptosis

The term *apoptosis* was first introduced by Kerr et al in 1972.[74] Hallmarks of apoptosis include chromatin condensation, nuclear segmentation, cytoplasmic shrinkage, blebbing, and formation of apoptotic bodies (See Table 4). The possibility that ischemic brain cells may also die by apoptotic mechanisms was not proposed until 1993. [75-77] In fact, ischemic cell death was considered the classical prototype for necrotic cell death based mainly on morphological criteria. *Necrosis* describes swollen cells with disruption of subcellular organelles, and is also called accidental cell death . [78] (see also Table 4). Now convincing biochemical, histochemical, molecular and genetic data favor *apoptosis* (or a closely related mechanism) as an additional mechanism of cell death in models of cerebral ischemia. There are numerous reports demonstrating oligonucleosomal DNA fragmentation either by gel electrophoresis (DNA laddering) or by the TUNEL in situ technique (terminal deoxynucleotidyl transferase mediated dUTP-biotin nick-end labeling). [75,77,79-82] Following focal ischemia apoptosis is predominant in the ischemic border-zone.[75,83,84] Following more severe insults TUNEL positive cells appear earlier than after mild insults.[19] DNA laddering and TUNEL staining, however, are not pathognomonic for apoptosis. Most neuropathologists accept electron microscopy criteria as the gold standard. Van Lookeren Campagne and Gill[85] found 1996 no ultrastructural evidence for apoptosis after MCA occlusion in the rat.

It is generally accepted that *new protein synthesis* is needed for the initiation of apoptosis.[74] In fact, the protein synthesis inhibitor cycloheximide protects cells and reduces injury after focal ischemia.[18,19,75] In general, cycloheximide suppresses synthesis of all proteins, including those that protect cells. Hence, is has remained elusive by inhibition of which downstream products neuroprotection of cycloheximide is mediated.

MOLECULAR MECHANISMS

During development, apoptosis is triggered by evolutionally highly conserved signals, and is regulated by a balance between death promoting and death inhibiting factors.[86-89] In the nematode *Caenorrhabditis elegans* the death promoting ced-3 and ced-4 genes and the death inhibiting gene ced-9 were identified.[90-93] As mammalian homologues for the ced-3 gene the so-called *caspases* (*c*: cystein proteinase; *-aspase*: cleavage after aspartate residues) were identified (formerly also known as interleukin-1β family).[94,95] These proteases consist of at least 12 familiy members and are important executioners of apoptosis. The most important caspase for apoptosis execution is caspase-3. Caspase-3 as all caspases is activated by proteolytic cleavage (posttranslational modification) and in turn cleaves itself multiple (>30) substrates. Known caspase-3 substrates are for example endonuclease (CAD plus inhibitor ICAD), lamin, spectrin, huntingtin, gelsolin, poly(ADP-ribose)polymerase (PARP, see above). Cleavage of these caspase substrates is thought to mediate the downstream events during apoptosis (beyond the point of no-return). Until recently , it had remained unclear how caspase-3 is activated during cerebral ischemia. Generally, two pathways exist: type I apoptosis via Fas/TNF-receptors, activation of intracellular death recepors and activation of caspase-8; type II apoptosis via release of cytochrome c from mitochondria, formation of the so-called apoptosome (which is a complex of cytochrome c, apoptosis activation factor and pro-caspase-9).[96] Hence, after their respective cleavage either caspase-8 (type I) or caspase-9 (type II) activate caspase-3. Recent data favours the idea that caspase-3 is activated during cerebral ischemia predominantly via release of cytochrome c into the cytosol.[28]

EXPERIMENTAL EVIDENCE FOR CASPASE-MEDIATED CELL DEATH FOLLOWING CEREBRAL ISCHEMIA

Both caspase-1 and -3 become activated in the ischemic territory following focal cerebral ischemia,[19,97] an event that precedes markers of morphological damage and TUNEL staining by hours. Animals with deletion of caspase-1 or caspase—3 have smaller ischemic lesions.[98-100] Following a general rule that severe insults lead to necrotic cell death while after milder insults apoptosis is prominent, caspase activation was analyzed following short durations of MCA occlusion. Following 30 min MCAo caspase-3 was not activated until 9-12 hours after ischemia onset, while after 2 h MCAo it was active within minutes after reperfusion. More than 50% of all neurons within the ischemic tissue double stain for activated caspase-3 and TUNEL.[20,97]

CASPASE INHIBITION PROTECTS FROM CEREBRAL ISCHEMIA

Small oligopeptides (tri- or tetrapeptides) that mimic the cleavage site of a protease substrate can be used as relatively selective caspase inhibitors.[101,102] For example YVAD.cmk (acetyl-tyr-val-ala-asp.chloromethylketone) is relatively selective for caspase-1 (Ki 0.76 nM) while zDEVD.fmk (N-benzyloxycarbonal.asp-glu-val-asp-fluoromethylketone) is more selective for caspase-3 (Ki <0.1 nM).[103,104] Intracerebroventricular administration of these inhibitors blocks the activation of capase-1 and—3, respectivelyand significantly protects from injury following middle cerebral artery occlusion and global ischemia.[19,20,100,105,106] Following mild insults the treatment window was extended to 9 hr after the onset of ischemia.[19,20] Interestingly, caspase inhibitors act synergistically with anti-excitotoxic drugs such as the glutamate receptor blocker MK-801.[107] Hence, it seems feasible that both anti-excitotoxic and anti-apoptotic drugs may be given in a neuroprotective cocktail ; low doses may limit potential toxicity from either drug alone. Of note, non-peptide caspase inhibitors are being developed that can be administered systemically.

CONCLUSION

Cell death following cerebral ischemia is mediated by a complex pathophysiologic interaction of different mechanisms. As outlined in this Chapter, *excitotoxicity, peri-infarct depolarization, inflammation,* and *apoptosis* seem to be the most relevant mediators and are promising targets for neuroprotective intervention.. After all, however, stroke is a vascular disease. Reperfusion of an occluded artery (e.g., by rt-PA) is still the most important early intervention and should be combined with neuroprotective strategies.

REFERENCES

1. Ginsberg MD. Animal models of global and focal cerebral ischemia. In: Welsh KMA, Caplan LC, Reis DJ, Siesjo BK, Weir B, eds. Primer on Cerebrovascular Disease. San Diego: Academic Press, 1997:124-126.
2. F ssler R, Martin K, Forsberg E et al. Knockout mice: how to make them and why. The immonological approach. Int Arch Allergy Immunol 1995; 196:323-334.
3. Majzoub JA, Muglia LJ. Knockout mice. N Engl J Med 1996; 334:904-907.
4. Rubin EM, Barsh GS. Biological insights through genomics: mouse to man. J Clin Invest 1996; 97:275-280.
5. Thomas KR. The knockout mouse: six yeras old and growing stronger. Am J Respir Cell Mol Biol 1995; 12:461-463.
6. Pulsinelli WA, Brierley JB, Plum F. Temporal profile of neuronal damage in a model of transient forebrain ischemia. Ann Neurol 1982; 11:491-498.
7. Fujii M, Hara H, Meng W et al. Strain-related differences in susceptibility to transient forebrain ischemia in SV-129 and CC57/balck/6 mice. Stroke 1997; 28:1805-1811.
8. Tamura A, Graham DI, McCulloch J et al. Focal cerebral ischemia in the rat: 1. Description of technique and early neuropathological consequences following middle cerebral artery occlusion. J Cereb Blood Flow Metab 1:53-60.
9. Tamura A, Graham DI, McCulloch J et al. Focal cerebral ischemia in the rat. 2. Regional cerebral bloow flow determined by [14C]iodoantopyrine autoradiography following middle cerebral artery occlusion. J Cereb Blood Flow Metab 1:61-69.
10. Belayev L, Busto R, Zhao W et al. HU-211, a novel noncompetitive N-methyl-D aspartate antagonist, improves neurological deficit and reduces infarct volume after reversible focal cerebral ischemia in the rat. Stroke 1995; 26:2313-2320.
11. Brint S, Jacewicz M, Kiessling M et al. Focal brain ischemia in the rat: methods for reproducible neocortical infarction using tandem occlusion of the distal middle cerebral artery and ipsilateral common carotid arteries. J Cereb Blood Flow Metab 1988; 8:474-485.

12. Hara H, Huang PL, Panahian N et al. Reduced brain edema and infarction volume in mice lacking the neuronal isoform of nitric oxide synthase after transient MCA occlusion. J Cereb Blood Flow Metab 1996; 16:605-611.
13. Hallenbeck JM, Dutka AJ. Background review and current concept of reperfusion injury. Arch Neurol 1990; 47:1245-1254.
14. Ames A III, Wright RL, Kowada M et al. Cerebral ischemia II. the no-reflow phenomenon. Am J Pathol 1968; 52:437-453.
15. Murakami K, Kondo T, Chan PH. Reperfusion following focal cerebral ischemia alters distribution of neuronal cells with DNA fragmentation in mice. Brain Res 1997; 751:160-164.
16. Leist M, Vollbracht C, K hnle S et al. Caspase-mediated apoptosis in neuronal excitotoxicity triggered by nitric oxide. Mol Med 1997; 3:750-764.
17. Bonfoco E, Krainc D, Ankarcrona M et al. Apoptosis and necrosis: Two distinct events induced, respectively, by mild and intense insults with N-methyl-D-aspartate or nitric oxide/ superoxide in cortical cell cultures. Proc Natl Acad Sci USA 1995; 92:72162-72166.
18. Du C, Hu R, Csernansky CA, Hsu C et al. Very delayed infarction after mild focal cerebral ischemia: a role for apoptosis? J Cereb Blood Flow Met 1996; 16:195-201.
19. Endres M, Namura S, Shimizu-Sasamata M et al. Attenuation of delayed neuronal death after mild focal cerebral ischemia in mice by inhibitors of the caspase family. J Cereb Blood Flow Metab 1998; 18:238-247.
20. Fink K, Zhu J, Namura S et al. Prolonged therapeutic window for ischemic brain damage caused by delayed caspase activation. J Cereb Blood Flow Metab 1998; 18:1071-1076.
21. Dawson VL, Kizushi VM, Huang PL et al. Resistance to neurotoxicity in cortical cultures from neuronal nitric oxide synthase-deficient mice. J Neurosci 1996: 16:2479-2487.
22. Rothman S. Synaptic release of excitatory amino acid neurotransmitter mediates anoxic neuronal death. J Neurosci 1984; 4:1884-1891.
23. Harms H, Wiegand F, Megow D et al. Acute treatment of hypertension increases infarct size in spontaneously hypertensive rats. Neuroreport 2000; 11:355-359.
24. Dirnagl U, Iadecola C, Moskowitz MA. Pathobiology of ischaemic stroke. An integrated view. Trends Neurosci 1999; 22:391-397.
25. Martin RL, Lloyd HG, Cowan AI. The early events of oxygen and glucose deprivation: Setting the scence for neuronal death? Trends Neurosci 1994; 17:251-2578.
26. Choi DW. Ischemia-induced neuronal apoptosis. Curr Opinion Neurobiol 1996; 8:667-672.
27. Chen H, Chopp MN, Vande Linde MQV et al. The effects of post-ischemic hypothermia on the neuronal injury and brain metabolism after forebrain ischemia in the rat. J Neurol Sci 1992; 107:191-198.
28. Furukawa K, Fu W, Li Y et al. The actin-severing protein gelsolin modulates calcium channel and NMDA receptor activities and vulnerability to excitotoxicity in hippocampal neurons. J Neurosci 1997; 17:8178-8186.
29. Endres M, Fink KB, Zhu J et al. Neuroprotective effects of gelsolin during murine stroke. J Clin Invest 1999; 103:347-354.
30. Beckman JS, Koppenol WH. Nitric oxide, superoxide, and peroxynitrite: the good, the bad, and ugly. Am J Physiol 1997; 271:C1424-1437.
31. Iadecola C. Bright and dark sides of nitric oxide in ischemic brain injury. Trends Neurosci 1997; 1997; 20:132-139.
32. Dalkara T, Moskowitz MA. The complex role ot nitric oxide in the pathophysiology of focal cerebral ischemia. Brain Pathol 1994; 4:49-57.
33. Zhang J, Dawson VL, Dawson TM. Nitric oxide activation of poly(ADP-ribose)synthetease in neurotoxicity. Science 1994; 263:687-689.
34. Szabo C, Zingarelli B, O«Connor M et al. DNA strand breakage, activation of poly(ADP-ribose)synthetase, and cellular energy depletion are involved in the cytotoxicity of macrophages and smooth muscle cells exposed to peroxynitrite. Proc Natl Acad Sci 1996; 93:1753-1758.

35. Endres M, Wang Z-Q, Namura S et al. Ischemic brain injury is mediated by the activation of poly(ADP-ribose)polymerase. J Cereb Blod Flow Metab 1997; 17:1143-1151.
36. Eliasson MJL, Sampei K, Mandir AS et al. Poly(ADP-ribose)polymerase gene disruption renders mice resistant to cerebral ischemia. Nature Med 1997; 3:1089-1095.
37. Endres M, Scott GS, Namura S et al. Role of perosynitrite and neuronal nitric oxide synthase in the activation of poly(ADP-ribose)synthetase, against peroxynitrite-induced glial damage and stroke development. Neurosci Lett 1998; 248:41-44.
38. Huang PL, Huang Z, Mashimo H et al. Hypertension in mice lacking the gene for endothelial nitric oxide snythase. Nature 1995; 377:239-242.
39. Huang Z, Huang PL, Ma J et al. Enlarged infarcts in endothelial nitric oxide synthase knockout mice are attenuated by nitro-L-arginine. J Cereb. Blood Flow Metab 1996; 16:981-987.
40. Endres M, Laufs U, Huang Z. Stroke protection by 3-hydroxy-3 methylglutaryl (HMG)-CoA reductase inhibitors mediated by endothelial nitric oxide snythase. Proc Natl Acad Sci USA 1998; 95:8880-8885.
41. Laufs U, Gertz K, Huang PL et al. Atorvastatin upregulates type III nitric oxide synthase in thrombocytes, decreases platelet activation, and protects from stroke in normocholesterolemic mice. Stroke 31:2437-2449.
42. Dugan LL, Choi DW. Excitotoxicity, free radicals, and cell membrane changes. Ann Neurol 1994; 35 Suppl:S17-21.
43. Kristian T, Siesjo BK. Calcium in ischemic cell death. Stroke 1998; 29:705-718.
44. Fujimura M, Morita-Fujimura Y, Murakami K et al. Cytosolic redristribution of cytochrome c after transient focal cerebral ischemia in rats. J Cereb Blood Flow Metab 1998; 18:1239-1247.
45. Siesj BK, Kristian T, Katsura. In: Ginsberg MD, Bogousslavsky J, Eds. Cerebrovascular Disease. Blackwell Science, 1998:1-13.
46. Hossmann KA. Periinfarct depolarizations. Cerebrovasc Brain Metab Rev 1996; 8:195-208.
47. Obrenovitch TP. The ischaemic penumbra: Twenty years on. Cerebrovasc Brain Metab Rev 1995; 7:297-323.
48. Astrup J,Siesjo BK, Symon L. Thresholds in cerebral ischemia The ischemic penumbra. Stroke 1981; 12:723-725.
49. Furlan M, Marchal G, Viader F et al. Spontaneous neurological recovery after stroke and the fate of the ischemic penumbra. Ann Neurol 1996; 40:216-226.
50. Read SJ, Hirano T, Abbott DF et al. Identifying hypoxic tissue after acute ischemic stroke using PET and 18F-fluoromisonidazole. Neurology 1998; 51:1617-1621.
51. Kaufmann AM, Firlik AD, Fukui MB et al. Ischemic core and penumbra in human stroke. Stroke 1999; 30:93-99.
52. Deutsch SI, Huntzinger JA, Rosse RB et al. The role of excitatory amino acids and intraneuronal calcium in the acute intoxicational effects of ethanol. Clin Neuropharmacol 1989; 12:1-13.
53. Olney JW, Labruyere J, Price MT. Pathological changes induced in cerebrocortical neurons by phencyclidine and related drugs. Science 1989; 244:1360-1362.
54. Turski L, Huth A, Sheardown M et al. ZK20075: a phosphonate quinoxalinedione AMPA antagonist for neuroprotection in stroke and trauma. Proc Natl Acad Sci USA 1998; 95:10960-10965.
55. Bond A, O«Neill MJ, Hicks et al. Neuroprotective effects of a systemically active group II metabotropic glutatmate receptor agonist LY354740 in a gerbil model of global ischemia. Neuroreport 1998; 9:1191-1193.
56. Nedergaard M, Hansen AJ. Characterization of cortical depolarizations evoked in focal cerebral ischemia. J Cereb Blood Flow Metab 1993; 13:568-574.
57. Haring HP, Berg EL, Tsurushita N et al. E-selectin apperas in nonischemic tissue during experimental focal ischemia. Stroke 1996; 27:1386-1391.

58. Mies G, Iijima T, Hossmann KA. Correlation between peri-infarct DC shift and ischaemic neuronal damage in rat. NeuroReport 1993; 4:709-711.
59. Iijima T, Mies G, Hossmann KA. Repeated negative DC deflections in rat cortex following middle cerebral artery occlusion are abolished by MK-801: Effect on volume of ischemic injury. J Cereb Blood Flow Metab 1992; 12:727-733.
60. Back T, Hirsch JG, Szabo K et al. Failure to demonstrate peri-infarct depolarizations by repetitive MR diffusion imaging in acute human stroke. Stroke 2000; 31:2901-2906.
61. O Neill LA, Kaltschmidt . NF-kappa B: A crucial transcription factor for glial and neuronal cell function. Trends Neurosci 1997; 20:252-258.
62. Ruscher K, Isaev N, Trendelenburg G et al. Induction of hypoxia inducible factor 1 by oxygen glucose deprivation is attenuated by hypoxic preconditioning in rat cultured neurons. Neurosci Lett 1998; 254:117-120.
63. Iadecola C, Zhan F, Casey R et al. Delayed reduction of ischemic brain injury and neurological deficits in mice lacking the inducible nitric oxide synthase gene. J Neurosci 1997; 17:9157-9164.
64. Lindsberg PJ, Carpen O, Paetau A et al. Endothelial ICAM-1 expression associated with inflammatory cell response in human ischemic stroke. Circulation 1996, 94:939-945.
65. Pirttil TR, Kauppinen RA. Recovery of intracellular pH in cortical brain slices following anoxia studied by nuclear magnetic resonance spectroscopy: Role of lactate removal, extracellular sodium and sodium/hydrogen exchange. Neurosci 1992; 47:155-164.
66. Gong C, Qin Z, Betz AL et al. Cellular localization of tumor necrosis factor alpha following focal cerebral ischemia in mice. Brain Res 1998; 801:1-8.
67. Feuerstein GZ, Wang X, Barone FC. Pathophysiology, diagnosis and management. In: Ginsberg MD, Bogousslavsky J, Eds. Cerebrovascular Disease. Malden: Blackwell Science, 1998:507-531.
68. Chopp M, Li Y, Jiang N et al. Antibodies against adhesion molecules reduce apoptosis after transient middle cerebral artery occlusion in rat brain. J Cereb Blood Flow Metab 1996; 16:578-584.
69. Dereski MO, Chopp M, Knight et al. the heterogenous evolution of focal ischemic neuronal damage in the rat. Acta Neuropathologica Berl 1993; 85:327-333.
70. Baird AE, Benfield A, Schlaug G et al. Enlargment of human cerebral ischemic lesion volumes measured by diffusion-weighted magnetic resonance imaging. Ann Neurol 1997; 41:581-589.
71. Iadecola C, Niwa K, Nogawa S et al. Reduced susceptibility to ischemic brain injury and N-methyl-D-aspartate-mediated neurotoxicity in cyclooxygenase-2-deficient mice. Proc Natl Acad Sci USA 2001; 98:1294-1299.
72. Barone FC Arvin B, White RF et al. Tumor necrosis factor-alpha. a mediator of focal ischemic brain injury. Stroke 1997; 28:1233-1244.
73. Bruce AJ, Boling W, Kindy MS et al. Altered neuronal and microglial responses to excitotoxic and ischemic brain injury in mice lacking TNF receptors. Nat Med 1996; 2:788-794.
74. Kerr JF, Wyllie AH, Currie AR. Apoptosis: a basic biological phenomenen with wide-ranging implication in tissue kinetics. Br J Cancer 1972; 26:239.
75. Linnik MD, Zobrist RH, Hatfield MD. Evidence supporting a role for programmed cell death in focal cerebral ischemia in rats. Stroke 1993; 24:2002-2009.
76. Tominaga T, Kure S, Narisawa K et al. Endonuclease activation following focal ischemic injury in the rat. Brain Res 1993; 608:21-26.
77. MacManus JP, Buchan AM, Hill IE et al. Global ischemia can cause DNA fragmentation indicative of apoptosis in rat brain. Neurosci Lett 1993; 164:89-92.
78. Majno G, Joris I. Apoptosis, oncosis, and necrosis. An overview of cell death. Am J Pathol 1995; 146:3-15.
79. MacManus JP, Hill IE, Huang ZG et al. DNA damage consistent with apoptosis in transient focal ischaemic cortex. NeuroReport 1994; 5:493-496.

80. MacManus JP, Hill IE, Preston E et al. Differences in DNA fragmentation following transient cerebral or decapitation ischemia in rats. J Cereb Blood Flow Metab 1995; 15:728-737.
81. Charriaut-Marlangue C, Margaill I, Plotkine M et al. Early endonuclease activation following reversible focal ischemia in the rat brain. J Cereb Blood Flow Metab 1995; 15:385-388.
82. Charriaut-Marlangue C, Margaill I, Represa A et al. Apoptosis and necrosis after reversible focal ischemia: an in situ DNA fragmentation analysis. J Cereb Blood Flow Met 1996; 16:186-194.
83. Li Y, Chopp M, Jiang N et al. In situ detection of DNA fragmentation after focal cerebral ischemia in mice. Mol Brain Res 1995; 28:164-168.
84. Li Y, Chopp M, Jiang N et al. Induction of DNA fragmentation after 10 to 120 min of focal cerebral ischemia in rats. Stroke 1995; 26:1252-1258.
85. van Lookeren Campagne M, Gill R. Ultrastructural morphological changes are not characteristic of apoptotic cell death following focal cerebral ischaemia in the rat. Neurosci Lett 1996; 213:111-114.
86. Ameisen JC. The origin of programmed cell death. Science1996; 27:1278-1279.
87. Bredesen DE (1995) Neural apoptosis. Ann Neurol 1995; 38:839-851.
88. Hockenberry D. Defining apoptosis. Am J Pathol 1995; 146:16-19.
89. Raff MC, Barres BA, Burne JF et al. Programmed cell death and the control of cell survival: Lessons from the nervous system. Science 1993; 262:695-700.
90. Chinaiyan Am, O Rourke K, Lane Bret al. Interaction of CED-4 with CED-3 and CED-9: A molecular framework for cell death.Science 1997; 275:1122-1126.
91. Ellis HM, Horvitz HR. Genetic control of programmmed cell death in the nematode *C. elegans.* Cell 1986; 44:817-829.
92. Ellis HM, Yuan J, Horvitz HR. Mechanisms and functions of cell death. Annu Rev Cell Biol 1991; 7:663-698.
93. Wu DG, Wallen WD, Nunez G. Interaction and regulation of subcellular localization of CED-4 by CED-9. Science 1997; 275:1126-1129.
94. Yuan J, Shaham S, Ledoux S et al.The *C. elegans* cell death gene ced-3 encodes a protein similar to mammalian interleukin-1 beta-converting enzyme. Cell 1993; 75:641-652.
95. Alnemri ES, Livingston DJ, Nicholson DW et al. Human ICE/CED-3 protease nomenclature Cell 1996; 87:1712.
96. Redd CJ. Cytochrome c: Can t live it Can t live without it. Cell 1997; 91:559-562.
97. Namura S, Zhu J, Fink K et al. Activation and cleavage of caspase-3 in apoptosis induced by experimental cerebral ischemia. J Neurosci 1998; 18:3659-3668.
98. Friedlander RM, Gagliardini V, Hara H et al. Expression of a dominant negative mutant of interleukin-1 beta converting enzyme in transgenic mice prevents neuronal death induced by trophic factor withdrawal and ischemic brain injury. J Exp Med 1997; 185:933-940.
99. Schielke GP, Yang GY, Shivers BD et al. Reduced ischemic brain injury in interleukin-1 beta converting enzyme-deficient mice. J Cereb Blood Flow Metab 1998; 18:180-185.
100. Hara H, Friedlander RM, Gagliardini V et al. Inhibition of interleukin 1beta converting enzyme family proteases reduces ischemic and excitotoxic neuronal damage. Proc Natl Acad Sci USA 1997: 94:2007-2012.
101. Shaw E. Cysteinyl proteinases and their selective inactivation. Adv Enzymol Relat Areas Mol Biol 1990; 63:271-347.
102. Sarin A, Wu ML, Henkart P. Different interleukin-1 beta converting enzyme family protease requirements for the apoptotic death of T lymphocytes triggered by diverse stimuli. J Exp Med 1996; 184:2445-2450.
103. Nicholson WD, Ali A, Thornberry NA et al. Identification and inhibition of the ICE/CED-3 protease necessary for mammalian apoptosis, Nature 1995; 376:37-43.
104. Nicholson WD, Thornberry NA. Caspases: killer proteases. Trends Biochem Sci 1997; 22:299-306.

105. Chen J, Nagayama T, Jin K et al. Induction of caspase-3-like protease may mediate delayed neuronal death in the hippocampus after transient cerebral ischemia. J Neurosci 1998; 18:4914-4928.
106. Cheng Y, Deskmukh M, D-Costa A et al. Caspase inhibitor affords neuroprotection with delayed administration in a rat model of neonatal hypoxic-ischemic brain injury. J Clin Invest 1998; 101:1992-1999.
107. Ma J, Endres M, Moskowitz MA. Synergistic effects of caspase inhibitors and MK-801 in brain injury after transient focal cerebral ischaemia in mice. Br J Pharmacol 1998; 124:756-762.
108. Minamisawa H, Mellergard P, Smith ML et al. Preservation of brain temperature during ischemia in rats. J Cereb Blood Flow Metab 1990; 10:365-374.
109. Busto R, Globus MYT, Dietrich D et al. Small differences in intraischemic brain temperature critically determine the extent of ischemic neuronal injury. J Cereb Blood Flow Metab 1987; 7:729-738.
110. Crumrine RC, LaManna JC. Regional cerebral metabolites, blood flow, plasma volume, and mean transit time in total cerebral ischemia in the rat. J Cereb Blood Flow Metab 1991; 11:272-282.

NEUROPROTECTIVE STRATEGIES IN ALZHEIMER'S DISEASE

Christian Behl*

INTRODUCTION

Alzheimer s disease (AD) is a deadly progressive neurodegenerative disorder. Various aspects of the biochemistry of the AD-associated amyloid β protein (Aβ) are well-described and understood. The deposition of Aβ in the AD brain is one crucial hallmark of AD pathology. Consequently, the processes of the generation of Aβ are *the* main targets for novel avenues of prevention and treatment. Besides the increasing amount of genetic information on familial forms of AD including those types caused by mutations in the AD genes APP, PS1 and PS2, various genetic and non-genetic risk factors have been described. Worldwide efforts to block Aβ generation or to block its deposition by immunization are of great interest but are still at a more or less experimental stage. In addition, it may be argued that the sporadic and strictly age-related forms of AD develop because the overall homeostasis of neuroprotection is disturbed rendering the nerve cells more vulnerable for exogenous insults including Aβ. Therefore, it is also wise to study neuroprotective factors and to decipher how such factors, including neurotrophins and the female sex hormone estrogen, transduce their signal leading to nerve cell protection. The identification of potential protective genes that are regulated by well known neuroprotective signals may elucidate novel targets for preventive approaches. A general and subtle neuroprotection may stabilize nerve cell survival and may leave the neurons more resistant to AD-associated pathogenetic insults.

Alzheimer s disease (AD) is the major cause of dementia in the elderly in developed countries. AD accounts for more than 70% of all late-onset cases of

*Max Planck Institute of Psychiatry, Kraepelinstrasse 2-10, 80804 Munich, Germany.

dementia. In the USA alone AD claims more than 100,000 lives per year. Given the increased life expectancy in the industrialized nations a further increase in the total number of people suffering from AD is expected. AD is a deadly disease and there is currently no cure. AD can be a familial disease that is inherited in an autosomal dominant fashion but the majority of AD cases are strictly age-related. The major clinical characteristic of AD is the slowly progressing global decline in cognitive function. Furthermore, AD patients suffer from general deterioration, they lose their memory and, therefore, their personality. Even almost 100 years after the German psychiatrist Alois Alzheimer first described his observation of a neurodegenerative disease of the brain, which was later called Alzheimer s disease ,[1] the diagnosis definite Alzheimer s disease still depends on a combination of evaluations, including a general neuropsychiatric anamnesis and neuropsychological tests. Only post mortally can the diagnosis AD be made on the basis of histopathologic characteristics, proteins and markers that are found in the brain of the diseased patient. The most prominent AD proteins are the amlyoid β protein (Aβ) and the tau protein. Basic research mainly focuses on the role of these two proteins in AD pathogenesis. Insoluble Aβ deposits are found extracellularly in the senile plaques of AD brain and massively phosphorylated tau-proteins inside the cells build up neurofibrillary tangles . The biochemistry of Aβ is known to be centrally affected in the various familial forms of AD and mounting evidence points towards Aβ as the target protein of AD.

Understanding the basic mechanisms of AD is, of course, the prerequisite for a possible therapy of this devastating mental disorder, but in addition to biochemical and molecular approaches it is necessary to develop methods for early diagnosis of the disease. In the future novel and more sensitive brain imaging techniques may help to better determine AD in the living patient early in the course of the disease. Indeed, the great dilemma of why there is no early diagnosis of AD is still due to the lack of a valid and reliable disease-marker. Considering that the majority of AD cases have a preclinical phase of up to decades and that the disease is slowly progressing before first obvious clinical disease symptoms, a marker that would indicate the start of AD (beginning of AD) would be of invaluable help for a possible prevention and later also for therapeutic approaches.

WHAT IS THE CAUSE OF AD?

The investigation of post mortem brain tissue of AD patients has detected the major histopathological hallmarks of this neurodegenerative disorder. Those are:

1. synaptic alterations,
2. nerve cell loss primarily in cerebral cortex, hippocampus and amygdala,
3. deposition of extracellular amorphous amyloid β protein (Aβ),
4. intracellular precipitation of hyperphosphorylated tau-protein.

The extracellular deposits of Aβ-aggregates build up the senile plaques . Hyperphosphorylated tau-protein forms the neurofibrillary tangles in the nerve cells (for review see refs. 2-4). Despite worldwide efforts for many years the exact pathogenesis of AD is still obscure and consequently current therapeutic approaches are hypothesis-driven rather than causal. The main hypotheses of AD pathogenesis are the following:

The acetylcholine-deficiency-hypothesis is based on the fact that in AD a massive decrease of the neurotransmitter acetylcholine (ACh) is found.[5] ACh is well acknowledged as the main mediator of cognitive functions and is, therefore, the central neurotransmitter necessary for cognition and memory. The demonstration of substantial abnormalities in the cholinergic system in the brain of AD patients initiated the discussion about ACh replacement strategy. Similar to the approach that is still used to treat Parkinson s disease, which can be described as a dopamine-deficiency disease, where L-dopamine replacement is used, the aim in AD is to stabilize and boost cholinergic function. This can be reached at various levels. Some approaches aim at the prevention of the degradation of this neurotransmitter in the synaptic cleft maintaining therefore constant ACh-levels. ACh is physiologically degraded by the enzyme ACh-esterase. Developed AD drugs block the activity of ACh-esterases and are therefore called ACh-esterase-inhibitors. Indeed, ACh-esterase blockers are frequently used and various formulations from different pharmaceutical companies are available (for review see ref. 6).

The arthritis-of-the-brain-hypothesis of AD concentrates on the investigation of the impact of inflammation on the AD disease process.[7] Again, the main support of this hypothesis is derived basically from histopathology. AD brain tissue clearly shows a massive accumulation of inflammatory components, such as complement and activated microglia (for review see ref. 8). Treatment of AD patients with anti-inflammatory drugs has been proposed and various clinical trials using such drugs have been already performed, others are still in progress. Two initial preliminary clinical studies were rather optimistic in suggesting that anti-inflammatory drugs can decrease the rate of cognitive decline in AD. Follow-up studies further suggested that non-steroidal anti-inflammatory drugs (NSAIDs) cause a delay in the onset or slow down the progression of AD. The most prominent activity of NSAIDs is the inhibition of cyclooxygenases (COX) and lipooxygenases, which may, ultimately, lead to the inhibition of prostaglandin synthesis and also to a block of ROS formation. The enzyme COX-2 is induced in response to the exposure of neurons to toxic concentrations of glutamate, suggesting a possible role for COX inhibitors in neuroprotection by counteracting the events following a glutamate insult. As a consequence of the growing body of evidence supporting a major role of inflammation during AD pathogenesis, additional anti-inflammatory drugs are tested or are planned to be tested in clinical AD trials in the future. Potential novel therapies will include several drugs used in the treatment of inflammatory diseases. These may also include glucocorticoids that are in use for the treatment of rheumatic diseases as well as anti-malaria drugs (for review see refs. 6, 9, 10).

The energy-depletion-hypothesis of AD underlines the central importance of a decreased glucose metabolism and, therefore, a decreased energy supply in the brain for AD pathogenesis. Indeed, normal aging of mammalian brain is associated with various metabolic changes, including variations in the neuronal insulin receptor, the

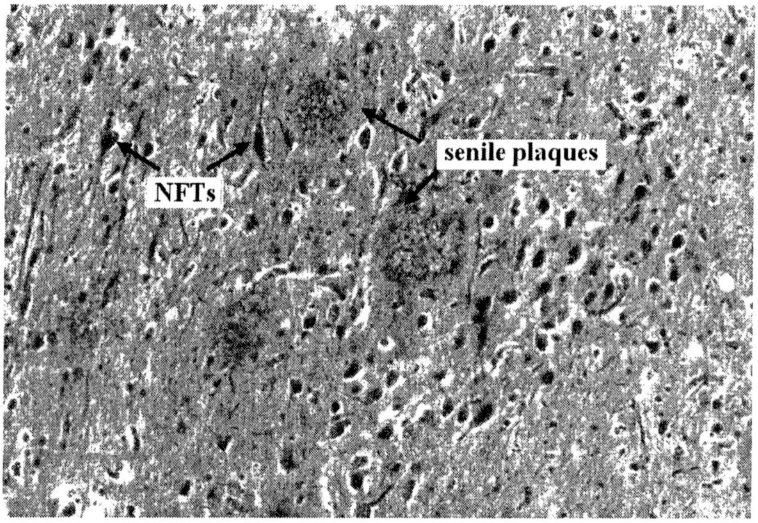

Figure 1. Senile plaques and neurofibrillary tangles as hallmarks of AD pathology. Extracellular deposition of Aβ build up the senile plaques. Hyperphosphorylated tau-protein comprise intracellular neurofibrillary tangles (NFTs). Senile plaques and NFTs are the overt pathological hallmarks in AD brain tissue.

desensitization of the neuronal insulin receptor by circulating cortisol and receptor dysfunction subsequent to changes in membrane structure and function. These changes may directly account for the aberrations in glucose/energy metabolism (for review see ref. 11). Furthermore, this hypothesis is driven by the fact that the majority of AD cases are late onset forms with unknown etiology.

Like the free radical theory of aging, the oxidative-stress-hypothesis of AD has been proposed (for review see refs. 12,13). Aging, in general, is associated with degenerative processes, including the degeneration of cells and tissues which can result in diseases, such as cancer, cardiovascular failure and neurodegenerative disorders. The free radical theory of aging, as previously mentioned, suggests that oxidative damage due to the accumulation of oxidative stress is causative for the degeneration of cells and tissues.[14]

The histopathological and the experimental evidence that support the oxidative stress hypothesis of AD is increasing (for review see ref. 15) and includes the following key findings in neuronal tissues:

- increased levels of highly oxidized proteins, DNA, and lipids,
- high amount of ions that can drive free radical generation,
- occurrence of advanced glycation endproducts,
- Aβ induces oxidative stress.

Most interesting were the findings that aggregates of Aβ can indirectly and directly induce oxidative stress in vitro(for review see refs. 13, 17) (Fig. 2)[16] Based on all this experimental evidence the oxidative stress hypothesis has been put to the test in various clinical trials. Indeed, in 1997 some beneficial effects of the lipophilic antioxidant α-tocopherol (vitamin E) for the treatment of moderately severe affected AD patients have been demonstrated.[18]

The amyloid-cascade-hypothesis summarizes the central role of the processing and the regulation of the amyloid β precursor protein and is still the most popular hypothesis of AD pathogenesis. The deposition of Aβ, the peptide which varies in size from 39 to 43 amino acids, in the senile plaques, has been already recognized by Alois Alzheimer and is believed by many neurobiologists to be the crucial step in AD pathogenesis (for review see refs. 3, 19). Processing of APP can lead to the generation of Aβ aggregates that have been shown to have neurotoxic activities under certain conditions in vitro and in vivo (for review see refs. 20-21). Besides direct toxic effects, such as the induction of oxidative stress in nerve cells, Aβ can also lead to secondary toxic effects via the attraction of inflammatory mediators and activation of inflammatory cells.

The tau-hypothesis of AD is fueled by the fact that intracellular neurofibrillary tangles which consist of hyperphosphorylated tau, a microtubule-associated protein of a molecular between 50 to 64 kilodalton, are a major hallmark of AD pathology (for review see ref. 22). Abundant tau-positive neurofibrillary lesions are found in AD brain tissue but the filamentous tau pathology is also central to a number of other dementing disorders, including Pick s disease, progressive supranuclear palsy corticobasal degeneration and familial frontotemporal dementia and Parkinsonism linked to chromosome 17 (FTDP-17). The discovery of mutations in the tau gene has firmly established the relevance of tau pathology for the neurodegenerative process (for review see ref. 23). Although for quite some time the researchers in favor of the tau-hypothesis (tau-oists) and those more in favor of the amyloid-cascade-hypothesis (A β-aptists) strictly followed their line of research, more and more links that combine Aβ and tau pathology have been defined (for review see ref. 24). Nevertheless, the amyloid-hypothesis of AD is strongly supported by genetic and other experimental evidence and still the major focus of AD basic research.

AD GENETICS AND BIOCHEMISTRY

A major breakthrough in AD research was the identification of gene mutations that are directly associated with the disease. For a decade it is clear that AD can be caused by autosomal dominant mutations in three separate genes:

- APP,
- presenilin 1 (PS1),
- presenilin 2 (PS2).

Inherited familial forms of AD (FAD) account for approximately 4-8% of all AD cases. Although these familial cases are the minority and the majority (92-96%)

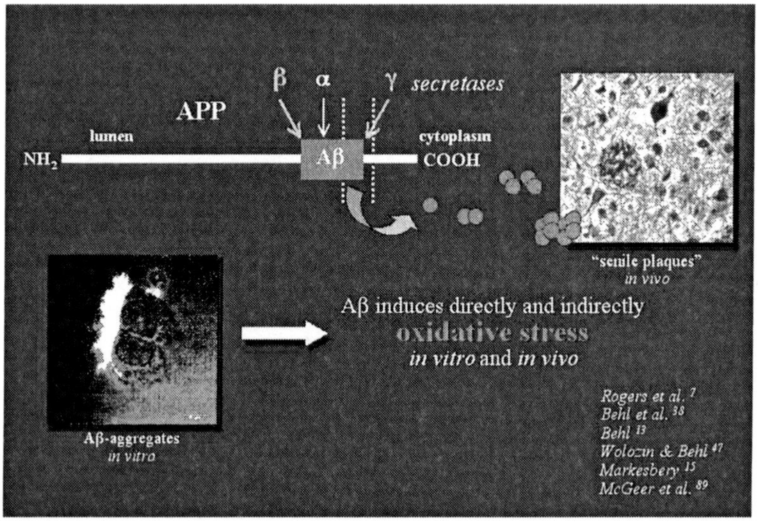

Figure 2. APP processing and oxidative stress in AD pathology. The processing of APP can be performed by β- and γ-secretases, which leads to the formation of Aβ capable of forming high molecular protein aggregates. In addition, the processing of APP by α-secretases is possible. α-secretases cut the Aβ sequences and prevent the formation of Aβ and its aggregates.

are sporadic forms without a clear genetic cause, the pathology of both types (familial and sporadic) is identical (for review see ref. 25).

Mutations in PS1 are the most frequent, APP mutations are the genetic basis for a small but significant percentage of AD cases and PS2 mutations are very rare. The most important difference between familial and sporadic cases is the age of onset of the disease. While almost all FAD cases start before the age of 60, sporadic AD is age-dependent. About 3-4% of the people age 65-74, up to 20% of the population aged 75-84 and up to 48% of the elderly population at an age over 85 are affected by AD.[26]

Since AD patients carrying or lacking gene mutations in APP, PS1 or PS2 develop a similar clinical (e.g., cognitive impairments) and histopathological (Aβ-, tau-pathology) pattern, these genes appear to be the basis for common pathological mechanisms. APP, PS1 and PS2 are the most studied proteins in the search for the cause of AD. Nevertheless, some authors stress that despite the common biochemistry, AD may be caused by a variety of different initial pathological processes and only late in the disease process funnel into the APP/PS1/PS2 pathway. In addition, a collection of risk factors for the disease have been described that increase the incidence of AD and which may also be studied properly with respect to their interaction with the downstream disease process. Before taking a closer look into such risk factors a short overview of the AD genes is given.

APP

The gene for APP is located on chromosome 21 which is involved in Down's syndrome (trisomy 21). Down's syndrome patients suffer from dementia of the AD-type very early in their life. This intriguing link between AD and two distinct neurodegenerative diseases with similar clinical characteristics puts the focus on APP. APP is a transmembrane protein containing a large extracellular region, a single membrane spanning domain and a small cytoplasmic tail (see Fig. 2). APP is expressed in neurons and in glial cells but the different splicing forms of APP show a more distinct pattern of expression. APP-695 is the most common form of the three splice variants of APP (APP-695, APP-751, APP-770) in neurons (for review see refs. 3, 19). Although the exact functions of APP in neurons is not well understood, its predominant localization in synapses suggest a role in basic synaptic signaling. Moreover, APP has been found to exert potent neurotrophic and neuroprotective activities partly by poorly understood mechanisms (see refs. 27-28).

APP is the precursor molecule for Aβ, the protein that is deposited as high molecular protein aggregates in senile plaques of AD brain. The biochemical pathway of the processing of Aβ is perhaps one of the most intensively studied molecular mechanisms in neurobiology. Aβ is released from APP through proteolysis. Responsible for the proteolytic activities are three types of secretases, α-, β-, γ-secretases (see Fig. 2). While approximately 90% of Aβ in brain tissue is Aβ 40 (40 amino acid long Aβ), about 10% make up Aβ 42/43. Aβ 42/43 is much more fibrillogenic and therefore readily induces Aβ aggregate formation (for review see 19).

Within the transmembrane region of APP there are cleavage sites for the three secretases. The APP processing that is considered to be normal is carried out by the α-secretase which cuts APP in the middle of the Aβ-sequence. Therefore α-processing precludes the formation of Aβ. Processing carried out by the β- and the γ-secretase produces an excess of Aβ (see Fig. 2). APP mutations increase this type of pathological processing, which leads to an overproduction of Aβ, the prerequisite for aggregate formation and deposition in the brain. Various types of APP mutations have been found so far (Swedish-mutation , London-mutation etc.) which all lead to a manifold induction of Aβ 42 production (for review see refs. 19, 30).

Presenilins

PS1 and PS2 are encoded by different genes on chromosome 14 and 1, respectively. Like APP, presenilins are predominantly expressed in neurons. Mutations in PS1, which make up the majority of AD mutations, enhance the generation of Aβ 42, suggesting that PS1 affects the γ-processing of APP. Indeed, it is believed by some investigators that PS1 itself could be the γ-secretase. Presenilin protein has multiple membrane-spanning domains and two highly conserved aspartate residues may account for an aspartyl protease activity. Additional biochemical evidence such as the direct binding of γ-secretase inhibitors to PS1 and PS2 supports this working hypothesis that PS1 could be the γ-secretase (for review see ref. 31).

In addition to γ-secretase the β-secretase is central to the biochemical processing of APP leading to the release of Aβ. After more than ten years of intensive research, the same protein has been identified as β-secretase by four independent laboratories. The β-secretase was named β-amyloid-cleaving-enzyme, BACE. BACE has just recently been cloned and the inhibition of BACE by pharmaceutical compounds is believed to be one main route towards the development of an AD drug. Already a second BACE, BACE 2, which is highly homologous to BACE has been cloned. The BACEs might belong to a new family of aspartyl proteases because, in contrast to other known aspartyl proteases, BACE is membrane-bound and is also insensitive to the protease inhibitor pepstatin. The BACE gene has been mapped to chromosome 11 and other than the APP and the PS genes has not yet been associated genetically with AD. Interestingly, the gene encoding BACE 2, similar to the APP gene, maps to the Down's syndrome region of chromosome 21 (for review see ref. 32).

Excellent reviews on the biochemistry of APP processing are available in the literature (for review see refs. 3, 19). The discovery of BACE was, indeed, a major step forward. Currently, the detailed studies on the structure and the regulation of these enzymes, which are crucial for Aβ formation, are going on. And it is the hope that this enzyme is the long awaited specific target for the prevention and treatment of AD. Inhibitors of BACE would prevent the formation of Aβ aggregates and, therefore the build-up of senile plaques.

Recent evidence demonstrated that members of the ADAM family (*A Disintegrin And Metalloprotease*) act as α-secretases. Indeed, for ADAM 10 the basal and the protein kinase C-stimulated α-secretase activity and other α-secretase characteristics have been described.[33] The enhancement and stimulation of α-secretase activity and therefore an increased generation and secretion of soluble APP (sAPPα) might be beneficial for the treatment and prevention of AD for several reasons. First, α-cleavage precludes the generation of Aβ. Second, sAPPα has been demonstrated to have neurotrophic and direct neuroprotective effects (e.g., Schubert and Behl, ref 27). Various enhancers of α-processing of APP have been described including also the female sex hormone estrogen.[34,35]

NEUROTOXICOLOGY OF Aβ

On the basis of genetic evidence and the biochemical analysis of APP processing, the overproduction and the deposition of Aβ are believed to be *the* central step in AD pathogenesis (amyloid cascade hypothesis). Important pathophysiological support for this view comes also from cell biology. In 1989 Bruce Yankner and colleagues found that Aβ may have toxic effects on neurons. This finding launched many investigations and, indeed, the direct toxicity of Aβ has been shown in a variety of cellular in vitro paradigms and also in certain in vivo models (for review see ref. 20). Intense investigations revealed different kinds of mechanisms that describe how Aβ induces nerve cell death. First of all, it is well acknowledged that Aβ needs to be aggregated to exert toxic activities. Aβ aggregates directly interact with the cell membrane. It was proposed that this interaction is not random, but rather mediated by the receptor for advanced glycation endproducts (RAGE).[36] This may apply for certain cell types but is definitively not the sine qua non prerequisite for the appearance of Aβ toxicity.[37] Nevertheless, it is possible that the precursor forms of the high molecular aggregates of Aβ, the protofibrils, can interact with RAGE. Mark Mattson s group nicely showed that Aβ renders nerve cells more vulnerable to secondary insults, such as glutamate toxicity, and that Aβ disturbs the intracellular Ca^{2+}-homeostasis. In addition, it became more and more clear that Aβ also may interact also with various neurotransmitter receptors and also could challenge neuronal cells at the level of neurotransmission (for review see ref. 17).

We demonstrated in 1994 that Aβ aggregates interact with the nerve cell membrane and induce a sequence of events leading to the accumulation of reactive oxygen species and peroxides inside the cells. Aβ induces oxidative stress in nerve cells which leads to the oxidation of the non-saturated carbohydrate side chains of membrane lipids and, ultimately, to cell lysis.[38] Various additional studies by different labs have supported the view that oxidative stress may be central to the Aβ-driven neurodegeneration and together with the histopathological evidence showing that oxidation end products are indeed present in the AD brain tissue, the oxidative stress hypothesis of AD is well acknowledged. Consequently, clinical trials employing free radical scavengers such as vitamin E (α-tocopherol) have been performed.[18] Various investigators have established an Aβ toxicity also in vivo (for review see ref. 20), which appears to be highly dependent on the particular model conditions used. Yankner s group showed convincingly that injected Aβ aggregates can induce nerve cell death in old primates,[39] which shows that it may not be simply the presence of Aβ in nervous tissue that induces toxicity. Rather certain preconditions and an increased neuronal vulnerability is necessary, as occurs in the aged brain. Consistent with that are findings showing that the preincubation of cultured neurons with glucocorticoids increase the vulnerability of the overall Aβ toxicity.[40]

In addition to this directly induced oxidative stress, Aβ can also indirectly generate an oxidative environment via the induction of a local immune response. Indeed, histopathology clearly demonstrates cellular and soluble mediators of inflammation in situ and the molecular pathways have been studied in vitro and ultimately led to the definition of the arthritis of the brain hypothesis of AD (see above).

In summary, there is good evidence that Aβ can develop neurotoxic effects under certain conditions. Independent of the particular cell death pathways that are induced by Aβ aggregates, oxidative stress appears to be a major player at various stages. There is still some debate whether Aβ may induce apoptosis or necrosis in situ but it is very likely that both pathways overlap and may occur side by side (for review see ref. 41). Despite this overwhelming set of data on the Aβ neurotoxicity, caution is proposed by the neuropathologists since the extent of cognitive decline in AD does not really correlate with the number of senile plaques. Here, the tau pathology shows a much more consistent correlation (for review see refs. 42-43).

AD RISK FACTORS

The cognitive defects that are the major clinical manifestations in AD are a consequence of dysfunctional synaptic activity and, ultimately, nerve cell loss. Since most AD cases are age-dependent, it is of great importance to study those factors and preconditions that may decrease the fitness and increase the vulnerability of nerve cells with regard to exogenous challenges. When such vulnerability or risk factors are present, the incidence of AD is increased. But in contrast to the above-mentioned mutations that dominantly lead to early onset AD, risk factors for AD increase the basic likelihood of getting the disease. In many cases risk factors influence the central process of AD pathogenesis including APP and tau biochemistry or render nerve cells more vulnerable to AD-associated neurodegeneration. AD risk factors are numerous but two groups can be defined, genetic risk factors and non-genetic risk factors.

Genetic factors that are AD risk factors are in general much more difficult to relate to AD than the APP or PS1 mutations since they are not genetically penetrant. Many more genetic risk factors will be identified in the future, but two such factors are the apolipoprotein E (ApoE) and interleukin-1 (IL-1).

ApoE is a serum protein that is a major player in the metabolism, storage and transport of cholesterol. ApoE is polymorphic and is encoded by three different alleles, named ApoE2, -3, -4. ApoE3 is the most frequently expressed. The exciting finding was that the prevalence of individuals having either one or two copies of the ApoE4 alleles in an AD population is significantly greater than in a control population.[44] Interestingly, there is a dose-dependant relationship between Aβ and ApoE4 and a link has been found between ApoE and oxidative stress. IL-1 on the other hand is a well known interleukin with a central role in inflammation. The establishment of IL-1 as AD risk factor is not as well established, but an association between AD and common population polymorphism in IL-1A and IL-1B genes has been shown (for review see ref. 25). Additional genetic susceptibility factors have been reported including also α2-macroglobulin or angiotensin I converting enzyme genes. Indeed, except for a rather small percentage of AD cases with a clear, dominant inheritance pattern, the genetic component of the majority of cases consists of a network of complex interactions of genetic susceptibility factors and very likely non-genetic and environmental factors (for review see ref. 45).

Non-genetic risk factors for AD represent a variety of preconditions including trauma and increased age-associated cholesterol-levels (hypercholesteriemia) and more general factors such as low education. Here, the recent discussion on the role of cholesterol in AD should be presented briefly. AD pathogenesis has many links to cholesterol metabolism. First, epidemiology shows that patients with elevated cholesterol levels have an increased risk for AD. Second, the above mentioned AD risk factor ApoE4 is involved in cholesterol metabolism and ApoE4 polymorphism results in elevated cholesterol levels. Third, the putative risk factor α2-macroglobulin binds to the same receptor as APOE4, the lipoprotein receptor-related protein, and plays a central role in cellular cholesterol uptake. Fourth, cholesterol, the basic constituent of all cellular membranes, has been shown to directly affect Aβ generation: deprivation of cells in cholesterol suppresses generation.[46] This cholesterol-hypothesis was investigated by a recent cross-sectional analysis of various populations of patients that received statins, inhibitors of 3-hydroxy-3-methylglutaryl coenzyme A reductase, the central enzyme of cholesterol biosynthesis. Indeed, a lower prevalence of diagnosed probable AD was found in patients taking two different statins (lovastatin and pravastatin).[47] While this is an exciting finding that supports the view that high cholesterol is indeed a risk factor for AD, no causative mechanisms so far have been proposed that demonstarte a reduction in cholesterol levels interferes with the central pathways of AD pathogenesis. Nevertheless, experimental in vitro and in vivo studies attempted to fix such a connection to the Aβ processing. Indeed, a study using mainly the statin, simvastatin, demonstrated that (1) simvastatin reduces intracellular and secretory neuronal Aβ 42 and Aβ 40 levels in vitro and (2) that administration of simvastatin to guinea pigs strongly reduces cerebral Aβ levels.[48] Another very recent in vitro study showed that low cholesterol stimulates the non-amyloidogenic pathway by its effect on the α-secretase ADAM10.[49]

MOUSE MODELS OF AD

Before the advent of knock-out and transgenic technologies, various types of animal models were developed that aimed to mimic and reproduce specific effects of AD pathology. Those models employed Aβ injections or depletion of certain brain regions in ACh through lesions. These models were rather limited and in most cases represented acute insults. They did not mirror the slowly progressive degenerative processes of AD. Today, human familial gene mutations can be inserted into the germline of mice or target gene expression can be specifically prevented leading to transgenic and knockout animals (in most cases mice), respectively. Although genetically modified animal models also suffer from certain limitations and ignoring that, perhaps, such a complex disease as AD cannot be reproduced in animals to the full extent, these mouse models are very valuable tools to study single steps of the disease process and novel therapeutic approaches.

Since various excellent reviews exist that elegantly cover the topic of AD k.o. and transgenic mice, here only a short summary of the currently available mouse models is given in Table 1 (for review: Emilien et al[50]). Since APP, presenilins and the tau protein appear to be the major AD players, these proteins are the prime targets of genetic engineering in vivo. APP was the first gene that was overexpressed in vivo, especially in the familial mutant APP forms using different types of promoters to boost the expression. Although certain characteristics of AD pathology were found, including changes in synapses (loss of synaptophysin staining) and modest loss of neurons, none of the mouse models showed neurodegeneration to an extent seen in the human AD tissue. While plaques could be created in APP transgenic mice, it was not possible to establish neurofibrillary tangles. These problems point to species differences and even mouse strain differences in the development of certain AD characteristics need to be considered. Another view is that rodents simply do not live long enough to develop these AD hallmarks. This supports the view that AD is indeed a multifactorial disorder that only occurs after a combined accumulation of various genetic and nongenetic risk factors. Besides neuropathology, in mouse models it is difficult to target behaviorial aspects of the disease such as changes in cognition and memory. In consequence of the fact that presenilins have been discovered as AD genes only some years ago and that the large number of PS1 mutations that are involved in familial AD, the PS1-transgenes have not had quite that impact on AD research as APP transgenic models had so far. Again, the main hallmark of AD, a massive loss of neurons, could not be found in PS1-transgenic mice. Even the PS1-APP double mutant does not develop significant neurodegeneration in the typical AD target regions (e.g., hippocampus, prefrontal cortex).

Various tau-mice have been created. Recently , mainly transgenic mice that overexpress a tau transgene derived from a human PAC that contains the coding sequence, intronic regions, and regulatory regions of the human gene were generated. It was found that the various isoforms of human tau are represented in the transgenic mouse brain at the mRNA and protein levels and human tau is distributed in neurites and at synapses, but is absent from cell bodies. A comparison between the genomic tau mice and mice that overexpress a tau cDNA transgene shows that overall, the distribution of tau is similar in the two lines, but human tau is located in the somatodendritic compartment of many neurons in the cDNA mice . Interestingly, tau-immunoreactive axonal swellings were found in the spinal cords of the cDNA mice , which correlated with a hind-limb abnormality. On the other hand, neuropathology was essentially normal in the genomic mice up to 8 months of age.[51]

CLINICAL AD THERAPY

The current therapy of AD that is used in patients on a daily basis is exclusively hypothesis-driven and not causal. Indeed, there are only a limited number of drugs currently available. All these compounds are ACh-esterase inhibitors, which have only a modest effect on the progression of AD. Although stabilization of the neurotransmitter ACh is an obvious approach and may help to some extent to keep up the ACh-driven neurotransmission necessary for complex processes such as cognition and memory, these drugs are not the long awaited breakthrough (for review see ref. 6). The drop in the ACh levels in AD is a consequence of disease processes that starts upstream in the neuron. It is believed by many researchers and especially by the pharmaceutical companies that targeting the main AD hallmarks, such as Aβ formation, Aβ deposition and, perhaps its clearance from the tissue will lead to the development of *the* AD drug of the future. In addition the prevention of NFTs and most importantly the stabilization of the neuronal homeostasis and survival are prime targets. Various drug developments are going on at the experimental levels and some of these therapeutic approaches will be pointed out shortly below.

EXPERIMENTAL AD THERAPIES

Vaccination

The idea of this recent approach is that the immunization with Aβ induces an immune response. This immune response tackles Aβ generated during the AD pathogenesis and facilitates clearance of Aβ. And indeed, Schenk et al first reported that Aβ deposition as plaques was successfully prevented in APP transgenic mice carrying a mutant APP gene after immunization of the animal with Aβ 42.[53] Interestingly, the immunization was successful when applied before Aβ accumulation as well as after Aβ deposition. Various aspects of this initial work were reproduced by two independent investigations. In addition, they showed that age-related behavioral deficits which are present in non-immunized animals were significantly attenuated by immunization.[54,55] These latter reports were successful not only at the neuropathological level (effect on Aβ deposition) but most importantly also at the behavioral level. First pilot studies in humans are going on and these future results can be awaited with great excitement.

Secretase Inhibitors

Based on the detailed knowledge of the APP processing including the cloning of BACE and the data that are already accumulated concerning γ-secretase activity one major therapeutic approach is the inhibition of the pathological processing of APP leading to Aβ generation. Although various experimental inhibitors are tested in in vitro paradigms, some drawbacks have to be considered, such as the accessibility of the enzymes and as with all inhibitors the specificity to avoid the interruption of physiological processes. In any case, in the future such inhibitors also will be tested in transgenic mice with an increased Aβ load in the brain tissue.

Anti-Inflammatory Drugs

As mentioned earlier, this concept is based on the fact that there is a strong inflammatory reaction in AD brain and that patients on chronic therapy with NSAID showed significantly lowered AD risk. Aspirin, ibuprofen and naproxen may be candidates to be tested. Employing ibuprofen in APP transgenic TG2576 mice, a significant reduction in the concentration of IL-1b in the brain, a reduced gliosis and a significant reduction in the overall Aβ deposition was found.[56] Due to possible side-effects in long-term treatment of these NSAID inhibitors of the enzyme cyclooxygenase-2 (COX-2) that are used for the treatment of arthritis and are known to have fewer side effects may be a better choice in the future.

Anti-Oxidants

Evidence is constantly accumulating that the brains of patients suffering from AD are challenged by severe constant oxidative stress. As pointed out earlier, this increased oxidative burden is a result of Aβ deposition, inflammation, a disturbed intracellular calcium homeostasis, or mitochondrial dysfunction. Among the drugs and formulations used so far for AD treatment are vitamin E (α-tocopherol), idebenone, estrogen, and extracts of *Gingko biloba*. Other antioxidatively acting new structures that are under clinical development include also so-called spin trapping agents such as a-phenyl-t-butylnitrone.

The most prominent of all antioxidants is the lipophilic free radical scavenger vitamin E. Vitamin E prevents the oxidative damage induced by Aβ and delays memory deficits in various animal models. A placebo-controlled, clinical trial of vitamin E in patients with moderately advanced Alzheimer disease was conducted by the Alzheimer s Disease Cooperative Study. Subjects in the vitamin E group were treated with 2000 IU vitamin E per day. The results indicated that vitamin E may slow functional deterioration leading to nursing home placement. New clinical trials are planned that will examine whether vitamin E can delay or prevent a clinical diagnosis of Alzheimer's disease in elderly persons with mild cognitive impairment that is frequently discussed as a first stage of AD (for review see ref. 57). Pharmacologists may ask whether vitamin E, which is a rather large molecule can in fact enter the brain via the blood-brain-barrier and may call for smaller antioxidant compounds with improved pharmacodynamic properties. Indeed, developments in that direction are under way and the authors lab just introduced a small phenolic compound with a high antioxidant neuroprotective potential.[58] 2,4,6-Trimethylphenol is currently being tested in various acute neurodegenerative conditions that are associated with oxidative stress including animal models of stroke and of Parkinson s disease. Whether such a lead compound may also act as neuroprotectant in AD will be investigated in the future.

The 2-year efficacy and safety of idebenone were studied in a prospective, randomized, double-blind multicentre study in patients with mild to moderate AD. During the placebo-controlled period (the first year of treatment), idebenone displayed statistically significant dose-dependent improvement with respect to various parameters. There was no evidence for a loss of efficacy during the second year of treatment, as a further improvement of most efficacy variables was found in the second year in comparison to the results at the 12 months visit. A dose effect relationship was maintained throughout the second year of treatment. This indicates that idebenone exerts its beneficial therapeutic effects on the course of the disease by slowing down its progression. Safety and tolerability of idebenone were good and similar to placebo.[59]

Based on many experimental data that present a neuroprotective activity for extracts of the plant *Ginkgo biloba* (e.g., Guidetti et al[60]), such plant formulations were also used to treat AD. While additional trials are ongoing some memory enhancing and otherwise beneficial effects have been reported.[61,62]

Neurotrophic Factors

This short overview of potential but up to now only experimental AD therapies should mention the use of neurotrophic factors. The well-studied nerve growth factor (NGF) is the prototype of a neurotrophin and is intimately related to the maintenance of the cholinergic function in the brain. Cholinergic neurons of the forebrain are the only cells in the adult brain that express high levels of the low-affinity p75 receptors for NGF. Of course, NGF may have various trophic functions in the brain which may lead to untoward side effects. Therefore, not NGF but rather certain NGF-mimetic compounds with selected activities and drugs that increase the expression and activity of certain neurotrophins is a strategy that is currently followed.

AD PREVENTION

On the basis of the experimental and clinical data, various preventive compounds can be considered in addition to antioxidants and anti-inflammatory drugs. Bearing in mind that the metabolism of cholesterol is of importance for AD pathogenesis, statins may act as preventive compounds. The prevention of an age-associated rise in cholesterol levels can be reached by various statin formulations. Also nutritional factors need to be investigated concerning their potential neuroprotective activities. Indeed, several studies have shown the existence of a correlation between cognitive skills and the serum concentration of folate, vitamin B12, vitamin B6, and homocysteine (for review see ref. 63). Of course, all the molecular and biochemical effects of all these factors with respect to AD-associated pathological processes are hard to determine and they may serve more general protective functions in the bodies physiological homeostasis.

In addition, the replacement of hormones whose levels decline during aging has to be included in this discussion. One main target hormone is the female sex hormone estrogen. Women more frequently suffer from sporadic AD compared to men and, at least in part, the massive drop in estrogen levels during menopause is believed to account for this effect. Since estrogen is a general neuroprotective factor with a wide range of neuronal effects, this is easy to conceive and will be pointed out below. Last but not least, psychosocial approaches should be mentioned, such as brain-jogging and an enhanced social and familial interaction.

By concentrating on the elucidation of the molecular signaling steps of AD-associated neurodegeneration, novel potential pharmaceutical targets for prevention and therapy will be elucidated. On the other hand it is clear that nerve cell death is the very final step in the pathology of AD, and many pathological alterations which affect neuronal function mainly at the synaptic level start years before cell death. Consequently, the cause of AD may not be described exclusively as a disease of nerve cell death but rather as a long-term impairment of neuronal function including synaptic neurotransmission. As mentioned above (see also Table 1) one central problem of AD research is the lack of good animal models. The first so-called

Alzheimer-mouse (The Games mouse) was generated as a mouse transgenic for the Swedish AD mutation driven by a platelet-derived growth factor promoter.[64] This mouse is characterized by an age-related increased deposition of Aβ and by synaptic loss in certain areas. The investigation of synaptic changes induced by various toxic stimuli, including Aβ, and also inflammatory mechanisms may lead to a better understanding of the molecular processes in AD pathogenesis. So far, transgenic mouse models have failed to represent the full spectrum of the pathological changes seen in human AD.

Is AD a Disease of Lost Neuroprotection?

But still another point needs to be seriously considered. The observed age-related decline in various neurotrophic and neuroprotective stimuli may render neurons more vulnerable to functional changes and ultimately to cell death. With respect to oxidative stress a decline in antioxidant defense systems might be of importance. Therefore, AD could be also defined as a disease of lacking trophic input (Loss-of-Protection-Hypothesis). Indeed, intrinsic neuronal protective mechanisms decline over time and leave behind neurons and their synaptic connections less well stabilized and less protected. One prominent example is the drop of the level of the female sex hormone estrogen after the menopause that is directly associated with an increased incidence of AD.[65] Consequently, supplementation through estrogen-replacement-therapy (ERT) reduces the AD risk significantly in females (see refs. 66-67). Therefore, ERT may be seen as a valuable approach for AD prevention (for review see ref. 68). The female sex hormone estrogen, has been shown to be a neuroactive and neuroprotective factor at various cellular levels. In addition to its hormone receptor-dependent neurotrophic activities such as the beneficial effects on neuronal morphology and synaptic function (for review see ref. 69), estrogen is also a powerful phenolic antioxidant with neuroprotective activities against oxidative apoptosis (for review see refs. 72-73).[70,71]

Several lines of experimental in vitro and in vivo evidence support the conclusion that the female sex hormone estrogen (17β-estradiol, the physiologically active compound) acts as neuroactive and neuroprotective compound (for review see ref. 73). Estrogen is a steroidal hormone and may exert its neuroprotective activities via binding to and activation of intracellular steroid receptors, which act as transcription factors. In the brain, where estrogen receptors (ERs) are expressed, the classical mode of estrogen action induces neuroprotection through a variety of processes such as an enhanced expression of nerve growth factor receptors or through the stimulation of an increased synaptic density and neuronal sprouting (for review see ref. 69). Interestingly, besides these classical delayed genomic activities various non-genomic and rather rapid interactions of E2 within the cells may take part in neuroprotective actions. Such rapid activities can act independently from the classical ER activation and are believed to be mediated in part via estrogen cross talk with various intracellular signaling pathways. Examples of such signaling pathways are cyclic-adenosine-mono-phosphate, G proteins, kinases, the redox-sensitive transcription factor NF-κB, and the mitogen-activated protein kinase (MAPK) signaling pathway (for review see ref. 74). In summary, the possibilities of interactions of estrogen with nerve cells are immense, and it is hard to pin down the actual mechanisms that are directly responsible for the neuroprotective activity.

In addition to the direct neuroprotective activities that are dependent or independent from ERs, another exciting link between the female sex hormone and AD exists. Estrogen affects APP processing. It has been shown that long-term incubation of cultured clonal and primary neurons with E2 enhances the α-processing of APP, therefore, increases the release of sAPPa and, correspondingly, decreases the formation of $A\beta_{1\text{-}42}$.[34] This dramatic effect of estrogen on the processing of APP also occurs in vivo. From these exciting data it can be speculated that the cessation of ovarian estrogen production in postmenopausal women might facilitate the formation and deposition of Aβ by increasing the local concentrations of $A\beta_{1\text{-}40}$ and $A\beta_{1\text{-}42}$ in brain. The modulation of Aβ processing and metabolism may be one way by which ERT prevents or delays the onset of AD in postmenopausal women.[75] Recently, we provided a direct link between two observations: (1) estrogen activates MAPK-signal pathways and (2) estrogen modulates APP processing leading to the enhancement of the α-secretory pathway. We found that in neuronal cells estrogen activates the phosphorylation of the MAP-kinases ERK1 and ERK2 within minutes and that this activation mediates the estrogen-driven α-processing of APP.[35]

Also the levels for another hormone, corticotropin-releasing-hormone (CRH), drop during aging. CRH is *the* central player of the stress response and of the hypothalamic-pituitary-adrenal-system (for review see ref. 76). We recently found that the activation of the CRH-R1 by its ligand CRH directly protects neurons against oxidative apoptosis and that this protection is associated with an increased release of sAPPa.[77] Therefore, as with estrogen we suggest that CRH is an intrinsic neuroprotective hormone that while present gives a permanent trophic input to the cells. A loss of this intrinsic system may lead to an enhanced sensitivity of neuronal cells to exogenous insults. Consistent with this view is the finding that CRH has cognition-enhancing effects and the increase of endogenous levels of CRH has been already proposed as potential therapeutic approach for AD (for review see ref. 78).

In conclusion, in the search for novel neuroprotective pathways, genes and pharmaceutical targets, attention should also be given on the elucidation of intrinsic mechanisms that protect neurons against death. By the investigation of the downstream genetic programs that are induced (or repressed) by neurotrophic factors such as estrogen and/or CRH, novel target genes of neuroprotection will be identified. With the advent of molecular high-throughput techniques to determine the mRNA expression profiles of cells employing DNA-array/DNA-chips, the expression patterns induced by neuroprotective signals will be identified. Such genes are then potential pharmaceutical targets for AD prevention.

Finally, intracellular signal pathways that affect nerve cell survival need to be studied with respect to their function in the development and prevention of AD. Here only the redox-sensitive transcription factor NF-κB should be mentioned that has been demonstrated to mediate the resistance of neuronal cells against oxidative apoptosis.[79] The identification of NF-κB-dependent gene expression will identify resistance genes which may then, ultimately, be regulated in their expression with pharmacological drugs. The admittedly idealistic idea is that a drug can specifically increase the expression of intrinsic protective factors that make the nerve cells more stable and prevents insults. Indeed, the analysis of genes that are up-regulated and/or down-regulated in cell types in vitro or brain regions in vivo that are selectively resistant against AD-associated neurodegenation can identify such resistance genes. The expression of the gene for the flavin-adenine-dinucleotide-dependent oxidoreductase *seladin-1* is downregulated in vulnerable regions and up-regulated in resistant areas of the brain. It is concluded that the decreased expression of seladin-1 in specific neurons may be a cause for selective vulnerability in AD.[80]

FINAL REMARKS

The incidence of AD will increase in the future. This means more and more people that suffer from this disease need full-time care, especially in the late stages of the disease. Besides the fatal consequences for the patient, this effort will be an enormous personal burden for the caregivers and a financial burden for the society. So what is needed so desperately are (1) markers that specifically detect this devastating disease early in its development or even before onset of the disease and (2) drugs for prevention and therapy. Immense research efforts in the last decades have identified crucial steps in the pathogenesis of AD. The biochemistry and the biology of AD target proteins such as APP, the presenilins, and BACE are studied and step by step their modes of action are understood. Clinically, we are still at the level of improving the neurotransmission by ACh in using ACh-esterase inhibitors. But many promising therapeutic approaches that are currently still more or less at the experimental stage are on the horizon including vaccination and the inhibition of γ- and β-secretase activity. Moreover, in consequence of many epidemiological studies various genetic and non-genetic risk factors have been identified which then in turn lead to novel preventive approaches. One good example is the risk factor high cholesterol and the preventive drug statins. More general preventive compounds include antioxidants, and the development of novel antioxidants when compared to vitamin E improved pharmacology and pharmacokinetics will improve the in vivo activity of these compounds. Finally, the identification and the functional analysis of target genes that are regulated by physiological modulators of neuroprotection such as neurotrophins and estrogen will bring even more potential targets for prevention into discussion. Ideally, all these research efforts in the search for AD therapy and prevention should be followed side-by-side with the same intensity, since we have to face the fact that due to the possible multifactorial origin of AD we will not be able to solve the AD problem with a single drug but rather with a combined pharmacological approach.

REFERENCES

1. Alzheimer A. ber eine eigenartige Erkrankung der Hirnrinde. Allg Zeitschr Psychiat Psych-Gericht Med 1907; 64:146-148.
2. Masters CL, Beyreuther K. Alzheimer s disease. BMJ 1998; 316: 446-448.
3. Selkoe DJ (1999) Translating cell biology into therapeutic advances in Alzheimer s disease. Nature 399: A23-A31.
4. Haass C, Mandelkow E. Proteolysis by presenilins and the renaissance of tau. Trends Cell Biol 1999; 9:241-244.
5. Whitehouse PJ, Price DL, Struble RG et al. Alzheimer s disease and senile dementia: Loss of neurons in the basal forebrain. Science 1982; 215:1237-1239.
6. Cacabelos R, Alvarez A, Lombardi A et al. Pharmacological treatment of Alzheimer disease: From psychotropic drugs and cholinesterase inhibitors to pharmacogenomics. Drugs Today 2000; 36:415-499.
7. Rogers J, Cooper NR, Webster S, Schultz J, McGeer PL, Styren SD et al. Complement activation by β-amyloid in Alzheimer disease. Proc Natl Acad Sci USA 1992; 89:10016-10020.
8. McGeer EG, McGeer PL. Brain inflammation in Alzheimer disease and the therapeutic implications. Curr Pharmac Design 1999; 5: 821-836.
9. Breitner JCS. The role of antiinflammatory drugs in the prevention and treatment of Alzheimer s disease. Annu Rev Med 1996; 47:401-411.
10. Eikelenboom P, Rozemuller AJM, Hoozemans JJM et al. Neuroinflammation and Alzheimer disease: Clinical and therapeutic implications. Alz Dis Ass Dis 2000; 14(Suppl 1):S54-S61.
11. Hoyer S. Brain glucose and energy metabolism abnormalities in sporadic Alzheimer disease. Causes and consequences: An update. Exp Gerontol 2000; 35(9-10 SI):1363-1372.
12. Beal MF. Aging, energy, and oxidative stress in neurodegenerative disorders. Ann Neurol 1995; 38:357-366.
13. Behl C. Alzheimer s disease and oxidative stress: Implications for novel therapeutic approaches. Prog Neurobiol 1999; 57:301-323.
14. Harman D. Aging: Phenomena and theories. Ann NY Acad Sci 1998; 854:1-7.
15. Markesbery WR. The role of oxidative stress in Alzheimer disease. Arch Neurol 1999; 56:1449-1452.
16. Behl C, Davies J, Cole GM et al. Vitamin E protects nerve cells from amyloid b protein toxicity. Biochem Bioph Res Co 1992; 186:944-950.
17. Mattson MP. Cellular actions of beta-amyloid precursor protein and its soluble and fibrillogenic derivatives. Physiol Rev 1997; 77(4):1081-1132.
18. Sano M, Ernesto C, Thomas RG et al. A controlled trial of selegiline, alpha-tocopherol, or both as treatment for Alzheimer s disease. N Engl J Med 1997; 336:1216-1222.
19. Selkoe DJ. Alzheimer s disease: Genes, proteins, and therapy. Physiol Rev 2001; 81(2):741-766.
20. Yankner BA. Mechanisms of neuronal degeneration in Alzheimers disease. Neuron 1996; 16:921-932.
21. Behl C. Amyloid beta-protein toxicity and oxidative stress in Alzheimer's disease. Cell Tiss Res 1997; 290:471-480.
22. Mandelkow E, Mandelkow EM. Microtubules and microtubule-associated proteins. Curr Opin Cell Biol 1995; 7(1):72-81.
23. Spillantini MG, Goedert M. Tau protein pathology in neurodegenerative diseases Trends Neurosci 1998; 21(10):428-433.
24. Chen M, Fernandez HL. Where do Alzheimer s plaques and tangles come from? Aging-induced protein degradation inefficiency. Front Biosci 2001; 6:E1-E11.
25. Chapman PF, Falinska AM, Knevett SG et al. Genes, models and Alzheimer s disease. Trends Genet 2001; 17(5):254-261.

26. Campion D, Dumanchin C, Hannequin D et al. Early-onset autosomal dominant Alzheimer disease: Prevalence, genetic heterogeneity, and mutation spectrum. Am J Hum Genet 1999; 65(3):664-670.
27. Schubert D, Behl C. The expression of amyloid beta protein precursor protects nerve cells from beta-amyloid and glutamate toxicity and alters their interaction with the extracellular matrix. Brain Res 1993; 629:275-282.
28. Mattson MP, Cheng B, Culwell AR et al. Evidence for excitoprotective and intraneuronal calcium-regulating roles for secreted forms of the beta-amyloid precursor protein. Neuron 1993; 10:243-254.
30. Hardy J. Amyloid, the presenilins and Alzheimer's disease. Trends Neurosci 1997; 20(4):154-159.
31. Wolfe MS, Haass C. The role of presenilins in gamma-secretase activity. J Biol Chem 2001; 276(8):5413-5416.
32. Skovronsky DM, Lee VMY. Beta-secretase revealed: Starting gate for race to novel therapies for Alzheimer s disease. Trends Pharmacol Sci 2000; 21(5):161-163.
33. Lammich S, Kojro E, Postina R et al. Constitutive and regulated alpha-secretase cleavage of Alzheimer s amyloid precursor protein by a disintegrin metalloprotease. Proc Natl Acad Sci USA 1999; 96(7):3922-3927.
34. Xu H, Gouras GK, Greenfield JP et al. Estrogen reduces neuronal generation of Alzheimer beta-amyloid peptides. Nat Med 1998; 4:447-451.
35. Manthey D, Heck S, Engert S et al. Estrogen induces a rapid secretion of amyloid β precursor protein via the mitogen-activated protein kinase pathway. Eur J Biochem 2001; 268:1-8.
36. Yan SD, Chen X, Fu J et al. Rage and amyloid-beta peptide neurotoxicity in Alzheimer's disease. Nature 1996; 382(6593):685-691.
37. Liu Y, Dargusch R, Schubert D. Beta amyloid toxicity does not require rage protein. Biochem Biophys Res Co 1997; 237(1):37-40.
38. Behl C, Davies JB, Lesley R et al. Hydrogen peroxide mediates amyloid β protein toxicity. Cell 1994; 77:817-827.
39. Geula C, Wu CK, Saroff D et al. Aging renders the brain vulnerable to amyloid beta-protein neurotoxicity. Nature Med 1998; 4(7):827-831.
40. Behl C, Lezoualch F, Trapp T et a. Glucocorticoids enhance oxidative stress-induced cell death in hippocampal neurons in vitro. Endocrinol 1997; 138(1):101-106.
41. Behl C. Apoptosis and Alzheimer s disease. J Neur Transm 2000; 107:1325-1344.
42. Terry RD. The pathogenesis of alzheimer disease An alternative to the amyloid hypothesis. J Neuropathol Exp Neurol 1996; 55(10):1023-1025.
43. Terry RD, Katzman R, Bick KL. Alzheimer Disease. New York: Raven Press, 1994.
44. Rebeck GW, Reiter JS, Strickland DK et al. Apolipoprotein E in sporadic Alzheimer s disease: Allelic variation and receptor interactions. Neuron 1993; 11(4):575-580.
45. Richard F, Amouyel P. Genetic susceptibility factors for Alzheimer s disease. Eur J Pharmacol 2001; 412(1):1-12.
46. Simons M, Keller P, Destrooper B et al. Cholesterol depletion inhibits the generation of beta-amyloid in hippocampal neurons. Proc Natl Acad Sci USA 1998; 95(11):6460-6464.
47. Wolozin B, Kellman W, Rousseau P et al. Decreased prevalence of Alzheimer disease associated with 3-hydroxy-3-methyglutaryl coenzyme A reductase inhibitors. Arch Neurol 2000; 57(10):1439-1443.
48. Fassbender K, Simons M, Bergmann C et al. Simvastatin strongly reduces levels of Alzheimer s disease beta-amyloid peptides A beta 42 and A beta 40 in vitro and in vivo. Proc Natl Acad Sci USA 2001; 98(10):5856-5861.
49. Kojro E, Gimpl G, Lammich S et al. Low cholesterol stimulates the nonamyloidogenic pathway by its effect on the alpha-secretase ADAM 10. Proc Natl Acad Sci USA 2001; 98(10):5815-5820.
50. Emilien G, Maloteaux JM, Beyreuther K et al. Alzheimer disease Mouse models pave the way for therapeutic opportunities. Arch Neurol 2000; 57(2):176-181.

51. Duff K, Knight H, Refolo LM et al. Characterization of pathology in transgenic mice over-expressing human genomic and cDNA tau transgenes. Neurobiol Dis 2000; 7(2):87-98.
52. Emilien G, Beyreuther K, Masters CL et al. Prospects for pharmacological intervention in Alzheimer disease. Arch Neurol 2000; 57(4):454-459.
53. Schenk D, Barbour R, Dunn W et al. Immunization with amyloid-beta attenuates Alzheimer disease-like pathology in the PDAPP mouse. Nature 1999; 400(6740):173-177.
54. Janus C, Pearson J, McLaurin J et al. A beta peptide immunization reduces behavioural impairment and plaques in a model of Alzheimer s disease. Nature 2000; 408(6815):979-982.
55. Morgan D, Diamond DM, Gottschall PE et al. A beta peptide vaccination prevents memory loss in an animal model of Alzheimer s disease. Nature 2000; 408(6815):982-985.
56. Lim GP, Yang F, Chu T et al. Ibuprofen suppresses plaque pathology and inflammation in a mouse model for Alzheimer s disease. J Neurosci 2000; 20(15):5709-5714.
57. Grundman M. Vitamin E and Alzheimer disease: The basis for additional clinical trials. Am J Clin Nutr 2000; 71(2):630S-636S.
58. Moosmann B, Behl C. The antioxidant neuroprotective effects of estrogens and phenolic compounds are independent from their estrogenic properties. Proc Natl Acad Sci USA 1999; 96:8867-8872.
59. Gutzmann H, Hadler D. Sustained efficacy and safety of idebenone in the treatment of Alzheimer s disease: update on a 2-year double-blind multicentre study. J Neur Transm 1998; (Suppl 54):301-310.
60. Guidetti C, Paracchini S, Lucchini S et al. Prevention of neuronal cell damage induced by oxidative stress in-vitro: Effect of different Ginkgo biloba extracts. J Pharm Pharmacol 2001; 53(3):387-392.
61. Oken BS, Storzbach DM, Kaye JA. The efficacy of ginkgo biloba on cognitive function in Alzheimer-disease. Arch Neurol 1998; 55(11):1409-1415.
62. Wesnes KA, Ward T, McGinty A et al. The memory enhancing effects of a Ginkgo biloba/ Panax ginseng combination in healthy middle-aged volunteers. Psychopharmacol 2000; 152(4):353-361.
63. Nourhashemi F, Gillette-Guyonnet S, Andrieu S et al. Alzheimer disease: Protective factors. Am J Clin Nutr 2000; 71(2):643S-649S.
64. Games D, Adams D, Alessandrini R et al. Alzheimer-type neuropathology in transgenic mice overexpressing v717f beta-amyloid precursor protein. Nature 1995: 373(6514):523-527.
65. Simpkins JW, Singh M, Bishop J. The potential role for estrogen replacement therapy in the treatment of the cognitive decline and neurodegeneration associated with Alzheimer s disease. Neurobiol Aging 1994; 15(Suppl 2):S195-S197.
66. Henderson VW, Paganini-Hill A, Emanuel CK et al. Estrogen replacement therapy in older women. Arch Neurol 1994; 51:896-900.
67. Tang MX, Jacobs D, Stern Y et al Effect of oestrogen during menopause on risk and age of onset of Alzheimer s disease. Lancet 1996; 348:429-432.
68. Henderson VW. Oestrogens and dementia. In: Novartis Found Symp, ed. 2000; 230:254-265.
69. McEwen BS, Alves SE. Estrogen actions in the central nervous system. Endocr Rev 1999; 20:279-307.
70. Behl C, Widmann M, Trapp T et al. 17-β estradiol protects neurons from oxidative stress-induced cell death in vitro. Biochem Biophys Res Com 1995; 216:473-482.
71. Green PS, Bishop J, Simpkins JW. 17 alpha-estradiol exerts neuroprotective effects on SK-N-SH cells. J Neurosci 1997; 17:511-515.
72. Behl C, Holsboer F. The female sex hormone oestrogen as a neuroprotectant. Trends Pharmacol Sci 1999; 20(11):441-444.
73. Behl C. Estrogen Mystery drug for the brain? The neuroprotective activities of the female sex hormone. Wien, NewYork: Springer, 2001.
74. Green PS, Simpkins JW. Neuroprotective effects of estrogens: Potential mechanisms of action. Intern J Develop Neurosci 2000; 18:347-358.

75. Petanceska SS, Nagy V, Frail D, Gandy S. Ovariectomy and 17 beta-estradiol modulate the levels of Alzheimer s amyloid beta peptides in brain. Neurology 2000; 54:2212-2217.
76. Holsboer F, Barden N. Antidepressants and hypothalamic-pituitary-adrenocortical regulation. Endocrine Rev 1996; 17:187-205
77. Lezoualc h F, Engert S, Berning B et al. Corticotropin-releasing hormone mediated neuroprotection against oxidative stress is associated with the increased release of non amyloidogenic amyloid beta precursor protein and with the suppression of nuclear factor-kappa B. Mol Endocrinol 2000; 14(1):147-159.
78. Behan DP, Grigoriadis DE, Lovenberg T et al. Neurobiology of corticotropin releasing factor (CRH) receptors and CRH-binding protein: Implications for the treatment of CNS disorders. Mol Psych 1996; 1:265-277.
79. Lezoualch F, Sagara Y, Holsboer F et al. High constitutive nf-kappa-b activity mediates resistance to oxidative stress in neuronal cells. J Neurosci 1998; 18(9):3224-3232
80. Greeve I, Hermans-Borgmeyer I, Brellinger C et al. The human DIMINUTO/DWARF1 homolog seladin-1 confers resistance to Alzheimer s disease-associated neurodegeneration and oxidative stress. J Neurosci 2000; 20(19):7345-7352.
81. Borchelt DR, Ratovitski T, Vanlare J et al. Accelerated amyloid deposition in the brains of transgenic mice coexpressing mutant presenilin 1 and amyloid precursor proteins. Neuron 1997; 19(4):939-945.
82. Destrooper B, Saftig P, Craessaerts K et al. Deficiency of presenilin-1 inhibits the normal cleavage of amyloid precursor protein. Nature 1998; 391(6665):387-390.
83. Duff K, Eckman C, Zehr C et al. Increased amyloid-beta-42(43) in brains of mice expressing mutant presenilin 1. Nature 1996; 383(6602):710-713.
84. Holcomb L, Gordon MN, Mcgowan E et al. Accelerated alzheimer-type phenotype in transgenic mice carrying both mutant amyloid precursor protein and presenilin 1 transgenes. Nature Med 1998; 4(1):97-100.
85. Hsiao K, Chapman P, Nilsen S et al. Correlative memory deficits, a-beta elevation, and amyloid plaques in transgenic mice. Science 1996; 274(5284):99-102.
86. Sturchler-Pierrat C, Abramowski D, Duke M et al. Two amyloid precursor protein transgenic mouse models with alzheimer disease-like pathology. Proc Natl Acad Sci USA 1997; 4(24):13287-13292.
87. Zheng H, Jiang MH, Trumbauer ME et al. Beta-amyloid precursor protein-deficient mice show reactive gliosis and decreased locomotor activity. Cell 1995; 81(4):525-31.
88. Shen J, Bronson RT, Chen DF, Xia WM, Selkoe DJ, Tonegawa S. Skeletal and CNS defects in presenilin-1-deficient mice. Cell 1997; 89(4):629-639.
89. McGeer PL, McGeer EG, Yasojima K. Alzheimer disease and neuroinflammation. J Neural Transm Suppl 2000; 59:53-57.

INDEX

A1 receptors 249-252, 255, 256, 258, 261, 263, 264, 268
A2A receptors 249, 250, 253, 257, 258, 260-264, 266-268
A3 receptors 249, 258, 267-269
ACE 136
Acetylcholine 255, 256, 260, 362, 477
ACh 477, 485, 486, 492
ACh-esterase 477, 486, 492
Acidosis 8, 459, 464
Active zones 199
Activin 129, 342-345, 353-355, 357, 365
Adeno-associated viral vector 432, 437, 447
Adenosine 209, 249-258, 260, 261, 263, 265-270, 402, 490
Adenovirus 56, 135, 322, 423, 436
Adenylate cyclase 12, 198, 199, 209, 234
AIDS 8, 318, 425
AMPA 9-11, 13, 162, 163, 210, 232, 341, 460, 462, 463, 470
Amplicon vectors 435
Amygdala 5, 10, 239, 264, 476
Amyloid 8, 24, 50, 53-55, 206, 207, 210, 211, 214, 240, 290, 341, 363, 475, 476, 479, 482, 485
Amyloid precursor protein 240
Amyotrophic lateral sclerosis (ALS) 8, 22, 53, 55, 62, 65, 66, 170, 225, 236, 303, 305, 314, 317, 336, 382
Androgen receptor 289
Angelman syndrome 55
Angiogenesis 134, 354, 376, 377, 379-382, 384-388
Antioxidant 17, 19, 268, 291, 341, 479, 488, 489, 490, 492
Antisense oligonucleotide 56, 62, 134, 136, 204, 233, 286, 287, 387, 433
AP-1 44, 51
Apaf-1 23, 60, 65, 66, 292, 294
ApoE 484
Apolipoprotein E 484
Apoptosis 5, 7, 9, 19-25, 42-46, 48, 50-60, 62-67, 91, 98, 127, 188, 190, 191, 205, 206, 264, 265, 282, 292, 294, 309, 313, 314, 342, 345, 363, 364, 382, 384, 455, 457, 459, 465-468, 483, 490-492
APP 240, 475, 479, 480-482, 484-487, 491, 492
Arachidonic acid 13, 16, 126, 237
Aspartate 3, 4, 10, 55, 64, 67, 99, 202, 228, 233, 242, 254, 255, 261, 270, 283, 339, 407, 409, 411, 462, 467, 481
Astrocytes 7, 20, 42, 88, 93, 96, 98-100, 104, 115, 116, 119-124, 130, 132, 134, 136, 209, 210, 212, 225, 227, 234, 237, 240, 241, 283, 295, 336-340, 359, 360, 375, 378, 379, 381, 383, 442, 456, 461
Astroglia 99, 225, 227, 234, 236, 240, 241, 269, 340, 358, 359, 364
Ataxia telangiectasia mutated 57, 66
Ataxin-1 289
Ataxin-3 289
ATM 46, 57, 58, 66
Axotomy 24, 46, 50, 62, 316, 364

BACE 482, 487, 492
BAPTA 186

Bax 21, 23, 24, 46, 47, 49, 52, 56-58, 60-63, 119, 342
Blood brain barrier (BBB) 88-90, 118, 120, 126, 127, 130, 131, 136, 176, 202, 203, 339, 397, 398, 400-410, 412, 415, 416, 420, 422-427, 431
Bcl-2 23, 43, 46, 47, 60-63, 66, 100, 119, 189, 191, 292-294
BDNF 49, 50, 102, 129, 130, 270, 303, 305, 306, 309-311, 315-320, 322, 397-400, 409, 411-423, 427, 441
Bicuculline 339
Bone morphogenetic proteins (BMPs) 353-355, 365, 366
Bradykinesia 207
Brainstem 317, 337

CA1 6, 57, 62, 98, 99, 101, 163, 165, 167, 172, 176, 188, 208, 211, 236, 239, 240, 257, 258, 264, 265, 283, 288, 291, 294, 337, 340, 414-417, 425, 457, 458, 460
CA3 6, 98, 185, 187, 188, 253, 257, 258, 264, 265, 283, 337, 343, 344, 414, 425, 433
CAG 240, 289
Calbindin 184, 192, 193, 341
Calpains 66
Calreticulin 187, 189
Calretinin 184, 185, 192
Caspase 7, 9, 13, 21-5, 42-49, 52, 56-67, 91, 99, 119, 161, 190, 292, 294, 313, 314, 324, 342, 381, 462, 467, 468
Cell adhesion 87, 88, 89, 90, 93, 120, 127, 377
Cerebellum 165, 184, 198, 227, 241, 250, 337, 457
Cerebral blood flow (CBF) 127, 342, 406-408, 410, 456, 461, 462, 464
Cerebral cortex 18, 55, 65, 90-93, 95-97, 99, 100, 135, 165, 167, 172, 175, 207, 213, 236, 257, 261, 267, 270, 314, 315, 337, 358, 359, 377, 404, 425, 457, 458, 476, 486
Cerebral ischemia 7, 9, 43, 45, 50, 59, 67, 87, 90, 91, 93-96, 98-102, 176, 186, 208, 236, 282, 283, 285, 287-289, 291, 294, 295, 315, 380-382, 385, 387, 388, 414, 455-468
Cerebrospinal fluid 119, 125, 128, 137, 227, 233, 339, 444
cGMP-phosphodiesterase 198
Chaperones 282, 289, 358
Chemokines 87-89, 93, 95, 118, 124-127, 131, 465
Chimeric peptides 397. 402, 407, 412, 413, 416, 419, 420
Chloride channels 230
Cholesterol 484, 485, 489, 492
Chromaffin cells 359, 360, 362, 442
CNTF 129, 130, 317, 323, 361, 362, 398, 400
Cognition 98, 304, 477, 486, 491
Connexin 340

COX 13, 127, 465, 466, 477, 487
CREB 49, 50, 55
CRH 491
Cyclooxygenase 13, 16, 460, 465, 477, 487
Cyclosporin A 14, 19, 137, 191
Cystic fibrosis 436, 437
Cystine knot 354, 356, 366
Cytochrome c 13, 22, 23, 24, 46, 49, 56, 57, 60, 62, 63, 65, 190, 191, 292, 294, 462, 467
Cytomegalovirus (CMV) 287, 440, 443
Cytoskeleton 13, 21, 235, 462

Dantrolene 189
Death receptor 42, 43, 45, 63, 64, 67, 292
Dementia 8, 43, 67, 475, 476, 479, 481
Dentate gyrus 188, 250, 342, 359
Dihydropyridines 175
DISC 42, 43, 45, 64, 67
DNA-chips 491
DNQX 163, 165, 207
Domoic acid 9, 10
Dopaminergic neurons 59, 62, 211, 241, 316, 319, 343, 355, 361, 363-365, 381

Endoplasmic reticulum 13, 14, 17, 24, 183, 187, 188, 282, 335
Endothelial cells 20, 89, 90, 93, 96, 99, 126, 127, 129, 135, 210, 269, 282, 283, 285, 320, 338, 376-378, 380, 382-384, 386, 387, 401, 456, 465
eNOS 20, 91, 209, 250-58, 260-263, 265-270, 342, 386, 461, 462
Ependyma 118, 121, 122, 311, 399, 400
Epilepsy 5-8, 197, 204-206, 214, 236, 239, 240, 281, 282, 303, 320
EPSCs 10, 11, 170, 231
ERK 47-51, 67, 312, 386, 387
Estrogen 475, 482, 488, 489, 490, 491, 492
Extracellular matrix 89, 116, 120, 121, 131, 335, 338, 340, 354, 377, 434

Fas 24, 42-46, 51, 57, 63, 65, 67, 127, 129, 198, 227, 231, 232, 281, 282, 294, 303, 304, 308, 311, 318, 338-340, 345, 467
Fetal tissue grafts 441
FGF1 335-337, 340
FGF2 335-345
FGF5 337, 340
FGF7 336, 337, 340
FGF8 337, 340
FGF9 335, 337, 340

Fibroblasts 58, 115, 116, 120-123, 130, 131, 134, 135, 441, 442
Fibrosis 115, 116, 119, 121, 133, 135, 136, 436, 437
Follistatin 343, 344
Forskolin 209
Free radicals 18, 57, 89, 118, 240, 249, 261, 457

G-protein 49, 205, 235, 462
GABA 168, 170, 204-206, 208, 211, 214, 227, 239, 257, 260, 266, 319, 339
Gap junctions 120, 340
GDNF 49, 129, 130, 317, 323, 354-357, 361-366, 399, 400
Geldanamycin 290, 291
Gene therapy 322, 423, 425, 426, 432, 436, 441, 443-445
Gene transfer 135, 281, 282, 286, 287, 323, 365, 382, 433-437, 441, 443-445
GFAP 88, 99, 119, 120, 121, 440, 442
Glia 8, 20, 21, 43, 49, 67, 87-96, 98-100, 102-104, 115-127, 129-131, 133-135, 197-199, 205, 207-209, 212-214, 225, 227, 228, 231, 233, 234, 236-238, 240-242, 250, 258, 260, 262, 263, 267-269, 282, 283, 286, 287, 291, 295, 310, 317, 318, 320, 324, 336-338, 340, 343, 353-355, 358-360, 362-364, 366, 381-383, 398-400, 425, 427, 430, 434, 456, 457, 461, 465, 468, 477, 481
Glial fibrillary protein 119
Glucocorticoids 122, 360, 477, 483
Glucose 7, 49, 91, 162, 165, 171, 176, 205, 208, 210, 227, 235, 282, 286, 287, 291, 293, 294, 379, 381, 402, 417, 422, 455, 458-460, 463, 464, 477, 478
Glutamate 3-14, 16, 20, 23-25, 50, 52, 55, 56, 91, 121, 162, 165, 168, 170-173, 176, 186, 191, 197-199, 201-205, 207-211, 213-215, 225-228, 230-242, 250, 254-256, 259-263, 265, 266, 267-269, 285, 286, 291, 341, 363, 365, 381, 407, 409, 411, 457, 460, 462-464, 468, 477, 483
Glutathione 291, 341, 357, 463
Glycine 8, 11, 199, 255
Granule cells 52, 185, 204, 207, 211, 283, 342, 360
Growth cone 115, 116, 123, 124, 131, 133, 134, 338

Heat shock protein 47, 281, 282, 285, 295, 296
Heme oxygenase-1 (HO-1) 282. 291, 294
Hemorrhage 116, 122, 125, 137, 320
Hepatic encephalopathy 8
Herpes simplex virus 286, 287, 342, 423, 435
HIF-1 379, 380, 385
HIF-2 379, 380, 385
High altitude brain edema 385
Hippocampus 5, 9, 10, 98, 165, 167, 171, 185, 186, 188, 191, 192, 198, 206, 209, 213, 227, 239, 250, 255, 257, 258, 260-262, 264, 265, 267, 268, 270, 283, 288, 315, 336, 337, 339, 343, 345, 359, 377, 416, 425, 443, 457, 458, 476, 486
HIV 43, 67, 210, 363, 432, 443, 444
Homer 204, 205
Hsp70 281-283, 285-295
HSV 286-288, 290, 423, 435, 440, 445
Huntingtin 55, 215, 290, 467
Hypercapnia 459
Hypoxia 6, 7, 22, 24, 25, 49, 50, 54-56, 93, 100, 126, 127, 162, 163, 165, 167, 171, 172, 176, 208, 236, 256, 291, 379, 380, 382, 384, 385, 387, 388, 459, 464

ICAM 90, 93, 99, 127, 465
Idebenone 488
IGF-1 129
IGF-2 129
Interleukin-1 (IL-1) 52, 87, 88, 91, 96, 97, 99, 100-102, 104, 119, 126, 127, 129, 130, 132, 136, 263, 268, 269, 465, 467, 484, 487
Inhibin 343, 353, 354, 355
iNOS 20, 91-93, 97, 99, 129, 461, 462, 465, 466
Insulin 49, 125, 127-129, 234, 235, 306, 307, 312, 317, 398, 400, 402, 404, 409, 427, 477, 478
Integrin family 90
Interferon 91, 127, 268, 464
Ionotropic glutamate receptors 4, 10, 12, 162, 165, 207, 341

JNK 24, 25, 44, 47, 48, 50-52, 54, 62, 67, 292, 294, 313

Kainic acid (KA) 9, 10, 51, 55, 186, 239, 240, 253, 255, 264, 266, 283, 285, 288, 289, 291, 296, 343, 344
Kindling 239, 315, 320
Kynurenic acid 262

L-dopamine 477
LacZ 287
Lamotrigine 170-172
Lathyrism 9, 10
Lifarizine 171, 173
Lipid peroxidation 20, 161, 240, 269
Liposomes 95, 242, 424, 425, 431, 433, 434, 445
LTP 102, 319

Malaria 477
Memory 49, 100, 304, 319, 320, 476, 477, 486, 488
Meninges 115, 116, 118, 130, 133, 359
Mercury 241
Metabotropic glutamate receptors 12, 173, 197, 199, 201, 215, 225, 463
Metalloproteinase 116, 123
Metamphetamine 207
Mibefradil 175
Microglia 20, 43, 87-90, 92-96, 98-100, 102-104, 115, 118, 119, 121, 122, 124-127, 129-131, 133, 250, 283, 295, 320, 338, 461, 465, 477
Midbrain 337, 343, 355, 361, 363, 365
Middle cerebral artery 55, 90, 172, 186, 191, 208, 283, 288, 290, 342, 385, 397, 416, 417, 419, 420-422, 457, 459, 461, 467
Mitochondria 8, 13-19, 22-24, 41-43, 46, 47, 49, 52, 56, 57, 60, 61, 63, 66, 67, 161, 162, 168, 173, 176, 183, 184, 189-192, 214, 253, 261, 265, 282, 286, 292, 341, 462, 467, 488
Mitogen-activated protein kinase (MAPK) 48, 51, 52, 313, 365, 491
MK-801 97, 99, 100, 171, 172, 207, 463, 468
MMP 123, 124, 127, 131, 132, 134
Modal gating 165, 167, 178
Motoneurons 43, 65, 67, 102, 170, 305, 310, 316, 317, 323, 360, 361, 364, 365
MPTP 50, 56, 59, 62, 185, 207, 241, 253, 316
Myelin 88, 94, 102, 103, 119, 121, 123, 124, 131, 170, 313, 317, 318, 423, 434

NAAG 202, 208, 209
Nanocontainer 424
Neurofibrillary tangles 476, 478, 479, 485
Neurogenesis 70, 337, 338, 342, 345
Neuropilins 375, 377, 378, 381-383, 387
Neurotrophins 89, 102, 119, 120, 123, 124, 130, 131, 303-324, 354, 356, 361, 397, 398, 400, 409, 410, 413, 441, 475, 489, 492
NGF 44, 46, 48, 49, 51, 53, 56, 62, 129, 130, 234, 305, 306, 308-312, 315-318, 320, 321, 323, 340, 356, 361, 398, 400, 409, 410, 411, 441, 442, 489
Nifedipine 175, 176
Nitric oxide (NO) 8, 13, 14, 19-21, 23, 49, 52, 87, 91, 92, 96, 98, 129, 130, 167, 209, 213, 250-258, 260, 261, 263, 265-270, 292, 342, 386, 457, 460-462, 465
Nitric oxide synthase 8, 13, 14, 20, 21, 87, 91, 292, 460
NMDA 6, 7, 8, 10-17, 19, 20, 22-25, 55, 59, 67, 87, 97, 99, 100, 162, 163, 165, 168, 171, 172, 173, 186, 198, 201, 204-211, 213, 231, 241, 250, 253, 255, 256, 258, 261, 262, 270, 339, 341, 343, 363, 365, 460, 462-464
nNOS 14, 20, 91, 460-462

p38 MAPK 48, 51, 52
p53 25, 45, 46, 47, 49, 51, 54-59, 62, 313, 438
p75 42, 45, 46, 48, 67, 98, 282, 305, 308-313, 315, 316, 317, 320, 322, 489
PACAP 234
PARP 20, 57, 59, 67, 461, 467
Parvalbumin 184, 192
Pegylation 410-415, 423, 424
Permeability transition pore 23, 47, 60, 63, 190
Phenytoin 167, 169-173
Photothrombosis 90, 91, 95, 97
PI3K 47-50, 52, 67
PKC 167, 204, 235, 386, 387
Plaques 240, 476, 478, 479, 481-483, 485, 487
Polyglutamine 240, 289-291
Prefrontal cortex 486
Procaspase-9 65, 292, 294
Progressive supranuclear palsy 479
Presenilin 1 (PS1) 475, 479-482, 484, 486
Presenilin 2 (PS2) 268, 475, 479-482
Purkinje cell 184, 192, 227, 231, 232, 250, 457

Radial glia 358, 359
Reactive oxygen species (ROS) 13, 16, 17, 19, 20, 24, 46, 161, 189, 190, 207, 237, 238, 247, 261, 313, 341, 345, 462, 477, 483
Receptor for advanced glycation endproducts (RAGE) 48, 483
Retina 3, 25, 48, 102, 123, 137, 198, 199, 226, 227, 311, 359, 381, 443, 444
Retroviral vector 286, 432, 433, 437, 438, 440-444
Rheumatic diseases 384, 477
Riluzole 169-172, 214
Ryanodine 187

Scar 9, 88, 99, 103, 104, 115, 116, 118-124, 128, 130-137, 270, 338, 340, 364
Schwann cells 320, 358, 359, 382
Secretase 480-482, 485, 487, 492
SERCA 187-189
Serotonin 319, 362
Sipatrigine 173
Smads 357, 358
SOD 53, 56, 62, 63, 237, 238, 291

Spinobulbar muscular atrophy 289
Spinocerebellar ataxia 290
Sprouting 103, 131, 319, 362, 384, 490
Src kinase 387
Statins 462, 485, 489, 492
Status epilepticus 5, 225
Stem cells 323, 324, 337, 444
Striatum 165, 206, 207, 209, 214, 236, 240, 250, 257, 260, 266, 267, 283, 288-290, 337, 365, 443, 457, 458
Substantia gelatinosa 227
Substantia nigra 185, 193
Superior colliculus 227
Superoxide dismutase 238, 269, 291, 341
SV40 440, 443
Synantocytes 115, 120, 121, 123
Synapse 121, 170, 198, 230, 232, 234, 239, 304, 317, 319, 362, 381, 481, 485, 486

T cells 45, 51, 59, 87, 89, 93-95, 98, 99, 102, 103, 119, 127, 130, 176, 209, 281, 282, 290, 291, 295, 303, 310, 317, 320, 321, 323, 324, 353, 354, 358, 434, 437, 438, 440, 443, 444, 465, 466
Tau-protein 476, 478
Tenascin-C 120
Tetrodotoxin 5, 120, 163
Thalamus 4, 165, 198, 264, 319, 337, 425
Thapsigargin 24, 188, 189
Theophylline 255, 257, 270
TNF 24, 42-46, 48, 52, 64, 65, 67, 87, 91, 96-101, 126-128, 132, 136, 263, 264, 268, 269, 305, 308, 309, 323, 379, 466, 467
TNF-related apoptosis-inducing ligand (TRAIL) 43-45, 64, 67
Transferrin 322, 397, 402-404, 407, 409, 415, 416, 423, 445
Transgene 286, 287, 290, 323, 339, 431, 432, 433, 434, 435, 436, 437, 440, 441, 442, 443, 444, 445, 486
Trinucleotide repeat 215, 289
Trisomy 21 481
TrkA 307, 310-313, 315, 318
TrkB 310-313, 315, 316, 319, 413-416
TrkC 310, 312, 313, 318
TRP 187
TTX 163, 165, 168, 170, 171
Tyrosine kinase 25, 49, 305, 307, 309, 311-313, 336, 355, 357, 364, 375, 377, 382

VEGF 120, 127, 130, 375-388
VIP 406-410
Vitamin E 479, 483, 488, 492

Ziconotide 176